About Island Press

Island Press is the only nonprofit organization in the United States whose principal purpose is the publication of books on environmental issues and natural resource management. We provide solutions-oriented information to professionals, public officials, business and community leaders, and concerned citizens who are shaping responses to environmental problems.

In 2004, Island Press celebrates its twentieth anniversary as the leading provider of timely and practical books that take a multidisciplinary approach to critical environmental concerns. Our growing list of titles reflects our commitment to bringing the best of an expanding body of literature to the environmental community throughout North America and the world.

Support for Island Press is provided by the Agua Fund, Brainerd Foundation, Geraldine R. Dodge Foundation, Doris Duke Charitable Foundation, Educational Foundation of America, The Ford Foundation, The George Gund Foundation, The William and Flora Hewlett Foundation, Henry Luce Foundation, The John D. and Catherine T. MacArthur Foundation, The Andrew W. Mellon Foundation, The Curtis and Edith Munson Foundation, National Environmental Trust, National Fish and Wildlife Foundation, The New-Land Foundation, Oak Foundation, The Overbrook Foundation, The David and Lucile Packard Foundation, The Pew Charitable Trusts, The Rockefeller Foundation, The Winslow Foundation, and other generous donors.

The opinions expressed in this book are those of the author(s) and do not necessarily reflect the views of these foundations.

About World Wildlife Fund

Known worldwide by its panda logo, World Wildlife Fund is dedicated to protecting the world's wildlife and the rich biological diversity that we all need to survive. The leading privately supported international conservation organization in the world, WWF has invested in more than 13,000 projects in 157 countries and has more than 1 million members in the United States.

WWF directs its conservation efforts toward three global goals: protecting endangered species, saving endangered species, and addressing global threats. From working to save the giant panda, tiger, and rhino to helping establish and manage parks and reserves worldwide, WWF has been a conservation leader for forty years. Visit the WWF Web site at www.worldwildlife.org or write to us.

About African Wildlife Foundation

Founded in 1961, the African Wildlife Foundation (AWF) is the leading conservation organization focused solely on the African continent. AWF designs and implements conservation strategies that are based on science and are compatible with human benefit. Since its inception AWF has ensured the continued existence of some of Africa's most rare species. AWF has also invested in training hundreds of African individuals who have gone on to play critical roles in conservation. AWF has pioneered the use of community conservation to demonstrate that wildlife can be conserved while human wellbeing is also improved. To learn more about AWF, visit www.awf.org.

Terrestrial Ecoregions
of Africa and Madagascar

Terrestrial Ecoregions of Africa and Madagascar

A Conservation Assessment

Neil Burgess

Jennifer D'Amico Hales

Emma Underwood

Eric Dinerstein

David Olson

Illanga Itoua

Jan Schipper

Taylor Ricketts

Kate Newman

World Wildlife Fund
United States

ISLAND PRESS

Washington · Covelo · London

Library of Congress Cataloging-in-Publication data.

Terrestrial ecoregions of Africa and Madagascar :
a conservation assessment / Neil Burgess . . . [et al].
     p.   cm.
   Includes bibliographical references and index.
   1. Biological diversity conservation—Africa.   2. Biotic
communities—Africa.   I. Burgess, Neil D.
QH77.A35T47   2004
333.95'16'096—dc22   2004002139

Printed on recycled, acid-free paper

Design by Bea Hartman, BookMatters

Manufactured in the United States of America
10 9 8 7 6 5 4 3 2 1

This book is dedicated to the memory of **Henri Nsanjama**, one of Africa's foremost conservation leaders and the former WWF-US vice president and senior advisor for the continent. Henri was killed in a car accident in Washington, D.C. in July 2000.

Henri was born in rural Malawi, and his career was one of steadfast dedication to conservation. He rapidly became a respected member of the conservation community in Malawi and eventually more widely throughout Africa. Immediately before joining WWF in November 1990, Henri headed the Department of National Parks and Wildlife in Malawi, where he had previously served as deputy director, wildlife control officer, and park warden. During his distinguished career in conservation, Henri chaired the Standing Committee of the Convention on International Trade in Endangered Species (CITES), served as Africa region representative for CITES, chaired a United Nations Food and Agriculture Organization working party on national parks and wildlife management, and coordinated wildlife issues for the Southern African Development Community.

Throughout his 10 years with WWF, Henri was an inspirational ambassador for conservation with the American public and our partners in Africa. He was at the forefront of efforts to include women in conservation and increase their educational opportunities, and he also coauthored *Voices from Africa: Local Perspectives on Conservation.*

This dedication celebrates Henri's lifelong contribution to global conservation efforts, in particular his tireless support of fellow African conservationists and the miraculous natural world they work to save. His smile and his example of dedication are still with us. May he rest in peace.

Contents

Appendixes

List of Special Essays

List of Figures

List of Tables

List of Boxes

Acronyms

AWF	African Wildlife Foundation	ICDP	Integrated Conservation and Development Project
CABS	Center for Applied Biodiversity Science (at Conservation International)	IMF	International Monetary Fund
CAMPFIRE	Communal Areas Management Programme for Indigenous Resources Project	IUCN	The World Conservation Union
		UNDP	United Nations Development Programme
CBD	Convention on Biological Diversity	UNEP	United Nations Environment Programme
CI	Conservation International		
CITES	Convention on International Trade in Endangered Species	UNESCO	United Nations Educational, Scientific and Cultural Organization
CPD	Centers of Plant Diversity (WWF and IUCN)	USAID	United States Agency for International Development
DFID	Department for International Development	USGS	United States Geological Survey
EBA	Endemic Bird Area	WCMC	World Conservation Monitoring Centre
GEF	Global Environment Facility	WRI	World Resources Institute
GIS	geographic information system	WWF	World Wide Fund for Nature
GTZ	Deutsche Gesellschaft für Technische Zusammenarbeit	WWF-US	World Wildlife Fund (USA)

Preface

African forests and other habitats provide a wealth of natural resources to their human inhabitants who have lived in close harmony with the landscape for millions of years. Many Africans still rely on an intimate understanding of the natural world to survive. Knowing when to hunt for certain species, plant crops, gather medicinal plants, cross open deserts, or burn off the local vegetation are part of the vast oral traditions of natural history codified by African cultures. In some ethnic groups, the wealth of natural history knowledge is now concentrated among elderly members while in others, the local lore is still passed down from one generation to the next. Although there is no objective way to measure "natural history literacy" in human populations, Africans rank high among the peoples of the world in their understanding of the rhythms and elements of the natural world.

This book examines the natural history of Africa and its islands, not by country, but by ecoregions. Ecoregions can be defined as large geographic units that share a characteristic set of species, habitats, environmental conditions, and ecological dynamics. While most Africans know well the ecoregion in which they live, this book adds the scientific underpinnings to this understanding through detailed and rigorous analyses about the biological features of these areas and the threats facing them. As a biologist, I appreciate the hard work that has gone into gathering and analyzing a unique set of biological and satellite data to provide the most up-to-date portrait of conservation potential in the region. As a conservationist, I am encouraged by the ambitious efforts to undertake large-scale conservation across the continent, from the rainforests of West Africa and Central Africa, to the greater Miombo, to the unique and precious fynbos of South Africa. As an African, I am delighted in the examination of certain myths that others from outside the region maintain that stymie efforts to advance conservation and sustainable development.

In the new millennium, some of the traditional foreign aid donors have shifted emphasis from strict conservation projects to poverty alleviation efforts. The message that we must get across to decision-makers is how closely conservation and poverty alleviation in Africa are linked. One only has to look at some of the desert ecoregions of Somalia or the dry plateau of Madagascar where extensive conversion and degradation of natural habitats has created an ecological trap that makes poverty alleviation extremely costly to achieve. We must do a better job of accounting for the ecological services that natural habitats provide, the subsidies from nature that provide the basis for a sustainable future.

If we can do nothing else, this generation needs to preserve options for how future generations can prosper in greater harmony with nature. The opportunities are there. Vast areas of Africa are still largely intact and other ecoregions would recover quickly with attention paid to restoration. To leave the Earth better than we found it should be the motto of every concerned citizen of the world. To leave an Africa better than we found it, we must embrace the visionary social and political ideas of Nelson Mandela and Kwame Nkrumah. And we must also embrace the ecological visions of success for African ecoregions outlined in this book and in other efforts now under way across the continent. To do less is to deny those who will come after us the extraordinary biological wealth that sustains the human spirit and underpins our vital march to a sustainable future.

Dr. Yaa Ntiamoa-Baidu, Director,
Africa and Madagascar Programme, WWF International
Avenue du Mont Blanc, 1196 CH Gland, Switzerland

Acknowledgments

Many people and organizations provided invaluable help to produce this book. Assistance involved providing and checking data, drafting and checking texts, sending reprints, and allowing the use of completed analyses and graphics. We apologize in advance to anyone who has been accidentally omitted.

A number of organizations and individuals assisted in the compilation of species data used in this book. Foremost among them are Jon Fjeldså, Carsten Rahbek, Louis Hansen, and Steffan Galster of the Zoological Museum of the University of Copenhagen, who helped to assemble a database of species by ecoregions for amphibians, snakes, small mammals, and birds in sub-Saharan Africa. The BIOMAPS project team at the Botanical Institute, University of Bonn (Gerold Kier, Jens Mutke, Wolfgang Küper, Alexander Herold-Mergl, Ulrike Lessnow, Holger Kreft, and Wilhelm Barthlott), allowed us to use their database of vascular plant distributions in Africa, as did Jon Lovett of the University of York in the United Kingdom. Donald Broadley of the Biodiversity Foundation for Africa reviewed data for the reptiles of Africa, Madagascar, and the smaller islands. Pamela Beresford of the American Museum of Natural History, New York, reviewed the bird database. Final data compilation into an ecoregion database was done at the Conservation Science Program of WWF-US. Despite the efforts of these people and organizations, we acknowledge that errors remain in the final database used in this book and accept responsibility for them.

The World Conservation Monitoring Centre (WCMC) in Cambridge, England, kindly supplied protected area data from the United Nations List of Protected Areas (March 2002). South Africa protected area data were kindly provided by the Council of Scientific and Industrial Research (CSIR, Environmentek).

The WWF-US Conservation Science Program completed the analysis presented here. Andrew Balmford of the Conservation Biology Group, Cambridge University, and Thomas Brooks of the Center for Applied Biodiversity Science of Conservation International provided analytical advice.

We thank the World Resources Institute (WRI) for providing human population data projected to the year 2025 and for providing digital data showing the distribution of Frontier Forests in Africa. Conservation International (CI) provided a copy of the CI biodiversity hotspots map for Africa and provided data on the CI tropical wilderness areas. Jon Lovett of the University of York provided a map of priority areas for plants in Africa, and the Zoological Museum of the University of Copenhagen provided a comparable map for vertebrates. WWF and IUCN provided a map of centers of plant diversity in Africa, and BirdLife International provided a map of the endemic bird areas in Africa. Generalized rainfall, temperature, topographic, and seasonality data were extracted from the Centre for Resource and Environmental Studies (CRES) at the Australian National University (1995) (also see Hutchinson et al. 1996).

Many people contributed to the overall process of completing the book. From organizations outside WWF these were Benabid Abdelmalek, Ndiaye Abdoulaye, Gaston Achoundong, Adelewale Adeleke, Marcellin Agnagna, Ndinga Assitou, Luciano Asumu, Conrad Aveling, Mohammed Bakkar, Phoebe Barnard, Wilhelm Barthlott, Simon Bearder, Marie Béatrice, Henk Beentje, Abdelmalek Benabid,

Belinga Hyrseinte Bengono, Salima Benhouhou, Pamela Beresford, Nora Berrahmouni, Inogwabini Bila-Isia, Michael Bingham, Sonia Blaney, Markus Borner, Pouka Boumso, Henri Bourobou Bourobou, Pierre Claver Bouzanga, Rauri Bowie, Frans Breteler, Donald Broadley, Thomas Brooks, Michael Brown, Marius Burger, Paul Burgess, Rosalyn Doggett Burgess, Antje Burke, John Burlison, Tom Butyinski, Neil Caithness, Alison Cameron, Bizantino Edelmiro Castano, Anthony Cheke, Patrice Christy, Tony Clarke, Jean-Gael Collomb, Marc Colyn, Fabio Corsi, Richard Cowling, Joel Cracraft, Helen Crowley, David Cumming, Henrique Pinto Da Costa, Tim Davenport, Richard Dean, Carlo De Filipi, Gamassa Deo, Paul Desanker, Jaime De Urioste, Jean-Pierre D'Huart, Samba Diallo, Luc Dimanche, Tim Dodman, Nike Doggart, Séraphin Dondyas, Charles Doumenge, Robert Dowsett, Françoise Dowsett-Lemaire, Bob Drewes, Marc Dubernet, Jean-Marc Duplantier, Chris Duvall, Jane Dwasi, Edem Edem, Atanga Ekobo, Lyndon Estes, Richard Estes, Marc Fenn, José María Fernández-Palacios, Brian Fisher, Lincoln Fishpool, Jon Fjeldså, Roger Fotso, Alliette Frank, Ib Friis, Jean-Marc Froment, Steffan Galster, Steve Gartlan, Angus Gascoigne, Annie Gautier, Meg Gawler, Karen Goldberg, Bronwen Golder, Catherine Graham, Elise Granek, Leticia Greyling, Michael Hales, Philip Hall, Louis A. Hansen, David Happold, John Hart, Terese Hart, John Hatton, Frank Hawkins, Cornelis J. Hazevoet, Inga Hedberg, Olov Hedberg, Charlotte Heijnis, Alexander Herold-Mergl, John Hirsh, Carl Hopkins, David Hoyles, Brian Huntley, Ratsirarson Joelisoa, Véronique Joiris, André Kamdem-Toham, Vital Katembo, Illisa Kelman, Gerold Kier, Jonathan Kingdon, Donovan Kirkwood, Luhunu Kitsidikiti, Rebecca Kormos, Guy Koumba Koumba, Holger Kreft, Thomas Kristensen, Wolfgang Küper, Sally Lahm, Dayton Lambert, Olivier Langrand, Marc Languy, Annette Lanjouw, Nadine Laporte, Francis Lauginie, Aiah Lebbie, James Ledesma, Jean Lejoly, Ulrike Lessnow, Maria Jose Bethencourt Linares, Rob Little, Jon Lovett, Pete Lowry, Chris Magin, Fiona Maisels, Jean Maley, Emile Mamfoumbi-Kombila, Victor Mamonekene, John Mauremootoo, Jean-Daniel Mbega, Malembe Mbo, Jeremy J. Midgley, Mussavu Mikala, Susan Minnemeyer, Zéphirin Mogba, Joslin Moore, Marie Pauline Moussavou, Trinto Mugangu, Hans Peter Müller, Jens Mutke, Herman Mwageni, Nna Jeanne Ndong, Norbert Ngami, Paul Robinson Ngnegueu, Dominique Ngongba, Martin Nicoll, Yaa Ntiamoa-Baidu, Omer Ntougou, John Oates, Thierry Oberdorff, Emmanuel Auquo Obot, Sue O'Brien, Danyelle O'Hara, Silvio Olivieri, Mari Omland, Richard Onouviet, Gordon Orians, Richard Oslisly, Steven A. Osofsky, Pete Phillipson, Mark Pidgeon, Shirley M. Pierce, Jacques Pierre, John Pilgrim, Andrew Plumptre, Derek Pomeroy, Isabelle Porteous, Paul Posso, John Poynton, Carsten Rahbek, Nanie Ratsifandrihamanana, Tony Rebelo, Gordon McGregor Reed, Pedro Regato, Gay Reinhartz, Winnie Roberts, Alan Rodgers, Karen Ross, Mary Rowen, Roger Safford, Michael Samways, Ivan Scales, Arne Schiøtz, Uli Schlieven, Jameson Seyani, Colleen Seymour, James Shambaugh, Rob Scott-Shaw, Rob Simmons, Ousmane Sissoko, Jean Claude Soh, Robert Solem, Laurent Some, Marc Sosef, Amy Spriggs, Mark Stanley-Price, Malcolm Starkey, Lisa Steel, Frank Stennmans, Peter Stephenson, Melanie Stiassny, Bob Strom, Nicodème Tchamou, Guy Teugels, Marc Thibault, David Thomas, Duncan Thomas, Jo Thompson, Mats Thulin, Jonathan Timberlake, Norbert Toe, Sylvia Tognetti, Theodore Trefon, Jaime A. de Urioste, Rudy van der Elst, Jean-Pierre Vande Weghe, Simon Van Noort, Renaat Van Rompaey, Amy Vedder, Suzanne Vetter, Toon Vissers, Jan Vlok, Van Wallach, Peter Walsh, Ross Wanless, Jan Lodewijk R. Were, Lee White, Chris Wild, David Wilkie, Paul Williams, Philip Winter, Mary Wisz, Eddy Wymenga, Derek Yalden, and Mahmoud Zahran.

From WWF-US: Robin Abell, Tom Allnutt, Christine Burdette, John Burrows, Richard Carroll, Lauriane Cayet, Eric Dinerstein, Tim Green, Ken Kassem, John Lamoreaux, Colby Loucks, Meghan McKnight, Miranda Mockrin, John Morrison, Melissa Moye, Dawn Murphy, Kate Newman, the late Henry Nsanjama, Sheila O'Connor, Judy Oglethorpe, Sue Palminteri, Dawn Pointer, George Powell, Emily Rowan, Meseret Taye, Michele Thieme, Wes Wettengel, and Eric Wickramanayake. Holly Strand from WWF-US read the entire book before its completion.

From the wider WWF–Africa and Madagascar Program: Adelewale Adeleke, Nora Berrahmouni, Allard Blom, Brigitte Carr-Dirick, Anthony Cunningham, Holly Dublin, Mark Fenn, Steve Gartlan, Steve Goodman, André Kamdem-Toham, Olivier Langrand, Rob Little, Ian MacDonald, John Newby, Yaa Ntiamoa-Baidu, Achille Raselimanana, Pedro Regato, and Paul Siegel.

The final draft was reviewed by the regional representatives of the WWF network in the Africa and Madagascar Program: Sam Kanyamibwa (Eastern Africa), Harrison Kojwang (Southern Africa), Jean-Paul Paddack (Madagascar), Laurent Somé (Central Africa), Souleymane Zeba (West Africa), and Pedro Regato (Mediterranean Programme). Peter J. Stephenson from WWF International in Switzerland also reviewed an earlier version of the text and provided many helpful comments.

We would like to give special thanks to WWF South Africa for hosting the initial WWF African Ecoregions Workshop in Cape Town (August 1998). This workshop was the foundation for this effort, and we have tried to follow the recommendations provided by that group in this book. In addition to the many experts present in South Africa, we would like to thank the experts at the Congo Basin Ecoregion Workshop (March 2000) and all others who have worked with the WWF-US Conservation Science Program since then.

Finally, we would like to thank the African Wildlife Foundation through funds from the Directorate-General for International Cooperation (DGIS) of the Netherlands, U.S. Agency for International Development (USAID), and the WWF Endangered Spaces Program for their generous financial support.

This publication was made possible through support from the Office of Environment, Bureau for Economic Growth, Agriculture, and Trade, USAID, under the terms of Award No. LAG-A-00-99-00048-00. The opinions expressed herein are those of the authors and do not necessarily reflect the views of the U.S. Agency for International Development.

This publication was also made possible through support from the DGIS under the terms of Activity No. WW204107. The opinions expressed herein are those of the authors and do not necessarily reflect the views of the Directorate-General for International Collaboration, the Netherlands.

Introduction

When most people think of wild nature, they envision Africa. African wildlife and wildlands exert a powerful grip on the global psyche. The sources of such inspiration are numerous, from the vast migrations of large mammals across the East African savanna to the unparalleled floral diversity of South Africa to the ancient mammal faunas of Madagascar; African ecoregions harbor some of the most spectacular biological diversity on the planet.

Unfortunately, the species and habitats so beautifully captured by wildlife photographers and television programs are under siege from a wide range of threats, including expansion of agriculture, human population growth, logging, hunting, civil unrest, and intentional burning. Nearly every habitat in Africa suffers from some degree of degradation. Meanwhile, the resources of governments and international organizations are far too limited to address these threats across the continent. Thus, there is an urgent and practical need to set priorities for conservation investment.

The existing literature documenting African biodiversity is rich in descriptive accounts, but this is the first attempt to set conservation priorities at the continental scale, using data from multiple groups of animals and plants, incorporating nonspecies biological values, and including a quantitative assessment of threat. As part of a global effort to identify the areas where conservation measures are needed most urgently, WWF has assembled teams of scientists to conduct assessments for the different regions of the world. The next

seven chapters address the major targets and issues surrounding biodiversity conservation for Africa and its islands.

Chapter 1 traces the biological and physical history of the African region and summarizes the main threats to its biological values. We outline the long history of conservation planning in the region, highlight the important role played by African societies in protecting natural resources, and illustrate the development of conservation management within the framework of politically defined boundaries (countries).

Chapter 2 outlines the biological and geographic framework used in our assessment. We use divisions in the distribution of animals and plants across Africa to define ecoregions: geographic areas that share the majority of their species and ecological processes. These 119 ecoregions are used as the basis of our conservation assessment in later chapters. Ecoregions are further grouped in habitat-based biomes, such as tropical moist forests, savanna-woodlands, and deserts, allowing our prioritization efforts to represent the full variation of habitats across the continent.

Chapter 3 describes the methods used in this continental analysis of African ecoregions. First, the chapter explains our analysis of biodiversity value, where data on endemic species and species richness are analyzed within biomes and across different taxonomic groups and where we incorporate nonspecies biological values (e.g., migrations). Second, we introduce the conservation status analysis, where protected area data, habitat extent and fragmentation, species and habitat threat data, and human population data are used to identify threats and opportunities in each ecoregion. Third, we outline the methods used to integrate biological value and threat to develop an assessment of the priority for conservation investment.

Chapter 4 presents the biodiversity priorities across the ecoregions of Africa. We show how mainland Africa contains a number of important centers of endemism (where species with small geographic ranges coincide). These centers are found mainly in the forested mountains across the tropical belt of Africa but are also in the drier habitats of the Cape of South Africa and the Horn of Africa in the northeast. Effective conservation in these areas would prevent the extinction of the majority of African species. We also show how mainland Africa is globally important for its spectacular concentrations of large mammals in both the savanna-woodland and Congo rainforest regions. Although most of the world has already lost its large mammal fauna, much still remains in Africa. The areas of importance for endemic species and those important for concentrations of large mammals show little overlap, offering clear choices for agencies interested in either the conservation of endemism hotspots or the conservation of large, intact ecosystems supporting concentrations of widespread species. On the offshore islands, especially Madagascar, biological uniqueness surpasses that of the African mainland. Besides the large numbers of endemic species, many of the islands contain evolutionary lineages of fauna and flora that have long since gone extinct on the mainland. Thus, these islands provide a glimpse back at how the world might have looked millions of years ago. Some of Africa's biological treasures occur in obvious places, well known to laypeople and scientists alike, such as the Serengeti grasslands and the Congo Basin forests. But many of our ecoregions of global importance are less familiar and often are overlooked by popular conservation thinking, such as the Horn of Africa, Angola scarp, Ethiopian Highlands, East African mountains, and Zambezian flooded grasslands.

Conservation decisions cannot be based only on biological values. Chapter 5 presents the results of the conservation status analysis of Africa, examining the current and future threats for each ecoregion. We show how the most important threats to the habitats and species across Africa come from conversion of natural habitat to agriculture by an expanding human population, from the overhunting of large mammals to provide bushmeat, and from the lack of conservation resources associated with the poverty of most African nations. We also show that most African biomes re-

main underprotected by official wildlife reserves: in the best-protected biome (savanna-woodlands) approximately 5 percent of the total area is conserved. There is a clear need to increase protection by creating additional conservation areas, particularly in the forest, montane grasslands, desert, and Mediterranean scrub biomes. Finally, the extinction risk of different regions of Africa is evaluated, showing how threats of extinction are concentrated in the montane regions of tropical Africa, on the offshore islands, and in the fynbos habitats and grasslands of South Africa.

Chapter 6 integrates the results of chapters 4 and 5 to establish priorities for conservation intervention. This integration shows that thirty-two ecoregions are of global biological importance for their concentration of endemic species but face high rates of threat. The majority of these ecoregions occur in tropical forests, particularly those on mountains. However, the focus on forests is not exclusive, and priority threatened ecoregions fall in eight of the nine biomes across Africa. These ecoregions should be a focus for conservation efforts aiming to prevent species extinction in coming years. We also identify twenty-nine ecoregions of global conservation importance primarily for their large intact habitats and concentrations of large mammals. These ecoregions are found mainly in the savanna-woodland biome, but examples lie in seven of the nine African biomes.

Chapter 6 also explores the validity of common myths and assumptions about African conservation. The first myth is that biologists disagree on where the most important areas for conservation are located in Africa. Biologists have been rapidly converging on a common set of priorities in the past few years. The second conservation myth is that postcolonial African governments oppose the creation of protected areas. On the contrary, African governments have created more parks and reserves since independence than during the colonial period, and the rate of preservation remains higher in Africa than in South America or southeast Asia. We conclude that there is a strong interest in developing protected area networks across the continent, at both national and local levels. The third myth we examine is that conservation in Africa is unaffordable. We outline the costs of conservation in protected areas and show that it is affordable given the political will of African nations and the support of the international community. The fourth myth is that there is too much conflict in Africa to achieve conservation. We show that this is largely erroneous: across this huge continent there are many stable regions, and even in countries that have faced war, conservation work often has been able to continue. However, we also demonstrate that conflict can bring devastating consequences to conservation efforts in war-torn nations. Intriguingly, there is a correlation between rates of conflict and the level of poverty and quality of governance in a given country. In our fifth myth we examine the consequences of disease for conservation efforts. The people, livestock, and wildlife of Africa have long been affected by diseases such as malaria and sleeping sickness, altering the demographic profile of the continent. In recent years the rise of HIV and AIDS in Africa has begun to have a dramatic effect on the human resources of the continent, including professional conservationists, wildlife and forestry staff, and scientists. The sixth myth is that rampant poaching eventually will cause the extirpation of elephant and rhinoceros from most areas of Africa; conservation efforts have shown that this is not the case if we provide adequate protection and habitat for these species. Finally, we address a widely perpetrated myth that conservation in Africa is impossible until people and countries achieve a certain level of wealth. The continued development of a protected area network and the efforts of conservation staff to maintain areas even under conditions of war and sometimes without being paid for years provides a simple but clear message of national and personal commitment to conservation even in the worst possible conditions.

Finally in chapter 7 we outline our vision for implementing bold conservation action across African ecoregions. We show that several different governments, international conservation agencies, and local peoples have embraced the concepts of large-scale conservation and have embarked on their own programs in different parts of Africa. Methods for developing a conservation plan for a single ecoregion are outlined, using relevant examples from Africa. Such plans normally identify a suite of biologically important landscapes or sites where projects should focus. They also outline a range of overarching policy and legal issues that must be addressed at a national level, or even outside Africa, far away from the area of impact.

Many mechanisms are available for conservation action in the landscapes and sites identified as priorities. These include land purchase in countries with appropriate land tenure systems, purchase of concessions in forests for conservation instead of logging, enhanced use of hunting and community wildlife management schemes to conserve populations of large mammals and their habitats, targeted improvement of protected area networks to fill gaps in the coverage of African biodiversity, expansion of the transboundary reserve concept from southern Africa to other parts of the continent, and the use of community conservation approaches in the important lands outside officially protected areas.

Funding the implementation of an ecoregional conservation plan is also discussed. Ideas such as linking ecoregion conservation programs to major debt restructuring or to major economic assistance are illustrated, and the links between ecoregion conservation and the efforts of African nations to alleviate poverty are stressed. Trust fund mechanisms, donor funding, and ways to use the tax system to fund ecoregional conservation in the long term are also presented.

The more important conclusions of this assessment include the following:

- A quantitative confirmation of the theory of "Island Africa" as outlined in 1989 by Jonathan Kingdon. We show the importance of offshore islands and montane habitat islands for species conservation and quantify levels of endemism, extinction risk, and the degree to which their habitats are protected or threatened.

- An overlap of geographic areas chosen as biodiversity priorities by various conservation nongovernment organizations (NGOs) shows a high level of agreement over the most important areas across the continent. The level of consensus between the different schemes has been increasing in recent years.

- For mainland Africa, the most important regions for protecting endemic species and preventing mass extinction are the mountains of the Cameroon-Nigeria border region; the mountains of the border regions between Uganda, Rwanda, and Burundi and the eastern Democratic Republic of Congo; the mountain chain running from southwest Kenya through eastern Tanzania; and the lowland habitats of the Cape Floral Kingdom in South Africa.

- Almost all offshore islands around Africa are critical for the conservation of endemic species and the prevention of extinction. This is particularly true for Madagascar, the Seychelles, the Comoros, the Mascarenes, and the Canary Islands.

- The most important regions in mainland Africa for the conservation of large, intact habitats and the populations of large mammals they support are the Congo Basin forests of Central Africa; the Miombo, Mopane, and *Acacia* woodlands of eastern and southern Africa; and the large flooded grasslands of the Sudd in Sudan and in eastern and southern Africa. Although the Sahara Desert is an extensive area of intact habitat, most of its large animals, especially desert-adapted antelopes, have been removed by people over the past century.

- The development of the protected area network across Africa remains an important conservation goal. Most biomes have less than 5 percent of their area in conservation areas. The Mediterranean scrub biome of South Africa and parts of North Africa has less than

1 percent of its area under conservation management. Well-designed and targeted reserve design programs will remain an important element of conservation in Africa for the foreseeable future.

- The importance of forest reserves managed by government forest departments as critical sites for biodiversity conservation has been widely overlooked. In eastern and southern Africa, much of the montane forest, containing numerous endemic and threatened species, is found in forest reserves. Working with forest departments to safeguard these areas and to improve their formal recognition as protected areas by conservation agencies is an important goal in Africa.

- Throughout Africa there remains a close relationship between the livelihoods of people and their use of natural resources. This fundamental link has been recognized by the many agencies and governments in Africa that support community approaches to conservation in the region. The World Summit on Sustainable Development in South Africa in 2002 reinforced this link and provided a platform to move community conservation approaches forward.

- Ecoregional conservation planning processes are under way in twenty-three priority ecoregions across Africa. National or regional governments lead some of these initiatives (e.g., in South Africa), whereas in other areas conservation NGOs are working in partnership with governments. WWF has organized or participated in ecoregion conservation planning exercises in the Fynbos, the Congo Basin forests, the Albertine Rift Mountains, the East African coastal forests, the Miombo Woodlands, and the Madagascar Spiny Thickets. Conservation International is involved with conservation planning across the Upper Guinea forest ecoregions and is working closely with South

African specialists in the Succulent Karoo, Fynbos, and succulent thicket regions of South Africa. Several agencies are also working with the government to develop plans for the important ecoregions on Madagascar. Many biodiversity visions and conservation action programs have already been written, and attempts to raise the funds to implement the plans are ongoing. In the Congo Basin, for example, up to $53 million was pledged in 2002 by the U.S. government for the implementation of an ecoregional action plan by a consortium of agencies. Funding has also been raised to implement the Fynbos and Succulent Karoo conservation plans in South Africa.

To provide more coverage of some of the key issues facing conservationists in the region today, we have commissioned twenty-one essays by regional experts. These are placed strategically in the different chapters to illustrate and clarify key themes. The nine appendixes include summary data used in the analyses, specific analytical methods, and a thorough text description for each of the 119 terrestrial ecoregions of Africa.

This book is part of a series that has already produced assessments of North and South America and southeast Asia (Dinerstein et al. 1995; Ricketts et al. 1999; Abell et al. 2000; Wikramanayake et al. 2002) and will eventually encompass all ecoregions covering the terrestrial and freshwater habitats of the world. The books are produced to assist conservationists, NGOs, and governments in setting strategic priorities at a large scale for conservation efforts across continents, and particularly to facilitate decisions on where to direct scarce conservation resources.

CHAPTER 1

The Forces That Shaped Wild Africa

To much of the world, Africa is perceived as a vast continent of open savannas teeming with wild and dangerous animals. Indeed, this postcard image, typified by the Serengeti plains of northern Tanzania, is the most visible image of the continent and the one most often encountered by visitors from other countries. The East African plains are indeed spectacular, yet they account for a small part of this magnificent continent. Africa contains huge contrasts of scenery and biology: the glaciers of Mount Kenya and the snow-capped peaks of Kilimanjaro, the dense, dark, and wet forests of the Congo Basin, and the sand deserts of the Sahara and Namib are just a few examples. The wondrous diversity is further enhanced by the offshore islands: Madagascar with its multitude of unique lemurs, small mammals, and peculiar plants, and the giant tortoises and flightless birds that were once characteristic of the smaller islands such as the Seychelles and Mascarenes (WCMC 1992; Groombridge and Jenkins 2000).

Unfortunately, Africa normally draws international attention in the environmental arena, not for its spectacular biodiversity values but because of its regular environmental tragedies. Periodic droughts and floods reduce crop yields and raise the specter of famine for rural people. Hungry people are forced to strip the land or hunt for anything that can be eaten or sold. One bad year per decade can wipe out any environmental gains over the other 9 years. However, the natural resources of Africa are threatened not only by climatic factors interacting with the needs of people. Numerous other threats are found, varying in intensity and with location. Poaching of elephants and rhinoceros to feed foreign demand for their tusks and horns has decimated these species across large parts of the continent. Commercial logging has already denuded the forests of West Africa and is now moving into the Congo Basin. Rural people light millions of fires every year to facilitate farming and other life-supporting activities, but in many habitats these fires alter the species composition and reduce tree cover. Large-scale commercial agriculture is starting to remove the wild habitats of eastern Africa after already achieving this across more fertile regions of southern Africa. As the human population continues to grow, there will be an increasing need for additional farmland to produce staple foods and for land for building.

In this first chapter we take a step back to define the features that shaped wild Africa before the widespread land-use changes of the past 100–200 years. We introduce the complex set of evolutionary and ecological forces that are responsible for the distribution and concentration of Africa's biodiversity today. We then move to the modern era to review the contemporary threats to African biodiversity that will permanently alter African wildlife unless addressed in the coming years. Finally, we trace the history of conservation planning in Africa and indicate how this study complements and furthers existing conservation efforts.

The Unique Natural Heritage of Africa

To understand the biological patterns of species and habitats that make Africa unique, it is first necessary to understand the geological, climatic, anthropologic, and evolutionary forces that have shaped its landscape. We outline each of these factors and indicate sources for more detailed information.

Ancient Geology and Geomorphology

Africa is a remnant of the ancient Southern Hemisphere continent of Gondwanaland, which started to break up around 290 million years ago (Hallam 1994; Adams et al. 1996). Some of the geological formations in Africa date back more than 3,500 million years, and much of the continent sits on rocks that are more than 700 million years old. Over millions of years, these rocks have weathered to produce soils with leached nutrients and lower agricultural potential than those in the United States or Europe. Because the African mainland is derived directly from Gondwanaland, it retains some elements of an ancient flora and fauna, especially in areas where more recently evolved species have not outcompeted the Gondwanaland relicts (in parts of the Cape Fynbos flora of South Africa, for example).

The ancient African continent is huge: the land areas of Europe, the United States, China, India, Argentina, and New Zealand fit within its borders (Reader 1998). Throughout the last couple of hundred million years this large landmass has remained approximately the same size, largely isolated from other landmasses, and has very slowly drifted north (Adams et al. 1996; Bromage and Schrenk 1999). Groups of African animals and plants evolved during this vast span of time, and their living relatives make up the current flora and fauna.

Some particularly enigmatic pieces of Africa's geological puzzle are the Indian Ocean islands of Madagascar and the Seychelles. Madagascar is another fragment of Gondwanaland that may have been isolated from mainland Africa for as long as 160 million years (Rabinowitz et al. 1983; Lourenço 1996). Safe in their isolation, primitive faunal groups survived on Madagascar even when they had become extinct on the African mainland. A few mainland plants still exhibit these ancient links; for example, some species of mosses and liverworts are shared between the mountains of eastern Africa and Madagascar (Pócs 1998). The Seychelles represent a small fragment of continental crust (a microcontinent) isolated for many millions of years (Plummer and Belle 1995), which explains their remarkable biological differences from Madagascar and mainland Africa.

Africa and Madagascar were geologically stable over the 80 million years between the late Cretaceous and the middle Tertiary periods. The land eroded into a flat plateau over vast regions of the continent, and some of these plateaus remain today (e.g., the Central African Plateau). Tectonic activity began afresh about 30 million years ago, uplifting many of the current mountain ranges, rerouting rivers, and accelerating erosion. A gradual northward movement and rotation of Africa into southern Europe produced major faults in the crust of eastern and south-central Africa, manifested as the famous Rift Valley system (Adams et al. 1996; Meijer and Wortel 1999). New volcanoes formed close to these rifts, and today a chain of large volcanoes runs from Ethiopia to Tanzania (Griffiths 1993). Tectonic movements continued throughout the Tertiary period, causing the repeated inundation of coastal eastern Africa (Kent et al. 1971; Griffiths 1993) and uplifting the Atlas Mountains of northwest Africa (Meijer and Wortel 1999). Crustal movements continue, especially in North Africa and along active volcanic lines such as those of the Rift Valley and the mountains of Cameroon (Burke 2001). In all areas where mountains are uplifted or volcanoes emerge there are enhanced opportunities for new species to evolve, a process that continues today.

A Changing Climate

The climate of Africa and its islands has varied dramatically over tens of millions of years. The slow northern drift and rotation of Africa into Europe has resulted in the southern migration of the tropical belt down the face of Africa, promoting the gradual drying of North Africa (Axelrod and Raven 1978; Bromage and Schrenk 1999).

Over the past 1–2 million years, there have been dramatic swings in the African climate associated with development of ice age cycles in the earth's climate (Bonefille et al. 1990; Gasse et al. 1990; Livingstone 1990; Fredoux 1994). The alternating warm (interglacial) and cold (glacial) periods are manifested across Africa as periods of higher and lower rainfall, lasting thousands of years. Plant communities respond to these climatic changes by expanding or contracting their ranges, and their characteristic fauna changes its distribution pattern accordingly. For example, the forest cover of the Congo Basin and its margins (e.g., Uganda) expanded during wetter interglacials (Jolly et al. 1997, 1998; Maley and Brenac 1998), only to shrink again in drier glacials (Jolly et al. 1997; Hoelzmann et al. 1998; Elenga et al. 2000). During the driest periods, the forest may even have shrunk to small refuges in areas of highest rainfall. The former refuges are now thought to contain particularly high levels of endemism and species richness (Diamond and Hamilton 1980; Mayr and O'Hara 1986; White 1993). The eastern and western margins of the Congo Basin exemplify this: both ends display higher levels of richness and endemism than the intervening central Congo Basin, which is thought to have lost much of its forest cover during the peak of glacial drying (Maley 1991, 1996).

The end of the last glacial period (approximately 10,000 years ago) saw an increase in rainfall over much of Africa and the subsequent expansion of forest cover (Bennet 1990; Maley 1996). As recently as 6,000 years ago there were lakes in the Sahara, and savanna woodland habitats were found across much of what is dry desert today (Clausen and Gayler 1997; Jolly et al. 2000). At that time the Sahara region supported abundant animal life, including rhinoceros and elephants (Kingdon 1989; Le Houérou 1992). Beginning about 5,000 years ago the climate became drier and forests contracted, Saharan lakes dried up, and the desert expanded. Loss of tropical forest and concurrent invasion by savanna woodland habitats culminated around 2,500 years ago when the central part of the Congo may have been savanna woodland (Maley 1996; Maley and Brenac 1998). Starting around 2,000 years ago a wetter climatic period allowed forest to re-expand into many savanna areas, and there is evidence from West and Central Africa that forest expansion might still continue in some places (Dale 1954; Leach and Fairhead 1998; Maley and Brenac 1998). However, in most parts of Africa, intensive tree cutting and conversion of forest to agriculture counteracts any forest expansion caused by increased rainfall (FAO 1999, 2001b).

In addition to major climatic shifts over thousands of years, weather patterns across much of Africa vary on a decade-to-decade and annual basis (Nicholson 1994). Drought years are followed by flood years. Wetter periods can last for a decade or two, followed by another drought period. These fluctuations over periods of years or decades have significant effects on the distribution of vegetation and associated animal species. For example, the boundary between the Sahara Desert and the Sahel has repeatedly shifted by hundreds of kilometers over the past few decades (Erhlich and Lambin 1996; Lambin and Erhlich 1996).

Current climatic conditions and climatic variability, in combination with topographic features of the continent, have given rise to a series of terrestrial habitat islands in the African continent. As with the offshore oceanic islands, the habitat islands in mainland Africa are isolated by a surrounding sea of inhospitable habitat (Kingdon 1989). A good example is the Afromontane forests scattered in areas of higher elevation and rainfall. Each of these forest islands is isolated from similar habitats by a sea of hotter, drier land, a situation that might have persisted for millions of years. The island analogy also holds true for other vegetation communities in Africa (e.g., mangroves, hypersaline vegetation, fynbos, Afromontane moorlands); their species are adapted to the relevant local climatic and vegetation conditions and are unable to cope with significant variations. Specialist species persist in these habitat islands even if they become extinct elsewhere, again paralleling the situation on islands that are often rich in ancient relics.

A Window on Evolution

Current patterns of species richness and endemism across Africa and its islands are influenced by evolutionary events, some that occurred millions of years ago and others that are still happening today.

The presence of many distantly related animals or plants in a small area indicates that the products of millions of years of evolution have become isolated (relictualized) there, whereas swarms of closely related species indicate a recent (or current) burst of speciation. The African flora and fauna contain many ancient elements that can be interpreted as relict from past eras. The lemurs of Madagascar are relict on this island; their relatives on mainland Africa were driven to extinction more than 10 million years ago by newly evolved mammal groups. On mainland Africa there are endemic species that suggest ancient connections to Madagascar (Pócs 1998), Australia (Cowling et al. 1997b), or South America (Goldblatt 1990). The South American and Australian affinities are particularly marked among the invertebrate fauna of the Cape Floral Kingdom in South Africa (examples include freshwater crustaceans [Phreatoicidea, Paramelitidae], harvestmen [Triaenonychidae], flies [genera: *Pachybates, Trichantha,* and *Peringueyomina*], Megaloptera, Dermaptera, bugs of the tribe Cephalelini, caddis flies [order: Trichoptera], and various beetles, notably stag beetles [family: Lucanidae] of the genus *Colophon* [Stuckenberg 1962]). These species are evidence of links back to the ancient Southern Hemisphere landmass of Gondwanaland. The Cape Floral Kingdom is also a center for relict Gondwanaland plant lineages, with five endemic plant families and 160 endemic genera (Goldblatt and Manning 2000).

Among plants, the gymnosperm *Welwitschia mirabilis* of the Namib and Kaokoveld deserts of southern Africa is perhaps the most spectacular relict plant of the region, as it is the sole living member of a family of plants otherwise extinct for millions of years. Monotypic endemic families with an ancient history also occur on the offshore islands around Africa, such as on the Seychelles (Medusagynaceae) and Madagascar (Didiereaceae). Relict animal species on the African mainland include the newly discovered Udzungwa partridge (*Xenoperdix udzungwensis*) in the Udzungwa Mountains of Tanzania (Dinesen et al. 1994) and the Congo peacock (*Afropavo congensis*) in the Congo Basin forests. Both of these have their strongest affinities with groups of species found mainly in Asia. Additional examples of plant and animal species with relict distributions are outlined in Craw et al. (1999).

Genetic analyses of African species are starting to shed light on their evolutionary history. It is well known that the Cretaceous extinction episode wiped out the dinosaurs, which were replaced by mammals and birds during a rapid burst of speciation in the early Tertiary period (Feduccia 1995). Genetic analyses are also demonstrating that many African bird groups date back more than 30 million years (Fjeldså 2000; Fjeldså et al. 2000; Roy et al. 2001). Analysis of geographic patterns of genetic age in African birds shows that the species in the rainforest areas of Central Africa are mostly fairly ancient but that a greater number of more recently evolved species occupy savanna habitats away from the main forest block (Fjeldså 1994; Fjeldså and Lovett 1997). On Madagascar, recent genetic research on forest birds indicates that many evolved from a single group of ancestors that arrived on the island between 9 and 17 million years ago (Cibois et al. 2001) and speciated to fill available niches. Other research indicates that the lemurs are derived from a single group of primates that colonized Madagascar 54 million years ago (Yoder et al. 1996). Insectivorous mammals (tenrecs) colonized the island around the same time (Douady et al. 2002) and also underwent a huge adaptive radiation; the carnivores of this island also date from a single radiation of an African colonist (Yoder et al. 2003).

Africa also contains groups that have undergone a recent burst of speciation into many closely related forms, a process that in some places is continuing. Two areas of Africa stand out in particular. The first is in South Africa close to the Cape. The fynbos habitats of this region contain thirteen genera that possess more than 100 species each. One genus alone, *Erica,* has 658 closely related species (Goldblatt and Manning 2000). The majority of these species are believed to have evolved in the past million years. The situation is similar in the adjacent Succulent Karoo, with several genera containing numerous closely related species (*Ruschia,* 136 spp.; *Conophytum,* 85 spp.; *Euphorbia,* 77 spp.; *Othonna,* 61 spp.; and *Drosanthemum,* 55 spp.) (Hilton-Taylor 1996). The Afroalpine heathland habitats on the volcanic mountains of East Africa also contain many examples of rapid evolu-

tion. Plants such as *Lobelia* and *Dendrosenecio* colonized the different volcanoes and rapidly speciated in habitats that are less than a million years old, and invertebrates have followed the same trends.

The Cradle of Humanity

Humans and their hominid ancestors have been part of the African landscape for up to 5 million years (Bromage and Schrenk 1999). The evolution of the human species alongside the natural habitats and species of Africa has proceeded for longer than in any other part of the world and may be important in understanding why the megafauna has survived in Africa when it has been eliminated almost everywhere else.

Early hominids probably had little influence on the distribution of habitats and species in Africa because they were predominantly hunter-gatherers, and their populations remained small. Direct evidence of how hominids and modern humans (*Homo sapiens*) might have changed the African landscape is scarce. However, it seems inconceivable that they had no effect over millions of years, considering the discovery of fire (Phillips 1965), tools, pastoralism, and settled agriculture (often involving imported crop plants from other parts of the world) (Sutton 1990).

Over millennia, human populations slowly increased in Africa, and civilizations emerged (Connah 1987; Iliffe 1995). Two of the practices that may have altered the African landscape are the use of fire and changing of the grazing regimes using imported cattle (Hanotte et al. 2002). Heavy grazing of savanna woodlands by cattle, coupled with burning early in the growing season, gives a competitive advantage to woody vegetation. The resulting thicket is denser than the original woodland. On the other hand, burning late in the dry season gives competitive advantage to grass and results in a grassland. Furthermore, the exclusion of fire in some habitats allows the spread of thicket and forest (see Trapnell 1959; Swaine 1992), indicating that some savanna is maintained by fires set by humans.

Hominids and humans left Africa a few hundred thousand years ago and gradually colonized the rest of the world (Bromage and Schrenk 1999; Groombridge and Jenkins 2000). Malayo-Polynesians and Africans colonized Madagascar approximately 2,000 years ago. Since then this island has experienced large-scale human alteration of habitat, the conversion of huge areas of forest to savanna, and numerous species extinctions (du Puy and Moat 1996; Lourenço 1996; McPhee and Marx 1997; Mittermeier et al. 1999; Goodman and Benstead 2003).

On many of the smaller islands around Africa, human colonization is even more recent, sometimes a matter of only a few hundred years. Habitat degradation and species extinction followed closely after human colonization of the smaller Indian Ocean Islands and the introduction of predators (e.g., on the Mascarene Islands [Mourer-Chauviré et al. 1999], Comoro Islands [Sinclair and Langrand 1998], and the Seychelles [Stoddart 1984]). Extinction also followed human colonization of the islands in the Atlantic Ocean, although these were less dramatic than in the Pacific, largely because there were fewer specialized endemics.

The extinction of species seen on the offshore islands around Africa seems to have been avoided on the African mainland (Martin and Klein 1984). This is probably because of the long period of co-evolution between hominids and African species, as opposed to the sudden arrival of people bringing predators, disease, invasive species, slash-and-burn agriculture, and hunting to the offshore islands. However, it is also possible that there was a gradual loss of species from the African mainland, but over a period of a few million years it is hard to distinguish human-induced extinction from that caused by natural events.

Centuries of Biological Study

Traditional African cultures have a deep knowledge of their natural environment and the species living in it, sometimes accumulated over tens of thousands of years, as among the San Bushmen of southern Africa. Traditional naturalists are able to recognize plants and animals and know the location of a species in a particular season, its breeding and dietary habits, whether it is edible, and whether it poses any threat. Traditional herbalists use this knowledge and have developed a healthcare system based on medicinal plants (Hedberg et al. 1982, 1983a, 1983b; Akerele et al. 1991; Iwu 1993; Schlage et al. 1999). Herbal healers continue to play an important part in African culture, and in rural areas they still supply more than 80 percent of medicines (Grifo and Rosenthal 1997). Unfortunately, knowledge of medicinal plants and other aspects of the natural environment was never written down in traditional societies, so human memory and lifespan limit its scope (Akerele et al. 1991; Marshall 1998).

European colonization of Africa introduced the Linnaean system of classifying plants and animals and documenting their distribution in a written format. All of the colonial powers (i.e., the British, French, Belgian, German, Italian, Spanish, and Portuguese colonies) in the newly created African countries in the late 1800s sent naturalists to collect, describe, and publish information on their species. Nearly all specimens of flora and fauna were sent back to museums in their respective European colonial countries. Through this process a vast collection of specimens, and consequently vast information resources on the natural history of Africa, were amassed outside the continent. After the end of colonialism African nations continued to collect material on their flora and fauna, often in partnership with the former colonial powers. In some countries a detailed understanding of the distribution of plants and animals has been compiled in Africa, but in many other countries the bulk of expertise and information remains in the hands of the former colonial power and the relevant research institutions.

The long history of scientific research in Africa has resulted in a significant amount of natural history information that perhaps surpasses that of South America or southeast Asia. However, a number of African regions are virtually unknown biologically because of poor access and long-term insecurity. These include much of the Congo Basin, northern Mozambique, most of Angola, and parts of the Horn of Africa. These areas are focal regions for additional scientific research and can be expected to yield new species of animals and plants.

Although Africa has a diverse flora and fauna, the available data indicate that it is less rich than South America or southeast Asia (Richards 1973). There are a number of reasons why Africa may have fewer species than these two other tropical continental areas. One is that Africa contains large areas of very dry habitat, particularly the huge Sahara Desert, and desert biomes typically are poorer in species than moister areas. Africa was also particularly affected by the desiccation events of the last ice ages, possibly causing the extinction of numerous species. Another explanation is that over the past few million years tropical Asia and South America have been much more dynamic geologically than Africa. The Andean mountain chain has risen over the past 10 million years in South America, and in Asia the complex movements of small pieces of crust have formed new islands and mountain ranges, forcing contact between different floras and faunas. These major tectonic events stimulate the evolution of new species. In similar areas of Africa—the young mountains of East Africa, the newly formed winter rainfall regime of South Africa, and young fold mountains such as the Atlas of North Africa—there are also newly evolved species. Africa may simply have fewer drivers of speciation than other tropical regions, and as a consequence much of the remaining flora and fauna is old, with some spectacular relict elements, especially on the offshore islands.

Challenges to Conserving Africa

Increasing Threats to Species

INCREASING EXTINCTION RISK

Lists of species threatened with extinction provide a yardstick against which the status of the world's species can be measured (Hilton-Taylor 2000). For vertebrates this number has increased globally from 2,580 in 1994 to 2,779 in 2002, an increase of 8 percent (www.redlist.org). Several studies indicate that we are in, or will soon enter, a major species extinction event (see Pimm et al. 1995; Pimm and Brooks 1999; Pimm and Raven 2000). Current extinction rates are estimated to be proceeding at 1,000 times the natural rate and could accelerate if remnant habitat patches supporting many unique species are removed (BirdLife International 2000; Groombridge and Jenkins 2000; Hilton-Taylor 2000).

Africa has already seen species become extinct; for example, the islands of Madagascar, Mascarenes, and Seychelles all lost species soon after the first arrival of humans. The risk of further extinction is heightened across the region as human population concentrates close to the areas of highest biodiversity value, perhaps selecting them for the same combinations of climate and soil (Balmford et al. 2001a, 2001b; Moore et al. 2002).

LOSS OF SPECIES RANGE

Many African species, even if not yet threatened by extinction, are also suffering from huge contractions in their range. Large savanna mammals have been lost from extensive areas of Africa (Stuart et al. 1990; East 1999). Starting from the European colonization of South Africa they have continued to disappear as human populations have grown across the continent and the demand for natural resources has escalated. In West Africa, for example, once-common savanna mammals such as giraffe, elephant, black rhinoceros, lion, and African wild dog are confined to a few (or even a single) protected areas. Although such catastrophic contractions of range have not yet occurred in most of eastern and southern Africa, there is significant competition for land between wildlife and humans in many areas.

BUSHMEAT HUNTING

A century ago there was no significant bushmeat trade in Africa. People hunted and consumed animals locally. Today, tens of millions of Africans living in rural areas rely on bushmeat for more than 80 percent of their protein (Robinson and Bennett 2000a, 2000b). A thriving trade in bushmeat from rural areas to cities has also developed, particularly in West and Central Africa (Bowen-Jones and Pendry 1999; Robinson et al. 1999; Bakarr et al. 2001b) but also increasingly in eastern and southern Africa (Barnett 2000). The commercialized bushmeat trade in West and Central Africa often is associated with the logging industry, which provides facilities for hunters and a transport system to the towns (Wilkie et al. 1992).

Bushmeat hunting has been an important conservation issue in West Africa (especially for primates) for a number of years (Oates 1986). Recently, however, it has assumed crisis proportions (Bushmeat Crisis Task Force 2001). It threatens the survival of vulnerable species throughout West and Central Africa, and as edible species are hunted, an empty forest is left behind (Redford 1992; Robinson and Bennett 2000b; Bennett and Robinson 2001). African forest species are particularly vulnerable to overhunting because the forests support a low animal biomass, estimated around 2,500 kg of meat per square kilometer, which can sustainably support about 1 person per square kilometer. In contrast, animal biomass is at least ten times higher in the savanna woodlands regions of eastern and southern Africa, so these areas can supply meat to far more people (Robinson and Bennett 2000b). Robinson and Bennett (essay 1.1) elaborate on the bushmeat problem in Africa.

INVASIVE SPECIES

Across the world it has been estimated that the mean percentage of alien plants is around 12 percent of a given flora, and on islands it is closer to 28 percent (Vitousek et al. 1997a). Invasive alien plants have pushed native flora and fauna close to extinction on the endemic-rich offshore islands of the Mascarenes, Seychelles, Cape Verde, and even Madagascar (Cronk and Fuller 2002). The fynbos habitats of South Africa are also greatly threatened by a variety of invasive plants (MacDonald 1994; Higgins et al. 1999). Introduced animals can also have disastrous effects on vulnerable island fauna and flora, especially if these islands were previously predator-free. The introduction of rats and cats onto small offshore islands has caused the extinction of island birds, reptiles, and invertebrates over the past century and dramatically reduced populations of many other species (WCMC 1992).

Increasing Threats to Habitats

HABITAT LOSS

Large areas of habitat have already been converted for human use in Africa, although the continent still contains huge areas of natural or near-natural vegetation. Current economic patterns indicate that most of the next generation of Africans will continue to live subsistence lifestyles, farming for food and deriving their fuelwood, meat, medicines, and building materials from natural resources. Given population doubling within 25 years, Africa is predicted to undergo tremendous agricultural expansion. An additional 30 percent of the remaining forests and natural woodlands are set to disappear in the next 30 years (Tilman et al. 2001).

Moist tropical forest is one of the major biomes in Africa. Across the globe it is estimated that tropical forests have declined in area by 50 percent over the past 2,000 years (Groombridge and Jenkins 2000; FAO 1999, 2001b; WRI 2001). This global habitat loss is continuing and may even be accelerating (Vitousek et al. 1997b; Cincotta and Engleman 2000; Woodroffe 2000). In Africa, the current loss of tropical forest is around 0.85 million hectares annually (around 0.43 percent per annum), or about half the rate of loss found in Asia (Achard et al. 2002). Some researchers dispute these rates of decline in Africa (e.g., Fairhead and Leach 1998a, 1998b; Lomborg 2001), citing the extremely dynamic nature of African forest and the fact that in some areas forest cover has expanded in the past century (Menaut et al. 1995; Maley 1996, 2001). Although there is evidence of historical forest expansion and recolonization in some places (Davies 1987), the current trend across the bulk of Africa is for rapid forest loss to agriculture and logging (FAO 1999, 2001b).

HABITAT FRAGMENTATION AND DISTURBANCE

As habitats are fragmented and levels of disturbance increase, some elements of biodiversity are lost. Human disturbance to African forests may reduce the number of habitat specialists they can support. For example, in the montane forests of East Africa, disturbed forests support lower numbers of the narrowly endemic species that characterize these habitats (Fjeldså 1999; Lovett et al. 2001; Newmark 2002). In other habitats with a high diversity of endemic plants, such as fynbos in South Africa, disturbance in the form of fire is needed to maintain diversity. But there is a threshold of burning that, if exceeded, results in declining diversity and the loss of specialized species (Cowling and Richardson 1995).

UNCONTROLLED LOGGING

One of the primary causes of forest loss in Africa has been logging. The logging industry has been extracting high-value timber from tropical forests and woodlands since the 19th century (Forests Monitor 2001). In the colonial era, African forests were regarded mainly as a source of income from logging or a source of new land for commercial farms. As a result, large areas of forest were destroyed, and the forests of southern, eastern, and northern Africa were stripped of their timber resources (see Poore 1989; Sayer et al. 1992; Whitmore and Sayer 1992). Where timber resources remained in West and Central Africa, commercial logging operations have continued to operate until today and have the potential to destroy the last African forest wildernesses (Bryant et al. 1997; Forests Monitor 2001; Minnemeyer 2002).

In countries where the forest has been heavily depleted or is inaccessible (e.g., East Africa), mechanized commercial logging has been replaced by informally organized pitsawing. Small mobile teams cut valuable timber trees and carry the sawn timber from the forest on their heads. There are almost no statistics on the scale of these operations, but in eastern Africa they cover almost all forests, even those more than 10 km from any road. As with mechanized logging, pitsawing opens up the forest and facilitates hunting and colonization for agriculture.

INCREASING RATES OF FIRE

Fossil charcoal provides evidence that fires have occurred naturally across the world for hundreds of millions of years (Stewart 1983), started by lightning, rotting vegetation, or volcanic eruptions. Many African habitats have evolved to survive regular burning (see White 1983; Frost and Robertson 1987; Desanker et al. 1997), and some (e.g., the fynbos of South Africa) are ecologically dependent on fire (Cowling and Richardson 1995).

Humans have been setting light to the African savanna for at least the past 1 million years (Bird and Cali 1998), particularly in eastern and southern Africa, which have the longest proven history of human habitation (Groombridge and Jenkins 2000). People traditionally burned African savanna to drive game animals to places where they could be easily killed, to stimulate the new growth of grasses for domestic grazing animals, and to clear an area to facilitate farming (Hall 1984). No other part of the world has this ancient history of enhanced burning, which may have caused significant vegetation changes over hundreds of thousands, or perhaps even millions, of years (Phillips 1965; Thompson 1975). Today, millions of fires are lit every year across Africa. Observation and experimental work indicate that the high rates of fire maintain some types of savanna-woodland in areas that would change to dense woodland or even to forest if fire were excluded (Swynnerton 1917; Trapnell 1959; Swaine 1992; Swaine et al. 1992). Too frequent burning also changes the vegetation composition in many other habitats (White 1983).

OVERHARVESTING OF FUELWOOD

Throughout much of Africa the fuel source for cooking and heating is firewood collected from natural vegetation close to home (Leach and Mearns 1988). This is often converted to charcoal, which is easier, lighter, and more convenient to transport. More than 75 percent of fuel in most African countries comes from wood; in rural areas this is closer to 100 percent (FAO 2001b; WRI 2001). These practices are sustainable in areas where population densities are low and where trees grow quickly. However, this is not the case in many areas, especially close to larger towns, in agricultural areas with high population pressure, and in dry areas where trees grow slowly. In these areas, cutting trees and bushes for fuel seriously degrades the landscape.

The introduction of exotic trees, such as species of *Eucalyptus* and *Pinus,* is one way to meet the demand for fuelwood in agricultural areas. In agricultural areas tree cover may increase above a certain human population density (Tiffen et al. 1994). However, the trees are mainly planted exotic species that are useful for building, for firewood, and for sale but support extremely low biodiversity and do not include any rare or threatened species (Pomeroy and Dranzoa 1998).

OVERGRAZING OF SENSITIVE HABITATS

In the past wild large mammal species occurred across much of Africa, and their grazing and browsing helped create the current vegetation. For example, when elephant populations achieve high densities they reduce the tree cover and convert areas to grassland (Laws 1970; Ben-Shahar 1993; Cumming et al. 1997).

In many drier parts of Africa human pastoralists and their cattle and goats have replaced the wild animals, and these in turn influence the vegetation. Some believe that pastoralists overstock and overgraze the land, bringing about its degradation in the long term (e.g., Pratt and Gwynne 1977). Hardin's theory of the tragedy of the commons (Hardin 1968) predicts a personal profit motive driving each individual to pasture more (privately owned) livestock on finite areas of (communally owned) rangeland. However, more recent studies indicate that pastoralism is causing habitat change rather than decay (Hesse and Trench 2000). African pastoralism is also regarded as much better for conservation than conversion to agriculture or urban areas. The Masai and other traditional pastoralists in Africa often help preserve biodiversity in their rangelands (Anderson and Grove 1987; Homewood and Rodgers 1991; Galaty and Bonte 1992; Homewood 1994; Brussard et al. 1994; Homewood and Brockington 1999).

Root Causes of Species and Habitat Loss in Africa

We have outlined some of the immediate threats to species and habitats in Africa. However, there are some deeper-seated causes of these threats, often called root cause threats. Those most affecting conservation in Africa are outlined in brief in this section.

INCREASING HUMAN POPULATIONS

For millions of years the human population of Africa was low and increased slowly (Reader 1998; AAAS 2000). Early hunter-gatherer peoples had comparatively little impact on habitats and species. However, the discovery of fire and later the development of agriculture facilitated the first large-scale transformations of the landscape. Initially agriculture in Africa used local species, but the importation of crop plant species from other parts of the world provided the food security that allowed human populations to expand (Sutton 1990). Many regions were densely settled when Europeans first colonized Africa (Pakenham 1991). African populations declined in the early colonial period (the early-middle nineteenth century to the early twentieth century), mainly because of the introduction of diseases but also as a consequence of wars and slave trading (Kjekshus 1977; Davies 1979; Sutton 1990; Pakenham 1991). In some parts of Africa formerly farmed areas reverted to bush, and wild animal populations rebounded.

Since the early twentieth century human populations have grown rapidly across Africa. Between 1975 and 1999 growth rates were 2.8 percent per annum across sub-Saharan Africa, the highest regional rate in the world (UNDP 2001). African population growth rates have fallen to 2.4 percent per annum, but this is still the highest regional rate in the world and gives a population doubling time of around 25 years. By 2015 there will be at least 866 million people in sub-Saharan Africa, even taking into consideration the effects of HIV/AIDS (UNDP 2001). There is little sign that these high population growth rates will slow appreciably in the coming decades, so demands for natural products and for farmland undoubtedly will increase dramatically.

CONTINUING POVERTY

Sub-Saharan Africa is consistently portrayed as the least developed and most impoverished area of the world, with the poorest countries located across the tropical belt (UNDP 2001). Across Africa, the growth rate of the per capita gross domestic product (GDP) was negative (−1.0 percent) between 1975 and 1999 and slightly less negative from 1990 to 1999 (−0.4 percent). This is the only continent with such a long-term negative GDP. The Structural Adjustment Programmes sponsored by the World Bank and International Monetary Fund have failed to improve the economic status of most countries and may have increased poverty in some countries (Ross 1996; Kaufmann et al. 1999; Wood et al. 2000a). The levels of debt in sub-Saharan African nations have also soared, from US$6,000 million in 1970 to US$330,000 million in 1999, falling slightly since then. Much of the national budget is spent on servicing these debts, leaving little money available for conservation or any other activity. In some African countries there have been signs of economic recovery in the past few years (UNDP 2001) and some attempts to reduce the debt burdens of African nations (GEO3 2002). It is hoped that these trends will continue and allow conservation to be better funded across a broader range of African countries.

DEBILITATING DISEASES

Africa has many serious diseases of humans and animals, which have been linked to widespread poverty (Snow et al. 1999; Sachs and Malaney 2002). For example, the most malaria-infested countries in the world are in sub-Saharan Africa (Sachs and Malaney 2002; Rogers et al. 2002). Large areas of Africa also remain affected by sleeping sickness (carried by the tsetse fly) (Rogers 1990, 2000), which causes gradual debilitation and eventual death if not treated. It has rendered regions of the continent uninhabitable. More recently there has also been a rapid spread of HIV/AIDS throughout the continent. This combination of life-threatening disease means that people are often working below their optimum, and with reduced life expectancy trained people often die before achieving their full potential in terms of national development or conservation achievements.

Climate Change

Much has been written about current and predicted global climatic change (e.g., IPCC 2001). In Africa, it is predicted that by 2050 the Sahara and semiarid parts of southern Africa may be as much as 1.6°C warmer, and equatorial countries (Cameroon, Uganda, and Kenya) might be about 1.4°C warmer (IPCC 2001). Rainfall is predicted to decline in already arid areas and increase in the tropical parts of the continent. In South Africa, modeling of the effect of climate change indicates significant shifts in biome distribution, with potentially disastrous consequences for the several thousand endemic plants of the Cape Floral Kingdom (Hannah et al. 2002). If similar intense climate changes and habitat shifts occur elsewhere in Africa, there may be a signifi-cant loss of biodiversity regardless of efforts made to protect habitats and species in reserves or parks.

Conservation Planning in Africa

Traditional Methods of Conservation Planning

Conservation planning is not new in the African context. For thousands of years the peoples of Africa developed methods of conserving resources in the African landscape. In fact, resource use planning was an everyday part of life; rules and regulations were developed to enforce controls. For example, hunting and fishing regulations that designated closed seasons when hunting and fishing were banned allowed target species to maintain their population levels and provide a sustainable source of food for people. Other forms of resource management included regulations against the killing of pregnant females. Habitat around water sources was protected to prevent water supplies from drying up, steep areas were left unfarmed to prevent landslides, strips of natural vegetation were retained to provide a ready source of natural materials close to the village, and medicinal plants were retained when farmland was being cleared (Campbell and Luckert 2002; Cunningham 2002; Laird 2002).

Planning activities were undertaken at the village or tribal level and generally revolved around providing food and environmental services for the village and its people, but environmental planning was also undertaken for traditional religious reasons. For example, the white-necked picathartes (*Picathartes gymnocephalus*) is a bizarre-looking bird that breeds on overhanging rocks in the Upper Guinea rainforest of West Africa. This species is sacred to the Mende people of Sierra Leone and Liberia, so they protect not only the bird but also its nesting rocks and some of the surrounding forest habitat. Throughout Africa similar rules exist to prohibit the hunting of different kinds of species and to actively protect the places where they live.

Protected areas were also a feature of African cultures long before the European colonialists arrived at the end of the nineteenth century and started to develop their own systems of government. Burial groves, sacred mountains, lakes, forests, and trees for the ancestors to inhabit were commonly recognized by different African communities. Protection was strict; people daring to break taboos faced severe penalties. Every chief and village had protected areas. Because information on these areas was not written down, formal statistics are only now becoming available (e.g., for Ghana: Hawthorne and Abu-Juam 1995; Kenya: Spear 1978; Tanzania: Hawthorne 1993a; Mwikomeke et al. 1998). Millions of small traditionally protected areas are scattered throughout village lands across Africa. Undoubtedly, they have made an important contribution to biodiversity conservation.

In many places these traditional systems declined throughout the colonial period, eroded by centralized government power, the universal adoption of schools for education, and the promotion of global religions such as Islam and Christianity. More recently a general process of westernization and globalization has further eroded these traditional beliefs.

Conservation Planning by African Nations

The nations and governments of Africa were established by colonial powers during the "scramble for Africa" in the nineteenth century (Pakenham 1991). Colonial governments set about establishing their own systems of planning and management of natural resources, typically within four major divisions: wildlife, forestry, fisheries, and agriculture. Wildlife departments were concerned with protecting large tracts of "wilderness" to maintain populations of large mammals for hunting and for recreational purposes. The focus was on savanna woodland and grassland species—especially elephant, black and white rhinoceros, buffalo, and lion—with much less attention paid to biological diversity in general. Forestry and fishery departments aimed to supply materials to the colonial government in Africa and the colonial power back in Europe. Forest reserves were established, but they were mainly for resource exploitation or water catchment conservation; the biological diversity of these systems was not considered (Rodgers 1993). Agriculture departments aimed to open up the bush for agricultural use and often undertook wildlife control programs in which millions of wild animals were killed.

After independence in the 1960s and 1970s the majority of African nations continued the wildlife, forestry, fishery, and agriculture divisions established by colonial administrations. The general rationale for these divisions remained similar to that during the colonial period but aimed to facilitate national development (Anderson and Groves 1987). Over the past 30 years these departments have become skilled at undertaking conservation planning work in their own countries, and some, such as South Africa, are now world leaders in this discipline (Cowling et al. 2003; Younge and Fowkes 2003). In other African nations where most information on the species and habitats was exported to colonial powers, efforts have been made over the past decade to repatriate knowledge to Africa.

A number of intergovernment processes and conventions have been put in place in the past 20 years to help governments manage their biological heritage. The Convention on Biological Diversity (CBD), established in 1992, is the most important of these. A number of publications were prepared to help governments implement this convention (McNeeley et al. 1990; Stuart et al. 1990; WCMC 1992; Reid et al. 1993). The focus of the convention's work has been at the level of the national government. For example, the CBD requires countries to produce and then implement their own Biodiversity Strategy and Action Plans. Organizations that assist countries in implementing the CBD (e.g., the United Nations Environment Programme, the World Conservation Union, and the Global Environment Facility) have largely continued to focus on countries, as have bilateral donor agencies (e.g., the U.S. Agency for International Development). Most African countries have ratified the CBD, and together with agencies supporting its implementation, this convention has achieved much for conservation in Africa over the past decade.

Moving from National Planning to Planning within Biological Regions

Although country-level planning is logistically convenient, countries do not necessarily coincide with the natural distribution of species and communities. This means that country-level planning alone risks missing the most important areas for biodiversity conservation if they happen to be shared by two or more nations. Country-level planning also does not permit overall analyses of the distribution of biodiversity in the framework of natural ecological divisions of the world, so it is hard to recognize the regions of global conservation priority. Increasingly over the past 10 years biodiversity values have been cataloged in terms of natural biogeographic units of land, independent of national borders. The earliest of these efforts stemmed from attempts to understand the biogeographic units of Africa and its islands, such as the classic work of Frank White on patterns of plant endemism and plant communities across Africa and its offshore islands (White 1983).

Scientific efforts to analyze natural areas of particular importance for biodiversity conservation also intensified before the CBD in 1992. For example, Norman Myers published an assessment of the location of global hotspots of plant endemism (Myers 1988, 1990), and the International Council for Bird Preservation (now BirdLife International) published a comprehensive assessment of endemic bird areas of the world (ICBP 1992). These publications sparked further efforts to describe biodiversity priorities in Africa (e.g.,

WWF and IUCN 1994; Mittermeier et al. 1998, 1999; Olson and Dinerstein 1998, 2002; Stattersfield et al. 1998; Brooks et al. 2001a). These efforts spawned vigorous debate on the kinds of biodiversity data that should be gathered, how analyses should be conducted, the appropriate scale of analyses, the best taxonomic groups for illustrating biodiversity priority, and many other topics (Freitag et al. 1996; van Jaarsveld et al. 1998b; Williams 1998). Hundreds of scientific papers relevant to establishing biodiversity priorities have been published in the last few years (many for Africa reviewed in Brooks et al. 2001a; Burgess et al. 2002b).

The critical question of how to translate the results to conservation on the ground continues to receive less attention, however. Herein lies the challenge for all conservationists and academic scientists fascinated by Africa's unique biological features. In the coming chapters we outline our own set of conservation priorities in Africa and the major mechanisms that we believe can promote conservation in the coming years. The WWF network and others in Africa are already tackling many of these issues, and this book provides a link between conservation analysis and conservation practice.

ESSAY 1.1

A Monkey in the Hand Is Worth Two in the Bush? The Commercial Wildlife Trade in Africa

John G. Robinson and Elizabeth L. Bennett

Until recently, we could comfort ourselves that the very inaccessibility of the forests of Central Africa would protect their flora and fauna and that establishing representative protected areas throughout the region would institutionalize that inaccessibility. This would compensate for the deficiencies in conservation efforts and reserve management (Oates 1999) and protect Africa's rainforest species. We can no longer delude ourselves that this will occur. Social and economic turmoil has swept the region (Fimbel and Fimbel 1997; Hart and Hart 1997; Draulans and Van Krunkelsven 2002), and forests have rapidly been opened up across the continent in the rush to extract resources (Robinson et al. 1999). Across the landscape, both inside and outside parks and reserves, people are harvesting wild species at ever-increasing rates. A voracious appetite for almost anything that is large enough to be eaten, potent enough to be turned into medicine, and lucrative enough to be sold is stripping wildlife from wild areas, leaving empty forest shells and an unnatural quiet.

Ever since they first inhabited Africa's forests some 40,000 years ago, humans have hunted wild animals. Wild meat has been the major source of animal protein for forest people throughout the region and continues to be today. For example, subsistence hunters in Kenya's Arabuko-Sokoke reserve (372 km^2) harvested 130,000 kg of wild meat per year (FitzGibbon et al. 1996, 2000), and in the 1,250-km^2 Korup National Park, west-ern Cameroon, 271,000 kg was harvested annually (Infield 1988). In the past, the hunted species must have been able to withstand such pressure, but wildlife populations are now being depleted. What has changed? First, Africa's rural population has grown and expanded, causing fragmentation of the forest and increased hunting pressures on its wildlife. In addition to the steady loss of natural forest has been a dramatic opening up of the remaining forest, especially by commercial timber operations. With access into and out of the forest, forest-dwelling people have increasingly entered the cash economy, and immigrants have flooded in. Hunting patterns have changed concomitantly. Local demand for animal protein increased, traditional hunting practices and taboos were lost, efficient modern technologies such as shotguns and wire snares were increasingly used, and, perhaps most of all, hunting was increasingly commercialized.

The distinction between subsistence and commercial hunting is rarely clear for forest-dwelling people. Selling of wildlife is an easy source of cash for rural people, with the money being used not only for essential and desired commodities but also for technologies to increase the efficiency of the hunt itself, thereby further increasing harvest rates. At the consumer end, the availability of new wildlife products creates and stimulates new markets, and the increased consumer spending power fosters the demand for wildlife meat and other products such as pets and traditional animal-based medicines.

Large-scale, capitalized logging enterprises contribute to the commercialization of the wildlife trade by introducing roads into remote forest areas, allowing hunters in and meat to be transported out for sale in towns (Robinson et al. 1999). It causes the hunting patterns of local communities to change. In northern Congo, for example, communities near logging roads increase their hunting when logging arrives because they can sell wild meat; in two such villages, per capita harvest rates were 3.2 and 6.5 times higher than in communities far from roads, and up to 75 percent of the meat by weight was sold (Auzel and Wilkie 2000). Moreover, the logging workers hunt for their own consumption: a single logging camp of 648 people in northern Congo harvested in 1 year some 8,251 individual mammals, with a total biomass of 124 tons (Auzel and Wilkie 2000).

The scale of today's wildlife harvest from African forests is difficult to assess accurately, but it is vast. For example, 12,974 mammal carcasses were sold in the course of a year in the market at Malabo, the largest town on Bioko Island, West Africa, equivalent to 112,000 kg of dressed meat (Juste et al. 1995). And it is not just ungulates that are sold: in Bioko, seven species of primates and two rodents also were sold. Elsewhere in Africa, the trade concentrates on duikers and monkeys but also includes a wide range of species, from elephants and great apes, through carnivores, aardvarks, and crocodiles down to flying foxes, elephant shrews, porcupines, and cane rats.

In Central Africa, rural people consume less than 0.15 kg of wild meat per person per day, and urban people consume perhaps a tenth as much (Wilkie and Carpenter 1998), although this might be a conservative estimate. These figures indicate that total wild meat consumption in the countries of the Congo Basin probably exceeds 1 million metric tons (Wilkie and Carpenter 1998). Harvests of this magnitude are not sustainable. If 65 percent of live weight is edible meat, and a million metric tons of wild meat is being consumed each year, then some 830 kg of animals are being harvested annually from each square kilometer of Central African forest. The maximum sustainable production of wild animals from these forests does not exceed 150 kg/km^2 (Robinson and Bennett 2000a), so wildlife is being strip-mined out of the forest. Hunting has reduced duiker populations in Central Africa by about 43 percent (Hart 2000; Noss 2000); in Bioko primates have been reduced by 90 percent in some areas and to local extinction in others (Fa 2000); and in Kenya, squirrels have been reduced by 66 percent and large ungulates to such low levels that it is no longer worthwhile hunting them (FitzGibbon et al. 2000). The eclectic nature of the hunt means that populations of a wide range of mammal, bird, and reptile species are affected. Sustained heavy hunting is extirpating vulnerable species, especially large-bodied species with low intrinsic rates of natural increase. As the species preferred in trade are extirpated from local areas, hunters shift to less preferred species.

The extirpation of hunted species has wider implications for the forests themselves. The species preferred by hunters generally are large-bodied, typically fruit eaters and herbivorous browsers. These species often play keystone roles in forest ecology as pollinators, seed dispersers, and seed predators and make up the majority of the vertebrate biomass. Their reduction or extirpation produces cascading effects through the biological community. The forests of Africa are becoming increasingly devoid of wildlife, with the concomitant breakdown of ecological processes, a phenomenon known as the empty forest syndrome (Redford 1992).

If the wildlife of Africa's forests is to be conserved, the trade in wildlife must be managed. Management must include targeted trade bans, with proper legal, administrative, and logistical mechanisms for enforcing them, and education programs at all levels of society. Timber companies and forest concessionaires have an important role to play in these efforts because they are often the only institutional presence in many rural areas, and it is their workers who hunt and their trucks that transport hunters in and meat out (Robinson et al. 1999). Some success is being achieved with individual companies, although such efforts are few and far between. Local forest people are doing much of the hunting and selling of wildlife and often have few alternative commodities that they can sell. They are also the first to suffer when the wild species on which they depend for food and cash no longer exist, so animal protein substitutes and other economic activities must be provided for rural peoples. As long as wildlife is seen as an open-access commodity to be harvested and traded, however, species populations will continue to wink out, diversity will continue to erode, and Africa's forests will become increasingly silent.

CHAPTER 2

Beyond Political Boundaries: A Conservation Map of Africa

More than fifty countries can be distinguished on the familiar mapped representation of Africa (figure 2.1). Not as common in atlases, ecological maps of Africa are strikingly different from maps showing only countries. Africa contains a plethora of distinctive habitats supporting unique assemblages of species, which are distributed according to their preferred ecological conditions. For example, the tropical heart of Africa is cloaked in a vast wet rainforest, with outliers extending along the West African coast and to the mountains and coasts of eastern Africa. Rainforests are found in thirty-six African countries but show little correspondence to national boundaries. Drier areas support huge expanses of savanna-woodlands with seasonal rainfall: green and lush during the wet season but often burnt and blackened during the dry season. Spreading in a vast arc around the central Congo Basin forests, savanna-woodlands also show little correspondence to national boundaries. Elsewhere, deserts, Afroalpine moorlands, mangroves, swamp wetlands, and Mediterranean scrub are found where conditions allow.

A map of biological units of land is much more than an academic issue; it is of fundamental importance to conservation planning and to understanding the types of conservation intervention that will work best in a given area. Some regional biological histories, such as those of small offshore islands, have produced a flora and fauna that are adapted to small patches of habitat and are highly intolerant of invasive competitors or other forms of disturbance. Other histories, such as those of the fire-climax savanna-woodlands of mainland Africa, are much more tolerant of different kinds of disturbance, and their species are adapted to dynamic habitat patches across a huge scale. We tried to consider these differences as we developed our framework for the biological divisions of Africa to use in creating a strategy for conservation investment across the continent.

In this chapter we focus a new lens on the African continent and create a map of the natural ecoregions of Africa. This map draws heavily on previous research that has classified Africa into ecological units. We then outline two broad classification systems for the natural world, within which we group the natural regions of Africa. The frameworks established in this chapter are used throughout the book to define and set conservation priorities and to interpret the conservation needs within the priorities.

Ecoregions

Ecoregions are large units of land or water that contain a distinct assemblage of species, habitats, and processes, and whose boundaries attempt to depict the original extent of natural communities before major land-use change (Dinerstein et al. 1995; Olson et al. 2001). We have defined ecoregions to span all of mainland Africa and its various offshore islands (figure 2.2). Although our ecoregions are new, they follow an ancient tradition of classifying natural

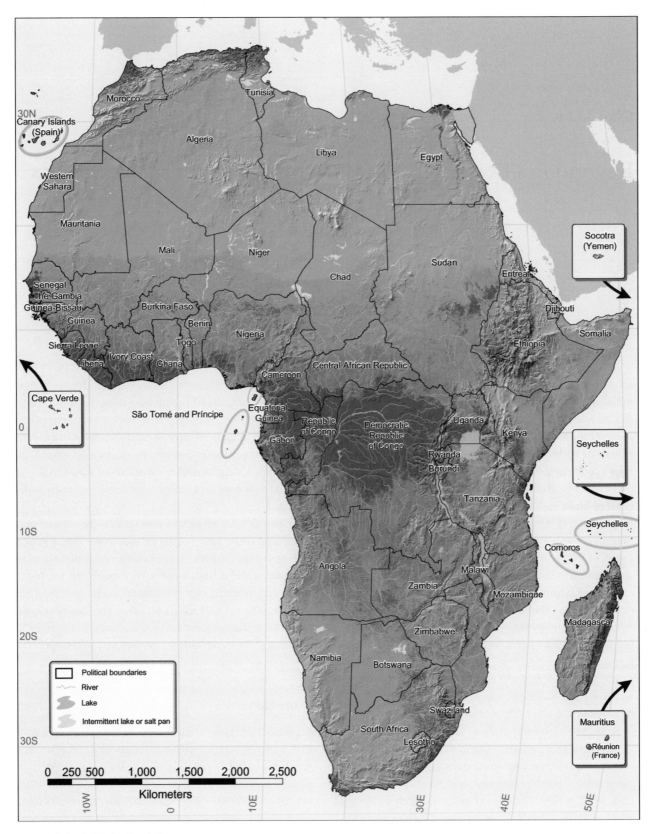

FIGURE 2.1. Political units of Africa.

units of land. Traditional African societies had complex classification systems for the surrounding landscape that recognized each habitat in terms of its values (which could include spiritual values) and the species it contained. Several hundred years ago explorers traversing the continent also realized that there was a great variation in habitat and species composition of the region, and they started to develop maps of the changing habitats. Explorers' accounts were gradually formalized by scientists into maps of the biological divisions of the world, a process that continues to the present day.

WWF uses ecoregions as a framework to establish conservation priorities and to undertake targeted programs of conservation action on the ground across large areas of land (chapter 7). We believe that ecoregions are preferable to countries as units for large-scale conservation planning because they share species, habitats, and ecological processes (Wright et al. 1998). Basing conservation planning efforts on political units risks ignoring important natural regions that extend beyond national borders. One of the best African examples of this is the Serengeti ecosystem of northern Tanzania and Kenya. Each year, 1.3 million wildebeest, 200,000 plains zebra, and 400,000 Thomson's gazelle undergo seasonal migrations from the Serengeti plains in northern Tanzania into the adjacent Masai Mara Reserve of Kenya. By early summer the herds return to Tanzania (figure 2.3). Rather than stop at the border between Kenya and Tanzania, the huge herds follow the rains in search of green forage. It is the seasonal rainfall that drives the ecological process of migration and structures this community, which can be mapped in two separate ecoregions. The long-term conservation of the Serengeti ecosystem requires that it be managed as a transboundary system within these ecoregions rather than as two separate national units. There are numerous other examples of divisions in natural biogeographic units by political boundaries; indeed, eighty-five African ecoregions cross country boundaries.

In this book we recognize 119 terrestrial ecoregions across Africa and its islands (figure 2.2; Olson et al. 2001). These ecoregions are based mostly on previously proposed biological divisions. For example, the majority of the terrestrial ecoregions of mainland Africa are derived from *The Vegetation of Africa* (White 1983). Similarly, the terrestrial ecoregions of Madagascar are based mainly on the *Bioclimates of Madagascar* (Cornet 1974). An initial map of ecoregions was subjected to a process of comment and expert review starting in 1998. This led to a modification of the

boundaries from the vegetation units outlined by White (1983) and also led to the subdivision of several of White's phytogeographic regions. For example, the Afromontane archipelago-like regional center of endemism of White (1983) was broken down from a single phytochorion into fifteen separate ecoregions. Division into smaller ecoregions was also achieved for all other phytochoria, based mainly on patterns of plant and animal endemism.

We placed distinct island groups (Seychelles, Aldabra, Cape Verde, Mascarenes, and Comoros) into separate ecoregions. In the case of the Canary Islands, most of the islands were placed into a single ecoregion, but the islands closest to the mainland of Africa were grouped with the Atlantic Coastal Desert ecoregion [92], with which they share affinities. However, the coastal islands of East Africa (Zanzibar, Pemba, Mafia, and others) were regarded as too similar to the Northern Zanzibar–Inhambane Coastal Forest Mosaic ecoregion [20] to warrant separation from it. A full description of each ecoregion is provided in appendix I.

The essays by Williams and Margules (essay 2.1) and by Lovett and Taplin (essay 2.2) present an alternative to ecoregion-based biogeographic classification using a grid-based approach.

Grouping Ecoregions into Broader Frameworks

Biomes

The palette of ecoregions that color the ecological map of Africa can be simplified into groupings that reflect shared vegetation composition. This larger-scale vegetation framework is important background to many of the analyses we present later in this book, so we outline it in some detail here.

Vegetation types with similar characteristics are grouped together as habitats, and the broadest global habitat categories are called biomes. WWF has developed a simple system of biomes as its framework for mapping habitats across the world (Olson and Dinerstein 1998; Olson et al. 2001). Each biome contains one dominant habitat. For example, all the warm, wet forests with limited seasonality and dominant broadleaf evergreen trees are grouped as tropical moist forests, whereas all tropical areas with greater seasonality and deciduous trees are grouped as savanna-woodlands.

Of the fourteen biomes identified by WWF (Olson et al. 2001), nine are found in Africa and its islands (table 2.1, figure 2.4). The area of land occupied by each of these biomes

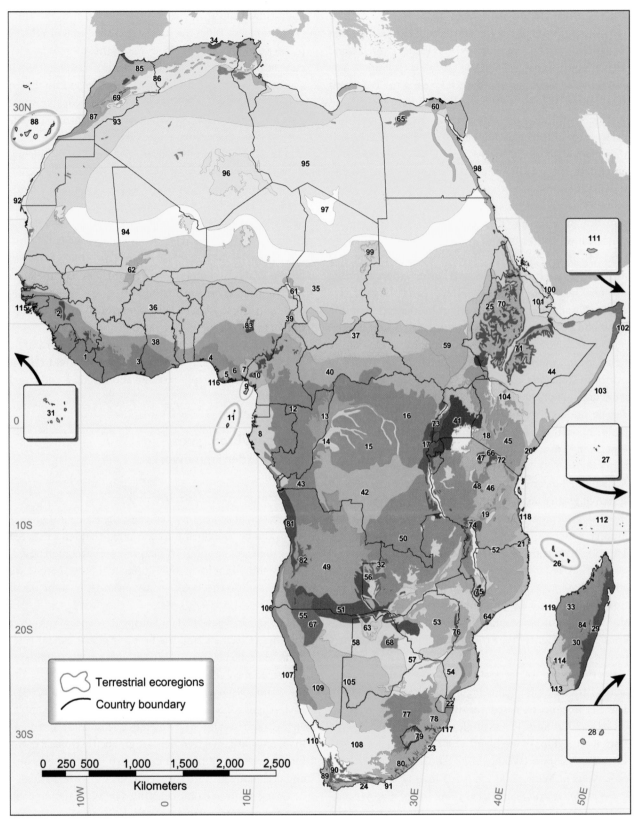

FIGURE 2.2. Terrestrial ecoregions of Africa.

Tropical and subtropical moist broadleaf forests

- 1 Western Guinean Lowland Forests
- 2 Guinean Montane Forests
- 3 Eastern Guinean Forests
- 4 Nigerian Lowland Forests
- 5 Niger Delta Swamp Forests
- 6 Cross-Niger Transition Forests
- 7 Cross-Sanaga-Bioko Coastal Forests
- 8 Atlantic Equatorial Coastal Forests
- 9 Mount Cameroon and Bioko Montane Forests
- 10 Cameroon Highlands Forests
- 11 São Tomé, Príncipe, and Annobon Moist Lowland Forests
- 12 Northwestern Congolian Lowland Forests
- 13 Western Congolian Swamp Forests
- 14 Eastern Congolian Swamp Forests
- 15 Central Congolian Lowland Forests
- 16 Northeastern Congolian Lowland Forests
- 17 Albertine Rift Montane Forests
- 18 East African Montane Forests
- 19 Eastern Arc Forests
- 20 Northern Zanzibar–Inhambane Coastal Forest Mosaic
- 21 Southern Zanzibar–Inhambane Coastal Forest Mosaic
- 22 Maputaland Coastal Forest Mosaic
- 23 KwaZulu-Cape Coastal Forest Mosaic
- 24 Knysna-Amatole Montane Forests
- 25 Ethiopian Lower Montane Forests, Woodlands, and Bushlands
- 26 Comoros Forests
- 27 Granitic Seychelles Forests
- 28 Mascarene Forests
- 29 Madagascar Humid Forests
- 30 Madagascar Subhumid Forests

Tropical and subtropical dry broadleaf forests

- 31 Cape Verde Islands Dry Forests
- 32 Zambezian *Cryptosepalum* Dry Forests
- 33 Madagascar Dry Deciduous Forests

Temperate coniferous forests

- 34 Mediterranean Conifer and Mixed Forests

Tropical and subtropical grasslands, savannas, shrublands, and woodlands

- 35 Sahelian *Acacia* Savanna
- 36 West Sudanian Savanna
- 37 East Sudanian Savanna
- 38 Guinean Forest-Savanna Mosaic
- 39 Mandara Plateau Mosaic
- 40 Northern Congolian Forest-Savanna Mosaic
- 41 Victoria Basin Forest-Savanna Mosaic
- 42 Southern Congolian Forest-Savanna Mosaic
- 43 Western Congolian Forest-Savanna Mosaic
- 44 Somali *Acacia-Commiphora* Bushlands and Thickets
- 45 Northern *Acacia-Commiphora* Bushlands and Thickets
- 46 Southern *Acacia-Commiphora* Bushlands and Thickets
- 47 Serengeti Volcanic Grasslands
- 48 Itigi-Sumbu Thicket
- 49 Angolan Miombo Woodlands
- 50 Central Zambezian Miombo Woodlands
- 51 Zambezian *Baikiaea* Woodlands
- 52 Eastern Miombo Woodlands
- 53 Southern Miombo Woodlands
- 54 Zambezian and Mopane Woodlands
- 55 Angolan Mopane Woodlands
- 56 Western Zambezian Grasslands
- 57 Southern Africa Bushveld
- 58 Kalahari *Acacia* Woodlands

Flooded grasslands and savannas

- 59 Sudd Flooded Grasslands
- 60 Nile Delta Flooded Savanna
- 61 Lake Chad Flooded Savanna
- 62 Inner Niger Delta Flooded Savanna
- 63 Zambezian Flooded Grasslands
- 64 Zambezian Coastal Flooded Savanna
- 65 Saharan Halophytics
- 66 East African Halophytics
- 67 Etosha Pan Halophytics
- 68 Makgadikgadi Halophytics

Montane grasslands and shrublands

- 69 Mediterranean High Atlas Juniper Steppe
- 70 Ethiopian Upper Montane Forests, Woodlands, Bushlands, and Grasslands
- 71 Ethiopian Montane Moorlands
- 72 East African Montane Moorlands
- 73 Rwenzori-Virunga Montane Moorlands
- 74 Southern Rift Montane Forest-Grassland Mosaic
- 75 South Malawi Montane Forest-Grassland Mosaic
- 76 Eastern Zimbabwe Montane Forest-Grassland Mosaic
- 77 Highveld Grasslands
- 78 Drakensberg Montane Grasslands, Woodlands, and Forests
- 79 Drakensberg Alti-Montane Grasslands and Woodlands
- 80 Maputaland-Pondoland Bushland and Thickets
- 81 Angolan Scarp Savanna and Woodlands
- 82 Angolan Montane Forest-Grassland Mosaic
- 83 Jos Plateau Forest-Grassland Mosaic
- 84 Madagascar Ericoid Thickets

Mediterranean forests, woodlands, and scrub

- 85 Mediterranean Woodlands and Forests
- 86 Mediterranean Dry Woodlands and Steppe
- 87 Mediterranean *Acacia-Argania* Dry Woodlands and Succulent Thickets
- 88 Canary Islands Dry Woodlands and Forests
- 89 Lowland Fynbos and Renosterveld
- 90 Montane Fynbos and Renosterveld
- 91 Albany Thickets

Deserts and xeric shrublands

- 92 Atlantic Coastal Desert
- 93 North Saharan Steppe
- 94 South Saharan Steppe
- 95 Sahara Desert
- 96 West Saharan Montane Xeric Woodlands
- 97 Tibesti-Jebel Uweinat Montane Xeric Woodlands
- 98 Red Sea Coastal Desert
- 99 East Saharan Montane Xeric Woodlands
- 100 Eritrean Coastal Desert
- 101 Ethiopian Xeric Grasslands and Shrublands
- 102 Somali Montane Xeric Woodlands
- 103 Hobyo Grasslands and Shrublands
- 104 Masai Xeric Grasslands and Shrublands
- 105 Kalahari Xeric Savanna
- 106 Kaokoveld Desert
- 107 Namib Desert
- 108 Nama Karoo
- 109 Namib Escarpment Woodlands
- 110 Succulent Karoo
- 111 Socotra Island Xeric Shrublands
- 112 Aldabra Island Xeric Scrub
- 113 Madagascar Spiny Thickets
- 114 Madagascar Succulent Woodlands

Mangroves

- 115 Guinean Mangroves
- 116 Central African Mangroves
- 117 Southern Africa Mangroves
- 118 East African Mangroves
- 119 Madagascar Mangroves

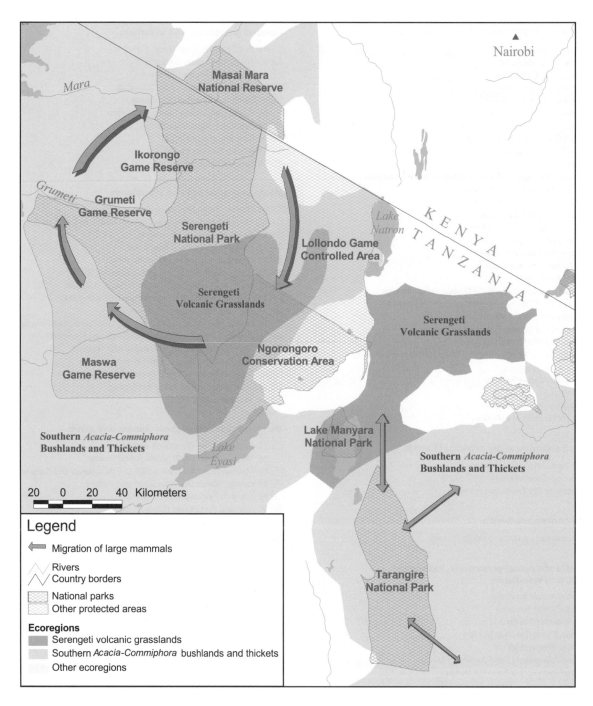

FIGURE 2.3. Wildebeest migration around the Serengeti.

varies greatly (table 2.1). The savanna-woodland biome occupies by far the largest area (more than 13 million km²), with the deserts and tropical moist forests also covering millions of square kilometers of land. These huge areas reflect the tropical position of Africa, where differences in rainfall largely control whether an area supports desert, savanna-woodland, or rainforest. In comparison, temperate coniferous forests are typical of cooler parts of the world and in Africa are confined to a small mountainous portion of North Africa, where they cover a little over 23,000 km². The nine biomes of Africa are described here, in order of descending area.

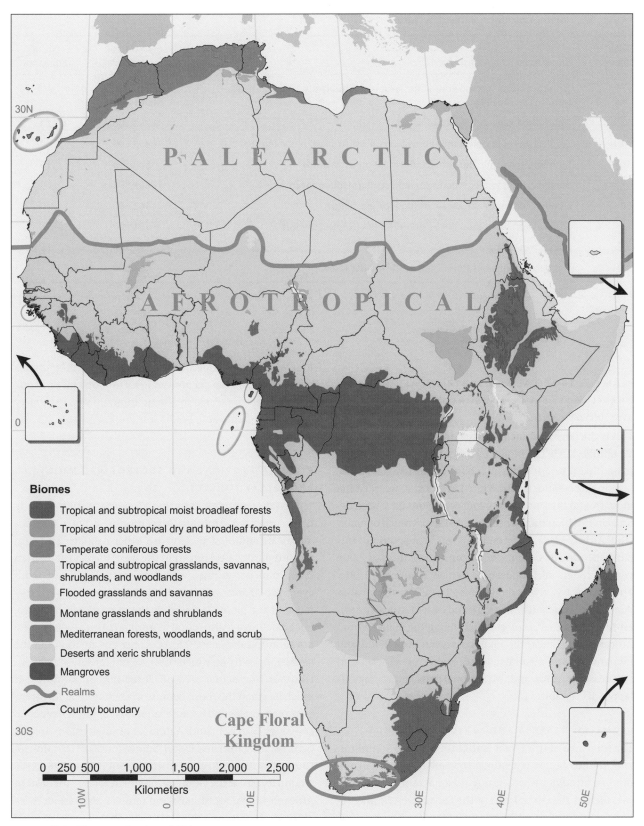

Biomes

- Tropical and subtropical moist broadleaf forests
- Tropical and subtropical dry and broadleaf forests
- Temperate coniferous forests
- Tropical and subtropical grasslands, savannas, shrublands, and woodlands
- Flooded grasslands and savannas
- Montane grasslands and shrublands
- Mediterranean forests, woodlands, and scrub
- Deserts and xeric shrublands
- Mangroves
- Realms
- Country boundary

PALEARCTIC

AFROTROPICAL

Cape Floral
Kingdom

0   250 500   1,000   1,500   2,000   2,500

Kilometers

30N

0

30S

10W   0   10E   30E   40E   50E

FIGURE 2.4. Biomes and realms of Africa.

TABLE 2.1. **Biomes of Africa, Ranked by Total Area.**

Biome Number	Biome Name (abbreviated name)	Approximate Biome Area (km²)	No. of Ecoregions
7	Tropical and subtropical grasslands, savannas, shrublands, and woodlands (savanna-woodlands)	13,981,100	24
13	Deserts and xeric shrublands (deserts)	9,753,200	23
1	Tropical and subtropical moist broadleaf forests (tropical moist forests)	3,485,500	30
10	Montane grasslands and shrublands (montane grasslands)	868,700	16
12	Mediterranean forests, woodlands, and scrub (Mediterranean scrub)	851,300	7
9	Flooded grasslands and savannas (flooded grasslands)	562,600	10
2	Tropical and subtropical dry broadleaf forests (tropical dry forests)	194,900	3
14	Mangroves (mangroves)	76,700	5
5	Temperate coniferous forests (temperate conifer forests)	23,100	1

BIOME 7: TROPICAL AND SUBTROPICAL GRASSLANDS, SAVANNAS, SHRUBLANDS, AND WOODLANDS

This is the largest biome in the region, covering more than 13 million km². The biome includes the areas that would naturally (in recent millennia) have supported grasslands, savanna-woodlands (sensu White 1983), forest, wet savanna mosaics, and thickets. Twenty-four ecoregions have been defined in this biome and are distributed across much of mainland Africa. Madagascar contains no ecoregions in this biome because its savannas were formed over the past 2,000 years through anthropogenic changes (Lourenço 1996). The biome has five subdivisions. These are the *Acacia*-dominated woodlands of western and eastern Africa, the forest-savanna mosaic ecoregions adjoining the main Central African forest block, the miombo savanna-woodland and mopane savanna-woodland ecoregions of eastern and southeastern Africa, and two seasonal grassland ecoregions (appendix A, table 2.2).

BIOME 13: DESERTS AND XERIC SHRUBLANDS

The total area of this biome is slightly less than 10 million km². Its distribution is controlled by rainfall—less than 500 mm per annum—with a strong seasonal profile. Some areas, such as the central part of the Sahara Desert, may receive no rainfall at all for years. Twenty-three ecoregions are found in this biome. Five subdivisions are recognized:

four where there is significant vegetation cover and rainfall close to 500 mm per annum (~4.9 million km²) and one where plants are scarce to absent and rainfall is extremely low and unpredictable (~4.87 million km²) (table 2.2 and appendix A).

BIOME 1: TROPICAL AND SUBTROPICAL MOIST BROADLEAF FORESTS

This biome comprises areas that support or would naturally have supported rainforest vegetation or a heterogeneous mosaic of rainforest mixed with other habitats. The rainforest typically is evergreen in the wettest areas (more than 2,000 mm per annum with low seasonality) and semi-evergreen in areas where rainfall is more seasonal. The majority of the biome is found in the tropical belt of Africa, but a small area is also found in southern Africa, where there is a subtropical climate. Thirty ecoregions are found in the biome (appendix A), which cover 3.5 million km² (table 2.1). However, the actual area of closed-canopy rainforest is much less than this, both because the biome includes forest-savanna mosaics and because large areas of forest have been converted into farmland. A number of subdivisions are recognized: the Upper and Lower Guinea forests of the Guineo-Congolian regional center of endemism (White 1983), the swamp forests of the Guineo-Congolian forests, the eastern African lowland forest-grassland mosaics, the various Afro-montane forests, and the forests distributed on islands off the coast of Africa (table 2.2).

TABLE 2.2. **Sub-Biome Areas in the Four Largest Biomes.**

Biome	Area (km²)	Biome	Area (km²)
Deserts and xeric shrublands		**Tropical and subtropical moist broadleaf forests**	
Desert	4,870,400	Afromontane forest	515,200
Island xeric woodland and shrubland	123,300	Eastern African lowland forest-grassland mosaics	327,100
North African xeric woodland and shrubland	3,146,000	Guineo-Congolian lowland moist forest	2,107,000
Northeastern African xeric woodland and shrubland	346,100	Guineo-Congolian swamp forest	235,700
Southern African xeric woodland and shrubland	1,267,400	Island moist forest	320,000
Tropical and subtropical grasslands, savannas, shrublands, and woodlands		**Montane grasslands and shrublands**	
Acacia savanna woodland	7,783,700	Alpine moorland	50,700
Guineo-Congolian forest-savanna mosaic	2,538,100	Montane forest-grassland mosaic	819,000
Miombo woodland	3,000,900		
Mopane woodland	606,800		
Seasonal grassland	52,000		

Biome 7: Herd of African elephants (*Loxodonta africana*) grazing in the savanna of Masai Mara National Reserve, Kenya. Photo credit: WWF-Canon/Martin Harvey

Biome 13: Dunes and an *Acacia* tree at Temet, Northern Aïr, Niger. Photo credit: WWF-Canon/John Newby

BIOME 10: MONTANE GRASSLANDS AND SHRUBLANDS

This biome is found in scattered locations throughout Africa, wherever there are suitable mountain areas over 800 m altitude (depending on location) with sufficient rainfall to develop grasslands and related habitats. Mountains with lower rainfall are grouped with the deserts, and those with higher rainfall are grouped into montane rainforests. Two subdivisions are recognized: the forest-grassland mosaics, where trees are found (c. 819,000 km^2), and upland heathland and Afroalpine areas, where trees are absent (~50,000 km^2) (table 2.2; appendix A). The sixteen ecoregions are regional units of these subdivisions.

BIOME 12: MEDITERRANEAN FORESTS, WOODLANDS, AND SCRUB

This biome is confined to areas with hot, dry summers and winter rainfall. The total area is about 851,000 km^2, mainly in North Africa. Seven ecoregions are found on the Canary Islands, in the Mediterranean Basin in North Africa, and in the Cape Region of South Africa. The vegetation ranges from mixed coniferous and broadleaf woodland to shorter vegetation made up of dwarf shrubs with an understory of herbs and grasslike plants, often with a large diversity of flowering bulbs.

BIOME 9: FLOODED GRASSLANDS AND SAVANNAS

Covering approximately 563,000 km^2, this biome includes seasonally flooded grasslands, swamps, and shallow lakes (both freshwater and saline). Major lakes such as Lake Tanganyika and Lake Victoria are excluded because they are described in a separate assessment of African freshwater ecoregions (Thieme et al. 2005). The two subdivisions recognized are freshwater swamp systems (~477,000 km^2) and slightly brackish to hypersaline swamp systems (~86,000 km^2) (appendix A). The ten ecoregions are regional variations of these two subdivisions.

BIOME 2: TROPICAL AND SUBTROPICAL DRY BROADLEAF FORESTS

Only one dry forest ecoregion is found on the African mainland, in Angola and Zambia. Two others occupy the Cape

Biome 1: Montane forest from Sanje Falls, Udzungwa Mountains National Park, Tanzania. Photo credit: WWF-Canon/Edward Parker

Biome 9: Marabou stork (*Leptoptilos crumeniferus*), Etosha National Park, Namibia. Photo credit: WWF-Canon/Martin Harvey

Biome 10: Giant lobelia (*Lobelia rhynchopetalum*), a giant herb growing at high altitudes of 3,000–4,500 m a.s.l., Bale Mountains National Park, Ethiopia. Photo credit: WWF-Canon/Martin Harvey

Biome 12: Fynbos vegetation. Cape Point, Republic of South Africa. Photo credit: WWF-Canon/Meg Gawler

Biome 14: Mangrove tree (Rhizophoraceae), Mafia Island, Tanzania. Photo credit: WWF-Canon/Edward Parker

Verde Islands and the western margin of Madagascar. In total, these three ecoregions cover ~195,000 km^2.

BIOME 14: MANGROVES

The mangroves of Africa and Madagascar are remarkably similar to each other. Two sub-biomes are recognized, following the split between East and West African mangroves outlined by White (1983). The five ecoregions delineated capture additional variation caused by climatic factors and marine biogeographic considerations, such as warm and cold currents and areas of upwelling, which influence the marine flora and fauna associated with the mangroves.

BIOME 5: TEMPERATE CONIFEROUS FORESTS

Temperate coniferous forests are absent from the Afrotropical realm and the offshore islands around Africa. They are confined to the North African mountains in the Palearctic realm. This biome is represented by one ecoregion that covers a total area of around 23,000 km^2 across Algeria, Morocco, and Tunisia.

Realms

Ecoregions define areas that share species, habitats, and ecological processes. In this regard they form medium-sized biogeographic units, which can be fitted into biogeographic systems that divide the distributions of plants and animals at the global scale. To illustrate the concept, the fauna of Africa south of the Sahara contains many shared species across a wide range of habitats. These species are not found in similar habitats in Europe or Asia, so sub-Saharan Africa is recognized as its own biogeographic region: the Afrotropical realm.

Attempts to classify the world into biogeographic realms began in the mid-nineteenth century when Alfred Russell Wallace proposed the first global system (Wallace 1876). For Africa he proposed two biogeographic divisions. From the Sahara Desert northward Wallace placed Africa in the Palearctic realm, together with Europe. South of the Sahara was placed in the Ethiopian realm, with three subdivisions: a vast region of woodland, savanna, and desert; the central rainforest region; and the southern region of diverse vegetation types (Wallace 1876). Madagascar and the other Indian Ocean islands were included in the Ethiopian realm but were regarded as distinct from the mainland. The divisions proposed by Wallace are still recognized today, despite many other attempts to subdivide Africa and its islands into biogeographic units (e.g., Allen 1878; Herbertson 1905; Gleason and Cronquist 1964; Good 1964; Cox and Moore 1993; Bisby 1995; Craw et al. 1999).

FAUNAL REALMS

Mainland Africa contains two faunal realms. The fauna of the Sahara Desert northward is placed with Europe in the Palearctic realm, and Africa south of the Sahara forms the Afrotropical realm (the Ethiopian realm of Wallace 1876) (figure 2.4). The Sahara Desert acts as a barrier to species movement between the Afrotropical and the Palearctic faunal realms. Except for the Canary Islands, which belong to the Palearctic realm, the islands around Africa are normally regarded as part of the Afrotropical realm. Madagascar is sometimes separated into its own faunal realm (Malagasy), based on the high rates of endemism and the presence of primitive mammal groups. Others recognize Madagascar and the smaller Indian Ocean islands as a separate subrealm (e.g., Pomeroy and Ssekabiira 1990) within the larger Afrotropical realm.

FLORAL REALMS

Three distinct floral realms fall within the African continent. The Paleotropical and Holarctic parallel the Afrotropical and Palearctic divisions recognized in the fauna. However, the exceptional levels of plant endemism in the Cape region of South Africa elevate that area as its own plant realm: the Cape Floral Kingdom (Cox and Moore 1993; Cowling and Richardson 1995) (figure 2.4). There is debate over the biogeographic classifications of the offshore islands, where the flora is dramatically different from that of mainland Africa. For example, more than 80 percent of the plant species of Madagascar and the Mascarenes are endemic to these islands (Cronk 1992, 1997; Lourenço 1996). The Indian Ocean islands often are regarded as a subrealm of the Paleotropical floral realm, as are the Canary Islands and other Macronesian islands off the western shore of Africa.

Other Classification Schemes

Several other schemes provide a biogeographic classification of Africa. Some of these schemes have been proposed for single groups of fauna, such as mammals (Balinsky 1962; Turpie and Crowe 1994), birds (Chapin 1932; Moreau 1966; Crowe and Crowe 1982; Crowe 1990; de Klerk et al. 2002a, 2002b), snakes (Hughes 1983), and amphibians (Poynton 1999). For plants there have also been a number of attempts to define larger- and smaller-scale biogeographic units across the continent (Monod 1957; de Laubenfels 1975; Brenan 1978; White 1983; Jürgens 1997). Many of the areas defined as biogeographic units for different taxon groups overlap wholly or partially with the ecoregions identified in this study.

Recently there have also been some attempts to look at biogeographic patterns in Africa below the scale of an ecoregion. Some have looked at biodiversity patterns within single ecoregions, such as the Fynbos of South Africa (Campbell 1985; Cowling and Richardson 1995; Cowling and Heijnis 2001), the Karoo of South Africa (Dean and Milton 1999), the eastern African coastal forests (Burgess and Clarke 2000), and the Eastern Arc forests (Lovett and Wasser 1993; Burgess et al. 1998d). Others have looked at fine-scale biogeographic divisions within countries (e.g., Lind and Morrison 1974; Gibbs Russel 1987; Huntley 1994; Low and Re-

belo 1996; Cowling et al. 1997b). More than half of the fine-scale biogeographic studies that exist for the African region focus on South Africa.

The challenge of this book is to take the framework of ecoregions, grouped within their larger habitat-based and biogeographic frameworks, and use them to establish priorities for conservation action across Africa. In chapter 3 we outline how we have used this biogeographic framework to devise methods for creating rigorous continental-scale conservation strategies and implementing them to achieve broad-scale conservation across the continent.

ESSAY 2.1

Grid Cells, Polygons, and Ecoregions in Biogeography and in Biodiversity Priority Area Selection

Paul Williams and Chris Margules

Biogeography can be seen as the study of the distribution patterns of species and other taxa in time and space. These patterns are one major component of biodiversity. Biodiversity priority areas are the areas toward which scarce conservation planning and management resources should be directed. For priority-setting, areas are compared with one another in terms of threats to persistence or urgency for conservation action and in terms of the particular valued assemblages of species or other taxa, sets of environmental classes, and landscape types that occur in them (Margules and Pressey 2000). Areas can be appropriate units for conservation action because biodiversity is a property of collections of things or of processes, which are often defined spatially by volume or, more loosely, by area. The consequence of this area dependency is that careful consideration must be given to the choice of areas to be compared in conservation prioritization.

Areas usually are represented using outlines on maps because maps are a very efficient way to communicate large amounts of information. Grid cells are regular repeating shapes, such as squares, although other shapes such as hexagons have been used (Scott et al. 1993). If the shapes of the area outlines are irregular, they are often called polygons. The earliest biogeographic analyses used complex polygons for biogeographic regions (Sclater 1858). It was not until a century later, with the establishment of the national biological recording schemes in the

United Kingdom, that grid systems came into widespread use (Perring and Walters 1962). Not long after, computers first became available, which facilitated using regular grid systems because it was easy to print symbols on a regular grid to represent maps. In due course, with the spread of modern geographic information system (GIS) software, the more complex graphics allowed a return to polygon mapping, using computers to handle the large amounts of data. It is now possible to acquire and store biological data that are precisely spatially referenced by using information from geographic positioning systems (GPSs). The choice of how to group and present data on maps can and should be based on what is appropriate for the questions being asked.

If the choice between polygons and regular arrays of grid cells depends on the aim of an analysis, what are the main factors influencing the choice? For the biogeographic analysis of patterns and processes, equal-area grid cells have advantages because of the effect of area extent on comparisons of, for example, species richness. For more applied analyses such as identifying priority areas for conservation, polygons can have advantages for representing realistic candidate areas for conservation management. However, ecoregions can perform several roles in biodiversity planning and priority-setting. For example, they can be used as polygons in their own right to set priorities for conservation action at global and continental scales, and they can then act as the spatial units within which

Using Equal-Area Grid Cells to Map Biogeographic Patterns

Why are equal-area grid cells often used in biogeography? If diversity is a property of collections of objects, then because larger areas are likely to have more individual organisms and more species, there is usually a strong positive relationship between species richness and area extent. The precise nature of this species-area relationship depends on many factors (Connor and McCoy 1979), including the kind of habitat, the history of the biogeographic region, and the shape of the landmass and its isolation, so that comparing samples from different-sized areas may not be straightforward. Consequently, for biogeographic analyses, controlling the area sampled by using equal-area grid cells can simplify comparisons by reducing at least one form of bias. It is then possible to compare levels of biodiversity on a per-unit-area basis, as when comparing the species richness of areas or their richness in narrowly endemic species. Polygons that vary greatly in area extent (such as entire ecoregions) can be compared for these properties only if generalized species-area relationships are assumed, and any important variations within them cannot then be studied. Furthermore, apparent patterns of diversity may depend critically on the grain size at which they are measured (Stoms 1994), so this should be made explicit.

Potential effects of variation in area extent and in area location on measures of species richness can be illustrated for sub-Saharan Africa using ecoregions and data for mammals and breeding birds (figure 2.5). These species distribution data were compiled by Neil Burgess, Carsten Rahbek, John Feldså, and Louis Hansen at the University of Copenhagen, in collaboration with the Natural History Museum (London), the University of Cambridge, and an international team of mammalogists and ornithologists (Burgess et al. 1998c). Note that the numbers of species are much higher on the scale of the ecoregion map (right) because there tend to be more species in the larger ecoregions than in the much smaller 1-degree grid cells (a consequence of the species-area relationship). At first glance, the broad patterns of relative richness appear similar, but there are some important discrepancies. The map of richness for grid cells (left) shows high richness in the Cameroon and Ethiopian highlands, whereas in these parts of the ecoregion map (right), richness appears lower. Generalized species-area relationships could be used to adjust the ecoregion-based richness estimates for effects of their relative extent, and this probably would increase the richness of the Ethiopian Highlands relative to surrounding

regions. However, this treatment is unlikely to solve the Cameroon problem because of the way grid cells were assigned to several different ecoregional groups. Of course, the cells could be reclassified into different groups, but the potential for this kind of problem to arise from the location of the ecoregion boundaries remains elsewhere on the map and in other biogeographic applications. For a discussion of some of the problems inherent in drawing regional boundary lines on maps and for some alternative approaches with examples for Africa, see Williams et al. (1999).

Using Polygons to Identify Priority Areas for Biodiversity

When selecting particular areas for conservation management, area extent is much less important than the conservation value or urgency for action of each area relative to its cost (however cost is most appropriately calculated). Therefore, candidate areas may be of any shape or size that is appropriate for management planning. Just two examples from the large Australian literature include selection among stream catchments, varying from 0.1 km^2 to more than 60 km^2 (Bedward et al. 1992), and selection among pastoral properties, varying from about 40 km^2 to about 520 km^2 (Pressey et al. 1994). One implication of this approach is that if an area were selected as a high priority, then it should be managed for conservation in its entirety, not just in some small parts; otherwise, some components of its value might be lost.

Whether grid cells or polygons are used, continent-wide studies of the most important areas for conserving biodiversity in Africa have, in practice, shown an encouraging degree of agreement (da Fonseca et al. 2000). For comparison with the ecoregion priority maps presented elsewhere in this book, figure 2.6 shows some important areas for biodiversity in sub-Saharan Africa based on a quantitative analysis of 1-degree grid cell data for mammals and birds from the Copenhagen database. Near–equal-area grid cells can be a useful starting point for continent-wide analysis when details of real local candidate areas (polygons) for conservation management are unavailable.

Using Ecoregions as Surrogates for Biodiversity

Measuring and mapping biodiversity are notoriously difficult because of the immense complexity of the problem. Consequently, people usually are obliged to resort to surrogates for the valued biodiversity that are easier to measure. Toward the end of the scale at which surrogates may be less precise but also less expensive to measure are approaches based on landscapes or ecosystems (Williams 1996).

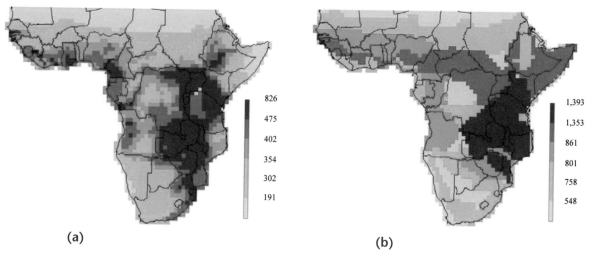

(a)   (b)

FIGURE 2.5. Distribution of species richness of 2,687 mammal and breeding bird species in sub-Saharan Africa (data described by Burgess et al. 1998c; Williams et al. 2000). (a) The map on the left shows the species richness of 1-degree grid cells. (b) The map on the right shows the species richness of ecoregions (1-degree grid cells are grouped into the largest intersecting ecoregion occupying more than 25 percent of each cell, and the total number of species per ecoregion is then plotted in all of the included cells). Each color class represents a consistent part (17 percent) of the frequency distribution of richness scores on the maps (except where constrained by large numbers of tied scores), with dark brown for maximum richness and light brown for minimum nonzero richness. National boundaries are plotted on the map as black lines. One-degree grid cells are approximately 105 km on each side.

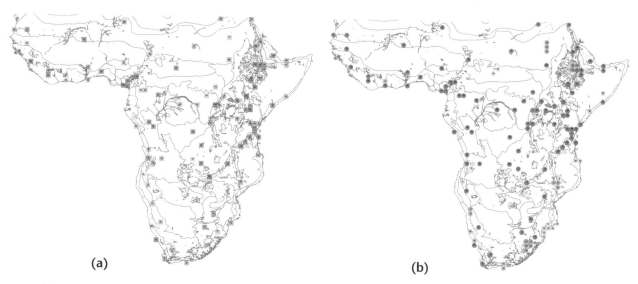

(a)   (b)

FIGURE 2.6. Distribution of the top 100 1-degree grid cells for seeking to represent the maximum number (97 percent) of the 2,687 mammal and breeding bird species in sub-Saharan Africa (data described by Burgess et al. 1998c; Williams et al. 2000) without viability or threat constraints. (a) Top 100 1-degree grid cells prioritized for biodiversity value (as a sequence of near–maximum coverage sets of increasing numbers of cells), in the sense that the redder cells contribute the most species, and orange, green, and blue cells add progressively fewer additional species. (b) All alternative areas that could be substituted to represent the same total number of species. There are no substitutes for the red areas; they are irreplaceable. The groups of orange, green, and blue cells are substitutable for one another, with few substitutes for the orange cells, more for the green cells, and still more for the blue cells. The numbers identify the ordered area choice from (a) for which these areas are alternatives. Ecoregion boundaries are plotted on the map as black lines. One-degree grid cells are approximately 105 km on each side.

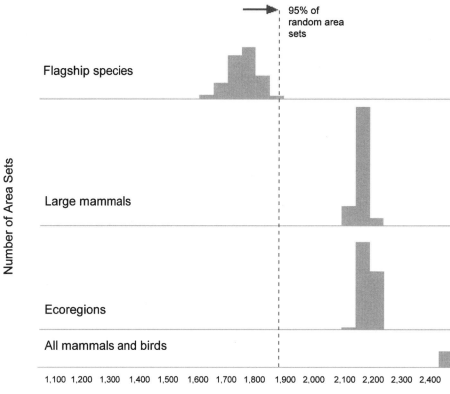

FIGURE 2.7. Frequency with which alternative sets of fifty 1-degree grid cells, chosen for maximum coverage of different biodiversity surrogates, represent different numbers of sub-Saharan mammal and breeding bird species (based on Williams et al. 2000). Surrogates used to make the selections were six flagship species (*Gorilla gorilla, Pan paniscus, Pan troglodytes, Loxodonta africana, Ceratotherium simum,* and *Diceros bicornis*), ninety-eight terrestrial ecoregions, the larger mammal species (224 species of Primates, Carnivora, Proboscidea, Perissodactyla, and Artiodactyla), and all 2,687 sub-Saharan mammal and breeding bird species. With the exception of the histogram at the bottom, which shows all alternative sets that represent the same number of surrogates, the scores are for a randomly drawn sample of 1,000 confirmed alternative sets. Scores to the left of the dotted line are within the range expected when choosing fifty grid cells at random.

Ecoregions might also be used as coarse-scale surrogates for biodiversity. For example, one recent analysis in sub-Saharan Africa sought to compare the representation of different groups of species achieved by using a variety of different surrogates (Williams et al. 2000). When choosing fifty areas among 1-degree grid cells, for example, figure 2.7 shows that maximizing the representation of the different ecoregions results in representing more mammals and birds than either choosing areas at random or choosing areas for the six flagship species. Maximizing the representation of ecoregions also performed just as well as maximizing the representation of all of the large mammal species (note that using knowledge of the distribu-

tion of all mammals and birds succeeded in representing more diversity of these groups, as would always be expected). The significance of this result is that information on ecoregions is much easier to obtain (e.g., from remote sensing) than accurate data on, for example, distribution patterns of mammals. Therefore ecoregions could provide a cost-effective surrogate for broader biodiversity in this kind of continent-wide priority area analysis when species-based data are unavailable or when there might be reasonable doubt that areas chosen to represent one group of species, such as mammals or birds, will also represent many other taxa (van Jaarsveld et al. 1998a; Howard et al. 1998).

Sub-Saharan African Phytogeography Patterns and Processes: History and Evolution in Conservation Priority-Setting

Jon C. Lovett and James Taplin

One of the challenges and opportunities for setting conservation priorities is that biodiversity is not randomly distributed but rather is clustered into centers of species richness and endemism. This is a challenge because we need to recognize and delimit these centers, but it is also an opportunity to target conservation efforts. This essay briefly describes patterns of species richness and endemism for sub-Saharan African plants and discusses them in the context of history and evolution. The patterns are related to phytogeographic classification systems and underlying ecological and evolutionary processes. An understanding of these processes is particularly important for the managers of biodiversity as they determine the limits to sustainable use.

Patterns

The distributions of African plants have been mapped over many years, principally in the *Distributiones Plantarum Africanum* series and in taxonomic monographs. We have reached a point at which this wealth of material can be synthesized to investigate continent-wide patterns of plant species richness and endemism (Linder 1998, 2001; Lovett et al. 2000b; La Ferla et al. 2002). A map of species richness based on the distributions of 3,563 plant species (9 percent of the sub-Saharan African flora) shows that richness is concentrated in southern Ghana, Cameroon, Gabon, the forests of the Congo Basin, the Albertine Rift, the Eastern Arc Mountains, and the South African Cape (figure 2.8). The top ten species richness hotspots together contain 40 percent of the African plant species in the sample. These hotspots are in Cameroon (Mount Cameroon and Bioko Montane Forests, Cameroon Highlands Forests, and Cross-Sanaga-Bioko Coastal Forest ecoregions), South Africa (Fynbos and Succulent Karoo ecoregions), the Democratic Republic of Congo (margins of the Albertine Rift Montane Forests and Northeastern Congolian Lowland Forests ecoregions), and Tanzania (Eastern Arc Forests, Northern Zanzibar–Inhambane Coastal Forest Mosaic, and Southern Rift Montane Forest-Grassland Mosaic ecoregions). Ecoregions that are rich in species also tend to have many restricted-range endemics (figure 2.9), with some

areas, such as the Cape and Eastern Arc, having exceptionally high numbers of geographically rare species. Both the Cape and Eastern Arc contain a high number of endemics, but they differ in that the Eastern Arc endemics occur in many different genera, whereas the Cape endemics belong to comparatively few genera. This difference can be attributed to markedly different evolutionary histories of the two floras.

History and Evolution

Distribution patterns of plants in sub-Saharan Africa show two important characteristics: some areas are much richer than others, and the reasons for species richness differ in different places. There are several possible explanations for differences in species richness. Some of the pattern can be explained by spatial constraints (Jetz and Rahbek 2001) in that richness is greatest in the middle of two boundaries. Ecology also plays an important role. Species richness is greater in areas of high rainfall (O'Brien 1993, 1998), and biodiversity hotspots often are on and around mountains, which offer a range of habitats over a short distance and also tend to be wet. History is likely to have played an important role in determining plant distribution patterns in sub-Saharan Africa. The African climate has fluctuated over time, having been cold and dry during the last glacial maximum (Bonnefille et al. 1990; Servant et al. 1993; Elenga et al. 2000). This may have reduced the area of forest to refugia where plants survived climate change but may also have caused extinctions in areas where forest loss has occurred. Interestingly, relictual taxa are found not only in rainforests but also in dry forest and arid woodlands (Lovett and Friis 1996) such as the Itigi thicket in Tanzania (Itigi-Sumbu Thicket ecoregion), the *Acacia-Commiphora* woodlands of northern Kenya and northeast Somalia (Somali *Acacia-Commiphora* Bushlands and Thickets ecoregion), and the dry *Juniperus* forest of eastern Ethiopia (Ethiopian Xeric Grasslands and Shrublands ecoregion). This suggests that a historical ecological explanation many be more complex than that proposed by the refugium model.

The complexity can in part be explained by in situ evolution of new plant species rather than range restriction through

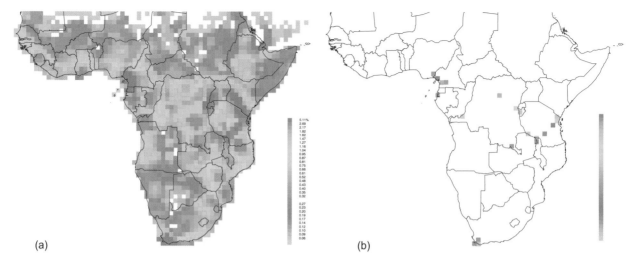

(a) (b)

FIGURE 2.8. **(a)** Smoothed species richness map of 3,563 plant species. **(b)** Top twenty species richness hotspots. Red indicates high species richness and blue low species richness. The 3,563 species are in 357 genera and 96 families and represent 9 percent of the sub-Saharan African flora. The data are biased by collecting intensity and availability of maps (La Ferla et al. 2002). A more complete dataset may show different patterns.

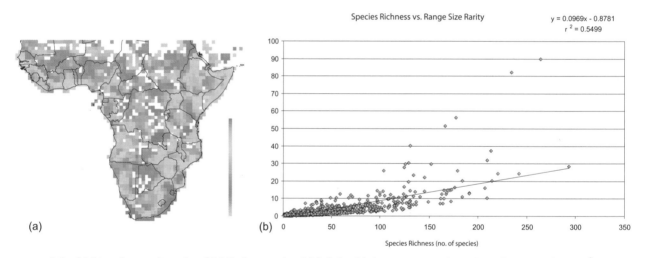

(a) (b)

FIGURE 2.9. **(a)** Map of range size rarity of 3,563 plant species. **(b)** Relationship between range size rarity and species richness ($r^2 = .55$). The outliers with high rarity scores are from the South African Cape.

climate change. Certain places have been subject to ecological conditions that encourage speciation, and these conditions are not the same in different places. In the Cape center of endemism, certain genera have radiated into many endemic species (Goldblatt 1978, 1997). The beginning of this burst of speciation has been dated to 7–8 m.y. B.P. (Richardson et al. 2001). It is thought that this radiation is facilitated by periodic fluctuations in climate and by fires (Cowling and Hilton-Taylor 1997). This contrasts with places such as the Eastern Arc of Tanzania, which contain many relictual taxa whose nearest relatives either are not in continental Africa or else occur west of

the arid corridor that separates the Eastern Arc from the larger Guineo-Congolian forests. The suggestion here is that the Eastern Arc has been environmentally stable over an exceptionally long period of time, and it is this stability that has led to both survival of relicts and evolution of new species (Lovett and Friis 1996; Fjeldså and Lovett 1997).

Phytogeographic Classification Systems

Phytogeographic classification systems are important in conservation management because they provide a framework within which priorities can be formulated. There has been a long history of development of African phytogeography (Friis 1998), and the most commonly used system is that of White (1983). For example, it was the basis for a review of Afrotropical protected areas (MacKinnon and MacKinnon 1986) and has been influential in the development of the ecoregion approach (e.g., Olson et al. 2001). However, White's system suffers from two main problems (Lovett 1998a). First, with the exception of the Cape Region, it does not recognize areas of high local endemism (biodiversity hotspots in the sense of Myers et al. 2000) because the system is designed to classify large regional centers of endemism. Second, it muddles similar vegetation types and distinct types. For example, the clearly distinct high-altitude Afroalpine region is included in a heterogenous Afromontane region, and in eastern Africa the upper elevation rainforests are separated into an Afromontane region when they are not distinct from midelevation and lowland forests in the Zanzibar-Inhambane region (Lovett et al. 2000a).

The ecoregion approach partly overcomes some of the problems by using endemism to drive the definition of regions of Africa at a smaller scale than that attempted by White (1983). Thus, the high-altitude Afroalpine region is recognized as distinctive and then subdivided into ecoregions based on geographic variation; for example, the East African Afroalpine is separated from that in Ethiopia. Moreover, the Afromontane forests of Africa are not grouped within a single phytochorion but instead are divided into a number of forested and forest-grassland mosaic ecoregions that are intended to reflect the natural divisions in the flora and fauna of the different mountain regions as closely as possible. Even the center of plant endemism in the Cape Floral Kingdom of South Africa is found in several ecore-

gions. However, one issue that the ecoregion approach has not been able to solve is that of gradations and transition zones. Consequently, sharp divisions have been proposed between systems that in practice intergrade, such as the Eastern Arc to lowland coastal forests of Tanzania. This is a general weakness of any mapping approach in which lines are placed on a map because the line indicates a hard boundary where actually none exists.

Processes and Conservation Management

Understanding the ecological processes underlying the evolutionary history of a flora is crucial to sustainable management (Lovett et al. 2000a). Moreover, it is no longer straightforward to protect large areas of land from human use, as was done in establishing Africa's famous national parks and game and forest reserves. Increasingly, policymakers are looking toward solutions that integrate conservation and development. Conservationists need to recognize the shaping of the land by human influence and combine the rights of access and productive land use with maintenance of plant and animal diversity. Biodiversity managers need to develop access and production systems that are compatible with the dynamics and resilience of the vegetation being conserved. For example, disturbance in the form of fire is important in maintaining the rich Cape flora; in its absence the fynbos is invaded by woody plants (Manders and Richardson 1992). In contrast, in the Eastern Arc, where the forests have evolved through a long period of environmental stability, management interventions such as logging result in a loss of endemic species and a lowering of species diversity (Lovett 1999).

CHAPTER 3

Setting Priorities: The Race against Time

The African continent has produced many strong figures in the conservation movement. Julius Nyerere, the first president of Tanzania, issued the famous Arusha Declaration in 1967, stating, "The survival of our wildlife is a matter of grave concern to all of us in Africa. These wild creatures amid the wild places they inhabit are not only important as a source of wonder and inspiration but are an integral part of our natural resources and of our future livelihood and well-being." The survival of many elements of wild Africa has resulted from the thinking of people such as Nyerere, both from within the African continent and from other parts of the world.

Until recently, African conservation focused on saving its charismatic megafauna. Efforts by Jane Goodall and Dian Fossey to save the great apes of Africa not only attracted converts to the cause of chimpanzees and gorillas but also inspired the career choices of many of today's conservationists. Today, even for first-time tourists to Africa, most of whom venture to the savanna game parks rather than the rainforests or deserts, saving the most charismatic large mammals of the plains—elephants, lions, cheetahs, rhinoceros—is the most obvious conservation need.

Over the past two decades many new themes have been developed to prioritize conservation work. These approaches have developed partly from an improved knowledge of the species, habitats, and important biological processes in the region. They have also developed from a sense of increasing threat to African nature, along with a general consensus that an extinction crisis looms in the coming decades (Gibbons et al. 2000; Groombridge and Jenkins 2000; Pimm and Raven 2000). Some of the most prominent conservation approaches used in Africa are outlined in this chapter.

The History of Setting Conservation Priorities

One of the earliest conservation priorities was to save Africa's large charismatic mammals. This idea remains attractive in Africa, not least because the diversity and abundance of large grazing and browsing herbivores is unparalleled on the planet (Kingdon 1997). A more recent rationale for focusing on large charismatic species is that they need huge protected areas, often linked by habitat corridors, to maintain viable populations. Proponents argue that conservation approaches that focus on the needs of these large "umbrella species" will also protect viable populations of less charismatic and smaller species such as spiders, snakes, and land snails (Sanderson et al. 2002b). These less charismatic species, which are no less important, receive less conservation support from the general public in Africa and elsewhere.

Other biologists argue that we should target most of our efforts to save the species most threatened with extinction, not the big charismatic species that are often more common and fairly well protected (gorillas and rhinos are notable exceptions). The development of the World Conservation Union's (IUCN's) Red Data Books into a systematic program of measuring the extinction risk of species facilitates this approach (Oldfield et al. 1998; Hilton-Taylor 2000). One problem is that although the Red Data Books are adequate for birds and mammals (and partly the trees), many other amphibian, reptile, plant, and invertebrate species are threatened but do not appear in these books (Possingham et al. 2002). In Africa, a focus on Red List species would tend to emphasize the biota of offshore islands and mountains (Cronk 1997; BirdLife International 2000). One risk of focusing all efforts on places important for threatened birds and mammals (and some trees) is the conservation of an unrepresentative set of areas. This misses mainland areas with narrowly distributed reptiles, invertebrates, and plants, which have distribution patterns different from those of birds and mammals.

An alternative line of conservation action emphasizes conserving the last vast tracts of tropical forests and other wilderness areas (Mittermeier et al. 2002). The wilderness area approach favors the conservation of large-scale ecological processes, arguing that unless the process is maintained, the species and habitat elements of biodiversity will inevitability decline (Smith et al. 1993; Cowling et al. 2003). An additional consideration is that by targeting areas where human populations are low, there is a greater chance of success than in regions with heavily converted habitats (Sanderson et al. 2002a). Others note that a focus on wilderness areas would fail to conserve numerous endemic African species that are found in regions with high human populations (Balmford et al. 2001a, 2001b; Burgess et al. 2002b).

Some taxonomists and evolutionary biologists believe that genetic difference should be used to develop conservation plans that aim to represent as much of the existing evolutionary variation as possible (Vane-Wright et al. 1991). Others take a forward-looking approach and suggest focusing conservation strategies around areas that are "speciation machines" to conserve the evolutionary potential of existing lineages (Cowling and Pressey 2001). Problems with these approaches arise from the incomplete genetic knowledge of the species of Africa and from an incomplete understanding of the speciation process. Moreover, the timescales for generating new species are likely to far exceed the time needed for a large proportion of African biodiversity to go extinct due to habitat destruction.

Finally, a minority of scientists feel that we cannot set priorities until we have completed inventories for large parts of unexplored or poorly explored regions of the continent or until the majority of the species have been named. In the ideal world this viewpoint is correct, but unfortunately the threats faced by African species and habitats often are so severe that conservationists need to act using the information that they have, even if it is imperfect.

Out of this complex stew of competing agendas emerge a host of questions. What should be our conservation targets for Africa, and what should we save first? The charismatic megafauna? The rarest and most threatened species? Island biotas? Rare habitats? Representative habitats? Ecological processes such as migrations, rainfall regimes, or predator-prey interactions that help maintain and create Africa's biodiversity? Large intact areas of habitat facing few human pressures? In the following pages we outline the approach we have taken to addressing these competing approaches, in which we have tried to develop an integrated prioritization method that considers all of these issues.

Analytical Approach

In this assessment we have attempted to take a logical approach to setting conservation priorities, one that draws on the best available data, satellite imagery, and expert opinion. Our approach seeks to integrate two different kinds of information: biological distinctiveness (i.e., biological value) and conservation status (i.e., level of conservation opportunity or threat). The method follows that outlined by Dinerstein et al. (1995) and developed further by Ricketts et al. (1999) and Wikramanayake et al. (2002). The approach also aims to ensure that biodiversity priorities are selected from each major biome and hence that there is full representation of biodiversity values in the final set of conservation proposals. Habitat representation is one of the key differences between the prioritization method outlined here and those presented in other schemes (Redford et al. 2003).

In August 1998 WWF convened a workshop in Cape Town, South Africa, to start developing an expert consensus for the methods and data to be used in this assessment. Workshop participants provided input to the development of the ecoregion map for Africa and also helped to compile data on species and threat values for ecoregions. As requested at the 1998 Cape Town workshop, over the past 5 years WWF has systematically gathered data on the biological values, threats, and opportunities for every ecoregion across Africa. Most of the data collected at the 1998 workshop have been superseded by quantified information based on primary data sources. The data-gathering process has involved contact with numerous scientists and conservationists working in Africa, from a wide spectrum of organizations based in Africa and elsewhere around the world.

The production of this book has also benefited from data compiled at several other workshops held in Africa. Foremost among these were the Upper Guinea Forests workshop in

Elmina, Ghana (December 1999), the Congo Basin workshop in Libreville, Gabon (March 2000), the Miombo Woodlands workshop in Harare, Zimbabwe (September 2001), the Lowland Coastal Forests of Eastern Africa workshop in Nairobi, Kenya (February 2002), and the Albertine Rift Mountains workshop in Kampala, Uganda (July 2002 and April 2003).

Biological Distinctiveness Index

The biological value of every ecoregion has been assessed using a number of different criteria (table 3.1; appendixes B and C). The Biological Distinctiveness Index (BDI), developed using these data, is the first main attribute we use to set conservation priorities among the ecoregions of Africa; the other is the Conservation Status Index (appendix D).

The six components of the Biological Distinctiveness Index are described here, grouped under the broad headings of species and nonspecies biological values.

Species Values

We analyzed both species endemism and species richness values of every ecoregion to develop a combined measure of importance for the species complement of every ecoregion (appendixes B and C).

SPECIES ENDEMISM

The most important measure of an ecoregion's distinctiveness is the number of endemic species it supports. For the purpose of this assessment, we define endemism in two ways: strict endemics are species that are confined to a single ecoregion, whereas near-endemics are species with restricted ranges that cross ecoregion boundaries. The criteria used to determine near-endemics were as follows:

- More than 75 percent of the species' global range is found only in one ecoregion or
- The species has a restricted global range (less than 50,000 km^2) but is found in more than one ecoregion. We used a 50,000-km^2 threshold to parallel that used for birds by BirdLife International (see ICBP 1992 and Stattersfield et al. 1998).

In the analysis we used near-endemism as our measure of endemism for four reasons. First, some ecoregions possess no strict endemics but do have near-endemics, whereas some have neither. Therefore, using near-endemics helps to separate ecoregions with some endemism values from those with none. Second, species that are near endemic can become de facto endemics in the near future if their habitat

TABLE 3.1. **Component Variables Used to Develop Biological Distinctiveness and Conservation Status Indices.**

Biological Distinctiveness Index	Conservation Status Index
Species endemism	Habitat loss
Species richness	Remaining habitat blocks
Unusual ecological or evolutionary processes	Degree of fragmentation
Rare habitat type	Degree of protection
Unique higher taxa	Future threat
Large intact ecosystems or wilderness	

is lost or other threats reduce their populations. Third, a high correlation exists between the strict endemic scores and near-endemic scores of the ecoregions in the study area (birds: $r = .841$, $p < .001$; mammals: $r = .845$, $p < .001$; amphibians: $r = .976$, $p < .001$; reptiles: $r = .896$, $p < .001$). Finally, if there is some debate about the boundary of an ecoregion, near-endemism is a more conservative approach to ensuring that the presence of a species in an area is considered in this analysis.

Bird, mammal, reptile, and amphibian strict and near-endemics were recorded from the literature and from expert review (appendix B). For invertebrates and plants we have used an expert assessment of levels of endemism developed at a workshop in Cape Town in 1998, ranked into four richness classes: very high, high, medium, or low within each biome.

Species with restricted ranges in Africa but wide global ranges were not included in the analysis of endemism. Such species occur most often on offshore islands and in the portion of the Palearctic realm in North Africa. Species that have been introduced into Africa were also excluded from the analysis.

SPECIES RICHNESS

For each ecoregion across Africa and neighboring islands we assessed species richness for vertebrates, invertebrates, and vascular plants. For more than 5,900 vertebrate species we assigned species to every ecoregion where they were reported to occur in literature and by experts (appendix B). Estimates of plant species richness were provided by the University of Bonn in Germany (Kier et al. in press), and invertebrate richness was based on relative classes provided by experts

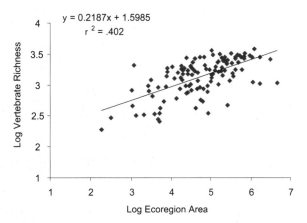

FIGURE 3.1. Relationship between ecoregion area and species richness in the 119 terrestrial ecoregions.

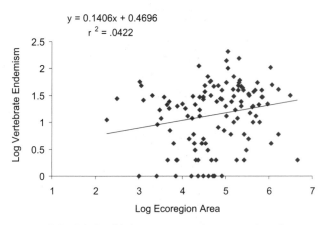

FIGURE 3.2. Relationship between ecoregion area and species endemism in the 119 terrestrial ecoregions.

at the Cape Town workshop. Invertebrate richness was ranked into four richness classes: very high, high, medium, or low within each biome.

Correcting Species Richness Data for Ecoregion Area. A simple summation of the number of species in an ecoregion provides one measure of relative biological importance. However, the largest ecoregion is more than 160,000 times larger than the smallest and therefore should support more species. Our data confirm the importance of the species-area relationship across African ecoregions (figure 3.1). To control for this effect, species richness per unit area ideally would be calculated for every ecoregion by sampling a standardized area in each. Such data are not available across Africa. An alternative is to use a mathematical approach to correct the richness score according to the area of each ecoregion, using a generalized equation developed from species-area studies from many different parts of the world (Rosenzweig 1995):

$$BVd = BV/A^z$$

where z is the species-area exponent, BV is the biological value in question, A is area (in square kilometers), and BVd is the biological value corrected for area. In our analysis, species richness was used as the biological value. We set $z = 0.2$ for mainland ecoregions and 0.25 for islands, which corresponds to empirical results for a wide variety of taxa and ecosystems (Rosenzweig 1995). For the purpose of this analysis, Madagascar was classified as mainland.

By using these equations we were able to generate residual richness scores from a regression analysis, which we then used to rank ecoregions in terms of their species richness with the effects of area removed.

A much weaker relationship exists in our data between ecoregion area and the number of endemic species (figure 3.2). This is probably because ecoregions often were delineated around areas of high endemism, such as a group of isolated mountaintops or small offshore islands. Because these relationships were not statistically significant, we did not correct endemism scores by ecoregion area.

Combining Endemism and Richness. To combine these values we first calculated the total number of near-endemic species of birds, mammals, reptiles, and amphibians in each ecoregion. For each taxon, endemism scores were then normalized to a range of 0–100 within each biome, allowing a relative comparison between taxa (appendix E). Biomes were used at this stage to adequately represent different habitats among the most important ecoregions for species. The expert-derived classes of plant and invertebrate endemism were also converted to numerical scores within biomes ("very high" translated to a score of 100, "high" to a score of 67, "medium" to a score of 33, and "low" to a score of 1). In this way, each taxonomic group contributes equally to further calculations, regardless of the number of species it contains. Although the total number of near-endemics would have offered a more precise reflection of the value of an ecoregion, our lack of quantitative data for plants and invertebrates prevented this approach. Simply summing the raw scores of endemism would also depress the relative importance of less speciose or lesser-known taxa (e.g., amphibians).

Next, the normalized endemism scores for each ecoregion were totaled, and the ranked list of ecoregions was divided into quartiles within the continent. The top 25 percent of ecoregions were assigned to the globally outstanding category, the next 25 percent to the regionally outstanding category, the next 25 percent to the bioregionally outstanding category, and the last 25 percent to the locally important category.

Finally, we adjusted these categories of endemism using the area-adjusted species richness calculations. Ecoregions that contained exceptional species richness for their size were identified (i.e., residual richness values, after controlling for area). The top 10 percent of ecoregions in terms of species richness within each biome were considered globally outstanding, regardless of their endemism ranking (appendix E). For example, a desert ecoregion that is regionally outstanding for endemism but contains exceptional species richness for its area would be elevated to globally outstanding for combined endemism and richness. If an ecoregion were not exceptionally rich in species (90 percent of the ecoregions in each biome), then its richness score would make no difference to its ranking in our combination of endemism and richness.

Many additional analyses could be performed on our richness and endemism data, and many different analytical approaches could be used. Some examples of possibilities are outlined in recent publications (Williams 1998; Margules and Pressey 2000; Williams et al. 1999, 2003; Balmford 2002). We present the raw species richness and endemism scores for each ecoregion, along with our calculations of biodiversity value based on this method (appendixes E and G).

Nonspecies Biological Values

Species data allow the recognition of one set of ecoregions of outstanding biological importance, but the use of species data alone does not fully describe the biological distinctiveness of an ecoregion. The WWF Conservation Science Program is guided by the four goals of conservation biology developed by Noss (1992). Only one of these principles relates specifically to species-based conservation targets (box 3.1), so we have tried to include criteria for measuring the other (nonspecies) biological attributes of ecoregions.

A current challenge in conservation biology is to develop means to measure these nonspecies characteristics of biodiversity in a way that is objective and can be analyzed simply (Cowling et al. 1999b, 2003; Mace et al. 2000). We have developed simple methods to analyze four nonspecies biological values across the ecoregions of Africa and its islands: ecological and evolutionary phenomena, globally rare habitat types, higher taxonomic uniqueness, and large intact ecosystems and wilderness (table 3.1; appendix C).

Initial assessments were completed by experts at the 1998

BOX 3.1. **The Four Goals of Conservation Biology (Noss 1992).**

- Represent, in a system of protected areas, all native ecosystem types and serial stages across their natural range of variation.

- Maintain viable populations of all native species in natural patterns of abundance and distribution.

- Maintain ecological and evolutionary processes, such as disturbance regimes, hydrological processes, nutrient cycles, and biotic interactions, including predation.

- Design and manage the system to respond to short-term and long-term environmental change and to maintain the evolutionary potential of lineages.

---

workshop in Cape Town, updated at several regional workshops, and finalized according to literature and expert opinion at WWF-US (appendix F). Each of the nonspecies attributes is described in this section.

UNUSUAL EVOLUTIONARY OR ECOLOGICAL PHENOMENA

This attribute attempts to quantify the distinctiveness of an ecoregion in terms of the evolutionary and ecological phenomena that can be conserved only in a limited number of areas (appendix C). Examples of evolutionary phenomena include adaptations to extreme environmental conditions among multiple taxa (e.g., the tops of some African mountains) or the evolution of species radiations (e.g., the fynbos in South Africa). Examples of ecological phenomena include the presence of intact vertebrate faunas or intact predator-prey interactions (an increasingly rare phenomena around the world), vast migrations of large vertebrates (also increasingly rare around the world), or extraordinary aggregations of breeding or nonbreeding vertebrates (such as bird breeding colonies or migratory stopover sites) (appendix C).

RARE HABITAT TYPE

Some habitats are so rare that there are only a few opportunities to achieve their conservation around the world. We used the results of a recent analysis to assess the global rarity of different habitats (Olson and Dinerstein 1998; Olson et al. 2001). These results were used to develop criteria for habitat rarity that we applied to each of the ecoregions of Africa, allowing the recognition of ecoregions globally important for their rare habitat type (appendix C).

HIGHER TAXONOMIC UNIQUENESS

Another measure of distinctiveness is uniqueness in terms of higher taxa, represented by endemic genera and families. Because genera and families are a higher level than species in the taxonomic hierarchy, narrowly endemic examples contribute more to an ecoregion's biotic distinctiveness than an endemic species. Such endemic higher taxa often represent relict or primitive groups of organisms that have become extinct over most of their former range. A good African example is the plant *Welwitchia mirabilis,* which is the only living member of its family. The criteria used to assess this attribute are presented in appendix C.

LARGE INTACT ECOSYSTEMS OR WILDERNESS

Intact ecosystems (or wilderness areas) often are the last places where ecosystem processes continue to function naturally. Intact ecosystem function is increasingly rare around the world, so we have identified the ecoregions that have the largest percentage of wilderness, following the recent literature (McCloskey and Spalding 1989; Bryant et al. 1997; Sanderson et al. 2002a; Mittermeier et al. 2002).

Integrating Species and Nonspecies Biological Values

To arrive at the final BDI for each ecoregion, we combined the species and nonspecies biological attributes of every ecoregion as follows. First, ecoregions were ranked into four classes of importance according to their species endemism and species richness values (appendix E). These classes were globally outstanding, regionally outstanding, bioregionally outstanding, and locally important. Second, we summed the nonspecies scores of every ecoregion (appendix F). Ecoregions scoring above 1.5 for nonspecies biological values (an arbitrary choice) were elevated to globally outstanding in the BDI, regardless of their values for species. For example, the Central Congolian Lowland Forest [15] is bioregionally outstanding based on its summed endemism and richness values but is elevated to globally outstanding in the final BDI when nonspecies values are considered.

The final ranking of ecoregions into four classes of importance, after species and nonspecies values had been considered, represented our final ranking for the BDI (appendix E). This information was used in later analyses.

Conservation Status Index

The Conservation Status Index (CSI) measures the degree of habitat alteration and habitat conservation of ecoregions and is the second major index we use to set priorities among the ecoregions of Africa (the other being the BDI).

Criteria for Assessing Current Conservation Status

The Current Conservation Status Index is an indication of current habitat integrity, developed using spatial data on the amount of habitat loss, presence of remaining large blocks, degree of habitat fragmentation, and percentage of protection (appendix D).

HABITAT LOSS

Habitat loss is known to be the most important threat to global biodiversity (Heywood and Watson 1995; Stattersfield et al. 1998; BirdLife International 2000; Groombridge and Jenkins 2000). Habitat loss reduces both the species and nonspecies biological values of ecoregions by removing the places where species can exist. We calculated habitat loss as the percentage of habitat that has been converted to farmland or urban areas, using geographic information system (GIS) analysis of a global landcover database (Hansen et al. 2000) combined with a global population database (Dobson et al. 2000) (see appendix D for details).

HABITAT BLOCKS

Large blocks of habitat can continue to support species and nonspecies values of an ecoregion even if there has been habitat loss across the region as a whole. Remaining habitat blocks for each ecoregion were calculated using GIS software and classified according to predetermined size thresholds (appendix D). The size thresholds for habitat blocks were tailored to each biome and aim to reflect the different scales at which ecological processes operate (Dinerstein et al. 1995; Ricketts et al. 1999; appendix D). For example, the species and ecological processes of small, naturally disjunct montane forest ecoregions have different area requirements from those of lowland forest ecoregions.

HABITAT FRAGMENTATION

As habitats are lost and habitat blocks are isolated from each other, fragmentation rates increase and species and nonspecies biological values decline. Small habitat fragments can remain important for maintaining populations of species with small area requirements. However, smaller areas are less capable of surviving large-scale disturbance such as fires and are less able to support large mammals with wider area requirements. Extensive fragmentation robs the landscape of the continuous habitat needed by some species and renders more sites susceptible to damaging disturbance

and invasion by non-native species. We have calculated fragmentation within each ecoregion as the edge-to-area ratio
of remaining habitat blocks (appendix D).

HABITAT PROTECTION

The area of habitat in protected areas provides an assessment
of the likelihood that species and nonspecies biological values might persist across an ecoregion. We calculated the percentage of each ecoregion that was protected using the Protected Areas Databases from the World Conservation
Monitoring Centre (WCMC 2002) and South Africa CSIR,
Environmentak (for South Africa protected area data only;
see appendix D). Only protected areas classified under IUCN
categories I–IV (IUCN 1998) were considered, although we
acknowledge that in some countries local and private parks
and government forest reserves provide protection for
species and nonspecies biological values that is not captured
in this analysis.

Development of the CSI

To calculate the CSI we used the results of the analysis of
habitat loss, habitat blocks, habitat fragmentation, and
habitat protection for every ecoregion. A weighted score was
then applied to each of these attributes, giving a point range
from 0 to 100 for the Current Conservation Status Index,
with higher values denoting a higher level of endangerment
(appendix D; table 3.2). Using this index, we assigned ecoregions to the five conservation status categories (critical, endangered, vulnerable, relatively stable, and relatively intact)
in a descending order of threat. The logic of these categories
is similar to that used in IUCN Red Data Books for threatened species (Mace and Lande 1991; Hilton-Taylor 2000).

Final (Future Threat-Modified) Conservation Status

Whereas the Current Conservation Status Index provides a
measure of the current status of each ecoregion, the Final
Conservation Status indicates where biodiversity might be
most threatened in the coming years. We have used two sets
of data to estimate future threats, one based on threat to
species and the other on threats to habitats.

The recently completed Red Data Book (Hilton-Taylor
2000) provides lists of species that are threatened with extinction in a number of different categories of threat. For
this index we used only the threatened bird and mammal
data because the majority of reptile and amphibian species
have not been assessed for their threats, and information
available for invertebrates is even sketchier. Assessments of

**TABLE 3.2. Relationship between Summed Scores for
Threat and the Conservation Status Index.**

Conservation Status Index	Summed Score for Threats*
Critical	80–100 points
Endangered	60–79 points
Vulnerable	40–59 points
Relatively stable	20–39 points
Relatively intact	0–19 points

*Habitat loss, habitat blocks, fragmentation, and habitat protection.

threat are available for trees (Oldfield et al. 1998; Walter and
Gillett 1998; Farjon and Page 1999), but the lack of detailed
species distribution data also prevented our use of these data.

A model of human population density in the year 2025
provides an indication of future habitat threat, assuming
that there is a positive relationship between human population density and habitat loss or degradation. We assigned
point values for both these attributes of future threat (appendix D).

The Final (Future Threat-Modified) Conservation Status
was calculated for every ecoregion by modifying Current
Conservation Status using the estimates of future threat. The
procedure was to elevate the ecoregions ranking highest (top
30 percent) for future threats (appendix D) by one level in
the CSI; for example, an endangered ecoregion scoring
highly for future threat would be elevated to critical.

Integrating Biological Distinctiveness and Conservation Status

The BDI and CSI describe the biological importance and
threat for every ecoregion. However, these indexes provide
no guidance on where conservation investment should be
targeted and therefore are not a complete prioritization system. For example, some biologically important ecoregions
might be so degraded that their biodiversity will decline
without intensive restoration efforts and protection of the
few remaining natural sites. Alternatively, some ecoregions
may offer cost-effective opportunities to maintain the ecological integrity of very large landscapes over the long term.
We have attempted to separate ecoregions into different
paths for conservation investment by integrating the results
of the BDI and CSI. To do this we used a matrix developed
by Dinerstein et al. (1995) and modified by Ricketts et al.
(1999). In this matrix, the biological distinctiveness cate-

TABLE 3.3. **Integration Matrix Used to Assign Ecoregions to Different Conservation Priority Classes.**

Biological Distinctiveness Index	Conservation Status Index				
	Critical	Endangered	Vulnerable	Relatively Stable	Relatively Intact
Globally outstanding	I	I	I	III	III
Regionally outstanding	II	II	II	III	III
Bioregionally outstanding	IV	IV	V	V	V
Nationally important	IV	IV	V	V	V

gories are arranged along the vertical axis, with the conservation status categories along the horizontal axis (table 3.3). Ecoregions are allotted to one of twenty cells in the matrix.

Each cell in the matrix is assigned to one of five classes (I–V) that reflect the nature and extent of the management activities likely to be needed for effective biodiversity conservation:

Class I: globally outstanding ecoregions that are highly threatened. Conservation actions in these ecoregions must be immediate to protect the remaining native species and habitats.

Class II: regionally outstanding ecoregions that are highly threatened. Conservation actions must be immediate to protect the remaining native species and habitats, but the overall biological value is lower than that of the Class I ecoregions.

Class III: ecoregions with globally or regionally outstanding biodiversity values that are not highly threatened at present. These ecoregions are some of the last places where large areas of intact habitat and associated species assemblages might be conserved.

Class IV: bioregionally outstanding and nationally important ecoregions that are highly threatened. Conservation actions are needed to protect the remaining native species and habitats, but the overall biological values are lower than those of Class II and much lower than those of Class I ecoregions.

Class V: bioregionally outstanding and nationally important ecoregions that are not particularly threatened at present. These ecoregions are some of the last

places where large areas of intact habitat and associated species assemblages might be conserved, but they are less important biologically than the Class III ecoregions.

This ranking of ecoregions into classes outlines an order of priority for addressing conservation needs by focusing on the most distinctive units of biodiversity when faced with limited resources. It does not imply that some ecoregions are unimportant. Biodiversity conservation is needed in every ecoregion to conserve the distribution range and genetic diversity of species and the distribution and extent of habitats and to maintain ecological processes. Every ecoregion also provides important ecosystem services that provide benefits to people all over Africa.

Because others might want to analyze our data for different purposes, data for each ecoregion are summarized in appendixes E and H. For example, other people may assign different values to the cells in our integration matrix and decide for themselves which classes of ecoregions they should focus their efforts on. To some the Class I ecoregions in the top left corner of the matrix might be regarded as too costly and difficult to achieve conservation. Others may decide that the Class I ecoregions are where the most effort should be expended to prevent mass extinction.

In the next four chapters we present the results of biological analyses, illustrate patterns in the distribution of threat, and outline where conservation should focus its attention in coming years.

CHAPTER 4

From Fynbos to the Mountains of the Moon: Ranking the Biological Value of Ecoregions

Africa contains natural treasures that offer glimpses into the past and reveal evolutionary processes still at work. The Cape Floristic Region of South Africa features plant groups with a history stretching back to the ancient continent of Gondwanaland. In the same part of South Africa, some plant genera have undergone a burst of speciation in the past million years, resulting in members of ancient plant lineages living side by side with newly evolved species. In the forests of West Africa the diversity of primates is among the highest in the world; even after 100 years of study the number of species is still subject to intense debate. In East Africa, some of the most unusual plants in the world are found above 3,500 m on the high volcanic peaks. Able to survive daily temperature differences of up to 40°C (from −20°C to +20°C), small ancestors evolved in isolation and over time attained the stature of small trees. Around the bases of the same volcanoes and stretching out across the wide East African plains reside the greatest concentrations and migrations of large wild herbivores in the world. They form the prey base for perhaps the largest and most intact assemblage of large predators on Earth.

An even greater divergence in flora and fauna is found on the islands offshore from mainland Africa. Madagascar contains groups of animals and plants that have long been extinct elsewhere in the world. For example, the diversity of primitive primates (lemurs) harks back to the fauna of millions of years ago. Until their extinction in the past 2,000 years, the giant ground birds of Madagascar would have also

been a unique feature of the island and a further link to ancient faunas that have gone extinct elsewhere. These spectacular ground birds and some larger species of ground-dwelling lemurs probably were hunted to extinction soon after the first arrival of people on the island (Martin and Klein 1984).

The distinctiveness of Africa's biological wealth is indisputable. Yet biodiversity varies across the landscape, with some areas more distinct or speciose than others. As described in chapter 3, we use a Biological Distinctiveness Index (BDI), which includes measures of species endemism, species richness, and biological processes (nonspecies values), to compare the biological priority of ecoregions across Africa and its islands. In this chapter we present the results of this analysis across Africa, in terms of sets of priority ecoregions for various measures of biological worth, brought together into a single ranked prioritization by the end of the chapter. Anthropogenic aspects are assessed within a Conservation Status Index, presented in chapter 5.

Richness

Vertebrate species richness is unevenly distributed across the ecoregions of Africa and offshore islands (figure 4.1), with the tropical forest and savanna-woodland ecoregions generally being the most speciose. Sixty-two of the 119 ecoregions contain more than 500 species of vertebrates

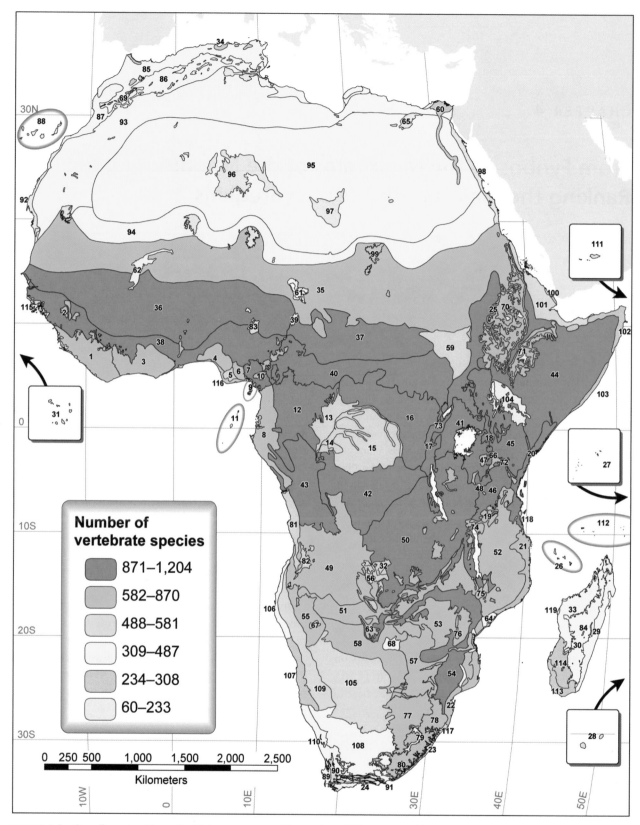

FIGURE 4.1. Overall species richness for birds, mammals, reptiles, and amphibians across African ecoregions. Classes are based on quantile groupings of data values.

(the combined total of all birds, mammals, reptiles, and amphibians), and four ecoregions (the Northern *Acacia-Commiphora* Bushlands and Thickets [45], Northern Congolian Forest-Savanna Mosaic [40], Albertine Rift Montane Forests [17], and Central Zambezian Miombo Woodlands [50]) contain more than 1,000 vertebrate species. Most of these exceptionally rich ecoregions are large savanna woodlands with inclusions of other habitats, such as wetlands. The other exceptionally rich area is the Albertine Rift [17], the largest mountain ecoregion in Africa, which covers an altitudinal and climatic gradient from tropical lowland rainforest to montane grasslands.

Birds

Bird species richness (figure 4.2a, appendix E) is highest in eastern Africa around the Albertine Rift Montane Forests [17], Victoria Basin Forest-Savanna Mosaic [41], East African Montane Forests [18], Northern Congolian Forest-Savanna Mosaic [40], and then into the *Acacia-Commiphora* Bushlands and Thickets [45, 46] and the Central Zambezian Miombo Woodlands [50]. The large size of these ecoregions, their high level of habitat heterogeneity, and their presence on a migratory flyway explains this pattern. The next highest band of species richness is found across the remainder of the tropical belt, with the exception of the western portion of the Upper Guinea forests and the center of the Congo Basin. The ecoregions of Madagascar and other offshore islands all have much lower bird species richness than mainland Africa. These results fit with other published assessments of the overall patterns of bird richness in Africa and its islands (e.g., Moreau 1966; Crowe and Crowe 1982; Pomeroy and Lewis 1987; Brooks et al. 2001a; de Klerk et al. 2002a, 2002b).

Mammals

Similar to bird richness patterns, the highest rates of species richness in mammals occur in eastern Africa, from the Albertine Rift Montane Forests [17], Victoria Basin Forest-Savanna Mosaic [41], *Acacia-Commiphora* Bushlands and Thickets [45, 46], down through to the Central Zambezian Miombo Woodlands [50] (figure 4.2b, appendix E). There is significant overlap between the ecoregions with the highest richness scores for birds and mammals (table 4.1). As with birds, all of the ecoregions in Madagascar and the offshore islands have much lower species richness than ecoregions in corresponding biomes on the mainland. Furthermore, dif-

Bearded vulture or lammergeier (*Gypaetus barbatus*), Ethiopia. Photo credit: WWF-Canon/Michel Gunther

Mountain gorilla (*Gorilla beringei beringei*), Virunga National Park, Democratic Republic of Congo. Photo credit: WWF-Canon/Martin Harvey

ferences in the size of ecoregions affect the species richness of mammals across ecoregions, with the large savanna ecoregions also supporting higher species richness through the many smaller-scale habitat features that they incorporate. The patterns found for mammals largely coincide with those observed in other studies (Turpie and Crowe 1994; Brooks et al. 2001a).

Reptiles

For reptiles, species richness is higher in some of the drier ecoregions in the Horn of Africa and southwest Africa than it is for other vertebrate groups. The ecoregions ranking highest in reptile richness are found in East Africa: the Northern Zanzibar–Inhambane Coastal Forest Mosaic [20], the Central Zambezian Miombo Woodlands [50], the So-

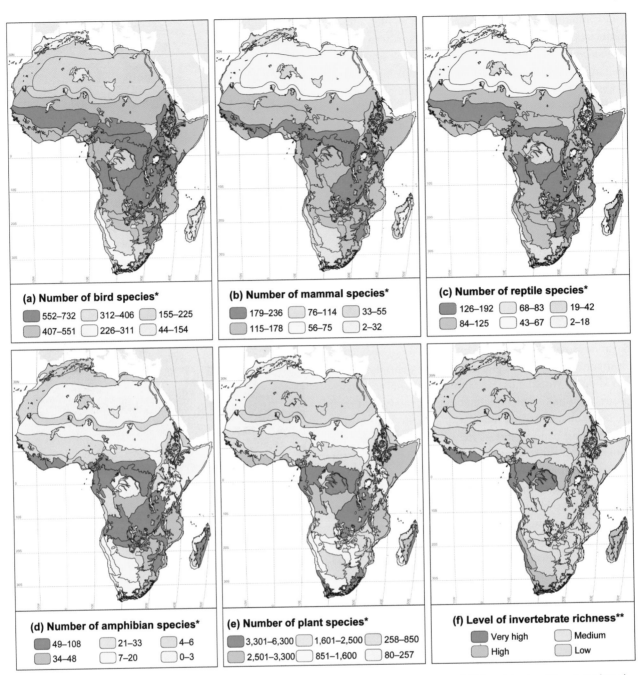

FIGURE 4.2. Species richness of (a) birds, (b) mammals, (c) reptiles, (d) amphibians, (e) plants, and (f) invertebrates. *Classes are based on quantile groupings of data values. **Classes are based on expert assessment of relative levels of richness within biomes.

mali *Acacia-Commiphora* Bushlands and Thickets [44], and the Zambezian and Mopane Woodlands [54] (figure 4.2c, appendix E). High reptile richness in the latter may result from the mixing of wetter forest and drier savanna communities, each with its own particular reptile assemblages. The next highest rates of richness are found across a wide range of ecoregions, including the forests of Madagascar, the forest-savanna mosaics around the Congo Basin, and the drier woodland-savanna ecoregions of western, eastern, and southern Africa. No other studies have looked at species richness patterns for reptiles across this region, although Brooks et al. (2001a) have presented results for snakes.

Biome	No. of Ecoregions	Birds	Mammals	Amphibians	Reptiles	Total
Tropical and subtropical moist broadleaf forests	30	386 (31.75)	119 (11.02)	45 (5.83)	83 (7.65)	633
Tropical and subtropical dry broadleaf forests	3	203 (92.9)	51 (25.48)	12 (5.29)	44 (20.26)	310
Temperate coniferous forests	1	218	60	3	27	308
Tropical and subtropical grasslands, savannas, shrublands, and woodlands	24	519 (20.18)	166 (7.47)	33 (3.70)	109 (8.28)	827
Flooded grasslands and savannas	10	276 (35.53)	75 (7.60)	9 (3.58)	30 (8.34)	390
Montane grasslands and shrublands	16	306 (38.7)	93 (10.12)	26 (3.32)	44 (6.9)	469
Mediterranean forests, woodlands, and scrub	7	245 (28.75)	69 (10.41)	13 (4.43)	54 (8.07)	381
Deserts and xeric shrublands	23	188 (19.77)	58 (5.85)	8 (1.04)	51 (6.44)	305
Mangroves	5	198 (19.73)	23 (4.92)	2 (0.20)	7 (1.32)	230

Amphibians

The distribution of amphibian richness is very different from that of reptiles. The highest levels of amphibian richness occur in the forests of coastal Nigeria, Cameroon, and Gabon, the lowland and montane forests of Madagascar, and the Central Zambezian Miombo Woodlands [50] and Eastern Arc Forests [19] (figure 4.2d, appendix E). Although most areas with high amphibian richness are wet forested areas, an exception to this general trend is the drier Central Zambezian Miombo Woodlands [50] of eastern to southern Africa. Its elevated richness can be partly explained by its huge size but also relates to the high habitat heterogeneity, particularly the presence of many small wetlands that are favored by amphibian species. The species richness patterns we present are similar to those reported elsewhere (Poynton and Broadley 1991; Poynton 1995, 1999; Schiøtz 1999; Brooks et al. 2001a).

Plants

The highest levels of plant species richness occur in the western coastal portions of the lowland rainforest in Cameroon and Gabon [7, 8], in the Madagascar Humid Forests [29], and in the Fynbos [89, 90] and Succulent Karoo [110] of South Africa (figure 4.2e, appendix E). The next highest level is found across the forest belt of tropical Africa, in the wetter parts of Madagascar, and in the majority of the coastal regions of South Africa. The exceptional richness of plants in

Golden frog (*Mantella aurantiaca*), southeastern Madagascar. Photo credit: WWF-Canon/Martin Harvey

Tropical rainforest flower on riverbank, moist forest in the western Congo Basin, Gabon. Photo credit: WWF-Canon/Martin Harvey

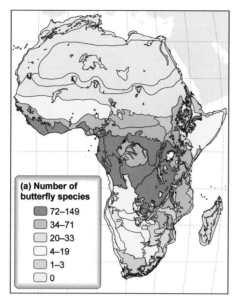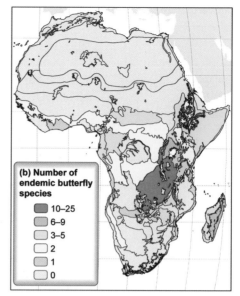

FIGURE 4.3. **(a)** Richness and **(b)** endemism for two subfamilies of butterflies: Acraeinae and Charaxinae (*N* = 426 species; data provided by Jacques Pierre, Muséum National d'Histoire Naturelle, Paris). Classes are based on quantile groupings of data values.

the Mediterranean scrub ecoregions of the southern and southwestern tip of Africa is well known (Hilton-Taylor 1987; Cowling et al. 1997a; Mittermeier et al. 1999) and is markedly different from the patterns seen for vertebrate animals. The high rates of plant richness in the lowland forests of Bight of Biafra and in eastern Africa are also well recognized (Brenan 1978; Lovett et al. 2000b; Kier and Barthlott 2001; Linder 2001; La Ferla et al. 2002; Mutke et al. 2001). In this area the high plant richness corresponds with a similar high species richness for vertebrate animals.

Invertebrates

Experts described areas of particularly high invertebrate species richness in tropical forests and some drier ecoregions (figure 4.2f). Among forests, highest richness occurs in the rainforests of the Upper Guinea and the western portion of the Congo Basin. The easternmost (humid) forests of Madagascar also have high invertebrate richness. In drier habitats, the Fynbos, Succulent Karoo, and Namib Desert are richest in invertebrates. To an appreciable extent these estimated patterns of invertebrate richness parallel those of plants, which is not surprising when the life histories of many invertebrates are closely tied to a few (or even a single) plant species. Similar patterns of species richness are reflected in quantitative data from two butterfly families (figure 4.3). In these examples, species richness is concentrated across the tropical belt,

especially in the Upper Guinea Forests, Albertine Rift, Congo Forests, and Central Zambezian Miombo Woodlands.

Distribution of Species Richness across Biomes

The African ecoregions fall in nine different biomes, which have different intrinsic levels of species richness. Average species richness across all groups of vertebrates varies from more than 820 species per ecoregion in the tropical and subtropical grasslands, savannas, shrublands, and woodlands down to around 230 species per ecoregion in the mangroves (and many of these are temporary visitors or are associated with mangrove margins). There is also wide variation in species richness between different taxonomic groups; for example, the tropical moist forest ecoregions have much higher amphibian species richness (forty-five species per ecoregion) than mangrove (two) or desert (eight) ecoregions (table 4.1).

The relationships between richness scores of four vertebrate taxa across the different biomes show a high degree of congruence, especially between mammals and birds ($r =$ .946, $p <$ .001), the two groups with the best data. When these relationships are analyzed separately for each biome, there are also numerous significant correlations (appendix G). The strength of these relationships results partly from area effects, with larger ecoregions tending to have more species in each taxonomic group. However, when species richness scores are corrected for the effects of area, the cross-taxon data are still significantly correlated. The implication is that similar factors drive species richness in different groups of organisms across Africa and its islands (also see Balmford et al. 2001a, 2001b; Jetz and Rahbek 2002).

A comparison of species richness across all ecoregions within a biome obscures the finer details of relative species richness between ecoregions within the same biomes. We ranked ecoregions within the specific biomes to provide a better understanding of richness patterns under similar environmental conditions (table 4.2). For the tropical moist forests, there is a gradual decline in richness from around 1,000 vertebrate species in the Albertine Rift Montane Forests [17] and Northern Zanzibar–Inhambane Coastal Forest Mosaic [20], which are large and heterogeneous, down to less than 100 vertebrates on the small offshore islands of the Granitic Seychelles [27] and Mascarenes [28]. The same progression of richness from large tropically located ecoregions with high richness to offshore island ecoregions with low species richness is also seen in the other biomes (table 4.2, appendix E).

Endemism

Across Africa, total numbers of animal and plant endemics vary widely between ecoregions. Some ecoregions contain hundreds of endemic species, whereas others have few or even none. Among the vertebrates, the highest numbers of endemics are found in the humid forests in eastern Madagascar and in the Albertine Rift Mountains on the mainland. Slightly lower levels of endemism are found in the Eastern Arc Mountains, coastal lowland forests of Kenya and Tanzania, savanna woodlands of the Horn of Africa, the Cameroon Highlands, drier forests of western Madagascar, the forests and woodlands of the Drakensberg, and the western Guinea forests (figure 4.4).

Birds

The highest concentrations of bird endemics are found in the Albertine Rift Montane Forests [17], Eastern Arc Forests [19], and Madagascar Humid Forests [29]. Slightly lower numbers of endemics are found in the Mount Cameroon and Bioko Montane Forests [9], the Cameroon Highlands Forests [10], various smaller offshore islands (Comoros [26], Mascarenes [28], and São Tomé, Príncipe, and Annobon Moist Lowland Forests [11]), and the Northern Zanzibar–Inhambane Coastal Forest Mosaic [20] (figure 4.5a). Among drier ecoregions the highest levels of bird endemism are in

Verreaux's sifaka (*Propithecus verreauxi*), Madagascar. Photo credit: WWF-Canon/Martin Harvey

Madagascar and the Horn of Africa. There is considerable agreement between this set of priority ecoregions for bird endemism, the regions defined as endemic bird areas by BirdLife International (Stattersfield et al. 1998), and those identified in other studies of bird endemism across Africa (Moreau 1966; Crowe and Crowe 1982; Pomeroy and Ssekabiira 1990; Brooks et al. 2001a; de Klerk et al. 2002a, 2002b).

Mammals

Mirroring the distribution of bird endemism, the highest number of mammal endemics occurs in the Albertine Rift Montane Forests [17] on the African mainland and the Madagascar Humid Forests [29] of eastern Madagascar (figure 4.5b). Following in importance are the Upper Guinea forests, the Kenyan Highlands, and the arid Horn of Africa. Of slightly lower importance are the Eastern Arc Mountains, the Northwestern and Northeastern Congolian lowland forests, and the Cameroon Mountains. Other sum-

TABLE 4.2. **Top Ten Ecoregions According to Vertebrate Richness in the Six Largest Biomes of Africa.**

Tropical and Subtropical Moist Broadleaf Forests	*Species Richness*	*Tropical and Subtropical Grasslands, Savannas, Shrublands, and Woodlands*	*Species Richness*
Albertine Rift Montane Forests [17]	1,149	Central Zambezian Miombo Woodlands [50]	1,205
Northern Zanzibar–Inhambane Coastal Forest Mosaic [20]	969	Northern Congolian Forest-Savanna Mosaic [40]	1,029
Northeastern Congolian Lowland Forests [16]	931	Northern *Acacia-Commiphora* Bushlands and Thickets [45]	1,012
Northwestern Congolian Lowland Forests [12]	909	Victoria Basin Forest-Savanna Mosaic [41]	988
East African Montane Forests [18]	886	Zambezian and Mopane Woodlands [54]	960
Cross-Sanaga-Bioko Coastal Forests [7]	885	East Sudanian Savanna [37]	937
Eastern Arc Forests [19]	870	Guinean Forest-Savanna Mosaic [38]	908
Atlantic Equatorial Coastal Forests [8]	866	Southern Congolian Forest-Savanna Mosaic [42]	899
Eastern Guinean Forests [3]	855	Somali *Acacia-Commiphora* Bushlands and Thickets [44]	895
Cameroon Highlands Forests	838	Western Congolian Forest-Savanna Mosaic [43]	889

Flooded Grasslands and Savannas	*Species Richness*	*Montane Grasslands and Shrublands*	*Species Richness*
Zambezian Flooded Grasslands [63]	754	Drakensberg Montane Grasslands, Woodlands, and Forests [78]	767
Sudd Flooded Grasslands [59]	525	Southern Rift Montane Forest-Grassland Mosaic [74]	738
Zambezian Coastal Flooded Savanna [64]	469	Ethiopian Upper Montane Forests, Woodlands, Bushlands, and Grasslands [70]	728
Lake Chad Flooded Savanna [61]	381	South Malawi Montane Forest-Grassland Mosaic [75]	616
Inner Niger Delta Flooded Savanna [62]	366	Highveld Grasslands [77]	609
Makgadikgadi Halophytics [68]	323	Eastern Zimbabwe Montane Forest-Grassland Mosaic [76]	595
East African Halophytics [66]	304	Angolan Scarp Savanna and Woodlands [81]	577
Etosha Pan Halophytics [67]	293	Maputaland-Pondoland Bushland and Thickets [80]	544
Saharan Halophytics [65]	182	Angolan Montane Forest-Grassland Mosaic [82]	540
		Ethiopian Montane Moorlands [71]	479

Mediterranean Forests, Woodlands, and Scrub	*Species Richness*	*Deserts and Xeric Shrublands*	*Species Richness*
Montane Fynbos and Renosterveld [90]	504	Kalahari Xeric Savanna [105]	551
Lowland Fynbos and Renosterveld [89]	466	Namib Escarpment Woodlands [109]	538
Mediterranean Woodlands and Forests [85]	441	Ethiopian Xeric Grasslands and Shrublands [101]	518
Albany Thickets [91]	417	Nama Karoo [108]	487
Mediterranean Dry Woodlands and Steppe [86]	374	Masai Xeric Grasslands and Shrublands [104]	463
Mediterranean *Acacia-Argania* Dry Woodlands and Succulent Thickets [87]	357	Mediterranean *Acacia-Argania* Dry Woodlands and Succulent Thickets [87]	357
Canary Islands Dry Woodlands and Forests [88]	109	Kaokoveld Desert [106]	366
		South Saharan Steppe [94]	350
		Sahara Desert [95]	337
		North Saharan Steppe [93]	327

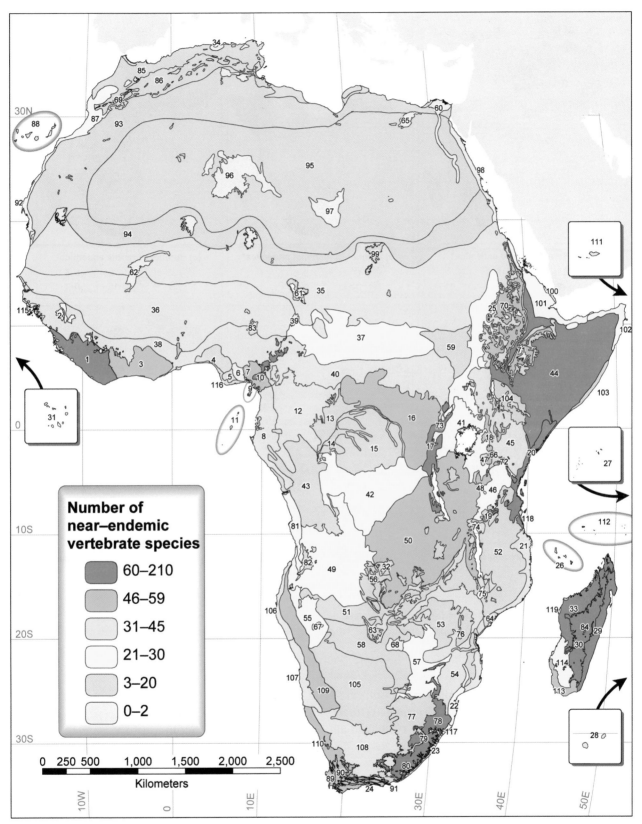

FIGURE 4.4. Combined near endemism for birds, mammals, reptiles, and amphibians across African ecoregions. Classes are based on quantile groupings of data values.

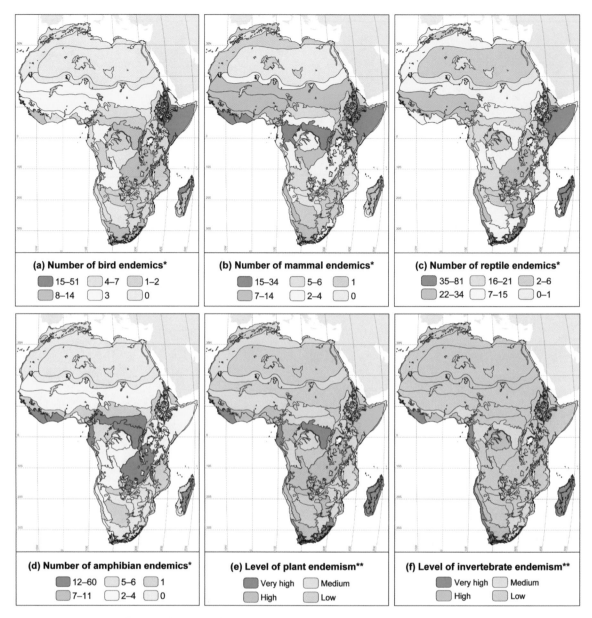

FIGURE 4.5. Species endemism of (a) birds, (b) mammals, (c) reptiles, (d) amphibians, (e) plants, and (f) invertebrates. *Classes are based on quantile groupings of data values. **Classes are based on expert assessment of relative levels of richness within biomes.

maries of the patterns of mammalian endemism across mainland Africa exist separately for large mammals (Turpie and Crowe 1994), primates (Hacker et al. 1998), and all species across sub-Saharan Africa (Brooks et al. 2001a). Regional assessments of mammalian endemism have also been undertaken in South Africa (Gelderblom and Bronner 1995; Gelderblom et al. 1995) and in the Congo Basin and West African forests (Happold 1996). In general our large-scale patterns of mammal endemism agree with the results of these previous studies.

Reptiles

The pattern of reptile endemism departs somewhat from that observed for birds and mammals, in particular in the Horn of Africa and southwest Africa (figure 4.5c). In the forests, the most important areas are the Madagascar Humid [29] and Sub-humid [30] Forests and the Northern Zanzibar–Inhambane Coastal Forest Mosaic [20]. In the drier areas, the most important area is the Somali *Acacia-Commiphora* Bushlands and Thickets [44] of the Horn of Africa, followed by the Succu-

Johnston's chameleon (*Chamaeleo johnstoni*), Albertine Rift Mountains. Photo credit: WWF-Canon/ Martin Harvey

Welwitschia (*Welwitschia mirabilis*), Namib Desert, Namibia. Photo credit: WWF-Canon/Frederick J. Weyerhaeuser

lent Karoo [110] and the Madagascar Spiny Thickets and Dry Forests [33, 113]. There are no previous assessments of the distribution of reptilian endemism across Africa and its islands, although Hughes (1983) and Brooks et al. (2001a) presented assessments of the snake faunas of Africa, and Lombard et al. (1995) looked at conservation priorities for the snakes of South Africa.

Amphibians

Amphibians are more confined to the wetter parts of Africa than other vertebrate groups, and endemism levels increase in wetter forest ecoregions, especially those on mountains. The Humid [29] and Subhumid [30] Forests of Madagascar have the highest levels of endemism (figure 4.5d). Somewhat lower levels are found in the Albertine Rift Montane Forests [17], Eastern Arc Forests [19], and Cameroon Highlands Forests [10] on the African mainland. Areas with slightly less importance include the Upper Guinea forests, the coastal and northeastern forests of the Congo Basin, and the Central Zambezian Miombo Woodlands [50]. Drier ecoregions are understandably poor in terms of endemic amphibians, with the only exception being of the Cape Region of South Africa, which has a highly adapted amphibian fauna. Poynton (1999) looked at broad patterns of endemism in the amphibians across Africa and its islands, and Brooks et al. (2001a) illustrated endemism patterns in sub-Saharan Africa. Our results agree with those of these previous studies.

Plants

Expert assessments of the distribution of plant endemism indicate the highest levels in the Fynbos [89, 90] and Suc-

culent Karoo [110] of South Africa (figure 4.5e). High rates of plant endemism are also found in lowland and montane forests across the tropical belt. Particularly important lowland forests are those of eastern Madagascar [29, 30], the western portion of the Upper Guinea rainforest, the coastal areas of Cameroon through Gabon [7, 8], the northeastern Congolian lowland forests [16], the northern portion of the eastern African coastal forest mosaic [20], and the KwaZulu coastal forests of South Africa [23]. In montane forests, the highest rates of endemism are in the Cameroon Highlands [10] and the Eastern Arc Mountains [19] of Kenya and Tanzania. The patterns of plant endemism we demonstrate are similar to those presented in a number of recent works (e.g., Lovett et al. 2000b; Kier and Barthlott 2001; Linder 2001; La Ferla et al. 2002). They also correspond to some of the older studies of plant endemism that presented their results in large biogeographical units (Brenan 1978; White 1983).

Invertebrates

As with species richness, the patterns of endemism we illustrate for invertebrates follow closely those of plants. The highest rates of invertebrate endemism are found in Madagascar, the western portion of the Upper Guinea forest, the coastal and montane forests of the western Congolian forests, the montane forests of East Africa, and the Fynbos, Succulent Karoo, and Namib Desert of South Africa and Namibia (figure 4.5f). These patterns are further illustrated using data from two subfamilies, Charaxinae and Acraeinae (figure 4.3). There are no published studies of the distribution of endemism in invertebrates across Africa with which to compare our results.

TABLE 4.3. **Mean Numbers of Endemic Vertebrate Species (and Standard Error) across Ecoregions in the Nine African Biomes.**

Biome	No. of Ecoregions	Birds	Mammals	Amphibians	Reptiles	Total
Tropical and subtropical moist broadleaf forests	30	13 (2.26)	9 (1.71)	12 (2.70)	18 (3.90)	52
Tropical and subtropical dry broadleaf forests	3	9 (4.67)	5 (4.51)	2 (1.53)	16 (9.94)	32
Temperate coniferous forests	1	1	0	0	1	2
Tropical and subtropical grasslands, savannas, shrublands, and woodlands	24	3 (0.69)	4 (0.97)	2 (0.69)	15 (2.98)	24
Flooded grasslands and savannas	10	1 (0.52)	0 (0.22)	0 (0.20)	1 (0.33)	2
Montane grasslands and shrublands	16	7 (1.14)	3 (1.06)	4 (0.72)	9 (2.02)	23
Mediterranean forests, woodlands, and scrub	7	6 (1.94)	3 (0.84)	3 (1.76)	13 (3.57)	25
Deserts and xeric shrublands	23	3 (0.76)	3 (0.62)	0 (0.18)	12 (2.65)	18
Mangroves	5	2 (1.25)	0 (0.00)	0 (0.00)	1 (0.40)	3

Distribution of Endemism across Biomes

Like species richness, levels of endemism vary widely according to biome (table 4.3). For example, the mangrove and flooded grassland biomes have fewer than three endemic (strict and near-endemic) vertebrate species per ecoregion, whereas the tropical moist forest biome supports more than fifty endemic vertebrates per ecoregion. For vertebrates, ecoregions in the Mediterranean scrub biome support only twenty-five endemic vertebrates per ecoregion, ranking this biome in third position out of the nine biomes. However, this ranking would rise to second place if numerical data on plant and invertebrate endemics were available, given the global importance of this biome for these groups.

Within the individual vertebrate groups there are significant differences in rates of endemism between biomes (appendix G), although there are fewer correlations between patterns of endemism in different taxonomic groups between biomes (table G.4) than was seen for species richness (table G.3). Endemism patterns between birds and reptiles show the lowest number of significant relationships, largely because the reptiles have high rates of endemism in the ancient deserts of the Horn of Africa and southwestern Africa that are not seen in birds. Many of the areas important for reptile endemism are also important for plants and invertebrate endemics, again indicating that there are some important differences in the distribution of different species groups across Africa and its islands.

Combining Richness and Endemism

The broad pattern of richness and endemism across Africa provides important insights for identifying conservation priorities. However, we seek to go beyond the simple illustration of patterns. Our method (chapter 3) aims to combine the species values of ecoregions objectively and to set species-based priorities within biomes. Furthermore, we seek to combine species values with nonspecies values to develop a single metric that captures the overall biological importance of every ecoregion.

To start this process we have combined the species richness and species endemism values of ecoregions, as described in chapter 3, to provide a single measure of their value for species. This combination was undertaken separately within each of the nine African biomes to ensure habitat representation in the set of ecoregions identified as priorities for species. Results show ecoregions ranked into four classes of importance in terms of their species assemblages (figure 4.6, box 4.1). There is a distinct tendency for ecoregions in the coastal margins of Africa, on offshore islands, or on mountains to rank highly in this analysis, regardless of biome. Globally outstanding ecoregions include most of the coastal region from Cameroon to South Africa, the western part of the Upper Guinea rainforest in Sierra Leone and Liberia, and all of Madagascar. In East Africa the coastal forest mosaic ecoregions and the major mountain ecoregions are globally outstanding for species, together with the Somali *Acacia-Commiphora* Bushlands and Thickets [44] of

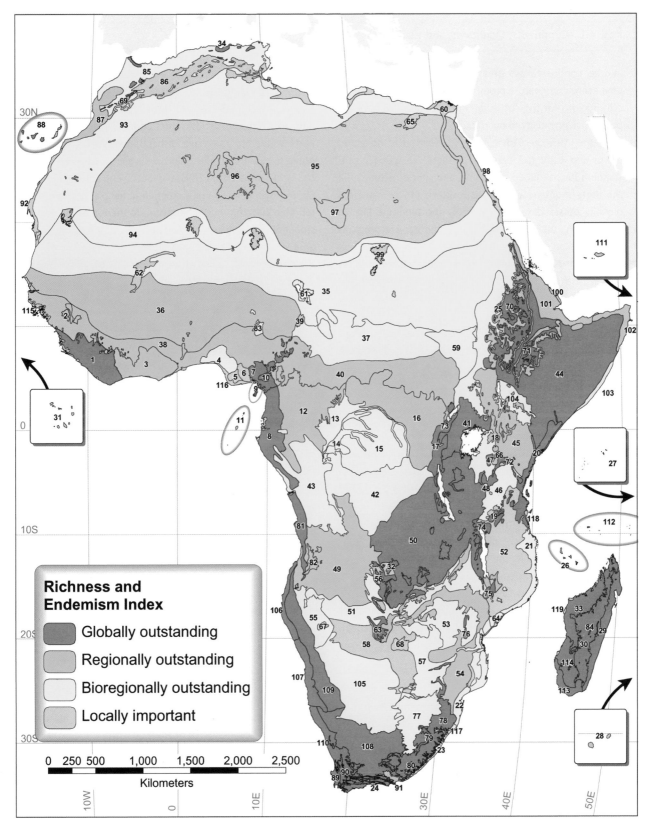

FIGURE 4.6. Richness and Endemism Index.

BOX 4.1. **Continents, Mountains, and Islands.**

In other continental assessments by WWF, islands and montane ecoregions feature prominently in the portfolio of priority regions for the conservation of species. We have explored this possibility in our data.

Of the 119 ecoregions in Africa, fourteen are island ecoregions, and eight (57 percent) of these are globally outstanding for combined endemism and species richness. Of the thirty montane ecoregions fourteen (46 percent) are globally outstanding, most notably the Albertine Rift. In comparison, the nonisland and nonmontane ecoregions contain fifteen globally outstanding ecoregions out of seventy-five (or 20 percent of lowland ecoregions). These results indicate that our priorities for species do favor islands and montane regions, even though we have used a representation approach to set priorities within biomes.

We were also interested in how the species-based priorities for this region would have changed if we had used different measures of endemism, which drives the majority of our results. In particular we wanted to know whether the balance between mainland lowland, mainland montane, and island ecoregions would have changed. To answer this question we have mapped the priority ecoregions for strict endemism (defined as species confined to a single ecoregion) and those for percentage endemism and then looked to see how they fell into the groups of lowland, montane, and island.

Using strict endemism changes the rank order of some of the most important ecoregions for vertebrates (figures 4.7 and 4.8). For example, the ten most important ecoregions defined using near-endemism contain four mainland lowland ecoregions, three mainland montane ecoregions, and three island ecoregions (all three on Madagascar). If we used strict endemism, then the top ten would include only two main-

land lowland ecoregions, the same three montane mainland ecoregions, and five island ecoregions (including two small island ecoregion chains: São Tomé, Princípe, and Annobon and the Comoros Islands). Strict endemism data for plants and invertebrates may have further tipped the balance of priorities toward the smaller islands. However, the Lowland Fynbos and Renosterveld [89] and Succulent Karoo [110] of South Africa have so many strict endemic plants and invertebrates that they would have risen to the top of the list for strict endemic species.

If we had used percentage endemism to establish priorities, this would have changed our results considerably. Many island ecoregions support fewer vertebrate species than lowland (mainland) and montane (mainland) ecoregions of similar area (figure 4.9a). They also tend to support more near-endemic and strict endemic species than montane and lowland ecoregions on the mainland (figure 4.9b–c). Because the offshore islands have few species and many endemic species, they dominate the list of important ecoregions in terms of percentage vertebrate endemism (figure 4.9d).

In conclusion, if we had used the number of strict endemics or percentage endemism in this assessment, even more of the islands and mountain ecoregions would have ranked at the top of our prioritized list of ecoregions for species values. By using near-endemism as the main driver of importance for species and by setting priorities within biomes, we have ensured that more lowland mainland ecoregions feature in the set of high-priority ecoregions than would have otherwise been the case. However, our approach has the disadvantage of leaving out some important islands and mountain ecoregions that have large numbers of endemic species that cannot be conserved anywhere else.

the Horn of Africa and the Central Zambezian Miombo Woodlands [50]. In North Africa, only the conifer woodlands emerge as globally outstanding for species values.

Capturing More Than Richness and Endemism: Ecological Features and Processes

The number of endemic species and level of species richness are important features for identifying the biological distinctiveness of an area. However, they fail to capture other

ecological features and processes that may distinguish an ecoregion as biologically distinct (Mace et al. 2000). Methodological details of how we quantify these nonspecies biological values are presented in chapter 3 (and appendix C), and results are outlined in this section.

Unusual Evolutionary Phenomena

This attribute is designed to capture ecoregions where there has been (and probably still is) rapid evolution of new taxa

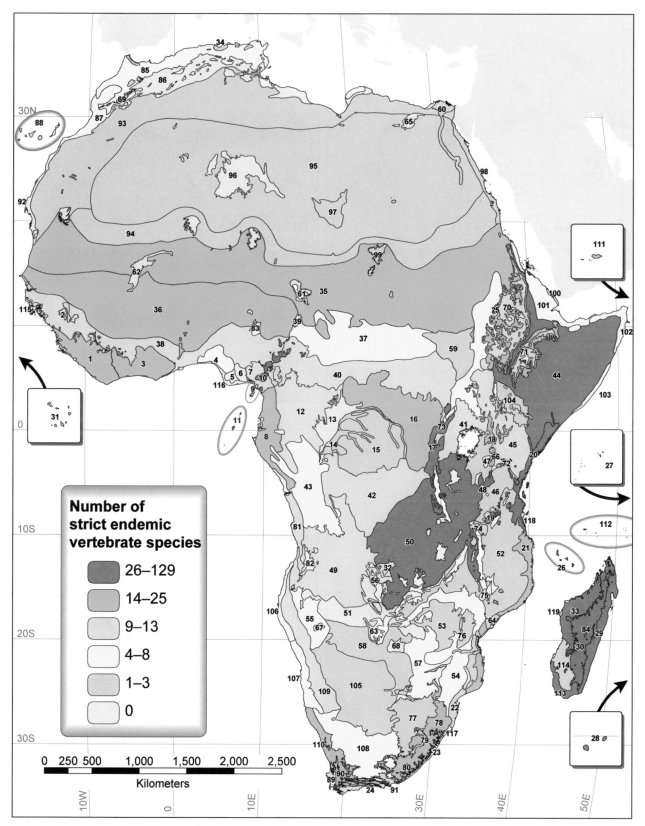

FIGURE 4.7. Combined strict endemism for birds, mammals, reptiles, and amphibians across African ecoregions. Classes are based on quantile groupings of data values.

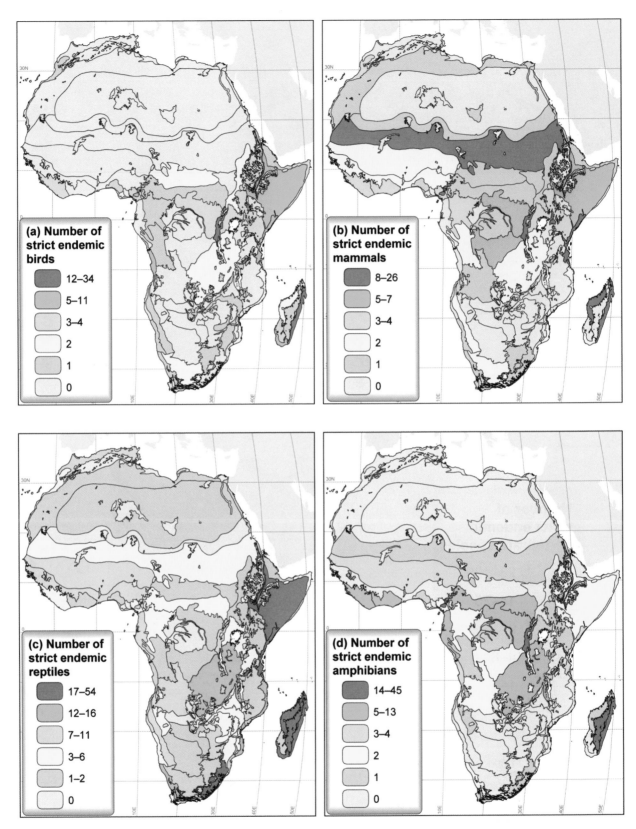

FIGURE 4.8. Numbers of strict endemic species for (a) birds, (b) mammals, (c) reptiles, and (d) amphibians. Classes are based on quantile groupings of data values.

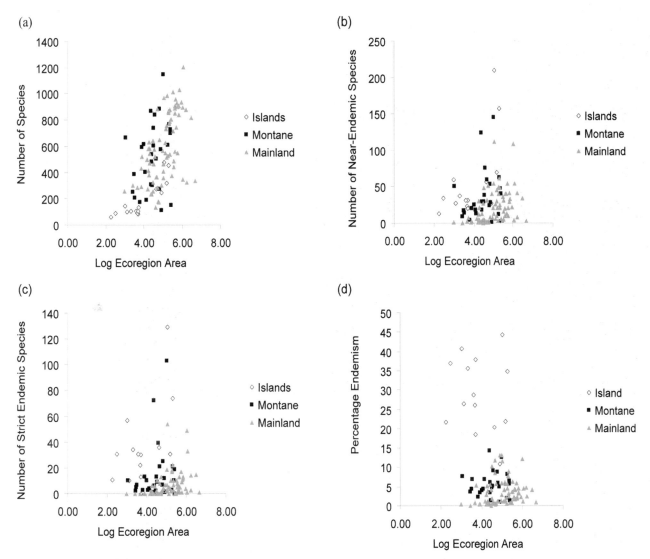

(a)

(b)

(c)

(d)

FIGURE 4.9. Differences in (a) species richness, (b) near endemism, (c) strict endemism, and (d) percentage endemism per unit area of ecoregions occupying offshore islands, mountains (mainland islands), and lowland continental ecoregions in Africa.

within closely related groups (figure 4.10a). For example, in the Fynbos more than 500 species in the genus *Erica* have evolved over the last million years. In the Succulent Karoo there are hundreds of newly evolved species in approximately thirty genera in the family Mesembryanthemaceae (Smith et al. 1998).

Other ecoregions where recent speciation is important are those covering the African moorland and Afroalpine area, particularly on the volcanic mountains in Kenya and Tanzania. In these areas—less than 2 million years old—species of giant *Lobelia* and *Dendrosenecio* have recently evolved as volcanic activity subsided and habitats became suitable for colonization. On these mountains speciation in plants has been paralleled by evolution of new species of

invertebrates. The large number of subspecies and races of birds and mammals on smaller offshore islands around the coast of Africa probably indicate sites of active speciation.

Unusual Ecological Phenomena

This attribute is designed to capture other biological phenomena such as major migrations of terrestrial vertebrates or birds (figure 4.10b). These are important ecological processes that would not be captured using other criteria of biological importance. The Serengeti plains, the Southern *Acacia-Commiphora* bushlands and thickets, and the Sudd swamp are all noted for their seasonal mammal migrations of global significance. Also, the entire Congo Basin, which

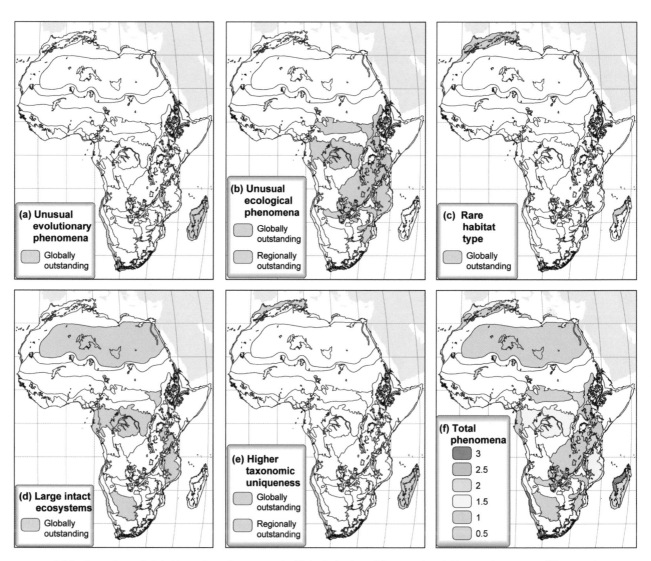

FIGURE 4.10. Ecoregions of global importance for nonspecies biological values: (a) unusual evolutionary phenomena, (b) unusual ecological phenomena, (c) rare habitat type, (d) large intact ecosystems, (e) higher taxonomic uniqueness, and (f) overlap of the various nonspecies biological attributes.

does not rank as important as a center of endemism, ranks as highly important for its large mammal migrations. Ecoregions ranking as globally important for possessing congregations of more than 1 million birds include the Inner Niger Delta, Seychelles, Aldabra, halophytic and flooded ecoregions of eastern Africa, and the coastal wetlands in the Atlantic coastal desert.

Rare Habitat Type

In the global context, Mediterranean climate ecoregions, tropical dry forests, and montane moorlands are the rarest habitat types on Africa and its islands (figure 4.10c; Olson

and Dinerstein 1998). Ecoregions with Mediterranean habitats are found primarily at the northern and southern ends of Africa, and those with montane moorlands are found on the tops of the highest tropical mountains in eastern Africa and Madagascar. Perhaps the best examples of tropical montane moorlands in the world are found on the volcanic peaks in Kenya and Tanzania and on the Rwenzori Mountains of Uganda. The tropical dry forests of Africa are found on the Cape Verde islands in the west, in central southern Africa, and on the western side of Madagascar. Of these, the Madagascar dry forests are the most important because they combine rare habitat with exceptional levels of species endemism.

Large Intact Ecosystems and Wilderness Areas

Ecoregions that contain huge areas of largely untransformed habitats, supporting a naturally functioning ecosystem, are becoming increasingly rare around the world (figure 4.10d). Africa supports several intact areas, including the Sahara Desert, the Sudd swamps, the Congo Basin, the Serengeti, parts of the Miombo and Mopane woodlands, the Kalahari Desert, and the Namib-Kaokoveld Desert. Conservation International and the Wildlife Conservation Society have also identified these areas as important wilderness regions worthy of conservation attention (Mittermeier et al. 2002; Sanderson et al. 2002a).

Higher Taxonomic Uniqueness

This attribute is designed to capture ecoregions that harbor relict flora and fauna that have an ancient evolutionary history but are now extinct over most of their range (figure 4.10e). These relicts most often occur as families or genera endemic to an ecoregion and typically supporting very few species (often a sole surviving example). Most notable in this regard are the ecoregions of Madagascar, which support an ancient fauna long since extinct on the African mainland. Relict families and genera are also found in the drier ecoregions of southwestern Africa from the Kaokoveld and Namib deserts to the Fynbos ecoregions. Many of the smaller offshore islands (Canaries, Mascarenes, Socotra, Seychelles) also contain relict families and numerous genera. Although not meeting our criteria (appendix C), some forest ecoregions also support a few endemic genera. Most notable among the forests are those of southeastern South Africa, the Eastern Arc Mountains of Tanzania, and the coastal lowland forests of Cameroon and Gabon.

By summing the presence of the five features just described, we illustrate the ecoregions of greatest importance for nonspecies biological values (figure 4.10f). Of particular importance are the island of Madagascar, coastal Namibia, and the Fynbos of South Africa. The next level of importance includes the Congo Basin forests, the *Acacia* and Miombo woodlands of eastern Africa, the Sudd swamps of southern Sudan, and the northwest African mountains.

Overall Patterns of Biological Distinctiveness

Bringing together the biological values of ecoregions in terms of species and nonspecies values allows us to propose a comprehensive set of conservation priorities for the African region. We aim to identify areas of exceptional endemism

Blue wildebeest (*Connochaetes taurinus*), Masai Mara National Reserve, Kenya. Photo credit: WWF-Canon/Martin Harvey

to prevent extinction while also identifying areas important for nonspecies biological values to address concerns over the maintenance of ecological processes in Africa. The results of these analyses show that collectively fifty-five ecoregions rank as globally outstanding for combined species and nonspecies values, seventeen as regionally outstanding, twenty-three as bioregionally outstanding, and twenty-four as locally important (table 4.4; figure 4.11).

By undertaking analyses in the framework of biomes, we also achieved habitat representation in the final set of priority ecoregions. Because analyses were completed within biomes, it is also relevant to present the results in terms of biomes to illustrate the distribution of the different ranks of importance across Africa (figure 4.12). The majority of the tropical forest ecoregions are globally important for their species and nonspecies biological values, except for those of southern Nigeria and parts of coastal eastern and southeastern Africa (figure 4.12a). The same is true of the

TABLE 4.4. Number of Ecoregions That Are Globally Outstanding for Species Endemism, Species Richness, or Nonspecies Values by Biome.

Biome	Endemism	Richness	Nonspecies Values	Total
Tropical and subtropical moist broadleaf forests	9	2	8	19
Tropical and subtropical dry broadleaf forests	1	1	0	2
Temperate coniferous forests	1	0	0	1
Tropical and subtropical grasslands, savannas, shrublands, and woodlands	2	2	2	6
Flooded grasslands and savannas	1	0	1	2
Montane grasslands and shrublands	7	2	3	12
Mediterranean forests, woodlands, and scrub	3	0	1	4
Deserts and xeric shrublands	7	0	2	9
Mangroves	0	0	0	0

montane grassland and forests ecoregions, which are mainly of the highest importance except for the Highveld grasslands of South Africa (figure 4.12c). In the savanna-woodlands biome the most important ecoregions are found in eastern Africa to the Horn of Africa and down through the miombo woodlands into Zambia (figure 4.12b). Further west and south the savanna-woodland ecoregions are of lower importance. In the desert biome, globally outstanding ecoregions are found in southwestern Africa, Madagascar, and the far north of the Horn of Africa, with much less important ecoregions covering the greater Sahara region (figure 4.12d). Although not mapped, the priority ecoregions in other biomes are widely distributed across Africa and its islands, illustrating that the conservation of the full range of African biodiversity requires efforts in a number of different areas.

Poorly Known Ecoregions

A common question asked of biologists undertaking priority-setting analyses is the degree to which available knowledge influences the results. It is certainly true that biological data are unevenly distributed across Africa. A number of ecoregions are poorly known in terms of species, such as the majority of the Congo Basin, the Southern Zanzibar–Inhambane Coastal Forest Mosaic [21], the Eastern Miombo Woodlands [52], the Sudd Flooded Grasslands [59], and the Angolan Scarp [81] (except for birds). Ecoregions with limited biological inventories may have depressed levels of endemism and species richness and therefore be underranked in these analyses.

Although we remain uncertain about the full diversity and distribution patterns of vertebrate animals and plants across Africa, the invertebrate fauna of large parts of the region remains at the frontier of taxonomic understanding. Intriguingly, the level of undescribed invertebrate species seems to vary between biomes and ecoregions. In the tropical forest biome an invertebrate taxonomist can discover up to 90 percent undescribed species at a newly visited site. The same species often are absent in the next forest patch only a few tens of kilometers away, where another suite of undescribed endemic species occurs. In other biomes the rates of undescribed species seem lower, although even this level of generalization is hampered by a lack of data.

We asked a number of experts to assess the rates at which new vertebrate species are being discovered in various taxonomic groups (see essays 4.1 and 4.2). Each essay shows that new species are still being regularly discovered and that more should be found in the future. The situation is similar for plants, where the rates of discovery of new plant species remain high, even in well-known areas such as the South African Fynbos. However, we predict that new inventories will only accentuate the geographic priorities already defined in this analysis.

In recent years computer modeling techniques have been developed to address the problems caused by the uneven distribution of biological data. Methods have been developed to predict species distribution patterns using existing (incomplete) distribution data, or by simply using environmental or biophysical factors. Essays by Boitani et al. (essay 4.3) and Underwood and Olson (essay 4.4) provide examples. Undoubtedly, as these types of models im-

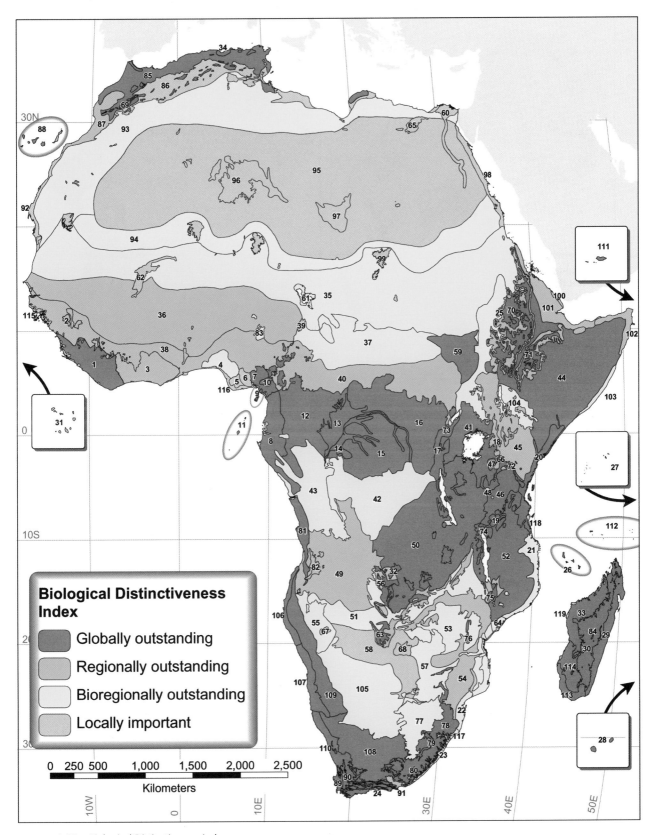

FIGURE 4.11. Biological Distinctiveness Index.

Biological Distinctiveness Index

- Globally outstanding
- Regionally outstanding
- Bioregionally outstanding
- Locally important

0 250 500 1,000 1,500 2,000 2,500

Kilometers

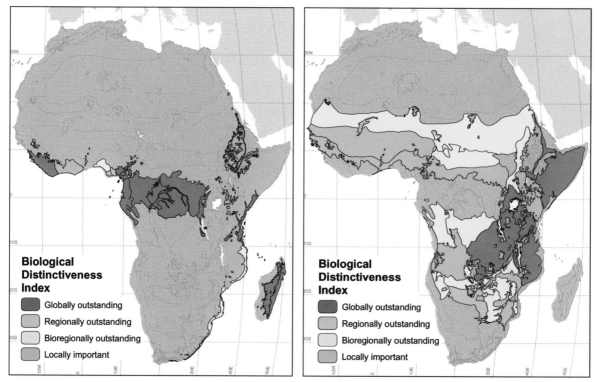

(a) Tropical and subtropical moist broadleaf forests biome.

(b) Tropical and subtropical grasslands, savannas, shrublands, and woodlands biome.

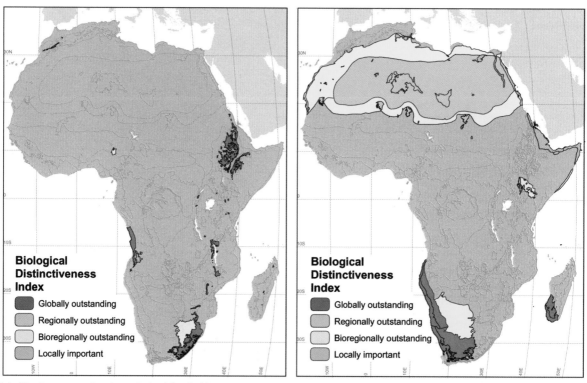

(c) Montane grasslands and shrublands biome.

(d) Deserts and xeric shrublands biome.

FIGURE 4.12. Priority ecoregions in terms of their Biological Distinctiveness Index, within different biomes: (a) tropical moist forests, (b) savanna-woodlands, (c) montane grasslands, and (d) deserts.

prove, they will produce better estimates of both current and historical biogeographic patterns in the distribution of species.

Conclusion

The combination of species and nonspecies biological values provides the first overall index of the biological importance of natural regions across Africa and its islands. If human disturbance were insignificant or occurred at equal intensity throughout Africa, we could focus exclusively on the most biologically unique areas. There would be no need to undertake the arduous task of assessing the degree of threat to every ecoregion in order to further assess where conservation action should focus. But the level of threat to biodiversity varies widely across the region. Some ecoregions are on the verge of complete conversion to human-dominated landscapes, whereas others are largely untouched. The next chapter develops our second filter, a calculation of threat and opportunity for every ecoregion that will allow us to decide where to focus conservation resources to best effect.

ESSAY 4.1

Recent Discoveries of Vertebrates on Mainland Africa and the Potential for Further Discoveries

Donald G. Broadley, Alan Channing, and Andrew Perkin

A common refrain among taxonomists is that our understanding of species and their distributions is too limited to set priorities for conservation action. Instead, we need more surveys, more funding for taxonomists to describe new species, and a greater effort to map and synthesize biogeographic data. There is certainly an urgent need to accelerate the naming of all taxa indigenous to the African continent and its surrounding islands. The authors of the first two essays in this chapter have dedicated their careers to achieving this goal. In this essay, three authors pool their knowledge of recent discoveries of terrestrial vertebrates, except birds, in mainland Africa. In essay 4.2, Goodman examines patterns of new discoveries of plants and animals in Madagascar. Two important questions emerge that relate to our confidence in setting continental and global-scale priorities. Is the rate of discovery of new species of vertebrates in mainland Africa all that extraordinary, suggesting a paucity in current knowledge? Are the ecoregions that are currently the richest in species and endemics the same areas where most of the new species occur or are expected to be found? Put another way, would adding recent discoveries to the lists of ecoregion endemics alter our rankings of biological distinctiveness? These essays conclude that the ecoregions where many new species await discovery are those already among the most diverse and of highest priority at continental and global scales.

Mammals

The mammalian fauna of Africa has been well studied, and most species were discovered long ago; for example, by 1900 around 80 percent of all known mammal species were already described. However, new mammal species continue to be found in Africa after intensive periods of field work, often backed up by comparative museum studies and genetic analysis (e.g., Bearder 1999; Honess and Bearder 1996; Schwartz 1996; Van Rompaey and Colyn 1998). For example, in the past 10 years seventeen new primate species and subspecies have been recognized from Africa (table 4.5). To compare with other parts of the world, ten species and subspecies of primates have been described from Madagascar, seven from Southeast Asia, ten from the Brazilian Amazon (seven of them marmosets), and three from the Brazilian Atlantic forest.

New species have also been discovered in other mammalian families, in particular in the rodents, shrews, and bats, but here we focus on new discoveries in the group of nocturnal primates popularly known as bush babies or galagos.

The galagos are secretive, cryptic primates with ancient ancestries and many adaptations to nocturnal life. They are often considered "lower," "primitive," and solitary, which reflects a misunderstanding of their biology and social behavior. Recent studies are transforming the way we understand this mammal

TABLE 4.5. **Species and Subspecies of Primate Recognized by Primatologists in the Last 10 Years.**

Monkeys	Nocturnal Primates
Northern talapoin, *Miopithecus ogouensis*	False potto, *Pseudopotto martini**
Ngotto moustached monkey, *Cercopithecus cephus ngottoensis*	Silver greater galago, *Otolemur monteiri*
	Mwera greater galago, *Otolemur sp. nov.**
Nigerian white-throated guenon, *Cercopithecus erythrogaster pococki*	Gabon Allen's galago, *Galago gabonensis*
	Makande Allen's galago, *Galago sp. nov.**
Semeliki black and white colobus, *Colobus badius semlikiensis*	Ukinga dwarf galago, *Galagoides sp. nov.**
Niger Delta red colobus, *Procolobus badius epieni*	Mozambique lesser galago *Galagoides granti*
	Malawi lesser galago, *Galagoides nyasae**
	Kalwe lesser galago, *Galagoides sp. nov.**
	Mt. Thyolo lesser galago, *Galagoides sp. nov.**
	Diani lesser galago, *Galagoides sp. nov.**
	Rondo dwarf galago, *Galagoides rondoensis*

*There still exist several discrepancies, which result from the use of different species and subspecies concepts (Oates et al. 2000; Groves 2001; Kingdon 1997) and from the characteristics and evidence used in differentiating the taxa. These six new species await further clarification and official description.

group and are leading to the discovery and naming of new species. Most galagos look very similar. They do not display obvious sexual dimorphism, do not have brightly colored pelages, and are color-blind. Instead, they rely heavily on auditory and olfactory senses to communicate and to recognize their mates. Different species occupy separate ecological niches, thus allowing several species to co-exist in the same habitat.

Traditional museum taxonomists described species on the basis of pelage and morphologic studies, often from a small number of specimens sampled across a wide geographic area or from different habitats. It is now emerging that the similar-looking species described in this way are unable to interbreed and differ consistently with respect to anatomy, biochemistry, physiology, and behavior. The most important taxonomic characters include those related to mate attraction, such as call, scent, facial patterns, and penile morphology, and those reflecting variation in ecological niches, such as limb proportions and nail and footpad structure. Existing galago taxonomy has been revised through these new criteria and especially through the results of intensive field work. In the field, where obtaining detailed observational data for nocturnal prosimians is difficult, species are most easily identified by their vocal repertoires. Vocalizations are highly important for selecting mates, rearing young, maintaining social groups, and avoiding predators.

In light of these new approaches, more galago species have been recognized. In 1986 most texts recognized only six galago species. Recent research has recognized as many as twenty-three species. Some of the latest discoveries of galago species have occurred in Tanzania. Three new species have recently been described: *Galagoides orinus, G. rondoensis,* and *G. granti.* Two of these species, *G. rondoensis* and *G. orinus,* were known from only one or two specimens collected in the 1920s from isolated forests in the Eastern Arc and coastal forests of eastern Tanzania. They were provisionally assigned as subspecies of *G. demidoff,* which occurs in the Congolian forests. The recent research has shown that they are very different from *G. demidoff.*

The Eastern Arc and coastal forests of East Africa remain potential areas for additional galago discoveries. For example, a distinct population of dwarf galago (*Galagoides sp. nov.*) has been found in the Taita Hills. It is a new record for that region and remains unidentified (Bytebier 2001). Moreover, *G. orinus,* which is distributed in several isolated populations in the remaining submontane and montane forests of the Eastern Arc Mountains of East Africa, displays interpopulation variation and may have taxonomic implications. Field surveys have also found possible new species in the Itombwe Mountains in the Albertine Rift system and the forests on the Cameroon-Nigeria border.

Museum studies are still providing new ideas for galago and other mammalian systematics (table 4.5). These studies largely take a phylogenetic approach to define a species, where certain levels of variation and differentiation in the pelage and morphology of museum specimens of have led to the naming of new species. These species have yet to be fully studied in the wild to determine whether they have different calls or habitat preferences, for example, but they provide a very useful pointer of where to look in Africa. They reveal that the highly threatened isolated forests of Malawi and northern Mozambique warrant further biodiversity surveys.

Many existing mammalian subspecies may also be elevated to species level as we grow to understand that they have diverged to a greater level than originally realized. Ultimately, the level of this type of species description—also known as species splitting—depends on how we define a species, and this is (and probably always will be) a contentious debate.

Further research on cryptic mammal taxa, which are generally small, arboreal, and often nocturnal (and often have entangled taxonomies), is likely to locate additional species in the future. The most likely locations for finding new mammal species remain the montane and lowland rainforest regions of tropical Africa, with the best sites probably in the Eastern Arc Mountains, the Albertine Rift Mountains, the Ethiopian Highlands, and the Cameroonian Highlands.

Amphibians

Amphibians are found in all habitats in Africa, from the oases of the Sahara, to the granite domes of the Namib Desert, to the lush rainforests of West Africa and the eastern Afromontane belt. Many parts of the continent remain unexplored for frogs and caecilians, and because of ongoing turmoil in many areas, it seems unlikely that a complete inventory will be available for decades.

Over the last 10 years or so, ten new species (about 10 percent of the known forms) have been discovered in southern Africa, with about the same number described from the rest of the continent. The low numbers of new species discovered is not a reflection of poor amphibian diversity but rather the result of few amphibian specialists. At the start of the year 2000, no more than five specialists were employed throughout the continent in African museums or universities. However, small numbers of graduate students from Africa and abroad and researchers from other continents complement these scientists.

New species are traditionally recognized by comparison to reference collections in museums. Amphibians, especially

frogs, have few body features that are useful for this purpose. Although closely related frog species can be easily distinguished in the field on the basis of a different advertisement call (produced by the male to attract the female) and biology, once preserved the same specimens might be impossible to distinguish.

For example, sand frogs of the genus *Tomopterna* occur from the southern tip of Africa through the arid areas of North Africa. They look alike, and recent publications recognize only a few widespread species. Field studies show that there are probably twice as many species, easily recognized in the field by the different calls and with different numbers of chromosomes.

Even well-explored areas, such as the mountains near Cape Town, continue to surprise. Over the last few years, three new species of moss frogs have been discovered on slopes in the Fynbos, one of the biologically best-known areas in Africa.

In East Africa the chain of old mountains known as the Eastern Arc has long been recognized as an important center of amphibian endemism. However, the last comprehensive amphibian survey of the Uluguru Mountains in Tanzania, to take but one example, was undertaken in 1926. A brief visit to the Ulugurus in early 2000 produced three species not found in 1926. This is an increase of 12 percent and indicates that the amphibian fauna of Africa is still incompletely known. It is to be expected that new species will be discovered in areas that have not been recently evaluated.

New species have been found in the Bale Mountains of Ethiopia, and similar habitats, though having the potential for many new species, are also very difficult to work in because of the steep dissected mountains and remain poorly explored.

As in other groups, new amphibian species are being discovered through molecular techniques such as DNA sequencing. Using this technique, cryptic species can be recognized. In contrast, the advertisement call of frogs is a valuable field tool for recognizing new species.

When considering which parts of Africa are most likely to produce new species, the question is perhaps best answered in reverse. The best-known areas, from an amphibian point of view, are parts of West Africa, such as Cameroon, parts of East Africa, such as Kenya, and much of southern Africa, such as Swaziland. New species undoubtedly will be found in all areas, including the countries that are best known at present. Taxonomists tend to revisit known interesting areas, and as long as they continue to find new species this is likely to continue.

In conclusion, Africa is rich in amphibian species, but difficult terrain and unstable political situations have hampered field explorations. Technological advances made over the last two decades, including sound analysis and routine DNA techniques, have made species identification simple. The shortage of suitably trained field personnel suggests that it will be a long time before all amphibian species have been discovered and named.

Reptiles

By far the largest number of new vertebrate species described over the past decade have been reptiles, totaling seventy-one new species. These include forty-four lizards, three amphisbaenians, and twenty-four snakes. Most of these are morphologically distinctive, but initial recognition has often been prompted by DNA studies. Many former subspecies have been recognized as good evolutionary species. These discoveries can be broken down as follows.

Lizards

A single new species of agamid lizard (*Uromastyx maliensis*) has been described from West Africa. Twelve new species of geckos include one from Egypt (*Tarentola mindiae*), two from Tanzanian coastal forests (*Lygodactylus broadleyi* and *L. kimhowelli*), three from northern South Africa (*Lygodactylus nigropunctatus, L. graniticolus,* and *L. waterbergensis*), and six from the arid southwestern regions of South Africa (*Pachydactylus kladaroderma, P. haackei, Phyllodactylus* [= *Goggia*] *hexaporus, P. hewitti, P. gemmulus,* and *P.* [= *Afrogecko*] *swartbergensis*).

Sixteen new skink species are all fossorial forms with limbs reduced or absent. Three come from Morocco (*Chalcides ghiarai, C. minutus,* and *C. pseudostriatus*), one from Cameroon (*Panaspis chriswildi*), one from the southern coastal desert of Angola (*Typhlacontias rudebecki*), six from islands off the East African coast (*Scelotes duttoni, S. insularis, Lygosoma lanceolatum, L. mafianum, Typhlosaurus bazarutoensis,* and *T. carolinensis*), one from coastal forest in southeastern Tanzania (*Scolecoseps litipoensis*), and four from KwaZulu-Natal (*Scelotes bourquini, S. fitzsimonsi, S. vestigifer,* and *Acontias poecilus*).

Eight cordylid lizards have been described from South Africa: *Cordylus oelofseni, C. cloetei, C. imkeae, C. aridus, Pseudocordylus nebulosus, Platysaurus monotropis, P. lebomboensis,* and *P. broadleyi.*

Seven new species of forest chameleon come from eastern Africa: *Chamaeleo conirostratum* from the Imatong Mountains of southern Sudan, *C. harennae* and *C. balebicornutus* from the Ethiopian Highlands, *C. marsabitensis* and *C. tremperi* from Kenya, *Rhampholeon uluguruensis* from the Eastern Arc Mountains of Tanzania, and *R. chapmanorum* from southern Malawi.

Amphisbaenids (Worm Lizards)

Three new amphisbaenians are all from southern Africa: *Zygaspis ferox, Z. kafuensis,* and *Monopeltis infuscata.*

Snakes

In the worm snakes, one new species of *Typhlops* (*T. debilis*) was described from the Central African Republic. A further nine worm snake species in the genus *Leptotyphlops* have also been named; three are from West Africa (*L. albiventer, L. broadleyi,* and *L. adleri*), one from Ethiopia (*L. parkeri*), one from Kenya (*L. drewesi*), one from coastal forests of Tanzania and Kenya (*L. macrops*), one from Mozambique (*L. pungwensis*), and two from South Africa (*L. sylvicolus* and *L. jacobseni*).

Eleven new species of colubrid snakes have also been named, including two from West Africa (*Dipsadoboa underwoodi* and *Rhamphiophis maradiensis*), one from the Congo Basin (*Philothamnus hughesi*), one from the Sudan (*Telescopus gezirae*), two from Ethiopia (*Pseudoboodon boehmei* and *Lycophidion taylori*), one from coastal forests of Tanzania (*Prosymna semifasciata*), two from Namibia (*Lycophidion namibianum* and *Coluber zebrina*), one from the Drakensburg of South Africa (*Montaspis gilvomaculata*), and one from the KwaZulu-Natal coast (*Lycophidion pygmaeum*).

One new species of elapid snake has been described from Somalia (*Elapsoidea broadleyi*), and the three vipers are from Cameroon (*Atheris broadleyi*), Uganda (*A. acuminata*), and the southwestern Cape (*Bitis rubida*).

Descriptions of many other new species from Africa are in preparation, including a new tortoise of the genus *Homopus* from Namibia, about eight new chameleons of the genus *Bradypodion* from South Africa and a *Rhampholeon* from Tanzania, five geckos of the genus *Afroedura* from South Africa, and three skinks of the genus *Mabuya* from Cameroon and Central African Republic. Two species of *Cordylus* will be described from northern Malawi and the Maasi plains of northern Tanzania and southern Kenya and Tanzania. Descriptions of seven more worm snakes (Leptotyphlopidae) are in the pipeline; these are from Libya, Central African Republic, Sudan, Ethiopia, Uganda, Kenya, and Tanzania. Two new colubrid snakes will be described: a small species of *Mehelya* from the Congo-Sudan border and a *Dromophis* from the north coast of Mozambique. A large Ethiopian viper of the genus *Bitis* awaits description.

Many regions of Africa remain poorly explored for reptiles. These include the Sudan, Ethiopia, the central Congo Basin, the Udzungwa Mountains of Tanzania, northern Mozambique, and the whole of Angola. All these areas undoubtedly harbor undescribed species, especially in isolated patches of forest. The regular use of pitfall traps in conjunction with plastic drift fences has resulted in vastly improved sampling of fossorial reptiles, especially those of the forest floor.

The majority of the recently described species are from regions where there have been active programs of field work: the southwestern Cape (University of Stellenbosch), the southwest arid region (Transvaal and Port Elizabeth Museums), the Tanzanian coastal forests (Frontier Tanzania), and the Bazaruto Archipelago off the southern Mozambique coast (Natural History Museum of Zimbabwe). Others are found during taxonomic revisions of genera or species groups, the relevant material often lying misidentified in museum collections for decades (e.g., many new species of *Leptotyphlops*). The major problem is the dearth of taxonomists—truly an endangered species. Indeed, people who are either retired or not employed as herpetologists now carry out much of the taxonomic research on African reptiles. The following references form the source of data for the new reptile species: Baha El Din (1997), Bauer et al. (1996, 1997), Böhme and Schmitz (1996), Bourquin (1991), Bourquin and Lambiris (1996), Branch (1997), Branch et al. (1995, 1996), Branch and Whiting (1997), Broadley (1990, 1991, 1994a, 1994b, 1994c, 1995a, 1995b, 1996, 1997, 1998), Broadley and Broadley (1997, 1999), Broadley and Hughes (1993), Broadley and Schätti (2000), Broadley and Wallach (1996, 1997a, 1997b), Caputo (1993), Caputo and Mellado (1992), Chirio and Ineich (1991), Haacke (1996, 1997), Hahn and Wallach (1998), Hallerman and Rödel (1995), Jacobsen (1992, 1994), Jakobsen (1996), Joger (1990), Joger and Lambert (1996), Lawson (1999), Mouton and van Wyk (1990, 1994, 1995), Necas (1994), Pasteur (1995), Rasmussen (1993), Rasmussen and Largen (1992), Tilbury (1991, 1992, 1998), Tilbury and Emmrich (1996), Trape and Roux-Estève (1990), and Wallach and Hahn (1997).

ESSAY 4.2

Measures of Plant and Land Vertebrate Biodiversity in Madagascar

Steven M. Goodman

Madagascar has received much conservation attention because of its extraordinary levels of endemism, unique ecoregions, and the plight of its remaining natural habitats. The island has been separated from the African mainland for more than 160 million years, long before the evolution of most modern land vertebrate families. Given this long period of isolation and the difficulty for organisms to disperse to the island across the 400 km of open water that separates it from Africa, Madagascar evolved a unique fauna and flora with a large percentage of biota found nowhere else in the world. The ancestors of the modern lineages that did cross this ocean faced different evolutionary pressures than on continental Africa and subsequently differentiated into some of the most remarkable radiations of unique species anywhere on Earth.

The last decade has seen a significant increase in number of researchers conducting biological inventories in poorly known areas of Madagascar, supplementing the taxonomic reevaluation of preserved specimens in the herbaria and museums of the world. This work has greatly modified estimates of the number of endemic forms occurring on the island and the importance of this biota in a global sense. Here I focus on aspects of the plant and vertebrate fauna of the island, but this is only a portion of its diversity, and the levels of species richness and discovery among invertebrates are equally, if not more, astounding.

During the taxonomic reassessment of the trees of Madagascar, Schatz (2000) recalculated estimates of the rates of endemism. Of the 4,220 species of trees and large shrubs that occur on the island, 96 percent were endemic to Madagascar (including the Comoro Islands). Furthermore, at the generic level, 161 of the 490 (33 percent) indigenous tree genera were endemic. Trees represent about 35 percent of the total flora of Madagascar, which is on the order of 12,000 species (Dejardin et al. 1973). Previous estimates of around 80 percent endemism for the plants of Madagascar may prove to be an underestimate. At a higher taxonomic level, eight plant families are endemic to the island: Asteropeiaceae, Didiereaceae, Didymelaceae, Kaliphoraceae, Melanophyllaceae, Physenaceae, Sarcolaenaceae, and Sphaerosepalaceae (Schatz et al. 2000).

The period between 1990 and 1999 saw a great increase in knowledge on the herpetofauna diversity and levels of endemism in Madagascar. In this decade, forty-six species and one subspecies of amphibian and fifty-six species and six subspecies of reptiles were described as new to science (Glaw and Vences 2000); this is far more than had been named from Madagascar in any previous decade. It represents an increase of about 25 percent for amphibians and 18 percent for reptiles reported from Madagascar. Most of these discoveries were based on field work and the collection of new material rather than reassessment of older specimens held in museum collections. Currently, 182 amphibian and 326 reptile species (excluding marine forms) have been named from the island, which show levels of endemism of 98 percent and 94 percent, respectively. These figures do not include the estimated forty-two to sixty-eight amphibian and twenty-four to thirty-four reptile species that have already been identified but are waiting to be described or resurrected from synonymy (Glaw and Vences 2000). These authors have estimated that as many as 300 amphibian species might occur on the island.

The avifauna of the island was thought to be reasonably well known, but recent ornithological work has resulted in the description of two new birds, one of which was placed in a new genus (Goodman et al. 1996, 1997). Furthermore, several species that were thought to be extinct have been rediscovered over the past few years; these include *Tyto soumagnei* and *Eutriorchis astur* (Thorstrom and René de Roland 1997, 2000). Several extant higher taxonomic groups of birds are endemic to Madagascar, including the families Mesitornithidae, Brachypteraciidae, Leptosomatidae (also occurring on the Comoros Islands), and Vangidae and the subfamilies Philipittinae and Couinae. The taxonomy of Malagasy birds is changing rapidly, largely as the result of molecular studies, but of the 270 or so species recorded from the island more than 50 percent are endemic.

Among mammals, new species and genera are still being uncovered at a high rate. The family Tenrecidae, which is probably endemic to the island, has been the focus of recent field and taxonomic research, and six species new to science have been described since 1992 (Goodman and Jenkins 1998; Jenkins 1988, 1993; Jenkins and Goodman 1999; Jenkins et al. 1996, 1997). The endemic subfamily of rodents on the island, the Nesomyinae, has seen an explosive increase in the number of recognized forms. For example, *Eliurus,* the most speciose genus in the subfamily, was listed by Honacki et al. (1982) as comprising two species; nine are now recognized, and several remain to be described (Carleton 1994; Carleton and Goodman

1998). Furthermore, two new genera of nesomyine rodents have been described in recent years: *Monticolomys* and *Voalavo* (Carleton and Goodman 1996, 1998).

Despite the wealth of information on the primates of the island, several new species have been described in the past 15 years. These include *Hapalemur simus* (Meier et al. 1987), *Propithecus tattersalli* (Simon 1988), and *Microcebus ravelobensis* (Zimmermann et al. 1998). A recent revision of western forms of *Microcebus* raised the number of described species from this region from three to seven species: *M. murinus, M. ravelobensis, M. myoxinus, M. griseorufus, M. tavaratra, M. sambiranensis,* and *M. berthae;* the fourth species was brought from synonymy, and the latter three species were new to science (Rasoloarison et al. 2000). Several other species of lemurs await description. Furthermore, lemurs have been rediscovered that were thought to be extremely rare and perhaps extinct (Meier and Albignac 1991). Also, a new species of carnivore, *Galidictis grandidieri,* was described from Madagascar during this same period (Wozencraft 1986). If one excludes two lemur species that were apparently introduced to the Comoro Islands, all of the native land mammals occurring on Madagascar are endemic to the island.

The Future

New field work on Madagascar is confirming its known global importance and adding to it. However, there are enormous problems for the protection of the dwindling number and size of forest blocks that hold this unique biota. Over the last few decades the forests of Madagascar have been dramatically reduced (Green and Sussman 1990; Nelson and Horning 1993). The current high level of habitat destruction, particularly in the lowlands and littoral forests of the east and dry deciduous forests of the west, is apparently unrelenting. Most of the habitat loss on Madagascar is associated with subsistence agriculture using slash-and-burn techniques or conversion of standing timber into charcoal. Compared with numerous other places in the world with a similar environmental plight, there is almost no commercial logging on the island. The fundamental problems underlying this ecological crisis are socioeconomic and demographic: an unbalanced economy and increasing human population. The further development of conservation projects and a national environmental consciousness over the next two decades will be paramount to preserve what remains of this truly remarkable flora and fauna.

ESSAY 4.3

Mapping African Mammal Distributions for Conservation: How to Get the Most from Limited Data

Luigi Boitani, Fabio Corsi, and Gabriella Reggiani

Of all environmental components, fauna is one of the most difficult to show on a map because of its mobility. At any given time, we can only have a certain probability of finding an individual of a given species at a given location. At the same time, detailed knowledge of faunal distribution patterns is one of the most important aspects in their conservation, in establishing and managing protected areas, and in environmental monitoring and management. Distribution patterns often are derived from literature sources and museum data that are decades old, but conservation action requires data on the current distribution of species, which can seldom be inferred by using available historical data. To solve these problems, we have started to develop computer models that predict the distribution of species by correlating details of the species' ecology to biophysical data on the environment. Our techniques are made possible through the use of models that combine geographic information system (GIS) techniques for handling large quantities of spatial information with the best data available on species-habitat relationships (Corsi et al. 2000).

We have compiled a databank on the distribution of 281 species of African large and medium-sized mammals in the orders Primates, Carnivora, Perissodactyla (except rhinos), Hyracoidea, Tubulidentata, Artiodactyla, Pholidota, and Lagomorpha and including seven species of large rodents (Boitani et al. 1999, http://www.gisbau.uniroma1.it/amd). The main goal of the project (African Mammal Databank [AMD], funded by the European Community) was to provide a continent-wide analysis of mammal distribution patterns for use in conservation-related activities.

Method for Developing the Models

The method used to develop the models combines two sets of data: the most recent polygons of the overall distribution of each species (the Extent of Occurrence [EO], sensu IUCN Species Survival Commission 1994) and information on the species-habitat relationship. The two sets of data are then com-

bined through a GIS analysis with the aim of refining the original polygon distribution to the areas characterized by environmental conditions as suitable for the species.

Polygon distributions and species-habitat relationships were collected and validated through the network of experts of the IUCN Species Survival Commission, and GIS environmental datasets were collected from various sources and agencies (i.e., the Digital Chart of the World, African Data Sampler, UN Food and Agriculture Organization, World Conservation Monitoring Centre, and U.S. Geological Survey).

Several ways exist to link the available environmental variables to the species' presence (Garshelis 2000) and subsequently to use this relationship to model species distribution with GIS (Corsi et al. 2000). We used two approaches in our modeling: a deductive approach and an inductive one (Stoms et al. 1992; Corsi et al. 2000).

The Categorical Discrete (CD) Model

This deductive approach describes the species' environmental preferences, as derived from existing knowledge (i.e., literature, specialists' knowledge), in terms of the environmental variables available in the GIS database.

Based on such available knowledge, one or more experts derive a ranking of suitability for each of the environmental layers analyzed and for each combination of the same layers. The experts' rankings are then used to classify the entire study area (the whole African continent in the AMD). To allow a simpler interpretation of the results, the combination of different scores is categorized into four suitability classes (suitable, moderately suitable, unsuitable, undefined).

The resulting map appears as a patchwork depicting more and less suitable areas. As a guide for interpreting the maps, the suitable and moderately suitable classes inside the original distribution polygon were interpreted as the true area of occupancy (AO) of the species (sensu Gaston 1991), and the areas outside the boundary of the original distribution polygon were interpreted as potential areas.

The Probabilistic Continuous (PC) Model

This inductive approach uses the information on species presence to build a function that is capable of ranking the entire study area according to a continuous suitability index. Probabilistic models are increasingly used in modeling species distributions (Knick and Dyer 1997; Corsi et al. 1999; Peterson et al. 1999) but have rarely been applied at the continental scale (Busby 1991; Walker 1990; Skidmore et al. 1996) and never in Africa.

The general method used for these models is based on a dataset of known locations of the species (known territories, observations, or radiolocations). These are used to define the species' ecological profile, which is obtained by selecting the characteristics of known locations from the available GIS environmental layers.

The characterization is then used to calculate the ecological distance of each defined portion within the study area from the species' preferred ecological conditions. This produces a model of the species' distribution, within which distance values can be appropriately scaled to represent environmental suitability for the species (Clark et al. 1993; Corsi et al. 1999).

To apply this method to the AMD project, a major modification was necessary because of the lack of a dataset comparable to that of the known locations. To build the species' ecological profile, we had to rely only on its distribution polygon. By assuming that, on average, most of the area in the polygon is occupied by the species, and by assuming a multivariate normal distribution of the environmental dataset used, we used the vector coverage of the average environmental conditions of the polygon, together with its variance-covariance matrix, as the ecological profile of the species (Garshelis 2000). Thus we could map the ecological variability in terms of ecological distance (using the Mahalanobis distance) from the average environmental conditions of the polygon, obtaining a description of the internal pattern of the extent of occurrence. Following these assumptions, we can expect the quality and reliability of the results to depend on the homogeneity of distribution of both the species and the environmental variables.

Model Validation

As purely theoretical speculation based on literature and GIS techniques, the resulting models are of limited use for effective conservation action. Therefore, a validation process was carried out in four countries: Botswana, Cameroon, Morocco, and Uganda. The countries were selected to maximize the ratio between representation of environmental conditions in the continent and financial cost. Each country was stratified according to White's vegetation maps, and each stratum was allocated randomly a number of sample points proportional to its extent, for a total of 427 sample points. In each point a team composed of one researcher of a local research institution (e.g., university, government agency, or nongovernment organization) and one of the AMD staff researchers compiled a complete checklist of all the species of interest to the AMD project based on direct observation, existing literature, and interviews of local populations.

Validation was applied only to the outputs of the CD models because the PC models were considered too preliminary to stand a validation procedure.

Results and Discussion

For each species considered by the AMP project, two separate maps depicting environmental suitability for the species were obtained by applying the two modeling methods just described. An example map is provided for *Canis adustus* (figure 4.13), giving the results of the CD model on the left and the PC model on the right.

All analyses were supported by the good performance of the models, which showed an average index of accordance, derived from ground-truthing, above 60 percent. Both types of models are useful tools to obtain adequately accurate maps of species distribution, especially to support decision making at a broad scale and in areas where alternative data are limited in quantity and poor in quality.

Besides making all the data available to the conservation community for further inclusion in biodiversity conservation analyses, the AMD project has outlined various applications of the models to conservation action planning. For instance, the resulting suitability maps have been compared with the existing network of protected areas to assess adequacy of conservation measures for each species analyzed, and a fragmentation assessment has been carried out on the AO to support IUCN Red List criteria. Furthermore, some preliminary analyses of biodiversity have been conducted by overlaying the resulting models to identify areas of species richness. Several interesting results were found. One example, shown in figure 4.14, shows the potential pitfall of using only the coarse-scale EOs to identify areas of higher species biodiversity. The curve built using the AOs, as derived from the models, significantly decreases species richness. This very simple result was obtained by taking into account the degraded environmental conditions that created a more complex patchwork of suitable and unsuitable habitat.

The weaknesses and caveats of GIS applications to animal distributions are well known to scientists (for a review see Corsi

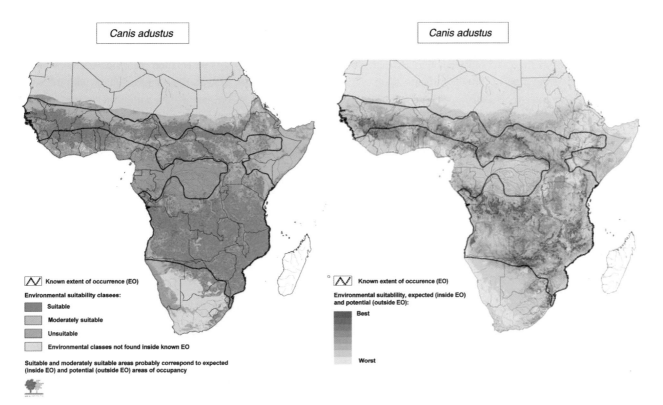

Canis adustus   *Canis adustus*

Known extent of occurrence (EO)

Environmental suitability clasees:

Suitable

Moderately suitable

Unsuitable

Environmental classes not found inside known EO

Suitable and moderately suitable areas probably correspond to expected
(inside EO) and potential (outside EO) areas of occupancy

Known extent of occurence (EO)

Environmental suitablility, expected (inside EO)
and potential (outside EO):

Best

Worst

FIGURE 4.13. Environmental suitability maps obtained from the categorical-discrete (CD, *left*) and the probabilistic-continuous (PC, *right*) models for *Canis adustus.* For the CD model, the suitability scale ranges from unsuitable (*red*) to suitable (*green*), with an intermediate moderately suitable class (*light green*). For the PC model suitability is shown in shades of green, with darker green being the most suitable.

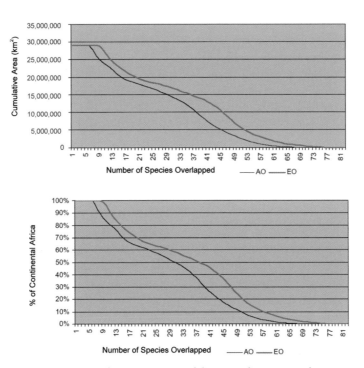

FIGURE 4.14. Cumulative percentage of the areas of co-presence for increasing numbers of species based on their extent of occupancy (EO) and areas of occurrence (AO).

et al. 2000) but less so to conservationists. The output of GIS models must be interpreted as an indication of existing trends and patterns, not as the exact representation of reality. The greatest danger is misunderstanding the true meaning of the models and the graphic representation of animal distribution, and hence the limitations of basing management action on insufficient ground data. For instance, the models use environmental datasets that have their own inherent errors, and the GIS can mask these by displaying final outputs in the form of appealing maps.

Despite these caveats, the AMD shows that the approach offers a valid way of formalizing existing information and using it for biodiversity conservation planning. Additionally, the approach creates a feedback process in which, with the addition of better environmental data and further insight into the ecology and biology of the species, the models can be constantly improved. Ultimately, even if the final results provide a distribution range with marginally enhanced resolution, this provides a significant improvement for ecologists and conservation planners.

ESSAY 4.4

Developing Predictive Models to Estimate Patterns of Biodiversity for Data-Poor Ecoregions: The Congo Basin as a Case Study

Emma Underwood and David Olson

The Congo Basin contains one of the last remaining large blocks of tropical forest on Earth. However, habitat loss, fragmentation, and only a small representation of forest in protected areas means there is a great need for a conservation strategy in the region. Unfortunately, much of the existing data on patterns of biodiversity across the Congo Basin are inappropriate for planning at the regional scale. Survey locations often are biased toward easily accessible areas and certain taxonomic groups and often are conducted at inappropriate spatial scales to provide valuable information. Therefore, making an objective conservation plan for this region is problematic. In this essay we explore some potential methods for modeling species richness and endemism patterns across the Congo Basin.

Method for Developing the Predictive Model

The ecological niche of a species or group of species is the conceptual basis of many predictive models. The fundamental ecological niche can be defined as the combination of ecological conditions in which the species is able to maintain populations (MacArthur 1972). When distributions are modeled at the regional scale, the types of data used to determine these niche dimensions often include temperature, precipitation, elevation, and vegetation (Townsend Peterson and Vieglais 2001).

One of the earliest and simplest methods to model species

distributions was to draw a boundary around the maximum extent of known species locations. More complicated algorithms soon expanded this approach and correlated known species occurrences with their environmental envelopes. Extrapolation from known locations using these dimensions provided the potential spatial distribution of the species across the landscape (e.g., a bioclimatic prediction system [BIOCLIM], Nix 1986). Other developments were made using statistical modeling techniques, such as logistic regression, classification and regression trees, and Bayesian statistics.

We developed a model for biodiversity patterns in the Congo Basin using a multiple regression approach. The model was based on research conducted in the neotropics that identified key biophysical factors as determining tree and liana species richness (Clinebell et al. 1995). Multiple regression analyses assessed the contribution of rainfall, soil, and elevation to explain the variance in species diversity. The key findings were that rainfall, seasonality, and soils are able to explain nearly 75 percent of the variance; essentially, species diversity is strongly positively correlated with total amount of rainfall and negatively correlated with rainfall seasonality. Our method assumes that the relationship between biophysical parameters and biodiversity distribution in the Afrotropics is the same as in the neotropics. The necessary biophysical data layers for the Congo Basin were compiled in a geographic information system (GIS) using a 5-km grid cell resolution. Data layers included

Predicted patterns of species richness   Predicted patterns of species endemism

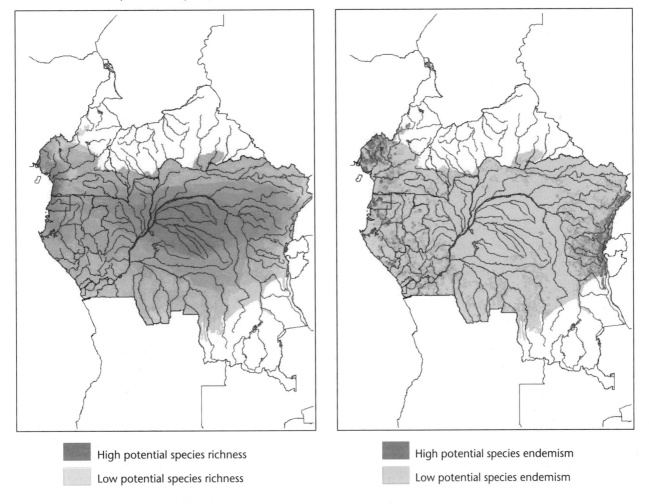

| | High potential species richness | | High potential species endemism |
| | Low potential species richness | | Low potential species endemism |

FIGURE 4.15. Predicted patterns of species richness and endemism across the Congo Basin.

precipitation (interpolated from 6,000 point locations in the region), elevation, and the number of dry months as an indicator of seasonality. Dry months are defined here as months with an average precipitation of less than 60 mm (a level at which precipitation was assumed to be less than the amount of evapotranspiration). The multiple regression equation derived from the analysis of the neotropics data was then applied to these data layers to produce a map of predicted species richness for the Congo Basin.

Predicted Patterns of Species Richness

Two areas of potentially high species richness emerged (figure 4.15). The first is an area in southwestern Cameroon, encompassed within the Atlantic Equatorial Coastal Forests ecoregion [8], particularly the Campo area. The second region is the extensive area of forest just south of the northern arc of the Congo

River, near the headwaters of the Bolombo River, which falls largely in the Central Congolian Lowland Forests [15].

Validating predictive models by conducting accuracy assessments is an essential part of distribution modeling. This highlights both errors of omission (where actual occupied areas are omitted) and errors of commission (where unoccupied areas are incorrectly included). In the absence of targeted field surveys for validation, we compared these results with the limited survey data from the region, which supported our predictions. However, we also used the presence and absence data for mammals, birds, frogs, and snakes from the 1- by 1-degree resolution distribution database of the Zoological Museum of the University of Copenhagen (figure 4.16). Although there was agreement in some areas, such as the coastal part of Cameroon, others showed large differences. Of particular interest is the central Congolian area, which is predicted to have high species richness by the model but has low species richness according to

Recorded patterns of species richness

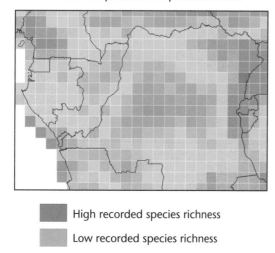

■ High recorded species richness

▨ Low recorded species richness

Recorded patterns of species endemism

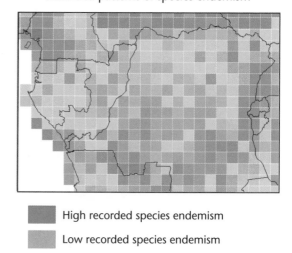

■ High recorded species endemism

▨ Low recorded species endemism

FIGURE 4.16. Recorded patterns of species richness and endemism of vertebrates across the Congo Basin (data from Zoological Museum, University of Copenhagen, Denmark).

available biological data. These differences might be explained in several ways:

- The Congo Basin is undersampled. With additional sampling we may find patterns similar to those predicted by the model.

- Modeled patterns of species richness would converge with the recorded data if the latter included plant and invertebrate data.

- Nutrient-poor soils or other unusual biophysical features in the central Congo Basin may account for generally low richness and thereby invalidate the predictive model.

- Historical factors, such as drier epochs and the incursions of savannas into the central Congo Basin, may have influenced the observed richness patterns, whereas the predictive model is based on current bioclimatic attributes.

Although it is reassuring to have a degree of overlap between predicted and recorded patterns of species distributions, one of the most valuable contributions of the models is to highlight these discrepancies. These areas of contradiction can then provide a framework for stratifying field work and further investigation.

Predicted Patterns of Species Endemism

We also developed a preliminary model to predict broad patterns of endemism. The results of the species richness analysis serve as a first filter because extremely species-rich forests of-

ten harbor endemic species at a subregional and local scale. This result was then weighted with information on topographic complexity, which was calculated using the standard deviation between adjacent 5-km cells of elevation across the Congo Basin. Topographic complexity is important because mountain ranges and inselbergs can promote allopatry and local specialization and therefore are likely to support local endemics and high beta diversity.

Several areas are predicted to have high levels of endemism (figure 4.15). These include the isolated ranges of the Congolian coastal forests, the Monts de Cristal and the Chaillu Massif in Gabon, and the foothills of the Albertine Rift Mountains on the eastern margin of the Congo Basin. Biological surveys previously conducted in these areas have located high numbers of narrow-range endemic species (figure 4.16), providing some degree of confidence in the results.

Although targeted field surveys must be conducted to evaluate and refine the models, these initial results indicate that predictive models can provide useful guidance on potential species patterns across the basin. Ideally, models developed in the future will also be based on correlations between known species locations and environmental parameters from studies conducted in the Afrotropics.

Shortcomings of Predictive Modeling

Regional modeling techniques, such as described here, have two shortcomings in terms of practical interpretation and implementation. First, their premise is to define the fundamental niche of the species, although in actuality it might be the realized niche that is critical. The realized niche incorporates more complex factors, such as species interactions, which are inherently difficult to quantify. Also, because of the regional scale of the input data, modeling does not take account of fine-scale factors, such as the microhabitat, that obviously affect the local distribution of species. Second, logistic regression techniques aim to predict the probability of a "yes" versus "no" in the independent (i.e., species distribution) variable (Skidmore et al. 1996), which is a concept that that is more appropriate in dealing with continuous data, such as elevation, as opposed to categorical data, such as vegetation type. However, the recent emergence of more sophisticated genetic algorithms, such as the genetic algorithm for rule set prediction (Stockwell and Noble 1991), have great potential because they combine a number of modeling rules, thus overriding the disadvantages of using each method individually.

Utility of Models in Conservation Planning

Predictive models herald tremendous hope for biodiversity conservation planning, which will no doubt increase as the technical capacity for modeling organismal biology also grows (Townsend Peterson and Vieglais 2001). However, the use and value of predictive models rest on two factors. First, results must be adequately ground-truthed to provide an assessment of accuracy. Second, the conservation community, from biologists to conservation policy analysts, need to embrace predictive models as useful tools in ecoregion planning, not just for species distributions at the regional scale but also to tackle other conservation challenges, such as predicting the distribution of invasive species.

CHAPTER 5

The Status of Wild Africa

In comparison with most other parts of the world (Western Europe, North America, and southeast Asia) the biodiversity values of Africa are still fairly intact and seemingly subject to less critical threats. However, this is no cause for complacency. Human populations are increasing rapidly through some of the highest birthrates in the world, governments are impoverished so that many protected areas and forest reserves go unmanaged, and unplanned privatization or commercialization of communal or government resources is converting intact areas into plantations, logging concessions, and large-scale farms. On top of this, insecurity and civil unrest affect many countries, reducing their capacity to undertake effective conservation work, and every year climatic extremes leave hundreds of thousands of people either without rain, or flooded, and consequently short of food. More environmental damage occurs in years with exceptionally poor climatic conditions that occur over decades of normal weather conditions as people are forced to strip the available resources or migrate to new and unfamiliar areas in order to survive. The unpredictability of climate; local, national, and regional economies; politics; governance; and security are at the root of many conservation challenges in Africa today.

In this chapter we look at the current and future-projected conservation status of ecoregions across Africa. Our conclusions provide challenges for conservationists working in Africa, and addressing these challenges forms the focus of the last two chapters of this book.

Current Status of Habitats

With increasing populations and continued poverty in many African nations, pressures on natural resources continue to rise, often to the detriment of remaining biodiversity. To better understand where to focus conservation attention, we need to understand the patterns and trends in the intensity of human land use. Here, we outline the conservation status of ecoregions as a measure of current and predicted land-use changes as they affect biodiversity. The methods used to measure the status of habitats in African ecoregions are outlined in chapter 3.

Habitat Loss

Some broad patterns of habitat loss are evident across Africa (Figure 5.1a). By biome, the highest average percentages of habitat conversion are found in the temperate coniferous forests of North Africa, mangroves, the montane grasslands, and the Mediterranean scrub (figure 5.2). Not surprisingly, the least conversion has occurred in the desert biome because it is unsuitable for agriculture or habitation. Approximately one-half of all ecoregions have more than 50 percent habitat loss (appendix H).

Ranking ecoregions by the percentage of converted habitat is a good indicator of where natural habitats are under the greatest pressure (table 5.1). Ecoregions where more than 95 percent of their area is already converted to

Forest being burnt to create new agricultural land in Madagascar. Photo credit: WWF-Canon/John Newby

Maize plantation, Udzungwa Mountains, Tanzania. Photo credit: WWF-Canon/John Newby

farmland or urban areas include the Mandara Plateau Mosaic [39], Cross-Niger Transition Forests [6], Jos Plateau Forest-Grassland Mosaic [83], and Nigerian Lowland Forests [4]. Nine other ecoregions have more than 80 percent habitat loss, including the Lowland Fynbos and Renosterveld [89] and the forests and grasslands of the Ethiopian Highlands [70, 71] (figure 5.1a).

Remaining Habitat Blocks

Although percentage habitat loss is a general indicator of the relative condition of an ecoregion, it does not identify the number and size of remaining intact habitat blocks that may still allow the persistence of biodiversity (figure 5.1b). A number of ecoregions possess few remaining intact habitat blocks. In North Africa these include the Nile Delta Flooded Savanna [60], the Mediterranean High Atlas Juniper Steppe [69], and the Mediterranean Conifer and Mixed Forests [34]. In West Africa habitat is no longer found as large blocks in the Guinean Montane Forests [2], the Eastern Guinean Forests [3], and the lowland forest ecoregions of southern Nigeria. In eastern Africa, few large habitat blocks remain in the Albertine Rift Montane Forests [17], the East African Montane Forests [18], and the Eastern Arc Forests [19], although these ecoregions naturally occurred as a number of different blocks. Further south the entire coastal strip from Mozambique to Cape Town lacks large habitat blocks.

Levels of Habitat Fragmentation

Whereas the size and number of remaining habitat blocks provide one measure of intactness, habitat fragmentation corresponds to the intensity of edge effects and can indicate the chances for habitat persistence (figure 5.1c). The

TABLE 5.1. **Top Twenty Ecoregions with the Highest Percentage of Habitat Loss.**

Ecoregion	% Habitat Loss
Cross-Niger Transition Forests [6]	98
Mandara Plateau Mosaic [39]	98
Nigerian Lowland Forests [4]	96
Jos Plateau Forest-Grassland Mosaic [83]	96
Eastern Guinean Forests [3]	90
Ethiopian Upper Montane Forests, Woodlands, Bushlands, and Grasslands [70]	89
South Malawi Montane Forest-Grassland Mosaic [75]	89
Niger Delta Swamp Forests [5]	88
Comoros Forests [26]	85
KwaZulu-Cape Coastal Forest Mosaic [23]	84
Lowland Fynbos and Renosterveld [89]	84
Victoria Basin Forest-Savanna Mosaic [41]	83
Ethiopian Montane Moorlands [71]	81
Southern Zanzibar–Inhambane Coastal Forest Mosaic [21]	79
Guinean Forest-Savanna Mosaic [38]	78
Zambezian Coastal Flooded Savanna [64]	78
Southern Africa Mangroves [117]	78
Mediterranean Conifer and Mixed Forests [34]	78
Mediterranean Woodlands and Forests [85]	77
Southern Rift Montane Forest-Grassland Mosaic [74]	77

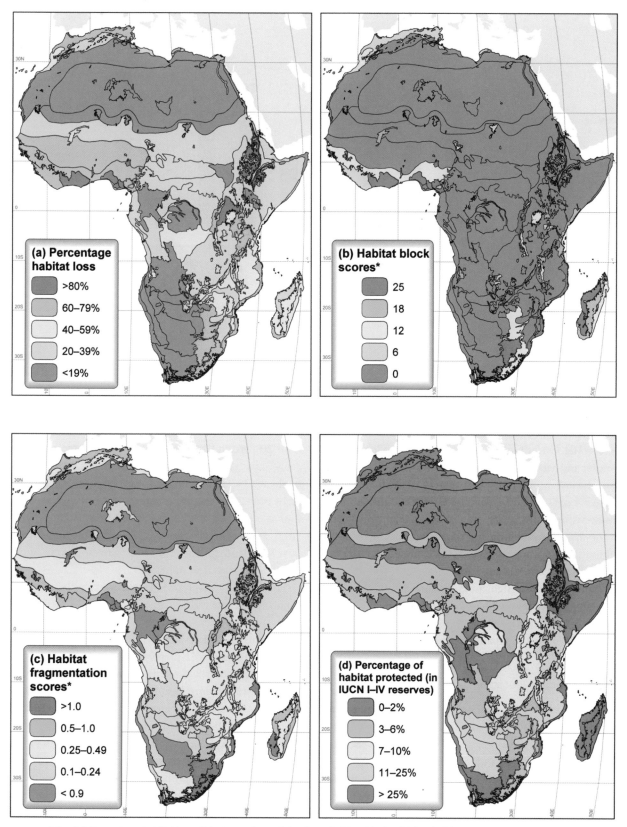

FIGURE 5.1. **(a)** Percentage habitat loss, **(b)** status of remaining large habitat blocks, **(c)** level of fragmentation, measured by edge:area ratio of habitat blocks, and **(d)** percentage of habitat protected in IUCN category I–IV protected areas. *See appendix D for scoring criteria.

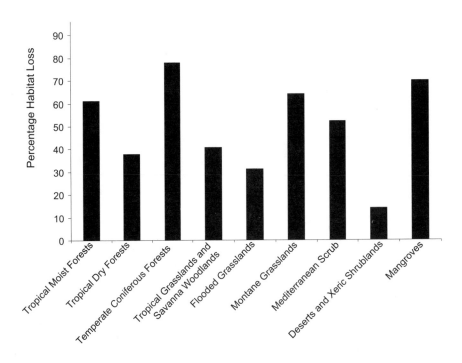

FIGURE 5.2. Percentage of biome converted to cropland or urban areas.

most fragmented ecoregions in North Africa are the Mediterranean High Atlas Juniper Steppe [69], the Mediterranean Conifer and Mixed Forests [34], and the Nile Delta Flooded Savanna [60]. Most of the forest areas of West Africa, except the Western Guinean Lowland Forests [1] and the Guinean Forest-Savanna Mosaic [38], are also highly fragmented. One of the major causes of habitat fragmentation in the tropical forests is logging, an issue that is explored in more detail in essay 5.1. In eastern Africa, all the major upland areas (Ethiopia, Kenya Highlands, Albertine Rift, Eastern Arc) are highly fragmented, with much of this fragmentation being natural. In southern Africa the fragmentation scores tend to increase toward the coastal regions.

Levels of Habitat Protection

There is great variation in the degree to which ecoregions are protected by IUCN category I–IV protected areas (Strict Nature Reserve/Wilderness Area, National Park, Natural Monument, Habitat/Species Management Area). The African regions under the best strict protection tend to be the savanna-woodland habitats, particularly those of eastern and southern Africa (figure 5.1d). The least protected areas are found in North Africa, Madagascar, drier parts of South Africa, and the most heavily deforested parts of West and

East Africa. Of the 119 ecoregions in Africa, 110 have less than 20 percent of their area officially protected, and 89 ecoregions have less than the 10 percent target suggested by the IUCN (MacKinnon and MacKinnon 1986). However, eight ecoregions are more than 20 percent protected, and four ecoregions—Kaokoveld Desert [106], Namib Desert [107], Etosha Pan Halophytics [67], and East African Montane Moorlands [72]—have more than 40 percent of their area under protection (table 5.2).

Some of the least-protected ecoregions are also those with high biodiversity values, such as Mount Cameroon and Bioko [9] (0 percent protected), Eastern Arc Forests [19] (1 percent), Succulent Karoo [110] (2 percent), Ethiopian Montane Forests [25] (2 percent), Lowland Fynbos and Renosterveld [89] (3 percent), Western Guinean Lowland Forests [1] (3 percent), East African Montane Forests [18] (5 percent), Albertine Rift Montane Forests [17] (7 percent), and Northern Zanzibar–Inhambane Coastal Forest Mosaic [20] (7 percent). These are also all areas of high human population density, where reserves tend to be small (Harcourt et al. 2001). The lack of protection in these high-biodiversity regions of Africa has been noted previously (Balmford et al. 1992, 2001a, 2001b).

Finally, although we believe that the broad patterns of protection outlined in figure 5.1d are true, the results are somewhat distorted by the protected area data considered. In particular, forest reserves, wildlife management areas, and private nature reserves are not coded as IUCN I–IV protected areas and thus are excluded from this analysis, yet they function as important biodiversity conservation areas. For example, the globally important Eastern Arc Forests [19] ecoregion has only one protected area in IUCN categories I–IV (the Udzungwa Mountains National Park), but almost all the remaining forest lies in more than 100 catchment forest reserves managed by the central government and where no extractive use is permitted. If we had included catchment forest reserves in the calculation of percentage habitat protected, the degree of protection in this ecoregion would have increased several times. Essay 6.2 provides further detail on the development of the protected area network in Africa.

Current Threat Patterns

The Current Conservation Status Index summarizes habitat loss, remaining habitat blocks, habitat fragmentation, and protected area status across African ecoregions (figure 5.3). Ecoregions are ranked into five levels of threat. We outline the ecoregions in the two categories of highest threat (critically threatened and endangered) and those in the lowest two categories (relatively stable and relatively intact).

Critically Threatened Ecoregions

The ecoregions regarded as critically threatened contain a combination of high levels of habitat loss, few remaining habitat blocks, high rates of habitat fragmentation, and little protection (figure 5.3). The Nile Delta Flooded Savanna [60] of North Africa is critically threatened, mainly because it has supported a high population density for more than 5,000 years and only tiny fragments of natural habitat remain. In northwest Africa, the Mediterranean Conifer and Mixed Forests [34] is also critically threatened because of a combination of extreme habitat loss and little protection.

A number of tropical forest ecoregions rank as critically threatened. In West Africa these are the Eastern Guinean Forests [3], Nigerian Lowland Forests [4], Mandara Plateau [39], Niger Delta Swamp Forests [5], Cross-Niger Transition Forests [6], and Guinean Mangroves [115]. In East Africa, the Eastern Arc Mountains [19], Southern Zanzibar–Inhambane Coastal Forest Mosaic [21], Maputaland Coastal Forest Mosaic [22], and KwaZulu-Cape Coastal Forest Mosaic [23] are critically threatened because of low levels of protection, high rates of conversion, and fragmentation. A number of the forested offshore islands are also regarded

TABLE 5.2. **Ranking of Ecoregions with More Than 10 Percent of Their Area in IUCN Category I–IV Protected Areas.**

Ecoregion	Percentage in IUCN I–IV Protected Areas
Etosha Pan Halophytics [67]	92
Namib Desert [107]	55
East African Montane Moorlands [72]	47
Kaokoveld Desert [106]	44
Serengeti Volcanic Grasslands [47]	32
West Saharan Montane Xeric Woodlands [96]	32
Rwenzori-Virunga Montane Moorlands [73]	28
Southern *Acacia-Commiphora* Bushlands and Thickets [46]	16
Montane Fynbos and Renosterveld [90]	14
Western Zambezian Grasslands [56]	13
Zambezian *Baikiaea* Woodlands [51]	13
Atlantic Coastal Desert [92]	13
Ethiopian Montane Moorlands [71]	13
Itigi-Sumbu Thicket [48]	13
Zambezian and Mopane Woodlands [54]	12
Makgadikgadi Halophytics [68]	12
Zambezian Flooded Grasslands	12
Angolan Mopane Woodlands [55]	12
Southern Africa Mangroves [117]	12
Madagascar Ericoid Thickets [84]	12
Eastern Miombo Woodlands [52]	12
Angolan Scarp Savanna and Woodlands [81]	12
Kalahari *Acacia* Woodlands [58]	11
Kalahari Xeric Savanna [105]	11
Northern *Acacia-Commiphora* Bushlands and Thickets [45]	11
Namib Escarpment Woodlands [109]	10
Maputaland Coastal Forest Mosaic [22]	10
East Sudanian Savanna [37]	10
Cross-Sanaga-Bioko Coastal Forests [7]	10

Source: Protected area data for 2002 kindly supplied by UNEP–World Conservation Monitoring Centre.

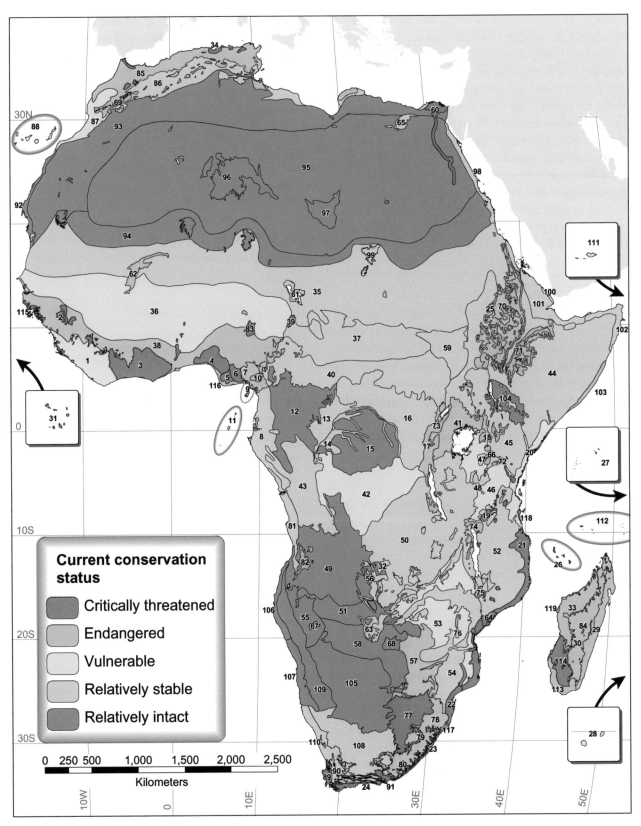

FIGURE 5.3. Current conservation status.

Current conservation status

- Critically threatened
- Endangered
- Vulnerable
- Relatively stable
- Relatively intact

as critically threatened, such as the Comoros Forests [26], Granitic Seychelles Forests [27], and Cape Verde Islands Dry Forests [31].

Some grassland and Mediterranean habitat ecoregions are also assessed as critically threatened, such as the High-veld Grasslands [77] of South Africa, which have been converted mostly to intensive cropland. The Jos Plateau Forest-Grassland Mosaic [83] in Nigeria is also heavily converted to smallholder farmland. The same is true of the Lowland Fynbos and Renosterveld [89] of South Africa, which has been converted to both agriculture and urban areas.

Endangered Ecoregions

In West and Central Africa the Guinean Forest-Savanna Mosaic [38], together with a number of the forests around the Nigeria-Cameroon border, are considered endangered. The relevant forested ecoregions are the Cameroon Highlands Forests [10], the Mount Cameroon and Bioko Montane Forests [9], the Cross-Sanaga-Bioko Coastal Forests [7], and the Central African Mangroves [116].

In East Africa, the Victoria Basin Forest-Savanna Mosaic [41], the Albertine Rift Montane Forests [17], and the East African Montane Forests [18] are also endangered. In all of these areas there has been considerable forest loss over millennia (Howard 1991). The montane forests and grasslands of the Ethiopian Highlands [70, 71] also fall into this category, primarily because of their high population density and conversion of habitats to farmland over millennia.

In southern Africa, the Southern Africa Bushveld [57], Drakensburg Montane Grasslands, Woodlands, and Forests [78], and Angolan Montane Forest-Grassland Mosaic [82] are all endangered. Almost all of Madagascar is regarded as endangered. The majority of the Mediterranean habitat ecoregions of northwestern Africa also fall in this category because of extensive habitat loss and low levels of protection.

Relatively Stable Ecoregions

In this category, habitats and biodiversity are intact over significant parts of the ecoregion. The ecoregions in the Sahel region of western Africa, the miombo and mopane woodlands of eastern and southern Africa, the Congo Basin forests, the Nama Karoo of South Africa, and the Horn of Africa fall mainly into this category. Although the name of the category implies little change, we believe that many of these ecoregions are changing rapidly, and the term *near-threatened* might be more appropriate. Very often the area of untouched

habitat is declining, and species are becoming ever more confined to protected areas. In particular, many of these ecoregions have retained their habitat, but the large mammal fauna has been severely affected by hunting.

Relatively Intact Ecoregions

The desert and semi-desert habitats of southern Africa, the Sahara, northern Kenya, and the central and western forests of the Congo Basin are regarded as relatively intact. To some extent the inclusion of most of the semi-desert areas in this category is misleading because these are pastoral areas, where little habitat transformation has taken place but where the habitats may be seriously degraded though overgrazing. The large mammal fauna may also be almost entirely removed through overhunting and replaced by domestic livestock, significantly reducing the importance of the area for conservation. The two large saline depressions in southern Africa (Etosha Pan Halophytics [67] and Makgadikgadi Halophytics [68]) that fall in this category are not suitable for agriculture and are found mainly in large protected areas. However, even these areas are not without threat because they are vulnerable to future changes in their water regimes (where water is diverted for human uses elsewhere). Moreover, logging concessions often cover the ecoregions of the Congo Basin that are regarded as relatively intact, so the future of these areas is also uncertain.

How Will the Patterns of Threat Change in the Future?

Our assessment of the current conservation status of ecoregions is a valuable measure of the threats facing habitats and species at this time. Yet a number of factors, such as changes in sociopolitical and economic status, infrastructure development, climate change, and population growth trends may quickly accelerate the levels of fragmentation and habitat loss. To complete a comprehensive assessment of future threats is a daunting task. Here we use likely future habitat loss and the number of species threatened by extinction to indicate ecoregions where future biodiversity decline is likely to occur across Africa.

Future Threats to Habitats: Increasing Population Densities

Human population densities are highly variable across Africa. In 1995, average population densities of more than

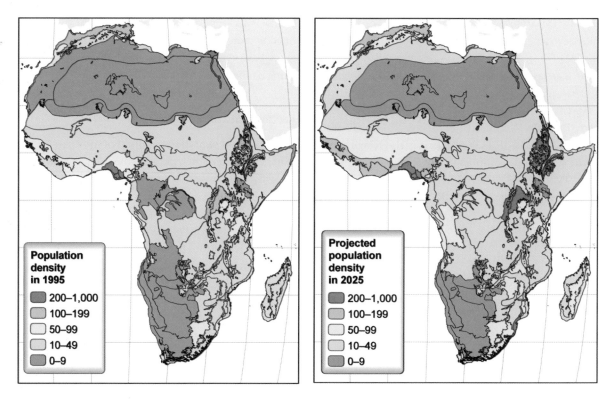

FIGURE 5.4. Estimated population density (people per square kilometer) by ecoregion in 1995 and 2025 (2025 data kindly provided by the World Resources Institute; CIESIN, Columbia University, IFPRI, and WRI 2000).

200 people/km^2 existed along the Nile River in North Africa, in southern Nigeria, and around Mount Mulanje in Malawi (figure 5.4). Average densities of more than 100 people/km^2 were found in the Kenyan Highlands, in northwest Africa, around Lake Victoria, along the Albertine Rift, in the Ethiopian mountains, and in parts of coastal South Africa. Based on projected population growth rate estimates from the World Resources Institute, by the year 2025 densities of more than 200 people/km^2 also are likely to occur around Lake Victoria and the Albertine Rift Mountains, throughout the Ethiopian Highlands, and more widely in coastal South Africa. Populations will also expand on Madagascar, throughout West Africa, and in coastal eastern and southern Africa. Extreme threats to habitats can also be predicted near many large capital cities, such as Kinshasa and Brazzaville, Lagos, Nairobi, and Dar es Salaam, associated with a flow of natural resources into the urban areas.

Future Threats to Species: An Index of Extinction Risk

A number of ecoregions possess high numbers of birds and mammals threatened with extinction. These areas are con-

centrated in the tropical belt of Africa, on Madagascar, and in southeastern South Africa (figure 5.5). Five of the top ten ecoregions in terms of their threatened bird and mammal fauna are montane regions, supporting mainly tropical montane forest and grassland. Four of these, the Eastern Arc Forests [19], Albertine Rift Montane Forests [17], Madagascar Subhumid Forests [30], and East African Montane Forests [18] possess particularly endangered faunas for these two taxonomic groups (table 5.3). If other groups of animals and plants were included, we are confident that the ecoregions of the Cape Floristic Region, including the Lowland and Montane Fynbos and Renosterveld [89, 90] and the Succulent Karoo [110] would also feature areas with heavily threatened species assemblages (mainly for plants).

Which Ecoregions Will Become More Threatened in the Near Future?

By combining information on the increasing risk of habitat loss (indicated by human population growth) and the increasing risk of species extinction (indicated by Red Data Book lists), we can determine possible future changes to the conservation status of African ecoregions. We have used

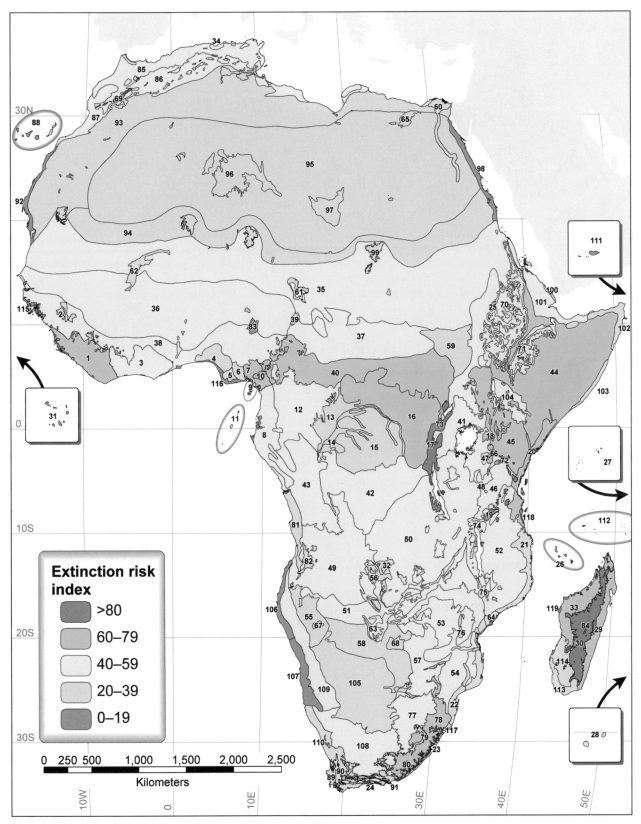

FIGURE 5.5. Future species threat measured by an extinction risk index. Based on the IUCN Red List for birds and mammals (Hilton-Taylor 2000).

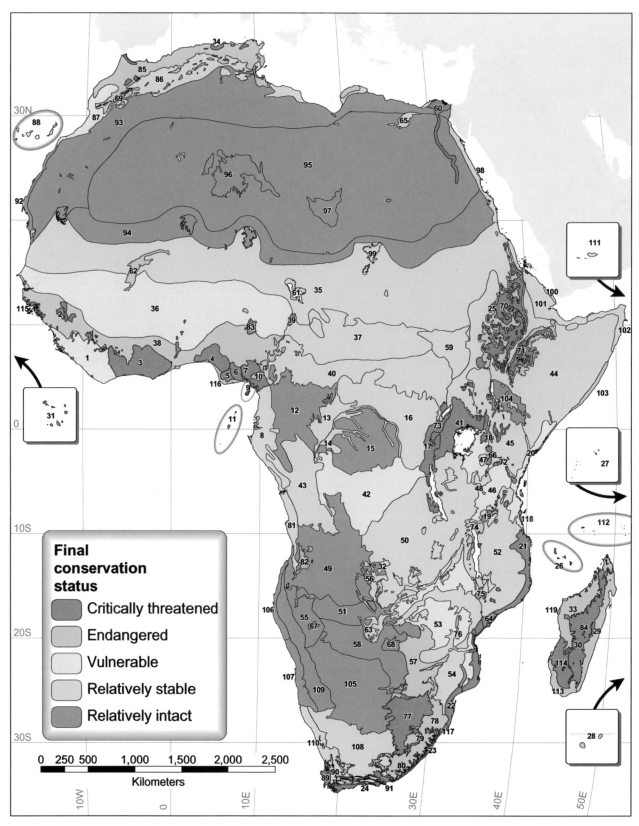

Final conservation status

- Critically threatened
- Endangered
- Vulnerable
- Relatively stable
- Relatively intact

0 250 500 1,000 1,500 2,000 2,500

Kilometers

FIGURE 5.6. Final (future threat-modified) conservation status.

Two female Dorcas gazelles (*Gazella dorcas,* VU) on a sand dune in Aïr, Niger. Photo credit: WWF-Canon/John Newby

TABLE 5.3. **Top Ten Ecoregions Ranked by Extinction Risk Score.**

Ecoregion	Extinction Index
Eastern Arc Forests [19]	90
Albertine Rift Montane Forests [17]	80
Madagascar Subhumid Forests [30]	80
East African Montane Forests [18]	72
Northern Congolian Forest-Savanna Mosaic [40]	70
Cameroon Highlands Forests [10]	70
Madagascar Humid Forests [29]	67
Northeastern Congolian Lowland Forests [16]	65
Cross-Sanaga-Bioko Coastal Forests [7]	64
Somali *Acacia-Commiphora* Bushlands and Thickets [44]	63

Source: Data from Hilton-Taylor (2000).

these data to elevate some ecoregions to higher levels of threat (see chapter 3 Appendix D), resulting in the Final (future threat-modified) Conservation Status Index (figure 5.6). Some of the ecoregions becoming more imperiled are the Albertine Rift Montane Forests [17], Victoria Basin Forest-Savanna Mosaic [41], Ethiopian Highlands [70, 71], Mount Cameroon and Bioko Montane Forests [9], Cross-Sanaga-Bioko Coastal Forests [7], São Tomé, Príncipe, and Annobon Moist Lowland Forests [11], East African Montane Forests [18], and Madagascar Subhumid Forests [30] (figure 5.6). The conservation status of other ecoregions remains unchanged when future threats are considered.

A serious challenge to conservationists is to prevent these ecoregions from losing so much of their habitat that their threatened species become extinct (Brooks et al. 2002; essay 5.2). To prevent an extinction crisis in Africa we must tackle some of the root causes of habitat loss and species endangerment and develop bold ways to enhance conservation efforts across the region in terms of both scope and impact. In the next two chapters we identify the best opportunities for doing so and explore some of the approaches we have at our disposal to achieve this goal.

ESSAY 5.1

Drivers of Forest Loss in West and Central Africa

Susan Minnemeyer and Liz Selig

In Africa, rainforests are found in West Africa close to the Atlantic coast and in the equatorial zone of Central Africa. The Dahomey Gap of Benin and Togo, a stretch of savanna 300 km wide, separates the rainforests of West and Central Africa into two distinct biogeographic regions (Martin 1991).

Both West and Central African forests have globally important levels of species richness and endemism. In addition to their contribution to global biodiversity, African forests store a sig-

nificant portion of the earth's carbon found in live vegetation and are among the globe's largest terrestrial carbon sinks (Matthews et al. 2000). The loss of Africa's rainforests probably would influence local (temperature, precipitation) and global (monsoons, storm systems) weather patterns. It is believed that coastal deforestation in West Africa may already contribute to regional droughts (Boahene 1998; Zheng and Eltahir 1998; Gaston 2000).

Deforestation: Comparing and Contrasting the Situations in West and Central Africa

West Africa

The majority of West Africa's rainforests occur in Liberia, Ghana, and Côte d'Ivoire, with smaller forest areas found in Guinea, Sierra Leone, and Togo. West African forests have been declining in area for decades, particularly from the 1950s onward, and much of what remains is degraded and fragmented (Sayer et al. 1992; Myers et al. 2000). In 1980, the United Nations Food and Agriculture Organization (FAO) reported that Côte d'Ivoire had the highest annual rate of deforestation in the tropics, more than ten times the average rate (Fairhead and Leach 1998a, 1998b). Today deforestation rates are half of what they were in the 1980s but at 3.1 percent per year are still among the world's highest (FAO 2001b). Root causes of deforestation in West Africa include a long history of commercial logging, high population density, rapid population growth, poverty, armed conflict, and population displacement.

Commercial logging has been a major activity in West Africa for decades. Typically, governments own the forests, and logging concessions are granted to international logging companies. In theory concessions are granted to raise revenue for national development. In practice, though, they are often distributed as favors for political party functionaries, government officials, and multinationals (Boahene 1998). Logging companies operating in West Africa are primarily European and Lebanese companies (ITTO 1999) that have little incentive to extract wood products sustainably, the jobs produced are often short term, and most of the wealth generated tends to leave the country.

Timber exports are greatest from Ghana and Côte d'Ivoire, with a 1998 value of US$206 and US$229 million, respectively (FAO 1998). These countries are among the world's top exporters of tropical sawnwood and veneer (ITTO 1999). Although tropical forests are selectively logged, with only the most valuable tree species removed, half or more of the remaining trees can be damaged (Frumhoff 1995; Sponsel et al. 1996). Additional indirect impacts result from the opening of previously inaccessible forest areas by logging roads, bringing hunting, settlement, and human-induced fires (Sponsel et al. 1996).

Another key factor in West Africa's deforestation is population growth and density, which are both well above the world average; see table 5.4 (Rudel and Roper 1997; Fairhead and Leach 1998a, 1998b; Cincotta et al. 2000; FAO 2001b). Population growth can be a strong driver of environmental degradation in developing countries, where it increases demand for fuelwood, water, food (including wild foods such as bushmeat), and forest conversion to subsistence cropland and commercial plantations. Because 90 percent of energy supplies in tropical Africa come from fuelwood or charcoal (Boahene 1998), an increasing population has a direct effect on the demand for wood.

Armed conflict in Sierra Leone, Guinea, and Liberia has also resulted in the loss of some of the most intact remaining forest areas. The mass migration of local populations resulting from the conflict has led to the deforestation of large areas around refugee camps in Guinea. Internal displacement of people in Liberia and Sierra Leone has also led to forest loss. Proceeds from commercial logging may also be used to support conflicts. Timber extraction in Liberia was reportedly used to support Sierra Leone's Revolutionary United Front (Global Witness 2001).

Central Africa

In contrast to West Africa, deforestation has been much less extensive in Central Africa. This is mainly because of past difficulties of accessing the remote interior forests. Some areas of Central Africa also have higher-elevation forests (Cameroon, Gabon) or extensive swamp forests (Congo, Democratic Republic of the Congo), which are particularly inaccessible for most logging equipment. Improvements in technology for logging and roadbuilding have made previously remote forests accessible, and the depletion of the forests of West Africa and Asia means that logging companies are focusing on new areas, such as Central Africa. The area of forest allocated under logging concessions has increased rapidly; for example, two-thirds of Gabon's forests and three-fourths of Cameroon's forests fall within concessions (Collomb et al. 2000, Bikie et al. 2000).

Currently, Gabon, Cameroon, and the Democratic Republic of the Congo rank among the world's top tropical log exporters (ITTO 1999). Gabon and Cameroon alone produced US$492.4 million and US$539.4 million in timber exports in 1998 (FAOSTAT 1998). Unlike West African countries that are shifting to value-added wood products, Gabon and Cameroon primarily export industrial roundwood. As in West Africa, the companies operating in Central African countries are predominantly European (Collomb et al. 2000; Bikie et al. 2000; ITTO 1999). The expatriate origin of logging companies, the export of raw logs, and local corruption mean that most of the value of the trees bypasses local communities.

The effects of commercial logging on biodiversity go far beyond the removal of selected tree species. Although selective logging in Central Africa rarely leads directly to deforestation

TABLE 5.4. **Population and Forest Statistics.**

Country	Population Density (n/km²)	Annual Population Change (%)	Annual Deforestation Rate (%)	Value of Timber Exports (million $US)
West Africa				
Côte d'Ivoire	45.7	1.8	−3.1	228.9
Ghana	86.5	2.7	−1.7	206.4
Guinea	30.0	0.8	−0.5	14.9
Liberia	30.4	8.6	−2.0	6.8
Sierra Leone	65.9	3.0	−2.9	0.8
Togo	83.0	2.7	−3.4	4.5
West Africa average	56.9	3.3	−2.3	
Central Africa				
Cameroon	31.6	2.7	−0.9	539.4
Central African Republic	5.7	1.9	−0.1	24.1
Congo	8.4	2.8	−0.1	75.1
Democratic Republic of the Congo	22.2	2.6	−0.4	54.1
Equatorial Guinea	15.8	2.5	−0.6	63.3
Gabon	4.6	2.6	0.0	492.4
Central Africa average	14.7	2.5	−0.4	
Global average	45.8	1.3	−0.2	

Source: FAO (2000).

because pressures to clear land for agriculture remain low, bushmeat hunting can have a devastating impact on wildlife populations (Frumhoff 1995). As logging companies move into previously inaccessible areas, which often have high wildlife population densities, hunting rates increase greatly to supplement workers' incomes. Logging company truck drivers routinely transport bushmeat to towns where it commands higher prices (Amman and Pierce 1995).

Armed conflict is also driving forest loss in central Africa. In particular, there has been extensive deforestation around refugee camps in the eastern Democratic Republic of the Congo, resulting in large part from fuelwood gathering by displaced populations (Draulans and Van Krunkelsven 2002). Conversely, conflict in the Congo, Democratic Republic of the Congo, and Central African Republic has prevented some commercial logging in the region because companies are reluctant to invest amid political unrest. It is no coincidence that the central African countries with the largest exports of commercial wood products, Gabon and Cameroon, are also the region's most stable. Conflict reduces the direct impacts of logging but increases factors such as poverty, landlessness, and migration that indirectly lead to deforestation.

Habitat Loss and Extinction Threat to African Birds and Mammals

Thomas Brooks

Habitats are being converted to human-dominated landscapes across Africa. Only about 35 percent of the continent's tropical forests remain (Sayer et al. 1992), and the dry lands are coming under increasing pressure for agriculture (Glantz 1994). The continent's biodiversity is renowned both for its richness and uniqueness (Kingdon 1989); for example, sub-Saharan Africa holds about a fifth of the world's mammals, birds, snakes, and amphibians (Burgess et al. 1998c). Therefore, a critical management priority is to assess the impacts of the continuing loss of natural habitats on this biodiversity.

To answer this question, we need to describe the relationship between the size of an area (*A*) and the number of species (*S*) that it holds. Small areas have fewer species than larger ones, according to a well-documented relationship, the species-area curve. Empirically, this relationship approximates a power function of the form $S \propto A^z$ (Preston 1962). Furthermore, habitat islands contain fewer species than equivalent areas of continuous habitat, such that *z* is approximately 0.15 in subsets of continuous habitat and approximately 0.25 in fragmented habitats (Rosenzweig 1995). Given the conversion of a known area of continuous habitat to fragments, we can use this relationship to make predictions about the number of species likely to be lost and the time over which these extinctions will occur (figure 5.7). Recent studies of African forest primates on continental (5,000,000 km^2) to national (up to 1,000,000 km^2) scales, on large mammals in East African savanna reserves (up to 20,000 km^2), and on birds in East African forest fragments (up to 100 km^2) make and test such predictions. In this essay I summarize these studies and synthesize their key conclusions.

At the coarsest scale, Cowlishaw (1999) considered threat to African forest primates. He first used forest cover data and the fragmented habitat *z*-value to show that even though no species have become extinct in recent years, African countries have a median "extinction debt" of four species expected to be lost because of deforestation to date. He then made similar predictions of "extinction debt" for endemic species on regional and continental scales and found that these were very similar

to the numbers of species listed as threatened "with a high probability of extinction in the wild in the medium-term future" (IUCN 2001: 14). Finally, he used the 2,017-km^2 land bridge island of Bioko as a model for the likely rate of loss. He suggests that forest fragments the size of the island, assuming that it held all of the species in adjacent Cameroon when separated from the mainland, lose about two-thirds of their species over the first 10,000 years after insularization and most much faster than that.

Soulé et al. (1979) sought to predict how rapidly mammal extinctions should occur in nineteen isolated savanna reserves in Kenya, Uganda, and Tanzania. They first used data on the number of species on the Malay Peninsula, its area, and the continuous habitat *z*-value to define a species-area curve for the region. They used this to estimate how many large mammal species occurred on each of the land bridge islands of the Greater Sundas, in southeast Asia, before their isolation from the peninsula by rising sea levels over the last 10,000 years. They then calculated the species loss over this period and hence, for each island, the coefficient (*k*) describing the exponential decay in species richness. Next, they used the consistent relationship between island size and *k* to estimate *k* for the East African reserves and finally used these values to predict numbers of large mammal extinctions from the reserves at given future dates. Thus, they predicted that small reserves would lose two-thirds of their species over the next 500 years and that even large reserves would lose one-third. Newmark (1996) compared these predictions with documented extinctions from six Tanzanian reserves over the last 35–83 years and found that the model predicted large mammal extinctions accurately.

At the finest scale of individual forest patches, as small as 1,000 m^2, Newmark (1991b) investigated the effects of habitat fragmentation on understory bird communities in the East Usambara Mountains of Tanzania. He found that local extinctions had already occurred to such an extent that the number of species surviving in the forest patches was strongly related to their areas. Furthermore, he demonstrated that these ex-

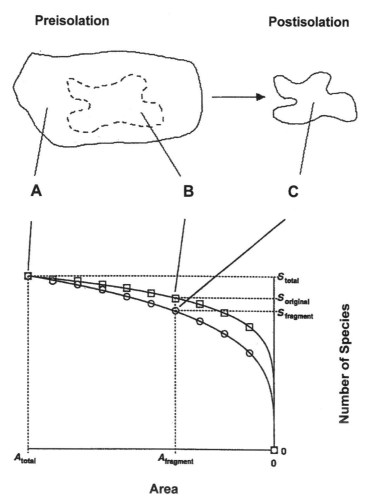

Preisolation

Postisolation

A   **B**   **C**

Area

FIGURE 5.7. Typical species-area relationships. Larger areas (*A*) have more species than smaller ones (*B, C*), and areas that have been long isolated— such as habitat fragments—have proportionately fewer species (*C*) than do equal-sized areas that are nested in continuous habitat (*B*). If a habitat is fragmented, shrinking in size from A_{total} to $A_{fragment}$, the number of species found in the area $A_{fragment}$ will therefore decline from $S_{original}$ to $S_{fragment}$. $S_{original}$, the number of species in a subset of continuous habitat, can be estimated using the equation $S_{original} = S_{total} ((A_{fragment}/A_{total})^{0.15}$, and $S_{fragment}$, the number of species in a habitat fragment, can be estimated using the equation $S_{fragment} = S_{total} ((A_{fragment}/A_{total})^{0.25}$. At a time *t* after the fragmentation event, these variables are related to each other by the equation $(S_{now} - S_{fragment})/(S_{original} - S_{fragment}) = e^{(-k.t)}$, where S_{now} is the number of species surviving at time *t*, and *k* is a decay constant describing the exponential loss of species from the fragment (Brooks et al. 1999).

tinctions were nonrandom, with rare forest interior species being the most susceptible. At a similar scale, in Kakamega forest in western Kenya, Brooks et al. (1999) used the areas of the surviving fragments, historical data on the entire forest avifauna, and bird survey data from each fragment to estimate the original (using the continuous habitat *z*-value) and predicted final

(using the fragmented habitat *z*-value) avifauna of each fragment (figure 5.7). We then reconstructed the forest's history to date the fragmentation and thus were able to estimate *k* and hence the half-life that characterized the decay of each fragment's avifauna (figure 5.7). The fragments had lost half of the species that they stood to lose by 23–80 years after their isolation. Many of the extinctions were of species that required a (now-broken) elevational gradient for migration between the wet and dry seasons.

This body of work provides several key lessons:

- The extent of habitat loss accurately predicts the number of threatened species in an area and the number of extinctions that the area will eventually suffer.
- The length of this time lag is scale dependent, varying from thousands of years for the entire continent down to decades at the scale of forest fragments.
- Rare species and those with specialized habitat requirements (e.g., deep forest, elevational gradients) tend to be the most vulnerable to extinction.

The key conclusion is that although habitat loss has already led to many local extinctions across Africa, the time lag between habitat loss and species loss means that we still have a narrow window in which to implement conservation measures (Pimm and Brooks 1999). The nature of the species most likely to become extinct first suggests that these measures must include not only strengthening the protection of already-reserved areas but also striving to reestablish corridors of natural habitat between reserves (Newmark 1993). Finally, it must be kept in mind that direct threats (e.g., from bushmeat hunting) often act in synergy with habitat loss (Wilkie et al. 1992) and that these must also be countered if extinctions are to be averted in Africa.

CHAPTER 6

Priorities, Realities, and the Race to Save Wild Africa

Although we have identified both the biologically most important ecoregions of Africa and the patterns of threat and opportunity, this does not provide a sufficient basis for conservation investment in the region. Faced with a range of globally important ecoregions for species and nonspecies biological values and varying levels of threat where should we focus our limited conservation budget? Where should we try to persuade governments to set aside additional reserves or change patterns of land allocation? And where should we approach large donor agencies to provide the funding to prevent the extinction of species and the loss of irreplaceable features of wild Africa, such as wilderness habitats with intact animal and plant assemblages?

This chapter offers answers to these questions. In the first part we synthesize the results of previous analyses to set priorities for conservation investment among Africa's 119 ecoregions. We use the results to

- identify the ecoregions that deserve greater emphasis because of their unique biological values and high level of threat
- highlight the ecoregions of outstanding diversity that contain large blocks of mostly intact natural habitat facing moderate to low threats, where cost-effective investments can be made now to safeguard biodiversity over the long term

In the second part of the chapter we examine in detail many of the issues that have been presented as reasons why conservation cannot be achieved in Africa. Our aim is to test

whether these myths and assumptions are true and whether they pose an insurmountable obstacle to achieving conservation in Africa. In many cases the myths are shown to be erroneous, and the prospects for conservation progress in Africa are at least as good as anywhere else in the tropical world, and we like to believe that in many cases they are far better.

Where to Act First to Conserve Africa

Many ecoregions are clearly outstanding in terms of their biological importance (chapter 4) but are under different levels of current and impending threat (chapter 5). We used a matrix to integrate the biological distinctiveness and conservation status scores for different ecoregions to develop sets of short- and long-term conservation priorities for conservation investment in coming years. This matrix places ecoregions in five different classes of conservation priority.

Class I are ecoregions with globally important biodiversity values that are highly threatened.

Class II are ecoregions with regionally outstanding biodiversity values that are highly threatened.

Class III are ecoregions with globally or regionally outstanding biodiversity values that still offer opportunities for large-scale conservation interventions.

Class IV are ecoregions with regionally or nationally important biodiversity values that are under threat.

Class V are ecoregions with regionally or nationally important biodiversity values where opportunities for large-scale conservation interventions exist.

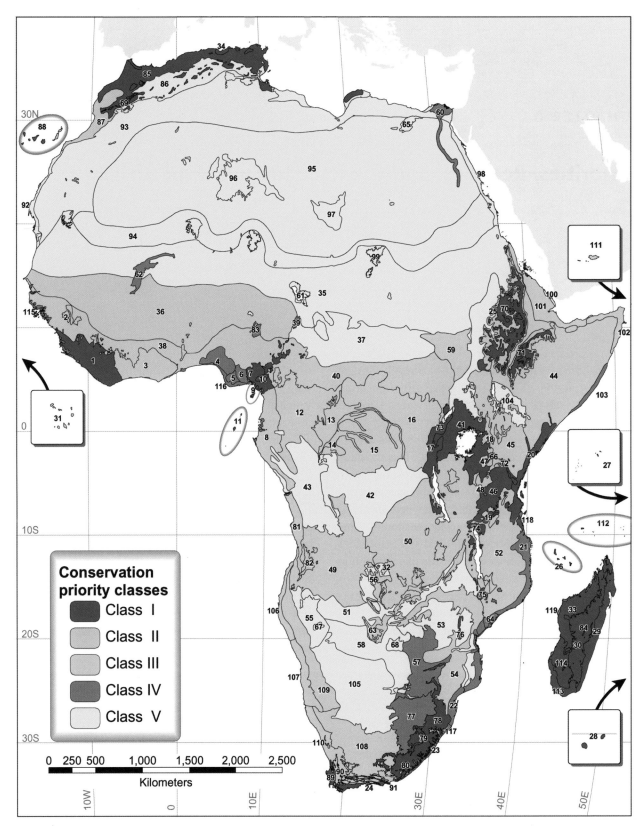

FIGURE 6.1. Distribution of conservation priority classes across Africa.

The distribution of these conservation priority classes provides a rapid overview of priority regions for conservation investment across Africa (figure 6.1). Details of their spatial distribution are outlined here.

Class I. The set of highly threatened and biologically important ecoregions is scattered across different parts of Africa. All of the large offshore island of Madagascar falls in Class I, as do the Indian Ocean Islands of Seychelles, Comoros, and Mascarenes. On the African mainland, the lowland and mountain ecoregions of northwest Africa fall into this class, and further south, the western portion of the Upper Guinea Forest and the Cameroon Highlands and associated lowland forests rank here. In eastern Africa, almost all the montane forest and forest-grassland mosaic ecoregions fall into Class I, as do the forest mosaic habitats around Lake Victoria and along the eastern African coastline. The Serengeti Volcanic Grasslands [47] and the Southern *Acacia-Commiphora* Bushlands and Thickets [46] represent eastern African examples of the savanna-woodland biome that rank as Class I priorities. In southern Africa, the Class I ecoregions are again dominated by montane ecoregions, particularly those associated with the Drakensberg Mountains but also including the Mediterranean habitats of the Fynbos in the Cape region of South Africa.

Class II. These ecoregions tend to be located in regions adjacent to clusters of Class I ecoregions. For example, in northwest Africa the only Mediterranean ecoregion that is not in Class I falls into Class II. Further south, large parts of the forest and forest-savanna mosaic of West Africa are in this category, excluding the biologically less valuable forest ecoregions of southern Nigeria. Highland areas of Ethiopia, Kenya, and Tanzania that were not in Class I also fall in Class II, as do some of the marginal areas of the Drakensberg and Cape Floral Kingdom of South Africa. The Cape Verde Islands also fall into this class.

Class III. These ecoregions generally occupy the large, less threatened but biologically important forest, savanna-woodland, wetland, and desert ecoregions of Africa. Most of the Horn of Africa, the miombo and mopane woodlands of eastern and southern Africa, the Nama Karoo and coastal desert regions of southwest Africa, and the Congo Basin forests are captured here. The much younger Sahara Desert does not feature in this set of ecoregions because it has much lower biological values than the ancient dry areas of northeast and southwest Africa. Significantly, the Angola Scarp ecoregion also falls into this category, the only montane region in that part of Africa to achieve such a high ranking for conservation investment. The scarp forests themselves are the most valuable part of this ecoregion, and because these occupy a tiny

part of the ecoregion that are significantly threatened, they might, on their own, be better assigned to Class I importance.

Class IV. The Class IV ecoregions are scattered across the biomes of Africa, exhibiting no obvious pattern. The biologically poor forests of southern Nigeria fall in this class, as do parts of the coastal habitats of Mozambique, the inland plateau areas of South Africa, and the Nile Delta of Egypt. Most of these ecoregions are heavily degraded and are not exceptional in biological importance. An exception might be the forest mosaic of coastal Mozambique, which has been poorly explored biologically but is contiguous with more important forest mosaics to the north and south.

Class V. This class covers the greater Sahara region of Africa, the Kalahari and some of the drier woodlands of southern Africa, and the forest-savanna mosaic habitats north and south of the main Congo Basin forests. Many of these areas are transitional in nature between forest and woodland or between woodland and desert and therefore do not support a unique flora and fauna. Others have been recently derived because of climatic desiccation (e.g., most of the Sahara).

Although conservation is important in every ecoregion, throughout this chapter we give highest prominence to the Class I and Class III ecoregions because of their globally outstanding biodiversity values. Of the 119 ecoregions of Africa, 32 fall into Class I (27%) and 29 into Class III (24%), just over half of the total. This analysis also brings to the fore a fundamental issue for global conservation and for priorities at the continental scale. The Class I and Class III ecoregions of highest priority tend to be located in different types of habitats and at different scales. For example in the montane grassland, Mediterranean scrub, and temperate coniferous forest biomes, there are many Class I ecoregions where conservation tends to be costly and urgent and involves restoration. Conversely, in the savanna-woodland, flooded grassland, and desert biomes there are many Class III ecoregions where opportunities remain for cost-effective conservation across huge areas.

Some conservationists argue that we should practice triage and write off Class I ecoregions. The reasoning is that the species and natural habitats in these highly threatened ecoregions will be too costly to protect over the long term. This argument would suggest that a better strategy is to invest the limited resources available to conservation in the more intact Class III ecoregions, where the odds of long-term conservation are better. Our view is that it is too early to write off any Class I ecoregion in Africa. Many of the endemic species in these areas require small patches of habitat and have the potential to avoid extinction if we are able to retain modest amounts of habitat in well-targeted reserve

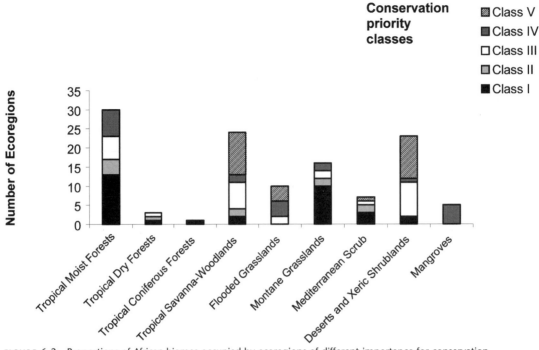

FIGURE 6.2. Proportions of African biomes occupied by ecoregions of different importance for conservation intervention.

networks or communally managed or private lands. We are not yet in the era of triage but of expanding the funding base and the number of stakeholders engaged in conservation to save as much as possible.

Distribution of Conservation Priority Classes across Biomes

The conservation priority classes are spread across all the biomes of Africa, indicating that our aim of representing the biological distinctiveness of each of the biomes of Africa has been quite successful (figure 6.2). Most biomes support at least one globally outstanding Class I or Class III ecoregion. The major exception is mangroves, which are somewhat transitional between terrestrial and marine habitats and hence difficult to rank using our terrestrially focused methods.

Distribution of Conservation Priority by Area and Location

The two largest biomes (savanna-woodlands and deserts) contain large areas of ecoregions that fall in the lowest rank of conservation priority (Class V) (figure 6.3). This is mainly because many ecoregions in these biomes contain widespread species and few endemics and lack multiple values

for nonspecies biological features. The same biomes, together with the tropical moist forests, also contain large areas of Class III ecoregions, which have high biological values but are unthreatened, where biodiversity is spread across huge areas of land. In comparison, the ecoregions falling in Classes IV, II, and I occupy much smaller areas of the various African biomes (figure 6.3).

Most of the Class I global priority ecoregions are small and occupy islands and montane regions of Africa (figure 6.4a). Few of the Class I ecoregions are found in lowland habitats on the African mainland (figure 6.4a). This distribution reflects the importance of these ecoregions primarily for concentrations of endemic species threatened by loss of the small remaining habitat patches. In contrast to the Class I ecoregions, most Class III priorities are found in lowland regions of the African mainland, where they tend to be very large (figure 6.4b). This reflects their importance primarily for ecological phenomena such as migrations, intact habitats and wilderness areas, and, to a lesser extent, species richness and endemism.

Comparison with Other Continents

The World Wildlife Fund (WWF-US) has now completed continental-scale ecoregional conservation assessments for

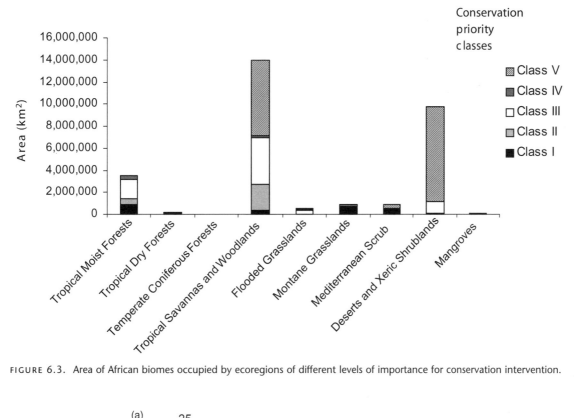

FIGURE 6.3. Area of African biomes occupied by ecoregions of different levels of importance for conservation intervention.

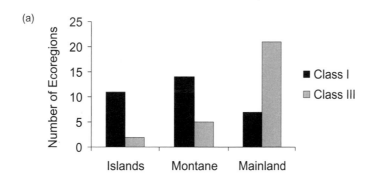

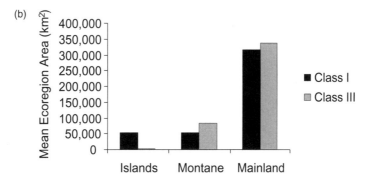

FIGURE 6.4. Distribution of Class I and Class III ecoregions across islands, montane areas, and mainland areas of Africa: (a) number of Class I and III ecoregions; (b) area of Class I and III ecoregions.

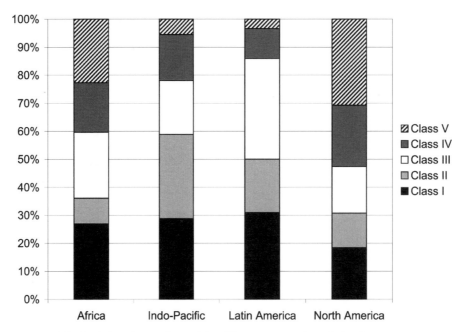

FIGURE 6.5. Breakdown by continent of the different conservation priority classes.

Latin America, North America, and the Indo-Pacific (Dinerstein et al. 1995; Ricketts et al. 1999; Wikramanayake et al. 2002). A comparison of the distribution of conservation priority classes between these regions shows that the percentage of Class I ecoregions in Africa is slightly less than for Latin America and the Indo-Pacific Region, the two other major tropical regions of the world (figure 6.5). This agrees with general assessments that the biodiversity importance of Africa is somewhat less than that of Asia or South America and that habitats in Africa are not yet as threatened as they are elsewhere in the tropics. However, Africa has more Class I ecoregions than North America, probably because of the lower density of endemic species across temperate and boreal portions of the world, including much of North America.

The percentage of Class III ecoregions in Africa is slightly more than in the Indo-Pacific and North American regions and somewhat less than in Latin America (figure 6.5). This confirms the greater loss of intact habitat from much of Asia and North America, aside from the boreal region of North America, and the forests on the island of Papua New Guinea. In South America the Amazonian forest is still intact over vast areas, leading in part to the higher percentage of Class III ecoregions in that region.

At the other end of the scale for conservation investment, the largest number Class V ecoregions of lowest biological importance is found in North America. This is presumably because temperate and boreal habitats have an intrinsically lower level of biological importance compared with tropical

and Mediterranean habitats. Some African ecoregions also fall into this category of low importance, mainly mangroves and some deserts, particularly the recently formed Sahara.

Portfolio of the Most Important Ecoregions

One of the complaints lodged against conservationists is that they offer little guidance to decision-makers because they view every place as a priority for conservation action. We go against this trend and offer clear priorities for conservation intervention across large regions of the mainland of Africa and its offshore islands. Our top priorities are the ecoregions falling in Class I and Class III in the integration matrix.

Class I: Globally Important and Highly Threatened Ecoregions

The conservation of these biologically valuable but highly threatened ecoregions will entail great expenditure because they are also often found in regions of high human population density where the standard approaches to conservation—large reserves and habitat corridors—will be problematic to implement. However, small reserves do exist in most of the Class I ecoregions, and these provide the core areas for conservation focus.

Our integration matrix places thirty-two of the African ecoregions in this class (table 6.1). Of these, 41 percent are from the tropical and subtropical moist broadleaf forests,

TABLE 6.1. **Class I Ecoregions.**

Ecoregion Name	Area (km²)	Biome*
Western Guinean Lowland Forests [1]	205,100	1
Cross-Sanaga-Bioko Coastal Forests [7]	52,100	1
Mount Cameroon and Bioko Montane Forests [9]	1,100	1
Cameroon Highlands Forests [10]	38,000	1
São Tomé, Príncipe, and Annobon Moist Lowland Forests [11]	1,000	1
Albertine Rift Montane Forests [17]	103,900	1
Eastern Arc Forests [19]	23,700	1
Northern Zanzibar-Inhambane Coastal Forest Mosaic [20]	112,600	1
Comoros Forests [26]	2,100	1
Granitic Seychelles Forests [27]	300	1
Mascarene Forests [28]	4,900	1
Madagascar Humid Forests [29]	112,100	1
Madagascar Subhumid Forests [30]	199,600	1
Madagascar Dry Deciduous Forests [33]	152,100	2
Mediterranean Conifer and Mixed Forests [34]	723,100	5
Victoria Basin Forest-Savanna Mosaic [41]	165,800	7
Southern *Acacia-Commiphora* Bushlands and Thickets [46]	227,800	7
Mediterranean High Atlas Juniper Steppe [69]	6,300	10
Ethiopian Upper Montane Forests, Woodlands, Bushlands, and Grasslands [70]	245,400	10
Ethiopian Montane Moorlands [71]	25,200	10
Rwenzori-Virunga Montane Moorlands [73]	2,700	10
Southern Rift Montane Forest-Grassland Mosaic [74]	33,500	10
South Malawi Montane Forest-Grassland Mosaic [75]	10,200	10
Eastern Zimbabwe Montane Forest-Grassland Mosaic [76]	7,800	10
Drakensberg Montane Grasslands, Woodlands, and Forests [78]	202,200	10
Drakensberg Alti-Montane Grasslands and Woodlands [79]	11,900	10
Madagascar Ericoid Thickets [84]	1,300	10
Mediterranean Woodlands and Forests [85]	358,300	12
Canary Islands Dry Woodlands and Forests [88]	5,000	12
Lowland Fynbos and Renosterveld [89]	32,800	12
Madagascar Spiny Thickets [113]	43,400	13
Madagascar Succulent Woodlands [114]	79,700	13

*1, tropical and subtropical moist broadleaf forests; 2, tropical and subtropical dry broadleaf forests; 5, temperate coniferous forests; 7, tropical and subtropical grasslands, savannas, shrublands, and woodlands; 9, flooded grasslands and savannas; 10, montane grasslands and shrublands; 12, Mediterranean forests, woodlands, and scrub; 13, deserts and xeric shrublands; 14, mangroves.

(a) By number of ecoregions

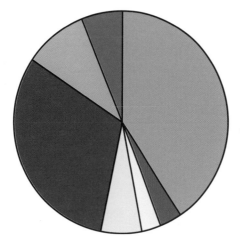

(b) By ecoregion area

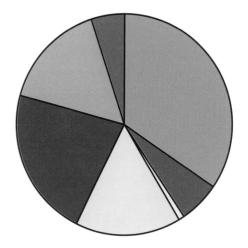

□ Tropical and subtropical
   Moist Broadleaf Forests

■ Tropical and subtropical dry
   broadleaf forests

□ Temperate coniferous forests

□ Tropical and subtropical
   grasslands, savannas,
   shrublands, and woodlands

■ Montane grasslands and
   shrublands

■ Mediterranean forests,
   woodlands, and scrub

■ Deserts and xeric shrublands

FIGURE 6.6. Proportion of globally important and threatened ecoregions (Class I ecoregions) in each biome by **(a)** number of ecoregions and **(b)** area.

more than in any other biome (table 6.1; figure 6.6). This is followed by 31 percent in the montane grasslands and shrublands biome and 9 percent in the Mediterranean forest, woodlands, and scrub biome. Table 6.1 summarizes the ecoregions that fall in Class I; full text descriptions of each ecoregion can be found in appendix I.

Class III: Globally Important but Less Threatened Ecoregions

Conservation of these biologically valuable but less threatened ecoregions faces fewer challenges than that of the Class I ecoregions. In Class III ecoregions large areas of habitat remain, human population densities are low, and there is

often a well-developed reserve network. Large reserves and habitat corridors linking these reserves can still be considered in these areas, and opportunities to engage local communities and private landowners in conservation-based management are good.

Our integration matrix indicates that 29 of the 119 ecoregions fall in the Class III category of priority (table 6.2). Most of these ecoregions fall in the tropical and subtropical grasslands, savannas, shrublands, and woodlands biome and the deserts and xeric shrublands biome (30 percent each), followed by the tropical and subtropical moist broadleaf forests (21 percent) (figure 6.7). Table 6.2 summarizes the ecoregions that fall in Class III; full text descriptions of each ecoregion can be found in appendix I.

Class I ecoregion: Spiny thicket of *Allaudia procera,* west of Fort Dauphin, southeastern Madagascar. Photo credit: WWF-Canon/Gerald S. Cubitt

Class II ecoregion: *Ruschia* ssp. (Mesembryanthemaceae family), Namaqualand, Republic of South Africa. Photo credit: WWF-Canon/Martin Harvey

(a) By number of ecoregions

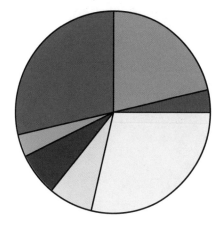

(b) By ecoregion area

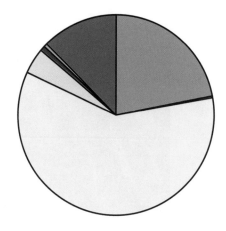

- ◻ (gray) Tropical and subtropical Moist Broadleaf Forests

- ◼ Tropical and subtropical dry broadleaf forests

- ◻ Tropical and subtropical grasslands, savannas, shrublands, and woodlands

- ◻ Flooded grasslands and savannas

- ◼ Montane grasslands and shrublands

- ◻ Mediterranean forests, woodlands, and scrub

- ◼ Deserts and xeric shrublands

FIGURE 6.7. Proportion of globally important and less threatened ecoregions (Class III ecoregions) in each biome by (a) number of ecoregions and (b) area.

Myths and Realities of African Conservation

Conservation in Africa is hindered by a number of long-standing myths and assumptions that are often unfounded or at best distort the true situation. Here we try to dispel some of the conservation myths about Africa and shed light on emerging developments in the conservation arena.

Myth 1: Biologists Cannot Agree on the Most Important Areas for Conservation in Africa

The basis for this myth is unclear, but it reached a sufficient level of controversy to prompt a commentary in the journal *Nature* on the harmful effects of discord and duplication of effort among conservation organizations in setting biodiversity priorities (Mace et al. 2000). Perhaps the myth re-

sulted from the promotion of African conservation in picturesque East Africa, with its large populations of plains ungulates against a backdrop of Mount Kilimanjaro. Other analyses have promoted regions rich in endemics as the best conservation targets, but endemism-focused approaches have ignored the globally important concentrations of large mammals (WWF and IUCN 1994; Stattersfield et al. 1998; Mittermeier et al. 1999; Myers et al. 2000). These differences reflect the fact that different approaches began with different conservation targets (Redford et al. 2003). Focusing on intact vertebrate assemblages will not yield the same set of priority areas as analyses of centers of plant endemism.

We explore this issue further in essay 6.1. Our conclusion is that despite the differences of approach, there is an overwhelming emerging consensus on the important re-

TABLE 6.2. **Class III Ecoregions.**

Ecoregion Name	Area (km^2)	Biome*
Atlantic Equatorial Coastal Forests [8]	189,700	1
Northwestern Congolian Lowland Forests [12]	434,100	1
Western Congolian Swamp Forests [13]	128,600	1
Eastern Congolian Swamp Forests [14]	92,700	1
Central Congolian Lowland Forests [15]	414,800	1
Northeastern Congolian Lowland Forests [16]	533,500	1
Zambezian *Cryptosepalum* Dry Forests [32]	38,200	2
Northern Congolian Forest-Savanna Mosaic [40]	708,100	7
Somali *Acacia-Commiphora* Bushlands and Thickets [44]	1,053,900	7
Northern *Acacia-Commiphora* Bushlands and Thickets [45]	326,000	7
Serengeti Volcanic Grasslands [47]	18,000	7
Angolan Miombo Woodlands [49]	660,100	7
Central Zambezian Miombo Woodlands [50]	1,184,200	7
Eastern Miombo Woodlands [52]	483,900	7
Zambezian and Mopane Woodlands [54]	473,300	7
Sudd Flooded Grasslands [59]	179,700	9
Zambezian Flooded Grasslands [63]	153,500	9
East African Montane Moorlands [72]	3,300	10
Angolan Scarp Savanna and Woodlands [81]	74,400	10
Montane Fynbos and Renosterveld [90]	45,800	12
Ethiopian Xeric Grasslands and Shrublands [101]	153,100	13
Somali Montane Xeric Woodlands [102]	62,600	13
Kaokoveld Desert [106]	45,700	13
Namib Desert [107]	80,900	13
Nama Karoo [108]	351,100	13
Namib Escarpment Woodlands [109]	225,500	13
Succulent Karoo [110]	102,700	13
Socotra Island Xeric Shrublands [111]	3,800	13
Aldabra Island Xeric Scrub [112]	200	13

*1, tropical and subtropical moist broadleaf forests; 2, tropical and subtropical dry broadleaf forests; 5, temperate coniferous forests; 7, tropical and subtropical grasslands, savannas, shrublands, and woodlands; 9, flooded grasslands and savannas; 10, montane grasslands and shrublands; 12, Mediterranean forests, woodlands, and scrub; 13, deserts and xeric shrublands; 14, mangroves.

gions to conserve in Africa. Agreement is generally based on two broad areas of biological value. First, a focus on endemic species as the best way to prioritize regions in terms of their species values will develop one set of priority regions across Africa. Many groups of vertebrates (birds, mammals, amphibians, snakes, chameleons) have similar patterns of endemism, and analysis of all of these groups, or one individually, will indicate similar priority areas. Looking at patterns in a few other groups of vertebrates (lizards), plants, and invertebrates will reveal some additional areas of importance for endemism but will also confirm the areas prioritized by the other groups. Second, various approaches

Priorities, Realities, and the Race to Save Wild Africa 107

that use nonspecies ecological values to define priorities (e.g., migrations, wilderness areas) have all identified similar areas of importance in the region (Sanderson et al. 2002a, Mittermeier et al. 2002). The results of the analyses presented earlier in this book also fit into this general pattern.

The increasing concordance between priority area schemes derived by different organizations using somewhat different approaches is encouraging. It enables us to produce maps showing large-scale conservation priorities across Africa and its islands. If such maps could be agreed on collectively by all the different organizations interested in global level biodiversity analysis, this would be a major step toward reaching a unified voice on what parts of the world are the most important for conservation investment.

Myth 2: African Nations Are Not Interested in Biodiversity Conservation

The creation and gazetting of strict protected areas is a frequently used measure of the commitment by governments to biodiversity conservation. Protected areas are widely recognized as the cornerstone of conservation, and a representative network of reserves is a widely accepted goal for a regional, national, or ecoregional conservation strategy. A common assumption is that most African protected areas were established during colonial rule, and African leaders should want to use natural resources to provide funds to bring their nations out of poverty. If this were true, then the creation of new protected areas should have declined greatly over the past 30 years.

Essay 6.2 shows that the opposite is generally true. Since independence, many African countries have increased levels of protection. Despite severe economic problems and often declining standards of living, African leaders and their governments have decided to set aside additional areas of land as national parks, game reserves, forest reserves, and other protected designations. Many countries have also upgraded the status of existing reserves to stricter levels of protection, making their unsustainable use more difficult. These efforts reflect a genuine commitment and interest in conservation of natural resources among a wide range of African governments. They are also linked to African nations' attempts to climb out of poverty by encouraging ecotourism and providing opportunities for the sustainable use and income generation from wildlife and forests.

Myth 3: Conservation Is an Expensive Investment in Africa

Africa is the poorest continent on the planet (see chapter 1). Thus, it is not surprising that in most countries the protected area network is underfunded (James et al. 1999; Wilkie et al. 2001; Balmford et al. 2000). In essay 6.3, Moore et al. investigate this issue within the ecoregional framework and calculate indicative costs of protected area management across ecoregions. They estimate that it would cost US$631 million per year to manage 10 percent of the 118 (of 119) African ecoregions, or around 0.1 percent of the combined gross national income of all of Africa. The cost distribution is not even across all ecoregions. Some ecoregions will cost far more to conserve because of their high population densities, settled agriculture, higher land prices, and other factors. Nevertheless, the world could easily afford to assist Africa with this effort if the will to do so existed.

The economic value of conserving these areas is placed into context by analyses that consider the value of ecosystem services and the costs of not conserving ecosystems (outlined by Costanza et al. 1997; Balmford et al. 2002, 2003). Under these approaches, conservation is an inexpensive way to ensure that important and valuable ecosystem services provided by the remaining natural habitats of Africa persist. Given the current stage of development in Africa, heavily based on the use of natural resources, these services must be maintained for economic progress to occur.

Myth 4: Conservation in the Most Biologically Important Regions of Africa Is Futile Because of the Intensity of Armed Conflict

Since 1970 more than 30 wars have been fought in Africa, and in recent years it has been the continent most affected by conflict. In regions with active wars, conservation work is often difficult and dangerous to implement (Appleton 1997; Shambaugh et al. 2001). Essay 6.4 explores relationships between conflict and biodiversity conservation across the ecoregions of Africa. This analysis shows that in 2002, of the thirty-two Class I priority ecoregions, only two (6.3 percent) were so seriously affected by ongoing wars that conservation efforts were all but impossible. This figure contrasts with how Africa is typically perceived. Five (16 percent) Class I ecoregions have serious security problems, and 14 (44 percent) have some security problems that prevent conservation efforts in at least part of the ecoregion. In total, 65 percent of the Class I ecoregions have significant security problems. Furthermore, the analysis indicates that conflict is not necessarily linked to the presence of high-value minerals or timber, as is often stated. A stronger correlation is with the poverty of nations and the effectiveness of their governments: where countries are poor and mismanaged, wars are much more likely to occur.

A hopeful sign across Africa is that many of the most intractable conflicts have either ended recently or are showing signs of ending (e.g., Angola, Mozambique, Republic of the Congo, Sudan, parts of Somalia, Democratic Republic of the Congo, Sierra Leone, and Liberia). Unfortunately, some conflicts have started in recent years (e.g., the formerly stable Côte d'Ivoire). In the countries where conflict is ending, Africa is poised to undergo a dramatic experiment in conservation, one that never happened when Asian and Latin American conflicts finished. To what extent can border areas that were once the site of conflicts be turned into transboundary conservation areas? In southern Africa there are already several examples of these "peace parks." Some of these possibilities are explored later in this book, in essay 7.3.

Myth 5: The Spread of Human and Animal Diseases Will Overwhelm Conservation Efforts in Africa

Historically, disease in Africa has dramatically affected both people and their livestock. The prevalence of deadly strains of malaria is one reason why human populations of Africa remained low until recent centuries. The threat of sleeping sickness and the presence of its vector, the tsetse fly, kept vast areas of sub-Saharan Africa sparsely populated by people and livestock. Diseases, and their control, have also had significant impacts on wildlife populations. For example, the control of the tsetse fly poses a threat to biodiversity in some parts of Africa. In addition, many diseases have directly affected wildlife populations; for example, rinderpest devastated ungulate populations in Africa just over a hundred years ago. But will malaria, tsetse fly, and other diseases of animals and people overwhelm conservation efforts in Africa? Essay 6.5 explores the links between disease and conservation and examines how collaborative approaches between veterinary scientists, epidemiologists, conservation biologists, and others can advance conservation efforts in Africa.

The spread of HIV/AIDS among the human population of Africa poses a new set of challenges. Human capital is an essential component of development, and the loss of members of the small group of African conservationists to

HIV/AIDS is a frightening prospect. Essay 6.6 explores some of the ramifications of the AIDS pandemic on human capacity and on land and natural resources. But will HIV/AIDS overwhelm conservation in Africa? This may be one myth that shifts to a reality unless better efforts at prevention and treatment take hold in the region.

Myth 6: Rampant Poaching Will Eventually Cause the Extirpation of Elephant and Rhinoceros from Most Areas of Africa

Hunting elephants for their valuable tusks was one of the prime motivations for the colonization of Africa by European powers (Alpers 1975; Pakenham 1991; Reader 1998). The hunting has continued apace for more than a century, and over the past two decades conservationists have been trying to control elephant hunting to prevent the extinction of the species. There has been much success in these efforts, and despite problems it does not seem likely that the elephant will go extinct in the near future (see essay 6.7).

The damaging effects of elephant poaching pale in significance when compared with the threat faced by two species of African rhinoceros (black and white). Rhino horn fetches huge prices in the Middle East (Yemen) and the Far East, and the demand and willingness to pay have resulted in the decimation of rhino populations across most of Africa. In the 1950s Africa supported tens of thousands of rhinos across wide swaths of the continent. Intense poaching reduced the populations to critically low levels, and despite trade bans, poaching continues in a number of countries. The positive news is that strenuous conservation efforts, particularly in southern Africa, have allowed populations of black rhino to increase slowly in recent years, from a low of 2,450 in 1992 to around 3,100 in 2001. The population increase for white rhino has been more rapid, from 8,300 in 1992 to around 11,670 in 2001, with almost all the animals being found in South Africa. Thus, some of the slowest-breeding mammals on Earth can recover their populations if protected from poaching and provided with suitable habitat (Dinerstein 2003).

The survival of both rhinoceros species will require eternal vigilance until a substitute for the use of rhino horn is widely accepted, but even these species probably will not become extinct in the near future. However, some of the subspecies of white and black rhino may not survive, particularly the northern white rhino in the Democratic Republic of the Congo, and the western black rhino in Cameroon, which may be lost entirely in the next few years.

Myth 7: The Conservation Problems of Africa Will Be Solved Only When the Standard of Living Greatly Improves across the Continent

This viewpoint assumes that conservation awareness and effort increases only after societies have reached a sufficient standard of living (Lomborg 2001). Therefore, we must alleviate poverty before we attempt to encourage conservation efforts. Some situations may seem to support this position; efforts at protecting national parks in the wealthier southern African nations are more successful than those in poorer Central Africa. However, there are probably just as many counterexamples. For instance, many conservationists would consider the bushmeat trade to be one of the most serious conservation issues in forested parts of Africa today. Yet in most countries it is the increasingly affluent urban citizens who are driving the demand for wild meat and therefore contributing to the extirpation of large mammals from poor rural forest areas. Why have these affluent citizens not switched from bushmeat to domestic livestock and begun to promote the conservation of wild animals? The answer relates to complex social factors, including peoples' assessment that bushmeat tastes better than that of domestic animals.

The economic importance of conservation in Africa was well recognized at the recent World Summit on Sustainable Development in South Africa. In vast parts of sub-Saharan Africa, where soils are nutrient-poor and eroding and climatic extremes bring droughts and floods, deriving income from natural habitats and species may be the only avenue out of rural poverty. The use of these resources must be planned and made sustainable. In addition, the mountains of East Africa are the sources of almost all water flowing to the main cities in that region, where most of the people and industry are located. If these mountain forests are lost, the water supplies may dry, and the possibility for economic development will disappear as well. In conclusion, past and future economic development in Africa is solidly rooted in the natural resources of the continent. If they are squandered or overused, there is nothing left to fall back on.

This chapter has identified the most important regions for conservation investment across the African continent and its islands and has shown that conservation is both possible and essential for the long-term benefit of the people of the region. The next challenge is to scale up our conservation efforts to fully conserve the biological diversity of this continent and more fully link conservation approaches to those that achieve economic and social development for people. Instead of creating single new reserves, we need to improve entire protected area networks across ecoregions. We must progress beyond isolated integrated conservation and development projects to conservation landscapes that include a mix of land-use designations from strict core reserves to sustainable extraction zones. And we must shift from offsetting threats to biodiversity on a site-by-site basis to address overarching threats that affect multiple areas: logging, bushmeat hunting, expansion of unsustainable agriculture, and loss of watershed protection. These threats are bad for biodiversity conservation and are reducing the development potential of the region. Chapter 7 shows how WWF envisages operationalizing large-scale conservation initiatives in Africa.

Do Biodiversity Priority Schemes Agree in Africa?

Neil Burgess and Jennifer D'Amico Hales

A number of priority area analyses have been published over the past few years. This essay compares the results of the schemes and assesses the extent to which their results concur or are coming closer together. In that way we aim to address the question of whether biologists can agree on what is important to conserve in Africa.

We first assessed the published schemes and grouped them under two broad headings: those that focus on species as the units driving the prioritization method and those that use habitats or ecological processes (nonspecies values) to drive the method. We then assessed to test the degree of overlap or difference with what has been attempted previously (e.g., da Fonseca et al. 2000).

Species-Focused Schemes

Published prioritization schemes use birds, plants, and several groups of vertebrates to determine priority regions for species conservation across Africa.

Endemic Bird Areas

BirdLife International has described the endemic bird areas (EBAs) of Africa (ICBP 1992; Stattersfield et al. 1998). These are regions that possess at least two bird species wholly confined to an area of land not exceeding 50,000 km². There are twenty-two of these regions on mainland Africa and fourteen on the offshore islands.

Hotspots

Conservation International has defined a hotspot as an area of land possessing at least 1,250 plant species as strict endemics (1 percent of the presumed global total), where at least 75 percent of the original habitat has been lost (Mittermeier et al. 1999; Myers et al. 2000). The hotspots of continental Africa include the Upper Guinea forests of West Africa, the Eastern Arc Mountains and lowland coastal forest mosaic of eastern Africa, and the Mediterranean habitats of northern Africa and the Fyn-bos and Succulent Karoo in South Africa. Madagascar also forms part of a hotspot that extends to cover the Mascarene, Comoros, and Seychelles island groups.

Centers of Plant Diversity

These areas aim to describe the most important places for plant conservation across Africa and its islands (WWF and IUCN 1994). The centers of plant diversity are quite variable in size. In southern Africa they are comparable in size to the hotspots and include many of the same areas. In tropical and northern Africa they are at the scale of portions of ecoregions but focus on key sites.

Priority 1-Degree Squares Identified Using Complementarity

Scientists at the Zoological Museum of the University of Copenhagen (Denmark) and the University of York (England) have developed 1-degree maps of the distribution of a large number of species of fauna and flora across the Afrotropical region (sub-Saharan Africa). These projects used mathematical methods to select grid squares to best represent the distributions of birds, mammals, snakes, and amphibians (in the case of the Copenhagen vertebrate mapping project) or a subset of the Africa plants (in the case of the York plant mapping project) across sub-Saharan Africa. A mathematical principle for efficient area selection called complementarity is built into these analyses. This procedure first selects the area that contains the most species, then that which contributes the greatest number of species not yet represented in the selection (i.e., the most complementary species), and so on, until all species are represented in the smallest number of areas possible. Such methods have been widely recognized as adding significant efficiency to conservation planning projects (Williams 1998; Margules et al. 1988; Kershaw et al. 1994; Muriuki et al. 1997; Pressey et al. 1997; Margules and Pressey 2000; Howard et al. 1998). However, they need considerable data on species distributions to function effectively.

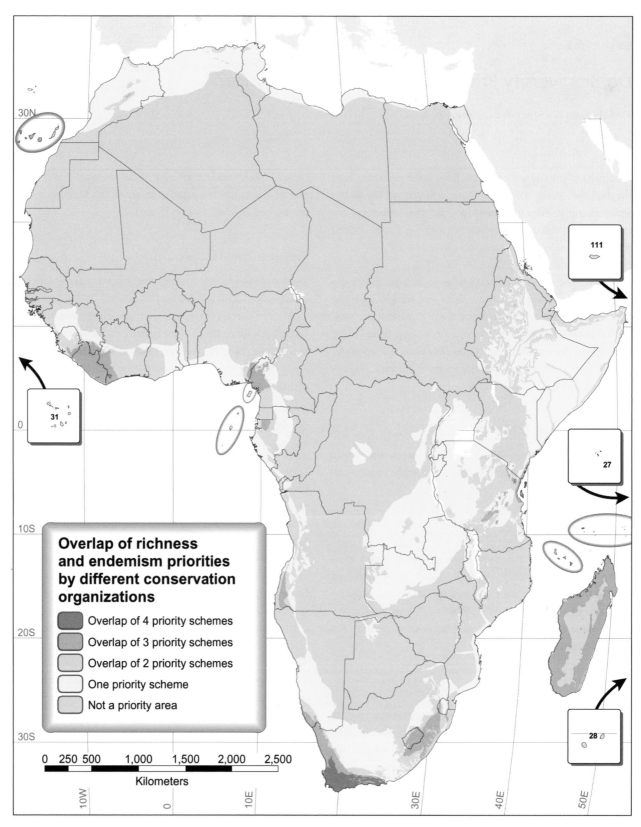

Overlap of richness and endemism priorities by different conservation organizations

- Overlap of 4 priority schemes
- Overlap of 3 priority schemes
- Overlap of 2 priority schemes
- One priority scheme
- Not a priority area

0 250 500 1,000 1,500 2,000 2,500
Kilometers

FIGURE 6.8. Overlap of different prioritization schemes based on species: endemic bird areas (Stattersfield et al. 1998), hotspots (Mittermeier et al. 1999), centers of plant diversity (WWF and IUCN 1994), and WWF ecoregions that are globally outstanding for species (this book).

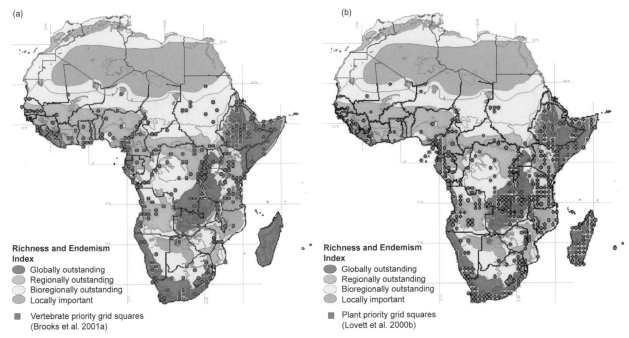

(a)

Richness and Endemism Index
- ● Globally outstanding
- ● Regionally outstanding
- ○ Bioregionally outstanding
- ● Locally important
- ■ Vertebrate priority grid squares (Brooks et al. 2001a)

(b)

Richness and Endemism Index
- ● Globally outstanding
- ● Regionally outstanding
- ○ Bioregionally outstanding
- ● Locally important
- ■ Plant priority grid squares (Lovett et al. 2000b)

FIGURE 6.9. Overlap of priorities identified using computer databases and the priority ecoregions. (a) Overlay of priority grid squares identified for vertebrates (Brooks et al. 2001a) and the WWF the ecoregions of global importance for species conservation. (b) Overlay of priority grid squares identified for plants (Lovett et al. 2000b) and the ecoregions of global importance for species conservation.

Agreement or No Agreement?

Despite differences in targets (Redford et al. 2003), there are broad agreements over large-scale conservation priorities in Africa. To illustrate the extent of spatial agreement over species priorities, we have overlapped EBAs (Stattersfield et al. 1998), hotspots (Mittermeier et al. 1999), and centers of plant diversity (WWF and IUCN 1994) with our globally important ecoregions for endemism and species richness (figure 6.8). The Cape Floristic Region in South Africa rises as a clear priority for all four prioritization schemes. This is followed by the western Upper Guinea lowland forest, Cameroon Mountains, part of the Kaokoveld, Tongoland-Maputaland, Eastern Arc Mountains, and the coastal forests of Kenya and Tanzania. These areas are a clear set of top priorities for conservation investment in terms of the narrowly endemic species of Africa. If we expand the analysis to include Madagascar and the nearby islands (Comoros, Mascarenes, and Seychelles), then there is agreement that the majority of Madagascar and at least the Mascarenes and the Seychelles are of exceptional biological importance at the global scale.

The degree of agreement between these schemes will be further heightened when Conservation International publishes the result of a re-evaluation of African hotspots. Since the initial hotspot analysis (Mittermeier et al. 1998), Conservation International has identified several potential additional hotspots

in Africa, including the Albertine Rift, the Horn of Africa, the highlands of Ethiopia, and Maputaland-Pondoland in eastern South Africa (Brooks et al. 2002). These new hotspots are all in regions of high biological priority according to our results and those of others. By identifying the Albertine Rift in their revised analysis, Conservation International will make this region an additional top priority confirmed by all four analyses.

We have also overlapped the areas of importance identified for 1-degree grids across Africa (Lovett et al. 2000b; Brooks et al. 2001a) with our most important ecoregions for combined endemism and species richness (figure 6.9). This analysis shows that many of the priority grid cells for plants and vertebrates (birds, mammals, amphibians, and snakes) fall in the priority ecoregions we have identified. To some extent the 1-degree analysis indicates the most important pieces of these ecoregions and thereby facilitates the process of refining priorities from larger ecoregions to smaller landscapes and sites.

Comparison of Schemes That Focus More on Habitat Criteria

A number of schemes have been published that use intact habitat and low human densities as their major measure of conservation importance, although some also incorporate species values.

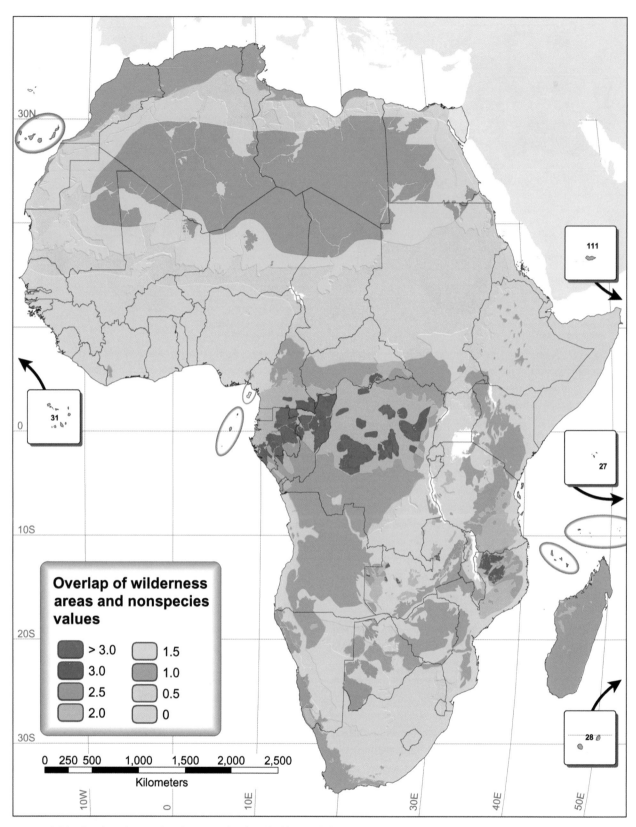

FIGURE 6.10. Overlap of ecoregions important for their "wild intact habitat" and nonspecies biological values with frontier forests (scored as 1) (Bryant et al. 1997), high-biodiversity wilderness areas (scored as 1) and low-biodiversity wilderness areas (scored as 0.5) (Mittermeier et al. 2002), and the "last of the wild" areas (scored as 1) (Sanderson et al. 2002a). Higher scores reflect greater condor-dance among authors.

Tropical Wilderness Areas

Conservation International has identified major tropical wilderness areas as regions where more than 75 percent of the original habitat remains intact (Mittermeier et al. 1998). The aim of this approach was to complement the hotspot approach where more than 75 percent of the habitat was destroyed and more than 1,500 endemic plants occurred. Since the first analysis, Conservation International has identified additional wilderness areas with intact habitat (Mittermeier et al. 2002). The revised approach identifies the Congo forests of Central Africa and the miombo-mopane woodlands and grasslands of southern Africa as high-biodiversity wilderness areas with at least 70 percent of their habitat intact and at least 1,500 endemic plants. Several other wilderness areas were identified that have lower rates of importance for endemic plant species: the Sudd Swamps of Sudan, the Okavango Delta of Botswana, the Serengeti ecosystem in Tanzania and Kenya, and the Sahara and Namib deserts (Mittermeier et al. 2002).

"The Last of the Wild"

The Wildlife Conservation Society has recently identified wilderness areas as those that have been least influenced by the human footprint (measured as population density, land transformation, road networks, and power infrastructure; Sanderson et al. 2002a). In Africa parts of the Sahara Desert, Horn of Africa, Sudd Swamp, Congo Basin forests, miombo woodlands, Kalahari Desert, and Namib Desert were classified as wildest.

Frontier Forests

The World Resources Institute's frontier forests are areas of tropical forest that have not been logged or subjected to other kinds of disturbance and therefore represent core wilderness areas within the tropical forest habitat (Bryant et al. 1997). In Africa, the frontier forests fall mainly in the Congo Basin. They are small areas that form parts of different ecoregions, but as core wilderness areas they generally contain the most intact areas biologically.

Agreement or No Agreement?

We overlapped the high-biodiversity wilderness areas and other wilderness areas of lower importance for biodiversity con-

servation (Mittermeier et al. 2002), "the last of the wild" (Sanderson et al. 2002a), and frontier forests (Bryant et al. 1997) onto our ecoregions of highest importance for their nonspecies values (figure 6.10). This approach shows broad agreement over the importance of the Congo Basin forests as a conservation target but also shows that the savanna-woodland habitats of eastern and southern Africa are important conservation targets for nonspecies biological features. Several of the desert regions are priorities for the conservation of their wilderness features.

The method we have used in this book provides the only data on a broader range of nonbiological features important to conservation. Further work on this issue would be justified to refine the approach we have taken and to better assess the level of agreement over conservation priorities for nonspecies biological attributes.

Conclusion

We believe that these analyses show that the degree of agreement between biologists over what is important in Africa is already strong and is getting stronger as analyses are refined and repeated using updated data.

The results of recent species-based assessments all point to the same areas as being of the highest priority. Although there is some difference in priority regions identified for plants and animals (resulting from real biological variation), there is a strong indication that biologists agree over what is really important in Africa.

The results of recent analyses aiming to identify areas of biological importance for nonspecies biological criteria (mainly intact wilderness areas) are also pointing at the same areas across Africa. As these analyses are refined, the degree of agreement seems to be increasing. Although further work to develop quantified criteria for nonspecies biological values is needed, the current combined map shows that biologists already agree over what is important.

Moving to finer-scale priority-setting approaches and looking for agreement is more problematic than at the largest scale. However, some analyses being developed indicate that some smaller-scale approaches, such as the important bird area approach (Fishpool and Evans 2001), will capture much of the diversity in other groups (Brooks et al. 2001b). This will be an area of further exploration and debate in coming years.

African Protected Areas: Supporting Myths or Living Truths?

Colby Loucks, Neil Burgess, and Eric Dinerstein

Protected areas are a key element of conservation efforts around the world. Some believe that they will become the last line of defense for the landscapes and wildlife that many people associate with Africa (Kramer et al. 1997), although others suggest that they are the legacy of colonial rule and have little relevance in contemporary land-use planning on a poor continent (Neumann 1996, 1998). Given such conflicting positions, it is not surprising that many myths have developed regarding protected areas in Africa, from the commitment to them by African nations, their location in romantic settings, and their importance in capturing areas of high biodiversity value. In this essay we address some of these myths and either refute or substantiate them by examining the relevant facts. Essay 6.3 explores another myth: that a comprehensive network of protected areas would be too expensive to maintain.

Myth: Africa's Parks and Game Reserves Are a Colonial Era Phenomenon, and as Countries Gained Independence They Gave Up on Establishing Protected Areas. This perception is largely untrue. Using the IUCN and UNEP (2003) protected area database for Africa we analyzed the creation of protected areas from 1895 to 2002. We limited our analysis to protected areas assigned as IUCN categories I–IV because these classes of reserves prohibit resource extraction. This cutoff is consistent with most assessments of protected area networks worldwide.

Colonial rule in Africa persisted through the mid-1960s. In 1951, Egypt became the first African nation since Liberia to gain independence, a gap of more than 100 years. Over the next 15 years, thirty-three countries across the continent also gained independence. Much has been written on the alienation of Africans from their land during the colonial period, when reserves often were said to have been established more to provide hunting and recreational opportunities for the colonial leaders than for the good of the country (Neumann 1996, 1998). However, both in terms of number and area, protected area establishment has been greater in the postcolonial period (1967–2002) than during the colonial period (figure 6.11). Through 1967 there were 226 reserves established, covering approximately 606,000 km². Since 1967 an additional 379 re-

serves covering 774,149 km² have been gazetted. However, the rate of reserve establishment began to wane in 1974. At that time a major increase in the global price of oil became problematic for many African economies. The slowdown in park creation under such conditions is understandable: locating reserve sites, mapping boundaries, gazettement, and establishing management regimes are expensive. If funds are not available, then parks remain only a dream, despite the commitment of dedicated African conservationists.

There was another surge in reserve gazettement in the mid-1980s, followed by a decline in declaration of new reserves that continued through 2002.

During this time, a large number of armed conflicts affected large parts of Africa (see essay 6.4). These conflicts have made the declaration and management of protected areas impossible in several countries and ecoregions. Even when government capacity has survived the effects of war, protected area agencies have generally been reduced to scarcely operational shells. Declaring new reserves under such a situation is not a high national priority.

To some extent the continued declaration of protected areas reflects an understanding that natural resources have formed the backbone of African economies for centuries. A current example is wildlife tourism and tourist hunting, which is very profitable to some countries. This is undoubtedly part of the answer, but the degree of commitment to protected areas goes beyond this simple economic explanation. Much stronger environmental nongovernment organizations in many countries, the increased use of environmental impact assessments for major projects, and the existence of national environment action plans and biodiversity conservation strategies through the Convention on Biological Diversity provide other reasons. In addition to protected area development there has also been numerous community-based natural resource management programs across Africa outside protected areas, including experiments with low-impact logging in logging concessions. All of these developments show a strong and increasing interest by African nations in conservation, particularly the link between conserving the environment and promoting development.

Current initiatives to expand the official protected area cov-

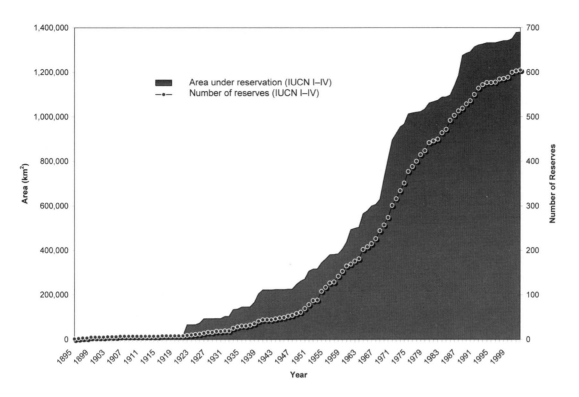

FIGURE 6.11. Reserve establishment in Africa, in numbers of reserves and area (square kilometers), from 1895 to 2002. Data from IUCN and UNEP version 6 (2003).

erage are under way in the Congo Basin (for example, thirteen new national parks were declared in Gabon in 2002). Similar efforts are being undertaken in South Africa, especially in the Fynbos and Karoo habitats. And the chance for transnational parks in the wake of armed conflict is an encouraging legacy. New community-based wildlife and forest management areas are also being established across many African countries.

However, it is not all good news. The Kenyan government recently announced its intention to degazette sections of several forest reserves for allocation to local farmers; although much of this was prevented by public outcry, it could have resulted in significant loss of forest, biodiversity, and water catchment values. Similar threats to established protected areas also occur in other countries.

Myth: Africa's Protected Areas Are a Reflection of Postcard Savanna Images, and the Grasslands of Eastern Africa Are Overrepresented in the Protected Area System of the Continent. The data to support or refute this assumption are mixed. We analyzed protected area establishment in the six dominant biomes, which contain 99.7 percent of IUCN I–IV protected areas in Africa. In decreasing area of protection, these biomes are tropical and subtropical grasslands, savannas, shrub-

lands, and woodlands (hereafter savanna-woodlands); deserts and xeric shrublands (deserts); tropical and subtropical moist broadleaf forests (rainforests); flooded grasslands and savannas (flooded grasslands); Mediterranean forests, woodlands, and scrub (Mediterranean scrub); and montane grasslands and shrublands (montane grasslands) (table 6.3).

Like biodiversity, Africa's protected areas are not randomly distributed; they are clumped together. Reserve establishment in Africa has been focused predominantly on the savanna-woodlands biome, in terms of total number of reserves established, total area, and percentage of the biome covered (figure 6.12, table 6.3). Despite the prolonged attention given toward gazetting savanna-woodland protected areas, the current reserve system still protects only 6.4 percent of the biome (table 6.3). The five other biomes are even less protected in terms of percentage and area, with less than 2 percent of the extremely diverse Mediterranean scrub biome being protected as of 2002.

Although savanna-woodland protected areas continue to dominate Africa's landmass, few new protected areas have been gazetted in this biome since 1974 (figure 6.12). However, the rate of protection in other biomes has been increasing. In the late 1960s and early 1970s and again in the mid-1980s rainforests experienced large increases in protection. In the mid- to

TABLE 6.3. **Protected Area Status (IUCN Categories I–IV) by Biome.**

Biome	Total Biome Area (km^2)	Number of Reserves	Reserve Area (km^2)	Percentage Biome in Reserves
Tropical and subtropical grasslands, savannas, shrublands, and woodlands	13,981,400	340	897,942	6.42
Deserts and xeric shrublands	9,753,200	59	237,490	2.43
Tropical and subtropical moist broadleaf forests	3,485,500	136	218,378	6.27
Flooded grasslands and savannas	562,600	6	22,717	4.04
Mediterranean forests, woodlands, and scrub	851,300	103	13,860	1.63
Montane grasslands and shrublands	868,700	98	11,791	1.36
Tropical and subtropical dry broadleaf forests	194,900	10	3,461	1.78
Temperate coniferous forests	23,100	3	63	0.27
Mangroves	76,700	2	3	0.00
Total	29,797,400	757	1,405,705	

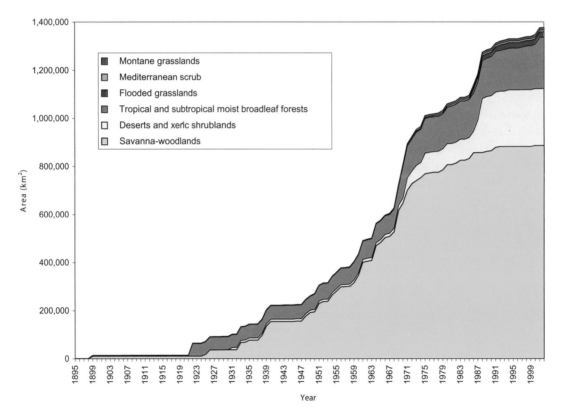

FIGURE 6.12. Gazettement of protected areas (IUCN Class I–IV) by year for the six largest biomes, containing 99.7 percent of the protected areas in Africa. Data from IUCN and UNEP version 6 (2003). Data are from protected areas designated with IUCN status, date of establishment, and area.

late 1980s a majority of the protected areas in the desert biome were gazetted. In the next three most protected biomes (flooded grasslands, Mediterranean scrub, and montane grasslands) the majority of protected areas were established after 1974.

Most of the protected areas in the Mediterranean scrub are found in southern Africa. North African countries declared few protected areas in their Mediterranean habitats through 2002. This is particularly true in the Nile Delta region of Egypt, which has been densely populated by people for thousands of years, as have northern Morocco, Algeria, and Tunisia. Although large expanses of Mediterranean habitat have also been converted to agriculture in South Africa, this country is making ambitious attempts to expand protected area coverage across large areas (e.g., the conservation areas proposed in the Cape Action Plan for the Environment strategy [CSIR 2000a, 2000b] and Cape Peninsula National Park). For deserts, the opposite is true. Most of the large increases in desert protection come from three large protected areas recently declared in the Sahara Desert of North Africa, which account for approximately 46 percent of the total desert area protected.

Myth: The Current IUCN I–IV Protected Areas Capture the Most Diverse and Biologically Important Areas. In the colonial and early postcolonial era many reserves were established as hunting or game reserves conserving the full spectrum of savanna mammal species, including the top predators (Rodgers 1993). Subsequently, many additional reserves have been established for the conservation of other charismatic mammal species, such as Gombe Stream National Park for chimpanzees (*Pan troglodytes*) and Bwindi Impenetrable National Park for mountain gorillas (*Gorilla gorilla*). Only recently have principles of conserving the full range of species diversity in Africa been applied to reserve networks (Cowling et al. 2003; Rebelo 1994), and outside South Africa most of these studies are theoretical and have not yet been implemented (Brooks et al. 2001a).

Forest Reserves: The Wildcard of Protected Area Biodiversity Conservation in Africa

Many African countries have large numbers of legally gazetted forest reserves managed by forestry (and not wildlife) departments. These reserves are not classified as protected conservation areas in the current IUCN system of classification and therefore fall outside the official protected area network. Throughout Africa there are more than 4,300 forest reserves that comprise approximately 616,700 km^2, or an additional 50 percent of the land area currently found in IUCN I–IV category reserves.

In some ecoregions, forest reserves constitute most habitat protection. For example, the Eastern Arc Forests [19] ecoregion of Tanzania and Kenya is globally outstanding for its biodiversity, specifically its endemism (Myers et al. 1999). Little protection is afforded to this ecoregion's biodiversity from IUCN I–IV protected areas. However, more than 3,300 km^2, or 14 percent of the ecoregion, is contained in forest reserves, and these reserves contain up to 90 percent of the remaining forest, which is where the bulk of the biodiversity is found (Burgess et al. 1998d, 2002a). Although these reserves may protect the biodiversity now, their future remains uncertain and depends largely on the management decisions of the government Forestry Division. If forest reserves remain unrecognized as protected areas and governments fail to take into account biodiversity, then species may go the way of Miss Waldron's red colobus monkey (*Procolobus badius waldroni*). This species' primary habitat was contained in the forest reserves of Ghana and Ivory Coast, but it was intensively hunted in these reserves, and its forest habitat was cleared and was declared extinct in 2000 (Oates et al. 2000).

Trends in the creation of forest reserves can also be compiled from available statistics, such as those from some administrative regions of Tanzania (figure 6.13). These data show how forest reserves were created during the colonial period of government and the early part of the postcolonial government. As with IUCN I–IV protected areas, the rate of establishment of forest reserves slowed markedly in the early 1970s and remained at a steady but low level until 1989, which is the last year for which we have data. Some reserves have been declared since that time, but numbers are small and other reserves have been abandoned because of low capacity in the Forest Division.

Conclusion

Our analyses have shown that general statements relating to the creation of Africa's protected areas may be true, but preconceived ideas begin to break down if a problem is analyzed more carefully. This is true in the case of protected area establishment after colonial rule, protection of habitat and species across all of Africa's biomes, and the biodiversity protection afforded by forest reserves throughout Africa. However, we have also identified room for improvement in terms of increased protection of all biomes, notably all those that are not savannas, and the possibility of working with governments to conserve habitat found in forest reserves or areas adjacent to protected areas. Importantly, we have not looked at management effectiveness of the existing reserves; many may be gazetted and then not managed adequately. A large project would be needed

(a)

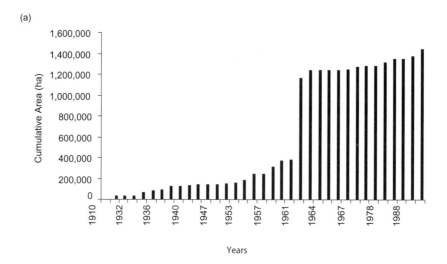

(b)

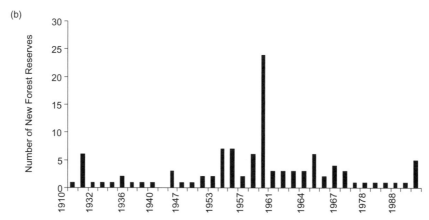

FIGURE 6.13. Annual increase in (a) total area and (b) number of forest reserves in Arusha, Coast, Tabora, and Iringa Regions of Tanzania, 1910–1989. Data from the Forest and Beekeeping Division, Dar es Salaam.

to determine how well protected areas are managed across the region and which ecoregions are therefore best protected by good management.

Sustainable use programs, community-based natural resource management programs, experiments with low-impact logging, awareness-raising and education, policy reform, and improved legislation reveal that both governments and African peoples are interested in conservation as they make clear the link between natural resource conservation (often in protected areas) and their own development needs.

Factoring Costs into Conservation Planning

Joslin Moore, Andrew Balmford, Tom Allnutt, and Neil Burgess

Financial resources for conservation are highly limited, driving the need to prioritize conservation efforts. Often, prioritization is driven purely by the biological value of the regions, with high priority being assigned to areas with high species richness or high endemism. However, the costs of achieving conservation also vary greatly between regions (Balmford et al. 2000, 2003). Thus, cost efficiency could be an important determinant of the effectiveness of conservation (Ando et al. 1998; James et al. 1999). Past studies suggest that cost is unlikely to correlate with biological value in any simple way, so we can expect that factoring costs into conservation prioritization will alter conservation priorities (Balmford et al. 2000; Williams et al. 2003). Here we test this assumption and examine the potential for incorporating costs into the prioritization of African ecoregions.

Estimating the Management Cost of Effective Conservation

Costs of conservation were derived from the findings of Balmford et al. (2003). They explored variation in the annual management costs of 139 area-based conservation programs (spanning thirty-seven nations and all continents except Antarctica) that were already successful in meeting conservation targets or for which managers were able to estimate the extra resources needed to make them fully effective. They found that these costs could be predicted from local socioeconomic information such as gross national income (GNI) and purchasing power parity (PPP, a relative measure of the local value of a currency and increases with the buying power of a U.S. dollar) and the opportunities for economies of scale (project area). Here, we improved their model slightly, using these three variables to explain 83 percent of the variation in the management costs per unit area of the 139 sites. The estimated recurrent management cost of individual areas (usually but not always reserves) was given by

$$\log (\text{annual cost, US\$/km}^2/\text{yr}) = 1.765 - 0.299 \log (\text{area, km}^2) + 1.014 \log (\text{PPP}) + 0.53 \log (\text{GNI/km}^2) - 0.771 \times \log (\text{area}) \times \log (\text{PPP}).$$

Thus, costs per unit area decrease with total reserve size and increase with GNI. The relationship between conservation cost and PPP is positive for very small protected areas but rapidly becomes negative for larger reserves.

We used this relationship to estimate the recurrent annual management costs for the conservation of 10 percent of the total area of each of 118 ecoregions in Africa (the Socotra Islands have no economic data). We scaled up from costs of management of a single reserve to 10 percent of an ecoregion as follows. We assumed that the existing distribution of reserve sizes in a given ecoregion was a guide to likely reserve size distributions in the ecoregion in the future. Then we calculated the cost of existing reserves in each ecoregion and the area currently conserved and used this to scale the estimate of cost to that needed if 10 percent of each ecoregion were protected, assuming that the distribution of reserve sizes remains similar over time. However, because for some ecoregions protected area data are poor, we used as our templates of eventual reserve size distribution the amalgamated distribution for all ecoregions in the same sub-biome (described in chapter 2). The Canary Islands Dry Woodlands and Forests [88] ecoregion had no associated protected area information and was the only representative of its sub-biome, so here we used the reserve size distribution of the Island Mediterranean Forest sub-biome instead.

How Much Would Effective Conservation Cost?

Our estimates suggest that around $630 million/yr is needed to effectively manage 10 percent of all 118 African ecoregions considered in this analysis (table 6.4). This represents 0.1 percent of total GNI for all of Africa. Between ecoregions, costs vary widely. Conservation in three ecoregions—the Mascarene forests, the Canary Islands dry woodlands and forests, and the Granitic Seychelles forests—were each estimated to cost more than $10,000/km^2/yr, with another nineteen each costing more than $1,000/km^2/yr. These twenty-two ecoregions account for 54 percent of the total cost for effective conservation in Africa. Conservation is most expensive in ecoregions that are small and are dominated by island, Mediterranean, grassland, and mangrove systems (table 6.4).

In contrast, effective conservation of some ecoregions is estimated to be much cheaper, most notably the vast tracts of

TABLE 6.4. **Management Costs for the Twenty-Six Sub-Biome Groups of Ecoregions.**

Sub-Biome	Ecoregions*	Area (km²)	Cost/km²/yr (US$)	Total Cost per Year (US$)
Island Mediterranean forest	1	4,974	12,172	6,054,757
Temperate coniferous broadleaf forest	3	673,582	2,840	191,274,191
Mediterranean scrub	4	195,730	2,792	54,652,536
East African mangroves	3	22,272	1,811	4,034,303
West African mangroves	2	54,448	1,252	6,814,820
Montane forest-grassland mosaic	10	818,048	1,022	83,572,672
Island moist forest	6	320,049	589	18,861,529
Alpine moorland	6	50,634	574	2,908,927
Island xeric woodland and shrubland	2	123,091	408	5,028,275
Afromontane forest	8	515,090	364	18,734,953
Southern African xeric woodland and shrubland	4	679,496	331	22,481,593
Eastern African lowland forest-grassland mosaics	4	307,607	310	9,526,376
Saline wetland	4	94,152	195	1,831,310
Island dry forest	2	156,642	194	3,033,769
African dry forest	1	38,228	182	696,765
Freshwater wetland	6	468,612	162	7,592,836
Guineo-Congolian swamp forest	3	235,786	142	3,347,345
Seasonal grassland	2	52,045	122	635,443
Guineo-Congolian lowland moist forest	9	2,107,023	116	24,385,720
Mopane woodland	2	606,778	105	6,396,890
Desert	6	4,870,446	98	47,645,160
Guineo-Congolian forest-savanna mosaic	6	2,538,201	92	23,415,050
Northeastern African xeric woodland and shrubland	5	930,407	91	8,490,966
Acacia savanna woodland	9	7,783,526	74	57,682,580
Miombo woodland	5	3,000,883	47	13,971,180
North African xeric woodland and shrubland	5	3,146,019	25	7,818,837
Africa (all sub-biomes combined)	118	29,793,773	212	630,888,783

*The number of ecoregions in the study that are classified as belonging to that sub-biome (see chapter 2).

arid lands in the Sahara and on the west coast of southern Africa and the lowland tropical forests and swamps of the Congo Basin. We estimate that reserves in these ecoregions typically cost less than $200/km²/yr (table 6.4).

If we consider total cost rather than cost per unit area, then just eleven ecoregions would cost more than $10 million/yr for effective conservation. Together, these eleven ecoregions account for more than 60 percent of the total estimate for all of Africa. These include some smaller Mediterranean and montane grass-land ecoregions but also some vast ecoregions that are also expensive; for example, the Sahelian *Acacia* savanna and the Nama Karoo are both fairly cheap per unit area but are so big that to conserve 10 percent of them would constitute a major investment.

Do Costs Correspond to Biological Value?

If the cost of conservation were negatively correlated with biological value, then incorporating cost into priority setting would be unnecessary: areas selected because of their high biological value would also be selected by virtue of their low cost. However, this is not so. We examined the patterns of correlation between our cost estimates and the species distribution database using Spearman correlation coefficients. The estimated total annual cost of conserving 10 percent of an ecoregion is positively correlated with both the ecoregion's species richness (r_s = .42, p < .001, N = 118) and the number of strict endemics (r_s = .33, p < .001, N = 118). Much of this pattern is driven by variation in area between ecoregions because species richness and total cost both rise with increasing ecoregion size (species richness, r_s = .6, p < .001, N = 118; cost, r_s = .61, p < .001, N = 118).

If we control for area, then the pattern revealed is rather different. Cost per square kilometer per year is negatively correlated with ecoregion area (r_s = −.53, p < .001, N = 118) because larger ecoregions tend to have larger reserves with lower management costs per unit area. In addition, taking area into account in the species richness and endemism scores (see chapter 3) means that cost and species richness are no longer correlated (r_s = .00, NS, N = 118), but cost does still correlate with the number of strict endemics (r_s = .28, p < .01, N = 118). The correlation with endemism is important because it indicates that we will be unable to meet our conservation goals if we focus only on cheap ecoregions; some expensive ecoregions must be incorporated if we want to include the ranges of endemic species in the reserve system. These findings are similar to those made in other studies, which also found that cost (or a proxy of cost) often co-varies with endemism (Balmford and Long 1994; Fjeldså and Rahbek 1998; Balmford et al. 2001a, 2001b).

Implications for Priority-Setting

Incorporating cost has the potential to increase the cost-effectiveness of conservation and to alter the priority attached to conserving different kinds of ecosystems. We tested this potential by comparing cumulative species representation and the associated total cost of reserves for ecoregion prioritization schemes that excluded and included cost. We used four simple ranking systems: richness (species richness of all vertebrates in the WWF-US database), endemism (number of strict endemics in each ecoregion), richness:cost ratio, and endemism:cost ratio. We also incorporated a complementarity approach by finding optimal combinations of ecoregions that maximized the number of species represented for a given cost. Complementarity seeks the set of areas that together represent as many species as possible, avoiding overlap. This contrasts with the ranking schemes, which do not explicitly consider overlap between ecoregions. The optimization problem was solved using the linear optimizing program Cplex (ILOG 1997–2000). Finally, we compared our results with those found when ecoregions were selected entirely at random.

The different ranking systems display a great deal of variation in their cost-effectiveness (figure 6.14). For example, if ecoregions were ranked by species richness, an annual budget of US$50 million would support the conservation of ten ecoregions that between them contain 2,916 vertebrate species (50 percent of the vertebrate species in the database). This is not much more effective than choosing ecoregions at random (figure 6.14). Similarly, the simple endemism scheme identifies that an annual budget of $50 million/yr should be focused on fifteen ecoregions that combined represent 3,597 vertebrates (62 percent of the total).

Schemes that incorporate cost perform better. If ecoregions are ranked by their richness:cost ratio, then $50 million/yr would support the conservation of twenty-five ecoregions that together contain 4,196 species, or 72 percent of the species in the database. The endemism:cost ratio ranking scheme indicates that thirty-five ecoregions containing 4,431 vertebrates (76 percent) could be conserved with a $50 million/yr budget. Incorporating complementarity into the approach can make further gains in cost-effectiveness. With a $50 million/yr budget, the optimization scheme based on complementarity identifies fifty-one ecoregions that together represent 4,851 (83 percent) of vertebrates in the database.

Including cost also alters the kinds of ecosystems likely to be ranked highly. If areas are ranked simply by species richness, then ecoregions with large areas (and hence high numbers of species) tend to be ranked highly. In contrast, if cost is incorporated into the ranking system, then attention is drawn to many smaller ecosystems—notably, many arid and semi-arid systems—that are not particularly species rich but are cheap to manage for conservation.

Conclusions and Caveats

This study illustrates how including costs in priority-setting can radically improve the cost-effectiveness of a conservation strategy and consequently do more to achieve conservation goals for a fixed budget. However, there are a number of caveats.

Our estimates focus only on on-site management costs of reserves. They do not incorporate costs of reserve establishment, which may be substantial, especially if the area is associated with conflicting high-income land uses such as in the lowland fynbos of South Africa (Frazee et al. 2003). They also do not deal with the off-site costs of dealing with wide-ranging threats such as climate change and altered fire or hydrological regimes. In addition, management cost estimates are based on

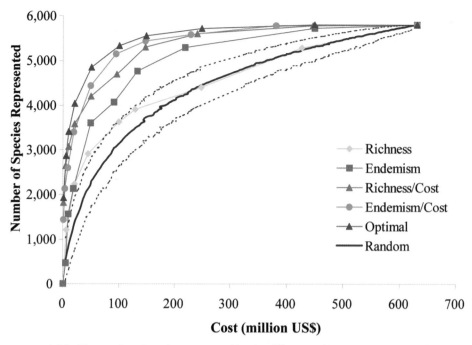

FIGURE 6.14. The number of species represented by the different ranking systems compared with the cost of conservation of the areas selected. The thick brown curve gives the mean number of species represented and corresponding cost when 1–118 ecoregions were chosen randomly 1,000 times. The dashed brown lines indicate the upper and lower 95 percent confidence intervals. With the exception of the random curves, the lines are interpolations between the symbols and should be interpreted accordingly.

current land uses and may not hold as reserve coverage increases (potentially increasing both the establishment and opportunity costs of reservation). We have also assumed for present purposes that conservation must be achieved through further reservation, yet of course effective conservation might be best served by a variety of approaches to area-based conservation (Frazee et al. 2003; Pence 2003). Nevertheless, despite these limitations, our estimate provides a reasonable measure of the relative costs of conservation in each ecoregion and therefore is a useful tool for informing conservation planning.

A second caveat is that this study examines prioritization of ecoregions, assuming that one expends all effort on a single ecoregion before moving onto the next. Clearly, this is not the case in reality where efforts are spread through a number of ecoregions. Using prioritization approaches to apportion effort between ecoregions would be a useful task and probably would increase the effectiveness of conservation planning.

However, it would necessitate more fine-grained data than we had available, so we could not consider it in this study.

Finally, this analysis focuses only on species representation and cost. Other factors are also important in a conservation strategy, including persistence, conserving ecological processes, and addressing threats. If we were undertaking a more comprehensive study, we would also address these issues. However, none of these caveats negates the importance of economics in determining efficient conservation strategies. By incorporating costs, we can achieve more for our conservation dollar and consequently make greater gains for conservation overall.

Acknowledgments

We would like to thank the Computer Science Department of the University of Copenhagen for kindly giving us access to the Cplex software. Results of this analysis have been published by Moore et al. (2004).

Armed Conflict and the African Environment

James Shambaugh, Nina Jackson, Rachel Gylee, Judy Oglethorpe, Rebecca Kormos,
Andrew Balmford, Jennifer D'Amico Hales, Sam Kanyamibwa, Neil Burgess,
Tommy Garnett, and Malik Doka Morjan

The tropical regions of sub-Saharan Africa encompass some of the planet's most outstanding biodiversity. Unfortunately, these regions also include many countries affected by armed conflicts. Since 1990, there have been fifty-six major armed conflicts throughout the world, of which nineteen were in Africa, more than in any other region (Gurr et al. 2000; Sollenberg and Wallensteen 2001). These conflicts, affecting 54 percent of African ecoregions, present serious challenges for biodiversity conservation. For example, the Democratic Republic of the Congo (DRC), which encompasses more than half of Africa's remaining forests in seven ecoregions, has been plagued by political instability and conflict since the early 1990s (Wolfire et al. 1998).

Armed conflicts cause untold suffering and enormous loss of human life, fragmenting societies and shattering economies—impacts that last long after hostilities end. They also wreak devastation on biodiversity and the natural resources on which people depend. For example, Uganda lost all of its rhinos during the internal conflicts of the Obote and Amin regimes of the 1970s, and populations of other large mammal species were greatly reduced by both the army and local populations hunting wild animals for meat. Conflict has also affected wildlife populations in protected areas in other conflict zones. In the Kahuzi-Biega National Park in DRC, war and the increased rates of poaching for ivory reduced the elephant population by 95 percent (Plumptre et al. 2001) and eastern lowland gorillas (*Gorilla gorilla graueri*) by about half (Appleton 1997). Assessments in the Marromeu Reserve and Gorongosa National Park in Mozambique revealed that populations of buffalo, elephant, hippo, waterbuck, and other species declined by at least 90 percent between the 1970s and 1994, during a period of civil warfare (Hatton et al. 2001). Armed conflict also has more general effects on natural resources. For example, more than two-thirds of Rwanda's Akagera National Park was severely degraded when returning refugees occupied the park (Plumptre et al. 2001); warfare in the Virungas region created a humanitarian crisis, which in turn led to serious pollution damaging aquatic biodiversity (Kalpers 2001); and Mozambique and Liberia both saw large-scale depletion of their natural resources through logging, which was most intense in the immediate aftermath of armed conflict (Hatton et al. 2001; Global Witness 2001).

Armed conflict and its associated activities also lead to a range of social, economic, and political impacts interrelated with the environment. Socially, armed conflict disrupts local communities and livelihoods by degrading the environment or reducing access to natural resources. Economically, conflict can disrupt established trade channels, decrease government revenues, impoverish local communities, and drive unsustainable and illicit extraction of natural resources (Draulans and Van Krunkelsven 2002). Politically, the absence of effective government control during periods of conflict creates a vacuum of authority that is often exploited by warring parties and other powerful interests at the expense of local communities. Many of these environmental impacts can in turn contribute to future armed conflict, thus creating a vicious circle (Homer-Dixon et al. 1993).

These impacts affect the environment both indirectly and directly. Direct impacts include local extinction of species through poaching for trade in arms and food and the overexploitation of natural resources by refugees. Indirectly, armed conflicts hinder the good governance that creates the conditions essential for effective long-term conservation. In addition, a conflict's impacts can extend to neighboring, nonwarring states (e.g., through migration of refugees, movements of government and rebel forces, illicit trade in natural resources and arms, and loss of tourism revenues). Finally, the greatest threats to the environment often occur during the transition from war to peace, after cessation of hostilities but before effective management of natural resources has resumed. In this immediate postwar period, natural resources often are exploited illegally by entrepreneurial operators while government is still weak. Resources may also be extracted unsustainably to pay off war debts and finance postconflict reconstruction.

Impacts on Biodiversity

Although the impacts of armed conflict on certain wildlife populations have been well documented, the impacts on overall biodiversity are less clear. The latter often are difficult to quantify be-

TABLE 6.5. **Five Globally Outstanding Ecoregions with the Greatest Proportion of Their Area in Countries in Conflict.**

Ecoregion	Country in Conflict	Percentage of Ecoregion That Falls in a Country in Conflict
Central Congolian Lowland Forests [15]	DRC	100
Eastern Congolian Swamp Forests [14]	DRC	100
Sudd Flooded Grasslands [59]	Sudan	100
Northeastern Congolian Lowland Forests [16]	DRC	94
Albertine Rift Montane Forests [17]	DRC	62

DRC, Democratic Republic of the Congo.

cause many areas lack a preconflict baseline for comparison. Several of the globally outstanding ecoregions in Africa are facing high levels of conflict (table 6.5), but the reasons for these conflicts and the factors driving them are not clear. More often than not, impact assessments are based on anecdotal information.

We have looked for correlations between areas of armed conflict and potential causes of those conflicts. The potential drivers of conflict we considered were human population density, cultural diversity, biological value, habitat degradation, and socioeconomic attributes such as national gross domestic product (GDP) and other indices of poverty. We chose these variables based on previous studies. For example, it has already been shown that across sub-Saharan Africa, high human population density correlates positively with vertebrate and plant species richness (Balmford et al. 2001a) and that cultural diversity correlates with vertebrate species richness (Moore et al. 2002). Both high population density and cultural diversity are thought to drive conflicts (Byers 1991; McNeely, 2000), so it may be expected that areas of higher biodiversity may be more prone to conflict. Ecological factors such as biodiversity are often cited as driving and catalyzing modern conflicts in Africa because with renewable resources becoming increasingly scarce, violence can erupt (Suliman 1999). Various socioeconomic factors are also commonly cited as driving conflicts (McNeely 2000; Homer-Dixon et al. 1993; Suliman 1999). For example, it has been stated that the presence of valuable minerals may drive a conflict as different groups struggle to control the resource for their own profit or to fund political struggle (Global Witness 2001). Moreover, if population densities exceed the carrying capacity of a region, this might lead to conflicts over land (Byers 1991) and other scarce resources (McNeely 2000). Finally, various studies have proposed a link between poverty, the breakdown of good governance,[1] and the incidence of armed conflict (Suliman 1999; Shambaugh et al. 2001).

An analysis at the University of Cambridge (England) by Rachel Gylee and Nina Jackson used African ecoregion and national data to provide quantitative tests of the various hypotheses on factors that correlate with conflicts across Africa. For this study, armed conflicts were classified into three categories according to intensity, defined as follows:

- Minor armed conflict: at least 25 battle-related deaths in a year and fewer than 1,000 battle-related deaths during the conflict
- Intermediate armed conflict: at least 25 but fewer than 1,000 battle-related deaths in a year and an accumulated total of at least 1,000 deaths during the conflict
- War: at least 1,000 deaths during the conflict

A weighting system was used to calculate the total conflict score of a country. Each year of a minor armed conflict scored one point, each year of an intermediate conflict scored two points, and each year of a war scored three points. For each country the total number of conflicts and their intensity and duration were calculated. Using this scoring system we calculated a total of 168 conflicts spanning 469 conflict years over the period 1946 to 2001.

Among African biomes, the tropical and subtropical moist broadleaf forests biome has the largest percentage of its area affected by conflicts, followed by mangroves and the tropical and subtropical grasslands, savannas, shrublands, and woodlands (table 6.6). Smaller areas of the deserts, Mediterranean habitats, and tropical dry forests are also affected by conflict.

However, when analyzed in more detail, none of the three major biome types showed a significant difference in mean conflict scores or mean area-independent conflict scores ($F(2, 116) = 0.78$, $p > .05$; $F(2, 116) = 0.13$, $p > .05$, respectively). There was also no significant difference between mean conflict scores and area-independent conflict scores when the nine biomes

TABLE 6.6. **Percentage of Biome in Areas Directly or Indirectly Affected by Armed Conflict in the Last 10 Years.**

Biomes	Area under Conflict (km²)	Percentage of Biome under Conflict
Tropical and subtropical moist broadleaf forests	2,171,610	62
Mangroves	43,373	61
Tropical and subtropical grasslands, savannas, shrublands, and woodlands	7,607,924	54
Temperate coniferous forests	9,418	41
Flooded grasslands and savannas	228,733	41
Montane grasslands and shrublands	350,094	40
Deserts and xeric shrublands	3,482,986	36
Mediterranean forests, woodlands, and scrub	293,844	35
Tropical and subtropical dry broadleaf forests	3,107	2

Calculations based on the area of countries directly or indirectly affected by conflict intersected with biomes.

were considered separately ($F(8, 110) = 1.51, p > .05$; $F(8, 110) = 0.13, p > .05$, respectively).

There was also no correlation between conflict levels and the majority of potential socioeconomic drivers of conflict. For example, ecoregions with minerals were not found to have a greater level of conflict than ecoregions with no minerals ($t(114) = -1.92, p > .05$). Certain countries have conflicts linked to illegal diamond or gold exploitation (e.g., Sierra Leone or DRC), but other mineral-rich but largely peaceful countries (e.g., Botswana, South Africa, and Namibia) counterbalance them. The area-corrected human population of a country did not correlate with the conflict score of that country, indicating that population density was not a driver of conflict ($r = .048, n = 119, p > .05$). Finally, there was no correlation between any of the factors used to measure biological degradation (see chapter 5) and the conflict score. Thus, although ecological degradation often is cited as being correlated with conflicts as a consequence or a driver (Suliman 1999; Homer-Dixon et al. 1993; Shambaugh et al. 2001), across the whole of Africa this is not the case.

However, one significant correlation did emerge between potential socioeconomic drivers and the levels of conflict across the whole of Africa. The wealth of a nation, as measured by per capita GDP, is weakly negatively correlated with its conflict score ($r = -.329, n = 119, p < .05$; figure 6.15). This means that the poorest countries are the most likely to have wars. Economic collapse and intense poverty often are linked to poor governance, which is also essential for sustainable management of natural resources (Shambaugh et al. 2001). Unfortunately,

Africa is the world's poorest continent, and has been getting poorer over the past 30 years (World Bank 2002). It is also the continent most afflicted by poor governance (Transparency International 2002).

These results are similar to those of Population Action International (R. Cincotta, pers. comm., 2003), which found strong correlations between the likelihood of nations going into conflict and demographic factors such as high infant mortality and a large youth population. These factors were themselves strongly linked to poverty. Population Action International also found some correlation between conflict and poor governance. Further analysis could incorporate demographic factors, corruption scales, and governance indexes into the model, factors that in addition to conflict data are a sign of the stability of a country. Moreover, future studies could also try to assess whether poverty is the primary cause of conflict or whether poverty is one of the results of conflict that leads to a downward spiral of greater poverty and greater chances for future conflict.

How Can Conservation Organizations Respond?

Conservation organizations working in areas affected by armed conflict face many challenges to the successful achievement of their mission to conserve biodiversity. At least some of these challenges can be overcome with a structured, strategic, and practical approach encompassing all institutional levels and during all phases of conflict (Shambaugh et al. 2001).

To begin, conservation organizations need to improve their

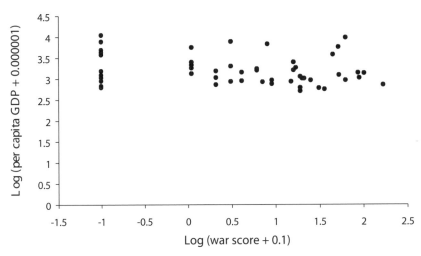

FIGURE 6.15. Correlation between the conflict score with per capita gross domestic product (GDP) across the countries of Africa.

capacity to maintain a presence during conflict; indeed, recent research has shown that protected areas suffered least when conservation groups maintained even a minimal presence during conflict (Plumptre et al. 2001). Building the technical and management capacity of junior and local staff in peacetime, maintaining good communication systems, and implementing appropriate safety and security measures are important steps in this direction. Conservation organizations should also develop working relationships with local nongovernment organizations and organizations from other sectors (e.g., development and relief) to facilitate closer collaboration before and during armed conflict. During conflicts and movements of refugees, conservation organizations can work with staff from humanitarian organizations to incorporate environmental aspects into planning and to minimize impacts of humanitarian operations by providing local environmental information (e.g., location of protected area boundaries, carrying capacity of the local natural resource base).

In addition, conservation organizations can help to strengthen governance, particularly after a conflict, by building capacity of local and national civil society organizations and supporting government rehabilitation processes, including policy implementation, which take the environment and sustainable use of natural resources into account (Kanyamibwa and Chantereau 2000).

Working at an ecoregion scale as opposed to site level offers numerous opportunities for responding effectively to armed conflict. For example, an ecoregion presence can facilitate the relocation of personnel, equipment, and offices within a region (and sometimes across international borders) when certain areas become too dangerous. At the same time, conservation organizations can help prevent brain drain by usefully employing staff in more stable parts of an ecoregion until an end to hostilities allows them to return. Working at the landscape level can also help conservation organizations create networks with several populations of valued species, to avoid reliance on individual sites and to provide fallback options in the event of conflict. In addition, corridors can provide the opportunity for species to take refuge in safe areas while safeguarding source populations for repopulating depleted areas.

Finally, conservation organizations should strive to resume activities as soon as possible after a conflict ends and actively promote integration of conservation with policy, planning, relief, and development activities. Collaboration with these sectors and the private sector is critically important for maintaining influence over postconflict redevelopment and for minimizing adverse environmental impacts as large-scale economic activities resume. This is crucial particularly in Africa, not only for ensuring the survival of threatened species and vulnerable ecological processes but also for conserving the natural resources that play a vital role in the livelihoods of millions of rural Africans.

Note

1. Governance is defined here as the exercise of authority through formal and informal traditions and institutions for the common good. Governance encompasses the process of selecting, monitoring, and replacing governments. It also includes the capacity to formulate and implement sound policies and the respect of citizens and the state for the institutions that govern economic and social interactions between them. All governance data obtained from the World Bank Institute Web site, accessed on July 18, 2002: http://www.worldbank.org/wbi/governance/govdata2001.htm.

Wildlife Health in Africa: Implications for Conservation in the Decades Ahead

Jennifer D'Amico Hales, Steven A. Osofsky, and David H. M. Cumming

The fundamental goals of conservation include maintaining ecological and evolutionary processes, viable populations, and blocks of natural habitat large enough to be resilient to large-scale disturbances (Noss 1992). Conservationists strive to achieve these goals through landscape-level planning. Yet the role that parasites and disease play in ecosystem dynamics by their impact on the size and distribution of species populations, and hence community structure, often is overlooked when such plans are being developed. Diseases, particularly introduced diseases, have had large impacts on wildlife populations throughout Africa. Notable examples include the rinderpest pandemic in ungulates and outbreaks of canine distemper virus and rabies in African wild dogs (*Lycaon pictus*), Ethiopian wolves (*Canis simensis*), and lions (*Panthera leo*) (Plowright 1982; Roelke-Parker et al. 1996; Sillero-Zubiri et al. 1996; Murray et al. 1999; Woodroffe 1999). Similarly, livestock disease control measures (e.g., for tsetse fly and foot-and-mouth disease) have opened large areas of wild land to unsustainable subsistence agriculture and subsequent loss of wildlife and habitats. Although these issues have been a focus of the veterinary community for decades, only in recent years has the broader conservation community recognized their importance (Osofsky et al. 2000; Deem et al. 2001).

Widespread anthropogenic changes to the environment (i.e., land cover change, habitat fragmentation and degradation) have amplified the role of disease as a regulating agent (Deem et al. 2001). Stresses resulting from edge effects, the loss of genetic diversity, overcrowding, and more extensive contact with domestic stock may increase a species' susceptibility to disease outbreaks (Lafferty and Gerber 2002) that may further stress populations and compound other threats. Because many protected areas have become "islands" surrounded by altered landscapes, epidemics can easily facilitate the extinction of a host species, particularly those with restricted ranges or small populations.

Anthropogenic changes to ecosystems and changes in human behavior also increase contact between humans, domestic animals, and wildlife, resulting in the greater likelihood of pathogen transfer and the emergence of new disease vectors

or changes to the ecology of existing diseases. For example, HIV has moved from primates to humans (Hahn et al. 2000). Nonhuman primates are also threatened by diseases from humans, such as measles in mountain gorillas and polio in chimpanzees (Daszak et al. 2000). However, some of the most severe impacts on wildlife populations have come from domestic animals. Introduced species often are the source of epidemics; for example, cattle introduced rinderpest to African ungulates in the late 1800s (Lafferty and Gerber 2002). The African wild dog (*Lycaon pictus*) also suffered critical declines and local extinctions caused by rabies and canine distemper viruses related to the presence of domestic dogs (Alexander and Appel 1994).

We are also learning more about other factors that can increase susceptibility to parasites and disease, such as invasive species, pollution, poor nutrition, resource exploitation, and climate change. Global climate change is already being linked to shifting distributions of disease vectors (Norris 2001; Epstein et al. 1998). For example, annual increases in temperature are associated with the expansion of malaria in the Usambara Mountains of the Eastern Arc (Matola et al. 1987) and in the Kenyan Highlands (Some 1994). Extreme weather patterns (Epstein et al. 1998) can also increase the vulnerability of range-restricted species to disease outbreaks.

Should epidemics and wildlife health issues be addressed in a conservation and management context? Maximizing the viability of species and communities is a common goal for both conservationists and animal health specialists, particularly for small, threatened populations, for which the risk of extinction is greatest. Yet conservationists and livestock specialists have often worked in opposition to each other. In the past, most disease control efforts were aimed at allowing the expansion of domestic livestock, often at the expense of biodiversity conservation and the long-term maintenance of environmental goods and services. Early efforts to control tsetse fly and the use of game fencing to manage foot-and-mouth disease severely affected wildlife populations over time.

Too often, decisions focused on single resources have had multiple adverse resource consequences. Examples include the control of foot-and-mouth disease to support a subsidized beef export market in Botswana and the control of tsetse fly in the Zambezi Valley in Zimbabwe. In Botswana, inappropriately sited fences decimated major wildlife populations and preempted sustainable wildlife tourism options (Pearce 1993). In Zimbabwe, subsistence farmers rapidly migrated into marginal areas cleared of tsetse fly, where they overwhelmed the autochthonous culture, displaced a rich wildlife resource, and developed an area that now depends on food aid in most years (Cumming and Lynam 1997). These kinds of decisions are starting to change as epidemiologically based tools, coupled with more appropriate economic analyses and natural resource accounting, gain wider acceptance in land-use planning and policy, wildlife and livestock management, and monitoring (Osofsky et al. 2000).

In recent years large-scale planning efforts across many parts of Africa have done much to advance the conservation of biodiversity. Yet as conservation landscapes are mapped and refined, we still know little about the factors needed to ensure the persistence of ecosystems and species. Corridors and buffer zones often are used to increase the functional size of areas. However, they can also facilitate the transfer of diseases into populations not already in contact with other pathogens and their hosts (Woodroffe 1999). Applying epidemiological approaches in planning can mitigate such risks. In most cases, however, conservationists lack even the most rudimentary baseline knowledge of what diseases already exist in species and populations of conservation interest. Obtaining such baseline information is the only way to begin monitoring the health of species and spaces.

Epidemiological modeling is increasingly being used by conservationists to identify the desirable size and structure of host populations to help reduce the likelihood of extinction. For example, new disease risk models have helped identify the amount of resources, habitat patch size, and viable population size that might be targeted to help ensure the survival of the critically endangered Ethiopian wolf (Haydon et al. 2002). Disease monitoring and surveillance, with the help of local communities, can also prevent epidemics in wildlife populations (Karesh et al. 2002). In places such as the Horn of Africa, the last known reservoir of rinderpest, local pastoralist communities working with veterinary authorities are able to identify the presence of rinderpest in their cattle and help mitigate its spread to wildlife populations.

There is a growing need for collaboration between veterinary scientists, epidemiologists, conservation biologists, economists, and a range of other disciplines in landscape and reserve planning and at the interface between agricultural lands and, for example, protected areas. Only a multidisciplinary approach is likely to prove successful in efforts to mitigate disease (Karesh et al. 2002). With many species now restricted to small areas surrounded by agricultural and urban landscapes, data on the presence, susceptibility, and transfer of pathogens should be evaluated in planning protected area networks, corridors, and multiple-use areas.

Collaboration between the conservation and health fields has already begun in academic consortia. Recently, scientists from two international animal health associations (one wildlife focused, the other agriculturally oriented) committed to the Pilanesberg Resolution, a call for more integrated approaches among health scientists and other disciplines to address wildlife and livestock health and the concomitant impacts on human livelihoods (http://www.wildlifedisease.org/includes/Documents/resolution.html; Karesh et al. 2002). Collaborations such as this present an opportunity for the environmental community to engage other disciplines in wildlife conservation in the context of development. As ecosystems are increasingly altered and the ecology of diseases changes, a more holistic understanding of the links between ecosystem integrity and ecological health will be important, if not essential, to the long-term persistence of biodiversity in the decades ahead.

ESSAY 6.6

HIV/AIDS, and Ecoregion Conservation

Jane Dwasi and Judy Oglethorpe

The HIV/AIDS pandemic is having huge impacts on society in Africa, causing great human suffering and increasing poverty in many countries. Perhaps less well known are its severe impacts on biodiversity and natural resources. Impacts occur through the loss of human capacity for conservation and management and through changes in land and resource use.

The scale of the pandemic is immense. In sub-Saharan Africa, more than 28 million people are living with HIV/AIDS, more than half the global number (UNAIDS 2000, 2001). People with HIV/AIDS in Africa usually die over a period of 2 to 10 years. The rates of HIV infection, prevalence of HIV/AIDS, and AIDS deaths are increasing in all countries, with the exception of Uganda, where the rate of new HIV infections has started falling (Uganda Ministry of Health 2001). Table 6.7 shows statistics for Kenya, Tanzania, and South Africa.

Unlike on other continents, most of the affected people in Africa are between 15 and 49 years of age and are often the most educated and the most socially and economically active (UNAIDS/WHO 2000; Whiteside and Sunter 2000; Dwasi 2002a, 2002b; Muraah and Kiarie 2001). Among this cohort, AIDS has become the leading cause of death. HIV in Africa is transmitted primarily heterosexually, so the epidemic has serious implications for families. HIV/AIDS has reduced life expectancy in many countries and created an imbalanced population structure, with widows, orphans, and the elderly forming the majority (NASCOP 1997; Whiteside and Sunter 2000; Muraah and Kiarie 2001).

There are severe economic impacts at individual, household, organization, and national levels. Medical and funeral expenditures have increased markedly. The loss of wages affects families very badly (lifetime loss per person in Kenya is estimated at US$7,643.20; Republic of Kenya 1997). Governments increasingly have to spend scarce funds on prevention, treatment, and mitigation. In many rural villages, households have had to sell natural and agricultural resources including land, crops, and livestock to pay for HIV/AIDS-related costs.

The pandemic is affecting all sectors, including biodiversity conservation and natural resource management. One of the most visible impacts is on the workforce. Environmental organizations are losing personnel, particularly trained and skilled people in leadership positions (who therefore have higher salaries and greater mobility through work-related travel, two important factors in HIV transmission). In one South African conservation department the number of deaths per year in the workforce has tripled compared with that before the pandemic. Long periods of illness and frequent deaths among staff are compromising planning and implementation of conservation activities and natural resource management. For example, some protected areas in South Africa are experiencing increased poaching as a result of reduced management capacity. Conservation organizations are also using scarce funds to pay for their staffs' medical and funeral expenses and terminal benefits.

Rural communities implementing community-based natural resource management activities are also vulnerable. For example, at Wuparu, a community conservancy in Caprivi, Namibia, AIDS deaths reduced the number of community game guards, and it is now very difficult to keep wildlife away from crops. In Nyamabale, Uganda, deaths of key conservation and natural resource management leaders have slowed the adoption of new conservation ideas and practices by community members. Without good leadership it becomes more difficult to empower and build rural communities' capacity to manage their local resources sustainably.

Environmental impacts caused by changes in livelihood practices of rural households are less clear. Some households are adopting coping strategies that increase dependence on natural resources for food, medicines, and shelter as agricultural production and income fall with the loss of labor. Other households have been wiped out, reducing the area of land under cultivation and resulting in the regeneration of natural vegetation (Rugalema et al. 1999). In seriously affected areas such as Rakai in Uganda and parts of Nyanza in Kenya, whole farming households have disappeared or left only young orphans who cannot farm. Indigenous knowledge of sustainable natural resource management is being lost.

In addition, there are many and varied examples of how the response to HIV/AIDS has increased natural resource exploita-

TABLE 6.7. **HIV/AIDS Statistics from Kenya, Tanzania, and South Africa.**

Country	Number of People Living with HIV/AIDS	Number of AIDS Deaths in 1999	Number of AIDS Orphans	National Population
Kenya	2,100,000	180,000	730,000	29,549,000
Tanzania	1,300,000	140,000	1,100,000	32,793,000
South Africa	2,900,000	250,000	420,000	39,900,000

Source: UNAIDS and WHO (2000).

tion. For instance, pressure on medicinal plants used to treat side-effects of AIDS is greatly increasing, and many are being harvested unsustainably. In KwaZulu-Natal, sea turtle eggs are being collected in the mistaken belief that they cure HIV/AIDS. In Kenya, logging has increased in many forests to make coffins (Dwasi 2002b).

There is still a large stigma associated with the disease. Victims and their families often are shunned by society and discriminated against in employment. Many countries and organizations continue to avoid discussion and actions that would limit disease spread and impacts. The insurance industry has complicated matters by requiring HIV testing before issuing medical coverage; many companies deny coverage to people infected with HIV and refuse payments to families in cases of AIDS deaths. Consequently, people do not want to undergo HIV testing, which is an important step in the fight against HIV/AIDS. People who know they are HIV positive often conceal it and strive to act normally, even when engaging in sexual activities. These problems promote the spread of HIV infection and defeat efforts to prevent or mitigate impacts.

The severe reduction of human capacity in government departments, communities, nongovernment organizations, universities, and the private sector will have a serious impact on ecoregion conservation in Africa. Expertise of many different kinds is needed for ecoregion development: biology, social sciences, economics, agriculture, development, traditional knowledge, planning, policy-making, politics, leadership, facilitation, management, implementation, training, negotiation, monitoring, and evaluation. Successful ecoregion conservation entails skilled people working at many different scales, from the broad ecoregion level to the national, landscape, and local levels. Capacity is already low in Africa: trained and experienced personnel are scarce and extremely valuable. Loss of these people leaves a big gap in a region or country's capacity for ecoregion conservation. It is likely that the HIV pandemic will delay and weaken development of ecoregion conservation efforts at

a time when a large, multidisciplinary effort with visionary leadership is needed.

What Can the Conservation Sector Do?

There is need for urgent action to reduce the impacts of HIV/AIDS. So far, the environmental sector has responded slowly. A few organizations and projects have developed coping strategies, such as the KwaZulu-Natal Department of Nature Conservation and the WWF East Africa Regional Program Office. Others are promoting awareness (USAID 2000a, 2000b) and researching impacts to guide future action (Dwasi 2002a, 2002b).

Awareness must be raised in conservation organizations, and experiences of existing coping strategies must be shared to encourage others to take action. In individual organizations, internal audits should assess HIV/AIDS impacts as a first step to developing a strategy. HIV/AIDS components should be incorporated into conservation programs, including counseling and education. Infected staff can benefit from nutritional and other medical advice. Organizational policy on health coverage and terms and conditions for affected employees should be clearly stated. Treatment with HIV/AIDS drugs should be considered if they become more widely available. The need for openness is very important. Collaboration with other sectors including health, population, and development is essential. Training of conservation managers and community natural resource workers is urgently needed to equip them to deal with the problem, and the health sector can help. HIV/AIDS funds should be leveraged and distributed to African conservation organizations. Although the conservation sector cannot tackle the HIV/AIDS pandemic itself, it can do its part to stem the loss of capacity and mitigate impacts on biodiversity.

Similarly, in communities managing natural resources HIV/AIDS awareness should be promoted and appropriate coping strategies developed to help limit spread of HIV and reduce impacts on natural resources and biodiversity. Indigenous knowledge of sustainable resource management practices should be preserved.

African governments should be encouraged to exercise strong and committed leadership in the fight against HIV/AIDS. Countries in Africa and elsewhere that are just starting to experience the pandemic can learn from those such as Uganda where a policy of open discussion has brought about significant behavioral change and a drop in new infection rates.

In all this work, timely and open communication is essential. The pandemic is already having devastating effects on people and biodiversity, but in many areas the worst impacts are yet to come.

ESSAY 6.7

The Future for Elephants in Africa

Peter J. Stephenson

The African elephant was among the first recognized flagship species, providing a rallying point to raise awareness and stimulate action and funding for broader conservation efforts (Leader-Williams and Dublin 2000). The conservation focus has been mainly on issues surrounding the killing (legal and illegal) of elephants for their ivory. However, elephants also serve as keystone or umbrella species, playing a pivotal role in ecosystem function and helping to maintain suitable habitats for myriad other species. Its conservation emphasizes many of the ideas of landscape design being promoted in Africa. It can be argued that a future for elephants therefore means a future for much of the biodiversity in Africa.

The Past

In Africa, people have always lived alongside elephants, and they have always exploited them for meat, hides, and ivory. One elephant provides a lot of meat, and African cave paintings indicate that they were hunted for this reason thousands of years ago. The earliest ivory carvings are more than 27,000 years old, showing that people have long appreciated the decorative and artistic value of ivory. However, because human populations across Africa were small until around 2,000 years ago (Reader 1998), hunting probably had no significant effect initially on elephant numbers or distribution. The decline in elephant numbers began with the huge demand for ivory created by the European colonization of Africa and the introduction of large-scale agriculture in countries such as South Africa (Hall-Martin 1992). Controlling the supply of valuable ivory from Africa to Europe was also a major incentive for the division on the continent into colonial-managed countries in the 1800s (Pakenham 1991).

Through much of the twentieth century the systematic hunting of African elephants for their ivory (both legal and increasingly illegal) continued to decimate populations. Between 1950 and 1985 there was a steady increase in ivory exported from Africa, up from 200 tons to more than 1,000 tons per annum (Convention on International Trade in Endangered Species [CITES] records). Elephants were hit particularly hard in the 1980s when an estimated 100,000 individuals were being killed per year and up to 80 percent of herds were lost in some regions (Alers et al. 1992; Cobb and Western 1989; Western 1989; WWF 1997, 1998). Most of the ivory was sold to the Far East, but after Japan the United States was the largest single importer, with a retail ivory trade worth US$100 million per year (Thomsen 1988). The sharp decline in elephant numbers in Africa caused an international outcry. North Americans and Europeans, some of who had never even seen a wild elephant and whose nations had contributed to the decline of the species in the first place, added their voices to an international demand to put an end to the killing.

In 1989 many importing countries imposed their own legislation to stop in the importation of raw ivory, and in the CITES meeting of 1989 the African elephant was placed in appendix I of CITES, preventing international trade in ivory and other elephant products. This ban was imposed in an attempt to cut off supply to the markets. In Africa antipoaching efforts were augmented where the means were available. Although poaching never completely stopped, elephant numbers recovered in many countries. However, the problems have not gone away. Although the use of ivory in Europe and North America has largely ceased, ivory is still in demand in the Far East; for example, ivory seals, or *hankos,* are still prized in Japan. The continued demand for ivory, as well as for wild meat, has maintained a hunting pressure on African elephants to the present day.

The Present

There may have been several million African elephants before the early twentieth century (Milner-Gulland and Beddington 1993); current numbers are estimated at 301,000–488,000 (Barnes et al. 1999). However, it is difficult to estimate exactly how many African elephants existed in the past or remain today. Many African countries do not make regular counts; indeed, population estimates for more than one-third of the range states (country where elephants occur) are derived mainly from guesswork (Barnes et al. 1999). Forest elephants beneath the canopy in Central and West Africa are particularly difficult to census with the typical aerial methods and can be estimated only using dung surveys (Barnes 1993). Despite problems with obtaining accurate counts, however, trends are apparent.

African elephants have less room to live in than ever before. People demand the same resources (fertile soils and water) as elephants (Parker and Graham 1989), and an expanding human population is converting more and more land into agriculture and settlement (Myers 1993); desertification further reduces available habitat (Cumming et al. 1990). Less than 20 percent of elephant habitat is under protection, and many elephant herds are now confined to isolated protected areas. When elephants try to follow traditional migration corridors through what was once forest or savanna, they are confronted with roads, fields, and villages. Many of the crops planted by people are attractive to elephants, and because an elephant can eat more than 200 kg of food a day, even a small herd can decimate a farmer's annual production during one night's foraging. Inevitably conflict arises, and the elephant always loses. As a result, human-elephant conflict (HEC) is one of the biggest issues facing elephant conservationists today.

The largest African elephant populations (and many of those increasing in size) are closely associated with the Central and Eastern Miombo and *Baikiaea* Woodlands [50–52] ecoregions, which stretch across key southern and eastern range states such as Namibia, Botswana, Zimbabwe, Zambia, Mozambique, and Tanzania. Other key ecoregions for elephants in East Africa include the East Africa *Acacia* Savannas and East African Coastal Forests; however, the coastal forests are very fragmented, and their small elephant populations are under severe threat. Most of the elephants surviving in West Africa are found in the remaining fragments of the Guinean Moist Forest Ecoregion. These animals may be among the most threatened by habitat loss and range contraction: elephant range in West Africa was estimated to have shrunk 93 percent between 1900

and 1984 (Roth and Douglas-Hamilton 1991), and encroachment continues. Many of the forests in Central Africa that cover elephant range have not been adequately surveyed, so no accurate, up-to-date population estimates are available for some ecoregions even though they must provide very important habitat for forest elephants. Key ecoregions for which population data are deficient include the Central Congo Basin Moist Forests in the Democratic Republic of Congo and the Western Congo Basin Moist Forests and Congolian Coastal Forests where they extend in to Gabon and the Republic of Congo (Barnes et al. 1999).

The illegal killing of elephants for ivory and meat continues. Several African states have thriving domestic ivory markets (Martin and Stiles 2000), yet many of these markets fall in countries that cannot possibly support their internal trades with ivory from their own diminished elephant populations. In many instances, the sources of ivory must be from illicit transborder trade routes, yet the movement of ivory into and out of these domestic markets, and the link with the illegal international trade, is poorly understood, further hampering law enforcement operations. Few range states have the resources to adequately monitor their elephant populations, the rates and levels of poaching, or the ivory trade, so it is difficult to make informed management decisions. Even if the information were available for adaptive management, the capacity of many states to act is inadequate. Developing countries rarely treat wildlife conservation as a high national priority, and many of the people responsible for protecting elephants are not given adequate funding to do their job. Many protected areas are woefully short of finances: some reserves in West Africa have an operating budget of less than US$5 per square kilometer per annum, and park guards in some range states are paid less than US$0.20 per month (Barnes 1999; Stephenson and Newby 1997).

So what can be done to turn the tide? Is there a future for elephants in Africa? Can they ever live alongside people in stable numbers? Can countries ever have the capacity and resources to guarantee the long-term survival of their elephants?

The Future

The future is difficult to predict on a continent as volatile as Africa. However, if the African elephant range states are committed to conserving their wildlife, if they put in place policies and legislation to assist elephant management, and if they are supported technically and financially by the international conservation community, it is possible to see a future for elephants. In this future, we can imagine an African continent criss-crossed with a network of landscapes in which humans and elephants live alongside each other. This vision can be achieved if existing initiatives aimed at conserving the elephant are replicated and expanded. Three issues in particular must be addressed over the coming decades.

Management

Range states need to know their elephant population levels in order to make informed management decisions. Priorities for surveys are the Central African forests and the priority ecoregions in the Congo Basin, for which no accurate data are currently available. Wildlife managers also need to know the rate of legal and illegal killing of elephants. This may be made possible by a system currently being put in place by CITES at the request of African governments. When established, Monitoring the Illegal Killing of Elephants (MIKE) will ensure that wildlife managers are using standardized and statistically validated protocols for collecting and analyzing poaching data. MIKE also aims to ensure regular population counts. The related Elephant Trade Information System (ETIS) will help track the stockpiles and movements of ivory and other elephant products. On the basis of better information, countries will need to develop and implement national elephant management strategies and work with their neighbors to develop subregional perspectives. Some countries such as Ghana and Cameroon have already developed strategies; West African countries produced a common sub-regional strategy. Other such plans must be developed and implemented so that elephant management is based on clear goals determined by key stakeholders.

Different countries will continue to have different approaches to elephant management. Some southern African countries will continue to maintain large, healthy populations and will continue to call for the right to sell elephant products. Others, mostly in eastern and southern Africa, will continue to promote tourism based on game viewing. It has been estimated that elephants may be worth US$25 million a year to Kenya's tourism industry (Brown and Henry 1993), and elephant viewing remains a high priority for many visitors to Africa (Goodwin and Leader-Williams 2000). Other countries that do not have the infrastructure for such large-scale ecotourism will prefer to allow limited trophy hunting. There is unlikely to be consensus between countries on the most suitable approach because of variations in socioeconomic conditions and in the size and distribution of national elephant populations. However, increased dialogue could lead to improved understanding and mutual respect. Dialogues between countries possessing elephant populations started in recent years and should continue to increase in frequency. Exchange visits could also be promoted to allow wildlife managers to share lessons and experiences.

Conflict Resolution

In Europe and North America most large mammals have been locally extirpated. Initiatives to reintroduce or increase populations of large carnivores have met with resistance from some local people, who are often uncomfortable living near potentially dangerous or destructive animals (Kellert et al. 1996). Needless to say, many Africans have similar reservations about living close to elephants. Efforts must be doubled to help mitigate HEC and to empower people living near the animals to make informed decisions on the choices available to mitigate or minimize the risk of conflict. Tools have been developed to monitor and mitigate conflict (Hoare 1999, 2001) and are being tested at a number of project sites (Stephenson 2002). These systems are simple, but choosing and implementing methods for mitigating HEC (Hoare 2001) takes some level of training and materials. Long-term solutions will entail appropriate land-use plans by communities throughout the elephant range. Funding will be needed to support such HEC mitigation, and if developed countries want African states with fewer financial resources to protect their own large wildlife, they should help foot the bill.

Protected Areas and Landscapes

Many protected areas are isolated islands of habitat surrounded by a sea of human agriculture and development. If that situation remains, particularly in West Africa, elephants will survive only in small, isolated populations in these protected areas (Barnes 1999). Over the coming decade elephant conservation must be reoriented toward conserving and managing broader landscapes in priority ecoregions. Using large, multiuse landscapes as a strategy to conserve mammals is increasingly advocated to take into account the integrity and function of ecosystems and other elements of biodiversity (Noss et al. 1996; Entwistle and Dunstone 2000; Linnell et al. 2000). Following the ecoregion approach, elephant landscape conservation must address threats on a large scale and beyond political boundaries. Some parts of the landscapes should be community-managed land where local people benefit in some way from the habitat and its wildlife. These landscapes must be established in the next 15 years, before the potential connections are severed (Newmark 1996).

The development and management of elephant landscapes is a huge challenge because it will entail cross-sectoral (and often transboundary) land-use planning involving different government and nongovernment agencies as well as local communities. Viable landscapes could be developed if existing initiatives in elephant range are scaled up. At the Yaoundé summit in 1999, Central African governments committed to protect and sustainably manage their forests and to work on transboundary initiatives. Since then, progress has been made toward establishing the Sangha Trinational Park between Cameroon, the Central African Republic, and the Republic of Congo. A transborder conservation initiative is also under way to link the forests of Dja (Cameroon), Minkebe (Gabon), and Odzala (Congo). A joint conservation plan is under development that will ensure protection of core areas and sustainable forest management and conservation-friendly land uses in surrounding zones.

Similar transnational conservation areas have been established in southern Africa (Hanks 2000). More efforts must be initiated along similar lines, taking into account regional and sub-regional, rather than just national, conservation priorities. Certainly the Congo Basin is a large enough wilderness area that existing transboundary initiatives could be built on to establish other new megaparks. There is much potential to build on existing protected areas in the Miombo and *Baikiaea* Woodlands, and opportunities also exist in West Africa to link remaining blocks of Guinean Moist Forest that traverse national boundaries (Parren et al. 2002). In every case, efforts must be made to involve key stakeholders such as private enterprise and local communities. Community involvement in wildlife management has had some successes in countries such as Zimbabwe and Namibia and could be expanded to other range states to increase incentives for conservation.

Key to the success of these landscape initiatives will be the provision of innovative sustainable funding mechanisms that ensure that resources are available for viable and effective management, especially in core protected areas. The use of trust funds is not new, but such long-term, large-scale approaches must be revitalized to incorporate lessons learned from pilot schemes.

What Next?

The challenge now is for African states and the international conservation community to mobilize adequate resources to scale up and replicate existing initiatives in order to ensure that whole landscapes are conserved in priority ecoregions. With elephants, more than with most other species, the scale of the efforts is important, and the scale is large.

CHAPTER 7

The Next Horizon: Large-Scale Conservation in Africa

The fact that conservation biologists have converged on identifying the most important areas for biodiversity conservation across Africa is an encouraging milestone (da Fonseca et al. 2000; essay 6.1). The challenge for conservationists is to turn this common set of large-scale continental priorities into conservation achievements at finer geographic scales. This process involves the development of spatially explicit and quantified plans defining what is needed to sustain biodiversity over the long term. For WWF, these plans are developed in the form of a biodiversity vision according to a set of standards that have been already defined (Dinerstein et al. 2000) but are being constantly updated. Biodiversity visions are based on a fundamental but rarely asked question: from the perspective of conservation biologists, what would successful conservation for an ecoregion look like 30 to 50 years hence? To answer this question, we must understand what biological features must be saved now and what features that were once prominent must be restored over the coming decades. The biodiversity vision also asks the hard question: if we cannot save everything everywhere, then what should we save? For African ecoregions, the biodiversity vision must also balance biodiversity conservation with the needs of a growing population and their demands on natural resources. This is even more difficult because human population density often is concentrated in areas rich in both species and endemics (Balmford et al. 2001a, 2001b; Moore et al. 2002). Compromises between the needs of people and those of biodiversity conservation will remain a major concern and challenge for the foreseeable future.

Typically the biodiversity vision will identify a set of priority sites or larger landscapes within an ecoregion that, if effectively conserved, would maintain the biological values of that area forever. Parallel sets of actions are also identified at the policy and legal levels that are needed to mitigate or remove threats to critical landscapes. In some ecoregions, the most important interventions may be policy related (e.g., to maintain water flow into a wetland threatened by upstream dams, such as in the Lake Chad Flooded Savanna [61] ecoregion in the Sahel).

This chapter outlines the WWF approach to conservation planning within ecoregions to develop a biodiversity vision and the implementation approach: the Ecoregion Action Program. Ways in which these plans can be made operational in the context of Africa are also outlined, as are the main approaches for implementing the plans and financing them. Examples are provided from the work of WWF (and other nongovernment organizations [NGOs] and government bodies) in the region. We conclude the chapter with some remarks on how we see the future of conservation in Africa, based on three different trajectories of natural resource use.

Ecoregion-Scale Planning

The analysis in chapter 6 provides a portfolio of ecoregions that are globally important and in need of conservation

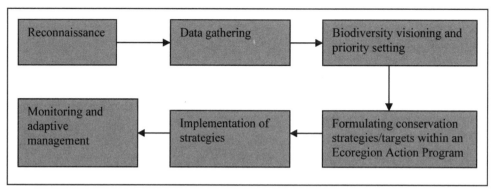

FIGURE 7.1. The WWF ecoregion planning process.

investment. This section outlines the methods that WWF and others use to separate the large priority ecoregions into smaller, more tangible conservation targets for implementation.

WWF Ecoregion Planning Approach

The WWF ecoregion planning process has several distinct stages: reconnaissance, data gathering, biodiversity visioning and priority setting, strategy formulation, implementation, and monitoring and learning (Dinerstein et al. 2000) (figure 7.1). Currently eight terrestrial ecoregions are the focus of WWF efforts in Africa. The involvement of WWF in these varies; for some WWF is the leading agency working with relevant stakeholders in the region, and for others WWF is a member of a consortium developing the plan. Elsewhere, other agencies, such as Conservation International, are developing and implementing similar plans. Ecoregions where ecoregion plans are being developed or implemented in Africa are as follows:

- The Miombo woodlands of eastern and southern Africa [49–56]
- The Congo Basin lowland forests of Central Africa [7, 9, 12–16]
- The dry and spiny forests of southern Madagascar [33, 113, 114]
- The Fynbos shrublands of South Africa [89, 90]
- The eastern African coastal lowland forests from Somalia south to Mozambique [20, 21]
- The Albertine Rift mountains of Central Africa from Uganda in the north to Democratic Republic of Congo in the south [17]
- The Upper Guinea lowland rainforests of Guinea in the east to Togo in the west [1–3]
- The humid forests of eastern Madagascar [29]

RECONNAISSANCE

The reconnaissance stage includes a preliminary overview of biological, socioeconomic, policy, legal, and institutional characteristics of the ecoregion. This stage provides the essential background for ecoregions that are poorly known to the organizations developing conservation plans or that span political boundaries where information has not previously been synthesized.

DATA GATHERING

The data-gathering stage assembles information that will be needed to develop a conservation vision and conservation strategy for the ecoregion. In general this includes species distribution information for key taxonomic groups, information on habitats across the ecoregion, and information on a number of different social issues (e.g., population, infrastructure, protected areas). Ideally this information should be assembled in a geographic information system (GIS) so that further analyses can be performed.

VISIONING AND BIODIVERSITY PRIORITY SETTING

The biodiversity vision stage is of great importance in the development of the conservation plan for an ecoregion. It presents a view of what conservation success might look like in the future in terms of species, habitat representation, and ecological processes. Often this boils down to maps of preferred land-use patterns across a region and the identification of important conservation sites or landscapes. For ecoregions that span several countries, this is the stage at which a common understanding of the importance of the area and what each nation has to offer to the process can be developed. Visions are the outcome of carefully planned workshops. There are two broad approaches to these vision workshops. Details of the biodiversity visioning process are presented in Dinerstein et al. (2000) and WWF-US (2002).

A **goal** is the long-term conservation outcome that should be achieved. For example, in a forested landscape of southwest Cameroon, the goal might be, "Conserve sufficient connected habitat to maintain viable long-term populations of lowland gorilla, forest elephant, and forest buffalo."

Focal elements are the set of biological characteristics that make an area significant for conservation. Focal elements include species (such as elephants, endemic bird and plant species), habitats (cloud forest, wetlands, aquatic habitats), and processes (colonial nesting sites for birds). For example, the endemic birds of the Cameroon Highlands would be fo-

cal species, the montane forest would be the focal habitat, and the hydrological supply of water from forests to lowland areas would be a focal ecological process.

A **target** is the amount, type, and configuration of the land needed to conserve the focal elements. It might also contain, in the case of species, a determination of viable population levels. A target for mountain gorillas might be, "Conserve a minimum of 500 km^2 of interconnected forests ranging in altitude from 2,500 to 3,000 m and access to water." A full set of targets should maintain the focal elements of an area.

—*Colby Loucks*

Expert Workshop Visions. WWF has used expert workshops to develop biodiversity visions for ecoregions around the world, including Africa. Expert workshops have the advantage of quickly capturing information from many knowledgeable people in a format that can be readily used (maps and databases). Workshops also bring different stakeholders together into a process and achieve a degree of national support for the implementation of a jointly produced biodiversity vision (and later the conservation action plan). Example visions developed using these methods are found at http://www.worldwildlife.org/ecoregions.

Computer-Assisted Visions. Alternative approaches to developing a biodiversity vision use objective data, often supplemented by expert knowledge. There is extensive scientific literature on computer-based biological analyses for setting conservation priorities (summarized in Williams 1998; Margules and Pressey 2000; Balmford 2002; Cowling et al. 2003). The advantage of these kinds of analyses is that they are objective and repeatable and produce efficient results that can be measured. Disadvantages are that they can alienate local stakeholders by producing a technological solution; they fail to address the local political realities and therefore can be easily ignored by those making conservation decisions (Pressey and Cowling 2001; Jepson 2001). They also suffer from a lack of insight. A computer cannot estimate a potentially important area, but knowledgeable people can make such assessments, which can then be tested.

Across Africa and elsewhere, attempts are being made to combine the advantages of the expert and computer-assisted approaches to the development of biodiversity visions. They have already been used in the Cape Floral

Kingdom (Cowling et al. 2003) and are being developed for the humid and subhumid forests [29, 30] of eastern Madagascar.

FORMULATION OF CONSERVATION STRATEGIES AND TARGETS FOR AN ECOREGION ACTION PROGRAM
The biodiversity vision process highlights the biological values of the area (including important sites and broader landscapes) and defines spatial patterns in socioeconomic threats and opportunities. To implement the vision, a conservation strategy must be developed. This process includes developing conservation goals, identifying the focal biological elements (species, habitats, and biological processes), and setting the necessary targets to achieve the biodiversity vision for the area (box 7.1).

In many ways the logic of creating a conservation strategy for an ecoregion, or a part of an ecoregion, is similar to that of the logical framework approach (LFA), which has become an essential part of project planning and implementation in developing countries (Caldecott 1998; Margolius and Salafsky 1998). This method can be adapted to create an overall conservation strategy, containing biological targets, which can be budgeted and turned into conservation proposals for implementation. WWF calls the product of this process an Ecoregion Action Program (Golder 2004).

STRATEGY IMPLEMENTATION
Clearly, the conservation plans developed through this process must be implemented. Most of the Ecoregion Action Programs that WWF has developed in partnership with relevant national governments across Africa are only starting to be implemented. There is a lot to learn, and there will be both

successes and failures along the way. Examples of programs under implementation are presented later in this chapter.

MONITORING AND ADAPTIVE MANAGEMENT

Monitoring and evaluation schemes are being developed to measure the impacts of implementing biodiversity visions and Ecoregion Action Programs. Monitoring the conservation impact, rather than the conservation process (meetings, newsletters, procurement), is an important goal for these schemes. Again, the logical framework approach provides clues on how to develop a monitoring scheme that provides ways to create detailed indicators of success, which can then be measured. Methods such as threat reduction assessment (Salafsky and Margolius 1999) have been developed recently to look at how a project is reducing threats to the system, which is important because these methods respond more quickly to project interventions than the biological measures of habitats and species.

Other Approaches at the Ecoregional Scale

Two other international conservation NGOs (Conservation International and BirdLife International) have identified large geographic regions as biodiversity priorities (Stattersfield et al. 1998; Mittermeier et al. 1999, 2002). Conservation International has used similar methods to those outlined in figure 7.1 to develop plans for large-scale conservation programs, such as in the Upper Guinea rainforests of West Africa (Bakarr et al. 2001a). BirdLife International has not developed plans for conservation in its endemic bird areas (EBAs) but has instead focused on identifying important bird areas (IBAs), which are sites in countries where conservation action will be implemented (e.g., for Africa, Fishpool and Evans 2001; see essay 7.2).

The Wildlife Conservation Society and the African Wildlife Foundation have also identified important areas for conservation, typically at a scale smaller than ecoregions and larger than sites. These conservation landscapes form the basis of conservation planning and implementation for these two organizations in Africa (Sanderson et al. 2002b; http://www.awf.org/heartlands).

Governments and supporting donors (e.g., the Global Environment Facility [GEF]) have also funded the development of conservation plans for globally important regions for biodiversity, as in the Cape Floristic Region of South Africa (CSIR 2000a, 2000b). Further government interest in these approaches is being expressed in other countries, as in Kenya and Tanzania and in the nations of the Congo Basin.

Ecoregion Planning

Of the thirty-two Class I ecoregions in Africa (see chapter 6), eight (25 percent) have active conservation planning exercises at the present time. In addition, fourteen (48 percent) of the twenty-nine Class III ecoregions have an active ecoregion conservation program. These programs are coordinated by a variety of different agencies, including WWF, Conservation International, GEF, government departments, and, in the case of the Albertine Rift, a consortium of other conservation groups including the Wildlife Conservation Society, the International Gorilla Conservation Programme, the Dian Fossey Gorilla Conservation Fund, and the Albertine Rift Conservation Society. Importantly, all the NGO-facilitated conservation plans have engaged with relevant government partners. As these programs move to the implementation stages, working with governments and with local communities will be increasingly important. Although many of these programs are in the early stages of their work, two—the Fynbos of South Africa [89, 90] and the Madagascar Spiny Thickets [113]—have completed the planning process and moved to action. Others are starting to implement newly developed plans: the Upper Guinea forests, the Congo Basin, the Miombo woodlands, and the eastern African coastal forests. In this section we highlight the approaches to planning and implementation taken in three programs

The Cape Action Plan for the Environment in South Africa

The Cape Floral Kingdom of South Africa is one of the most important regions of the world in terms of biodiversity. Out of 9,000 plant species, more than 6,000 are found nowhere else on Earth. Numbers of endemics are also high in invertebrates, and there are some endemic birds, mammals, amphibians, reptiles, and fish. The habitats of the Cape are also highly degraded, especially the Lowland Fynbos and Renosterveld [89] ecoregion. There is also a large urban population in Cape Town and an important wine-growing and agricultural area in the region.

THE PLANNING PROCESS

The Cape Action Plan for the Environment (CAPE) was a project managed by WWF-South Africa in collaboration with numerous South African agencies and funded by the GEF. The aims of the CAPE project were to identify the priorities for conservation on the basis of biodiversity and threats, develop a long-term strategy and vision, draft a 5-year action plan with priority activities, and identify potential funding sources for these activities. To achieve these aims the CAPE process brought together provincial planners, local authorities, civil society groups, and conservation bodies to focus on conservation as a key theme in land zoning and development.

A shared GIS resource was developed during this period, managed by the Western Cape Nature Conservation Authority. This GIS system is based on a decision support program developed in Australia (C-Plan), which allows an interactive assessment of the conservation implications of a plan aiming to conserve species, representative areas, and ecosystem processes. This GIS analysis blended expert assessment with computer-based analytical methods looking at endemism, area requirements of species, and ecological processes (Cowling et al. 2003).

The CAPE process developed a vision and a goal as a framework for the program (box 7.2). Working from this vision, the Cape Action Plan identified conservation interventions and defined the roles and responsibilities of large numbers of different people and organizations (CSIR 2000a, 2000b; Younge and Fowkes 2003). The conservation interventions operate at a number of scales and include protecting irreplaceable species, representing habitats, ensuring connectivity along altitudinal transects, and working on policy and legal issues.

IMPLEMENTATION

Conservation has moved quickly. A coordinator was hired in May 2001 to move the agenda forward, and significant funds for implementation have been raised. For example, in 2002 around $3 million was pledged by the Critical Ecosystem Partnership Fund to support the Cape Action Plan. Partners in South Africa also seek support locally. Currently, forty-six projects have been funded that address the goals of the plan. Novel long-term financing mechanisms are also in the works, such as tax incentives for farmers to preserve areas of critical habitat that contain narrowly endemic species and are thus irreplaceable. Although there are many challenges, the conservation planning work undertaken in the cape provides many lessons that can be applied elsewhere in Africa (and globally). Full details of the scientific planning, the results, and conservation action programs were recently published in the journal *Biological Conservation* (issue 112) and in essay 7.1.

The Congo Basin Biodiversity Vision and the Yaoundé Summit

The forests of the Guinean-Congolian region contain the second largest contiguous forest block in the world, stretching

BOX 7.2. **The CAPE Vision and Goal.**

The CAPE Vision: We, the people of South Africa, as proud custodians of the Cape Floral Kingdom, will protect and share its full ecological, social, and economic benefits now and in the future.

The CAPE Goal: By the year 2020, the natural environment and biodiversity of the Cape Floral Kingdom will be effectively conserved, will be restored wherever appropriate, and will deliver significant benefits to the people of the region in a way that is embraced by local communities, endorsed by government, and recognized internationally.

from Uganda to the shores of the Gulf of Guinea. The area supports the highest species richness in Africa. There is much local endemism in the Atlantic coastal lowland forest to the west and the associated montane habitats of Monte Alén, Monte Cristal, Mont Doudou, and the Chaillu Massif. The foothill forests of the Albertine Rift Mountains in the east are also rich in narrowly endemic species of plants and animals.

However, perhaps the most important biological value of the Congo forests is the assemblage of large forest mammals and birds. This megafauna is the most diverse of any forest in the world and includes forest elephant, forest buffalo, gorilla, chimpanzee, bongo, mandrill, and hornbills. Comparable assemblages are almost gone from the forests of Asia and were never present in the South American Amazon forests.

THE PLANNING PROCESS

The heads of state of Central African nations met in 1999 to sign the Yaoundé Declaration, accelerating conservation planning in this region (box 7.3). The declaration committed the signatories to conserving 10 percent of the forests in protected areas and to ensuring that sustainable forest management was practiced throughout the region. After this declaration, a large expert workshop was organized in Gabon by WWF and collaborators. More than 160 scientists identified 168 areas of taxonomic importance, including 41 important areas for birds, 41 for mammals, 24 for invertebrates, 15 for reptiles and amphibians, and 42 for plants. These areas were synthesized, resulting in the identification of 77 important areas (covering 1,251,935 km^2) for biodiversity conservation. The important areas were later consolidated to create eleven priority landscapes for conservation intervention that form the core of the Congo Basin Forest Partnership (Kamdem-Toham et al. 2003a, 2003b; fig-

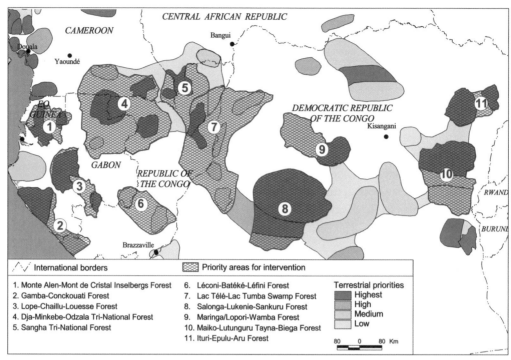

FIGURE 7.2. The Congo Basin Forest Partnership: priority landscapes for intervention.

Legend of the map:

- International borders
- Priority areas for intervention

1. Monte Alen-Mont de Cristal Inselbergs Forest
2. Gamba-Conckouati Forest
3. Lope-Chaillu-Louesse Forest
4. Dja-Minkebe-Odzala Tri-National Forest
5. Sangha Tri-National Forest
6. Léconi-Batéké-Léfini Forest
7. Lac Télé-Lac Tumba Swamp Forest
8. Salonga-Lukenie-Sankuru Forest
9. Maringa/Lopori-Wamba Forest
10. Maiko-Lutunguru Tayna-Biega Forest
11. Ituri-Epulu-Aru Forest

Terrestrial priorities
- Highest
- High
- Medium
- Low

80 0 80 Km

ure 7.2). These landscapes capture the species values of the region and are designed to maintain forest cover and the important biological processes of the region.

Governments in the region and the NGOs working there have adopted the biodiversity vision for the Guinean-Congolian Forest as the blueprint for conservation in the region (see http://www.worldwildlife.org/ecoregions/congo .htm).

There has been a great increase in conservation activity as a result of the Yaoundé Declaration (box 7.3) and the production of the biodiversity vision. Since 1999, a total of 40,607 km² of new protected areas have been declared or are being gazetted across the forests of the region, a 36 percent increase.

At the national level, the biological vision triggered reevaluation of protected area networks. In Gabon, thirteen new national parks covering 30,000 km² (10 percent of the country) were declared in 2003, and a similar process is under way in Cameroon. The Democratic Republic of the Congo and Central African Republic are planning similar reviews of their protected area networks.

IMPLEMENTATION

The biodiversity vision for this area has also been turned into an action plan (Kamdem-Toham 2002), a process involving the relevant stakeholders in the region. Part of the funding

BOX 7.3. **The Yaoundé Declaration.**

In March 1999, the heads of state of Central Africa signed the Yaoundé Declaration, which contains twelve commitments to forest conservation and sustainable forest management. At the core of the declaration is the recognition that protecting the forests entails a regional approach and coordinated policies. It marked a turning point in the political commitment to the environment in the region. The Democratic Republic of the Congo has since officially expressed its willingness to join the initial signatories: Cameroon, Central Africa Republic, Chad, Congo-Brazzaville, Equatorial Guinea, and Gabon.

Recognition by the UN General Assembly

The Yaoundé Declaration was recognized by the United Nations 54th General Assembly (Resolution 54/214) as a mechanism to ensure sustainable forest management and conservation in Central Africa. This resolution is titled "Conservation and Sustainable Development of Central African Forest Ecosystems," and it commends the declaration of Yaoundé and recommends its implementation by both the countries of the region and the international community as providing the framework for ensuring forest conservation and sustainable management in Central Africa.

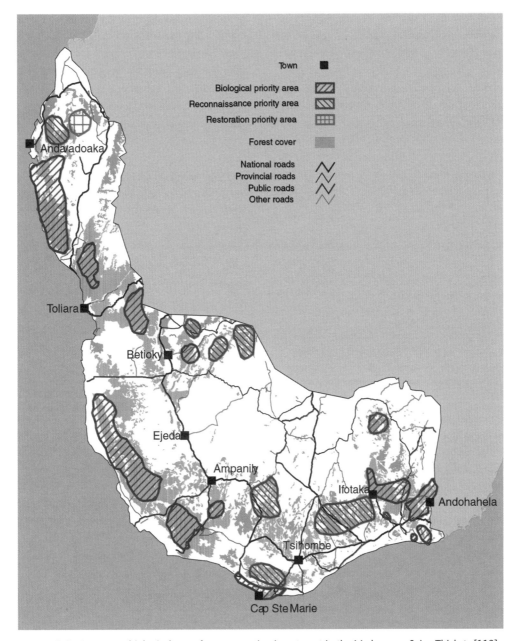

FIGURE 7.3. Important biological areas for conservation investment in the Madagascar Spiny Thickets [113].

needed to implement the plan has already been secured. At the World Summit on Sustainable Development in Johannesburg in September 2002, U.S. Secretary of State Colin Powell announced that the United States would support a Congo Basin Forest Partnership, an effort linking NGOs, industry, and governments to slow and even reverse deforestation in the Congo Basin (figure 7.2). This partnership will receive up to $53 million in the first 5 years. Although further funding is needed, this is a major step forward in a region where governments have been taking great strides,

albeit with little fanfare, in forest conservation over the past 3 years (Kamdem-Toham et al. 2003a, 2003b). Documents produced by WWF on the Congo Basin ecoregional planning process are available at http://www.worldwildlife.org/ecoregions.

The Spiny Thicket of Madagascar

The Madagascar Spiny Thickets [113] is an arid ecoregion located in southern and western Madagascar. The ecoregion supports a wealth of drought-adapted endemic species, including lemurs, tenrecs and other mammals, birds and reptiles, and many species of plants such as the spiny Dideraceae "trees."

Despite the area's exceptional endemism, less than 3 percent is represented in protected areas, which are too small to be self-sustaining. As much as 30 percent of the spiny thicket remains under natural vegetation, but this is quickly being degraded by clearing for shifting agriculture for subsistence or cash, for firewood and charcoal, and for livestock that are kept for ceremonial purposes.

THE PLANNING PROCESS

A working vision and action plan for this ecoregion were developed in Madagascar using available data and extensive discussions with scientists familiar with the ecoregion (WWF 2000, 2001). Because there were insufficient data on the distribution of individual species across the ecoregion, a habitat proxy was used as the basis for assessing the representation of biodiversity features in existing protected areas and for identifying areas without representation. First, the ecoregion was divided into five sub-regions on the basis of large-scale geologic formations or climatic factors. Experts confirmed the taxonomic uniqueness of each sub-region. The sub-regions were further divided into habitats using satellite imagery. The result was a map that serves as a proxy for the coarse-level distribution of biodiversity throughout the ecoregion (figure 7.3). Although the map is not purported to detail all habitats, it is assumed that representing each of the fifteen major habitat units or "representation units" in a system of protected natural areas would conserve the full array of biodiversity in the ecoregion.

Experts on Madagascar agreed that an area goal of 25 percent for each of these representation units would be a suitable target for long-term conservation. Maps were prepared to show how the proposed priority biodiversity conservation areas could be distributed given the current patterns of protected areas, remaining intact habitat, and an assessment of socioeconomic data relating to threats and conservation opportunities throughout the ecoregion.

IMPLEMENTATION

An Ecoregion Action Plan has been completed for this ecoregion (WWF 2001). Coordinators have been appointed and are responsible for developing programs around the most important areas identified while building consortia of national, regional, and local government and community participants. Because traditional power structures, based on local royalty and elders, are still an important component of southern Madagascar governance, special efforts were made to involve and strengthen these connections.

In two of the three priority areas, WWF is collaborating with the national park service to establish the new Mikea National Park and to extend the Tsimanampetsotsa National Park. In addition, there are ongoing efforts to develop regional parks and strengthen ecotourism infrastructure as an economic incentive for conserving habitat. Community-based natural resource management is being promoted to transfer management authority for habitat patches to local communities through traditional agreements or pacts (dina). Strategies are also being developed to reduce deforestation caused by the production of charcoal and fuelwood for urban markets.

Biodiversity Visions and Ecoregion Action Programs

Biodiversity visions and ecoregion action programs aiming to address conservation problems across large parts of Africa are currently operational in twenty-two (36 percent) of the sixty-one priority ecoregions for conservation investment across Africa (figure 7.4).

Implementation of these programs involves either direct interventions in landscapes or sites or indirect policy and legal work. In this section we outline ways in which these approaches are being used to achieve conservation gains in Africa and discuss the opportunities such approaches offer for increasing the effectiveness of conservation. We believe that these opportunities represent cause for optimism and hope for the future of biodiversity conservation in this region against a background of considerable pessimism (e.g., Vitousek et al. 1997b; Terborgh 1999; Gascon et al. 2000; Pimm et al. 2001).

Landscapes and Sites

Biodiversity visions typically identify core areas of habitat that must be conserved if valued biological features (species, habitats, and ecological processes) are to be retained. Some of these areas may already be parks and reserves, but in most cases there are additional areas outside these networks. Where park establishment is impossible, other approaches are needed to ensure the conservation of biological values. Working to develop habitat corridors between protected areas in high-biodiversity landscapes often is an important conservation strategy in these unprotected lands. Some of the most important implementation approaches are outlined in this section.

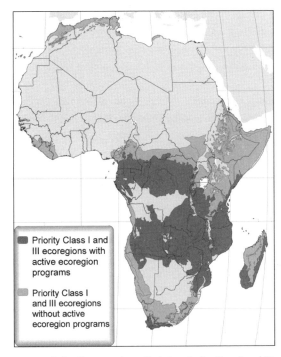

FIGURE 7.4. Conservation efforts in priority Class I and III ecoregions by WWF programs. (This map does not include a number of ecoregions where WWF is active at a local scale. WWF is also active in a number of freshwater and marine ecoregions not shown here.)

Legend:
- Priority Class I and III ecoregions with active ecoregion programs
- Priority Class I and III ecoregions without active ecoregion programs

IMPROVED PROTECTED AREA NETWORKS

One of the cornerstones of conservation across Africa is a functional protected area network that encompasses the distribution of the unique biodiversity values and is sufficiently funded to operate effectively. Concentrations of charismatic large mammals dictate much of the current constellation of parks and reserves in Africa (Goodwin and Leader-Williams 2000). Therefore, savanna-woodland habitats are well represented in the current reserve network of Africa, whereas other habitats are not nearly as well covered (essay 6.2). The main gaps in protection are in endemism-rich ecoregions, especially montane forests, Mediterranean scrub, and some of the dryland areas (particularly the Horn of Africa) (MacKinnon and MacKinnon 1986; Balmford et al. 2001a; de Klerk et al. 2004; Fjeldså et al. 2004). The same areas typically contain the highest numbers of threatened species (chapter 5).

The ecoregion action programs operating in Africa have investigated this issue in their own regions. In the Fynbos and Congo Basin biodiversity visions, a number of gaps in the protected area network were identified. By working with governments in the Congo Basin, conservationists have already filled some of these gaps (Kamdem-Toham et al.

2003b). For example, thirteen new national parks have been declared in Gabon. Protected area gap analyses and consequent gap filling could help direct biodiversity visions in other ecoregions. Such approaches are best suited to landscapes or sites with extremely high values of biological importance (which are often unique and irreplaceable); other mechanisms involving sustainable use are more appropriate for designing corridors to link these core areas together.

A number of quantitative approaches exist to design comprehensive reserve networks for countries or across entire regions (Williams 1998; Margules and Pressey 2000; Balmford 2002). Some broad-scale proposals for an idealized network of conservation sites across Africa include the large-scale work of the University of Copenhagen (Brooks et al. 2001a) and finer-scale studies in South Africa (e.g., essay 7.1; Rebelo 1994, 1997; Cowling et al. 2003). The Important Bird Areas Program of BirdLife International also provides a set of critical areas for bird conservation across the region, some of which are already protected (Fishpool and Evans 2001; essay 7.2).

TRANSBOUNDARY PROTECTED AREAS

Areas along national borders tend to be politically and economically neglected and have few people. They often contain valuable intact habitats, so there has been an increasing emphasis on transboundary protected areas in Africa (Cumming 1999; Sandwith et al. 2001; Van der Linde et al. 2001; http://www.peaceparks.org/; essay 7.3). Because borders often are areas of conflict in Africa, transboundary parks have come to symbolize opportunities to enhance peace and cooperation between nations (IUCN and WCPA 1997; Sandwith et al. 2001).

In southern Africa, seven large areas covering more than 200,000 km² are being promoted as ambitious transboundary conservation areas. These are the |Ai-|Ais/Richtersveld Transfrontier Conservation Park between South Africa and Namibia (5,921 km²), the Great Limpopo Transfrontier Park between South Africa, Mozambique, and Zimbabwe (35,000 km²), the Kgalagadi Transfrontier Park between Botswana and South Africa (37,991 km²), the Limpopo/Shashe Transfrontier Conservation Area between Botswana, South Africa, and Zimbabwe (4,872 km²), the Lubombo Transfrontier Conservation Area between South Africa, Swaziland, and Mozambique (4,195 km²), the Maloti-Drakensberg Transfrontier Conservation and Development Area between South Africa and Lesotho (13,000 km²), and an area from Lake Malawi/Nyasa to the Indian Ocean (100,000 km²) (http://www.peaceparks.org).

Similar transfrontier priority landscapes have also been

identified for the forests of Central Africa, such as Sangha Trinational Park between Cameroon, the Central African Republic, and the Republic of the Congo and the Transborder Conservation Initiative to link the forests of Dja (Cameroon), Minkebe (Gabon), and Odzala (Congo). Establishing these transfrontier reserves would result in significant progress in achieving the Congo Basin biodiversity vision (Kamdem-Toham et al. 2003b).

LAND PURCHASE

In areas where it is possible to purchase land for conservation, this is a secure method for conserving important landscapes or sites (Stein et al. 2000). Land purchase for conservation is common in the savanna-woodland habitats of southern Africa. This is because the land laws in southern Africa allow freehold purchase of land, whereas throughout tropical Africa land is either communally owned or vested with the state. Changes in the land laws of many eastern, central, and western African nations are now making land purchase possible (Wily and Mbaya 2001). For example, targeted land purchase might achieve much for the conservation of the small endemic-rich forest patches in coastal eastern Africa as identified in the recent WWF Ecoregion Action Program (Younge et al. 2002). Here, a priority area of biodiversity in the former Mkwaja Ranch was recently purchased by an Italian NGO and then donated to the Tanzanian government as an addition to the planned Sadaani National Park. Future land purchases of this type may become an important conservation strategy in tropical Africa.

CONSERVATION CONCESSIONS

In many African countries the government allocates areas of land (called concessions) to a company (generally foreign) that wants to use that area of land for commercial purposes. Two forms of concessions might be used to assist the conservation of landscapes and sites identified as priorities in a biodiversity vision.

Forest Concessions. Forest concessions are sold to logging companies for the purpose of logging the trees and selling them for profit. These concessions can cover huge areas and may operate for decades. In the Congo Basin, European companies hold concessions covering more than 11 million ha (Forests Monitor 2001), or 5.5 percent of the remaining forest. To achieve conservation across large areas of forest, conservation organizations have begun to bid on logging concessions, thereby preventing the felling of timber. Successful examples already exist in the Congo Basin, where such actions support the goals of the Congo Basin biodiversity vi-

sion (Elkan et al. 2002; Kamdem-Toham et al. 2003b). Although these concessions might be ecologically less pristine than core habitat, they can still function as habitat corridors for wide-ranging species such as the forest elephant.

Hunting Concessions. Hunting concessions are common in countries where hunting for sport, trophies, or subsistence is important. Companies or private individuals buy concessions and then sell rights to hunt animals to tourists (Leader-Williams et al. 1996; Hulme and Murphree 1999; Metcalfe 1999). Along similar lines are community wildlife harvesting strategies whereby communities own the rights to harvest certain numbers of animals for food or sell those rights to others, and in return they conserve habitats and the species that live there (CAMPFIRE 1997; Arnold 1998; McKean 2000; Roe and Jack 2001; Eves and Ruggerio 2002). Both of these approaches offer opportunities to conserve large landscapes with intact biodiversity values, such as are found in the miombo ecoregion complex of southern and eastern Africa (Gumbo et al. 2003). Here, huge areas of natural habitat are needed to maintain ecological processes and migrations of large megaherbivores such as elephants.

Hunting concessions are an important way to give natural habitats and their species a value to local populations, which makes it much more likely that they will survive over the long term. Hunting in game reserves and in the village lands surrounding them has also provided good conservation and revenue sources in some countries, such as in Tanzania (Leader-Williams et al. 1996; Leader-Williams and Dublin 2000).

COMMUNITY MANAGEMENT

There is much debate about whether local communities act as forces for conservation or threats to it. A variety of publications advocate the involvement of local people in the management of habitats, species, and protected areas (e.g., Leach and Mearns 1996; Borrini-Feyerabend and Buchan 1997; Hackel 1999; Hulme and Murphree 1999). Other publications focus on the need to maintain areas where people are excluded and where a centralized management focuses on the maintenance of biodiversity values (e.g., Kramer et al. 1997; Spinage 1996; 1998; Oates 1995, 1999; Attwell and Cotterill 2000; Bruner et al. 2001). Others have tried to understand where community management has worked and what the lessons are for conservation (e.g., Sanjayan et al. 1997; Larson et al. 1998; Wells et al. 1999; Newmark and Hough 2000; Adams and Hulme 2001).

Although this kind of debate is useful, it has often become polarized, caught up in political issues surrounding land ownership, colonial history, and the respective roles of elites and poor rural people. Because the people of Africa retain the closest association with natural resources of any continent globally, the community conservation paradigm is a potentially powerful way to achieve a biodiversity vision across large geographic regions. Here we outline some of the relevant issues.

Traditional societies often protect areas of natural habitat (Schrijver 1997; Berkes 1999; Posey 1999), and some traditional societies need large expanses of natural habitat to maintain their lifestyles. Examples include the pygmies of the Congo Basin (Stiles 1994; Gibson and Marke 1995), the Masai of eastern Africa (Homewood and Rodgers 1991), and some of the other nomadic and pastoral peoples across the drier areas of Africa (WWF and Terralingua 2000). Where the lifestyles of people coincide with conservation ideals for the management of large areas of natural habitat, maintaining lifestyles could help achieve the goals of a biodiversity vision.

The linkage of conservation and development ideals in a single project has become one of the most discussed ideas in conservation over the past 20 years. In particular, in poor rural Africa development agencies and their funding partners have warmly received project concepts that might achieve the double goal of wildlife conservation and improved human livelihoods. One of the main mechanisms for this purpose has been the integrated conservation and development project (ICDP). Hundreds of these projects have operated across Africa in the past two decades (Barrett and Arcese 1995; Pimbert and Pretty 1995; Alpert 1996; Newmark and Hough 2000; Hughes and Flintan 2001). They have met with mixed success and generally delivered less conservation and development than was originally expected (Oates 1999; Ite and Adams 2000; Dubois and Lowore 2000; Hughes and Flintan 2001; Jeanrenaud 2002). A particular criticism has been that they are an indirect way of achieving conservation (or development) and that it might be better simply to go for the most direct routes available—for example, by paying people to deliver certain types of conservation (Ferraro and Kiss 2002). However, development assistance agencies are increasing their emphasis on poverty alleviation, and many support conservation-related activities only if human development issues are included. Instead of advocating the cessation of ICDPs, we believe it is more productive to participate in the debate on how they can be better designed. Specifically, ICDPs should never be viewed as a replacement for strict nature reserves but rather as important components of a larger conservation landscape that contains strictly conserved areas and areas used by people. Essay 7.4 outlines how one organization is scaling up from an individual site-based ICDP to a series of linked interventions across a large landscape on the Bamenda Highlands, which covers a significant part of the Cameroon Highlands Forests [10] ecoregion.

Changes in the laws of a number of African countries allow villages a much greater opportunity to manage their own lands, including the development of local protected areas. Both village forest reserves and community-based wildlife management areas are now being promoted as appropriate conservation areas in many countries (Hasler 1999; IUCN Regional Office for Southern Africa and SASUSG 2000; Roe and Jack 2001). The approaches pioneered in southern Africa are now being tested in eastern (Barrow et al. 2000) and West and Central Africa (Abbot et al. 2000). Locally managed protected areas can greatly contribute to the maintenance of landscapes and sites across ecoregions and the species that live in them.

RESTORATION

Habitats across most of Africa are in better condition than those in Europe or North America, where restoration has become an important aspect of conservation action. However, in some areas habitat and species restoration is needed to implement biodiversity visions. For example, in the Fynbos of South Africa, the invasion of native habitats by foreign plant species seriously reduces overall biodiversity (CSIR 2000a, 2000b). Ambitious programs to eradicate the invasive plants have been implemented and are showing success. The program was motivated by human self-interest: the preservation of water supplies and the useful employment of large numbers of unemployed people. Other African examples include the restoration of appropriate fire-burning regimes in different types of savanna-woodland ecoregions, the restoration of sensitive montane heathland habitats after tourist damage on Kilimanjaro, and the restoration of mangrove habitats by replanting exercises. Although some of these interventions affect only small areas of ecoregions, others can have much broader significance.

Another class of restoration project is found on offshore islands such as the Mascarenes. These programs have aimed to prevent the extinction of plant and animal species through captive breeding, removal of predators and alien plants, and replanting or release of species (Jones and Hartley 1995; Jones et al. 1995). Despite immense challenges, some of these programs have successfully restored areas of native habitat and prevented rare species from going extinct. Some of the ecoregions placed in the critical category of threat by our analysis may need such intense levels of intervention.

Policy and Legal Mechanisms

Besides conserving landscapes and sites, we see policy and legal mechanisms as important strategies for achieving our biodiversity goals in Africa.

A FRAMEWORK FOR EXISTING INTERVENTIONS

Biodiversity visions help conservationists working in a region to recognize their individual contributions to the broader goal of conserving the species and nonspecies biological values of a particular ecoregion. Such an approach underpins the Cape Action Plan in South Africa, the Congo Basin Partnership in Central Africa, and the Spiny Forest Program in Madagascar. These plans have been successful in bringing together a wide range of different agencies working at a variety of spatial scales. This model is also proposed for a forthcoming GEF project in the Eastern Arc Forests [19], the Albertine Rift Montane Forests [17], and the Northern and Southern Zanzibar-Inhambane Coastal Forest Mosaics [20, 21]. In all three areas existing projects, government national parks and forest reserves, and development interventions will be linked through the development of a common conservation strategy that aims to ensure the long-term persistence of the biodiversity (and water, electricity, and natural resource supply) functions of these forests.

A LINK TO INTERNATIONAL CONVENTIONS AND TREATIES

Several international conventions and government agreements address the conservation of species, habitats, and ecological processes (table 7.1). A biodiversity vision for an ecoregion provides an entry point for discussions with international conventions on how the vision might help them achieve their own targets. For example, the Ramsar Convention encourages governments to designate important wetlands as Ramsar sites. If a suitable biodiversity vision were developed for the Zambezian Flooded Grasslands [63] ecoregion, this convention might provide the mechanism to connect with relevant governments to achieve that vision. There would be a synergy between the goals of the biodiversity vision and that of the Ramsar Convention, to which most governments have signed.

Some regional conventions and agreements in Africa affecting the environment include river basin management (e.g., Zambezi River System Agreement), conservation of the East African marine and coastal environment (Nairobi Convention), and Congo Basin forest conservation and management (Yaoundé Declaration). Biodiversity visions for these ecoregions can provide the framework for the work of

the convention or agreement to tackle conservation issues in their region of interest. The success of such an approach is outlined earlier for the Congo Basin forests. For details on other African conventions, see "Beyond Boundaries: Transboundary Natural Resource Management in Sub-Saharan Africa" on the Biodiversity Support Program Web site (http://www.bsponline.org).

A LINK TO NATIONAL POLICIES

Linking a biodiversity vision to the existing frameworks of policies and laws of existing nations increases its acceptance by government bodies. For example, incorporating the ecoregion actions in national forest policies, national forest programs, or other suitable national policies (Government of Tanzania 1998, 2001; Perrings and Lovett 2000) would institutionalize the program at the national level and this would facilitate implementation in forest ecoregions. Likewise, African governments have programs for wetlands, mangroves, and water conservation, increasing the opportunities to align ecoregional conservation with existing national processes.

The process of building the biodiversity vision into relevant national processes and laws is made easier when the ecoregion falls entirely within a single country (e.g., the Cape Fynbos of South Africa [89, 90]), where there is one government, one set of national policies, and one set of stakeholders. For ecoregions that cross one or more national borders, the situation becomes more complex. For example, the Northwestern Congolian Lowland Forests [12] ecoregion spans four different countries. Each has different sets of forest policies, laws, ethnic and colonial histories, and current attitudes toward conservation, and these considerations must be taken into account. Ownership of the vision and the process by the countries concerned is much more likely to ensure that it becomes part of their own development plans.

Funding a Biodiversity Vision

The cost of achieving the biodiversity vision for an ecoregion might seem expensive, certainly in comparison to enacting conservation at a few sites. However, if we wait 30 years before starting, it will be much more expensive. A good example is the globally important Everglades ecoregion in the United States. Had legislators enacted the restoration plan proposed by biologists three decades ago, the cost of revitalizing this area would have been $300 million. Today, the price tag for restoring about 11,000 km², or roughly half of the original ecoregion, exceeds $8 billion. When viewed

TABLE 7.1. International Conventions and Agreements as Tools for Implementing Biodiversity Visions in African Ecoregions.

Convention Name	Convention Scope
The Convention on Biological Diversity (1992)	The Convention on Biological Diversity's objectives are "the conservation of biological diversity, the sustainable use of its components and the fair and equitable sharing of the benefits arising out of the utilization of genetic resources."
The Ramsar Convention on Wetlands (1971)	Mission statement: "The Convention's mission is the conservation and wise use of wetlands by national action and international cooperation as a means to achieving sustainable development throughout the world."
The Convention on Migratory Species (Bonn Convention, 1979)	"Conservation and effective management of migratory species of wild animals require the concerted action of all States within the national jurisdictional boundaries of which such species spend any part of their life cycle."
The UN Framework Convention on Climate Change (1992)	This convention acknowledges that human activities have been substantially increasing the atmospheric concentrations of greenhouse gases, that these increases enhance the natural greenhouse effect, and that this will result on average in an additional warming of the earth's surface and atmosphere and may adversely affect natural ecosystems and humankind.
The World Heritage Convention (1972)	The United Nations Educational, Scientific and Cultural Organization (UNESCO) seeks to encourage the identification, protection and preservation of cultural and natural heritage around the world considered to be of outstanding value to humanity. This is embodied in an international treaty called the Convention Concerning the Protection of the World Cultural and Natural Heritage, adopted by UNESCO in 1972.
The UN Convention to Combat Desertification (1994)	A convention that was drawn up at the 1992 Rio Earth Summit. Signatory countries are required to undertake activities in arid or semi-arid areas that contribute to the sustainable use of those habitats and do not lead to further desertification.
The Convention on International Trade in Endangered Species (CITES, 1973)	The international wildlife trade, worth billions of dollars annually, has caused drastic declines in the numbers of many species of animals and plants. The scale of overexploitation for trade aroused such concern for the survival of species that an international treaty was drawn up in 1973 to protect wildlife against such overexploitation and to prevent international trade from threatening species with extinction.
International Tropical Timber Agreement (1983, revised 1994)	This agreement covers industrial tropical timber reforestation, forest management, sustainable use, and forest conservation policies. The 1994 revision incorporates sustainable development principles from the Forest Principles agreed at the UN Conference on Environment and Development in Rio de Janeiro in 1992.
General Agreement on Tariffs and Trade (GATT) (1947 and 1994)	GATT is administered through the World Trade Organization and aims to facilitate the progressive liberalization of world trade by reducing barriers to trade that take the form of tariffs on imported products. Because certain international environmental agreements and national environmental measures restrict trade in order to promote environmental objectives, there are strong debates about clashes of GATT with international environmental agreements and the consequent environmental impacts of GATT.
Africa Convention for the Conservation of Nature and Natural Resources (1968)	The convention focuses on sustainable use and conservation of soil, water, flora and fauna, and cooperation over the management of transboundary natural resources. Its secretariat is the Organization of African Unity.
Zambezi River System Agreement	This agreement establishes an action plan for the environmentally sound management of the Zambezi River System, covering eight countries in southern Africa.

from this perspective, financing a biodiversity vision, given all of the ecological services provided via conservation, may be a bargain we cannot ignore. Fortunately, a number of different sources are being tapped or designed to provide funding for large-scale conservation.

TAXATION SYSTEMS

Tax breaks and levies provide an important potential national mechanism to finance the implementation of a biodiversity vision. For example, the Eastern Arc Forests [19] and the East African Montane Forests [18] ecoregions are the sources of rivers that provide water to large towns and industry in the region. About 60 percent of the water supply in Tanzania comes from these forests, and some of the major rivers have hydroelectric schemes that generate around 50 percent of the countries' electricity. A tax levy on water users reinvested to the management of the forests and the social development of the people living around them would provide sustainable forest conservation in these ecoregions in perpetuity. The conservation of global biodiversity would be a side benefit of maintaining the forests for their water functions. Similar schemes operate successfully in Costa Rica and have helped reforest large areas of that country (Castro et al. 1998). A taxation system is also proposed as a mechanism to finance parts of the CAPE strategy in South Africa (Pence 2003).

TRUST FUNDS AND RELATED OPPORTUNITIES

Global financial markets may also provide a mechanism to generate funds from investments that allow conservation actions identified by the vision to continue in perpetuity (box 7.4). Such trust funds can be established at the scale of ecoregions (e.g., the Fynbos and the Eastern Arc Forests), landscapes (e.g., Bwindi Forest in Uganda), or sites (e.g., Gombe Stream in Tanzania) (GEF 1999). Donors are sought to contribute to the fund, which uses its assets to manage natural resources over the longer term (IPG 2000; Phillips 2000; Spergel 2001, 2002). Trust funds take a number of economic forms. The most common type is an endowment fund, in which the capital is invested and the interest (usually 5–10 percent annually) is used to fund conservation activities. Another popular model is the revolving fund, in which funds collected are spent each year, as when tourists pay a fee or tax to fund conservation activities.

The best-known trust fund in Africa is the Mgahinga and Bwindi Impenetrable Forest Conservation Trust. This fund was legally established in 1993, and the first $4.3 million was provided by GEF in 1994. Further contributions from

---

BOX 7.4. **Environmental (Conservation) Trust Funds.**

A trust fund can be broadly defined as money or other assets that can be used only for a specific objective or objectives, must be maintained separate from other financing sources, and is managed and controlled by an independent board of directors. Environmental trust funds have been established in more than forty countries around the world to provide long-term sustainable financing for the environment, including protected areas, species conservation, and grant funding for nongovernment organizations and community groups (Moye and Carr-Dirick 2002).

—*Melissa Moye and Brigitte Carr-Dirick*

---

the U.S. Agency for International Development and the Dutch government pushed the fund to more than $8.3 million by year 2000. The program provides funds for conservation management of the reserves and for local development projects. Support for local development has been critical to winning the support of people living around the protected area and the species of conservation concern it contains, especially the mountain gorilla (Hamilton et al. 2000).

Debt-for-nature swaps (box 7.5) also provide opportunities for countries and conservationists to link large-scale environmental gain to a reduction in their external debts (Kaiser and Lambert 1996; Restor 1997; Moye 2000). These are carried out when a country has a debt that it cannot finance. An external body purchases the debt in hard currency and then negotiates with the government to redeem the debt in local currency, which is then used for nature conservation. Although debts purchased around the world by international environmental organizations have reached $46.3 million and leveraged $128.7 million from debtor countries in local currency, the approach has not been used much in Africa outside Madagascar.

BILATERAL AID AND GOVERNMENT FINANCING

In Africa huge financial resources are invested every year by development assistance agencies from richer countries. Portions of these funds are available for environment and development projects, many of which are implementing component parts of biodiversity visions. Working with development agencies during the formulation of a vision and afterwards provides opportunities to realize the funds needed for conservation activities. Besides government donors, private foundations may also provide important funding opportunities.

Much of the current funding from governments and foundations focuses on poverty alleviation. Africa is the world's poorest continent, so the drive for poverty alleviation is most intense here. Therefore, it is no surprise that the World Summit on Sustainable Development (WSSD), a meeting 10 years after that in Rio de Janeiro launched the Convention on Biological Diversity, took place in Africa in 2002. Among the conclusions of the meeting were that Africa should be the focus of much development assistance in coming years and that the link between the conservation of natural resources and economic development in Africa is a close one. Natural resources will continue to drive economies in Africa, and the sustainable use of these resources therefore is of concern to all. A potentially important regional initiative was announced at the WSSD: the New Partnership for Africa's Development (NEPAD). This is a commitment by African leaders to the people of Africa. It provides a framework for sustainable development on the continent to be shared by all Africans. The international community has welcomed NEPAD and has pledged support (including significant financial support). Although also focused on poverty alleviation, NEPAD may provide significant opportunities for achieving biodiversity visions across broad parts of Africa.

Structural Adjustment Programs (SAPs) promoted by the International Monetary Fund (IMF) and World Bank are mechanisms that aim to assist the development of countries with a 1997 per capita gross domestic product of $925 or less (mainly countries in Africa). These programs have been heavily criticized by many environmentalists because they have caused a reduction in government budgets and capacity to manage natural resources across many African nations, particularly by dramatically reducing the number of government-employed conservation officers (Ross 1996; Wood et al. 2000a). In general, the early examples also failed economically (World Bank 1994). These programs have been restructured to become the Enhanced Structural Adjustment Facility (ESAF) (IMF 1999). By 1999 some $9 billion had been dispersed through SAPs and ESAF programs. Environment is not a core issue generally supported in an ESAF. However, as these programs are operationalized across Africa there are opportunities to incorporate environmental concerns, such as those outlined in large-scale ecoregion action plans, into the process.

Similarly, the Heavily Indebted Poor Countries (HIPC) Program is a multilateral debt forgiveness program that was

---

BOX 7.5 **Debt-for-Nature Swaps.**

A debt-for-nature swap involves the cancellation of external debt of a developing country in exchange for local currency funding for nature conservation and environmental protection in that country (Moye and Carr-Dirick 2002).
—*Melissa Moye and Brigitte Carr-Dirick*

BOX 7.6 **Heavily Indebted Poor Countries (HIPC) Initiative.**

Launched in 1996 and revised in 1999, the HIPC Initiative is an agreement by the international community to help poor countries with good policy performance to escape from unsustainable debt burdens by providing comprehensive debt relief. To qualify for debt relief, HIPC countries must adopt a poverty reduction strategy paper (PRSP) through a broad-based participatory process. PRSPs provide a framework for allocating debt relief savings generated by HIPC in the form of government budgetary resources and donor assistance. Although the initial focus of PRSPs was on the education and health sectors, there is growing recognition that the environment must be better integrated into PRSPs (Moye and Carr-Dirick 2002).
—*Melissa Moye and Brigitte Carr-Dirick*

---

initiated by the World Bank and IMF to cover more than thirty of the world's poorest countries (mainly in Africa) (box 7.6). HIPC was started in 1996 and was reformed in 1999. Although HIPC stresses the need for countries to address links between debt relief and the root causes of debt and poverty, few existing programs target environmental degradation as one of the causes of poverty. The exception in Africa is Madagascar, which has received significant funding for the environment through the HIPC process. Given that the total expected cost of the HIPC debt relief program is close to $55 billion (http://www.worldbank.org/hipc), working to incorporate environmental and conservation issues into the national poverty reduction strategy paper for the program could create major funding opportunities.

Another related potential funding approach relevant for forest ecoregions is that of carbon credits. The UN Framework Convention on Climate Change and the related Kyoto Protocol contain a clean development mechanism (box 7.7) that may provide a way to finance some forms of conservation. Huge plantations of trees are envisaged in some parts of the world as industrialized countries allow their own CO_2 emissions to continue by purchasing carbon credits in the form of tree plantations elsewhere in the world (Sandalow and Bowles 2001). This mechanism could also be operated like the concession system and might facilitate the conservation of large tracts of natural forest.

Finally, perhaps the least exploited source of funds for African conservation is the private sector. We expect that in certain ecoregions, such as in the Cape Fynbos, private sector financing will increase dramatically in the years to come.

Looking to the Future

Several million years ago, our human ancestors were endemic to a few African ecoregions. Whether it was the Sahelian *Acacia* Savanna [35], as recent discoveries in Chad might suggest, or in the *Acacia-Commiphora* Bushlands and Thickets of East Africa [44–46], the first humans exerted a tiny ecological footprint on the natural habitats of the continent. The dramatic increase in human population and its ability to dominate large parts of the world has changed this. In 2003, the human footprint is almost everywhere in Africa save for some of the most remote reaches of the Congo Basin (Sanderson et al. 2002a). A large fraction of these people are very poor and rely on natural resources to survive, including those from regions of exceptional importance for biodiversity conservation.

Using the model of the recent assessment of the ecoregions of Asia (Wikramanayake et al. 2002), we have asked ourselves at the end of this assessment how we see the future of biodiversity in Africa. Three scenarios can be envisaged, the first in which current declines in habitat area, quality, and species range and abundance continue to accelerate. In the second scenario, existing conservation efforts and development efforts of the nations of Africa produce some success, and the habitat and species attributes stabilize. In the third scenario conservation works for the benefit of nature and human development, and ecological development supports the economies of African nations into the future.

Scenario 1: catastrophic. If habitat loss (driven largely by population growth and agricultural expansion and aggravated by severe impacts of climate change) accelerates across Africa, large parts of the remaining areas of habitat (including those in protected areas) may be converted for human use. This will cause the loss of endemic species and biological phenomena such as large-scale animal migrations. The process will start first in regions of very high rural population pressure and around large towns; indeed, it has already started in some parts of Africa (e.g., Nigeria and parts of Kenya). As habitats are lost and decline to patches, the well-proven species-area relationship will also come into play, and first the largest mammal species and then ever-smaller species will start to decline in numbers and become extinct from the smaller habitat patches. If the species is found only in a few locations, the loss of area of its preferred habitat eventually will be enough to cause its extinction. Fragmentation of habitats will also make hunting of bushmeat easier, causing further declines in mammal numbers. Eventually, a situation close to that of Western Europe might prevail, where large-bodied species have been removed from everywhere except the most remote regions, many other species are reduced to remnant populations, and a few generalist species dominate over vast areas.

The trajectory for human development under this scenario cannot be easily predicted. In North America and Europe such an approach has created the conditions for wealth, but in Africa it may not be as favorable because the combination of poor soils, human disease, erratic rainfall, and low capacity makes the generation of wealth more difficult. In Africa, perhaps more than in any other continent, the loss of ecological services and the natural resource base on which millions of human livelihoods still depend entirely may result in increased livelihood insecurity, increased poverty, and, in some cases, civil unrest. Certainly the rapid increase of human population in Africa over the past decades has not been paralleled by increased living standards, which most cases indicate are declining (UNDP 1999; chapter 1).

Scenario 2: status quo. Under this scenario, the existing nationally protected areas of Africa are maintained, and the current areas of community-managed habitats are expanded. The conservation of natural habitats in these areas is expected to contribute to national development through ecotourism, the provision of essential services to the economy (e.g., water), and sustainable production of wild meat and wood for domestic use. Eventually many of the regions outside the parks and community-managed areas will be converted to farmland, and the isolation of some of the reserves will cause the local loss of large mammal species. Increased pressure on protected areas will result in degradation and in some cases loss. Large mammal migrations will decline almost to extinction. Poorly designed reserve networks will result in the extinction of some species in regions that are not already protected. The development scenario for humans is difficult to predict, but given current patterns there would be a general trend toward urbanization, and if trade barriers around the world are lowered and debt burdens are eased, then there might be moderate increases in African wealth.

Scenario 3: optimistic. Under this scenario reserve networks (managed by both government and local communities) will be designed collaboratively with relevant bodies who see these areas contributing to their livelihood and economic development. Both protective and extractive reserves will be set up and provide sustainable benefits to human populations. The value of wild animals as a food source will be recognized, and game ranching will become widespread and profitable, safeguarding the species and habitat diversity across large regions of Africa. Improved agricultural systems and better management practices also will allow an increase in food productivity without expanding to new land. The movement of people to urban areas will accelerate so that some regions become depopulated, promoting their use for wild animal ranching, tourism, and extensive production of valuable native timber. The peoples of Africa will become sufficiently wealthy that most poverty-related diseases will be eradicated, and active local tourism will develop, with both Africans and foreign tourists visiting the parks to enjoy African wilderness areas and their species. Economies will be changed from those that use natural resources to those with a more diverse economic structure.

We hope that the governments of the region will make decisions and commitments to embark on the road to the third scenario and that human development in Africa will follow a path that maintains natural values, in the best of African traditions. Twenty years from now we hope that the forests of West Africa and the Congo Basin contain representative networks of protected areas in a matrix of well-managed forest concessions providing habitat for species and wealth for the nations of the region. We also hope that the savanna-woodland habitats continue to provide wildlife spectacles in their national parks while the matrix between the parks contains community and privately managed game conservancies where habitats and their species survive and native animals provide income-generating opportunities to local populations. In the mountains of Africa we hope that the networks of forest reserves and community reserves can be maintained to continue to provide water flows to downstream users while providing critical habitats to the large numbers of endemic species. In the Mediterranean habitats, intense human pressures necessitate especially intense conservation efforts, particularly related to good design of reserve networks at the scale useful to the endemic species of these regions, which often need only tiny patches of land. Much progress toward this end has been made in South Africa, but similar programs are now needed in the Mediterranean regions of North Africa (especially the northwest). Across the deserts of Africa, the overhunting of large antelope species must be brought under control, or more of these species will become extinct in coming years. The next 50 years will reveal whether our hopes bear fruit. Africa, and the world, will be greatly diminished if we fail to leave room for the habitats, species, and biological processes so characteristic of wild Africa in this developing continent.

ESSAY 7.1

Planning for Persistence: Designing Conservation Areas for Diversification Processes in the Species-Rich Winter Rainfall Ecoregions of Southern Africa

Richard Cowling, Bob Pressey, and Mandy Lombard

The Cape Floral Kingdom (Mediterranean scrub) and the Succulent Karoo (desert) are two winter rainfall ecoregions located in Africa's southwestern tip. Both regions are extremely rich in endemic plant taxa and have experienced rapid and recent (post-Pliocene) ecological diversification of many plant lineages; there are numerous genera with large clusters of closely related species (flocks) that have subdivided habitats at a very fine scale. Although they occupy only 186,000 km^2, these two ecoregions are home to some 12,000 plant species, of which 80 percent are endemic.

The existing reserve systems in both ecoregions are grossly inadequate for achieving realistic targets for conserving their biodiversity. Although about 11 percent of the Cape Floral Kingdom is conserved in statutory reserves (state nature reserves and national parks), this system is strongly biased in favor of mountain habitats. Therefore, 26 percent of the mountainous area is conserved, compared with only 3 percent of the severely transformed and highly threatened lowlands (Cowling et al. 1999a). Moreover, of the eighty-eight broad habitat units (surrogates for vegetation types), only seventeen are reserved at levels that match or exceed reservation targets (set as hectares of intact habitat) proposed in a current study of conservation requirements (Cowling et al. 1999a). The existing protected area system in the Succulent Karoo is similarly inadequate. Only 2.1 percent of the Succulent Karoo is conserved in six statutory reserves (Lombard et al. 1999). Larger reserves (more than 10,000 ha) occur in only four of the Succulent Karoo's twelve bioregions.

Clearly, the reserve system in both ecoregions must be expanded in a systematic way to fulfill biodiversity representation targets (e.g., species, broad habitat units). However, the ultimate goal of conservation planning should be to design systems that enable biodiversity to persist in the face of natural and human-induced change. Therefore, the expansion of the reserve system must also accommodate the processes, both ecological and evolutionary, that sustain and generate biodiversity (e.g., migration of ungulates, maintenance of viable populations of large herbivores). Few if any of the existing reserves in the Cape Floral Kingdom and the Succulent Karoo effectively accommodate these processes. This essay outlines an approach that we are developing to design reserves that will conserve the evolutionary processes that have generated and will continue to generate biodiversity in this part of Africa.

Because conservation planning is a spatial exercise, the first stage in planning is to identify the spatial components of the evolutionary processes that need conservation. Examples are habitat gradients or geographic barriers that are associated with lineage turnover. On the basis of our present understanding of diversification processes in the Cape Floral Kingdom and Succulent Karoo, we identified many such spatial components that must be protected to maintain evolutionary processes. As examples of our approach, table 7.2 lists the spatial components of evolutionary processes in the Cape Floral Kingdom, and figure 7.5 shows their spatial location in the Succulent Karoo.

The second stage is to set explicit conservation targets for each of these components (see table 7.2 for the Cape Floral Kingdom). In the larger conservation plan for this ecoregion, now under way, these are combined with targets for representing biodiversity patterns and for maintaining various ecological processes.

The third stage is to design a system of conservation areas by selecting from areas that can potentially fulfill targets for the spatial components in table 7.2. Invariably, the options associated with area selection are constrained by several factors, including the need to incorporate the existing reserve system, to avoid excessively transformed areas, and to select, where possible, areas that also contribute to targets for biodiversity patterns and ecological processes. In many instances, such as in the lowland regions of the Cape Floral Kingdom, habitats are so extensively transformed that it is no longer possible to achieve desirable targets for species diversification; the evolutionary future of this ecoregion has already been severely compromised.

Our method for selecting areas for conservation is based largely on the idea of irreplaceability. This is an indication, now

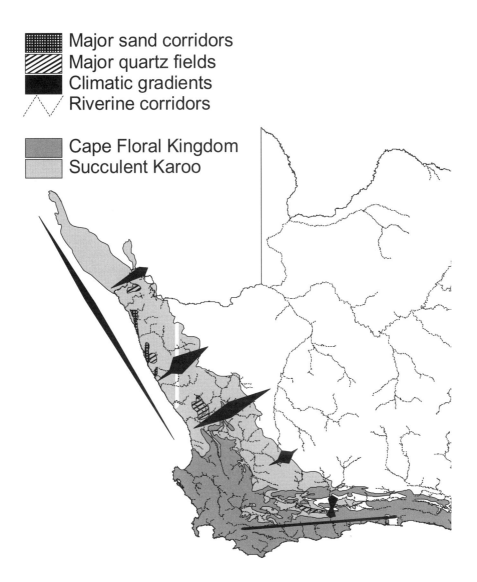

Major sand corridors
Major quartz fields
Climatic gradients
Riverine corridors

Cape Floral Kingdom
Succulent Karoo

FIGURE 7.5. Location of spatial components in the Succulent Karoo [110] required to sustain evolutionary processes. Quartz fields are associated with recent diversification of numerous succulent and bulb lineages and are recognized centers of plant endemism (Cowling et al. 1999b). Other components sustain evolutionary processes similar to those described for the Cape Floristic Kingdom in table 7.2. Thickness of lines indicating climatic gradients is proportional

implemented quantitatively in a software system called C-Plan (Pressey 1998), of the likelihood of any particular area being needed to achieve the conservation targets set for a region. Some areas are totally irreplaceable: if they are not given adequate protection, one or more targets will not be achieved. Other areas are replaceable to varying degrees; some have a few possible replacements, others many. A map of irreplaceability for pattern targets (e.g., hectares of each broad habitat unit) therefore is a map of the options available for achieving those targets, shown for each individual area within the region (e.g.,

watersheds, habitat fragments, tenure parcels). Measuring and mapping irreplaceability for pattern targets is not new (Ferrier et al. 2000; Margules and Pressey 2000), but doing the same for the types of process targets in table 7.2 is a new challenge for conservation planning. Therefore, part of our current work on the Cape Floral Kingdom involves some new research and development to meet this challenge. Once we have the analytical capabilities to map irreplaceability for each aspect of patterns and natural processes, we will develop and apply a protocol that combines these layers of planning options sequentially.

TABLE 7.2. **Spatial Components of Evolutionary Processes in the Cape Floral Kingdom (CFK).**

Spatial Component	Method of Identification	Target	Key Evolutionary Process Conserved
Juxtaposed edaphically different habitats	Planning units with particular combinations of broad habitat units (surrogates for vegetation types) that reflect strong edaphic contrasts (limestone and adjacent acidic substrata) known to be associated with plant diversification processes. Exclude unsuitable planning units based on fragmentation of native vegetation and lack of contiguity with other units.	At least one example of each specified combination of broad habitat units in each major climatic zone.	Ecological diversification of plant lineages in relation to fine-scale edaphic gradients
Entire sand movement corridors	Planning units containing three specific dune pioneer habitats. Exclude any corridors (sediment sources) with limited conservation potential of surrounding land (particularly in the sediment sink or downwind zones). Assume that stands of dense alien plants make corridors irrecoverable.	At least one entire corridor of each type.	Ecological diversification of plant lineages in relation to fine-scale edaphic gradients
Whole riverine corridors	Major rivers that link inland basins with coastal plains. Include untransformed corridors or parts of corridors.	All of any intact riverine corridor or the untransformed parts of other major corridors (five river systems, ten river corridors).	Migration and exchange between inland and coastal biotas
Gradients from uplands to coastal lowlands and interior basins	Planning units on the following interfaces of upland and lowland • coastal range–coastal plain • coastal range–interior basin • inland range–interior basin • inland range–Karoo basin that would allow construction of corridors between these landscapes.	At least one example of each gradient in each of the major climate zones. Gradient width should encompass at least one untransformed planning unit and maximize climatic heterogeneity.	Ecological diversification of plant and animal lineages in relation to steep environmental gradients

Macro-scale climatic gradients	Opportunities to complement gradients between lowlands and uplands (meso-scale) with macro-scale connectivity in two main directions: • north-south in the western CFK along the coastal forelands and inland mountains • east-west in the southern and eastern CFK along coastal forelands, coastal mountains, interior basins, and interior mountains.	Unbroken transects along all geographic gradients.	Geographic diversification of plant and animal lineages in relation to macroclimatic gradients
Mega-wilderness areas	Contiguous planning units that encompass ca 500,000 ha of untransformed habitat, transcend biome boundaries, and include all or part of a riverine corridor.	One in the northwestern, one in the southern, and one in the southeastern CFK.	Maintenance of all evolutionary processes, including predator-prey processes involving top predators
Transitions between major broad habitat unit categories and biome boundaries (see Cowling et al. 1999b)	Where possible, expand conservation areas to encompass these transitions.	As many transitions as possible.	Exchange between phylogenetically distinct biotas

The components are identified geographically and given quantitative targets for conservation planning. The term *planning units* refers to areas used in our current planning exercise as the preliminary building blocks of an expanded system of conservation areas. They are 16th degree grid cells, each covering about 4,000 ha. About 2,510 planning units cover the whole kingdom. Adapted from Cowling and Pressey (2001).

The fourth stage of our planning approach is implementation: the scheduling of conservation action, entailing choices in both space and time. In principle, priorities for implementation within a region should be guided by the irreplaceability of individual candidate areas and their vulnerability to threatening processes. In this way, scheduling of conservation action should minimize the extent to which conservation targets are compromised before areas are given adequate protection (Cowling et al. 1999b). However, when conservation targets deal with the representation of both patterns and processes, as is the case for our two ecoregions, there are no established ways to compare the relative risks of alternative approaches to implementation. For example, how should the outright loss of an extensively transformed and fragmented habitat be compared with the loss of a section of climatic gradient comprising adequately conserved habitat but essential for sustaining evolutionary processes? Resolving these conflicts is another major challenge for conservation planning and is another subject of ongoing research in our work in South Africa.

An important characteristic of the implementation stage is its duration. With limited conservation resources, full implementation of a regional conservation plan will take years, possibly decades. When Cape Nature Conservation and South African National Parks (also partners in the planning process) take over responsibility for implementing our conservation plan, managers will have to make day-to-day decisions about the fates of many candidate areas as pressures for development arise. These decisions will incrementally determine the fate of much of the biodiversity in the Cape Floral Kingdom and Succulent Karoo. We expect that information on the irreplaceabilities of candidate areas, for biodiversity patterns and processes, will help managers resolve conflicts between development and conservation, both by establishing the regional importance of particular areas and by indicating the availability of alternative areas.

Among the novel contributions of the conservation planning approach we are developing are the spatial identification of evolutionary drivers and the setting of explicit targets for these spatial components. Our study faces other challenges, particularly the difficult trade-offs between the representation of patterns and processes. In common with other conservation projects, it also faces trade-offs between biodiversity conservation needs and socioeconomic considerations. There are no easy answers for resolving these conflicts, nor can they be ignored.

We emphasize that the concepts and analytical techniques we are developing and applying in the Cape Floral Kingdom and Succulent Karoo are generally applicable. The big challenge for all ecoregions is to identify the spatial components of evolutionary processes and set targets for these. Biodiversity is being lost at an alarming rate in many parts of the world. The current focus on pattern representation in conservation planning will only temporarily slow the rate of extinction. Including explicit consideration of processes is vitally important in planning for evolutionary futures everywhere.

ESSAY 7.2

From Endemic Bird Areas to Important Bird Areas and Conservation Action at Sites across Africa

Lincoln Fishpool

BirdLife International's endemic bird area (EBA) analysis, begun in 1987, is a significant contribution to identifying priority areas for global biodiversity conservation using birds as indicators (ICBP 1992; Stattersfield et al. 1998). The analysis is based on the distributions of landbird species that have a total global breeding range estimated at less than 50,000 km^2. These birds are defined as being of restricted range. Places where the breeding ranges of two or more restricted-range species overlap are called EBAs, such that the boundary of the EBA entirely includes the complete breeding ranges of at least two restricted-range species. The number of restricted-range species confined to an individual EBA varies from two to eighty, and the size of EBAs may range from a few square kilometers (e.g., oceanic islands) to more than 600,000 km^2 that hold a large number of only partially overlapping species.

Some 2,561 species (more than 25 percent of all birds) have restricted ranges, and of these 2,451 are confined to one of 218 EBAs identified worldwide by Stattersfield et al. (1998). The remainder occur in secondary areas, generally places where only one restricted-range species occurs. Historically, some 20 percent of the world's birds were confined to EBAs whose area covered 2 percent of the earth's land surface. Today, as a result of habitat loss, the same 20 percent of birds are now limited to only 1 percent of the land. Some 50 percent of all restricted-range species are considered to be of global conservation concern.

Thirty-seven EBAs and twenty-seven secondary areas have been identified in Africa, holding 382 restricted-range species. Almost all of the islands around Africa are or contain EBAs, with five in Madagascar. On mainland Africa, the majority are located in the east and south of the continent, particularly in highland areas. The EBAs with the greatest numbers of restricted-range species are the Tanzania-Malawi Mountains (equivalent to three WWF ecoregions; thirty-seven species), the Albertine Rift Mountains (also thirty-seven species), and the Cameroon Mountains (two WWF ecoregions; twenty-nine species). However, the largest EBAs in Africa are the Upper Guinea forests (340,000 km^2) (three WWF ecoregions) and the Cameroon and Gabon lowlands (280,000 km^2) (three WWF ecoregions).

Full details of all EBAs, including descriptions, species lists, assessments of habitat types and loss, threats, and priorities for conservation action for EBAs, together with a discussion of congruence between EBAs and endemism in other groups, are given in Stattersfield et al. (1998). Many of the EBAs in Africa are also ranked as globally important ecoregions by WWF. Furthermore, it has been shown that, in addition to birds, EBAs capture the distributions of far more mammal, amphibian, and snake species than would be expected by chance, so they are a good overall conservation target for several groups of vertebrates (Burgess et al. 2002b).

Moving from Large-Scale Priority-Setting to Identifying Sites for Conservation Action

Many EBAs are too large for a site-based conservation approach. In addition, the majority of EBAs naturally contain areas of unsuitable habitat from which most or all restricted-range species are absent. Anthropogenic habitat modification means that this is true of much larger proportions of EBAs than was the case historically. Local conservation action therefore entails targeting representative key sites within the EBAs. BirdLife's Important Bird Areas (IBA) Programme provides the means of selecting priority sites within EBAs, and indeed elsewhere, for conservation action.

The purpose of this program is to conserve a global network of key sites for biodiversity. Put more formally, it seeks to identify, document, and promote the conservation and sustainable management of a network of sites that are important for the long-term viability of naturally occurring bird populations across the geographic range of bird species for which a site-based approach is appropriate. Sites are identified worldwide using standard, internationally recognized criteria, and the selection process ensures that, taken together, the sites form a network throughout the species' biogeographic distributions. The program aims to guide the implementation of national conservation strategies through the promotion and development of national protected area programs. It is also intended to assist the conservation activities of international organizations and to promote the implementation of global agreements and regional measures.

This global initiative began in Europe in the 1980s, and it has since made a significant contribution toward realizing a bird conservation strategy for the continent (Grimmett and Jones 1989; Heath and Evans 2000). More recently, an inventory of IBAs has been published for the African region (Fishpool and Evans 2001). To date, some 7,000 IBAs have been identified worldwide, a total that is expected to rise to 12,000 by 2004. The standard categories by which IBAs in Africa have been identified and that are being applied in other regions are as follows:

Globally threatened species: category A1. The site regularly holds significant numbers of a globally threatened species or other species of global conservation concern.

Restricted-range species: category A2. The site is known or thought to hold a significant component of a group of species whose breeding distributions define an EBA or a secondary area.

Biome-restricted assemblages: category A3. The site is known or thought to hold a significant component of the group of species whose distributions are largely or wholly confined to one biome.

Globally important congregations: category A4. The site may qualify on any one or more of the following criteria:

- The site is known or thought to hold, on a regular basis, 1 percent or more of a biogeographic population of a congregatory waterbird species.
- The site is known or thought to hold, on a regular basis, 1 percent or more of the global population of a congregatory seabird or terrestrial species.
- The site is known or thought to hold, on a regular basis, at least 20,000 waterbirds, or at least 10,000 pairs of seabird, of one or more species.
- The site is known or thought to be a "bottleneck site" where at least 20,000 pelicans (Pelecanidae), storks (Ciconiidae), raptors (Accipitriformes and Falconiformes), or cranes (Gruidae) pass regularly during spring or autumn migration.

These categories, explained in more detail in Fishpool et al. (1998) and Fishpool and Evans (2001), differ in what they seek

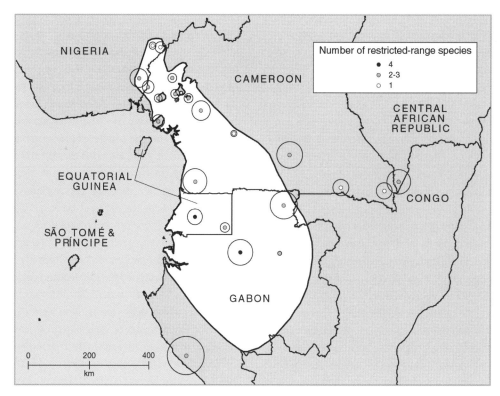

FIGURE 7.6. Network of important bird areas in the Cameroon and Gabon lowlands endemic bird area of the western Congo Basin lowland forest.

to do. A1 and A4 are essentially quantitative; the first selects sites at which significant numbers of individuals of species on the IUCN Red List (BirdLife International 2000) occur, whereas A4 identifies sites for more than threshold numbers of individuals considered vulnerable by virtue of their congregatory behavior at different points in their life cycle, such as at breeding colonies or when on migration. By contrast, sites are selected by A2 and A3 according to the species complements they hold such that, between them, they form a network that holds all or as many as possible of the species whose overlapping distributions define an EBA or the species that are confined to a biome. For these, complementarity analysis of species occurrence at sites is used to aid site selection. Figure 7.6 shows the network of IBAs that qualify under category A2 for the restricted-range species that occur in the Cameroon and Gabon lowlands EBA. Thus, the IBA approach is both site and species based. Ecological processes are not explicitly a target, although migration phenomena are captured in many sites identified under category A4.

In all, 1,228 IBAs have been identified in Africa, covering some 7 percent of the land area. Of these, 67 percent (824 sites) qualify under category A1, 39 percent (483 sites) for category A2, 64 percent (786 sites) for A3, and 40 percent (497 sites)

for category A4. The majority of sites therefore qualify for more than one category; only 32 percent (387 sites) qualify for one only, and 4 percent (50 sites) qualify for all four categories.

Many of the 1,228 IBAs are found in the highest biological priority Class I and Class III ecoregions identified here. In addition to their significance for birds, IBA sites in East Africa have been shown to be important for other vertebrate taxa (Brooks et al. 2001b), indicating that they might be an efficient conservation target for many different biological groups.

Implementation of Conservation Action at IBAs

The IBA Programme in Africa is now moving from site identification to conservation. For example, in 1998 BirdLife partner organizations in ten African countries secured funding from the Global Environment Facility, through the United Nations Development Programme, to turn their IBA inventories into national plans of monitoring, action, and advocacy. Implementation of this project, titled African NGO-Government Partnerships for Sustainable Biodiversity Action, has resulted in the development and operationalization of innovative participatory conservation concepts and techniques, including guidelines for prioritizing conservation action at IBAs, National Liaison Committees, National IBA Conservation Strategies, and Site Support Groups. This approach is now being extended to other countries.

The advantage of the IBA approach is that it identifies an objectively defined priority set of sites where conservation action must be concentrated. Because IBA identification and protection are undertaken in country by national BirdLife partner organizations or other bird conservation nongovernment organizations and scientists wherever possible, the approach also builds a trained national constituency capable of implementing the program indefinitely. In addition, there is significant scope for the involvement of local people in conservation activities at IBAs. This also builds an appreciation of the national and global importance of their home area and can foster local pride and the desire to see that these sites, and the biodiversity they support, persist in the long term.

ESSAY 7.3

Conservation across Political Boundaries

Judy Oglethorpe, Harry van der Linde, and Anada Tiéga

The borders of African countries rarely coincide with ecoregion boundaries. National borders were established mostly by European colonial powers in the highly political "scramble for Africa," with very little consideration of the geographic distribution of African peoples and even less consideration of ecological divisions of the landscape. Of the 119 terrestrial ecoregions in Africa and Madagascar, 85 (71 percent) occur in more than one country, mostly with direct connectivity across borders. Sixty-five percent of these transboundary ecoregions cover three or more countries. The West Sudanian Savanna [36] ecoregion spans thirteen countries, followed by the Sahelian *Acacia* Savanna [35] and the Guinean Forest-Savanna Mosaic [38] (twelve countries each) (figure 7.7). At a finer scale, national borders dissect many individual ecological processes, ecosystems, and home ranges of large mammal species. These include water catchments, mangroves in coastal border areas, and the home range of the Masai Mara-Serengeti wildebeest population, which moves between Kenya and Tanzania.

Ecoregions that cross national boundaries pose significant conservation challenges in terms of developing a coherent management strategy, but in the majority of areas the continuation of the same habitat and species assemblages across political borders provides opportunities for large-scale transboundary conservation. These opportunities range in scale from local collaboration over management of transboundary protected areas and cross-border community-based natural resource management to water catchment management, landscape and ecoregion conservation, and regional economic development that integrates sound natural resource management (van der Linde et al. 2001).

Many African conservationists are turning to transboundary collaboration in areas where single-country approaches to biodiversity conservation and natural resource management are insufficient (figure 7.8). Transboundary collaboration has existed for some years but in the conservation field has only recently started to operate over larger scales and at formal levels between governments. Recent examples of formal transboundary agreements are the new Kgalagadi Transfrontier Park agreement between Botswana and South Africa, which covers 38,000 km^2; the Greater Limpopo Transfrontier Park agreement between Mozambique, South Africa, and Zimbabwe, covering 35,771 km^2 (Greater Limpopo Transfrontier Park Joint Management Board 2002); and the Yaoundé Summit agreement between countries of the Congo Basin forests aiming to enhance conservation of the region's forests through development of transboundary parks and other actions. In 2001 there were thirty-five transborder protected area complexes in Africa involving thirty-four countries, including 148 individual protected areas (Zbicz in Sandwith et al. 2001).

Opportunities for Transboundary Ecoregion Conservation

Border areas often comprise large tracts of land with some of the most intact species assemblages and least disrupted ecological processes because they tend to be remote and undeveloped. Transboundary approaches have particular value for the conservation of species and ecological processes that need very large areas to survive, such as wide-ranging and migra-

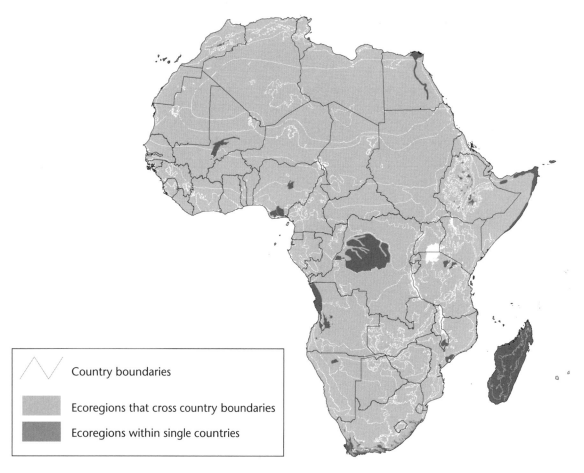

FIGURE 7.7. Area covered by transboundary ecoregions in Africa (including noncontiguous areas).

tory species (e.g., elephants). Sufficient land to achieve this conservation goal may not be available in a single country, but in a border region an allocation of land by each country can create a much larger conservation area that is more viable ecologically.

In addition to the benefits of large areas for species and migration processes, these large reserved areas may have sufficient resilience to withstand climate change. Maintaining large, intact, healthy areas with linkages across climatic, altitudinal, and other gradients should help to reduce anticipated impacts. The size of the area needed varies with local conditions including steepness of gradients and resilience of natural systems to change. Our understanding of climate change impacts and ways to mitigate them is still very incomplete but growing. It is clear that protection of natural systems from other human-

induced stresses such as habitat fragmentation and degradation, pollution, and invasive species will greatly enhance resilience.

In ecoregions that cross national borders, there may be opportunities to prioritize areas for conservation action, taking into account the ease of working in different countries. For example, in the Virunga Mountains on the borders of Uganda, Rwanda, and the Democratic Republic of the Congo, when conflict in one country prevented conservation work, efforts shifted to another country (Lanjouw et al. 2001). This approach allowed conservation to continue in this key area for mountain gorilla conservation despite the complex regional conflicts. Over the years transboundary collaboration between countries has been promoted despite conflict and breakdown of formal diplomatic relations, including training, communication, fire

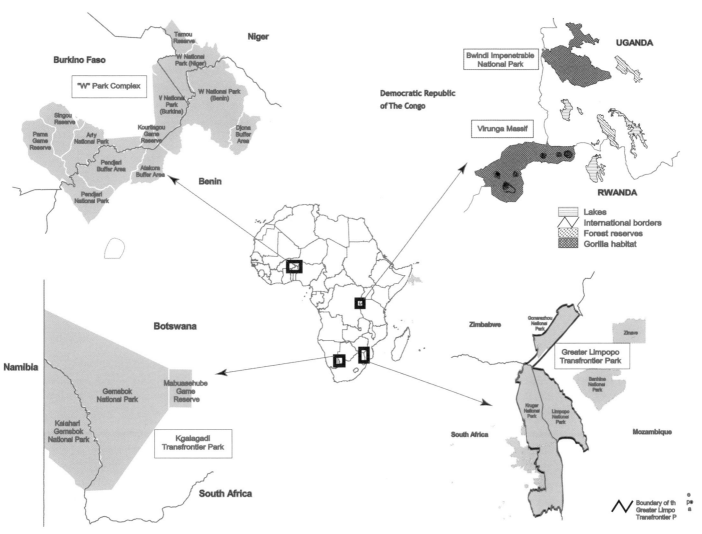

FIGURE 7.8. Examples of transboundary protected areas in Africa (Virunga-Bwindi region; Lanjouw et al. 2001).

control, antipoaching patrols, and monitoring. These conservation efforts succeeded: the mountain gorilla population in the Virungas has increased from 320 to 355 individuals in the past 10 years (Kalpers et al. 2003).

A collaborative transboundary approach can also help to improve the integration of conservation into national and regional economic development plans and promote long-term financial sustainability for conservation. In southern Africa transboundary natural resource management is being integrated into regional economic development through the Southern Africa Development Community (SADC). In this case, the main economic base is tourism. Transboundary collaboration ensuring conservation of the natural resource base and facilitating movement of tourists across borders is enabling the region to realize its potential for nature-based tourism, and many individual countries stand to benefit (SADC 1998; Griffin et al. 1999).

Transboundary collaboration can help reduce cross-border threats to biodiversity conservation and sustainable use of natural resources through collaboration to solve common problems. For example, the Central African Republic, the Republic of Congo, and Cameroon collaborate with joint antipoaching patrols in the Sangha River Trinational Initiative to control illegal hunting (Steel and Curran 2001). District officials in Uganda and Tanzania are discussing how to regulate the illegal cross-border timber trade occurring in the Minziro-Sango Bay Forest Ecosystem (Rodgers et al. 2001). Benin, Burkina Faso, and Niger are discussing how to tackle poaching, livestock, and water problems that affect shared resources in the "W" Park complex, including the elephant population (Magha et al. 2001).

Challenges for Ecoregion Conservation across Political Borders

Transboundary ecoregion conservation also presents many challenges. For example, it is sometimes difficult for bordering countries to reach a shared vision and set priorities for an ecoregion when each country has different national conservation policies and legislation or priorities and goals that do not correspond with what is most appropriate for the ecoregion as a whole. This is particularly so when land tenure systems and land uses, wealth, politics, religion, and other characteristics differ markedly across a border. Stakeholders have to be convinced that the benefits outweigh the transaction costs in order to collaborate across borders.

At the landscape level, biodiversity conservation and natural resource management systems on each side of the border may also not be compatible. For example, an upstream country may dam a transboundary river and seriously degrade wetland habitats, associated biodiversity, and fish yields downstream across its border. This occurred in the Zambezi Delta in Mozambique when the Kariba dam was built in Zimbabwe. Competition often occurs for shared resources. One country may overharvest migratory large mammal populations on its side of the border, adversely affecting the other country. Border fences built to demarcate boundaries and control movements of people and livestock can restrict movements of large mammals and cause their populations to decline (e.g., Kalahari mammal populations moving between Botswana and Namibia). In countries of the rainforest and woodland belt, illegal loggers sometimes cross borders to extract timber in neighboring countries.

Finally, local communities divided artificially by international borders may lose their traditional holistic approaches to natural resource management and be forced to use resources in an inappropriate and partitioned manner. This has occurred with some transhumant populations in drier zones of West Africa who traditionally ranged over large distances with their livestock, making use of seasonal rains and grazing to optimize land use. International boundaries have sometimes curtailed movements, resulting in overgrazing and rangeland degradation. Bilateral and multilateral agreements now exist to facilitate movements (Lycklama à Nijeholt et al. 2001).

Transboundary Lessons

Transboundary collaboration is not easy, and some hard-learned lessons are emerging. These include the importance of building trust, ensuring good communication, promoting participation and good governance, and finding win-win situations; all take time. A shared vision is essential not only across the border but within each country and entails a new way of thinking and planning. Good political will is a great asset. Not all activities have to be undertaken jointly, and different actors play different roles.

Institutional arrangements may have to be tailored to fit the vision. Formal agreements take time to negotiate and are not always necessary; it is best to start informally and involve higher levels only when necessary. The local level is very important and should not be neglected. Capacity is necessary on both sides of the border, and additional financial and personnel resources are needed for collaboration. Incorporation of conservation into general land-use planning on both sides of the border is valuable. Finally, there is no blueprint for transboundary collaboration or ecoregion management, so it is important to share experiences and collective learning to enable adaptive management. Transboundary lessons from Africa are outlined in more detail by van der Linde et al. (2001).

Strict Protected Areas or Integrated Conservation and Development Strategy: A Field-Based Perspective

David Thomas

Recent reviews of the effectiveness of integrated conservation and development projects (ICDPs) have fueled the debate between those favoring an approach that is inclusive of people and those arguing that conservation can be effectively achieved only through fences and fines ("fortress conservation"; see Oates 1995, 1999; Spinage 1996, 1998). However, rather than being mutually exclusive options, these two approaches define a continuum of possible strategies that can be tailored to the unique circumstances of each area. Protected areas need not exclude people from sharing in the benefits of conserved habitats (e.g., resource management or new tourism enterprise). And where local people are involved in decision-making this need not mean that protection is impossible; indeed, it may be their chosen land-use strategy.

The case for fortress conservation made in the polemic by Oates (1999) has been strongly criticized (e.g., Fairhead and Leach 2002). Some of the arguments being made for a more flexible, people-centered approach to conservation include the following:

- Protected areas may fail to conserve wildlife and have negative effects on food security, livelihoods, and culture of local people (Roe 2001).
- If protectionist approaches worked, fewer countries would be seeking alternatives (Child 1996).
- Markets can make an important contribution to the achievement of conservation goals (Hulme and Murphree 1999).
- Protected areas are expensive to establish and maintain (Roe 2001).
- Different approaches to conservation and natural resource management are suited to different wildlife, community, and institutional factors (Roe 2001; MacKinnon and Wardojo 2001; Adams and Hulme 2001; Abbot et al. 2001).
- Poverty alleviation is an increasing focus of global aid policy; conservation cannot avoid this and must therefore justify itself and work within these terms (MacKinnon 2001).

Protected areas still occupy only a fraction of the earth's surface, and biodiversity conservation globally will depend on effective management of natural resources beyond their boundaries. In many cases taking an exclusionist approach is inconceivable for a combination of economic, social, and political reasons, and conservation objectives will have to find resonance with the interests of local people. The following example from Cameroon examines a case in which a strict protected area approach to conservation was considered inappropriate, despite the site's high global priority. It also highlights some of the factors that have contributed to the success of a strategy that integrates conservation with development.

Starting with an ICDP at the Site Level

The Kilum-Ijim Forest in Cameroon extends over about 20,000 ha on the slopes of Mount Oku (3,011 m) and the adjoining Ijim Ridge and is the last significant remnant of Afromontane forest in West Africa (figure 7.9). Mount Oku lies in the Bamenda Highlands, part of the Cameroon Highlands Forests [10] ecoregion. The forest is a globally important center of endemism: fifteen bird species endemic to the Cameroon Mountains can be found at Kilum-Ijim, including Bannerman's turaco (*Tauraco bannermani*) and the banded wattle-eye (*Platysteira laticincta*). Both species are threatened, and the forest represents the only possibility of conserving viable populations of these two species (Stattersfield et al. 1998). Surveys of other taxa also demonstrate very high endemicity (e.g., Wild 1994; Hutterer and Fulling 1994; Cheek et al. 2000).

In the mid-1980s the Kilum-Ijim Forest was highly threatened by conversion to agriculture (Macleod 1987). Alternative conservation strategies debated at that time were either to advocate the establishment of a protected area controlled and managed by the state or to use community agreements focusing on the sustainable use of forest resources without creating an official protected area. After the passage of a new Forestry Law in 1994, the Kilum-Ijim Forest Project (which had begun in 1987) adopted the latter approach, based on management plans drawn up by the community and approved by the Ministry of Environment and Forests. The project has helped communities develop institutional capacity (e.g., management committees) and negotiate the complex legal process of forest registration and man-

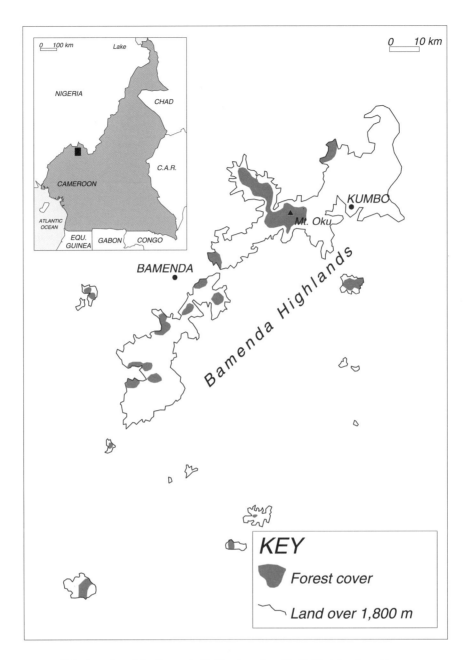

FIGURE 7.9. Location of the Bamenda Highlands region, North West Province, Cameroon, and approximate distribution of forest in 1987 (*dark green*). The black lines around the polygons indicate land above 1,800 m altitude.

agement plan approval. This has been combined with a development program to improve the productivity and sustainability of land use outside the forest and the production and value of products harvested from inside the forest. Although there is no protected area, the people-centered approach at Kilum-Ijim takes place in a legal context (provided by a Provincial Decree) that imposes state controls on land use (e.g., banning inappropriate activities such as burning and grazing in the forests).

Biodiversity monitoring indicates that biodiversity values inside the forest are being maintained. These achievements are significant considering the continued loss and degradation of other montane forests, including government forest reserves, over the same period (Yana Njabo and Languy 2000).

One of the major reasons for the success of this site-based ICDP is that the forest and its resources are the responsibility of the local chiefs (Fons) of Oku, Kom, and Nso, supported by their councils (the Kwifon). Despite weakening of traditional responsibilities, caused in part by assumption of management authority by the central government, the local chiefs still hold power. For example, when BirdLife International (then ICBP) began working at the Kilum-Ijim forest in 1987 one of the first steps was to help local communities demarcate a forest boundary beyond which no further clearance for agriculture would take place. Fifteen years later that boundary remains intact, and traditional authorities deal with the rare infringements promptly and effectively.

Scaling Up to the Landscape or Ecoregional Scale: The Bamenda Highlands Forest Project

The Kilum-Ijim Forest is the largest remaining block of montane forest in the Bamenda Highlands, but many scattered remnants exist elsewhere, varying in size from 100 to 16,000 ha (McKay 1994). The high biodiversity value of the region and the biological variation between individual forest blocks make conservation of as many patches as possible a high priority. There is also the possibility of linking individual patches in the future, should economic and political circumstances in Cameroon permit. As with the Kilum-Ijim Project, each of these forest patches is heavily used by the communities living around it. Building on the success of the Kilum-Ijim Project and responding to requests from communities in other parts of the Bamenda Highlands, the Bamenda Highlands Forest Project (BHFP) has scaled up the community-based approach to forest conservation to encompass the entire Bamenda Highlands ecoregion. Here too the focus is on a participatory approach that links forest conservation to local development and local values.

With so many individual forest patches as potential candidate sites for conservation and limited resources at its disposal, the BHFP decided that it would work only with communities and forests where there was strong support for forest conservation and sustainable management: communities should want to save their forests. After a communication and information drive, during which the project's objectives were publicized in local newspapers and on local radio, requests for support were invited. So far, forty communities have come forward requesting the project's assistance. Each community (and its forest) is then visited by project staff, who provide more details on what the project is able to offer, and through dialogue with the community and participatory appraisal methods the project is able to assess the level of commitment to forest conservation. This factor is then combined with the biodiversity significance of the individual forest (size, condition, position in relation to other forests) to arrive at a decision regarding the level of investment that the project will make. Although no community is denied support, the project's resources are focused on the communities or forests where there is the greatest community interest and the highest conservation value.

Forest conservation is not a soft option for communities. Project leaders made clear from the start that it will take commitment and investment from the community. The project's most significant input is to help develop capacity for sustainable forest management at community level (e.g., institution building, training, conflict resolution), and only limited capital support is provided (e.g., for boundary demarcation). The project also works closely with local nongovernment organizations (through secondments and partnerships), with the aim of increasing their capacity to support community forestry management after the project ends. Combined with the low-input approach, this will build sustainability into the project process.

Conclusion

Integrated conservation and development is a continuum of approaches, from buffer support around protected areas to community wildlife management in the absence of protected areas. The approach at Kilum-Ijim and the Bamenda Highlands has sought a position that accommodates the strengths and weaknesses of communities, traditional authorities, and government. It combines significant community control over forest resources with restrictions imposed by a government official in the framework of individual management plans approved and monitored by the Forest Department. Integrated conservation and development does not negate the need for rules and regulations, and finding the right balance and appropriate locus for responsibility and authority is key to success.

The conditions that led to the rejection of a strict protected area approach at Kilum-Ijim are to be found in various combinations throughout the developing world. Where globally important biodiversity is at stake, this demands innovative solutions that seek to engage local people in the conservation process rather than place them in opposition. The Kilum-Ijim Forest Project and Bamenda Highlands Forest Project provide an example of a successful ICDP that has been scaled up from a site level to address conservation at a programmatic, ecoregion scale.

Acknowledgments

The United Nations Development Programme through the Global Environment Facility, the Department for International Development (UK) through the Civil Society Challenge Fund and the Community Fund of the UK National Lottery Charities Board, have provided funds for the Kilum-Ijim Forest Project and Bamenda Highlands Forest Project.

African Heartlands: A Science-Based and Pragmatic Approach to Landscape-Level Conservation in Africa

Philip Muruthi

There is global concern over the rapid rate at which species are disappearing (Hilton-Taylor 2000), and various approaches have been developed to conserve biodiversity (Redford et al. 2003). Some conservation biologists advocate prioritization based on biodiversity hotspots (Myers et al. 2000), whereas others suggest that aggregate biodiversity levels are more important (Johnson 1995; Noss 1996; Dinerstein and Wikramanayake 1993; Olson and Dinerstein 1998). Although conservation approaches are diverse, they are not always incompatible when it comes to looking at conservation targets (Redford et al. 2003). Conservation biologists are converging on identifying the most important areas of biodiversity conservation in Africa (da Fonseca et al. 2000).

At the same time, governments in Africa are focused on national strategies to reduce poverty, combat the scourge of HIV and AIDS, and promote economic growth in the context of environmentally and socially sustainable development. African governments face the need to marshal scarce resources and to use any local assets that can provide an advantage in the competitive global environment. For the many parts of Africa that have been blessed with an abundant and globally significant natural heritage, wildlife and pristine habitats provide an important economic and environmental resource.

There is no universal agreement on what we are trying to save or how to do it. The challenge is to carve a conservation approach grounded in both science and practicality. This essay describes the African Wildlife Foundation (AWF) approach to landscape-level conservation, as embodied in its African Heartlands Program.

The African Heartlands Program

AWF's African Heartlands Program strives to conserve Africa's wildlife in large, cohesive conservation landscapes that are biologically important and have the scope to maintain healthy populations of wild species and natural ecological processes in perpetuity. The desirability of conserving large areas is an almost universally accepted principle in conservation biology. The African Heartlands aim to maintain the ecological integrity of the landscape over time. The program augments and strengthens the area under protection and manages the surrounding areas according to the needs of the native species, ecosystem processes, and local stakeholders. As demonstrated by increased species extinction rates in small, isolated parks (Dobson 1996; Woodroffe and Ginsberg 1998), protected areas are by themselves incomplete ecosystems incapable of conserving a great variety of biodiversity. AWF supports in situ conservation by linking existing protected areas with natural areas to form a contiguous landscape. These landscapes are biologically coherent and safeguard livelihoods (e.g., development of conservation enterprises) for local people.

The effects of fragmentation and habitat loss can have far-reaching effects on the flora and fauna, soil water resources, genetic and ecological processes and functions, and patterns of human ecology (Hobbs 1993; Forman 1998; Bennet 2003). These in turn can compromise people's livelihoods, particularly in Africa. Preventing habitat fragmentation and reduction is critical in maintaining the diversity of vegetation, increasing the likelihood of occurrence of rare or specialized habitats, maintaining species richness, and ensuring the sustainability of natural disturbance regimes. AWF's African Heartlands Program therefore strives to maintain and restore connectivity. Connectivity is crucial as key habitats become increasingly isolated from any wildlife that could move in from the outside when the areas around are clear-cut, overgrazed, or colonized. The land set aside to protect biodiversity is only a small fraction of the total area of natural habitat that is being converted to agriculture or harvested for timber.

AWF's concern for maintaining species and communities, habitats, and other entities is complemented by a concern for the ecological and evolutionary processes that brought these entities into being and will allow them to persist and evolve over time. African landscapes exhibit habitat heterogeneity and ecological gradients, aspects of natural variation that African Heartlands encourages. Ecosystem processes that depend on some vector for transmission through the landscape are most sensitive to isolation, and these include pollination, seed dispersal, and predator-prey relationships.

Current AWF Heartlands are located in central, eastern, and southern Africa (figure 7.10). The number of Heartlands will increase with time and resources to encompass other geopolitical areas and ecosystems of Africa.

Undertaking the African Heartlands Program

Using its Heartland Conservation Process (HCP;[1] AWF 2003), AWF first prioritizes and selects Heartlands, then plans and implements activities in these priority landscapes together with its multiple partners and adapts when necessary (http://www.awf.org/heartlands). AWF then considers the range of landscapes in Africa that merit its investment. Preselection draws on prioritization work at the continental level that has been undertaken by other organizations (ecoregions, Olson and Dinerstein 1998; biodiversity hotspots, Myers et al. 2000; important bird areas, Fishpool and Evans 2001; biosphere reserves, http://www.unesco.org/mab; heritage sites, http://www.unesco.org/heritage).

Using a variety of biological, ecological, social, and economic criteria, AWF identifies large landscapes of exceptional biological value where AWF can work over the long term to have significant conservation impact. AWF uses more than 40 years of accumulated experience of practicing conservation in Africa to identify large landscapes where field-level conservation can be practical and effective. After prioritization and selection, AWF uses the science-based HCP to work with partners to identify conservation targets, goals, threats, and opportunities and to design strategies and interventions. AWF and partners then implement activities in a Heartland and adapt when necessary. Analyses and syntheses of results are undertaken regularly through AWF's Program Impact Assessment (PIMA) system (http://www.awf.org). The HCP provides a useful framework for effective conservation in African Heartlands.

AWF and partners conduct systematic HCP analyses during which specific features of biodiversity are explicitly selected. These features of biodiversity are called conservation targets (Groves 2003). They include species, species assemblages, ecological communities, and systems.

Conservation targets drive landscape-scale conservation planning, including the process of identifying threats, developing strategies, measuring success, and approximately delineating the boundaries of a Heartland. The size of a specific Heartland is determined by combination of characteristics of conservation targets such as the ranging patterns of keystone species and the size of a watershed. AWF's African Heartlands approach incorporates time-tested species approaches while placing species into the context of a large landscape encompassing their ecological needs such as breeding, feeding, sea-sonal movements, and shelter. The Wildlife Conservation Society has a similar approach (landscape species approach, Sanderson et al. 2002b). Adequate protection of taxa with large home ranges can lead to successful protection of smaller organisms. In general, the ecology of conservation targets means that Heartlands straddle international borders to enhance landscape integrity.

There is no predetermined size for AWF's African Heartlands. AWF works at the scale of conservation landscapes, a size smaller than ecoregions and larger than sites (Poiani et al. 2000; Redford et al. 2003). These conservation landscapes form the basis for conservation planning and implementation (http://www.awf.org/heartlands; Redford et al. 2003; Groves 2003). Working at the landscape scale ensures that AWF is conserving an area large enough to sustain a majority of conservation targets and yet is of a manageable size for intervention strategies to be applied effectively.

AWF's initial planning horizon for work in a Heartland is 15 years, accommodating temporal scales beyond usual project funding cycles. This will allow the achievement of conservation goals and tracking of factors acting at larger spatial scales that may take longer to become apparent.

What Does AWF Do in a Heartland?

AWF works closely with a wide variety of partners—central and local government, private sector, communities, research organizations—to ensure that conservation targets and their environments persist in the long term. This is achieved by applying conservation strategies relating to land and habitat conservation, applied research, conservation enterprise, capacity building, and leadership.

Land and Habitat Conservation

AWF explores and applies appropriate mechanisms to bring land to conservation in each Heartland and country. AWF works with landowners to help them decide which lands will be reserved for wildlife and which lands will be used for farms, grazing, and tourist lodging and to bring other benefits to the landowners. For example, in the Masai Steppe Heartland, AWF and partners secured Manyara Ranch, an important habitat link between Tarangire and Manyara National Parks, through the Tanzania Land Conservation Trust (http://www.awf.org/succcess/manyara.php). In the Samburu and Kilimanjaro Heartlands, AWF is helping local communities to undertake participatory natural resource management planning and implementation, thus securing key areas for conservation and meeting the livelihood needs of the communities.

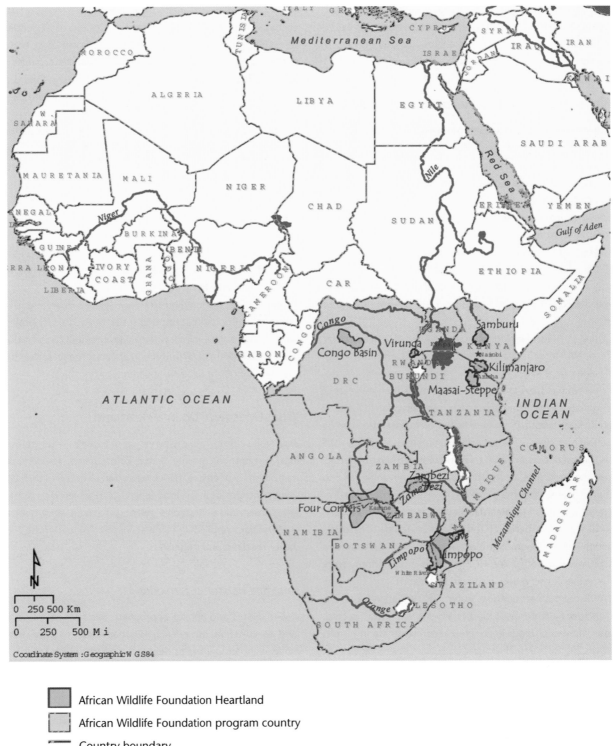

FIGURE 7.10. African Wildlife Foundation Heartlands (as of December 2003).

Legend:

African Wildlife Foundation Heartland

African Wildlife Foundation program country

Country boundary

Major lake or river

AWF works with protected area authorities in Heartlands to support protected area planning, management, enforcement, and monitoring. This has involved developing general management plans, building staff houses, providing safe drinking water, and improving visitor services given the often high dependency of protected areas on tourism revenue.

In all Heartlands, the conservation of wildlife movement corridors, habitat linkages, dry season refuges, and dispersal areas is an important strategy. Gap analysis is used to select areas of land to set aside as corridors. AWF studies the status, including use of these corridors by wildlife, as an essential part of their conservation. We facilitate land-use planning by landowners. In Kilimanjaro Heartland, for example, AWF helped formalize the Kitendeni corridor so that elephants and other species can move between Kilimanjaro Forest Reserve, Kilimanjaro National Park in Tanzania, and Amboseli National Park in Kenya. In each heartland we identify key corridors and habitats to be secured in the long term. Despite potential risks posed by corridors, it is undeniable that they can significantly enhance the conservation of protected areas.

Applied Research

Applied research into the status of conservation targets and threats to their conservation is an important component of AWF's work in Heartlands. Between one-third and one-half of conservation targets in the existing Heartlands are species or species assemblages. AWF undertakes multifaceted research, addressing ecological and socioeconomic issues, and uses the findings to inform overall conservation in the landscape. AWF's species research and management work is collaborative, involving AWF researchers and a variety of partners, consisting of individual researchers, institutions, landowners, communities, nongovernment organizations, and government agencies. AWF addresses significant interactions between humans and wildlife species. Negative interactions often lead to loss of property and deaths of wildlife and humans. Positive interactions between wildlife species and humans, if enhanced (e.g., through tourism), can lead to long-term coexistence. The threat of wildlife extinctions is real in Africa (Hilton-Taylor 2000), so AWF emphasizes conservation of endangered species and their habitats. Through our research theme "Essence of Africa: Species, Key Populations and Ecological Processes" we try to ensure that AWF protects different ecotypes of a species (such as savanna, forest, and desert elephants) and ecological phenomena such as migrations. Because of their important roles in natural ecosystems and the special challenges their conservation poses, AWF accords predators and their conservation special emphasis. The importance of disease as a threat to wildlife populations and to human livelihoods is addressed in a separate theme: disease and conservation.

Conservation Enterprises

AWF is demonstrating that conservation enterprises, though no panacea, are a useful strategy in extending the area of land under sound conservation. AWF has also pioneered a method for assessing the impact of conservation enterprises on local people and their livelihoods.

Conservation enterprises enable AWF to leverage additional land for wildlife and to help develop community-based business ventures that improve and safeguard livelihoods for local people. AWF has multidisciplinary enterprise support teams working in each Heartland, helping communities identify business opportunities, supporting community–private sector partnerships, helping to build marketing capacity and market opportunities, and enabling sound governance and good business management skills in community organizations. Much of AWF's conservation enterprise work has focused on the tourism sector, including the development of ecotourism lodges, tented camps, community campsites, fishing lodges, cultural bandas, and handicrafts. For example, AWF was asked to intervene in a court case involving a Masai Community and a South African operator. The new deal brokered by AWF formally recognizes community rights and has provided significant income to the community while protecting a key migration area. In another Heartland, AWF assisted a community owning key habitat between the Okavango and the Moremi Game Reserve to replace outdated tourism chalets that they had inherited from the government and create a new, competitive camp that will provide income. The community voluntarily created its main center of settlement far from this wildlife corridor to protect its revenue-generating potential. AWF also supports nontourism enterprise development including beekeeping and honey products, shade coffee, bottled water, and "wildlife" tea and works with local development organizations to help safeguard community livelihoods based on livestock and crops.

Capacity-Building and Leadership

AWF supports capacity-building at many levels. At the community level, AWF works with local leaders and members of village natural resource committees to gain the skills and experience they need to help manage land and conservation enterprises. AWF has found that community groups like to learn from each other; therefore, exchange visits in which two or more communities learn from the experiences, successes, and mistakes of each other are commonly arranged within and between Heartlands. The Ranger-Based Monitoring Program originally developed to protect mountain gorillas in the Virunga Heartland has been adapted and applied to the Samburu and Kilimanjaro Heartlands.

At a higher level, AWF is committed to developing and supporting Africa's future conservation leaders. This commitment led to the creation of the Charlotte Conservation Fellowship Program, which provides educational grants to Africans pursuing advanced degree studies in conservation-related fields under the assurance that they will return and apply their learning on the continent.

Measuring Success

AWF measures success by applying its PIMA system (AWF 2003). PIMA tracks the viability and status of conservation targets, the severity and scope of critical threats, and key socio-economic factors. PIMA was based on many other approaches, such as The Nature Conservancy's Measures of Success, and adapted for use in Africa.

How African Heartlands and Ecoregions Compare

As mentioned earlier, AWF works with WWF's ecoregions during the prioritization and selection stages of the HCP. Unlike WWF's Global 200 ecoregion approach, African Heartlands is not a strict prioritization approach. Ecoregion-based conservation works at larger scales than African Heartlands. Both approaches answer the questions of where and how to undertake conservation. Heartland conservation planning and ecoregional planning (Dinerstein et al. 2000) address landscape-level action and recognize multiple conservation targets ranging from species to ecosystems larger than Heartlands. There is much room for collaboration between these two approaches, which can complement each other. Informed collaboration will enhance long-term conservation of Africa's biodiversity.

Note

1. HCP has been adapted from The Nature Conservancy's Site Conservation Planning process (The Nature Conservancy 2000) with help from The Nature Conservancy.

APPENDIX A

Hierarchical Classification of Ecoregions in Terms of Biogeography and Habitats

	Biogeographic Framework		ECOREGION NO.	*Habitat Framework*	
REALM	BIOREGION	ECOREGION		BIOME	SUB-BIOME
Afrotropical	Western Africa and Sahel	Western Guinean Lowland Forests	1	Tropical and subtropical moist broadleaf forests	Guineo-Congolian lowland moist forest
Afrotropical	Western Africa and Sahel	Guinean Montane Forests	2	Tropical and subtropical moist broadleaf forests	Afromontane forest
Afrotropical	Western Africa and Sahel	Eastern Guinean Forests	3	Tropical and subtropical moist broadleaf forests	Guineo-Congolian lowland moist forest
Afrotropical	Western Africa and Sahel	Nigerian Lowland Forests	4	Tropical and subtropical moist broadleaf forests	Guineo-Congolian lowland moist forest
Afrotropical	Western Africa and Sahel	Niger Delta Swamp Forests	5	Tropical and subtropical moist broadleaf forests	Guineo-Congolian swamp forest
Afrotropical	Western Africa and Sahel	Cross-Niger Transition Forests	6	Tropical and subtropical moist broadleaf forests	Guineo-Congolian lowland moist forest
Afrotropical	Central Africa	Cross-Sanaga-Bioko Coastal Forests	7	Tropical and subtropical moist broadleaf forests	Guineo-Congolian lowland moist forest
Afrotropical	Central Africa	Atlantic Equatorial Coastal Forests	8	Tropical and subtropical moist broadleaf forests	Guineo-Congolian lowland moist forest
Afrotropical	Central Africa	Mount Cameroon and Bioko Montane Forests	9	Tropical and subtropical moist broadleaf forests	Afromontane forest
Afrotropical	Central Africa	Cameroon Highlands Forests	10	Tropical and subtropical moist broadleaf forests	Afromontane forest
Afrotropical	Central Africa	São Tomé, Príncipe, and Annobon Moist Lowland Forests	11	Tropical and subtropical moist broadleaf forests	Island moist forest
Afrotropical	Central Africa	Northwestern Congolian Lowland Forests	12	Tropical and subtropical moist broadleaf forests	Guineo-Congolian lowland moist forest
Afrotropical	Central Africa	Western Congolian Swamp Forests	13	Tropical and subtropical moist broadleaf forests	Guineo-Congolian swamp forest
Afrotropical	Central Africa	Eastern Congolian Swamp Forests	14	Tropical and subtropical moist broadleaf forests	Guineo-Congolian swamp forest
Afrotropical	Central Africa	Central Congolian Lowland Forests	15	Tropical and subtropical moist broadleaf forests	Guineo-Congolian lowland moist forest
Afrotropical	Central Africa	Northeastern Congolian Lowland Forests	16	Tropical and subtropical moist broadleaf forests	Guineo-Congolian lowland moist forest

	Biogeographic Framework		ECOREGION	Habitat Framework	
REALM	BIOREGION	ECOREGION	NO.	BIOME	SUB-BIOME
Afrotropical	Eastern and Southern Africa	Albertine Rift Montane Forests	17	Tropical and subtropical moist broadleaf forests	Afromontane forest
Afrotropical	Eastern and Southern Africa	East African Montane Forests	18	Tropical and subtropical moist broadleaf forests	Afromontane forest
Afrotropical	Eastern and Southern Africa	Eastern Arc Forests	19	Tropical and subtropical moist broadleaf forests	Afromontane forest
Afrotropical	Eastern and Southern Africa	Northern Zanzibar–Inhambane Coastal Forest Mosaic	20	Tropical and subtropical moist broadleaf forests	Eastern African lowland forest-grassland mosaic
Afrotropical	Eastern and Southern Africa	Southern Zanzibar–Inhambane Coastal Forest Mosaic	21	Tropical and subtropical moist broadleaf forests	Eastern African lowland forest-grassland mosaic
Afrotropical	Eastern and Southern Africa	Maputaland Coastal Forest Mosaic	22	Tropical and subtropical moist broadleaf forests	Eastern African lowland forest-grassland mosaic
Afrotropical Southern Africa	Eastern and Coastal Forest Mosaic	KwaZulu-Cape	23	Tropical and subtropical moist broadleaf forests	Eastern African lowland forest-grassland mosaic
Afrotropical	Eastern and Southern Africa	Knysna-Amatole Montane Forests	24	Tropical and subtropical moist broadleaf forests	Afromontane forest
Afrotropical	Horn of Africa	Ethiopian Lower Montane Forests, Woodlands, and Bushlands	25	Tropical and subtropical moist broadleaf forests	Afromontane forest
Afrotropical	Madagascar–Indian Ocean	Comoros Forests	26	Tropical and subtropical moist broadleaf forests	Island moist forest
Afrotropical	Madagascar–Indian Ocean	Granitic Seychelles Forests	27	Tropical and subtropical moist broadleaf forests	Island moist forest
Afrotropical	Madagascar–Indian Ocean	Mascarene Forests	28	Tropical and subtropical moist broadleaf forests	Island moist forest
Afrotropical	Madagascar–Indian Ocean	Madagascar Humid Forests	29	Tropical and subtropical moist broadleaf forests	Island moist forest
Afrotropical	Madagascar–Indian Ocean	Madagascar Subhumid Forests	30	Tropical and subtropical moist broadleaf forests	Island moist forest
Palearctic	African Palearctic	Cape Verde Islands Dry Forests	31	Tropical and subtropical dry and monsoon broadleaf forests	Island dry forest
Afrotropical	Eastern and Southern Africa	Zambezian *Cryptosepalum* Dry Forests	32	Tropical and subtropical dry and monsoon broadleaf forests	African dry forest
Afrotropical	Madagascar–Indian Ocean	Madagascar Dry Deciduous Forests	33	Tropical and subtropical dry and monsoon broadleaf forests	Island dry forest
Palearctic	African Palearctic	Mediterranean Conifer and Mixed Forests	34	Temperate coniferous forests	Temperate coniferous broadleaf forest
Afrotropical	Western Africa and Sahel	Sahelian *Acacia* Savanna	35	Tropical and subtropical grasslands, savannas, shrublands, and woodlands	*Acacia* savanna woodland
Afrotropical	Western Africa and Sahel	West Sudanian Savanna	36	Tropical and subtropical grasslands, savannas, shrublands, and woodlands	*Acacia* savanna woodland
Afrotropical	Western Africa and Sahel	East Sudanian Savanna	37	Tropical and subtropical grasslands, savannas, shrublands, and woodlands	*Acacia* savanna woodland
Afrotropical	Western Africa and Sahel	Guinean Forest-Savanna Mosaic	38	Tropical and subtropical grasslands, savannas, shrublands, and woodlands	Guineo-Congolian forest-savanna mosaic

	Biogeographic Framework		ECOREGION NO.	Habitat Framework	
REALM	BIOREGION	ECOREGION		BIOME	SUB-BIOME
Afrotropical	Western Africa and Sahel	Mandara Plateau Mosaic	39	Tropical and subtropical grasslands, savannas, shrublands, and woodlands	Guineo-Congolian forest-savanna mosaic
Afrotropical	Central Africa	Northern Congolian Forest-Savanna Mosaic	40	Tropical and subtropical grasslands, savannas, shrublands, and woodlands	Guineo-Congolian forest-savanna mosaic
Afrotropical	Eastern and Southern Africa	Victoria Basin Forest-Savanna Mosaic	41	Tropical and subtropical grasslands, savannas, shrublands, and woodlands	Guineo-Congolian forest-savanna mosaic
Afrotropical	Central Africa	Southern Congolian Forest-Savanna Mosaic	42	Tropical and subtropical grasslands, savannas, shrublands, and woodlands	Guineo-Congolian forest-savanna mosaic
Afrotropical	Central Africa	Western Congolian Forest-Savanna Mosaic	43	Tropical and subtropical grasslands, savannas, shrublands, and woodlands	Guineo-Congolian forest-savanna mosaic
Afrotropical	Horn of Africa	Somali *Acacia-Commiphora* Bushlands and Thickets	44	Tropical and subtropical grasslands, savannas, shrublands, and woodlands	*Acacia* savanna woodland
Afrotropical	Eastern and Southern Africa	Northern *Acacia-Commiphora* Bushlands and Thickets	45	Tropical and subtropical grasslands, savannas, shrublands, and woodlands	*Acacia* savanna woodland
Afrotropical	Eastern and Southern Africa	Southern *Acacia-Commiphora* Bushlands and Thickets	46	Tropical and subtropical grasslands, savannas, shrublands, and woodlands	*Acacia* savanna woodland
Afrotropical	Eastern and Southern Africa	Serengeti Volcanic Grasslands	47	Tropical and subtropical grasslands, savannas, shrublands, and woodlands	Seasonal grassland
Afrotropical	Eastern and Southern Africa	Itigi-Sumbu Thicket	48	Tropical and subtropical grasslands, savannas, shrublands, and woodlands	*Acacia* savanna woodland
Afrotropical	Eastern and Southern Africa	Angolan Miombo Woodlands	49	Tropical and subtropical grasslands, savannas, shrublands, and woodlands	Miombo woodland
Afrotropical	Eastern and Southern Africa	Central Zambezian Miombo Woodlands	50	Tropical and subtropical grasslands, savannas, shrublands, and woodlands	Miombo woodland
Afrotropical	Eastern and Southern Africa	Zambezian *Baikiaea* Woodlands	51	Tropical and subtropical grasslands, savannas, shrublands, and woodlands	Miombo woodland
Afrotropical	Eastern and Southern Africa	Eastern Miombo Woodlands	52	Tropical and subtropical grasslands, savannas, shrublands, and woodlands	Miombo woodland
Afrotropical	Eastern and Southern Africa	Southern Miombo Woodlands	53	Tropical and subtropical grasslands, savannas, shrublands, and woodlands	Miombo woodland
Afrotropical	Eastern and Southern Africa	Zambezian and Mopane Woodlands	54	Tropical and subtropical grasslands, savannas, shrublands, and woodlands	Mopane woodland
Afrotropical	Eastern and Southern Africa	Angolan Mopane Woodlands	55	Tropical and subtropical grasslands, savannas, shrublands, and woodlands	Mopane woodland
Afrotropical	Eastern and Southern Africa	Western Zambezian Grasslands	56	Tropical and subtropical grasslands, savannas, shrublands, and woodlands	Seasonal grassland

Biogeographic Framework			ECOREGION NO.	Habitat Framework	
REALM	BIOREGION	ECOREGION		BIOME	SUB-BIOME
Afrotropical	Eastern and Southern Africa	Southern Africa Bushveld	57	Tropical and subtropical grasslands, savannas, shrublands, and woodlands	*Acacia* savanna woodland
Afrotropical	Eastern and Southern Africa	Kalahari *Acacia* Woodlands	58	Tropical and subtropical grasslands, savannas, shrublands, and woodlands	*Acacia* savanna woodland
Afrotropical	Western Africa and Sahel	Sudd Flooded Grasslands	59	Flooded grasslands and savannas	Freshwater wetland
Palearctic	African Palearctic	Nile Delta Flooded Savanna	60	Flooded grasslands and savannas	Freshwater wetland
Afrotropical	Western Africa and Sahel	Lake Chad Flooded Savanna	61	Flooded grasslands and savannas	Freshwater wetland
Afrotropical	Western Africa and Sahel	Inner Niger Delta Flooded Savanna	62	Flooded grasslands and savannas	Freshwater wetland
Afrotropical	Eastern and Southern Africa	Zambezian Flooded Grasslands	63	Flooded grasslands and savannas	Freshwater wetland
Afrotropical	Eastern and Southern Africa	Zambezian Coastal Flooded Savanna	64	Flooded grasslands and savannas	Freshwater wetland
Palearctic	African Palearctic	Saharan Halophytics	65	Flooded grasslands and savannas	Saline wetland
Afrotropical	Eastern and Southern Africa	East African Halophytics	66	Flooded grasslands and savannas	Saline wetland
Afrotropical	Eastern and Southern Africa	Etosha Pan Halophytics	67	Flooded grasslands and savannas	Saline wetland
Afrotropical	Eastern and Southern Africa	Makgadikgadi Halophytics	68	Flooded grasslands and savannas	Saline wetland
Palearctic	African Palearctic	Mediterranean High Atlas Juniper Steppe	69	Montane grasslands and shrublands	Alpine moorland
Afrotropical	Horn of Africa	Ethiopian Upper Montane Forests, Woodlands, Bushlands, and Grasslands	70	Montane grasslands and shrublands	Montane forest-grassland mosaic
Afrotropical	Horn of Africa	Ethiopian Montane Moorlands	71	Montane grasslands and shrublands	Alpine moorland
Afrotropical	Eastern and Southern Africa	East African Montane Moorlands	72	Montane grasslands and shrublands	Alpine moorland
Afrotropical	Eastern and Southern Africa	Rwenzori-Virunga Montane Moorlands	73	Montane grasslands and shrublands	Alpine moorland
Afrotropical	Eastern and Southern Africa	Southern Rift Montane Forest-Grassland Mosaic	74	Montane grasslands and shrublands	Montane forest-grassland mosaic
Afrotropical	Eastern and Southern Africa	South Malawi Montane Forest-Grassland Mosaic	75	Montane grasslands and shrublands	Montane forest-grassland mosaic
Afrotropical	Eastern and Southern Africa	Eastern Zimbabwe Montane Forest-Grassland Mosaic	76	Montane grasslands and shrublands	Montane forest-grassland mosaic
Afrotropical	Eastern and Southern Africa	Highveld Grasslands	77	Montane grasslands and shrublands	Montane forest-grassland mosaic
Afrotropical	Eastern and Southern Africa	Drakensberg Montane Grasslands, Woodlands, and Forests	78	Montane grasslands and shrublands	Montane forest-grassland mosaic
Afrotropical	Eastern and Southern Africa	Drakensberg Alti-Montane Grasslands and Woodlands	79	Montane grasslands and shrublands	Alpine moorland

	Biogeographic Framework		ECOREGION NO.	Habitat Framework	
REALM	BIOREGION	ECOREGION		BIOME	SUB-BIOME
Afrotropical	Eastern and Southern Africa	Maputaland-Pondoland Bushland and Thickets	80	Montane grasslands and shrublands	Montane forest-grassland mosaic
Afrotropical	Eastern and Southern Africa	Angolan Scarp Savanna and Woodlands	81	Montane grasslands and shrublands	Montane forest-grassland mosaic
Afrotropical	Eastern and Southern Africa	Angolan Montane Forest-Grassland Mosaic	82	Montane grasslands and shrublands	Montane forest-grassland mosaic
Afrotropical	Western Africa and Sahel	Jos Plateau Forest-Grassland Mosaic	83	Montane grasslands and shrublands	Montane forest-grassland mosaic
Afrotropical	Madagascar–Indian Ocean	Madagascar Ericoid Thickets	84	Montane grasslands and shrublands	Alpine moorland
Palearctic	African Palearctic	Mediterranean Woodlands and Forests	85	Mediterranean forests, woodlands, and scrub	Mediterranean scrub
Palearctic	African Palearctic	Mediterranean Dry Woodlands and Steppe	86	Mediterranean forests, woodlands, and scrub	Mediterranean scrub
Palearctic	African Palearctic	Mediterranean *Acacia-Argania* Dry Woodlands and Succulent Thickets	87	Mediterranean forests, woodlands, and scrub	Mediterranean scrub
Palearctic	African Palearctic	Canary Islands Dry Woodlands and Forests	88	Mediterranean forests, woodlands, and scrub	Island Mediterranean forest
Cape	Cape Floristic Region	Lowland Fynbos and Renosterveld	89	Mediterranean forests, woodlands, and scrub	Mediterranean scrub
Cape	Cape Floristic Region	Montane Fynbos and Renosterveld	90	Mediterranean forests, woodlands, and scrub	Mediterranean scrub
Afrotropical	Eastern and Southern Africa	Albany Thickets	91	Mediterranean forests, woodlands, and scrub	Mediterranean scrub
Palearctic	African Palearctic	Atlantic Coastal Desert	92	Deserts and xeric shrublands	Desert
Palearctic	African Palearctic	North Saharan Steppe	93	Deserts and xeric shrublands	North African xeric woodland-shrubland
Palearctic	African Palearctic	South Saharan Steppe	94	Deserts and xeric shrublands	North African xeric woodland-shrubland
Palearctic	African Palearctic	Sahara Desert	95	Deserts and xeric shrublands	Desert
Palearctic	African Palearctic	West Saharan Montane Xeric Woodlands	96	Deserts and xeric shrublands	North African xeric woodland-shrubland
Palearctic	African Palearctic	Tibesti-Jebel Uweinat Montane Xeric Woodlands	97	Deserts and xeric shrublands	North African xeric woodland-shrubland
Palearctic	African Palearctic	Red Sea Coastal Desert	98	Deserts and xeric shrublands	Desert
Afrotropical	Western Africa and Sahel	East Saharan Montane Xeric Woodlands	99	Deserts and xeric shrublands	North African xeric woodland-shrubland
Afrotropical	Horn of Africa	Eritrean Coastal Desert	100	Deserts and xeric shrublands	Desert
Afrotropical	Horn of Africa	Ethiopian Xeric Grasslands and Shrublands	101	Deserts and xeric shrublands	Northeastern African xeric woodland-shrubland
Afrotropical	Horn of Africa	Somali Montane Xeric Woodlands	102	Deserts and xeric shrublands	Northeastern African xeric woodland-shrubland
Afrotropical	Horn of Africa	Hobyo Grasslands and Shrublands	103	Deserts and xeric shrublands	Northeastern African xeric woodland-shrubland
Afrotropical	Eastern and Southern Africa	Masai Xeric Grasslands and Shrublands	104	Deserts and xeric shrublands	Northeastern African xeric woodland-shrubland

Biogeographic Framework			ECOREGION NO.	Habitat Framework	
REALM	BIOREGION	ECOREGION		BIOME	SUB-BIOME
Afrotropical	Eastern and Southern Africa	Kalahari Xeric Savanna	105	Deserts and xeric shrublands	Southern African xeric woodland-shrubland
Afrotropical	Eastern and Southern Africa	Kaokoveld Desert	106	Deserts and xeric shrublands	Desert
Afrotropical	Eastern and Southern Africa	Namib Desert	107	Deserts and xeric shrublands	Desert
Afrotropical	Eastern and Southern Africa	Nama Karoo	108	Deserts and xeric shrublands	Southern African xeric woodland-shrubland
Afrotropical	Eastern and Southern Africa	Namib Escarpment Woodlands	109	Deserts and xeric shrublands	Southern African xeric woodland-shrubland
Cape	Cape Floristic Region	Succulent Karoo	110	Deserts and xeric shrublands	Southern African xeric woodland-shrubland
Afrotropical	Horn of Africa	Socotra Island Xeric Shrublands	111	Deserts and xeric shrublands	Northeastern African xeric woodland-shrubland
Afrotropical	Madagascar–Indian Ocean	Aldabra Island Xeric Scrub	112	Deserts and xeric shrublands	Island xeric woodland-shrubland
Afrotropical	Madagascar–Indian Ocean	Madagascar Spiny Thickets	113	Deserts and xeric shrublands	Island xeric woodland-shrubland
Afrotropical	Madagascar–Indian Ocean	Madagascar Succulent Woodlands	114	Deserts and xeric shrublands	Island xeric woodland-shrubland
Afrotropical	Western Africa and Sahel	Guinean Mangroves	115	Mangroves	West African mangroves
Afrotropical	Western Africa and Sahel	Central African Mangroves	116	Mangroves	West African mangroves
Afrotropical	Eastern and Southern Africa	Southern Africa Mangroves	117	Mangroves	East African mangroves
Afrotropical	Eastern and Southern Africa	East African Mangroves	118	Mangroves	East African mangroves
Afrotropical	Madagascar–Indian Ocean	Madagascar Mangroves	119	Mangroves	East African mangroves

APPENDIX B

Taxonomic Details and Sources of Species Distribution Data

General Approach to Taxonomy

For each vertebrate group, we first created a standardized list of species, following where possible the most recent standard taxonomic checklist or consulting with relevant experts. In cases where there has been a good taxonomic revision since the publication of the standard taxonomic list, we followed the more recent taxonomic treatment. For plants and invertebrates we did not develop a full species list but instead relied on estimates provided by the University of Bonn (plants) and experts (invertebrates).

Distribution Data

Species distribution data were compiled at the WWF-US Conservation Science Program in Washington, D.C., through cooperation with the Zoological Museum of the University of Copenhagen, Denmark, the BIOMAPS project at the University of Bonn, and various taxonomic experts. Attempts have been made to make the data as accurate as possible, but because the distributions of more than 5,000 species of vertebrates have been considered, some errors are inevitable.

Birds

Taxonomy

The taxonomic listing largely follows Sibley and Monroe (1990, 1993), with the addition of species described since that time. All terrestrial species and waterbirds that breed in the African Palearctic, Afrotropical region, or Madagascar and the other smaller offshore islands or that regularly visit this region as nonbreeding migrants (either from the Palearctic or Madagascar) were included, totaling 2,247 species. Pelagic seabird species that breed on a few of the offshore islands were excluded, as were vagrant species that have been only incidentally recorded in the region (Dowsett and Forbes-Watson 1993). The English bird names follow Thayer's Birder's Diary v. 2.5 CD-ROM (1997), based on Sibley and Monroe (1990, 1993).

Distribution Data

The main distributional sources for mainland Africa were the published volumes of the *Birds of Africa* (e.g., Fry et al. 1988; Keith et al. 1992; Urban et al. 1986, 1997; Brown et al. 1982). The distribution maps in these volumes, together with written habitat descriptions, were used to assign species to ecoregions. For species not published in the *Birds of Africa* by early 1999, distribution data within ecoregions were added at the Zoological Museum of the University of Copenhagen in Denmark, using a database with a resolution of 1- by 1-degree squares. The data sources used in the University of Copenhagen database are presented at http://www.zmuc.dk/commonweb/research/biodata .htm.

The distributions of birds in northern African ecoregions, on Madagascar, and on other offshore islands were taken from Ripley and Bond (1966), Penny (1974), Madge and Burn (1988), Langrand (1990), Dowsett and Dowsett-Lemaire (1993), Pérez del val et al. (1994), Peet and Atkinson (1994), Hazevoet (1995), Van Perlo (1995), Clarke and Collins (1996), Thayer's Birder's Diary (1997), Christy and Clarke (1998), Sinclair and Langrand (1998), and Stattersfield et al. (1998).

The ecoregional distribution of narrowly endemic birds was further checked against a number of publications (Hall and Moreau 1970; Snow 1978; Lewis and Pomeroy 1989; Harrison et al. 1997; Stattersfield et al. 1998). Dr. Pamela Beresford from the American Museum of Natural History completed a final review of the bird database, using the library resources at that museum. All these data were merged into a single database by WWF-US.

Mammals

Taxonomy

The mammal list follows Wilson and Reeder (1993), updated by the addition of species that have been recently described. The one major departure from Wilson and Reeder is in the Galagonideae, where we follow the species names, order, and distribution data presented by Kingdon (1997). In some ecoregion descriptions for West and Central Africa the updated species names for primates (Groves 2001) are followed, but these updates are not reflected in the database used for analysis. In total, distributions of 1,167 mammal species were assessed from the study region. The English mammal names follow Wilson and Cole (2000), except for some primates, for which we follow Groves (2001).

Distribution Data

For the larger mammal species, range maps and habitat descriptions were used to assign species to ecoregions (Haltenorth and Diller 1977; Skinner and Smithers 1990; Kingdon 1997; Boitani et al. 1999), with additional reference to *Walker's Mammals of the World* (Nowak and Paradiso 1999) and the database of the Zoological Museum of the University of Copenhagen (sub-Saharan Africa). For large mammal groups covered by an IUCN Species Action Plan (e.g., Oliver 1993; Oates 1986, 1998; East 1999; see also http://www.iucn.org), we also used these publications to refine the assignment of species to ecoregions.

For the smaller mammal species in sub-Saharan Africa—Chiroptera (bats), Rodentia (rodents), and Insectivora (shrews, otter shrews, golden moles, and hedgehogs)—data have been entered to ecoregions using the 1- by 1-degree resolution database of the Zoological Museum at the University of Copenhagen, Denmark. Their data sources are listed at http://www.zmuc.dk/commonweb/research/biodata.htm. Madagascar mammal distribution data came from a number of publications (Mittermeier et al. 1994; Garbutt 1999), and a similar process was used for North Africa (Le Berre 1989, 1990; Wilson and Reeder 1993; Kingdon 1997) and the smaller offshore islands (Wilson and Reeder 1993). All data were merged into a single database by WWF-US.

Reptiles

Taxonomy

The list of snakes is based on Welch (1982). Newly described species have been added, together with changes according to the current taxonomic interpretations of Dr. Jens Rasmussen of

the Zoological Museum in Copenhagen, Dr. Don Broadley of the Natural History Museum in Zimbabwe, and Van Wallach in the United States (Leptotyphlopidae only). The list of chelonians is based on Iverson (1992), and the list of other reptiles (e.g., lizards, skinks) is based on an unpublished list provided by Dr. Donald Broadley in Zimbabwe. In total, 1,763 species were assessed from the study area. The English names for reptiles follow the European Molecular Biology Laboratory (EMBL) reptile database (http://www.embl-heidelberg.de/~uetz/LivingReptiles.html).

Distribution Data

For the snakes in sub-Saharan Africa data come mainly from the databases of the Zoological Museum of the University of Copenhagen, Largen and Rasmussen (1993), Sprawls and Branch (1996), and Branch (1998). The data sources used in the University of Copenhagen database are presented at http://www.zmuc.dk/commonweb/research/biodata.htm. For chelonians, distribution data follow Iverson (1992), and for crocodiles they follow King and Burke (1989). For North Africa, data were compiled from Le Berre (1989) and Schleich et al. (1996). For the offshore islands and for Madagascar the data were compiled from Glaw and Vences (1992), Henkel and Schmidt (1995, 2000), the EMBL reptile database, Gardner (1986), and Showler (1994). Angus Gascoigne provided species lists for São Tomé and Príncipe. Once entered into ecoregion format, the initial data were checked and refined by Dr. Don Broadley in Zimbabwe and entered into a single database at WWF-US.

Amphibians

Taxonomy

The list of amphibians was based on the works of Frost (1999) and Duellman (1993). Taxonomic changes to these lists have been made using recently published papers. In total, 779 species were assessed for this project. The English names of amphibians follow Frost (1999).

Distribution Data

For Sub-Saharan Africa distribution data come from the 1- by 1-degree resolution distribution database of the Zoological Museum, updated using some recent references (e.g., Schiøtz 1999 [tree frogs]; Largen 2001 [Ethiopia]). The references used in the Copenhagen database are available at http://www.zmuc.dk/commonweb/research/biodata.htm. For North Africa data come from Le Berre (1989) and Borkin (1999), and for Madagascar and smaller offshore islands information was taken from Henkel and Schmidt (1995, 2000), Glaw and Vences (1992), Clarke and Collins (1996), Poynton (1999), and Duellman (1993, 1999). Angus Gascoigne provided species lists for São Tomé and Príncipe.

Plants

Estimates of plant species richness were provided by the BIO-MAPS project at the University of Bonn in Germany (Kier et al. in press). Where possible these estimates were compared against numerical data on richness of certain ecoregions contained in a number of scientific publications (WWF and IUCN 1994; Hazevoet 1995; Cowling and Pierce 1999a, 1999b; Cowling et al. 1999b; Mittermeier et al. 1999).

For plant endemism, estimates derived at a WWF workshop in 1998 in Cape Town were used to determine endemism within four classes. These assignations were verified against es-timates of plant endemism in a database developed at the University of York in the United Kingdom (Lovett et al. 2000b) and a number of other scientific publications (Hedberg 1986, 1997; White 1993; Huntley 1994; WWF and IUCN 1994).

Invertebrates

Invertebrate information on the continental scale was provided by experts at the workshop in Cape Town in August 1998. These data were placed into four broad classes to describe the level of richness and endemism ("very high," "high," "medium," and "low") within each biome.

APPENDIX C

Quantifying Nonspecies Values of Ecoregions

For this assessment we identified four features that indicate the nonspecies biological value of ecoregions. Scores were applied to each of these four features according to the criteria set out here. Scores were either 1 (global) or 0 (none), except for migrations and higher taxonomic uniqueness, where we used scores of 1 (global), 0.5 (regional), and 0 (none). Once every ecoregion was scored, we summed the values, and every ecoregion that scored more than an arbitrary 1.5 was elevated to globally outstanding in the Biological Distinctiveness Index, regardless of its species conservation values.

Unusual Evolutionary or Ecological Phenomena

Evolutionary Phenomena

EXTRAORDINARY ADAPTIVE RADIATIONS OF SPECIES

Criteria: An ecoregion scored 1 for adaptive radiations if

- It contains twenty or more species within a clear adaptive radiation in one or more genera in the same family.
- It contains extensive adaptive radiations (ten or more species displaying clear adaptive radiation to different resource niches) in genera from at least two different families.

Unusual Ecological Phenomena

INTACT LARGE VERTEBRATE ASSEMBLAGES

Criteria: An ecoregion scored 1 for this feature if

- It contains all of the largest carnivores, herbivores, and frugivores and other feeding guilds in that ecosystem, and

these species still fluctuate within natural ranges and play an important ecological role in the system.

Note: no more than three ecoregions were selected as globally outstanding for this feature per biome if the large carnivores, herbivores, and frugivores had widespread distributions throughout the biome.

MIGRATIONS OR CONGREGATIONS OF LARGE VERTEBRATES

Criteria: An ecoregion scored 1 for this feature if

- A migration of large terrestrial vertebrates occurs that exceeds 100 km in length, includes more than several thousand individuals, and is accompanied by the full complement of native large predators.
- Enormous aggregations (millions) of breeding or migratory birds occur.

Regionally important migrations were also recognized, and these scored 0.5.

Global Rarity of Habitat Types

Criteria: An ecoregion scored 1 for this feature if:

- It is one of the globally rare habitat types recognized by WWF, which in Africa are tropical dry forests (fifty-nine ecoregions worldwide), montane moorlands (eight ecoregions worldwide in three widely separated regions), and Mediterranean forests, woodlands, and scrub (thirty-nine ecoregions worldwide found in five distinct regions).

183

Higher Taxonomic Uniqueness

Criteria: An ecoregion scored 1 for this feature if

- The ecoregion contains one endemic family of plants or vertebrates.
- More than 30 percent of the genera of vascular plants or animals are estimated as endemic to the ecoregion.

An ecoregion was considered of regional importance (scoring 0.5) if there are more than five endemic genera, especially if these genera indicate a link back to groups more common millions of years in the past.

Ecoregions Harboring Examples of Large Relatively Intact Ecosystems

Criteria: An ecoregion scored 1 for this feature if it is a core part of a wilderness area defined by Conservation International (Mittermeier et al. 2002). These are areas at least 10,000 km^2 in size, at least 70 percent intact, and with 5 people/km^2 or fewer once cities and towns have been excluded. We refined this assessment of wilderness areas to fit as closely as possible to our ecoregional framework, also using the results of Sanderson et al. (2002a) for this purpose.

APPENDIX D

Development of the Conservation Status Index

Current Conservation Status

The Current Conservation Status Index of an ecoregion is derived using four different variables. The total possible score for this index is 100 points, made up of the following maximum values:

Habitat loss = 40 points

Habitat blocks = 25 points

Fragmentation = 20 points

Level of protection (IUCN I–IV) = 15 points

Habitat Loss

Habitat loss per ecoregion was estimated as the percentage of converted habitat derived from two sources. The first data source was the University of Maryland (UMD) Global Landcover data (Hansen et al. 2000; Hansen and Reed 2000, derived from advanced very high resolution radiometer imagery from 1992: http://gaia.umiacs.umd.edu:8811/landcover/). The "urban" and "cropland" land-cover classes were used to define areas of converted habitat. The second data source was population density, derived from the LandScan Global Population 1998 database, which depicts population density and total numbers of people, by 1-km grid cells (Dobson et al. 2000, http://www.ornl.gov/gist/projects/LandScan). A population density of ten or more people per square kilometer was used to define where population pressures start to negatively affect habitats, although figures between five and twenty people per square kilometer have been used elsewhere (Mittermeier et al. 1998; Hoare and Du Toit 1999; Eswaran et al. 2001). The Landscan data modeled human populations along all roads, including those in protected areas; we removed the population information from protected areas in IUCN categories I–IV, on the assumption that human population pressure in these protected areas should be negligible. Data from UMD and Landscan were combined to show the spatial distribution of habitat loss. Calculations of habitat loss from all ecoregions were

TABLE D.1. **Relationship between Percentage Habitat Loss and Points Contributed to the Current Conservation Index.**

Percentage of Habitat Loss	Contribution to Current Conservation Status Index (points)
≥80%	40
60–79%	30
40–59%	20
20–39%	10
0–19%	0

then scaled so that they contributed a maximum of 40 points to the Current Conservation Status Index (table D.1).

We recognize that our analysis measures only some of the biologically important changes in habitat. For example, semi-desert and desert habitats have experienced little habitat loss to agriculture or urbanization. In these areas the main land use is pastoralism, and although this may have degraded habitats and reduced their biological value, it cannot be measured using currently available land-use databases.

Habitat Blocks

Habitat blocks were calculated for every ecoregion, starting from the remaining habitat areas identified in the habitat loss analysis. Geographic information system technology was used to create buffer zones along all existing roads from the Digital Chart of the World (ESRI 1993, 1:1,000,000), updated for Central Africa using road data from the Central Africa Regional Program for the Environment (CARPE 1998, http://carpe.umd.edu/). Information was not available to tailor buffer distances to each biome. However, in forests, 80 percent deforestation has been shown to take place within 2 km of a road (Mertens and Lambin 1997), and in open habitats intense resource degradation can

Tropical and Subtropical Moist Broadleaf Forests Biome (Biome 1)
and Tropical and Subtropical Dry Broadleaf Forests Biome (Biome 2)

Point Value	Ecoregion Size >10,000 km²	Ecoregion Size (continuous habitat) <10,000 km²	Ecoregion Size (discontinuous habitat) <10,000 km²
0	>5,000 or ≥3 blocks >2,000	>3,000 or ≥3 blocks >1,000	≥2 blocks >500
6	>3,000 or ≥3 blocks >1,000	>1,000	≥2 blocks >300
12	>2,000	>500	≥2 blocks >200
18	>1,000	>250	100
25	None >1,000	None >250	None >100

Temperate Coniferous Forests Biome (Biome 5)

Point Value	Ecoregion Size >10,000 km²	Ecoregion Size <10,000 km²
0	>4,000 or ≥3 blocks >1,500	>2,500 or ≥3 blocks >800
6	>3,000 or ≥3 blocks >1,000	>800
12	>2,000	≥3 blocks >250
18	>1,000	>250
25	None >1,000	None >250

Tropical and Subtropical Grasslands, Savannas, Shrublands, and Woodlands Biome (Biome 7), Flooded
Grasslands and Savannas Biome (Biome 9), and Montane Grasslands and Shrublands Biome (Biome 10)

Point Value	Ecoregion Size >10,000 km²	Ecoregion Size <10,000 km²
0	>2,000 or ≥3 blocks >800	>1,000 or ≥3 blocks >500
6	>1,000	>500
12	>500	>250
18	>250	>100
25	None >250	None >100

(continued)

extend one day's walk from a road (~40 km). Therefore, we used a conservative buffer distance of 15 km, although buffers of 2–5 km from a road have been used in other studies (Minnemeyer 2002; Wilkie et al. 1998a, 1998b; Godoy et al. 2000). Intact habitat blocks were defined as those that had not been buffered, and the areas of all blocks within ecoregions were calculated.

Points were then assigned to each ecoregion according to the size distribution of the remaining habitat blocks, varied according to different biomes (table D.2). The habitat block scores were then scaled to contribute a maximum of 25 points to the Current Conservation Status Index, with less intact systems receiving a higher score than more intact systems (table D.2).

Habitat Fragmentation

This analysis is used to quantify the heterogeneity of the landscape. To calculate this ratio we took the total perimeter of remaining blocks divided by the total area of remaining blocks for each ecoregion (table D.3). The resulting fragmentation score contributes up to 20 points of the Current Conservation Status Index.

Habitat Protection

This calculation measures the degree of formal protection in each ecoregion. We overlaid polygons of IUCN I–IV protected areas onto ecoregions to calculate the percentage protected (protected area data from WCMC 2002; South Africa protected area data provided by and CSIR, Environmentek [South Africa]). Only protected areas coded IUCN I–IV were used because they have the highest levels of conservation protection, thus omitting all reserves coded IUCN V–VI and uncoded reserves such as forest reserves. Lower levels of protection were assigned higher point values, contributing a maximum of 15 points to the Current Conservation Status Index (table D.4).

Mediterranean Forests, Woodlands, and Scrub Biome (Biome 12)
and Deserts and Xeric Shrublands Biome (Biome 13)

Point Value	Ecoregion Size >10,000 km²	Ecoregion Size <10,000 km²
0	≥2 blocks >750 or ≥3 blocks >500	≥2 blocks >500 or ≥3 blocks >300
6	>750	>500
12	>500	>250
18	>250	>100
25	None >250	None >100

Mangroves Biome (Biome 14)

Point Value	Ecoregion Size >3,000 km²	Ecoregion Size 1,000–3,000 km²	Ecoregion Size <1,000 km²
0	>1,000 or ≥3 blocks >500	>750 or ≥3 blocks >500	90–100 percent
6	>500	>500	70–90 percent
12	>250	≥3 blocks >200	40–70 percent
18	>100	>75	10–40 percent
25	None >100	None >75	<10 blocks

The value ">500" means that the unit contains at least one habitat block greater than 500 km². The value "90 percent" means that the unit contains at least one habitat block that is 90 percent the size of the largest original unit. For a unit of any given size, the table should be read from top to bottom until a statement is reached that is true of the unit.

Calculating the Current Conservation Status Index

The Current Conservation Status Index was calculated by adding the scores obtained for habitat loss, habitat blocks, habitat fragmentation, and habitat protection. This index has a point range from 0 to 100, which is used to define five categories of Current Conservation Status:

80–100 points = critical
60–79 points = endangered
40–59 points = vulnerable
20–39 points = relatively stable
0–19 points = relatively intact

Based on WWF's experience from other terrestrial continental assessments and drawing on expert opinion, we have developed conservation scenarios for each of the five categories of Current Conservation Status:

Critical: Remaining intact habitat is restricted to isolated small fragments with low probabilities of persistence without conservation assistance. Some species are already extirpated (or extinct) because of habitat loss. Remaining habitat fragments fail to meet the minimum area requirements to maintain viable populations of other species and ecological processes over the long term. Top predators have been exterminated or nearly so. Land use in areas between remaining fragments is incompati-

TABLE D.3. **Points Assigned to the Degree of Fragmentation in Each Ecoregion.**

Edge:Area Ratio	Points
>1.0	20
0.50–1.0	15
0.25–0.49	10
0.10–0.24	5
0–0.09	0

TABLE D.4. **Points Assigned to Different Levels of Ecoregion Protection.**

Percentage of Ecoregion Protected	Points
0–2%	15
3–6%	12
7–10%	8
11–25%	4
>25%	0

TABLE D.5. Points Assigned to Population Density Scores Calculated for 2025.

Projected Population Density in 2025	Point Value
≥200 people/km^2	60
100–199 people/km^2	45
50–99 people/km^2	30
10–49 people/km^2	15
0–9 people/km^2	0

Source: World Resources Institute.

TABLE D.6. Points Assigned to Level of Extinction Threat to Species.

IUCN Weighted Scores	Point Value
≥80	40
60–79	30
40–59	20
20–39	10
0–19	0

ble with maintaining most native species and communities. The spread of alien species may be a serious ecological problem, particularly on islands.

Endangered: Remaining intact habitat is restricted to isolated fragments of varying size (a few larger blocks may be present) with medium to low probabilities of persistence without conservation assistance. Some species are already extirpated because of habitat loss. Top predators are greatly reduced in range and abundance, and large-scale ecological processes are altered. Land use in areas between remaining fragments is largely incompatible with maintaining most native species and communities.

Vulnerable: Remaining intact habitat occurs across a wide range of habitat blocks, including some that are large and likely to persist over the long term, especially if given adequate conservation attention. Some area-sensitive species have been extirpated from parts of the ecoregion or are declining, particularly top predators, larger primates, and hunted species. Land use in areas between remaining fragments sometimes is compatible with maintaining native species and communities.

Relatively stable: The natural habitats have been slightly altered in some areas, causing local species declines and disruption of ecosystem processes. Disturbed areas can be extensive but are still patchily distributed relative to the area of intact habitats. Ecological links between intact habitat blocks are still largely functional. Groups of species that are sensitive to human activities, such as top predators, larger primates, and ground-dwelling birds, are present but occur at densities below the natural range of variation.

Relatively intact: The natural habitats in an ecoregion are largely intact, with species, populations, and ecosystem processes occurring within their natural ranges of variation. Groups of species that are sensitive to human activities, such as top predators, larger primates, and ground-dwelling birds, occur at densities within the natural range of variation. The biota is able to disperse naturally across the landscape.

Final (Future Threat-Modified) Conservation Status

In order to assess where biodiversity might be increasingly threatened in the future, we modified our Current Conserva-

tion Status Index using estimates of future threat. Two sets of data were used to estimate future threat.

Future Threat to Habitats

Projected population density for the year 2025 was used as a proxy for future threat to habitats in each ecoregion. Projected population density calculations were provided by the World Resources Institute, based on original data from CIESIN et al. (2000) and UNDP (1999). Points were assigned according to the projected population density, with a maximum contribution of 60 points to the Final (Threat-Modified) Conservation Status Index (table D.5).

Future Threat to Species

The IUCN Red Data Book (Hilton-Taylor 2000) was used to assess the future threat to species, measured as their risk of becoming extinct. Data for birds and mammals were used because information is most complete for these taxa (Hilton-Taylor 2000), whereas it is only partially complete for amphibians (S. Stuart, pers. comm., 2002) and very incomplete for reptiles. The analysis included all species in each ecoregion that are considered vulnerable, endangered, or critically endangered. Weightings were assigned according to each species' status, with critical given a weighting of three, endangered given a weighting of two, and vulnerable given a score of one. The species scores were added to produce a future species threat score for each ecoregion, with a maximum contribution of 40 points to the Final (Future Threat-Modified) Conservation Status Index (table D.6).

The points for future threat to habitats (contributing up to 60 points) and future threat to species (contributing up to 40 points) were added to produce our Final (Threat-Modified) Conservation Status Index, rated between 0 and 100. Ecoregions scoring 70 or more were elevated one class from their Current Conservation Status. For example, an endangered ecoregion with a future threat score greater than 70 was elevated to critical, whereas the conservation status for ecoregions with moderate or low threat levels remained unchanged.

Biological Data Used for the Ecoregion Analysis

Ecoregion Number	Ecoregion Name	Area (km²)	Amphibian Richness	Amphibian Endemism	Amphibian Strict Endemism	Bird Richness	Bird Endemism	Bird Strict Endemism	Mammal Richness	Mammal Endemism	Mammal Strict Endemism
	*Tropical and subtropical moist broadleaf forests*										
1	Western Guinean Lowland Forests	205,100	57	17	13	454	11	3	177	21	3
2	Guinean Montane Forests	31,100	20	13	11	406	3	0	128	9	1
3	Eastern Guinean Forests	189,700	61	16	13	517	6	0	185	18	4
4	Nigerian Lowland Forests	67,300	35	2	1	407	2	1	134	3	1
5	Niger Delta Swamp Forests	14,400	3	0	0	333	1	0	101	2	0
6	Cross-Niger Transition Forests	20,700	29	0	0	373	0	0	112	1	0
7	Cross-Sanaga-Bioko Coastal Forests	52,100	108	9	2	485	13	1	163	11	4
8	Atlantic Equatorial Coastal Forests	189,700	93	13	9	484	7	0	169	4	0
9	Mount Cameroon and Bioko Montane Forests	1,100	81	12	5	403	23	3	126	11	2
10	Cameroon Highlands Forests	38,000	89	26	15	573	24	7	138	14	8
11	São Tomé, Príncipe, and Annobon Moist Lowland Forests	1,000	11	11	11	106	29	29	6	1	1
12	Northwestern Congolian Lowland Forests	434,100	75	6	2	525	3	1	186	15	4
13	Western Congolian Swamp Forests	128,600	19	1	0	321	2	0	132	0	0
14	Eastern Congolian Swamp Forests	92,700	13	1	1	352	3	0	123	1	1
15	Central Congolian Lowland Forests	414,800	18	2	1	342	1	0	121	4	1
16	Northeastern Congolian Lowland Forests	533,500	62	17	8	554	10	3	198	15	5
17	Albertine Rift Montane Forests	103,900	66	39	33	732	51	34	221	34	26
18	East African Montane Forests	65,500	17	4	2	621	14	4	186	19	9
19	Eastern Arc Forests	23,700	78	33	23	547	37	17	161	15	7
20	Northern Zanzibar–Inhambane Coastal Forest Mosaic	112,600	55	8	3	552	17	10	171	11	8
21	Southern Zanzibar–Inhambane Coastal Forest Mosaic	147,000	47	1	0	445	8	2	114	2	2
22	Maputaland Coastal Forest Mosaic	30,200	47	1	0	389	4	0	102	0	0
23	Kwazulu-Cape Coastal Forest Mosaic	17,800	40	6	0	373	7	0	80	3	1
24	Knysna-Amatole Montane Forests	3,100	24	5	3	272	5	0	52	3	1
25	Ethiopian Lower Montane Forests, Woodlands, and Bushlands	248,800	28	6	5	483	5	3	130	8	5
26	Comoros Forests	2,100	1	1	1	79	23	21	6	2	2
27	Granitic Seychelles Forests	300	5	4	4	54	11	11	2	1	0
28	Mascarene Forests	4,900	1	0	0	54	18	18	8	3	3
29	Madagascar Humid Forests	112,400	88	60	45	180	35	16	77	34	14
30	Madagascar Subhumid Forests	199,300	86	48	30	154	14	1	71	14	0

[a]Plant data provided by University of Bonn, Germany.

[b]VH = very high; H = high; M = medium; L = low (values based on expert estimates at the Africa ecoregion workshop, Cape Town, 1998)

[c]GO = globally outstanding; RO = regionally outstanding; BO = bioregionally outstanding; LI = locally important

[d]No = no globally outstanding phenomena; INT = intact vertebrate assemblages and habitat; MIG = globally outstanding migrations; RAD = globally outstanding species radiations; RAR = rare biome; UHT = unique higher taxa.

Reptile Richness	Reptile Endemism	Reptile Strict Endemism	Vertebrate Richness	Vertebrate Endemism	Vertebrate Strict Endemism	Plant Richness[a]	Plant Endemism[b]	Invertebrate Richness[b]	Invertebrate Endemism[b]	Endemism Index	Richness Index	Richness and Endemism Index[c]	Nonspecies Values[d]	Biological Distinctiveness Index[c]
87	12	3	775	61	22	3,300	VH	VH	VH	54	34	GO	No	GO
51	4	1	605	29	13	2,700	M	H	H	26	35	RO	No	RO
92	9	1	855	49	18	3,200	H	VH	H	39	37	RO	No	RO
74	1	1	650	8	4	3,000	H	H	H	25	35	BO	No	BO
63	3	0	500	6	0	1,500	L	H	M	7	32	LI	No	LI
69	1	0	583	2	0	2,100	L	H	M	6	38	LI	No	LI
129	18	6	885	51	13	4,000	H	H	H	38	61	GO	No	GO
120	16	8	866	40	17	6,000	VH	VH	VH	44	48	GO	INT	GO
55	4	0	665	50	10	3,300	M	VH	VH	39	94	GO	No	GO
38	12	9	838	76	39	3,300	H	H	VH	52	47	GO	No	GO
22	18	16	145	59	57	950	M	M	H	33	10	RO	RAD, UHT	GO
123	16	6	909	40	13	3,800	H	VH	M	30	36	RO	INT	GO
77	3	0	549	6	0	2,300	L	VH	H	13	25	BO	INT	GO
75	2	0	563	7	2	2,100	L	H	H	13	26	BO	INT	GO
77	3	0	558	10	2	3,500	M	VH	H	20	21	BO	INT	GO
117	12	1	931	54	17	3,500	VH	H	M	40	33	RO	INT	GO
130	21	10	1,149	145	103	3,200	H	H	H	71	55	GO	No	GO
62	16	10	886	53	25	4,000	L	H	VH	35	44	RO	No	RO
84	39	25	870	124	72	3,200	VH	H	VH	70	59	GO	No	GO
192	76	34	970	112	55	3,000	H	H	H	51	52	GO	No	GO
130	34	16	736	45	20	2,600	L	M	M	16	35	BO	No	BO
107	22	2	645	27	2	2,700	M	H	L	12	45	BO	No	BO
88	14	5	581	30	6	2,000	M	L	L	14	41	BO	No	BO
36	4	1	384	17	5	1,200	L	M	H	16	33	BO	No	BO
60	21	6	701	40	19	4,000	M	L	H	28	27	RO	No	RO
18	11	10	104	37	34	1,500	H	M	H	33	6	RO	RAD, UHT	GO
31	18	16	92	34	31	280	VH	H	VH	42	9	GO	MIG, UHT, RAD	GO
19	10	9	82	31	30	950	H	M	H	32	2	RO	RAD, UHT	GO
129	81	54	474	210	129	6,000	VH	VH	VH	95	42	GO	RAD, UHT	GO
140	81	43	451	157	74	3,200	VH	H	VH	75	31	GO	RAD, UHT	GO

Ecoregion Number	Ecoregion Name	Area (km²)	Amphibian Richness	Amphibian Endemism	Amphibian Strict Endemism	Bird Richness	Bird Endemism	Bird Strict Endemism	Mammal Richness	Mammal Endemism	Mammal Strict Endemism
Tropical and subtropical dry broadleaf forests											
31	Cape Verde Islands Dry Forests	4,600	2	1	0	66	11	11	5	0	0
32	Zambezian *Cryptosepalum* Dry Forests	38,200	14	0	0	380	0	0	93	1	0
33	Madagascar Dry Deciduous Forests	152,100	20	5	1	162	16	2	55	14	8
Temperate coniferous forests											
34	Mediterranean Conifer and Mixed Forests	23,100	3	0	0	218	1	1	60	0	0
Tropical and subtropical grasslands, savannas, shrublands, and woodlands											
35	Sahelian *Acacia* Savanna	3,053,600	16	2	1	540	3	0	151	14	8
36	West Sudanian Savanna	1,638,400	25	4	3	552	3	0	173	7	2
37	East Sudanian Savanna	917,600	31	1	0	607	5	2	178	5	1
38	Guinean Forest-Savanna Mosaic	673,600	33	1	0	551	3	0	199	7	2
39	Mandara Plateau Mosaic	7,500	10	0	0	319	0	0	84	0	0
40	Northern Congolian Forest-Savanna Mosaic	708,100	63	12	7	610	2	1	211	3	1
41	Victoria Basin Forest-Savanna Mosaic	165,800	27	2	1	623	5	3	219	9	3
42	Southern Congolian Forest-Savanna Mosaic	569,700	42	4	2	553	0	0	178	7	5
43	Western Congolian Forest-Savanna Mosaic	413,400	33	2	1	552	2	1	174	3	2
44	Somali *Acacia-Commiphora* Bushlands and Thickets	1,053,900	19	2	2	539	15	8	160	18	7
45	Northern *Acacia-Commiphora* Bushlands and Thickets	326,000	15	2	1	688	6	4	210	11	4
46	Southern *Acacia-Commiphora* Bushlands and Thickets	227,800	17	2	0	596	5	0	178	4	0
47	Serengeti Volcanic Grasslands	18,000	11	0	0	426	3	0	133	0	0
48	Itigi-Sumbu Thicket	7,800	12	0	0	359	0	0	107	0	0
49	Angolan Miombo Woodlands	660,100	52	3	2	531	3	0	170	5	1
50	Central Zambezian Miombo Woodlands	1,184,200	84	15	13	698	8	2	236	4	2
51	Zambezian *Baikiaea* Woodlands	264,400	50	1	0	507	2	0	161	0	0
52	Eastern Miombo Woodlands	483,900	41	1	0	456	1	0	145	0	0
53	Southern Miombo Woodlands	408,300	45	0	0	508	6	0	176	1	0
54	Zambezian and Mopane Woodlands	473,300	50	0	0	552	7	1	195	1	0
55	Angolan Mopane Woodlands	133,500	20	2	1	383	2	0	125	3	0
56	Western Zambezian Grasslands	34,000	29	0	0	366	0	0	135	1	0
57	Southern Africa Bushveld	223,100	37	0	0	471	1	0	148	1	0
58	Kalahari *Acacia* Woodlands	335,500	30	0	0	470	1	0	127	0	0

Reptile Richness	Reptile Endemism	Reptile Strict Endemism	Vertebrate Richness	Vertebrate Endemism	Vertebrate Strict Endemism	Plant Richness[a]	Plant Endemism[b]	Invertebrate Richness[b]	Invertebrate Endemism[b]	Endemism Index	Richness Index	Richness and Endemism Index[c]	Nonspecies Values[d]	Biological Distinctiveness Index[c]
15	11	11	88	23	22	257	M	M	M	30	0	RO	RAR	RO
34	2	0	521	3	0	1,100	L	H	L	2	79	GO	RAR	GO
83	35	20	320	70	31	2,100	VH	H	VH	100	72	GO	RAD, RAR, UHT	GO
27	1	0	308	2	1	2,300	M	M	VH	56	100	GO	RAR, UHT	GO
113	15	6	820	34	15	1,300	L	M	L	22	6	BO	No	BO
129	27	9	879	41	14	2,100	M	L	L	26	25	RO	No	RO
121	11	1	937	22	4	2,300	M	L	L	19	38	BO	No	BO
125	30	11	908	41	13	2,500	H	M	M	34	47	RO	No	RO
18	2	0	431	2	0	600	L	M	L	0	35	LI	No	LI
145	16	3	1,029	33	12	2,500	M	M	L	27	62	RO	No	RO
119	5	3	988	21	10	2,700	L	H	M	22	82	GO	No	GO
126	10	2	899	21	9	3,000	L	M	L	13	54	BO	No	BO
130	13	2	889	20	6	3,300	M	M	L	15	61	BO	No	BO
177	74	32	895	109	49	2,600	H	M	L	64	38	GO	No	GO
99	17	1	1,012	36	10	1,300	M	M	L	28	48	RO	MIG	RO
90	10	3	881	21	3	2,300	H	H	L	25	54	BO	MIG, INT	GO
64	2	0	634	5	0	1,200	L	H	L	4	66	GO	MIG, INT	GO
41	2	0	519	2	0	1,000	4	M	M	11	62	BO	No	BO
86	18	8	839	29	11	2,000	H	M	L	26	39	RO	No	RO
186	27	15	1,204	54	32	4,000	H	M	L	46	79	GO	No	GO
88	3	0	806	6	0	1,400	H	M	L	15	48	BO	No	BO
101	11	1	743	13	1	2,900	M	M	L	10	44	LI	INT	GO
123	13	3	852	20	3	2,200	M	M	L	16	53	BO	No	BO
163	32	4	960	40	5	2,500	M	M	M	27	65	RO	No	RO
101	19	4	629	26	5	1,100	L	M	L	11	36	BO	No	BO
49	3	0	579	4	0	800	L	L	L	1	50	LI	No	LI
135	23	7	791	25	7	1,500	H	M	L	18	53	BO	No	BO
91	4	1	718	5	1	1,600	L	M	L	2	33	LI	No	LI

Ecoregion Number	Ecoregion Name	Area (km²)	Amphibian Richness	Amphibian Endemism	Amphibian Strict Endemism	Bird Richness	Bird Endemism	Bird Strict Endemism	Mammal Richness	Mammal Endemism	Mammal Strict Endemism
Flooded grasslands and savannas											
59	Sudd Flooded Grasslands	179,700	7	0	0	395	0	0	99	1	1
60	Nile Delta Flooded Savanna	51,000	7	0	0	184	0	0	55	1	1
61	Lake Chad Flooded Savanna	18,800	3	0	0	293	2	0	69	2	0
62	Inner Niger Delta Flooded Savanna	46,000	4	0	0	288	2	0	58	0	0
63	Zambezian Flooded Grasslands	153,500	36	2	2	498	5	1	130	0	0
64	Zambezian Coastal Flooded Savanna	19,500	21	0	0	309	1	0	83	0	0
65	Saharan Halophytics	53,900	4	0	0	93	0	0	58	0	0
66	East African Halophytics	2,600	1	0	0	223	0	0	76	0	0
67	Etosha Pan Halophytics	7,200	1	0	0	217	0	0	64	0	0
68	Makgadikgadi Halophytics	30,400	1	0	0	256	0	0	55	0	0
Montane grasslands and shrublands											
69	Mediterranean High Atlas Juniper Steppe	6,300	4	0	0	106	0	0	35	0	0
70	Ethiopian Upper Montane Forests, Woodlands, Bushlands, and Grasslands	245,400	33	5	2	517	15	4	122	14	2
71	Ethiopian Montane Moorlands	25,200	29	5	2	348	7	0	85	14	4
72	East African Montane Moorlands	3,300	12	2	2	117	5	1	67	3	1
73	Rwenzori-Virunga Montane Moorlands	2,700	4	1	1	180	5	0	61	1	1
74	Southern Rift Montane Forest-Grassland Mosaic	33,500	48	5	1	485	15	1	159	4	0
75	South Malawi Montane Forest-Grassland Mosaic	10,200	34	6	3	412	5	1	133	1	0
76	Eastern Zimbabwe Montane Forest-Grassland Mosaic	7,800	23	6	4	400	3	2	115	2	2
77	Highveld Grasslands	186,200	29	0	0	397	4	1	115	1	1
78	Drakensberg Montane Grasslands, Woodlands, and Forests	202,200	47	6	1	450	10	0	152	2	1
79	Drakensberg Alti-Montane Grasslands and Woodlands	11,900	18	2	0	288	7	0	71	0	0
80	Maputaland-Pondoland Bushland and Thickets	19,500	38	8	0	351	5	0	92	2	0
81	Angolan Scarp Savanna and Woodlands	74,400	32	5	4	387	14	6	108	0	0
82	Angolan Montane Forest-Grassland Mosaic	25,500	34	2	1	355	5	2	107	3	0
83	Jos Plateau Forest-Grassland Mosaic	13,300	16	0	0	60	2	2	54	2	2
84	Madagascar Ericoid Thickets	1,300	22	10	5	44	4	0	17	5	1

Reptile Richness	Reptile Endemism	Reptile Strict Endemism	Vertebrate Richness	Vertebrate Endemism	Vertebrate Strict Endemism	Plant Richness[a]	Plant Endemism[b]	Invertebrate Richness[b]	Invertebrate Endemism[b]	Endemism Index	Richness Index	Richness and Endemism Index[c]	Nonspecies Values[d]	Biological Distinctiveness Index[c]
24	2	1	525	3	2	1,300	L	L	L	20	41	BO	MIG, INT	GO
42	0	0	288	1	1	800	L	L	L	9	32	LI	No	LI
16	0	0	381	4	0	600	L	L	L	24	40	BO	No	BO
16	0	0	366	2	0	500	L	L	L	7	26	LI	MIG	LI
90	3	1	754	10	4	1,400	L	L	M	56	88	GO	MIG	GO
56	2	1	469	3	1	1,100	L	L	L	15	85	BO	No	BO
27	1	0	182	1	0	100	L	L	L	6	9	LI	No	LI
4	0	0	304	0	0	80	L	L	L	0	41	LI	MIG	LI
11	1	0	293	1	0	80	L	L	L	6	28	LI	No	LI
11	0	0	323	0	0	80	L	L	L	0	15	LI	No	LI
27	4	3	172	4	3	1,000	H	M	VH	29	20	RO	RAR, UHT	GO
56	13	2	728	47	10	2,000	M	L	H	63	41	GO	No	GO
17	3	0	479	29	6	700	M	L	H	50	35	GO	RAD, RAR	GO
9	4	3	205	14	7	400	M	L	H	30	22	RO	RAD, RAR	GO
9	2	0	254	9	2	400	M	L	H	25	21	BO	RAD, RAR	GO
46	14	7	738	38	9	1,900	M	L	VH	57	79	GO	No	GO
37	13	6	616	25	10	1,900	M	L	H	38	85	GO	No	GO
57	9	5	595	20	13	1,500	M	L	H	35	82	GO	No	GO
68	7	1	609	12	3	1,900	M	L	M	19	41	BO	No	BO
118	44	19	767	62	21	3,700	VH	L	M	62	74	GO	No	GO
29	9	3	406	18	3	800	VH	L	H	42	38	GO	No	GO
63	6	1	544	21	1	2,200	L	M	M	29	76	RO	No	RO
50	9	2	577	28	12	1,200	H	L	M	43	42	GO	No	GO
44	8	3	540	18	6	2,300	H	L	M	32	67	RO	No	RO
61	9	0	191	13	4	1,300	M	H	H	24	38	BO	No	BO
19	8	4	102	27	10	800	VH	L	VH	63	34	GO	RAD, RAR, UHT	GO

Mediterranean forests, woodlands, and scrub

Ecoregion Number	Ecoregion Name	Area (km²)	Amphibian Richness	Amphibian Endemism	Amphibian Strict Endemism	Bird Richness	Bird Endemism	Bird Strict Endemism	Mammal Richness	Mammal Endemism	Mammal Strict Endemism
85	Mediterranean Woodlands and Forests	358,300	4	0	0	283	0	0	89	1	0
86	Mediterranean Dry Woodlands and Steppe	292,200	5	0	0	231	0	0	80	0	0
87	Mediterranean *Acacia-Argania* Dry Woodlands and Succulent Thickets	100,100	7	1	0	229	6	1	75	4	3
88	Canary Islands Dry Woodlands and Forests	5,000	2	1	0	88	8	4	9	3	2
89	Lowland Fynbos and Renosterveld	32,800	27	9	4	296	13	0	75	5	1
90	Montane Fynbos and Renosterveld	45,800	31	11	6	311	11	0	88	6	0
91	Albany Thickets	17,100	14	1	0	280	3	0	68	1	0

Deserts and xeric shrublands

Ecoregion Number	Ecoregion Name	Area (km²)	Amphibian Richness	Amphibian Endemism	Amphibian Strict Endemism	Bird Richness	Bird Endemism	Bird Strict Endemism	Mammal Richness	Mammal Endemism	Mammal Strict Endemism
92	Atlantic Coastal Desert	40,000	4	0	0	75	0	0	34	0	0
93	North Saharan Steppe	1,676,100	6	0	0	184	0	0	73	1	1
94	South Saharan Steppe	1,101,700	6	0	0	220	0	0	71	4	1
95	Sahara Desert	4,639,900	8	0	0	210	0	0	57	0	0
96	West Saharan Montane Xeric Woodlands	258,100	6	0	0	57	0	0	53	1	0
97	Tibesti–Jebel Uweinat Montane Xeric Woodlands	82,200	0	0	0	58	0	0	38	1	1
98	Red Sea Coastal Desert	59,300	6	0	0	87	0	0	46	1	1
99	East Saharan Montane Xeric Woodlands	27,900	2	0	0	232	1	0	64	3	2
100	Eritrean Coastal Desert	4,600	5	0	0	209	0	0	48	0	0
101	Ethiopian Xeric Grasslands and Shrublands	153,100	6	1	0	359	3	1	74	6	1
102	Somali Montane Xeric Woodlands	62,600	13	0	0	159	5	3	58	5	1
103	Hobyo Grasslands and Shrublands	25,600	15	0	0	147	3	2	54	4	2
104	Masai Xeric Grasslands and Shrublands	101,000	9	2	2	327	1	0	99	4	0
105	Kalahari Xeric Savanna	588,100	15	1	0	333	0	0	99	1	0
106	Kaokoveld Desert	45,700	16	1	0	233	2	0	55	2	0
107	Namib Desert	80,900	7	1	0	189	6	1	40	6	1
108	Nama Karoo	351,100	11	0	0	300	8	2	106	8	2
109	Namib Escarpment Woodlands	225,500	13	2	0	303	5	1	105	9	2
110	Succulent Karoo	102,700	15	3	2	225	5	1	75	7	1
111	Socotra Island Xeric Shrublands	3,800	0	0	0	78	6	6	2	0	0
112	Aldabra Island Xeric Scrub	200	0	0	0	44	6	4	6	0	0
113	Madagascar Spiny Thickets	43,300	7	0	0	151	14	8	35	4	2
114	Madagascar Succulent Woodlands	79,800	5	0	0	152	7	1	35	6	0

Reptile Richness	Reptile Endemism	Reptile Strict Endemism	Vertebrate Richness	Vertebrate Endemism	Vertebrate Strict Endemism	Plant Richness[a]	Plant Endemism[b]	Invertebrate Richness[b]	Invertebrate Endemism[b]	Endemism Index	Richness Index	Richness and Endemism Index[c]	Nonspecies Values[d]	Biological Distinctiveness Index[c]
65	15	8	441	16	8	1,500	M	L	L	16	32	BO	RAR, UHT	GO
58	3	0	374	3	0	1,200	L	M	L	0	28	LI	RAR	LI
46	10	3	357	21	7	1,600	H	M	M	41	37	RO	RAR	RO
10	8	7	109	20	13	1,130	VH	M	H	51	2	GO	RAD, RAR, UHT	GO
68	21	5	466	48	10	4,000	VH	VH	VH	88	91	GO	RAD, RAR, UHT	GO
74	31	15	504	59	21	5,000	VH	H	VH	97	98	GO	RAD, RAR, UHT	GO
55	7	2	417	12	2	1,700	H	L	M	27	73	RO	No	RO
19	1	0	132	1	0	300	L	L	L	1	19	LI	MIG	LI
64	7	2	327	8	3	1,150	M	L	L	11	23	BO	No	BO
53	0	0	350	4	1	500	M	L	L	13	23	BO	No	BO
62	3	1	337	3	1	500	L	L	L	2	16	LI	INT	LI
34	1	0	150	2	0	550	L	L	L	3	19	LI	INT	LI
15	0	0	111	1	1	580	L	L	L	2	12	LI	INT	LI
32	1	1	171	2	2	350	L	L	M	8	26	LI	No	LI
7	0	0	305	4	2	700	L	L	L	7	37	LI	No	LI
25	2	0	287	2	0	100	L	L	L	1	56	LI	No	LI
79	20	2	518	30	4	850	L	M	L	29	52	RO	No	RO
44	14	3	274	24	7	1,500	M	L	L	27	50	RO	No	RO
22	6	2	238	13	6	800	H	M	L	25	51	BO	No	BO
28	6	1	463	13	3	900	L	M	L	23	53	BO	No	BO
104	8	1	551	10	1	700	L	L	L	11	51	BO	INT	BO
62	26	8	366	31	8	500	M	VH	VH	45	61	GO	RAD, UHT	GO
67	27	5	303	40	7	1,000	L	VH	VH	52	42	GO	RAD, INT, UHT	GO
70	15	1	487	31	5	1,100	H	H	H	53	49	GO	No	GO
117	32	7	538	48	10	1,100	H	H	H	70	65	GO	No	GO
94	39	15	409	54	19	4,850	VH	H	VH	86	79	GO	RAD, UHT	GO
28	25	25	108	31	31	750	H	L	M	34	12	RO	RAD, UHT	GO
10	7	7	60	13	11	180	L	L	L	10	12	LI	MIG, RAD	GO
82	38	26	275	56	36	1,100	VH	M	VH	74	48	GO	RAD, UHT	GO
55	14	1	247	27	2	1,400	VH	H	VH	59	35	GO	RAD, UHT	GO

Ecoregion Number	Ecoregion Name	Area (km²)	Amphibian Richness	Amphibian Endemism	Amphibian Strict Endemism	Bird Richness	Bird Endemism	Bird Strict Endemism	Mammal Richness	Mammal Endemism	Mammal Strict Endemism
Mangroves											
115	Guinean Mangroves	23,500	2	0	0	191	0	0	32	0	0
116	Central African Mangroves	30,900	2	0	0	217	3	0	28	0	0
117	Southern Africa Mangroves	1,000	2	0	0	224	1	0	26	0	0
118	East African Mangroves	16,100	2	0	0	235	1	0	26	0	0
119	Madagascar Mangroves	5,200	1	0	0	125	7	0	4	0	0

Reptile Richness	Reptile Endemism	Reptile Strict Endemism	Vertebrate Richness	Vertebrate Endemism	Vertebrate Strict Endemism	Plant Richness[a]	Plant Endemism[b]	Invertebrate Richness[b]	Invertebrate Endemism[b]	Endemism Index	Richness Index	Richness and Endemism Index[c]	Nonspecies Values[d]	Biological Distinctiveness Index[c]
8	1	0	233	1	0	50	L	L	L	0	0	LI	No	LI
6	0	0	258	1	0	50	L	L	L	0	0	LI	No	LI
2	0	0	132	7	0	50	L	L	L	0	0	LI	No	LI

Nonspecies Biological Values of Ecoregions

(Only Ecoregions with Nonspecies Biological Values Are Included in the Table)

Tropical and subtropical moist broadleaf forests

Ecoregion No.	Ecoregion Name	Ecological Phenomena	Evolutionary Phenomena	Rare Habitat	Wilderness	Higher Taxonomic Uniqueness	Total Score	Comments
8	Atlantic Equatorial Coastal Forests	—	—	—	1	—	1.0	Mittermeier et al. (2002). Tropical wilderness area.
11	São Tomé, Príncipe, and Annobon Moist Lowland Forests	1	—	—	0.5	—	1.5	Endemic genera and gigantism and dwarfism in birds. Endemic genera of gastropods and other plant and invertebrate groups.
12	Northwestern Congolian Lowland Forests	0.5	—	—	1	—	1.5	Mittermeier et al. (2002). Tropical wilderness area. Migrations of forest taxa still occur.
13	Western Congolian Swamp Forests	0.5	—	—	1	—	1.5	Mittermeier et al. (2002). Tropical wilderness area. Migrations of forest taxa still occur.
14	Eastern Congolian Swamp Forests	0.5	—	—	1	—	1.5	Mittermeier et al. (2002). Tropical wilderness area. Migrations of forest taxa still occur.
15	Central Congolian Lowland Forests	0.5	—	—	1	—	1.5	Mittermeier et al. (2002). Tropical wilderness area. Migrations of forest taxa still occur.
16	Northeastern Congolian Lowland Forests	0.5	—	—	1	—	1.5	Mittermeier et al. (2002). Tropical wilderness area. Migrations of forest taxa still occur.
26	Comoros Forests	—	1	—	0.5	—	1.5	Endemic genera of plants and other taxa. Radiations of species in several genera.
27	Granitic Seychelles Forests	1	1	—	1	—	3.0	One endemic plant family (Medusagynaceae), one endemic amphibian family (Sooglossidae). Radiations of palms. Aride Island, 1,219,200 breeding seabirds; Desnoeufs Island, 1,080,000 breeding seabirds (Fishpool and Evans 2001).
28	Mascarene Forests	—	1	—	0.5	—	1.5	33% of plant species are endemic, with several endemic plant genera. Round Island keel scaled boa (Round Is.) and Round Island burrowing boa (Round Is., probably extinct) represent a unique family. Formally supported the endemic bird family Raphidae, but the dodo and Rodriquez solitaire are both extinct.
29	Madagascar Humid Forests	—	1	—	1	—	2.0	Representatives of Madagascar's eight endemic plant families (Asteropeiacea, Didiereacea, Didymelaceae, Diegodendraceae, Geosiridaceae, Humbertiaceae, Sphaerosepalaceae, and Sarcolaenaceae). Examples of many of Madagascar's 238 endemic plant genera. Extraordinary radiation of lemurs and tenrecs. One endemic bat, *Myzopoda aurita*, constitutes an endemic family (eastern forests). Members of three endemic bird families (mesites, asities, and ground rollers).
30	Madagascar Subhumid Forests	—	1	—	1	—	2.0	Representatives of Madagascar's eight endemic plant families (Asteropeiacea, Didiereacea, Didymelaceae, Diegodendraceae, Geosiridaceae, Humbertiaceae, Sphaerosepalaceae, and Sarcolaenaceae). Examples of some of Madagascar's 238 endemic plant genera. Extraordinary radiation of lemurs. Members of three endemic bird families (mesites, asities, and ground rollers).

Tropical and subtropical dry broadleaf forests

No.	Ecoregion							Description
31	Cape Verde Islands Dry Forests	—	—	1	—	—	1.0	Tropical dry forest is globally rare.
32	Zambezian *Cryptosepalum* Dry Forests	—	—	1	—	—	1.0	Tropical dry forest is globally rare.
33	Madagascar Dry Deciduous Forests	—	1	1	1	—	3.0	Members of Madagascar's eight endemic plant families (Asteropeiacea, Didiereaceae, Didymelaceae, Diegodendraceae, Geosiridaceae, Humbertiaceae, Sphaerosepalaceae, and Sarcolaenaceae). 238 endemic plant genera. Extraordinary radiation of lemurs. Members of three endemic bird families (mesites, asities, and ground rollers). Tropical dry forest is globally rare.

Temperate coniferous forests

No.	Ecoregion							Description
34	Mediterranean Conifer and Mixed Forests	—	—	1	—	0.5	1.5	Many Mediterranean Basin endemic plant genera, including groups with a Tertiary evolutionary history and those exhibiting more recent species radiations. Mediterranean habitats are globally rare.

Tropical and subtropical grasslands, savannas, shrublands, and woodlands

No.	Ecoregion							Description
37	East Sudanian Savanna	0.5	—	—	—	—	0.5	Seasonal elephant migration has been severely reduced by poaching, but annual migrations of korrigum (*Damaliscus lunatus*) still take place.
45	Northern *Acacia-Commiphora* Bushlands and Thickets	1	—	—	—	—	1.0	Wildebeest migration population of >1,000,000 individuals (East 1999). Lake Bogoria National Reserve, 1,092,240 waterbirds; Lake Nakuru National Park, 1,526,580 waterbirds (Fishpool and Evans 2001)
46	Southern *Acacia-Commiphora* Bushlands and Thickets	1	—	—	1	—	2.0	Migrations of elephant and other mammals still occur. Large blocks of intact habitat remain.
47	Serengeti Volcanic Grasslands	1	—	—	1	—	2.0	Serengeti National Park, Ngorongoro Conservation Area, Maswa Game Reserve, and Ikongoro and Loliondo game controlled areas together include what is likely to be the largest concentration of large mammals anywhere in the world. Wildebeest migration population exceeds 1,000,000 individuals (East 1999). Lake Manyara National Park, 2,294,349 waterbirds (Fishpool and Evans 2001). Mittermeier et al. (2002). Tropical wilderness area.
50	Central Zambezian Miombo Woodlands	0.5	—	—	—	—	0.5	Migrations of large mammals still occur in some places.
51	Zambezian *Baikiaea* Woodlands	0.5	—	—	—	—	0.5	Migrations of large mammals still occur in some places.
52	Eastern Miombo Woodlands	0.5	—	—	1	—	1.5	Mittermeier et al. (2002). Tropical wilderness area. Selous-Mikumi National Parks and surrounding areas harbor one of the largest remaining elephant populations. Migration still occurs between Selous in Tanzania and Niassa in Mozambique.
54	Zambezian and Mopane Woodlands	0.5	—	—	—	—	0.5	Migrations of large mammals still occur in some places.

Ecoregion No.	Ecoregion Name	Ecological Phenomena	Evolutionary Phenomena	Rare Habitat	Wilderness	Higher Taxonomic Uniqueness	Total Score	Comments
Flooded grasslands and savannas								
59	Sudd Flooded Grasslands	1	—	—	1	—	2.0	The kob (*Kobus kob*), tiang (*Damaliscus lunatus tiang*), and Mongalla gazelle (*Gazella thomsonii albonotata*) migrate in this ecoregion. An important migration route stopover for birds, contains at least 2,959,491 waterbirds (Fishpool and Evans 2001). Mittermeier et al. (2002). Tropical wilderness area.
61	Lake Chad Flooded Savanna	0.5	—	—	—	—	0.5	Important stopover for migratory waterbirds and passerines, with more than 1 million individuals in some years (Fishpool and Evans 2001).
62	Inner Niger Delta Flooded Savanna	1	—	—	—	—	1.0	Important stopover for millions of migratory passerines and waterbirds and the most important area in West Africa for overwintering Palearctic waterbirds. Lac Débo–Lac Oualado Débo, 1,011,086 waterbirds (Fishpool and Evans 2001).
63	Zambezian Flooded Grasslands	1	—	—	—	—	1.0	Lake Ngami, 1,569,400 waterbirds; Lake Chilwa and floodplain, 1,011,180 waterbirds (Fishpool and Evans 2001). Large mammal migrations.
66	East African Halophytics	1	—	—	—	—	1.0	Lake Natron and Engaruka basin, 1,500,000 waterbirds (Fishpool and Evans 2001).
Montane grasslands and shrublands								
69	Mediterranean High Atlas Juniper Steppe	—	—	1	—	0.5	1.5	Endemic plant genera showing Tertiary evolutionary histories. Globally rare habitat type.
71	Ethiopian Montane Moorlands	—	1	1	—	—	2.0	Species radiations in plant genera adapted to the harsh environmental conditions. Invertebrate radiations also responding to environmental pressures. Globally rare habitat.
72	East African Montane Moorlands	—	1	1	—	—	2.0	Species radiations in plant genera adapted to harsh environmental conditions, especially in giant lobelias and giant groundsels. Also invertebrate radiations responding to the same environmental pressures. Globally rare habitat.
73	Rwenzori-Virunga Montane Moorlands	—	1	1	—	—	2.0	Species radiations in plant genera adapted to the harsh environmental conditions. Also invertebrate radiations responding to the same environmental pressures. Globally rare habitat.
84	Madagascar Ericoid Thickets	—	1	1	—	1	3.0	Globally rare habitat. Contains examples of Madagascar endemic families and genera with ancient evolutionary history and species radiations.

Mediterranean forests, woodlands, and scrub

No.	Ecoregion								Notes
85	Mediterranean Woodlands and Forests	—	—	1	—	—	0.5	1.5	Plant radiations of Mediterranean endemic genera. Also contains Tertiary relict species. Globally rare habitat type.
86	Mediterranean Dry Woodlands and Steppe	—	—	1	—	—	—	1.0	Plant radiations of Mediterranean endemic genera. Also contains Tertiary relict species. Globally rare habitat type.
87	Mediterranean *Acacia-Argania* Dry Woodlands and Succulent Thickets	1	—	—	1.0	—	—	—	Plant radiations of Mediterranean endemic genera. Also contains Tertiary relict species. Globally rare habitat type.
88	Canary Islands Dry Woodlands and Forests	—	1	—	—	1	0.5	2.5	Plant species radiations of endemic genera (e.g., *Aeonium* (32 spp.), *Sonchus* (24 spp.), *Sideritis* (25 spp.). More than 20 endemic genera. Globally rare habitat.
89	Lowland Fynbos and Renosterveld	—	1	1	—	1	1	3.0	Seven endemic plant families in the Cape Floristic Region (Penaeaceae, 5 genera, 25 spp.; Stilbaceae, 5 genera, 12 spp.; Grubbiaceae, 2 genera, 5 spp.; Roridulaceae, 1 genus, 2 spp.; Geissolomataceae, 1 sp.; Retziaceae, 1 sp.; Bruniaceae, 12 genera, 75 spp.). Species radiations in plants and invertebrates. Globally rare habitat.
90	Montane Fynbos and Renosterveld	—	1	1	—	1	1	3.0	Seven endemic plant families in the Cape Floristic Region (Penaeaceae, 5 genera, 25 spp.; Stilbaceae, 5 genera, 12 spp.; Grubbiaceae, 2 genera, 5 spp.; Roridulaceae, 1 genus, 2 spp.; Geissolomataceae, 1 sp.; Retziaceae, 1 sp.; Bruniaceae, 12 genera, 75 spp.). Species radiations in plants and invertebrates. Globally rare habitat.

Deserts and xeric shrublands

No.	Ecoregion								Notes
92	Atlantic Coastal Desert	1	—	—	—	—	—	1.0	Banc d'Arguin National Park, 2,778,524 waterbirds (Fishpool and Evans 2001).
95	Sahara Desert	—	—	—	—	1	—	1.0	Mittermeier et al. (2002). Tropical wilderness areas.
96	West Saharan Montane Xeric Woodlands	—	—	—	—	1	—	1.0	Mittermeier et al. (2002). Tropical wilderness areas.
97	Tibesti-Jebel Uweinat Montane Xeric Woodlands	—	—	—	—	1	—	1.0	Mittermeier et al. (2002). Tropical wilderness areas.
105	Kalahari Xeric Savanna	—	—	—	—	1	—	1.0	Mittermeier et al. (2002). Tropical wilderness areas.
106	Kaokoveld Desert	—	1	—	—	—	1	2.0	One endemic monotypic plant family (Welwitschiaceae). Taxonomically isolated, monotypic genera. Plant radiations (e.g., *Lithops*, 40 spp.).
107	Namib Desert	—	1	—	—	1	1	3.0	One endemic monotypic plant family (Welwitschiaceae). Taxonomically isolated, monotypic genera. Plant radiations (e.g. *Lithops*, (40 spp). Mittermeier et al. (2002). Tropical wilderness areas.
110	Succulent Karoo	—	1	—	—	—	1	2.0	High succulent diversity of global importance. 35% plant endemics including numerous endemic genera. The plant family Mesembryanthemaceae is near endemic. Radiations in plants and invertebrates.
111	Socotra Island Xeric Shrublands	—	1	—	—	1	0.5	1.5	Endemic genera of plants and birds. Radiations in plants and invertebrates.
112	Aldabra Island Xeric Scrub	1	1	—	—	—	—	2.0	Cosmoledo Atoll, 3,302,850 breeding seabirds. Radiations of island birds. Giant tortoises.
113	Madagascar Spiny Thickets	—	1	—	—	—	1	2.0	Members of Madagascar endemic plant families, especially the near-endemic Didiereaceae. Numerous endemic plant genera. Radiation of lemurs. Representatives of endemic bird families (mesites, asities, and ground rollers).
114	Madagascar Succulent Woodlands	—	1	—	—	—	1	2.0	Members of Madagascar endemic plant families, especially the near-endemic Didiereaceae. Numerous endemic plant genera. Radiation of lemurs. Representatives of endemic bird families (mesites, asities, and ground rollers).

Ecoregions with none of these values are omitted.

APPENDIX G

Investigating Patterns of African Richness and Endemism

Testing Differences in Species Richness between Biomes

The levels of species richness, endemism, and strict endemism differ between African biomes (table G.1). To explore the differences between the species assemblages of biomes, we conducted post facto mean pairwise comparisons using the conservative Tukey test with a significance level of .05 (the temperate coniferous forests biome was excluded because it contains only one ecoregion). The tropical and subtropical grasslands, savannas, shrublands, and woodlands biome was significantly more speciose than all other biomes for bird and mammal species richness, and significantly more species-rich than desert, flooded, montane, and mangrove biomes for reptiles. For amphibian richness, moist forests were significantly more species rich than flooded grasslands, Mediterranean scrub, deserts, and mangroves. This differed somewhat for amphibian endemism, where moist forests had significantly higher values for endemism than savanna-woodlands, flooded and montane grasslands, and deserts. Moist forests also had significantly higher values than savanna-woodlands, flooded grasslands, and deserts for bird endemism, whereas for mammal endemism, moist forests had significantly higher endemism values than flooded grasslands, montane grasslands, deserts, and mangroves. Reptile endemism showed little significant variation among biomes.

These tests serve to highlight the most striking differences between biomes. They show that, on average, grasslands are far more speciose for birds, mammals, and reptiles, and moist forests are more speciose for amphibians than other biomes. However, moist forests exhibit higher levels of bird, mammal, and amphibian endemism than other biomes, whereas reptile endemism appears to be more evenly distributed among biomes. These results confirm that comparing ecoregions across all

TABLE G.1. **Results of 1-Way ANOVAs Testing for Differences in Species Richness, Endemism, and Strict Endemism between Biomes.**

Taxon	Richness	Endemism	Strict Endemism
Birds	12.125***	5.158***	3.149**
Mammals	14.832***	3.988***	2.688**
Reptiles	10.453***	2.113*	—
Amphibians	9.135***	4.858***	4.159***

$*p < .05$, $**p < .01$, $***p < .001$; no entry, $p > .05$.

biomes would have undervalued ecoregions within less speciose biomes (e.g., deserts and xeric shrublands and flooded grasslands and savannas).

Testing Differences in Species Richness in Relation to Ecoregion Area

We calculated the relationship between ecoregion area and species richness for birds, mammals, reptiles, amphibians, and plants. For all three biomes with more than twenty ecoregions there was a significant relationship between species richness and area (table G.2). Within tropical and subtropical moist broadleaf forests (thirty ecoregions) and tropical and subtropical grasslands, savannas, shrublands, and woodlands (twenty-four ecoregions), relationships were significant for all taxa. There was no significant relationship between area and amphibian richness and area and plant richness within deserts and xeric shrublands.

For the other biomes (with fewer than twenty ecoregions), across-ecoregion species-area relationships were rarely signifi-

TABLE G.2. Relationship between Ecoregion Area and Species Richness (Both Log-Transformed) for Biomes with More Than Twenty Ecoregions.

Biome	No. of Ecoregions	Bird Richness	Mammal Richness	Reptile Richness	Amphibian Richness	Plant Richness
Tropical and subtropical moist broadleaf forests	30	.416	.593	.569	.290	.594
Tropical and subtropical grasslands, savannas, shrublands, and woodlands	24	.579	.490	.710	.317	.498
Deserts and xeric shrublands	23	.248	.477	.367	ns	ns

All r^2 values shown are significant at $p < .05$, ns = not significant.

TABLE G.3. Results of Pearson Correlation Analyses of Richness Values across Taxonomic Groups in the Same Biome.

Biome	No. of Ecoregions	Bird-Mammal	Bird-Amphibian	Bird-Reptile	Mammal-Amphibian	Mammal-Reptile	Amphibian-Reptile
Tropical and subtropical moist broadleaf forests	30	.938	.475	.479	.527	.583	.630
Tropical and subtropical grasslands, savannas, shrublands, and woodlands	24	.930	.448	.708	.588	.730	.570
Flooded grasslands and savannas	10	.863	.710	ns	.823	.712	.967
Montane grasslands and shrublands	16	.922	.787	.523	.817	.612	.640
Mediterranean forests, woodlands, and scrub	7	.911	ns	.972	ns	.941	ns
Deserts and xeric shrublands	23	.832	.542	.618	.628	.568	.612
Mangroves	5	ns	.930	ns	.975	.902	.910

All r values are significant at $p < .05$; ns = not significant.

cant at $p < .05$. This does not mean that within each of these ecoregions, species richness is independent of area (merely that across a small sample of ecoregions, confounding effects masked significant area effects).

Testing for Cross-Taxon Congruence in Species Richness within Different African Biomes

If the patterns of species richness are the same for different vertebrate groups across the biomes of Africa, then one indicator group (e.g., birds) could provide equivalent answers to gathering data on all different groups of vertebrates. At larger scales and in the tropical forest biomes, in particular, other studies have demonstrated good congruence between the priority areas selected by one taxonomic group, and those selected by another (e.g., Howard et al. 1998; Burgess et al. 2000; Brooks et al. 2001b; Moore et al. 2003). Our analysis shows that in general there is a high degree of congruence in species richness scores between different taxonomic groups within the various biomes, with the strongest relationship between birds and mammals (table G.3). This indicates that in the context of

African ecoregions, using species richness data for one vertebrate group will be a good indicator of the patterns for other vertebrate groups.

Testing for Cross-Taxon Congruence in Species Endemism within Different African Biomes

If patterns of endemism across different vertebrate groups were the same, then the highest priority ecoregions selected for one vertebrate group (e.g., birds) would be equally important for all the other groups. To test this, we correlated species endemism by taxa within the various biomes. The results show that congruence across taxonomic groups is notably lower for endemism (table G.4) than it is for species richness (table G.3), suggesting that different factors influence endemism patterns in the various vertebrate groups. Within biomes, cross taxon congruence in vertebrate groups is strongest in the tropical moist forest biome, and weakest in the flooded grasslands, tropical dry forests and montane grasslands ecoregions (table G.4).

These results echo findings from other studies. Some have

TABLE G.4. **Results of Pearson Correlation Analyses of Endemism Values across Taxonomic Groups in the Same Biome.**

Biome	No. of Ecoregions	Bird-Mammal	Bird-Amphibian	Bird-Reptile	Mammal-Amphibian	Mammal-Reptile	Amphibian-Reptile
Tropical and subtropical moist broadleaf forests	30	.625	.667	.415	.788	.443	.669
Tropical and subtropical dry broadleaf forests	3	ns	ns	ns	ns	ns	.997
Tropical and subtropical grasslands, savannas, shrublands, and woodlands	24	.615	ns	.785	ns	.631	ns
Flooded grasslands and savannas	10	ns	.861	ns	ns	ns	.670
Montane grasslands and shrublands	16	ns	ns	ns	ns	ns	ns
Mediterranean forests, woodlands, and scrub	7	.926	.839	ns	.843	.803	.902
Deserts and xeric shrublands	23	.509	ns	.736	.513	.576	.566

All *r* values are significant at $p < .05$; *ns* = not significant.

looked at patterns of endemism in the African tropical moist forest biome at broad scales and have demonstrated good congruence between different taxonomic groups (Howard et al. 1998; Burgess et al. 2000). Others have looked at finer scales or across different biomes and have demonstrated poor cross-taxon congruence (Lombard 1995; Freitag et al. 1997; Lawton et al. 1998). The small number of ecoregions available for analysis in some biomes (e.g., in the tropical dry forests) may also be affecting the significance of the results. Because of a lack of data we have not been able to investigate cross-taxon endemism patterns among vertebrates, plants, and invertebrates. However, it is clear that these relationships are weak in the Mediterranean habitats, which have exceptional numbers of endemic plants and invertebrates, and much fewer endemic vertebrates. These results indicate that it will not be possible to use a single indicator taxa group to track the endemism priorities for all vertebrate groups (especially reptiles), nor for plants and invertebrates. Therefore, the additional effort required to develop multitaxon distributional databases is justified.

Data Used in the Conservation Status of Ecoregions

Tropical and subtropical moist broadleaf forests

Ecoregion Number	Ecoregion Name	Ecoregion Size (km²)	Habitat Loss Score	Habitat Blocks Score	Habitat Fragmentation Score	Habitat Protection Scored	Current Conservation Status	Future Threat to Habitat	Future Threat to Species	Future Threat	Final Conservation Status	Biological Distinctiveness Index	Conservation Priority Class
1	Western Guinean Lowland Forests	205,100	30	6	10	12	Vulnerable	30	30	60	Vulnerable	GO	I
2	Guinean Montane Forests	31,100	30	25	20	15	Critical	30	10	40	Critical	RO	II
3	Eastern Guinean Forests	189,700	40	25	15	15	Critical	45	20	65	Critical	RO	II
4	Nigerian Lowland Forests	67,300	40	25	20	15	Critical	60	10	70	Critical	BO	IV
5	Niger Delta Swamp Forests	14,400	40	25	20	15	Critical	60	10	70	Critical	LI	IV
6	Cross-Niger Transition Forests	20,700	40	25	20	15	Critical	60	10	70	Critical	LI	IV
7	Cross-Sanaga-Bioko Coastal Forests	52,100	30	25	10	8	Endangered	45	30	75	Critical	GO	I
8	Atlantic Equatorial Coastal Forests	189,700	10	0	5	8	Relatively stable	15	20	35	Relatively stable	GO	III
9	Mount Cameroon and Bioko Montane Forests	1,100	20	25	5	15	Endangered	60	20	80	Critical	GO	I
10	Cameroon Highlands Forests	38,000	30	18	10	12	Endangered	30	30	60	Endangered	GO	I
11	São Tomé, Príncipe, and Annobon Moist Lowland Forests	1,000	30	25	5	15	Endangered	60	10	70	Critical	GO	I
12	Northwestern Congolian Lowland Forests	434,100	0	0	0	12	Relatively intact	15	20	35	Relatively intact	GO	III
13	Western Congolian Swamp Forests	128,600	10	0	10	15	Relatively stable	15	10	25	Relatively stable	GO	III
14	Eastern Congolian Swamp Forests	92,700	10	0	10	15	Relatively stable	15	10	25	Relatively stable	GO	III
15	Central Congolian Lowland Forests	414,800	0	0	5	8	Relatively intact	15	10	25	Relatively intact	GO	III
16	Northeastern Congolian Lowland Forests	533,500	10	0	5	12	Relatively stable	15	30	45	Relatively stable	GO	III
17	Albertine Rift Montane Forests	103,900	30	25	15	8	Endangered	60	40	100	Critical	GO	I
18	East African Montane Forests	65,500	30	18	10	12	Endangered	45	30	75	Critical	RO	II
19	Eastern Arc Forests	23,700	30	25	15	15	Critical	30	40	70	Critical	GO	I
20	Northern Zanzibar–Inhambane Coastal Forest Mosaic	112,600	20	18	10	8	Vulnerable	30	30	60	Vulnerable	GO	I
21	Southern Zanzibar–Inhambane Coastal Forest Mosaic	147,000	30	25	20	15	Critical	30	10	40	Critical	BO	IV
22	Maputaland Coastal Forest Mosaic	30,200	30	25	20	8	Critical	60	10	70	Critical	BO	IV
23	Kwazulu-Cape Coastal Forest Mosaic	17,800	40	25	20	15	Critical	60	10	70	Critical	BO	IV

#	Ecoregion	Area											
24	Knysna-Amatole Montane Forests	3,100	20	25	5	12	Endangered	15	0	15	Endangered	BO	IV
25	Ethiopian Lower Montane Forests, Woodlands, and Bushlands	248,800	30	0	15	15	Endangered	45	20	65	Endangered	RO	II
26	Comoros Forests	2,100	40	25	20	15	Critical	60	10	70	Critical	GO	I
27	Granitic Seychelles Forests	300	30	25	20	15	Critical	60	0	60	Critical	GO	I
28	Mascarene Forests	4,900	30	0	10	15	Vulnerable	60	10	70	Endangered	GO	I
29	Madagascar Humid Forests	112,400	20	12	15	15	Endangered	30	30	60	Endangered	GO	I
30	Madagascar Subhumid Forests	199,300	20	18	10	15	Endangered	30	40	70	Critical	GO	I

Tropical and subtropical dry broadleaf forests

#	Ecoregion	Area											
31	Cape Verde Islands Dry Forests	4,600	30	25	15	15	Critical	45	0	45	Critical	RO	II
32	Zambezian *Cryptosepalum* Dry Forests	38,200	0	6	5	12	Relatively stable	0	10	10	Relatively stable	GO	III
33	Madagascar Dry Deciduous Forests	152,100	10	25	15	15	Endangered	15	30	45	Endangered	GO	I

Temperate coniferous forests

#	Ecoregion	Area											
34	Mediterranean Conifer and Mixed Forests	23,100	30	25	20	15	Critical	45	20	65	Critical	GO	I

Tropical and subtropical grasslands, savannas, shrublands, and woodlands

#	Ecoregion	Area											
35	Sahelian *Acacia* Savanna	3,053,600	10	0	5	15	Relatively stable	15	20	35	Relatively stable	BO	V
36	West Sudanian Savanna	1,638,300	30	0	10	12	Vulnerable	30	20	50	Vulnerable	RO	II
37	East Sudanian Savanna	917,600	10	0	5	8	Relatively stable	15	20	35	Relatively stable	BO	V
38	Guinean Forest-Savanna Mosaic	673,600	30	12	15	12	Endangered	45	20	65	Endangered	RO	II
39	Mandara Plateau Mosaic	7,500	40	25	20	15	Critical	45	10	55	Critical	LI	IV
40	Northern Congolian Forest-Savanna Mosaic	708,100	10	0	5	12	Relatively stable	15	30	45	Relatively stable	RO	III
41	Victoria Basin Forest-Savanna Mosaic	165,800	40	0	15	8	Endangered	60	20	80	Critical	GO	I
42	Southern Congolian Forest-Savanna Mosaic	569,700	20	0	10	15	Vulnerable	15	20	35	Vulnerable	BO	V
43	Western Congolian Forest-Savanna Mosaic	413,400	10	0	10	15	Relatively stable	30	20	50	Relatively stable	BO	V
44	Somali *Acacia-Commiphora* Bushlands and Thickets	1,053,900	10	0	5	15	Relatively stable	15	30	45	Relatively stable	GO	III
45	Northern *Acacia-Commiphora* Bushlands and Thickets	326,000	10	0	10	15	Relatively stable	15	30	45	Relatively stable	RO	III
46	Southern *Acacia-Commiphora* Bushlands and Thickets	227,800	30	0	10	4	Vulnerable	30	20	50	Vulnerable	GO	I
47	Serengeti Volcanic Grasslands	18,000	10	6	10	0	Relatively stable	15	20	35	Relatively stable	GO	III
48	Itigi-Sumbu Thicket	7,800	10	12	10	4	Relatively stable	15	10	25	Relatively stable	BO	V

Tropical and subtropical grasslands, savannas, shrublands, and woodlands (continued)

Ecoregion Number	Ecoregion Name	Ecoregion Size (km²)	Habitat Loss Score[a]	Habitat Blocks Score[b]	Habitat Fragmentation Score[c]	Habitat Protection Score[d]	Current Conservation Status	Future Threat to Habitat	Future Threat to Species[f]	Final Conservation Status	Biological Distinctiveness Index[g]	Conservation Priority Class
49	Angolan Miombo Woodlands	660,100	0	0	5	12	Relatively intact	15	20	Relatively intact	RO	III
50	Central Zambezian Miombo Woodlands	1,184,200	10	0	10	8	Relatively stable	15	20	Relatively stable	GO	III
51	Zambezian *Baikiaea* Woodlands	264,400	0	0	5	4	Relatively intact	0	20	Relatively intact	BO	V
52	Eastern Miombo Woodlands	483,900	20	0	10	4	Relatively stable	15	20	Relatively stable	GO	III
53	Southern Miombo Woodlands	408,300	20	0	10	12	Vulnerable	30	20	Vulnerable	BO	V
54	Zambezian and Mopane Woodlands	473,300	10	0	10	4	Relatively stable	15	20	Relatively stable	RO	III
55	Angolan Mopane Woodlands	133,500	0	0	5	4	Relatively intact	0	10	Relatively intact	BO	V
56	Western Zambezian Grasslands	34,000	0	0	5	4	Relatively intact	0	10	Relatively intact	LI	V
57	Southern Africa Bushveld	223,100	30	12	15	12	Endangered	15	20	Endangered	BO	IV
58	Kalahari *Acacia* Woodlands	335,500	0	0	5	4	Relatively intact	0	10	Relatively intact	LI	V

Flooded grasslands and savannas

Ecoregion Number	Ecoregion Name	Ecoregion Size (km²)	Habitat Loss Score[a]	Habitat Blocks Score[b]	Habitat Fragmentation Score[c]	Habitat Protection Score[d]	Current Conservation Status	Future Threat to Habitat	Future Threat to Species[f]	Final Conservation Status	Biological Distinctiveness Index[g]	Conservation Priority Class
59	Sudd Flooded Grasslands	179,700	0	0	5	15	Relatively stable	15	10	Relatively stable	GO	III
60	Nile Delta Flooded Savanna	51,000	30	25	20	15	Critical	60	10	Critical	LI	IV
61	Lake Chad Flooded Savanna	18,800	10	0	10	15	Relatively stable	15	10	Relatively stable	BO	V
62	Inner Niger Delta Flooded Savanna	46,000	20	18	20	12	Endangered	15	10	Endangered	LI	IV
63	Zambezian Flooded Grasslands	153,500	10	0	10	4	Relatively stable	15	20	Relatively stable	GO	III
64	Zambezian Coastal Flooded Savanna	19,500	30	18	20	12	Critical	15	10	Critical	BO	IV
65	Saharan Halophytics	53,900	0	0	5	15	Relatively stable	15	10	Relatively stable	LI	V
66	East African Halophytics	2,600	10	25	20	15	Endangered	15	10	Endangered	LI	IV
67	Etosha Pan Halophytics	7,200	0	0	0	0	Relatively intact	0	10	Relatively intact	LI	V
68	Makgadikgadi Halophytics	30,400	0	0	5	4	Relatively intact	0	10	Relatively intact	LI	V

Montane grasslands and shrublands

#	Ecoregion	Area					Status					Status				Code	
69	Mediterranean High Atlas Juniper Steppe	6,300	10	25	20	15	Endangered	15	20	0	15	Endangered	15	0	15	GO	I
70	Ethiopian Upper Montane Forests, Woodlands, Bushlands, and Grasslands	245,400	40	0	15	15	Endangered	60	15	20	80	Critical	60	20	80	GO	I
71	Ethiopian Montane Moorlands	25,200	40	6	15	4	Endangered	60	15	20	80	Critical	60	20	80	GO	I
72	East African Montane Moorlands	3,300	10	6	10	0	Relatively stable	45	10	10	55	Relatively stable	45	10	55	GO	III
73	Rwenzori-Virunga Montane Moorlands	2,700	20	25	20	0	Endangered	60	20	0	60	Endangered	60	0	60	GO	I
74	Southern Rift Montane Forest-Grassland Mosaic	33,500	30	18	10	8	Endangered	30	10	20	50	Endangered	30	20	50	GO	I
75	South Malawi Montane Forest-Grassland Mosaic	10,200	40	25	10	12	Critical	60	10	10	70	Critical	60	10	70	GO	I
76	Eastern Zimbabwe Montane Forest-Grassland Mosaic	7,800	20	25	15	12	Endangered	30	15	10	40	Endangered	30	10	40	GO	I
77	Highveld Grasslands	186,200	30	25	20	15	Critical	30	20	20	50	Critical	30	20	50	BO	IV
78	Drakensberg Montane Grasslands, Woodlands, and Forests	202,200	30	12	20	12	Endangered	30	20	30	60	Endangered	30	30	60	GO	I
79	Drakensberg Alti-Montane Grasslands and Woodlands	11,900	30	25	20	12	Critical	15	20	10	25	Critical	15	10	25	GO	I
80	Maputaland-Pondoland Bushland and Thickets	19,500	30	18	15	15	Endangered	30	15	20	50	Endangered	30	20	50	RO	II
81	Angolan Scarp Savanna and Woodlands	74,400	10	0	10	4	Relatively stable	30	10	10	40	Relatively stable	30	10	40	GO	III
82	Angolan Montane Forest-Grassland Mosaic	25,500	20	18	15	15	Endangered	30	15	10	40	Endangered	30	10	40	RO	II
83	Jos Plateau Forest-Grassland Mosaic	13,300	40	25	20	15	Critical	60	20	0	60	Critical	60	0	60	BO	IV
84	Madagascar Ericoid Thickets	1,300	30	12	15	4	Endangered	45	15	0	45	Endangered	45	0	45	GO	I

Mediterranean forests, woodlands, and scrub

#	Ecoregion	Area					Status					Status				Code	
85	Mediterranean Woodlands and Forests	358,300	30	6	15	15	Endangered	45	20	20	65	Endangered	45	20	65	GO	I
86	Mediterranean Dry Woodlands and Steppe	292,200	10	0	10	15	Relatively stable	30	20	20	50	Relatively stable	30	20	50	LI	V
87	Mediterranean *Acacia-Argania* Dry Woodlands and Succulent Thickets	100,100	20	12	10	15	Vulnerable	30	20	20	50	Vulnerable	30	20	50	RO	II
88	Canary Islands Dry Woodlands and Forestss	5,000	20	6	15	15	Vulnerable	60	15	0	60	Vulnerable	60	0	60	GO	I
89	Lowland Fynbos and Renosterveld	32,800	40	25	20	15	Critical	30	10	10	40	Critical	30	10	40	GO	I
90	Montane Fynbos and Renosterveld	45,800	0	6	15	4	Relatively stable	15	20	20	35	Relatively stable	15	20	35	GO	III
91	Albany Thickets	17,100	30	25	20	15	Critical	15	10	0	25	Critical	15	10	25	RO	II

Deserts and xeric shrublands

Ecoregion Number	Ecoregion Name	Ecoregion Size (km2)	Habitat Loss Score[a]	Habitat Blocks Score[b]	Habitat Fragmentation Score[c]	Habitat Protection Score[d]	Current Conservation Status	Future Threat to Habitat[e]	Future Threat to Species[f]	Final Conservation Status	Biological Distinctiveness Index[g]	Conservation Priority Class
92	Atlantic Coastal Desert	40,000	0	5	4	15	Relatively intact	0	15	Relatively intact	LI	V
93	North Saharan Steppe	1,676,100	0	0	15	15	Relatively intact	10	25	Relatively intact	BO	V
94	South Saharan Steppe	1,101,700	0	0	12	0	Relatively intact	10	10	Relatively intact	BO	V
95	Sahara Desert	4,639,900	0	0	15	0	Relatively intact	10	10	Relatively intact	LI	V
96	West Saharan Montane Xeric Woodlands	258,100	0	5	0	0	Relatively intact	10	10	Relatively intact	LI	V
97	Tibesti-Jebel Uweinat Montane Xeric Woodlands	82,200	0	0	15	0	Relatively intact	10	10	Relatively intact	LI	V
98	Red Sea Coastal Desert	59,300	0	5	15	0	Relatively stable	0	0	Relatively stable	LI	V
99	East Saharan Montane Xeric Woodlands	27,900	6	0	15	0	Relatively stable	10	10	Relatively stable	LI	V
100	Eritrean Coastal Desert	4,600	25	15	15	0	Vulnerable	10	10	Vulnerable	LI	V
101	Ethiopian Xeric Grasslands and Shrublands	153,100	0	10	12	15	Relatively stable	20	35	Relatively stable	RO	III
102	Somali Montane Xeric Woodlands	62,600	0	5	15	15	Relatively stable	10	25	Relatively stable	RO	III
103	Hobyo Grasslands and Shrublands	25,600	25	15	15	45	Endangered	10	55	Endangered	BO	IV
104	Masai Xeric Grasslands and Shrublands	101,000	0	5	12	0	Relatively intact	20	20	Relatively intact	BO	V
105	Kalahari Xeric Savanna	588,100	0	0	4	0	Relatively intact	10	10	Relatively intact	BO	V
106	Kaokoveld Desert	45,700	0	0	0	0	Relatively intact	0	0	Relatively intact	GO	III
107	Namib Desert	80,900	0	0	0	0	Relatively intact	0	0	Relatively intact	GO	III
108	Nama Karoo	351,100	0	10	15	0	Relatively stable	20	20	Relatively stable	GO	III
109	Namib Escarpment Woodlands	225,500	0	5	8	0	Relatively intact	20	20	Relatively intact	GO	III
110	Succulent Karoo	102,700	0	5	15	0	Relatively stable	10	10	Relatively stable	GO	III
111	Socotra Island Xeric Shrublands	3,800	6	10	15	15	Relatively stable	0	15	Relatively stable	GO	III
112	Aldabra Island Xeric Scrub	200	0	0	0	0	Relatively intact	0	0	Relatively intact	GO	III
113	Madagascar Spiny Thickets	43,300	18	15	15	15	Endangered	10	25	Endangered	GO	I
114	Madagascar Succulent Woodlands	79,800	25	20	15	15	Critical	10	25	Critical	GO	I

Mangroves

115	Guinean Mangroves	23,500	30	25	20	15	Critical	45	0	45	Critical	LI	IV
116	Central African Mangroves	30,900	30	0	15	15	Endangered	60	0	60	Endangered	LI	IV
117	Southern Africa Mangroves	1,000	30	25	15	4	Endangered	60	0	60	Endangered	LI	IV
118	East African Mangroves	16,100	30	18	20	15	Critical	60	0	60	Critical	LI	IV
119	Madagascar Mangroves	5,200	10	25	20	15	Endangered	15	0	15	Endangered	LI	IV

[a]See appendix D, table D.1.
[b]See appendix D, table D.2.
[c]See appendix D, table D.3.
[d]See appendix D, table D.4.
[e]See appendix D, table D.5.
[f]See appendix D, table D.6.
[g]GO = globally outstanding; RO = regionally outstanding; BO = bioregionally outstanding; LI = locally important.

APPENDIX I

Ecoregion Descriptions

Ecoregion Number: **1**

Ecoregion Name: **Western Guinean Lowland Forests**

Bioregion: **Western Africa and Sahel**

Biome: **Tropical and Subtropical Moist Broadleaf Forests**

Political Units: **Côte d'Ivoire, Guinea, Liberia, Sierra Leone**

Ecoregion Size: **205,100 km^2**

Biological Distinctiveness: **Globally Outstanding**

Conservation Status: **Vulnerable**

Conservation Assessment: **I**

Authors: *Aiah R. Lebbie, Paul Burgess*

Reviewed by: *Mohamed Bakarr, Lincoln Fishpool*

Location and General Description

The Western Guinean Lowland Forests [1] stretch from eastern Guinea, across Sierra Leone and Liberia, to the Sassandra River in southwestern Côte d'Ivoire. The topography is flat to undulating and is composed mainly of heavily eroded Precambrian basement rocks. Altitude ranges between 50 and 500 m, although there are a few isolated mountains of higher elevation. Several major rivers cross the ecoregion, including the Sewa, Mano, St. Paul, and Cavally. The soils are generally poor and heavily leached, except along river valleys and in inland swamps (Gwynne-Jones et al. 1977).

The ecoregion covers one of the wettest parts of West Africa: seasonal rains average 3,300 mm per year, and the Freetown Peninsula in Sierra Leone receives more than 5,000 mm of rainfall annually (Cole 1968). Rainfall peaks in July, and the dry season lasts from November to April (White 1983; Peters 1990). Seasonal temperatures range between 30°C and 33°C during the hot season and 12°C and 21°C during the cold season, when the Harmattan winds blow south from the Sahara Desert (Cole 1968).

The warm and humid climate has permitted the development of rainforest vegetation. White (1983) classifies the forest as part of the Upper Guinea block of the Guineo-Congolian regional center of endemism, with moist evergreen forest in the wetter areas and moist semi-deciduous further inland (Cole 1968; Vooren and Sayer 1992; Mayers et al. 1992; Lawson 1996). Swamp and riparian forests are embedded in the other forest types. Farmbush, the secondary growth derived from forest that follows slash-and-burn agriculture, is increasingly the dominant vegetation type in this ecoregion.

Outstanding or Distinctive Biodiversity Features

This ecoregion supports high species richness and large numbers of endemic species, with the patterns of endemism often stated to reflect the location of past forest refuges during times of heightened desiccation (Booth 1958; Moreau 1969; Grubb 1978; Hamilton 1981; Kingdon 1989; Happold 1996; Grubb et al. 1998).

More than 3,000 plant species occur, including at least 200 endemic species. The endemic liana family Dioncophyllaceae contains three monotypic genera (Guillaumet 1967; Jenkins and Hamilton 1992; WWF and IUCN 1994). In the vicinity of the Sassandra River in southwestern Côte d'Ivoire, seventy-two endemic plant species have been reported (Magenot 1955; Guillaumet 1967; Hall and Swaine 1981).

Two duikers, Jentink's duiker (*Cephalophus jentinki,* vulnerable [VU]) and zebra duiker (*Cephalophus zebra,* VU) are endemic, as are two carnivore species: the Liberian mongoose (*Liberiictis kuhni,* endangered [EN]) and Johnston's genet (*Genetta johnstoni,* data deficient [DD]) (Hayman 1958; Schlitter 1974; Taylor 1989, 1992; Hilton-Taylor 2000). Miller's striped mouse (*Hybomys planifrons*) is also strictly endemic, and there are fifteen near-endemic mammal species. Primates include endemic subspecies of the Diana monkey (*Cercopithecus diana diana,* EN), lesser spot-nosed monkey (*Cercopithecus petaurista petaurista*), and sooty mangabey (*Cercocebus torquatus atys*). Other threatened species include the Western chimpanzee (*Pan troglodytes verus,* EN), forest elephant (*Loxodonta africana cyclotis,* EN), and pygmy hippopotamus (*Hexaprotodon liberiensis,* VU).

Forest birds are diverse and include a number of endemic and rare species. Strictly endemic bird species are the Liberian greenbul (*Phyllastrephus leucolepis,* critical [CR]) and Gola malimbe (*Malimbus ballmanni,* EN). Near-endemic species are the white-necked picathartes (*Picathartes gymnocephalus,* VU), white-breasted guinea-fowl (*Agelastes meleagrides,* VU), rufous fishing-owl (*Scotopelia ussheri,* EN), brown-cheeked hornbill (*Ceratogymna cylindricus*), Turati's boubou (*Laniarius turatii*), western wattled cuckoo-shrike (*Campephaga lobata,* VU), rufous-winged illadopsis (*Illadopsis rufescens*), yellow-throated olive greenbul (*Criniger olivaceus,* VU), and white-eyed prinia (*Prinia leontica*) (Allport et al. 1989; Allport 1991; Jenkins and Hamilton 1992; Thompson 1993; Hilton-Taylor 2000; Fishpool and Evans 2001).

The herpetofauna is also diverse (Welch 1982) and contains a large number of endemic species. Strict endemics include Merlin's clawed frog (*Pseudhymenochirus merlini*), known only from Guinea and Sierra Leone (Chabanaud 1920; Menzies 1967), and the Freetown long-fingered frog (*Cardioglossa aureoli*), which is known only from the mountains close to Freetown in Sierra Leone. Other notable endemics include the Tai River frog (*Phrynobatrachus taiensis*), Liberian long-fingered frog (*Cardioglossa liberiensis*) and Ivory Coast toad (*Bufo danielae*) (Schiøtz 1964, 1967; Harcourt et al. 1992; Vooren and Sayer 1992). The reptile fauna includes three strictly endemic species: Los Archipelago worm lizard (*Cynisca leonina*), Benson's mabuya (*Mabuya bensonii*) and Liberia worm snake (*Typhlops leucostictus*).

There are numerous information gaps in the invertebrate fauna for this ecoregion, but several recent inventories in Sierra Leone led to the discovery of new species, especially among the order Coleoptera (*Euconnus* spp. and *Termitusodes* spp.) (Franciscolo 1982, 1994; Kistner 1986; Castellini 1990). New discoveries in the orders Lepidoptera and Diptera have also been made (Belcastro 1986; Munari 1994), with two endemic dragonfly species, *Argiagrion leoninum* and *Allorhizucha campioni,* also known from Sierra Leone (Stuart et al. 1990).

Status and Threats

Current Status

Human impacts on the vegetation have been severe and prolonged (Sowunmi 1986; Gillis 1988). Today's remaining forests could be described as late secondary stands (Voorhoeve 1965; Lebbie 2001). Côte d'Ivoire and Sierra Leone show the greatest level of forest fragmentation and loss, whereas Liberia still retains large forest blocks. High forest is most intact in national parks and forest reserves, even though civil wars have prevented management in some areas.

Taï Forest National Park in Côte d'Ivoire is the largest protected area at more than 3,300 km². Côte d'Ivoire also contains the Mont Peko National Park (340 km²) and Nzo Faunal Reserve (930 km²). In Liberia, Sapo National Park (1,292 km²) protects parts of this habitat, together with forest reserves such as Gola (2,070 km²), Kpelle (1,748 km²), Lorma National Forests (435 km²), Krahn-Bassa National Forest (5,140 km²), and Grebo National Forest (2,673 km²). The forest reserves are currently allocated for logging. Sierra Leone has declared one small area of forest as the Tiwai Island Wildlife Sanctuary (12 km²) but has much larger areas in forest reserves. These include Gola North and South, Western Area, Kangari Hills, Tama-Tonkoli Forest, Dodo Hills, Nimini Forest, and Geboi Hills forest reserves (Davies 1987; Davies and Birkenhager 1990; Harcourt et al. 1992). The Ziama and Diécké forest reserves in Guinea also support important forest areas. An important unprotected forest (approximately 50 km²) remains on the Kounounkan massif, southeast of Conakry in Guinea (Barnett et al. 1994).

Types and Severity of Threats

Anthropogenic pressures for farmland, timber, bushmeat, fuelwood, and mineral resources continue to cause the loss of high forest, especially outside reserves. Most high forest areas are now isolated in a sea of farmbush vegetation. The global demand for valuable hardwoods continues to drive logging operations, and logging roads provide access routes for farmers and hunters to the forest (Sayer et al. 1992).

The recent civil conflicts in Sierra Leone and Liberia led to uncontrolled logging, mining, and bushmeat hunting (Garnett and Utas 2000). In Liberia, an estimated 50,000 m³ of *Heritiera utilis* (Niangon) was exported in 1999 alone, from a total roundwood volume of 335,543 m³ exported by approximately twenty logging companies (Garnett and Utas 2000). During the civil war, logging activities also increased in the Western Area Forest Reserve in Sierra Leone, with refugees providing labor. The loggers selectively targeted two species, *Heritiera utilis* and *Terminalia ivorensis* (Lebbie 2001). Commercial collection of firewood and charcoal production are also a problem in the forests close to Freetown (Cline-Cole 1987). The favored species are *Phyllocosmus africanus, Parinari excelsa,* and *Xylopia quintasii* (Lebbie 2001).

Hunting for bushmeat threatens the survival of larger mammals in this ecoregion (Anstey 1991; Bakarr et al. 1999, 2001b). Antelopes, forest pigs, and primates dominate the bushmeat trade in urban areas, and grasscutter (*Thryonomys swinderianus*) and Gambian giant rat (*Cricetomys gambianus*) dominate in rural areas because they are readily available.

Mining is a locally intense and destructive practice in Sierra Leone. Mining of bauxite and titanium dioxide (rutile) in the southeast of Sierra Leone has resulted in forest loss, with the subsequent dredging leaving large water bodies polluted with heavy metals.

Justification for Ecoregion Delineation

The Western Guinean Lowland Forests [1] and adjacent Eastern Guinean Forests [3] comprise the lowland part of the Upper Guinea forest block of the Guineo-Congolian rainforest (White 1983). The two ecoregions are separated by the Sassandra River, which represents an important biogeographic boundary for primates, duikers, amphibians, and lizards. The northern limit of the Western Guinean Lowland Forests primarily follows the vegetation unit delimited by White (1983).

Ecoregion Number:	**2**
Ecoregion Name:	**Guinean Montane Forests**
Bioregion:	**Western Africa and Sahel**
Biome:	**Tropical and Subtropical Moist Broadleaf Forests**
Ecoregion Size:	**31,100 km²**
Political Units:	**Côte d'Ivoire, Guinea, Liberia, Sierra Leone**
Biological Distinctiveness:	**Regionally Outstanding**
Conservation Status:	**Critical**
Conservation Assessment:	**II**
Authors:	Aiah R. Lebbie, Paul Burgess
Reviewed by:	Mohamed Bakarr, Lincoln Fishpool

Location and General Description

The Guinean Montane Forests [2] consist of scattered mountains and high plateau areas across four West African countries, from Guinea in the west to Côte d'Ivoire in the east. Some landscapes rise precipitously (e.g., Loma Mountains and Tingi Hills in Sierra Leone and Mount Nimba on the border between Liberia, Guinea, and Côte d'Ivoire), whereas others, such as the Fouta Djallon in Guinea, a heavily eroded plateau with an elevation of 1,100 m (Morton 1986), rise more gently. Bintumani Peak on Loma Mountain (1,947 m) is the highest peak west of Mount Cameroon (Cole 1968; Atkinson et al. 1992). Several other mountains reach heights between 1,860 and 1,387 m but their peaks have been rounded by millions of years of erosion and weathering.

The mountains in this ecoregion are formed of Precambrian basement rocks (Morton 1986). Lithosols are the most common soil type on many of the mountains, although soil type may vary by aspect and degree of weathering. Soils are generally infertile, and some are rich in mineral deposits such as iron ore on Mount Nimba and the Fouta Djallon Plateau (Curry-Lindahl 1966; Morton 1986).

Most major rivers in West Africa originate in the Guinean Montane Forests. For example, the westernmost tributary of the Niger River originates in the Loma Mountains of Sierra Leone, and the Senegal and Gambia Rivers originate in the Fouta Djallon of Guinea. The Sewa River in Sierra Leone also has many of its tributaries arising from the Loma Mountains and Tingi Hills.

Annual rainfall ranges from 1,600 to 2,400 mm (Morton 1986), with wide variation between the southern side of the mountains, which face the Atlantic Ocean, and the northern slopes in the rainshadow. These leeward slopes are also subjected to the dry Harmattan winds blowing from the Sahara Desert. Temperatures vary widely on these mountain slopes, with maximum temperatures reaching 33°C and minimum temperatures below 10°C.

White (1983) classified these forests as part of the Afromontane archipelago-like regional center of endemism. The broad range in elevation, coupled with underlying geology and human influences, has given rise to different plant associations on these mountains. At mid-altitudes (above 1,000 m), forest often is shrouded in clouds, resulting in a proliferation of epiphytes. With increasing altitude on the highest mountains, forests change to grassland intermixed with bamboo, wetlands, and gallery forests. The vegetation of the drier, northern slopes is better adapted to desiccation than the vegetation of the southern slopes.

On the Fouta Djallon, agriculture and burning have transformed the *Parinari excelsa* forest into grassland (Adam 1958). The dominant flora of the grassland includes the genera *Anadelphia, Loudetia,* and *Tristachya* (Morton 1986). Grassland also occurs on Mount Nimba, where it is dominated by *Andropogon* and *Loudetia* (Curry-Lindahl 1966; Morton 1986), and is also found on Mount Loma. The forest habitats of Mount Nimba, above 800 m, are dominated by *Parinari excelsa, Gaertnera paniculata, Garcinia polyantha,* and *Syzygium staudtii* and a rich array of epiphytes (Jaeger et al. 1968; Johansson 1974; Morton 1986). On Mount Loma the submontane gallery forests contain *Parinari excelsa, Anthonotha macrophylla, Pseudospondias microcarpa, Amphimas pterocarpoides, Daniella thurifera, Terminalia ivorensis, Allanblackia floribunda,* and *Musanga cecropioides* (Cole 1968), along with tree ferns, *Cyathea camerooniana* and *Marattia fraxinea,* and the bamboo *Oxythenanthera abyssinica.*

Outstanding or Distinctive Biodiversity Features

The diversity and endemism of the flora and fauna of Mount Nimba are well known, but many other parts of this ecoregion remain poorly known biologically.

Within the ecoregion, thirty-five endemic plants including

eleven paleoendemics have been recorded (Schnell 1952; Cole 1967, 1974; Morton 1972; Jaeger and Adam 1975). Paleoendemics include *Borreria macrantha, Cyanotis lourensis, Droogmansia scaettaiana, Eriosema parviflorum, Eugenia pobeguinii, Hypolytrum cacuminum,* and *Kotschya lutea* (Cole 1974). Studies of the Loma Mountains have cataloged 1,576 plant species in 757 genera and 135 families (Jaeger et al. 1968; Johansson 1974). Nine plant species are endemic to this mountain: *Afrotrilepis jaegeri, Digitaria phaeotricha* var. *patens, Dissotis sessilis, Gladiolus leonensis, Ledermanniella jaegeri, Loudetia jaegeriana, Loxodera strigosa, Schizachyrium minutum* (*S. brevifolium*), and *Scleria monticola* (Jaeger 1983). The four endemic plant families of tropical Africa are also represented in the Loma Mountains by *Triphyophyllum peltatum* (Dioncophyllaceae), *Octoknema borealis* (Octoknemataceae), *Bersama abyssinica* (Melianthaceae), and *Napoleona leonensis* and *Napoleona vogelii* (Lecythidaceae).

The fauna of the ecoregion is also rich and includes a number of endemic species. Four mammals are either strict endemics or narrowly shared with the surrounding lowland habitats. These are the Mount Nimba otter shrew (*Micropotamogale lamottei,* EN), two species of white-toothed shrew (*Crocidura obscurior* and *C. nimbae*), and a species of leaf-nosed bat (*Hipposideros marisae,* VU) (Hilton-Taylor 2000). The western chimpanzee (*Pan troglodytes verus,* EN) also occurs in this ecoregion, with high densities reported from Mount Loma.

Avifaunal diversity is high and includes the endemic white-eyed prinia (*Prinia leontica,* VU), the poorly known yellow-footed honeyguide (*Melignomon eisentrauti,* DD), and the more widely distributed black-capped rufous warbler (*Bathmocercus cerviniventris*) (Colston and Curry-Lindahl 1986; Gatter 1997; Hilton-Taylor 2000; Fishpool and Evans 2001). The white-eyed prinia is found only in the gallery forests of the Guinea Highlands at 700–1,550 m (Stattersfield et al. 1998). The presence of the rare white-necked picathartes (*Picathartes gymnocephalus,* VU) has also been confirmed in the Loma Mountains, Mount Nimba (Thompson 1993), and Mount Peko (Conservation International, pers. comm., 2000). Nimba also contains birds such as rufous-naped lark (*Mirafra africana*) and stonechat (*Saxicola torquata*), which are confined as breeding species to montane grasslands in Africa.

The ecoregion is also important for endemic amphibians, including *Nimbaphrynoides occidentalis,* an endemic toad occurring in savannas on Mount Nimba (Curry-Lindahl 1966).

Several new species of insects in the family Coleoptera have been reported for both the Loma and the Nimba Mountains (Villiers 1965). For the Loma Mountains, these include *Promecolanguria lomensis, Barbaropus bintumanensis,* and *Barbaropus explanatus.* The species recorded on Mount Nimba include *Promecolanguria dimidiata, Promecolanguria pseudosulcicollis, Promecolanguria mimbana, Promecolanguria armata,* and *Barbaropus nigritus.*

Status and Threats

Current Status

Some of the mountain zones remain largely untouched, whereas others have been severely degraded and fragmented. Habitats on Mount Nimba have undergone marked fragmentation, as have those of the Fouta Djallon. In contrast, the Loma Mountains in Sierra Leone still had intact habitats in the early 1990s (Atkinson et al. 1992).

This ecoregion is not well protected. The Loma Mountains is a proposed national park under the authority of the Wildlife Conservation Branch of the Sierra Leone government. The Tingi Hills in Sierra Leone is a forest reserve. There are also numerous small forest reserves in the highland areas of Guinea and a large one at Ziama.

Despite the various designations of Mount Nimba as a strict nature reserve, World Heritage Site, and biosphere reserve, mining for iron ore was still occurring before the Liberian civil war. Although the war has stopped mining operations, future mining is likely (Garnett and Utas 2000). The Massif du Ziama is also a biosphere reserve and a World Heritage Site.

Types and Severity of Threats

Mining, fires, and deforestation are considered the principal threats in this ecoregion. Mount Nimba contains massive deposits of high-grade iron ore, making it the target of extensive mining operations in the past. On the Liberian side of Mount Nimba, a multinational mining operation has done enormous damage to the forests and streams (Curry-Lindahl 1966; Sayer et al. 1992), but the recent Liberian civil war has put a 10-year halt to mining operations. A high population density and human-made fires have caused deforestation in the Fouta Djallon (Allport 1991). Current information on the Loma Mountains and Tingi Hills in Sierra Leone is lacking because of the decade-long civil war, although the Loma Mountains were in good condition in the early 1990s (Atkinson et al. 1992). No data are available on the status and threats to the habitats of the other mountains in this ecoregion.

Justification for Ecoregion Delineation

White (1983) mapped montane areas in Sierra Leone, Liberia, and Côte d'Ivoire as "undifferentiated montane vegetation" and in Guinea as "mosaic of lowland rainforest, secondary grassland, and montane elements." This ecoregion follows these vegetation divisions, but the lower boundary has been delineated at 600 m elevation because at this elevation the vegetation is regarded as distinct from the surrounding lowlands (see Coe and Curry-Lindhal 1965).

Ecoregion Number:	**3**
Ecoregion Name:	**Eastern Guinean Forests**
Bioregion:	**Western Africa and Sahel**
Biome:	**Tropical and Subtropical Moist Broadleaf Forests**
Political Units:	**Benin, Côte d'Ivoire, Ghana, Togo**
Ecoregion Size:	**189,700 km²**
Biological Distinctiveness:	**Regionally Outstanding**
Conservation Status:	**Critical**
Conservation Assessment:	**II**
Authors:	*Aiah R. Lebbie, Paul Burgess*
Reviewed by:	*Mohammed Bakarr, Lincoln Fishpool*

Location and General Description

The Eastern Guinean Forests [3] extend from the east banks of the Sassandra River in western Côte d'Ivoire to the edge of Lake Volta in Ghana. There is a small extension east of Lake Volta in the Togo Hills. The dry lowland area on the eastern edge of the ecoregion is called the Dahomey Gap and is a major biogeographic barrier. Forest habitats extend to 08°N, where they gradually fade into a mosaic of forest patches and tall grasslands of the Guinean Forest-Savanna Mosaic [38].

Topography generally undulates between 50 and 300 m, with occasional inselbergs rising to more than 400 m. In the Togo Hills, there is more altitudinal variation, with the maximum elevation reaching 1,000 m (Sayer et al. 1992). The parent rock is mostly Precambrian basement. In Ghana, the soils in the moist semi-deciduous forest zone are fertile, whereas the soils in the evergreen forests are less fertile (Owusu et al. 1989).

Temperatures in the south range between 22°C and 34°C, whereas in the north temperatures are more extreme and can reach a maximum of 43°C and fall to 10°C on cold nights. The rainfall is seasonal with distinct wet and dry seasons, with a longer dry season than in the Western Guinean Lowland Forests [1]. In Benin and Togo, rainfall seldom exceeds 1,500 mm, but further to the west rainfall can average 2,500 mm per year. The rainfall declines inland, and the dry season becomes longer.

This ecoregion falls within the Upper Guinea forest block of the Guineo-Congolian regional center of endemism (White 1983). Moist evergreen forest in the extreme south grades into moist semi-evergreen forest further inland, which in turn becomes dry semi-evergreen forest in the northern parts of the ecoregion. Moist evergreen forest trees include *Entandrophragma utile, Khaya ivorensis,* and *Triplochiton scleroxylon* (Hall and Swaine 1981). The semi-deciduous forest is dominated by *Celtis* spp., *Mansonia altissima, Pterygota macrocarpa, Nesogordonia papaverifera, Sterculia rhinopetala,* and *Milicia excelsa* (Vooren and Sayer 1992).

Trees of the forest fragments in the Togo Hills include *Milicia excelsa, Triplochiton scleroxylon, Antiaris africana, Diospyros mespiliformis, Afzelia africana,* and *Ceiba pentandra* (Sayer et al. 1992).

Outstanding or Distinctive Biodiversity Features

This ecoregion has a rich flora and fauna, with many endemics. The Sassandra River separates the Western and Eastern Guinean Forests [1, 3] and has influenced faunal distribution and evolution. For example, the western and eastern subspecies of the sooty mangabey (*Cercocebus torquatus atys,* C. t. *lunulatus*) and Diana monkey (*Cercopithecus diana diana,* C. d. *roloway*) are separated by this river. Moreover, a subspecies of the king colobus, *Colobus polykomos vellerosus* (listed as *C. vellerosus* by Hilton-Taylor 2000) and Lowe's subspecies of Campbell's monkey, *Cercopithecus campbelli lowei,* are found only east of the Sassandra River.

Four small mammals are strictly endemic to this ecoregion: Wimmer's shrew (*Crocidura wimmeri,* EN), Ivory Coast rat (*Dephomys eburnea*), Cansdale's swamp rat (*Malacomys cansdalei*), and the Togo mouse (*Leimacomys buettneri,* CR). Other near-endemic mammals are Lowe's monkey (*Cercopithecus mona lowei*), lesser spot-nosed monkey (*Cercopithecus petaurista petaurista*), olive colobus (*Procolobus verus*), and royal antelope (*Neotragus pygmaeus*), as well as small rodents and shrews, including Kitamps rope squirrel (*Funisciurus substriatus*), western palm squirrel (*Epixerus ebi*), Gambian sun squirrel (*Heliosciurus punctatus*), *Oenomys ornatus,* and *Crocidura muricauda.* Most of these near-endemic species are found only in the Upper Guinea Forests [1–3]. The rare pygmy hippopotamus (*Hexaprotodon liberiensis,* VU) also occurs marginally in the western part of this ecoregion (Vooren and Sayer 1992). Small populations of forest elephant (*Loxodonta africana cyclotis,* EN) also occur, often isolated in unconnected forest patches. Species such as *Tieghemella heckelii* and *Balanites wilsoniana* appear to be dependent on forest elephants for regeneration (Voorhoeve 1965; Hall and Swaine 1981; Hawthorne and Parren 2000).

The ecoregion contains high bird species richness and shares several restricted-range species with the Western Guinean Lowland Forests [1] (Allport 1991). Near-endemic species are white-breasted guineafowl (*Agelastes meleagrides,* VU), western wattled cuckoo-shrike (*Campephaga lobata,* VU), brown-cheeked hornbill (*Ceratogymna cylindricus*), rufous-winged illadopsis (*Illadopsis rufescens*), copper-tailed glossy-starling (*Lamprotornis cupreocauda*), Sharpe's apalis (*Apalis sharpei*), black-capped rufous warbler (*Bathmocercus cerviniventris*), white-necked picathartes (*Picathartes gymnocephalus,* VU), rufous fishing-owl (*Scotopelia ussheri,* EN), yellow-bearded bulbul or yellow-throated olive greenbul (*Criniger olivaceus,* VU), and green-tailed bristlebill (*Bleda eximia,* VU) (Stattersfield et al. 1998; Fishpool and Evans 2001).

Although the herpetofauna is poorly known, there are high rates of amphibian endemism, including endemic tree frogs such as *Hyperolius bobirensis, Hyperolius laurenti,* and *Hyperolius viridigulosus* (Schiøtz 1999). Reptile endemics include the near-endemic *Cophoscincopus durus*. Larsen (1994) recorded 120 species of butterflies believed to be endemic to the West African forest ecoregions. He also documented a number of butterflies that are narrowly endemic to the Togo Hills, an area of complex biogeography.

Status and Threats

Current Status

What remains of the Eastern Guinean Forests is highly fragmented, largely because of human activities over hundreds if not thousands of years (Martin 1991). Most of the remaining large tracts of forest are in protected areas under the control of national government wildlife or forestry departments. The local population also protects numerous small forest patches as sacred groves. Protected sites cover barely 1 percent of the total ecoregion area. In Ghana, important protected areas include Kakum National Park, Bia National Park, Nini-Suhien National Park, Ankasa Game Production Area, Kayabobo National Park, and Agumatsa Wildlife Sanctuary. There are also a large number of forest reserves that are used for timber production (Hawthorne 1993b). In Côte d'Ivoire, the largest remaining areas of forested habitats are found in the southeastern region in the Marahoue National Park and along the Comoé River, and there are also a number of forest reserves. In Togo and Benin the forest area is much reduced from the turn of the century (Aubreville 1937), and remaining fragments comprise semi-evergreen or deciduous forest.

Types and Severity of Threats

Forests have been lost in this region primarily to slash-and-burn agriculture but also commercial logging and fuelwood collection in urban areas (Allport 1991). Fire associated with traditional agricultural practices is the greatest threat to semi-deciduous forests in Ghana (Hawthorne 1991). Since the 1970s, forest conversion for export production of coffee and especially cocoa have led to widespread forest loss in Côte d'Ivoire and Ghana.

Logging of natural forest for valuable hardwoods such as iroko (*Milicia excelsa*) and mahogany (*Khaya* and *Entandrophragma* spp.) has also contributed to the decline in forest area in all four countries in this ecoregion. Domestic demand for timber is moderately high, and there is a large export industry for processed and unprocessed logs, particular from Côte d'Ivoire and Ghana. Where forestry management authorities lack capacity, roads cut to access commercial hardwoods allow agriculturists to penetrate the forests. Commercial hunters also take advantage of the access provided by logging roads to travel further into the forest to harvest wildlife (Oates 1999). The impact of bushmeat hunting to meet the growing demand in urban and rural areas has the potential to cause local as well as global extinctions (Robinson and Bennett 2000b). One of the endemic

primates, Miss Waldron's red colobus (*Procolobus badius waldroni*), was recently declared extinct, primarily because of hunting (Oates et al. 2000).

In Ghana, heavily logged sites or previously burnt areas are at greatest risk of subsequent fires (Hawthorne 1991). The demand for fuelwood, particularly from burgeoning urban centers such as Abidjan and Accra, also continues to drive exploitation of remaining unprotected forests (Leach and Mearns 1988; Munslow et al. 1988; Nketiah et al. 1988; Owusu et al. 1989).

Justification for Ecoregion Delineation

This ecoregion comprises the eastern portion of the Upper Guinea forest block of the Guineo-Congolian regional center of endemism (White 1983). It is divided from the Western Guinean Lowland Forests [1] at the Sassandra River, which forms a significant biogeographic barrier. The forest boundaries follow the rainforest vegetation unit delineated by White (1983). The eastern margin is separated from the Nigerian Lowland Forests [4] by the dry habitats of the Dahomey Gap, which has long been recognized as a significant natural break in the Guineo-Congolian rainforest block. The finger of rainforest that extends into Togo is intended to encompass mounts Togo du Fazao and de l'Atakora using the 600-m elevation contour.

Ecoregion Number:	**4**
Ecoregion Name:	**Nigerian Lowland Forests**
Bioregion:	**Western Africa and Sahel**
Biome:	**Tropical and Subtropical Moist Broadleaf Forests**
Political Units:	**Nigeria, Benin**
Ecoregion Size:	**67,300 km²**
Biological Distinctiveness:	**Bioregionally Outstanding**
Conservation Status:	**Critical**
Conservation Assessment:	**IV**
Authors:	Jan Lodewijk R. Werre
Reviewed by:	John Oates

Location and General Description

The Nigerian Lowland Forests [4] are confined to a narrow band along the coast in southwestern Nigeria. The ecoregion extends from the eastern margin of the Dahomey Gap in Benin to the Niger River in the west and is more than 100 km wide at its widest extent. To the north, forest habitats grade into the Guinean Forest-Savanna Mosaic [38], and to the east the Niger River forms a boundary with the Cross-Niger Transition Forests [6].

Much of the ecoregion is situated on a gently undulating coastal plain of 150 m altitude. Two prominent scarps are also found, the first just north of the coastal swamps, with an average elevation of 160 m, and the second over 250 m that extends from Aiyetoro in the west and past Ore in the east. The underlying rocks are Precambrian basement, which are exposed in the north but covered by Tertiary-aged sediments further south. Soils are well drained and moderately to strongly leached ferrasols (Barbour et al. 1982).

The distribution of vegetation in West Africa depends mainly on the climate, which becomes increasingly dry further inland from the coast. Typical mean annual rainfall varies from 2,000 to 2,500 mm in the rainforest zone near the coast to 1,500–2,000 mm in the mixed deciduous forest zone further north. The distribution of wet and dry months is uniform throughout the ecoregion, with a 3-month dry season from December to February. A number of sizable rivers drain this ecoregion. From west to east the most important of these rivers are the Ogun, the Oshun, the Oni with its tributary the Shasha or Omo, the Owenna, and the Osse.

Three vegetation zones cross this ecoregion: the rainforest zone in the south, the mixed deciduous forest zone, and then the parkland zone further north. The first two are climax systems, but the parkland zone probably is anthropogenic and maintained by annual bush fires (Richards 1939). The rainforest is dominated by members of the Leguminosae (*Brachystegia* spp., *Cylicodiscus gabunensis*, *Gossweilerodendron balsamiferum*, *Piptadeniastrum africanum*) and Meliaceae (*Entandrophragma* spp., *Guarea* spp., *Khaya ivorensis*, *Lovoa trichilioides*) families (Richards 1939; Rosevear 1954; Jones 1955, 1956; White 1983; Sayer et al. 1992). In the drier northern portion of this ecoregion dominant trees belong to the Sterculiaceae (*Cola* spp., *Mansonia altissima*, *Nesogordonia papaverifera*, *Pterygota* spp., *Sterculia* spp., *Triplochiton scleroxylon*), Moraceae (*Antiaris africana*, *Ficus* spp., *Milicia excelsa*), and Ulmaceae (*Celtis* spp., *Holoptelea grandis*) families.

Outstanding or Distinctive Biodiversity Features

Despite the discrete biogeographic boundaries formed by the Niger River and the Dahomey Gap, levels of endemism are low. Floristically the ecoregion contains few strictly endemic plant species. However, some of the plant assemblages that contain both Upper and Lower Guinea plants are believed to be unique.

Five strictly endemic animal species are present. Ibadan malimbe (*Malimbus ibadanensis*, EN) occurs in the northernmost forest fringes in the parkland zone, only in the Ibadan area. The Benin genet (*Genetta bini*) was described by Rosevear (1974) from a single specimen from the Ohusu Game Reserve north of Benin and has not been found since (Oates and Anadu 1982; Happold 1987). However, a survey of the Niger Delta revealed the presence of the crested genet (*Genetta cristata*, EN), which also ranges to Cameroon, and this species may encompass the Benin genet (Powell 1995). The near-endemic white-throated guenon (*Cercopithecus erythrogaster*, EN) contains two endemic subspecies (ssp. *erythrogaster* and *pococki*). The endemic Nigeria crag gecko

(*Cnemaspis petrodroma*) and the Petter's toad (*Bufo perreti*) were collected in the early 1960s during an expedition in the Idanre Forest Reserve.

Other threatened animals include the African elephant (*Loxodonta africana*, EN) and chimpanzee (*Pan troglodytes*, EN) (Hilton-Taylor 2000). There may be a subspecific difference between chimpanzee populations in western and eastern Nigeria (Gonder et al. 1997). If substantiated, the subspecies of chimpanzee found in this ecoregion would be critically endangered. If the Benin genet were proven to be part of the same species as the crested genet, then it would be classified as endangered (Hilton-Taylor 2000). However, if the Benin genet were shown to be a separate species, it would be classified as endangered in its own right.

Status and Threats

Current Status

Archaeological evidence shows that the human population has been high for a long time. For example, in Okomu National Park layers of charcoal and pottery below the forest floor indicate that the present forest regenerated over the last 700 years (Jones 1955; White and Oates 1999). By the early twentieth century the Nigerian Lowland Forests [4] were fragmented into a series of disconnected blocks (Richards 1939). Most remaining forest patches were declared as forest reserves by the colonial administration. The last comprehensive survey of the reserves in the ecoregion was conducted in the early 1980s (Oates and Anadu 1982). Nearly all forest reserves visited were farmed, transformed into single-dominant plantations of exotic tree species (rubber [*Hevea braziliensis*], *Gmelina arborea*, and *Tectona grandis*), were heavily exploited for their remaining timber, and contained ample evidence of hunting.

Protected areas in this ecoregion have been created mainly from colonial forest reserves and include the Omo Biosphere Reserve; Akure-Ofosu, Ala, and Owo Strict Nature Reserves; Orle River Game Reserve; Ifon, Kwale, and Gilli-Gilli Game Reserves; and Okumu National Park. Omo's Strict Nature Reserve (46 km²) was established in 1949 and approved as a biosphere reserve in 1977. In the 1980s good forest remained in the core area, although an old logging road indicated past activities. In the early 1980s Okomu was identified as one of the largest and least disturbed forest reserves, with the largest surviving population of white-throated guenon (Oates and Anadu 1982). A 67-km² wildlife sanctuary was created in 1985, and in 1987 the area increased to 114 km². In May 1999 the Nigeria National Parks Service took over the site and apparently increased the protected area to 181 km². Ifon Game Reserve has no villages or roads, and because it lies on the boundary between this ecoregion and the Guinean Forest-Savanna Mosaic [38] it contains little valuable timber. As a result, it was last exploited in 1913 by selective felling. Gilli-Gilli Game Reserve lies 30 km southwest of Benin City. It has been extensively logged, and few large trees remain. Kwale Game Reserve was established in 1932 and covers only 3 km². In its present form it is unlikely to assist the survival of populations of large mammals (Oates and Anadu 1982).

Types and Severity of Threats

Human population density in this ecoregion is high; two decades ago large sections already had population densities of 250–400 people/km², with the remainder having population densities of 100–200 people/km² (Barbour et al. 1982). The population of Nigeria has risen rapidly since then. Farming, logging, and hunting are the most important human activities in the region. All forests in the ecoregion and many of the species they support therefore are highly threatened.

Justification for Ecoregion Delineation

The Nigerian Lowland Forests [4] ecoregion is based on the "lowland forest–drier type" vegetation unit of White (1983), with slight modifications based on advanced very-high-resolution radiometer [AVHRR] imagery (e.g., including the sliver of swamp forest that originally separated the lowland forest area from the mangrove) (Loveland et al. 2000). The Nigerian Lowland Forests [4] contain much lower rates of plant and animal endemism than other West and Central African lowland forests, but have some unique species. The Niger River Delta borders this forest unit on the west, and to the east are the drier habitats of the Dahomey Gap.

Ecoregion Number:	**5**
Ecoregion Name:	**Niger Delta Swamp Forests**
Bioregion:	**Western Africa and Sahel**
Biome:	**Tropical and Subtropical Moist Broadleaf Forests**
Political Units:	**Nigeria**
Ecoregion Size:	**14,400 km²**
Biological Distinctiveness:	**Locally Important**
Conservation Status:	**Critical**
Conservation Assessment:	**IV**
Authors:	Jan Lodewijk R. Werre
Reviewed by:	John Oates, Emmanuel Auquo Obot

Location and General Description

The Niger Delta Swamp Forests [5] form a triangle bordered by the town of Aboh on the Niger River at the northernmost tip, the Benin River along the western boundary, and the Imo River along the eastern boundary. The ecoregion is separated from the Atlantic Ocean by a band of mangroves.

The Niger Delta is the product of sedimentation since the upper Cretaceous, and its low relief is responsible for the meandering and frequent shifting of the Niger and its tributaries. The continuous movement of the delta's creeks has resulted in a mosaic of soil types. Remnants of old levees consist mostly of water-permeable sand and loam. The soils of the depressions behind them (backswamps) consist mostly of waterlogged clay covered by peat, and higher lying sections contain more silty loam (NEDECO 1961).

The climate of the Niger Delta is characterized by a long rainy season from March or April to October. Precipitation increases from the north of the delta (with an average of 2,500 mm) to the coastal area, where mean annual rainfall averages around 4,000 mm. The dry season peaks in January and February, but even during these months an average monthly mean of 150 mm rainfall falls in the delta. Relative humidity rarely dips below 60 percent and fluctuates between 90 and 100 percent for much of the year. The average annual temperature is approximately 28°C (Barbour et al. 1982).

The most important determinant of biological variation in the delta is its hydrology. In addition to precipitation, tidal movements and the Niger River flood determine the hydrological regime. The flood begins toward the end of the rainy season in August, peaks in October, and tapers off in December. The yearly rainfall determines some fluctuation in flow, but since 1968, after the completion of the Kainji Dam, the opening and closing of the dam sluices also has been important.

The swamp forest can be subdivided into three zones based on hydrological variation (Powell 1995). The first zone is the flooded forest. During the rainy season water levels slowly rise, eventually leading to complete inundation, generally from October to December. Some of the more common tree species include *Lophira alata*, *Pycnanthus angolensis*, *Ricinodendron heudelotii*, *Sacoglottis gabonensis*, *Uapaca* spp., *Hallea ledermannii*, *Albizia adianthifolia*, *Irvingia gabonensis*, *Klainedoxa gabonensis*, and *Treculia africana*. The second zone is the eastern delta flank, which is shrinking relative to the western flank (NEDECO 1961, 1966). The third zone is the central backswamp area, crossed by old creek levees. This area is not often flooded and is not influenced by the tides, but most of the forest soils are waterlogged. This forest is dominated by Euphorbiaceae (*Uapaca* spp., *Klaineanthus gaboniae*, *Macaranga* spp.), Annonaceae (*Xylopia* spp., *Hexalobus crispiflorus*), Guttiferae (*Symphonia globulifera*), Rubiaceae (*Hallea ledermannii*, *Rothmannia* spp.), Myristicaceae (*Coelocaryon preussii*, *Pycnanthus marchalianus*), and Ctenolophonaceae (*Ctenolophon englerianus*).

Outstanding or Distinctive Biodiversity Features

There is little information on the species composition of this ecoregion. Wildlife surveys in the delta were not conducted until the late 1980s (Oates and Anadu 1982; Oates 1989; Werre 1991, 2000; Powell 1993, 1995, 1997; Bocian 1998). A number of species that were not known from the delta and were also new for Nigeria were discovered in the 1990s (Powell 1995).

The Upper and Lower Guinean biota, which was considered separated by the Dahomey Gap, overlaps in the delta (Happold 1987; Powell 1995, 1997). The Delta's floral assemblage appears to be unique, although endemic plants are not known. The Delta is also regarded as a small center of endemism for animals (Grubb 1990).

There are no endemic animal species in this ecoregion, but there are two threatened endemic mammal subspecies, the Niger Delta red colobus (*Procolobus pennantii epieni*, EN) and the Niger Delta pygmy hippopotamus (*Hexaprotodon liberiensis heslopi*, CR). There are also two near-endemic monkey species: the white-throated guenon (*Cercopithecus erythrogaster*, EN) and Sclater's guenon (*Cercopithecus sclateri*, EN). A number of other species have been recently recorded that are either new to Nigeria or had not been observed west of the Cross River, such as black-fronted duiker (*Cephalophus nigrifons*), pygmy scaly-tailed flying squirrel (*Idiurus* sp.), and small green squirrel (*Paraxerus poensis*). Other threatened species found here are African elephant (*Loxodonta africana*, EN), chimpanzee (*Pan troglodytes*, EN), and crested genet (*Genetta cristata*, EN).

Status and Threats

Current Status

Although the Niger Delta is wedged between two of Africa's most densely populated ecoregions, for many years it escaped the habitat destruction typical of these areas because of its relative inaccessibility. Roads stop at the delta's boundaries, and further travel is by dugout canoe or motorized boats. The delta used to support low human population densities and little socioeconomic activity other than fishing, some farming in the drier sections, and the collection of forest products. Two events changed this. First, oil was discovered in the Niger Delta in the 1950s, and the associated activities (road and canal building) opened up large sections of remote delta habitat for exploitation. Second, in the 1950s abura (*Hallea ledermannii*) become the most important timber species in Nigeria after *Triplochiton scleroxylon*. Initially the abura came from the swamp forest between the Nigerian Lowland Forests [4] and the mangrove belt. After this was depleted loggers started to focus on the delta, where exploitation was facilitated by increasing oil exploration efforts. This species has been eradicated from the delta, and logging has shifted to other species that float, a necessity in an area where transportation is restricted to waterways.

There are no effective protected areas in the Delta. Three forest reserves exist: Upper Orashi, Nun River, and Lower Orashi, which comprise a total of 239 km². However, these are heavily exploited for their timber. Nine other forest reserves have been proposed. The only effective habitat protection is found in small sacred groves protected by communities. Community protec-

tion is also afforded to different animal species. In a number of lakes crocodiles receive protection, and in one area, Nembe, chimpanzees are protected (Bocian 1998).

Types and Severity of Threats

A growing human population, conflicts between different ethnic groups, national political instability, and unsustainable exploitation of natural resources all occur in this ecoregion.

The human population of the Niger Delta is growing rapidly, with the result that most of the natural resources (e.g., fish, timber) can no longer meet local needs. The inhabitants of the densely populated ecoregions next to the delta have depleted their natural resources even more and look to the delta to provide alternatives. People from as far as Ogoni, located more than 50 km to the east of the delta, come to the central area for part of the year to fish. Fish populations are now declining, and frozen marine fish is sold in the larger towns of the delta.

Justification for Ecoregion Delineation

The Niger Delta Swamp Forests [5] comprise the largest swamp forest habitat in Africa after the Congolian swamp forests. This ecoregion is based on the "swamp forest" vegetation mapped by White (1983). However, the southern boundary has been modified based on reference to a classified AVHRR 1-km satellite image of the continent (Loveland et al. 2000). The swamp forest is biologically distinct in that it harbors endemic mammal subspecies, *Procolobus badius epeini* and *Hexaprotodon liberiensis heslopi*.

Ecoregion Number:	**6**
Ecoregion Name:	**Cross-Niger Transition Forests**
Bioregion:	**Western Africa and Sahel**
Biome:	**Tropical and Subtropical Moist Broadleaf Forests**
Political Units:	**Nigeria**
Ecoregion Size:	**20,700 km²**
Biological Distinctiveness:	**Locally Important**
Conservation Status:	**Critical**
Conservation Assessment:	**IV**
Authors:	Jan Lodewijk R. Werre
Reviewed by:	John Oates

Location and General Description

The Cross-Niger Transition Forests [6] are found in southern Nigeria between the Niger River to the west and Cross River to the east. Most of the ecoregion's relief is low and undulating, with few notable topographic features. Bedrock dating to the Upper Cretaceous underlies the northern part of the ecoregion, whereas to the south the deposits are from the Tertiary (Anonymous 1954; Buchanan and Pugh 1955). All soils in the ecoregion are strongly leached and highly weathered ferrasols (Barbour et al. 1982).

Mean annual rainfall is somewhat higher than in the Nigerian Lowland Forests [4] to the west, ranging from 2,000 to 2,500 mm near the coast and 1,500 to 2,000 mm further north. The northern margins receive 1,250–1,500 mm rain per annum (Barbour et al. 1982). Although there is variation in the amount of rainfall, the seasonal distribution of wet and dry months is uniform, with the dry season lasting from December to February. Other than the Niger and Cross Rivers only two large rivers, the Imo and the Kwa Ibo, drain this ecoregion. The high human population densities found in the ecoregion long predate colonial times. Population densities are between 100 and 500 people/km^2.

This ecoregion harbors species typical of the Upper Guinea Forest Region to the west and the Cross-Sanaga-Bioko Coastal Forests [7] to the east (Schiøtz 1967; Oates 1989). It is therefore transitional between Upper Guinea and Lower Guinea blocks of the Guineo-Congolian regional center of endemism (White 1983). As with the Nigerian Lowland Forests [4] ecoregion, three vegetation zones cross the ecoregion: the rainforest zone to the south, the mixed deciduous forest zone farther north, and the parkland zone still farther north. These vegetation zones reflect decreasing rainfall farther inland. The wetter forest would have become dominated by members of the Leguminosae (*Brachystegia* spp., *Cylicodiscus gabunensis*, *Gossweilerodendron balsamiferum*, *Piptadeniastrum africanum*) and the Meliaceae families (*Entandrophragma* spp., *Guarea* spp., *Khaya ivorensis*, *Lovoa trichilioides*) (Sayer et al. 1992). The drier northern sections of this ecoregion probably were once dominated by Sterculiaceae (*Cola* spp., *Mansonia altissima*, *Nesogordonia papaverifera*, *Pterygota* spp., *Sterculia* spp., *Triplochiton scleroxylon*), Moraceae (*Antiaris africana*, *Ficus* spp., *Milicia excelsa*), and Ulmaceae (*Celtis* spp., *Holoptelea grandis*). Today, cultivation and fire have destroyed much of the rainforest and other natural habitats. Regular burning favors grasses and fire-hardy, gnarled trees. Common grasses include *Andropogon gayanus*, *A. schirensis*, and *A. tectorum*, together with the trees *Annona senegalensis*, *Afzelia africana*, and the palm *Borassus aethiopum* (White 1983).

Distinctive Biodiversity Features

The Cross-Niger Transition Forests [6] ecoregion has extremely low rates of endemism for a tropical forest. There are only two near-endemic species, Sclater's guenon (*Cercopithecus sclateri*, EN) and the crested chameleon (*Chamaeleo cristatus*). There are no endemic amphibians. The near-endemic bird Anambra waxbill (*Estrilda poliopareia*, VU) is considered to be typical of the Cross-Niger region. Although data are harder to find for plants,

few species of restricted distribution seem to be found here, and strict endemics are either absent or extremely few in number (White 1983).

Happold (1987) compared the mammal faunas east and west of the Niger River and found that 34 percent of the ninety-seven rainforest species are confined to one side. More recent surveys in the Niger Delta (Powell 1997; Werre 2000) have indicated that the true number is a bit lower, but the Niger River remains a formidable zoogeographic barrier. The Cross River, which forms this ecoregion's eastern boundary, is much smaller than the Niger River and less of a barrier. The absence in this ecoregion of a number of species found in the Cross-Sanaga-Bioko Coastal Forests [7] therefore may be more an artifact of the high level of deforestation in this ecoregion than a genuine biogeographic feature.

Status and Threats

Current Status

Archeological evidence indicates that high human population densities are not a recent phenomenon in this ecoregion. A rich archeological record, dating as far back as the ninth century A.D., shows that levels of human activity were already considerable at that time (Shaw 1977; Barbour et al. 1982). During colonial times so little forest remained in this area that logging companies did not take much of an interest, and forestry departments did little to establish forest reserves.

Wildlife has been heavily depleted in this ecoregion, to the extent that bats and frogs, animals that are generally avoided in Africa, have become part of the local diet. The few remaining mammal species that thrive in farmland are sold for high prices; greater cane-rat (*Thryonomys swinderianus*) sells for more than the average Nigerian earns in 2 weeks. The remaining native animal populations are restricted to narrow bands of riverine forest, but hunting pressure here is intense, and small-scale logging and agricultural conversion also threaten these forests.

No significant sections of forest remain in this ecoregion, although some habitat remains in forest reserves: Anambra (194 km^2), Mamu River (70 km^2), Osomari (115 km^2), Akpaka (296 km^2), and Stubbs Creek (~80 km^2). These reserves have been converted mostly to plantations of exotic tree species. Stubbs Creek Forest Reserve contains one of the few remaining larger blocks of natural forest and may be the only remaining opportunity for significant forest conservation in this ecoregion. The other intact forest patches are mainly in traditionally protected sacred groves, which are generally small.

Types and Severity of Threats

This ecoregion has a long history of high-density human settlement. Conversion to agriculture and depletion of the native fauna for bushmeat is long-standing, and little habitat and few animals survive. Anthropogenic fires have also altered and destroyed native vegetation.

The high level of deforestation is the most likely reason for the absence of large mammals. Reports from the 1940s indicate that large animals were already extremely scarce (Marchant 1949). Subsequent surveys by Oates (1989) supported these findings, confirming that larger mammals were extremely rare and that hunting pressure was intense. Sclater's guenon, though endangered, still occurs in the ecoregion. Some communities protect forest patches for traditional reasons, and the populations of Sclater's guenon associated with these groves (Oates et al. 1992), although these traditions are starting to lapse (Oates 1996).

Justification for Ecoregion Delineation

Located between the Cross River and the Niger River, this ecoregion contains remnant forests with low species richness and low rates of endemism relative to adjacent ecoregions. The biota is transitional between the Upper Guinean and Lower Guinean forest blocks of the Guineo-Congolian regional center of endemism (White 1983).

Ecoregion Number:	**7**
Ecoregion Name:	**Cross-Sanaga-Bioko Coastal Forests**
Bioregion:	**Central Africa**
Biome:	**Tropical and Subtropical Moist Broadleaf Forests**
Political Units:	**Cameroon, Equatorial Guinea, Nigeria**
Ecoregion Size:	**52,100 km²**
Biological Distinctiveness:	**Globally Outstanding**
Conservation Status:	**Critical**
Conservation Assessment:	**I**
Authors:	*Allard Blom, Jan Schipper*
Reviewed by:	*Steve Gartlan, John Oates*

Location and General Description

The Cross-Sanaga-Bioko Coastal Forests [7] ecoregion extends from the left bank of the Cross River in southeastern Nigeria, follows the coast as far south as the Sanaga River in Cameroon, and extends inland up to 300 km. It also includes the lowland forests of the Island of Bioko. The area is one of low topographic relief at the eastern and western margins but increasingly rugged topography in the foothills of the Nigerian-Cameroon Mountains.

This is one of the wettest areas in tropical Africa. In the southwestern foothills of Mount Cameroon and on southwest Bioko,

rainfall can exceed 10,000 mm per annum with little seasonal variation. Away from the montane influence rainfall averages 3,000 mm per annum along the coast and falls to around 2,000 mm inland. Humidity is always high, rarely dropping below 90 percent. Temperatures range from a maximum between 27°C and 33°C to a minimum between 15°C and 21°C, with minor seasonal differences. A number of river systems drain this ecoregion into the Atlantic Ocean. The Cross River forms the northern boundary of the ecoregion, and the Sanaga River forms its southern boundary. Other important rivers include the Meme, Wouri, Kwa, Ndian, and Nyong rivers.

Much of the ecoregion lies on Precambrian rocks. These basement rocks have been weathered for millions of years and are overlain by thick and heavily leached ferrasols. Because Mount Cameroon and Bioko are volcanoes, adjacent parts of the ecoregion have rocks and soils derived from lava and ash. Bioko was separated from the mainland some 12,000 years ago as sea levels rose at the end of the last ice age.

This lowland forest ecoregion falls within the Lower Guinea block of the Guineo-Congolian regional center of endemism (White 1983). The principal vegetation is coastal evergreen rainforest, with mixed moist semi-evergreen rain forest further inland (White 1983). The flora here shares affinities with other forest types in the Lower and Upper Guinea blocks of the Guineo-Congolian lowland forest (White 1979) and is dominated by species from the plant families Annonaceae, Leguminosae, Euphorbiaceae, Rubiaceae, and Sterculiaceae.

Outstanding or Distinctive Biodiversity Features

There is exceptional species richness in the rainforests of this ecoregion. Combined with the Atlantic Equatorial Coastal Forests [8] to the south, these two ecoregions support about 50 percent of the 7,000–8,000 plants endemic to tropical West Africa (Cheek et al. 1994). At least 200 plant species are endemic. This ecoregion is a center of diversity for the genera *Cola* (Sterculiaceae), *Diospyros* (Ebenaceae), *Dorstenia* (Moraceae), and *Garcinia* (Guttiferae). Endemic trees in the lowland forests include *Deinbollia angustifolia, D. saligna,* and *Medusandra richardsiana.* There are also endemic families and genera of plants, indicating a long evolutionary past (Cheek et al. 1994).

This ecoregion also has high vertebrate species richness and contains the highest numbers of forest-restricted birds and mammals in Africa (Burgess et al. 2000). The lowland forests are of particular importance for the conservation of primates, including the strictly endemic Preuss's red colobus (*Procolobus pennantii preussi,* EN) and the near-endemic red-eared monkey (*Cercopithecus erythrotis,* VU), crowned guenon (*Cercopithecus pogonias,* VU), drill (*Mandrillus leucophaeus,* EN), pallid needle-clawed galago (*Euoticus pallidus* [ssp. *pallidus*] on Bioko, EN), and Pennant's red colobus (*Procolobus pennanti pennanti,* EN) of Bioko. The Cross River population of lowland gorilla is a highly threatened strictly endemic subspecies, *Gorilla gorilla* diehli (CR) (Hilton-Taylor 2000; Sarmiento and Oates 2000). This area also supports the threatened chimpanzee subspecies *Pan troglodytes vellerosus* (EN). Two small mammal species are strictly endemic: the Bibundi bat (*Chalinolobus egeria*) and the pitch shrew (*Crocidura picea,* CR). Near-endemic small mammals include the long-footed shrew (*Crocidura crenata*) and Eisentraut's mouse shrew (*Myosorex eisentrauti,* EN). Korup National Park in Cameroon holds one of the priority populations of forest elephants (*Loxodonta africana cyclotis,* EN) in Africa (AECCG 1991). In some parts of the ecoregion, forest is sufficiently large and interconnected that migrations of forest elephant still occur.

Although no strictly endemic bird species occur, three of the six species characteristic of the Cameroon and Gabon lowlands endemic bird area are found: forest swallow (*Hirundo fuliginosa*), grey-necked rockfowl (*Picathartes oreas*), and Rachel's malimbe (*Malimbus racheliae*) (Stattersfield et al. 1998).

The herpetofauna is highly diverse; Korup alone contains 174 species of reptiles and amphibians. Among the reptiles, the forest chameleon *Chamaeleo camerunensis* and two worm lizards (*Cynisca schaeferi* and *C. gansi*) are strictly endemic. The amphibian fauna is also exceptionally diverse but contains few known strict endemics, including Schneider's banana frog (*Afrixalus schneideri*), Dizangue reed frog (*Hyperolius bopeleti*), and Werner's river frog (*Phrynobatrachus werneri*).

Status and Threats

Current Status

Despite much deforestation, extensive areas of rainforest remain, particularly in the border region between Cameroon and Nigeria (Kamdem-Toham et al. 2003b). A number of forest blocks remain around Korup National Park, which itself covers 1,259 km^2. The Cross River National Park in Nigeria (4,000 km^2) is made up of the Oban and Okwango divisions, with the Afi River Forest Reserve nearby. The Takamanda Forest Reserve in Cameroon is contiguous with the Okwangwo division of the Cross River Park. Part of the Gashaka Gumti National Park in Nigeria also contains lowland forest (estimated at 200 km^2).

Types and Severity of Threat

Over the past 100–200 years, commercial logging and plantation agriculture have been the main causes of deforestation on the mainland, followed by subsistence agriculture, which often occurs after the logging has opened up an area. Lowland forest habitats on Bioko have also been lost through conversion to plantations and farming activities, except in the southern sector, where they are inaccessible because of rugged topography (Fa 1991). Despite these losses much more forest remains than in the adjacent Cross-Niger Transition Forests [6]. Although the original forest is now fragmented in many areas, some large

habitat blocks remain, and in the border region between Nigeria and Cameroon these blocks are still connected.

A major threat to the fauna of the area is the overhunting of larger mammal species for bushmeat (Bowen-Jones and Pendry 1999). In some areas, this trade is fully commercialized and supplies protein to major towns. In other areas, certain species such as lowland gorillas are hunted for their religious, magical, and supposed medicinal properties. The wildlife trade is also a cause of species depletion of reptiles. Another threat is the pressure to establish rubber, wood pulp, oil, and palm plantations in the forest zone of Nigeria.

Although more then 10 percent of the ecoregion is officially protected in national parks, in reality these parks do not adequately protect fauna and flora because of low staffing, inadequate budgets, and lack of political will. Some larger mammal species in the Korup and Cross River National Parks are severely threatened from hunting, and populations of elephant, drill, and red colobus have been reduced.

Justification for Ecoregion Delineation

This forest ecoregion is a part of the Guineo-Congolian regional center of endemism (White 1983). The biogeographic barriers of the Sanaga River in Cameroon and the Cross River in Nigeria define the mainland boundaries of this ecoregion. These rivers are particularly important barriers for primates (e.g., drill, *Mandrillus leucophaeus* and red-eared guenon, *Cercopithicus eurythrotis*) and amphibians (e.g., Dizangue reed frog, *Hyperolius bopeleti*). The Bioko lowland forest shares close biological affinities with the adjacent mainland forests because it was connected to the mainland during lower sea levels during the last ice age and thus is included in this ecoregion.

Ecoregion Number: **8**

Ecoregion Name: **Atlantic Equatorial Coastal Forests**

Bioregion: **Central Africa**

Biome: **Tropical and Subtropical Moist Broadleaf Forests**

Political Units: **Angola, Cameroon, Republic of the Congo, Democratic Republic of the Congo, Equatorial Guinea, Gabon**

Ecoregion Size: **189,700 km²**

Biological Distinctiveness: **Globally Outstanding**

Conservation Status: **Relatively Stable**

Conservation Assessment: **III**

Authors: *Allard Blom, Jan Schipper*

Reviewed by: *Steve Gartlan, Marc Thibault, Sonia Blaney, André Kamdem-Toham*

Location and General Description

The Atlantic Equatorial Coastal Forests [8] extend from the Sanaga River in west central Cameroon south through Equatorial Guinea into the coastal and inland areas of Gabon, the Republic of the Congo, and the Cabinda Province of Angola. Along the coast the ecoregion ends in the extreme west of the Democratic Republic of the Congo (DRC), just north of the mouth of the Congo River. The southern 400 km of the ecoregion comprises a tongue of forest inland of the coastal plain and surrounded by the Western Congolian Forest-Savanna Mosaic [43].

Low undulating hills and plains characterize the topography of the northern portion of the ecoregion. Altitude increases gradually inland up to 800 m. Coastal mountain ranges with altitudes above 1,000 m are found in the southern sector of the ecoregion, including Monte Alen, Monts de Cristal, and Monts Doudou. A number of important river systems cross the ecoregion. The Sanaga River forms the northern limit, and further south rivers include the Ogooué, the Nyanga, and the Kouilou. Precambrian basement rocks characterize the geology. The southern extension is delineated by the limits of the Precambrian rocks, with younger rocks on either side. Ferrasols characterize the northern part of the ecoregion, whereas cambisols and nitosols are found to the south.

The ecoregion lies in the wet tropics and receives high rainfall with limited seasonality. Rainfall varies from 2,000 mm per annum in the north to 1,200 mm in the southern sector. Temperatures range from an annual mean maximum between 24°C and 27°C to an annual mean minimum between 18°C and 21°C, and there is little seasonal variation. Humidity is high throughout the year, except in the south.

The forests of Gabon have few inhabitants, and population densities are generally one to ten people per square kilometer, except in logging camps. The forests contain forest-dwelling Bakola and Bagyeli people and other groups who practice subsistence hunting, farming, and fishing (Luling and Kenrick 1998). In the southern mountains, population densities are higher, with more than fifty people per square kilometer.

The ecoregion falls within the Lower Guinea Forest Block of the Guineo-Congolian regional center of endemism (White 1979, 1983). Coastal evergreen moist forests characterize the vegetation, with mixed semi-evergreen moist forests in the drier southern extension (White 1993). Strips of Guineo-Congolian edaphic grassland occurs along the coast, and montane forest and grassland occurs on mountain ranges (White 1983). The forests are further described in Fa (1991) for Equatorial Guinea, Gartlan (1989) for Cameroon, Hecketsweiler (1990) for Congo, Wilks (1990) for Gabon, and IUCN (1989), Sayer et al. (1992), Weber et al. (2001), and Kamdem-Toham et al. (2003b) for Central Africa in general.

Outstanding or Distinctive Biodiversity Features

This ecoregion supports exceptional species richness and has many endemic species. The endemics are concentrated in the mountains to the south, although there are also endemic species in the lowland forest.

This ecoregion and the Cross-Sanaga-Bioko Coastal Forests [7] to the north support about 50 percent of the 7,000–8,000 plants endemic to tropical West Africa (Cheek et al. 1994). More than 200 plant species are strictly endemic to this ecoregion. The site of highest importance for plant conservation is Monts de Cristal in Gabon, which has more than 3,000 plant species and 100 strict endemics (WWF and IUCN 1994). Another important montane site is the Monts Doudou, with more than 1,000 vascular species and 50 strict endemics.

This is a critical area for the conservation of large forest mammals. Globally important populations of gorilla (*Gorilla gorilla gorilla,* EN), chimpanzee (*Pan troglodytes,* EN) and forest elephant (*Loxodonta africana cyclotis,* EN) occur (Tutin and Fernandez 1984; Barnes 1987; Barnes et al. 1995; Said et al. 1995). Extensive areas of forest also remain, and because human populations generally are low, the fauna can be intact. Migrations of large forest animals still occur between this and adjacent ecoregions. Other important larger mammal populations include mandrill (*Mandrillus sphinx,* VU), black colobus (*Colobus satanas*), bongo (*Tragelaphus euryceros*), and forest buffalo (*Syncerus caffer nanus*) (Blom et al. 1990; Castroviejo Bolivar et al. 1990). Near-endemic mammals include the sun-tailed monkey (*Cercopithecus solatus,* VU) (Blom et al. 1992), long-footed shrew (*Crocidura crenata*), lesser Angolan epauletted fruit bat (*Epomophorus grandis,* DD), and African smoky mouse (*Heimyscus fumosus*) (Happold 1996).

The avian species richness is high. No strict endemic bird species occur, but restricted-range species include Verraux's batis (*Batis minima*), black-necked wattle-eye (*Platysteira chaly-bea*), forest swallow (*Hirundo fuliginosa*), Rachel's malimbe (*Malimbus racheliae*), Ursula's sunbird (*Nectarinia ursulae*), African river martin (*Pseudochelidon eurystomina*), Bates's weaver (*Ploceus batesi,* EN), and the Dja River warbler (*Bradypterus grandis,* VU) (BirdLife International 2000). Most of these species are shared only with the Northwestern Congolian Lowland Forests [12] and a few only with the Cross-Sanaga-Bioko Coastal Forests [7]. Other threatened birds are grey-necked rockfowl (*Picathartes oreas,* VU) and Loango weaver (*Ploceus subpersonatus,* VU) (Christy 2001a).

The reptiles and amphibians also have high species richness and contain endemic species. There are seven strict endemic amphibian species, including the Apouh night frog (*Astylosternus schioetzi*), Perret's shovelnose frog (*Hemisus perreti*), Gabon dwarf clawed frog (*Hymenochirus feae*), Ogowe River frog (*Phryno-batrachus ogoensis*), and Andre's clawed frog (*Xenopus andrei*). Some of the strictly endemic reptiles include the French Congo worm lizard (*Cynisca bifrontalis*), Haugh's worm lizard (*Cynisca haughi*), and Cameroon racer (*Poecilopholis cameronensis*).

Status and Threats

Current Status

Large areas of rainforest remain in southern Cameroon, Gabon, and the Republic of the Congo, but most have been selectively logged. Logging has been particularly severe in Equatorial Guinea (Fa 1991) and is widespread in Cameroon and Gabon. The area of untouched primary forest is small.

The total area under protection is 28,664 km², or 15.1 percent of the ecoregion. Protected areas in the northern part of the ecoregion are Campo-Maan National Park and Douala-Edéa Wildlife Reserve in Cameroon and the Monte Alén National Park in Equatorial Guinea. In Gabon, protected areas include part of the Lopé Reserve, the Gamba complex, and the Wonga-Wongué Presidential Reserve. An additional thirteen forest national parks are being established in Gabon. The Republic of the Congo contains the Conkouati and Dimonika-Mayombe Reserves.

Types and Severity of Threats

Logging concessions cover almost the entire ecoregion, including many protected areas. Extensive logging poses a serious threat to the continued existence of primary stands of rainforest. In the southern and northern areas of the forest, where human population density is higher (Cameroon, northern Equatorial Guinea, DRC, and Republic of the Congo), logging has opened up the forest, and agriculturists have colonized some areas. Because the human population is small in the central portion of this ecoregion, the threat from agricultural conversion is low.

The main threat to the larger mammals comes from hunting for bushmeat and fetishes (Bowen-Jones and Pendry 1999). There are also oil exploration and production facilities in Gabon that are favoring human settlements. The logging and oil industries facilitate hunting, poaching, and the bushmeat trade by providing markets, transport, and access to remote forests.

Elephants are poached throughout the ecoregion for their meat and ivory. The trade in African grey parrots (*Psittacus erithacus*), especially in Cameroon, is threatening the survival of this bird.

Justification for Ecoregion Delineation

The Atlantic Equatorial Coastal Forests [8] form part of White's Guineo-Congolian regional center of endemism (White 1983). The northern limit of the ecoregion is bordered by the Sanaga River, which represents a significant biophysical boundary for many vertebrates (e.g., *Mandrillus sphinx*). The ecoregion stretches approximately 200 km inland, although this is difficult to define precisely. The coastal littoral vegetation and the forests on the mountain ranges in the south of the ecoregion contain distinct biological features and may be separated into new ecoregions in the future.

Ecoregion Number:	**9**
Ecoregion Name:	**Mount Cameroon and Bioko Montane Forests**
Bioregion:	**Central Africa**
Biome:	**Tropical and Subtropical Moist Broadleaf Forests**
Political Units:	**Cameroon, Equatorial Guinea**
Ecoregion Size:	**1,100 km²**
Biological Distinctiveness:	**Globally Outstanding**
Conservation Status:	**Critical**
Conservation Assessment:	**I**
Authors:	*Allard Blom, Jan Schipper*
Reviewed by:	*Steve Gartlan, Françoise Dowsett-Lemaire, Robert Dowsett*

Location and General Description

The Mount Cameroon and Bioko Montane Forests [9] form part of a volcanic chain that extends from the Gulf of Guinea islands to the Cameroon Highlands. Rising to 4,095 m in elevation, Mount Cameroon is the tallest peak in the region. The ecoregion occupies the higher elevations of Mount Cameroon and Bioko, although on Mount Cameroon high rainfall and intense cloud cover depresses montane habitats to as low as 500 m elevation on the southwest side of the mountain.

Both Mount Cameroon and Bioko are recent volcanoes, and Mount Cameroon is still active, erupting in 1982 and 1999 (Fotso et al. 2001). Etinde, on the southwestern side of Mount Cameroon, is an older, nonactive volcanic peak. The volcanic history is apparent in the areas of rifting on the north and south slopes, the barren lava fields from the 1982 and 1999 eruptions, and the presence of collapse scars and cinder cone features.

The southwestern sides of Mount Cameroon and Bioko have an almost continuous wet rainy season, with rainfall reaching 10,000 mm per year in the lower tropical altitudes. Rainfall gradually declines at higher altitudes, with less than 2,000 mm falling at the summit of Mount Cameroon. There is also a significant rainshadow on the north and eastern sides of Mount Cameroon. At the base of these mountains, temperatures average 25.5–27°C and can reach between 32°C and 35°C in the hottest months (March and April). Temperatures decline at approximately 1°C per 150 m of elevation.

This ecoregion falls in the Afromontane archipelago-like regional center of endemism (White 1983). The ecoregion has close floristic affinities to the Guineo-Congolian forests and the Cameroon Highlands (White 1979, 1983), the montane forests and sub-alpine communities of the Rwenzori-Virunga Mountains, and forests of the Angola Scarp. A number of factors, such as elevation, aspect, and climate, influence the diversity of vegetation types in the ecoregion. On Bioko, three distinct areas of montane forest occur above 1,500 m: Pico Basilé (maximum elevation 3,011 m), Gran Caldera de Luba (2,261 m), and Pico Biao (2,009 m). The montane forest is characterized by *Schefflera abyssinica*, *Prunus africana*, and *Nuxia congesta*. On Mount Cameroon, montane forest extends from 1,675 m to between 2,200 and 2,400 m in elevation. This grades into montane scrub and ultimately into montane grasslands, sub-alpine communities, and rocky habitats at the highest elevations (Killick 1979). The well-drained slopes formed by porous lava flows on Mount Cameroon support meadows of tussock grass and sedge vegetation, including *Pennisetum monostigma*, *Swertia abyssinica*, *Myrica humilis*, and *Agauria salicifolia*. Sub-alpine grassland communities occur at about 2,000–2,200 m on Mount Cameroon and 2,500 m on Bioko (Castroviejo et al. 1986; Morton 1986). Characteristic species include *Andropogon amethystinus*, *Deschampsia mildbraedii*, *Agrostis mannii*, *Koeleria cristata*, and *Bulbostylis erratica*.

Outstanding or Distinctive Biodiversity Features

This ecoregion contains exceptional species diversity and endemism in its flora and fauna. High species richness results from the diversity of habitats found in a restricted geographic area, ranging from submontane and montane forests to sub-alpine grasslands.

At least forty-two plant species and three genera are strictly endemic, and another fifty species are near endemic to Mount Cameroon (Cheek et al. 1994; WWF and IUCN 1994; Cable and Cheek 1998). Twenty-nine of these near-endemic species are also found on the island of Bioko. Most of the endemics at higher altitudes are recently evolved, comprising nineteen of the near-endemics and nineteen of the strict endemics, whereas paleoendemic plants, including the three endemic genera, occur only in the foothills. A number of species that occur in nearby highlands are also missing on Mount Cameroon. Frequent volcanic activity at 20-year intervals may explain why Mount Cameroon lacks stands of *Arundinaria alpina* and *Podocarpus latifolius*, common in the Cameroon Highlands Forests [10].

More than 370 bird species have been recorded, including several endemics (Fotso et al. 2001). Two species are strictly endemic to Mount Cameroon: Mount Cameroon francolin (*Pternistis camerunensis*, EN) and Mount Cameroon speirops (*Speirops melanocephalus*, VU). Notable species on Bioko include *Nectarinia ursulae*, *Psalidoprocne fuliginosa*, *Picathartes oreas*, and the endemic Fernando Po speirops (*Speirops brunneus*, VU), which is confined to Basilé Peak (Pérez del Val 2001). Additional restricted-range montane species are shared by this ecoregion and the Cameroon Highlands Forests [10] (Stattersfield et al. 1998).

Mammals display moderate levels of diversity and endemism. The arrogant shrew (*Sylvisorex morio*, EN) is strictly endemic, whereas the Cameroon soft-furred mouse (*Praomys morio*, VU) is one near-endemic species that is confined to narrow altitudinal bands. Other species found in the ecoregion include red-eared monkey (*Cercopithecus erythrotis*, VU), Preuss's monkey (*Cercopithecus preussi*, EN), black colobus monkey (*Colobus satanas*, VU on Bioko), drill (*Mandrillus leucophaeus*, EN), Hun shrew (*Crocidura attila*, VU), Cameroon climbing mouse (*Dendromus oreas*, VU), and Eisentraut's mouse shrew (*Myosorex eisentrauti*, EN, only on Bioko) (Hilton-Taylor 2000; Dowsett-Lemaire and Dowsett 2001). There is an endemic subspecies of drill (*Mandrillus leucophaeus poensis*) on Bioko.

Although the herpetofauna is diverse, there is only one strictly endemic toad on Mount Cameroon, *Werneria preussi*. Other near-endemic species include the four-digit toad (*Didynamipus sjöstedti*), Cameroon mountain chameleon (*Chamaeleo montium*), Sjöstedt's five-toed skink (*Leptosiaphos gemmiventris*), and Tandy's smalltongue toad (*Werneria tandyi*). Mount Cameroon is also important for butterflies, including the endemic *Charaxes musakensis* (Fotso et al. 2001).

Status and Threats

Current Status

On the eastern side of Mount Cameroon, up to half of the forest cover has already been lost (Stattersfield et al. 1998). Less forest habitat has been lost on the southern slopes of Mount Cameroon. Because the soils of the region are fertile, further conversion of some forest to agriculture is regarded as inevitable. Much of the foothills of Mount Cameroon are the property of the Cameroon Development Corporation, a parastatal agroindustrial institution in the process of privatization. Forest reserves on the northwestern (Bambuko) and northeastern (Southern Bakundu) flanks of the mountain provide little protection and are becoming increasingly degraded. These reserves were established for timber production and are also encroached on for farming and subject to illegal logging and hunting pressures. Three further reserves have been proposed by the Mount Cameroon Project, a collaboration between the Government of Cameroon, the Department of International Development, Deutsche Gesellschaft für Technische Zusammenarbeit, and Global Environment Facility. However, despite much work over the last decade to gazette the reserves of Mabeta-Moliwe, Etinde, and Onge, there has been little substantive progress. In the Mokoko-Onge area, the Moko Wildlife Management Association is strongly opposed to further forest exploitation, instead preferring that this area be protected and managed by their community (F. Dowsett-Lemaire, pers. comm., 2002).

The montane forest of Bioko is protected by Basilé National Park (Parque Nacionale de Pico Basilé, 330 km²) and Luba Crater Scientific Reserve (Reserva Cientifica de la Caldera de Luba, 510 km²), which were both given legal status in 2000. Habitat loss on the mountains of Bioko is low, and the southern slopes of the mountain are almost completely undisturbed from sea level to the summit at 2,261 m. Fire is a threat to some montane forest and grassland, as it is on Mount Cameroon.

Types and Severity of Threat

An expanding human population seeking new agricultural land and bushmeat is the main source of threat to habitats of the ecoregion. Areas that receive slightly lower rainfall are in most demand and are most likely to be converted into agricultural lands. The collection of nontimber forest products, including firewood and pygeum (*Prunus africana*), occurs, although the killing of *Prunus* for its bark has largely been stopped. Oil palm, rubber, and banana plantations are found across much of the Mount Cameroon foothills. Environmental impact studies carried out by the Mount Cameroon Project may impose constraints on further plantation creation, but some expansion is inevitable. The fact that none of Mount Cameroon is found in a national park is a major threat to the future of this important area.

In some areas larger mammals have nearly disappeared (southern, eastern, and northern slopes of Mount Cameroon). The Mount Cameroon Project, trying to develop parameters for sustainable hunting, was unable to produce an assessment of sustainable offtake because mammal numbers were so low. Hunting pressure is also rising on Bioko. Because of its inaccessibility the southern part of Bioko has a largely intact mammal fauna. There is also an active trade in live animals from the ecoregion to supply the pet trade, including amphibians, reptiles, mammals, and birds.

Justification for Ecoregion Delineation

This ecoregion forms a part of the Afromontane archipelago-like regional center of endemism (White 1983). The ecoregion was delineated from White's (1983) "Afromontane undifferentiated montane woodland" unit. The Mount Cameroon and Bioko Montane Forests [9] ecoregion is distinct from the Cameroon Highlands Forests [10] because of the younger age of their volcanoes and the active nature of Mount Cameroon. There are also differences in the flora and fauna. The montane area (above 1,500 m elevation) of Bioko was included because it shares biological affinities with Mount Cameroon and was connected to the mainland during periods of lower sea levels in the last ice age.

Ecoregion Number:	**10**
Ecoregion Name:	**Cameroon Highlands Forests**
Bioregion:	**Central Africa**
Biome:	**Tropical and Subtropical Moist Broadleaf Forests**
Political Units:	**Cameroon, Nigeria**
Ecoregion Size:	**38,000 km²**
Biological Distinctiveness:	**Globally Outstanding**
Conservation Status:	**Endangered**
Conservation Assessment:	**I**
Authors:	*Allard Blom, Jan Schipper*
Reviewed by:	*Steve Gartlan, Fiona Maisels, David Thomas, Françoise Dowsett-Lemaire, Robert Dowsett*

Location and General Description

The Cameroon Highlands Forests [10] encompass the mountains of the border region between Nigeria and Cameroon, excluding Mount Cameroon. Included are the Rumpi Hills, Bakossi Mountains, Mount Nlonako, Mount Kupe, and Mount Manengouba, north to the Bamenda-Banso Highlands and with outliers northeast to the Mambila Plateau and northwest to the Obudu Plateau of Nigeria. Further northeast the ecoregion continues along the western flank of the Adamawa Plateau to Tchabal Gangdaba, with two small outliers further east (Stuart 1986; Gartlan 1989; Stattersfield et al. 1998).

The majority of the ecoregion occupies a range of extinct volcanoes, some of which still contain remnant craters, hot springs, and other signs of volcanism. The highest point is Mount Oku, at 3,011 m, in the Bamenda-Banso Highlands. Most of the remainder is below 2,600 m in elevation. The soils derived from the volcanic rocks are fertile and suitable for agriculture.

Although it is located in tropical Africa, the ecoregion's mean maximum temperatures are below 20°C because of the altitude. At the southern extremity, closer to the coast, rainfall is around 4,000 mm per annum, declining inland to 1,800 mm or less. Adequate rainfall, together with fertile soils, contributes to a high human population density. In parts of the Bamenda Highlands, there are up to 300 people/km².

These mountains fall in White's (1983) Afromontane archipelago-like regional center of endemism. Vegetation consists of submontane forests between 900 and 1,800 m altitude and above this a mixture of montane elements, including montane forests, montane grasslands, bamboo forests, and sub-alpine communities. Five tree species characterize the forested montane zone: *Nuxia congesta*, *Podocarpus latifolius*, *Prunus africana*, *Rapanea melanophloeos*, and *Syzygium guineense bamendae* (Letouzey 1985). The sub-alpine zone occurs above 2,800 m and is found only on Mount Oku in this ecoregion.

Outstanding or Distinctive Biodiversity Features

The Cameroon Highlands Forests [10] are of extremely high biological importance, with many endemic species. The forests have affinities with the highland forests of Angola (Kingdon 1989), and especially with forests in East Africa. Most trees in the montane forests of Cameroon are also found in the mountains of eastern Africa, such as *Alangium chinense*, *Albizia gummifera*, *Apodytes dimidiata*, *Cassipourea gummiflua*, *Croton macrostachyus*, *Ilex mitis*, *Olea capensis*, *Podocarpus latifolius*, *Polyscias fulva*, *Prunus africana*, *Schefflera abyssinica*, *Strombosia schefferi*, *Xymalos monospora*, and, at edges, *Agauria salicifolia*, *Maesa lanceolata lanceolata*, *Myrica humilis*, *Nuxia congesta*, *Pittosporum viridiflorum*, *Rapanea melanophloeos*, and *Scolopia zeyheri* (Dowsett-Lemaire 1989a).

Tree species diversity in these forests tends to be low, but the diversity of nonwoody plants such as grasses is high. Highest levels of tree endemism are found in the submontane region. There is also a significant endemic flora in the grasslands, heathlands, moorlands, and other nonforested habitats at higher elevations, such as around the summit of Mount Oku (Maisels et al. 2000). These nonforested habitats share elements with the high mountain plant communities of the Rwenzori-Virunga Montane Moorlands [73].

The ecoregion contains exceptional levels of avian endemism, including the strictly endemic Bamenda apalis (*Apalis bamendae*), Bangwa forest warbler (*Bradypterus bangwaensis*), white-throated mountain-babbler (*Kupeornis gilberti*, EN), banded wattle-eye (*Platysteira laticincta*, EN), Bannerman's weaver (*Ploceus bannermani*, VU), Bannerman's turaco (*Tauraco bannermani*, EN) and Mt. Kupe bushshrike (*Telophorus kupeensis*, EN) (Bowden and Andrews 1994; Stattersfield et al. 1998). An additional nine montane endemics are shared with the Mount Cameroon and Bioko Montane Forests [9]: mountain sawwing (*Psalidoprocne fuliginosa*), grey-throated greenbul (*Andropadus tephrolaemus*), Cameroon olive-greenbul (*Phyllastrephus poensis*), black-capped wood-

land-warbler (*Phylloscopus herberti*), green longtail (*Urolais epichlora*), white-tailed warbler (*Poliolais lopezi*), Cameroon sunbird (*Nectarinia oritis*), Ursula's sunbird (*Nectarinia ursulae*), and Fernando Po oliveback (*Nesocharis shelleyi*). There are an additional fourteen species in common with just Mount Cameroon and not Bioko, including Cameroon greenbul (*Andropadus montanus*), grey-headed greenbul (*Phyllastrephus poliocephalus*), yellow-breasted boubou (*Laniarius atroflavus*), green-breasted bushshrike (*Malaconotus gladiator*, VU), mountain robin-chat (*Cossypha isabellae*), and the race *Cisticola chubbi discolor* (sometimes considered a separate species *C. discolor*) (Dowsett 1989; Dowsett-Lemaire and Dowsett 1998, 2000; Fotso et al. 2001).

Small mammals strictly endemic to this region include Eisentraut's striped mouse (*Hybomys eisentrauti*, EN), Cooper's mountain squirrel (*Paraxerus cooperi*), Mount Oku mouse (*Lamottemys okuensis*, EN), Mittendorf's striped grass mouse (*Lemniscomys mittendorfi*, EN), Oku mouse shrew (*Myosorex okuensis*, VU), Rumpi mouse shrew (*M. rumpii*, CR), western vlei rat (*Otomys occidentalis*, EN), Hartwig's soft-furred mouse (*Praomys hartwigi*, EN), and Isabella's shrew (*Sylvisorex isabellae*, VU) (Hutterer and Fulling 1994). In addition to these smaller species, there is also a population of an endemic subspecies of lowland gorilla (*Gorilla gorilla diehli*, EN), several groups of drill (*Mandrillus leucophaeus*, EN) in Bakossi, and healthy populations of Preuss's red colobus (*Procolobus pennantii preussi*, EN) and chimpanzee (*Pan troglodytes*, EN).

Exceptionally high levels of endemism are found among amphibians, with nearly forty species that are strictly endemic (Stuart 1986): *Hylarana longipes*, *Petropedetes parkeri*, *P. perreti*, *Phrynobatrachus cricogaster*, *P. steindachneri*, *P. werneri*, *Phrynobatrachus* sp. (Oku), *Phrynodon* sp. 1 *sensu* Amiet (1975), *Phrynodon* sp. 2 *sensu* Amiet (1975), *Arthroleptis adolfifriedericii*, *Cardioglossa melanogaster*, *C. oreas*, *C. pulchra*, *C. schioetzi*, *C. trifasciata*, *C. venusta*, *Astylosternus nganhanus*, *A. perreti*, *A. montanus*, *A. rheophilus*, *Leptodactylodon axillaris*, *L. bicolor*, *L. boulengeri*, *L. erythrogaster*, *L. mertensi*, *L. polyacanthus punctiventris*, *L. perreti*, *Afrixalus lacteus*, *Hyperolius adametzi*, *H. riggenbachi* (including *hyeroglyphicus*, now considered conspecific, J.-L. Amiet, pers. comm., 2002), *Leptopelis nordequatorialis*, *Xenopus amieti*, *Xenopus* sp., *Bufo villiersi*, *Werneria bambutensis*, *W. tandyi*, and *Wolterstorffina mirei*. A new *Leptodactylodon* (*wildi*) was also described in 2000 and appears endemic to Bakossi (Amiet and Dowsett-Lemaire 2000). There are also a number of near-endemic amphibians that also occur on lower hills near Yaoundé or Mount Cameroon: *Leptodactylodon ornatus*, *Hyperolius koehleri*, *H. kuligae*, *Werneria mertensi*, and *Wolterstorffina parvipalmata*.

Among the reptiles, the following species are considered narrow endemics: *Atractaspis coalescens*, *Chamaeleo eisentrauti*, Pfeffer's chameleon (*C. pfefferi*), four-horned chameleon (*C. quadricornis*), *Cnemaspis gigas*, *Leptosiaphos ianthinoxantha*, angel's five-toed skink (*L. lepesmei*), and *Panaspis chriswildi*.

There is also a significant overlap between the flora and fauna of this ecoregion and that of the nearby Mount Cameroon and Bioko Montane Forests [9]. Fifty near-endemic plant species are shared between the two ecoregions, and similar affinities are seen in other taxa.

Status and Threats

Current Status

Montane forests above 1,800 m altitude formerly occurred over a much greater area of the Cameroon Highlands (Letouzey 1985). Today they are reduced to small patches, which are under threat from agricultural encroachment and burning by pastoralists (Stuart 1986; MacLeod 1987). The area around Mount Oku in the Bamenda-Banso Highlands supports around 100 km^2 of forest, and there are other patches in this region (Maisels and Forboseh 1999; Maisels et al. 2000; Thomas 1987). The Bakossi Mountains have at least 200 km^2 of midaltitude and montane forest above 1,000 m, and the lowland forest ("Western Bakossi") covers some 400 km^2. The Mount Nlonako Faunal Reserve contains a partial forest continuum from the montane section to the lower levels. Tchabal Mbabo, in the northern sector, contains almost 50 km^2 of virtually pristine montane forest (Thomas and Thomas 1996). In Nigeria, the biggest patch in the Gotel Mountains covers 46 km^2 (peak Gangirwal), and there are other patches. Montane forest remnants also remain in gullies of the Obudu Plateau and on the highland areas further to the north, such as the Mambila and Mana Plateaus and Gashaka Gumti; all are smaller than the Cameroon patches (Dowsett 1989; Sayer et al. 1992; Stattersfield et al. 1998).

This is one of the least-protected ecoregions in Africa. No part of this ecoregion is under formal protected status in Cameroon, although local traditional rulers still exert authority over land use. The main section of Bakossi (550 km^2) has been proposed as "protection forest," banning all logging. Kupe has been proposed as a strict nature reserve, and the boundaries of this reserve were successfully delineated with the participation of the local people in 2000–2001. Nlonako Mountain is a faunal reserve, which protects it from logging. The Gashaka Gumti National Park in Nigeria contains some montane forest, and some fragments remain at Obudu in the Okwangwo section of the Cross River National Park. Forest loss at Mount Oku/Ijim has been halted since 1986 by projects working with local chiefs and community members around the forest area (Maisels and Forboseh 1999; McKay and Coulthard 2000). However, large mammals are virtually extinct in Oku, forest succession is impeded by grazing of goats, and the remaining montane fragments are highly vulnerable to fires set by cattle grazers (Maisels et al. 2000).

Types and Severity of Threat

The natural habitats of this ecoregion are highly threatened and are still being lost through conversion to agriculture, unsustainable use of timber, fires from farmland, and collection of firewood and construction materials (Collar and Stuart 1988; Gartlan 1989; Alpert 1993). Because volcanic rock produces good soils, there remains considerable pressure to convert areas to farmland. Forest cover has declined by more than 50 percent since the 1960s through conversion for cultivation because of the fertile soils and reliable rainfall in this area (Stuart 1986; Collar and Stuart 1988; Alpert 1993).

The once common African cherry tree (*Prunus africana*) has suffered serious declines in the last two decades because of uncontrolled exploitation of its bark for pharmaceutical use. Firewood collection is also a major cause of degradation of forest patches, and hunting also threatens the remaining larger mammals.

Justification for Ecoregion Delineation

This ecoregion forms part of the Afromontane archipelago-like regional center of endemism (White 1983). Numerous endemic species characterize the ecoregion, which is delineated as montane areas above 900 m elevation. The Cameroon Highlands Forests [10] are distinct from Mount Cameroon and Bioko Montane Forests [9] because there are no active volcanoes, and the flora and faunas are distinct.

Ecoregion Number:	**11**
Ecoregion Name:	**São Tomé, Príncipe, and Annobon Moist Lowland Forests**
Bioregion:	**Central Africa**
Biome:	**Tropical and Subtropical Moist Broadleaf Forests**
Political Units:	**Democratic Republic of São Tomé and Príncipe, Equatorial Guinea**
Ecoregion Size:	**1,000 km^2**
Biological Distinctiveness:	**Globally Outstanding**
Conservation Status:	**Critical**
Conservation Assessment:	**I**
Author:	*Angus Gascoigne*
Reviewed by:	*None*

Location and General Description

The São Tomé, Príncipe, and Annobon Moist Lowland Forests [11] cover the three islands of Príncipe, São Tomé, and Annobon in the Gulf of Guinea, off the west coast of central Africa. These islands are part of a volcanic chain that continues inland as the Cameroon-Nigerian Mountains. The volcanoes date back to the Tertiary and are considered inactive. Príncipe is the closest to mainland Africa and has an area of 128 km². São Tomé lies further out and covers approximately 836 km². Annobon is the furthest from the coast, with an area of 17 km².

The volcanic plugs and mountainous parts of the islands rise up to 948 m altitude on Príncipe, 2,024 m on São Tomé, and 695 m on Annobon. The only flat land on these islands is found at the base of the volcanoes (Jones and Tye 1988). The volcanic soils derived from basalts and phonolites are fertile and have been used for plantation crops in the past.

All three islands are in the wet tropical belt. On São Tomé, annual rainfall ranges from 1,000 mm in the northeast to more than 4,000 mm in the southwest. Mean annual maximum temperatures range from 30°C to 33°C, and minimum temperatures range from 18°C to 21°C, with little seasonal variation and high humidity all year. On Príncipe, rainfall patterns are similar, whereas on the island of Annobon rainfall is somewhat less.

The islands were uninhabited in 1470–71, when the Portuguese discovered them. Colonization began in the early sixteenth century when São Tomé became the world's largest sugar producer and, after this crop's decline, the island grew into an important slave trading post. Annobon became a Spanish colony in the eighteenth century and now forms part of Equatorial Guinea. In the nineteenth century, coffee and cocoa plantations were established on São Tomé and Príncipe. After the closure of the estates the islands remained populated and became an independent country in 1975.

The islands were not mapped in White's (1983) *Phytogeographical Classification of Africa* but have many elements in common with the Lower Guinea Forest Block of the Guineo-Congolian regional center of endemism. The original vegetation of both islands comprised forests of various types, including lowland and montane forests on the wetter side of the islands and mossy forest at the highest altitudes. The rainshadow side of the islands supported a drier forest type, which has been extensively cleared for farmland and estates. Although forests were heavily damaged in the past, there is regeneration to secondary forest in some areas.

Outstanding or Distinctive Biodiversity Features

The biological values of the islands have been described in a number of publications (Collar and Stuart 1988; Jones and Tye 1988; Atkinson et al. 1991, 1993; Jones 1994). Endemism at the generic, specific, and subspecific levels is exceptionally high.

The flora contains thirty-seven endemic plant species on Príncipe, ninety-five on São Tomé (along with one endemic genus), and twenty on Annobon (Figueiredo 1994; WWF and IUCN 1994). Only sixteen of the region's endemic plants are shared by more than one island. The Pteridophyte flora and the families Rubiaceae, Orchidaceae, and Euphorbiaceae have high generic diversity and many endemics (Figueiredo 1994). Significant endemic radiations are found amongst several genera, such as *Begonia* and *Calvoa*.

These islands are highly important for bird conservation. There are twenty-eight endemic bird species on Príncipe and São Tomé, including three endemic genera, *Amaurocichla bocagii, Horizorhinus dohrni,* and *Neospiza concolor* (Christy 2001b). Four bird species were recently rediscovered after having been unobserved for more than 60 years: the dwarf olive ibis (*Bostrychia bocagei,* CR), Newton's fiscal (*Lanius newtoni,* CR), the São Tomé grosbeak (*Neospiza concolor,* CR), and the São Tomé short-tail (*Amaurocichla bocagii,* VU) (Hilton-Taylor 2000). Other threatened endemic species include the São Tomé olive pigeon (*Columba thomensis,* VU), São Tomé oriole (*Oriolus crassirostris,* VU), Annobon paradise-flycatcher (*Terpsiphone smithii,* VU), Annobon white-eye (*Zosterops griseovirescens,* VU), and Príncipe white-eye (*Zosterops ficedulinus,* VU) (Pérez del Val 2001). A number of endemic bird subspecies also exist.

The islands have few indigenous mammals. The shrew *Crocidura thomensis* (VU), found on São Tomé, is the only endemic terrestrial mammal. São Tomé has two endemic bat species, *Chaerephon tomensis* (VU) and *Myonycteris brachycephala* (EN). Three endemic bat subspecies are also found (Juste and Ibanez 1994).

Rates of endemism are also high in other taxonomic groups. Of the twenty-four reptile species recorded for these islands, only six are not endemic, and these were probably introduced by ships (D. Broadley, pers. comm., 2000). Among the Rhopalocera (Lepidoptera), there are thirteen single-island endemic species on São Tomé and six on Príncipe (Pyrcz 1992; Wojtusiak and Pyrcz 1997). Terrestrial gastropods show rates of endemism above 75 percent on all three islands, with several endemic genera and a monospecific endemic family, the São Tomé door snail (*Thyrophorella thomensis*) (Gascoigne 1994a).

Island adaptations such as gigantism and dwarfism occur. The São Tomé olive pigeon, the São Tomé giant sunbird (*Nectarinia thomensis,* VU), and giant begonias (*Begonia crateris* and *B. baccata*) are larger than similar continental species. The dwarf olive ibis is much smaller than other members of its genus. Significant species radiations also exist (e.g., in the gastropod genus *Bocageia* and the plant genus *Calvoa*).

Status and Threats

Current Status

There are around 40 km² of primary forest on Príncipe and 240 km² on São Tomé. Secondary forest is regenerating over large areas of old plantation. On Annobon, much of the forest, with the exception of the high peaks of Santa Mina and Quioveo, has been modified by humans but remains important for endemic species. Many endemic species have adapted to cocoa and coffee plantations because of the use of shade trees to protect these crops.

No protected areas currently exist on São Tomé and Príncipe. Proposals have been made to protect the remaining areas of primary forest on São Tomé and Príncipe as national parks, such as the Parques Naturais d'Ôbo, which would cover a total of 293 km². Although a law establishing procedures for the proclamation and management of a protected area system was passed in 1999, they have not yet been declared. The entire island of Annobon was recently ratified as a protected area.

Types and Severity of Threats

Large areas of forest were cleared on these islands for sugar, coffee, and cocoa plantations during colonial periods. Rainforest in the north of Príncipe was also severely modified during a campaign against sleeping sickness from 1911 to 1916. However, many endemic species adapted to the shade forest found in coffee and cocoa plantations. After the 1930s, and especially after independence in 1975, many plantations were abandoned, and there was some regeneration to secondary forest. Since the mid-1980s, land reforms have led to the development of market gardening and consequent land conversion from coffee and cocoa plantations. Agricultural practices on Annobon traditionally have been based on a forest agricultural system that was less damaging to biodiversity than large-scale plantations. However, agricultural encroachment into the primary montane forest zones of Pico Quioveo and Pico Santa Mina remains a danger.

On all three islands, a number of terrestrial mammals, both domestic and wild, have been introduced over the centuries (Dutton 1994). It is now impossible to evaluate the damage they have caused. Recent introductions of terrestrial gastropod species have been recorded on all three islands (Gascoigne 1994b), and other recent species introductions are likely.

Little direct exploitation of the endemic terrestrial wildlife occurs. Medicinal plant use is almost exclusively concerned with nonendemic species, and hunting is restricted to the introduced primate *Cercopithecus mona*, feral pigs, and the common nonendemic bat species *Eidolon helvum*. The endemic São Tomé green pigeon (*Treron australis virescens*) and the São Tomé olive pigeon (*Columbus thomensis*) are occasionally shot.

A cause for conservation concern is the capture of the African grey parrot (*Psittacus erithacus*) on the island of Príncipe for the international pet trade. It is not known whether the level of exploitation is sustainable (Juste 1996).

Justification for Ecoregion Delineation

The forested oceanic island ecosystems of São Tomé, Príncipe, and Annobon contain many endemic species. Given the size of the islands, their similar geological history and vegetation, and the fact that some of the endemic species range across a number of islands, they are regarded as a single ecoregion.

Ecoregion Number:	**12**
Ecoregion Name:	**Northwestern Congolian Lowland Forests**
Bioregion:	**Central Africa**
Biome:	**Tropical and Subtropical Moist Broadleaf Forests**
Political Units:	**Cameroon, Central African Republic, Republic of the Congo, Gabon**
Ecoregion Size:	**434,100 km²**
Biological Distinctiveness:	**Globally Outstanding**
Conservation Status:	**Relatively Intact**
Conservation Assessment:	**III**
Authors:	*Allard Blom, Jan Schipper*
Reviewed by:	*Steve Gartlan, Fiona Maisels, André Kamdem-Toham*

Location and General Description

The Northwestern Congolian Lowland Forests [12] stretch across four countries: Cameroon, Gabon, Republic of the Congo, and the Central African Republic (CAR). It is bordered to the north and south by forest-savanna mosaics, to the east by swamp forest, and the Atlantic Equatorial Coastal Forests [8] in the west.

Most of the ecoregion lies at altitudes between 300 and 800 m, with the highest elevations to the north and in the Chaillu Massif to the south. Mean annual rainfall ranges from 1,400 to 2,000 mm in the central portion, with most rain falling during two distinct wet seasons. Temperatures are tropical, with an annual mean maximum between 27°C and 30°C and an annual mean minimum between 18°C and 21°C. Humidity is high throughout the year.

The majority of the area overlies Precambrian bedrock, with pre-Cretaceous sediments to the north (Juo and Wilding 1994). In most places thick and heavily leached ferrasol overlies the bedrock, sometimes overlain by alluvial deposits.

The human population of the ecoregion is generally less than five people per square kilometer, and large areas of the forest in Gabon and Congo are almost devoid of people. Most forest people are pygmies from groups such as the BaAka, BaKa, and BaKola (Luling and Kenrick 1998), together with some Bantu cultivators. Population densities are higher around towns and cities such as Yaoundé and Bangui.

This ecoregion is a part of the Guineo-Congolian lowland rainforest in the Guineo-Congolian regional center of endemism (White 1983). Two types of forest are recognized: a mixed moist semi-evergreen forest and a single-dominant moist evergreen forest type (Letouzey 1968, 1985; White 1983). Characteristic species of the ecoregion include the large emergent trees

Entandrophragma congoense, Pentaclethra eetveldeana, Pericopsis elata, and *Gilbertiodendron dewevrei,* shrub species from the genus *Drypetes* (*D. calvescens* and *D. capillipes*), and various lianas and rattans. *Raffia* palms are abundant along river valleys.

Outstanding or Distinctive Biodiversity Features

Species richness is high throughout the ecoregion, although large areas of the forest remain biologically unknown (Kamdem-Toham et al. 2003b), as indicated by the recent discovery of new species of birds and small mammals (Ray and Hutterer 1996; Beresford and Cracraft 1999).

Knowledge of the flora has greatly improved over the last decade (Wilks 1990; Dowsett-Lemaire 1996; Lejoly 1996; White 1995; White et al. 2000). There are an estimated 7,151 vascular plants in Gabon, more than 3,600 in the CAR, 8,260 in Cameroon, and 6,000 in Congo (Stuart et al. 1990; Hecketsweiler 1990; Hecketsweiler et al. 1991; WCMC 1992). However, the number of endemic plant species is not known.

Mammalian richness is among the highest of any forest ecoregion in Africa, especially for primates. Dzanga-Sangha National Park in CAR alone contains 105 species of nonvolant mammals. This ecoregion harbors the highest numbers of gorillas (*Gorilla gorilla,* VU) and possibly chimpanzees (*Pan troglodytes,* EN) in the world (Fay and Agnagna 1992; Tutin and Fernandez 1984; Blom et al. 2001). It also supports large populations of forest elephant (*Loxodonta africana cyclotis,* EN) (Carroll 1988; Fay and Agnagna 1991; Alers et al. 1992; Barnes et al. 1995). In some parts of this ecoregion, such as Mouabale Ndoki National Park in Congo and Langoue in Gabon, the elephants remain largely undisturbed. Strictly endemic mammals include Dollman's tree mouse (*Prionomys batesi*), Remy's shrew (*Suncus remyi,* CR), and the shrew *Sylvisorex konganensis* (Ray and Hutterer 1996). Near-endemic species include sun-tailed monkey (*Cercopithecus solatus,* VU), black colobus (*Colobus satanas,* VU), western needle-clawed galago (*Euoticus elegantulus*), Glen's wattled bat (*Chalinolobus gleni*), forest horseshoe bat (*Rhinolophus silvestris*), and five shrew species (*Crocidura attila,* VU; *C. crenata; C. ludia,* VU; *C. manengubae; C. mutesae*). Another threatened species in these forests is the mandrill (*Mandrillus sphinx,* VU) (Blom et al. 1992).

The bird fauna is also diverse. Odzala National Park alone contains 442 species (Dowsett and Dowsett-Lemaire 1997; Dowsett-Lemaire 1997). The Trinational area of Nouabale-Ndoki National Park in the Republic of the Congo, Lobeke National Park in Cameroon, and Dzanga-Sangha National Park in the Central African Republic contain at least 428 species (Dowsett-Lemaire 1996, 1997; Christy 1999), including one recently discovered endemic forest robin, *Stiphornis sanghensis* (Beresford and Cracraft 1999). Restricted-range birds include Verreaux's batis (*Batis minima*), Rachel's malimbe (*Malimbus racheliae*), and forest swallow (*Hirundo fuliginosa*) (Stattersfield et al. 1998; Christy 2001a; Dowsett-Lemaire 2001; Fotso et al. 2001).

The species richness of amphibians and reptiles is also high. Among the amphibians are two endemic clawed frog species, *Xenopus boumbaensis* and *X. pygmaeus.* Strict and near-endemic reptiles include Fuhn's five-toed skink (*Leptosiaphos fuhni*), gray chameleon (*Chamaeleo chapini*), crested chameleon (*C. cristatus*), Cameroon stumptail chameleon (*Rhampholeon spectrum*), and Zenker's worm snake (*Typhlops zenkeri*).

Status and Threats

Current Status

This ecoregion contains large areas of forest and forms a part of one of the world's last remaining tropical forest wildernesses (Gartlan 1989; IUCN 1989; Wilks 1990; Mittermeier et al. 1998). About one-third of the forest is classified as "frontier forests" that are largely in their natural state (Bryant et al. 1997).

Many pristine areas of forest are located in protected areas, including Lobéké, Nouabale-Ndoki, Odzala, Dzanga, Ndoki, and Mbam Djerem. When other reserves such as the Dzangha-Sangha Special Reserve, Minkébé, Dja, Boumba-Bek, Nki, and Ngotto are also included, the total area under protection is 44,166 km², or roughly 10 percent of the ecoregion. One of the largest areas under protection is the Sangha Trinational protected area (10,650 km²), which combines the Nouabalé-Ndoki National Park (more than 4,000 km²) in northern Republic of the Congo, the Dzanga-Sangha complex in the CAR, and the Lobéké National Park in Cameroon. Although the forest around Ngotto in CAR currently has no official protected area status, the Forêt de Ngotto (730 km²) is in the final stages of gazettement.

Types and Severity of Threat

Most of the ecoregion has been allocated to forestry concessions, even in protected areas (Minnemeyer 2002). Logging in the region is selective, although there are concerns about sustainability of the logging operations (Sayer et al. 1992; Putz et al. 2000; Minnemeyer 2002). The major impacts of logging come indirectly from opening up the forest. Logging concessions provide access to the forest interior (the logging roads), a local market for bushmeat hunted from the forest (the logging camps), and a transport system to bring bushmeat to towns (the logging trucks) (Bennett and Robinson 2000; Wilkie et al. 2000; Wilkie and Laport 2001). Many forest animals are hunted for bushmeat, especially duikers (*Cephalophus* spp.). Large predators, such as crowned eagle (*Stephanoaetus coronatus*), leopard (*Panthera pardus*), and golden cat (*Profelis aurata,* VU), are also affected by bushmeat hunting because their prey animals are removed.

Some forest species are also hunted for trophies, fetishes, and the pet trade. Elephants are hunted for their meat and ivory, and gorillas and chimps for their magical properties. The trade in African grey parrots (*Psittacus erithacus*) is well developed, especially in Cameroon. Crocodiles and lizards are also collected for the pet trade (Thorbjarnarson 1992, 1999).

Justification for Ecoregion Delineation

This ecoregion forms part of the greater Guineo-Congolian regional center of endemism (White 1983). The northwestern limit of the ecoregion is the Sanaga River, a faunal boundary for such species as the golden angwantibo (*Arctocebus aureus*), white-bellied duiker (*Cephalophus leucogaster*), mandrill (*Mandrillus sphinx*), and elegant needle-clawed galago (*Euoticus elegantus*). The Oubangui River also represents a faunal boundary to the northeast. Other borders follow the "Guineo-Congolian wet and dry rainforest" delineated by White (1983). Small areas of swamp forest were subsumed in the ecoregion.

Ecoregion Number:	**13**
Ecoregion Name:	**Western Congolian Swamp Forests**
Bioregion:	**Central Africa**
Biome:	**Tropical and Subtropical Moist Broadleaf Forests**
Political Units:	**Republic of the Congo, Democratic Republic of the Congo, Central African Republic**
Ecoregion Size:	**128,600 km²**
Biological Distinctiveness:	**Globally Outstanding**
Conservation Status:	**Relatively Stable**
Conservation Assessment:	**III**
Authors:	*Allard Blom, Jan Schipper*
Reviewed by:	*Fiona Maisels, Jo Thompson, André Kamdem-Toham, Annette Lanjouw*

Location and General Description

The Western Congolian Swamp Forests [13] stretch from the eastern Republic of the Congo through to the western portion of the Democratic Republic of the Congo (DRC), and into the Central African Republic. This ecoregion lies on the western bank of the Congo River between the confluence of the Lualaba (Upper Congo) and the Lomami rivers to the confluence of the Lefini and Congo rivers. The Congo can be 15 km wide along this section and braided between a maze of alluvial islands.

The ecoregion is found in the Cuvette Congolaise, a sedimentary basin that straddles the equator. The topography is predominantly a featureless alluvial plain at an altitude of 380–450 m. In the recent geological past (the past four millennia), the area may have supported a lake over much of its extent. Also, during the driest periods associated with the Pleistocene ice ages,

climatic desiccation may have caused forest habitats to retreat to the river margins (Ngjelé 1988; Colyn 1991; Colyn et al. 1991b; Maley 1996).

This area falls in the wet tropics, with a mean annual rainfall around 1,800 mm per annum. Mean maximum temperatures are around 30°C, and mean minimum temperatures are 21–24° C. There is little seasonality, and humidity normally is high. In the wet season, the swamp forests are flooded to a depth of 0.5–1.0 m, and during the dry season they dry out again. The soils of the ecoregion are classified as gleysols because they are waterlogged throughout the year. The human population is low and typically involved with hunting and fishing activities in the forest and its rivers.

This ecoregion contains swamp forest, flooded grasslands, open wetlands, rivers, and some drier forest areas on slightly raised land. Swamp forests grow extensively along the meandering tributaries to the Congo River, especially along the Mangala, Giri, Likouala aux Herbes, Sangha, Likouola, and Kouyou rivers. Species such as *Guibourtia demeusei*, *Mitragyna* spp., *Symphonia globulifera*, *Entandrophragma palustre*, *Uapaca heudelotii*, *Sterculia subviolacea*, and *Alstonia congensis* characterize the swamp forests. Permanently flooded swamp regions host almost monospecific stands of *Raphia* palm. Levee forests occur on higher ground and are dominated by *Gilbertiodendron dewevrei*.

Outstanding or Distinctive Biodiversity Features

This ecoregion supports low levels of species richness and endemism. Available data indicate that faunal endemism is lower than in the Eastern Congolian Swamp Forests [14], but this may result from differences in the intensity of biological study. Together with the Eastern Congolian Swamp Forests [14] and the Central Congolian Lowland Forests [15], this area has been interpreted as a possible forest refuge during the drier climatic periods associated with ice ages (White 1993; Colyn et al. 1991b).

One of the primary values of this ecoregion is as an intact forest wilderness where most species populations fluctuate within natural ecological limits. Large tracts of intact forest remain because of the difficulty of logging in these swamps (Bryant et al. 1997; Forests Monitor 2001; Minnemeyer 2002).

Important mammal populations are found in this ecoregion, especially of western lowland gorilla (*Gorilla gorilla gorilla*, EN), chimpanzee (*Pan troglodytes*, EN), and forest elephant (*Loxodonta africana cyclotis*, EN) (Fay et al. 1989; Blake et al. 1995). Large numbers of forest buffalo (*Sycerus caffer nanus*) used to occur, but most have been hunted out. There may also be important seasonal migrations of elephants (S. Blake, pers. comm., 2002).

This ecoregion is separated from the Eastern Congolian Swamp Forests [14] by the Congo River, which forms an important biogeographic boundary. For example, crowned guenon (*Cercopithecus pogonias*, EN), moustached guenon (*C. cephus*), chimpanzee (*Pan troglodytes*, EN), agile mangabey (*Cer-*

cocebus agilis), gray-cheeked mangabey (*Lophocebus albigena albigena*), Guereza lowland colobus (*Colobus guereza*), potto (*Perodicticus potto edwardsi*), golden angwantibo (*Arctocebus aureus*), and western lowland gorilla occur only on the right bank of the Congo River. In comparison, Wolf's guenon (*Cercopithecus wolfi*), bonobo (*Pan paniscus,* EN), golden-bellied mangabey (*Cercocebus agilis chrysogaster*), black crested mangabey (*Lophocebus albigena aterrimus*), and dryas guenon (*Cercopithecus dryas*) occur only on the left bank of the Congo (Colyn et al. 1991b). The distribution of Demidoff's galago (*Galagoides demidoff*) subspecies follows a similar pattern, with *G. d. anomurus* and *G. d. murinus* found only on the right bank of the Congo and *G. d. phasma* found only on the left. Allen's swamp monkey (*Allenopithecus nigroviridis*) is found on both sides of the Congo River.

There is a low rate of species richness and endemism in other vertebrate groups. There are two near-endemic bird species: the African river-martin (*Pseudochelidon eurystomina*, DD) and the Congo martin (*Riparia congica*). There is one near-endemic amphibian, the Yambata river frog (*Phrynobatrachus giorgii*), and three near-endemic reptiles: gray chameleon (*Chamaeleo chapini*), Witte's beaked snake (*Rhinotyphlops wittei*), and *Gastropholis tropidopholis*.

Status and Threats

Current Status

Together, the Eastern and Western Congolian Swamp Forest ecoregions contain approximately 124,000 km^2 of swamp forest habitats (Sayer et al. 1992). The remote and difficult swamp forest habitat makes many human activities problematic, and habitats remain largely intact (Kamdem-Toham et al. 2003b).

The ecoregion contains one large Ramsar site (4,390 km^2) in the Republic of the Congo: Lac Télé-Likouala-aux-Herbes Community Reserve, which was gazetted in 1998. The reserve is located along the River Likouala-aux-Herbes and four of its major tributaries (Tanga, Mandoungouma, Bailly, and Batanga) as well as Lac Télé, which is the home of the mythical giant dinosaur-like animal called Mokele Mbembe.

Types and Severity of Threat

The habitats of this ecoregion are threatened by logging concessions in the Congo and DRC. The main threat to larger mammals is from hunting and poaching. Closer to the Congo and Ubangui rivers, poaching probably has already eliminated elephants. Across the ecoregion, numbers of roads have increased, facilitating hunting in areas that were once inaccessible. For example, along the northern and northwestern border of the Lac Télé-Likouala-aux-Herbes Community Reserve a road has been planned for transportation of wood from a logging concession located just outside the reserve.

Justification for Ecoregion Delineation

The boundaries of the Western Congolian Swamp Forests [13] largely follow those of White (1983). However, the sections of swamp forest that White extends along the Sangha, Dja, and Ngoko rivers were subsumed into the Northwestern Congolian Lowland Forests [12]. The entire swamp forest is floristically similar, so the division into two ecoregions is based on differences in the fauna. For example, chimpanzee and Western lowland gorilla are present only on the right bank of the Congo, and the bonobo is present only on the left bank.

Ecoregion Number:	**14**
Ecoregion Name:	**Eastern Congolian Swamp Forests**
Bioregion:	**Central Africa**
Biome:	**Tropical and Subtropical Moist Broadleaf Forests**
Political Units:	**Democratic Republic of the Congo**
Ecoregion Size:	**92,700 km^2**
Biological Distinctiveness:	**Globally Outstanding**
Conservation Status:	**Relatively Stable**
Conservation Assessment:	**III**
Authors:	*Allard Blom, Jan Schipper*
Reviewed by:	*Fiona Maisels, Jo Thompson, André Kamdem-Toham, Annette Lanjouw*

Location and General Description

The Eastern Congolian Swamp Forests [14], combined with the neighboring Western Congolian Swamp Forests [13], comprise one of the largest swamp forests in the world. The Eastern Congolian Swamp Forests are found on the left bank (facing downriver) of the Congo River and its tributaries, forming a large arc across the central portion of the Congo Basin.

The ecoregion is almost entirely flat and occurs between 350 and 400 m in elevation. It is part of the wet tropics, with mean annual rainfall greater than 2,000 mm and mean maximum temperatures above 30°C. Mean minimum temperatures range from 18°C to 21°C. There is little seasonality, and the humidity is high. Human population densities average around twelve people per square kilometer, concentrated in villages along the major river systems.

This ecoregion encompasses a number of the Congo River's largest tributaries. The most dramatic change in topography and the largest riparian barrier is the Stanley Falls, located near Kisangani. The most important tributaries and other water bodies are (from west to east) Lake Ntomba, Lake Tumba, the Ruki-

Momboyo and Ruki-Busira-Tshuapa-Lomela systems, the Lulonga-Maringa-Lopori system, and the Lomami system.

It is believed that in the past million years this swamp area has been both significantly drier and wetter than at present. An enormous lake covered the area for part of the period. It is also believed that at least some of the forest was lost and replaced by savanna-woodland during the height of ice age cold periods, which were marked as desiccation events in Africa (Ngjelé 1988; Kingdon 1989; Colyn 1991; Colyn et al. 1991b; White 1983, 1993). The ecoregion's soils are predominantly gleysols, indicating the waterlogging that the area experiences each year. These soils are found over a large area mapped as "basins and dunes" (Juo and Wilding 1994), which may indicate the past extent of climatic desiccation (Maley 1996, 2001).

White (1983) defined these forests as Guineo-Congolian swamp forest and riparian forest, part of the Guineo-Congolian regional center of endemism. The vegetation consists of a mosaic of open water, swamp forest, seasonally flooded forest, dryland forest, and seasonally inundated savannas, all of which are affected by the seasonal flooding of the Congo River and its major tributaries. The swamp forests are characterized by species such as *Guibourtia demeusei, Mitragyna* spp., *Symphonia globulifera, Entandrophragma palustre, Uapaca heudelotii, Sterculia subviolacea,* and *Alstonia congensis.* Permanently flooded swamp contains monospecific stands of *Raphia* palm. Levee forests occur on higher ground and support the trees *Gilbertiodendron dewevrei* and *Daniellia pynaertii.* Open areas are home to giant ground orchids (*Eulophia porphyroglossa*), and riverbanks often are lined with arrowroot (*Marantochloa* spp.).

Outstanding or Distinctive Biodiversity Features

The flora and fauna of this ecoregion are moderately species rich but contain few endemic species. The flora of the Eastern Congolian Swamp Forests [14] shares elements with both the Western Congolian Swamp Forests [13] to the northwest and the Central Congolian Lowland Forests [15] to the south but has very few endemics (White 1983).

In the mammals there is one strictly endemic rodent species, *Praomys mutoni.* Near-endemic small mammals include Allen's striped bat (*Chalinolobus alboguttatus*) and Muton's soft-furred mouse (*Praomys mutoni*). Large gaps in knowledge remain.

Avifaunal richness is moderately high, but there are no known endemics. Some Congo forest restricted bird species present here include the Congo sunbird (*Nectarinia congensis*), African river martin (*Pseudochelidon eurystomina,* DD), and Congo martin (*Riparia congica*) (Demy and Louette 2001).

Several amphibians and reptiles are near endemic, but only one is strictly endemic: the tiny wax frog (*Cryptothylax minutus*). Near-endemic reptiles include the gray chameleon (*Chamaeleo chapini*) and *Gastropholis tropidopholis.*

The Congo River presents a formidable biogeographic and dispersion barrier to many species, with the clearest examples found among the primates. For example, Angolan colobus (*Colobus angolensis*), Wolf's guenon (*Cercopithecus wolfi*), bonobo (*Pan paniscus,* EN), golden-bellied mangabey (*Cercocebus galeritus chrysogaster*), northern black mangabey (*Lophocebus atterimus atterimus*), southern talapoin (*Miopithecus talapoin*), and dryas guenon (*Cercopithecus dryas*) occur only on the left bank of the Congo (Colyn et al. 1991a). In comparison, crowned guenon (*Cercopithecus pogonias,* EN), chimpanzee (*Pan troglodytes,* EN), agile mangabey (*Cercocebus agilis*), and gray-cheeked mangabey (*Lophocebus albigena*) occur only on the right bank. The distribution of Demidoff's galago (*Galagoides demidoff*) subspecies follows a similar pattern, with *G. d. anomurus* and *G. d. murinus* found only on the right bank of the Congo and *G. d. phasma* found only on the left. Allen's swamp monkey (*Allenopithecus nigroviridis*) is found on both sides of the Congo River.

The presence of species restricted to the central portion of the Congo Basin, such as the bonobo (*Pan paniscus,* EN), suggests that forest cover did persist here during the dry periods of the ice age (Colyn et al. 1991a; Prigogine 1986). However, the fact that few narrowly endemic species occur when compared with other Congo Basin ecoregions indicates that the forests may have declined greatly in extent during these dry periods (White 1993; Kingdon 1997).

Status and Threats

Current Status

It is estimated that 124,000 km^2 of swamp forests remain in the Congo Basin (Sayer et al. 1992), with perhaps half in this ecoregion. Lomako Reserve, Lomami Lualaba Forest Reserve, and parts of Salonga National Park fall in this ecoregion. The Tumba area has also been proposed for protection, and priority areas for biodiversity conservation have been outlined (IUCN 1989; Doumenge 1990; Kamdem-Toham et al. 2003b).

Types and Severity of Threat

Logging and associated poaching are the major threats in this ecoregion because of the ease of access through the Congo River and its tributaries. The Service Permanent d'Inventaire d'Amenagement Forestier has noted that extensive areas along the left bank of the Congo River have been allocated as logging concessions (J. Thompson, pers. comm., 2002).

Hunting is a major threat. Larger species are hunted for bushmeat, elephants are hunted for ivory and meat, and bonobos are hunted for meat, fetishes, and the pet trade. The Congo River is a highly navigable waterway, making most of the area accessible to poachers (Lanjouw 1987). Anecdotal information suggests that elephants have disappeared from large areas (Bakarr et al. 2001b). Elephant hunting in DRC is extremely well organized and professional (Alers et al. 1992; A. Blom, pers. obs., 2000). Areas close to the Congo River and other major waterways may have also suffered reduction in other wildlife populations, including bonobos (Van Krunkelsven et al. 2000; Van Krunkelsven 2001).

Justification for Ecoregion Delineation

The boundaries of the Eastern Congolian Swamp Forests [14] roughly follow those mapped by White (1983). Although the entire swamp forest is floristically similar (albeit species poor), it was separated into an eastern and western section based on the significant ecological barrier that the Congo River presents to nonflying vertebrates. This is clearly shown by the primates of this ecoregion, especially by the presence of bonobo (*Pan paniscus*), which is absent in the Western Congolian Swamp Forest [13].

Ecoregion Number:	**15**
Ecoregion Name:	**Central Congolian Lowland Forests**
Bioregion:	**Central Africa**
Biome:	**Tropical and Subtropical Moist Broadleaf Forests**
Political Units:	**Democratic Republic of the Congo**
Ecoregion Size:	**414,800 km²**
Biological Distinctiveness:	**Globally Outstanding**
Conservation Status:	**Relatively Intact**
Conservation Assessment:	**III**
Authors:	Allard Blom, Jan Schipper
Reviewed by:	Jo Thompson, André Kamdem-Toham

Location and General Description

The Central Congolian Lowland Forests [15] lie in the central part of the Congo Basin, south of the wide arc formed by the Congo River. The river functions as a distribution barrier to many species, thereby isolating this lowland basin along its northern, eastern, and western limits. The headwaters of the Lopori, Maringa, Ikelemba, Tshuapa, Lomela, and Lokoro rivers lie in this ecoregion. Because of the flat topography of the area, most of these rivers flow slowly, with heavy sediment loads, and have numerous alluvial islands.

The ecoregion is low-lying, with the central portion around 400 m above sea level. The elevation and topographic complexity increase toward the southeast, where hills up to 800 m are found. The ecoregion has approximately 2,000 mm of rainfall annually. The mean maximum temperature is around 30°C in the central portion and falls to around 27°C along the south-

east margins. The mean minimum temperature is between 18°C and 21°C, with little seasonality and high humidity.

This ecoregion occupies an ancient part of the African landscape, eroded over millennia and subject to alluvial deposition in some places. Soils often are nutrient-poor oxisols developed over ancient dune fields believed to date from the Quaternary ice ages (Juo and Wilding 1994), or are more recent alluvial deposits from the many rivers. There may have been dramatic changes in the forests of this ecoregion over thousands of years. Part of the area is a proposed forest refuge where forest survived during the drier periods (last 12,000 years B.P.) associated with ice ages (Axelrod and Raven 1978; Louette 1984; Mayr and O'Hara 1986; Prigogine 1986; Colyn et al. 1991b; Maley 1991, 1996). Research indicates that the forest also shrank during an arid phase around 2,500 B.P. (Maley 2001). In wetter climatic periods, a large inland lake may have also submerged this ecoregion.

Human populations are small, and the highest densities are found along the rivers, where people fish, cultivate cassava, and hunt animals from the forest (Lanjouw 1987). Most areas have fewer than 5 people/km².

This ecoregion falls in the Guineo-Congolian regional center of endemism (White 1983, 1993) and supports a number of different vegetation types. The northwestern portion is a mosaic of seasonally inundated and permanent swamp forest with mixed terra firma moist evergreen and semi-evergreen rainforest along levees and other areas of relief. Evergreen forests are dominated by stands of *Gilbertiodendron dewevrei*. To the south the forests become drier and change to semi-evergreen rainforest and finally a mosaic of Lower Guinea rainforest and grasslands. Semi-deciduous forest is characterized by *Staudtia stipitata*, *Scorodophloeus zenkeri*, *Anonidium mannii*, and *Parinari glaberrimum*.

Outstanding or Distinctive Biodiversity Features

In common with most other forested ecoregions of the Congo Basin, this ecoregion is noted primarily for its size and habitat intactness, with populations of most species fluctuating within normal ecological limits.

The vegetation in this region is diverse, with estimates between 1,500 and 2,000 vascular plant species, of which about 10 percent are endemic (Ngjelé 1988; WWF and IUCN 1994). In terms of the fauna, scientists are just beginning to document the species composition (Thompson 2001, 2003). Available data indicate that species richness and endemism are markedly lower than in other moist forest ecoregions of the Congo Basin. This may be caused by a lack of biological investigation, by the river barriers (Grubb 2001), by the loss of forest in recent millennia through climatic desiccation (Maley 1991, 1996, 2001), or even through the past inundation of the region beneath a large lake.

Among mammals, there is only one strictly endemic species, the dryas guenon (*Cercopithecus dryas*). Other near-endemic mammals include the golden-bellied mangabey (*Cercocebus ga-leritus chrysogaster*), bonobo (*Pan paniscus,* EN), okapi (*Okapia johnstoni*), Allen's swamp monkey (*Allenopithecus nigroviridis*), Angolan cusimanse (*Crossarchus ansorgei*), and Wolf's guenon (*Cercopithecus wolfi*). The ecoregion also supports an important population of forest elephant (*Loxodonta africana cyclotis,* EN) and is especially noted for containing the world's largest bonobo populations (Kano et al. 1994; Thompson-Handler et al. 1995; Kortlandt 1996; Van Krunkelsven et al. 2000; Thompson 2003).

There are no known strict endemic species among the avi-fauna, but there are two near-endemics, the Congo peacock (*Afropavo congensis,* VU) and the yellow-legged malimbe (*Mal-imbus flavipes*) (Thompson 1996). The known herpetofauna also shows low rates of endemism. One amphibian is strictly en-demic to the ecoregion, the Gembe reed frog (*Hyperolius robus-tus*), and another is near endemic, the Lomami screeching frog (*Arthroleptis phrynoides*). There are no strictly endemic reptiles, but near-endemic reptiles include Vanderyst's work lizard (*Monopeltis vanderysti*), *Gastropholis tropidopholis, Mehelya lau-renti,* and *Polemon robustus.* This ecoregion also offers critical habitat to the African slender-snouted crocodile (*Crocodylus cataphractus*).

Status and Threats

Current Status

Much of the forest habitat of the ecoregion remains intact, even to the south, where the transition to savanna woodland is the result of climatic rather than anthropogenic effects (Sayer et al. 1992; Minnemeyer 2002). It is estimated that more than 100,000 km^2 of forest still exists.

There is one large protected area, Salonga National Park (36,500 km^2). One of the largest national parks in the world and the second largest tropical forest national park in the world, it is found in two parts. Several other sections of the ecoregion have been proposed for protection, most notably Lomako and Lomami-Lualaba (IUCN 1989; Kamdem-Toham et al. 2003b).

Types and Severity of Threat

The threats to the habitats in this ecoregion are low, and the conservation potential is high. Some forest has been lost to farming and logging; however, this does not pose a significant threat to habitats or species at this time. Poaching poses a seri-ous threat to the elephants of this ecoregion (Alers et al. 1992; Van Krunkelsven et al. 2000), and there is also some hunting of bonobo and capture for the live animal trade.

Justification for Ecoregion Delineation

This ecoregion is a part of the Guineo-Congolian regional cen-ter of endemism (White 1983). Located under the arc of the

Congo River, it is distinguished from neighboring ecoregions by its isolated position, lower rates of species richness and en-demism, and the presence of bonobo (*Pan paniscus*). The eco-region's northern limit is the Eastern Congolian Swamp Forests [14]. The southern boundary follows the line just north of the Fimi River, where White (1983) distinguished wetter lowland rainforest from drier lowland rainforest.

Ecoregion Number:	**16**
Ecoregion Name:	**Northeastern Congolian Lowland Forests**
Bioregion:	**Central Africa**
Biome:	**Tropical and Subtropical Moist Broadleaf Forests**
Political Units:	**Democratic Republic of the Congo, Central African Republic**
Ecoregion Size:	**533,500 km²**
Biological Distinctiveness:	**Globally Outstanding**
Conservation Status:	**Relatively Stable**
Conservation Assessment:	**III**
Authors:	*Allard Blom, Jan Schipper*
Reviewed by:	*John Hart, Terese Hart, André Kamdem-Toham*

Location and General Description

The Northeastern Congolian Lowland Forests [16] is a roughly triangular ecoregion located in the northeastern portion of the Democratic Republic of the Congo (DRC), extending to the southeastern portion of the Central African Republic (CAR). The northern margin is formed by the transition to savanna and woodland habitats, the eastern border is bounded by the Al-bertine Rift Montane Forests [17], and the southern and west-ern margins are delimited by the Congo River and its tributar-ies, primarily the Elila River.

The ecoregion declines in elevation from east to west, from the Albertine Rift Mountains (up to 1,500 m) toward the Congo River (around 600 m), with a chain of inselbergs east to west across the northern portions. Precambrian basement rocks un-derlie the eastern and northern margins of the ecoregion, but further south the rocks are of more recent origin, associated with sedimentation in the Congo Basin. In the northern forests soils are thick and heavily leached ferrasols. To the south the soils are primarily nitosols, which are less weathered and better for crop cultivation, and to the east volcanic material from the Al-bertine Rift volcanoes improves soil fertility.

The climate is humid and tropical. Rainfall averages 1,500–2,000 mm per annum, declining eastward towards the Alber-tine Rift Mountains, with a well-marked dry season from January to March. The mean annual maximum temperature is between 27°C and 33°C, and the mean minimum is between 15°C and 21°C, with variation dependent mostly on elevation. The human population is greatest in the east, close to the Albertine Rift, reaching densities of fifty people per square kilometer. Human populations are lowest in the center of the rainforest, falling to less than five people per square kilometer. Some areas, notably the Maïko region, have no permanent settlements.

In terms of the phytogeographic classification of White (1979, 1983), this entire ecoregion falls in the Guineo-Congolian regional center of endemism. Forest cover has fluctuated over geological time, with forest declines associated with climate desiccation events that occurred in the Pleistocene ice ages and also since the end of the last ice age (e.g., 2,000–2,500 B.P. (Hart et al. 1994). Currently most of this ecoregion is lowland moist forest, with some transitional submontane forest in the east, where it borders the Albertine Rift Valley and some drier transitional types in the north (White 1983). Tree species include *Julbernardia seretii, Cynometra alexandri,* and *Gilbertiodendron dewevrei,* and the latter dominates the forest in some areas (White 1983; Hart et al. 1989; Hart 1990).

Outstanding or Distinctive Biodiversity Features

The flora and fauna of this ecoregion are diverse and have large numbers of endemic species. Levels of endemism are highest on the lower slopes of the Albertine Rift Mountains. The central and western portions have fewer endemic species, which may reflect declining biological study further west. About 1,500 species of plants are known from the Ituri region (WWF and IUCN 1994), with recent collections including the endemic cycad *Encephalartos ituriensis* and a sapotaceous tree of a South American genus (J. Hart, pers. comm., 2002). The inselberg flora of the northern part of the ecoregion also contains plants with links to the mountains of eastern Africa (J. Hart, pers. comm., 2002).

Mammals display high levels of endemism. Strictly endemic species include the giant genet (*Genetta victoriae*), aquatic genet (*Osbornictis piscivora*), mountain shrew (*Sylvisorex oriundus,* VU), African foggy shrew (*Crocidura caliginea,* CR), Congo shrew (*C. congobelgica,* VU), and fuscous shrew (*C. polia,* CR). Ten additional species are considered near endemic, including okapi (*Okapia johnstoni*) (Hart and Hart 1988, 1989), the owl-faced monkey (*Cercopithecus hamlyni*), L'Hoest's monkey (*C. lhoesti*), the pied bat (*Chalinolobus superbus,* VU), Allen's striped bat (*Chalinolobus alboguttatus,* VU), Misonne's soft-furred mouse (*Praomys misonnei*), and Verschuren's swamp rat (*Malacomys verschureni*). This ecoregion also contains the most important population of the eastern lowland gorilla (*Gorilla beringei graueri,* EN).

The avifauna includes two strictly endemic species: Neumann's coucal (*Centropus neumanni*) and the golden-naped weaver (*Ploceus aureonucha,* EN). Near-endemic species include Nahan's francolin (*Pternistis nahani,* EN), Ituri batis (*Batis iturien-*

sis), Turner's eremomela (*Eremomela turneri,* EN), Congo peacock (*Afropavo congensis,* VU) (Hart and Upoki 1997), Sassi's greenbul (*Phyllastrephus lorenzi*), Bedford's paradise-flycatcher (*Terpsiphone bedfordi*), and, on the eastern margins with the Albertine Rift, Chapin's mountain-babbler (*Kupeornis chapini*).

The herpetofauna has not been well studied. There are a few range-restricted reptiles, including the strictly endemic Zaire dwarf gecko (*Lygodactylus depressus*). There are seven strictly endemic amphibian species: the olive shovelnose (*Hemisus olivaceus*), Kigulube reed frog (*Hyperolius diaphanus*), Kunungu reed frog (*H. schoutendeni*), Mertens's running frog (*Kassina mertensi*), Buta River frog (*Phrynobatrachus gastoni*), Christy's grassland frog (*Ptychadena christyi*), and Pangi Territory frog (*Rana amieti*).

Status and Threats

Current Status

Currently, no accurate data are available on the status of the habitats across the entire ecoregion (see Doumenge 1990; Sayer et al. 1992; Witte 1995; and Kamdem-Toham et al. 2003b for general information). Although the full extent of forest is unknown, it easily exceeds 100,000 km^2, and much of this is continuous. In the DRC, an important part of the Ituri forest is now protected in the Okapi Faunal Reserve. Other lowland forest areas are protected in the Kahuzi-Biega National Park, the Maïko National Park and the Yangambi Reserve and Rubi Tele Domaine de Chasse, the Maika Penge Reserve (near Isiro), and the Bili-Uere Domaine de Chasse (IUCN 1998). The total area under protection is roughly 31,000 km^2, representing around 6 percent of the ecoregion. Lowland forest of the Itombwe Massif is currently unprotected but of high conservation value.

The central part of the ecoregion contains extensive stands of forest with a low density of forest-dwelling Mbuti pygmies (Luling and Kenrick 1998) and associated agricultural peoples (Wilkie and Curran 1993; Wilkie et al. 1998a, 1998b; Wilkie and Finn 1990). Soils and agriculture potential are better in the east and south, especially in Kivu Province, and much of the original habitat has been lost.

Types and Severity of Threats

The main threats to this ecoregion are mining, logging, large-scale human population movements as a result of war, and the bushmeat and wildlife use that accompany these other threats. Gold, diamonds, and the rare metal coltan are mined. Mining has already seriously degraded many important forest areas, such as Maïko, Ituri, and the Kahuzi Biega lowlands (in the latter there are currently 15,000 miners).

Bushmeat hunting, often associated with illegal mining, is a major threat to the larger animals, even in the remote Ituri forest (Alers et al. 1992; Hall et al. 1997, 1998b; Hart and Hall 1996). Elephant poaching has also had a major impact in this region, even in protected areas such as the Kahuzi-Biega and Maïko national parks and Okapi (Ituri) Faunal Reserve.

Logging has occurred in several locations but was largely interrupted by the war in the DRC. Recently the logging industry has been moving in from Uganda, taking advantage of the relative security in the east (especially the Ituri region). Logging concessions are planned for much of this region.

The recent wars in Rwanda, Burundi, and eastern DRC have had serious impacts on protected area management capabilities and have led to the widespread migrations of refugees. In some places these refugees have cleared large areas of forest for subsistence agriculture, especially in the eastern sector (U. Klug, pers. comm., 2000).

Justification for Ecoregion Delineation

The Northeastern Congolian Lowland Forests [16] ecoregion is a part of the Guineo-Congolian regional center of endemism (White 1983) and is distinguished by high endemism. It is bound by the Uele River in the northeast, the Congo River and its tributaries (primarily the Elila River) in the south, and the Bomu River as it flows into the Oubangui River in the west. These rivers form distribution boundaries to some mammal species, such as the fishing genet (*Osbornictis piscivora*). The eastern flank of the ecoregion consists of transitional forest in the foothills of the Albertine Rift Mountains up to 1,500 m altitude; locating the exact boundary in the east is problematic.

Ecoregion Number: **17**

Ecoregion Name: **Albertine Rift Montane Forests**

Bioregion: **Eastern and Southern Africa**

Biome: **Tropical and Subtropical Moist Broadleaf Forests**

Political Units: **Democratic Republic of the Congo, Uganda, Rwanda, Burundi, and Tanzania**

Ecoregion Size: **103,900 km²**

Biological Distinctiveness: **Globally Outstanding**

Conservation Status: **Critical**

Conservation Assessment: **I**

Authors: *Rauri Bowie, Allard Blom*

Reviewed by: *Marc Languy, Derek Pomeroy, Andrew Plumptre, Tim Davenport*

Location and General Description

The Albertine Rift Montane Forests [17] ecoregion lies along a major mountain range running north to south across the equa-

tor close to the center of Africa. The range is not continuous but instead comprises a number of distinct montane areas. Starting in the north from the Lendu Plateau in the Democratic Republic of the Congo (DRC) (Bober et al. 2001), the range runs south through mountains on the border between Uganda and eastern DRC, through western Rwanda and Burundi, and to some isolated massifs in Tanzania and DRC on either side of Lake Tanganyika.

The Albertine Rift is formed from a combination of uplifted Precambrian basement rocks and recent volcanoes, associated with the development of Africa's Great Rift Valley. The Rift is now partly filled by lakes, including Lake Tanganyika, and is flanked by the uplifted mountain areas, especially on the western margin. The highest peaks in the Albertine Rift are found in the Rwenzori Mountains at 5,110 m. Above 3,500 m habitats are placed in the Rwenzori-Virunga Montane Moorlands [73] ecoregion.

The topographic complexity of the Albertine Rift has resulted in a diversity of climatic regimes. Although the Rift is located in the center of tropical Africa, the high mountains have an essentially temperate climate. Average rainfall across the range typically is between 1,200 and 2,200 mm per annum, although it reaches 3,000 mm per annum on the western slope of Rwenzori.

Much of this ecoregion, especially in Burundi and Rwanda, supports one of the highest rural human population densities in Africa (300–500 people/km²). Population densities are lower in the DRC, but human migrations have occurred in recent years, and some areas that used to have few people (e.g., Itombwe) may now have large populations.

The habitats of the ecoregion are dominated by montane rainforest (White 1983). In the west, the montane forest grades below about 1,500 m into submontane and lowland forests of the Guineo-Congolian rainforest, and in the east the forest grades into forest-savanna mosaic habitats. Details of the vegetation composition in the Albertine Rift Mountains are found in Lind and Morrison (1974), Langdale-Brown et al. (1964), White (1983), and in the Uganda Forest Department Biodiversity Reports for sixty-five forests, including sixteen in the Albertine Rift ecoregion (Howard and Davenport 1996).

Outstanding or Distinctive Biodiversity Features

This ecoregion contains exceptionally high species richness because of its central location in Africa, juxtaposition of habitats, and altitudinal zonation. Exceptionally large numbers of endemic animal species occur. Endemism is found at all altitudes and extends into the lower-altitude forests on the western margins (Prigogine 1985; Vande weghe 1988a, 1988b).

Among vertebrates, birds possess at least thirty-seven endemic species (Bober et al. 2001). Amphibians also possess exceptional endemism, with thirty-four strict endemic species (Plumptre et al. 2003). Endemism is concentrated in the reed frogs (*Hyper-

olius), screeching frogs (*Phrynobatrachus*), river frogs (*Arthroleptis*), and clawed toads (*Xenopus*). The endemic mammalian community contains thirty-four strictly endemic species (Plumptre et al. 2003). The endemic mammal fauna is dominated by small mammals, including the strictly endemic Ruwenzori otter shrew (*Micropotamogale ruwenzorii,* EN), one of only two species of the family Tenrecidae found on mainland Africa.

The ecoregion also harbors the mountain gorilla (*Gorilla beringei beringei,* CR), one of the rarest animals in Africa (Schaller 1963; Fossey 1983; Harcourt et al. 1983; Aveling and Harcourt 1984; Aveling and Aveling 1989; McNeilage 1996; Watts 1998a, 1998b, 1998c). Primates dominate the other threatened species, including a subspecies of owl-faced monkey (*Cercopithecus h. kahuziensis,* EN), golden monkey (*C. mitis kandti,* EN), eastern lowland gorilla (*Gorilla beringei graueri,* EN), and chimpanzee (*Pan troglodytes schweinfurthii,* EN) (Harcourt et al. 1983; Aveling and Harcourt 1984; Aveling and Aveling 1989; McNeilage 1996; Hall et al. 1998a, 1998b). The near-endemic Ruwenzori duiker (*Cephalophus rubidus,* EN) is also found in the upper parts of the Rwenzori range.

In comparison to the other vertebrate groups, the number of endemic reptiles is low, with sixteen strict endemics (Plumptre et al. 2003). These include four species of chameleons (*Chamaeleo* spp.) and four species of skinks in the genus *Leptosiaphos*. There are known to be 117 strict endemic butterflies in the Albertine Rift ecoregion, of which 25 are endemic to Tanzania, 23 to DRC, 21 to Uganda, and 2 to Rwanda (Plumptre et al. 2003). Invertebrates are heavily undercollected, as indicated by the new species found in the Rwenzoris (Salt 1987).

Botanical knowledge of the area is more fragmentary than for animals. The number of strictly endemic plants is currently estimated at about 567 species (Eilu et al. 2001). Some sites are better known for their plants; for example, the Bwindi Impenetrable Forest in Uganda supports an estimated 1,000 plant species, 8 of which are tree species found only locally (WWF and IUCN 1994).

Status and Threats

Current Status

Most parts of the Albertine Rift forests remain only in protected areas or in the most rugged and inaccessible areas. Elsewhere the landscape is dominated by farmland (Hamilton 1984; Howard 1991). In Uganda, areas of forest are protected in the Rwenzori, Bwindi-Impenetrable, Mgahinga Gorilla, Kibale, and Semliki National Parks. Other protected areas include Kalinzu, Kasyoha-Kitomi, Echuya, and Mafuga, although there is very little natural vegetation remaining in Mafuga and extreme habitat destruction in Echuya (Howard and Davenport 1996). In the DRC, what little forest remains on the Lendu Plateau is in forest reserves, and the Kahuzi-Biega National Park and the Virunga National Park protect the montane forests on the western side of Lake Kivu.

In Rwanda and Burundi, forests are protected in the Volcanoes National Park and in the Nyungwe National Park, the latter forming a transboundary protected area with Kibira National Park of Burundi. In Tanzania, the forests of Gombe Stream and the Mahale Mountains are protected as national parks, and other forest areas are found in forest reserves. Although it is difficult to give accurate overall statistics, roughly 13,500 km^2 is gazetted, representing around 14 percent of the ecoregion.

Forests further to the south in DRC are largely unprotected. The unprotected forests of the Itombwe Mountains used to be in good condition but are now being converted to farmland by refugees (Ilambu et al. 1999). Other unprotected forests in DRC include those of Mount Kabobo and the Marungu Highlands.

Types and Severity of Threat

The major threat to biodiversity conservation in this ecoregion is the clearance and fragmentation of forest habitats for agriculture by poor farmers. Other threats include hunting and poaching. Firewood collection is also a serious problem in many areas (WWF 1994), as are fire and the invasion of exotic species.

Populations of many other large mammal species, including elephant (*Loxodonta africana,* EN) and hippopotamus (*Hippopotamus amphibius*), have been decimated in the region's turbulent political past. This is especially the case in the Virunga National Park in DRC. A recent aerial survey in the northern sector has shown that fewer than 100 hippos survive along the Semliki, compared with more than 8,000 in 1959 and 1,000 in 1989, whereas more than 3,000 domestic cattle were also counted. People have also converted about 200 km^2 of the Virunga National Park for farming. More than 20,000 illegal fishers have also invaded the west coast of Lake Edward. In total, the number of people living in this park is now about 60,000.

The recent wars in Rwanda, Burundi, and DRC caused an overspill of refugees into western Uganda and Tanzania. The wars also prevent the effective management of some protected areas, thereby increasing problems of encroachment and poaching. In Rwanda the political situation has significantly improved in recent years. Large mining camps also occur in DRC forests, in particular looking for columbo-tantalite, an ore used for making semiconductors for electronic goods.

Justification for Ecoregion Delineation

This ecoregion contains the highest levels of faunal endemism in Africa. Plant endemism is markedly lower but still considerable. The ecoregion largely follows the "undifferentiated montane vegetation" unit of White (1983) around the 1,500-m contour. The Lendu Plateau and the boundary just north of the Rwenzori-Virunga Montane Moorlands [73] are expanded from the boundaries of White to better conform with BirdLife's endemic bird area (Stattersfield et al. 1998), and we have included some additional outliers in Tanzania.

Ecoregion Number: **18**

Ecoregion Name: **East African Montane Forests**

Bioregion: **Eastern and Southern Africa**

Biome: **Tropical and Subtropical Moist Broadleaf Forests**

Political Units: **Kenya, Tanzania, Uganda, Sudan**

Ecoregion Size: **65,500 km²**

Biological Distinctiveness: **Regionally Outstanding**

Conservation Status: **Critical**

Conservation Assessment: **II**

Authors: *Jan Schipper, Neil Burgess*

Reviewed by: *Derek Pomeroy, Marc Languy, Alan Rodgers*

Location and General Description

The East African Montane Forests [18] encompass moderate- to high-altitude habitats along a chain of isolated mountains flanking the Rift Valley. At higher altitudes the ecoregion grades into the Afromontane heathland-moorland and Afroalpine vegetation in the East African Montane Moorlands [72] ecoregion. The northern extent is at Mount Kinyeti in the Imatong Mountains in southern Sudan, Mount Moroto and Mount Elgon in Uganda, and the Aberdare Range to Mount Kenya in Kenya. Farther south, the ecoregion includes the forests of the Nguruman Scarp in southern Kenya and Mount Kilimanjaro, Mount Meru, and Ngorongoro Highlands in northern Tanzania. Outliers are also found in eastern Uganda (e.g., mounts Kadam and Napak), along the eastern Rift Valley of Kenya (mounts Kulal, Nyiru, and Bukkol), and in northern Tanzania as far south as the Marang forests (Mbulu and Hanang) (Baker and Baker 2002). Most of these mountains are volcanic, and some are still active.

The ecoregion occupies elevations between about 1,500 and 3,500 m altitude. The climate is temperate and seasonal, with night temperatures falling below 10°C in the cold season (coldest in July and August) and rising to above 30°C during the day in the warm season. At the higher elevations frosts are possible. Rainfall varies between 1,200 and 2,000 mm per annum, with a distinct wet (October–December and March–June) and dry (January–February and July–October) season. The climate of these mountains is wetter than the surrounding lowlands but has a pronounced rainshadow, with the eastern and southern faces being significantly wetter than the western and northern ones.

The cracking of the African plate system, resulting in the Rift Valley, also produced the volcanoes typical of this ecoregion, including Mount Kilimanjaro and Mount Meru in Tanzania, Mount Kenya in Kenya, and Mount Elgon in Uganda. Other less obviously volcanic ranges are the Ngong Hills, Aberdare Mountains, and Mau Range in Kenya. The volcanic baserock weathers to create fertile soils that are suitable for agriculture. Human population is high throughout the ecoregion and is rapidly increasing. Population density is 200–300 people/km² in many areas. In some cases human occupation extends within the borders of protected areas (Gathaara 1999; UNEP et al. 2002).

Submontane and montane forests would have once dominated the ecoregion, together with montane grassland, bamboo, and rocky habitats. Much of the original vegetation has been converted to farmland where it is not protected, especially at lower altitudes. The forest habitats contain a diverse assemblage of tree species, including the timber species *Ocotea usambarensis, Juniperus procera, Podocarpus falcatus, P. latifolius, Nuxia congesta,* and *Newtonia buchanii.* At higher altitudes the bamboo *Arundinarium alpina* may become dominant, with scattered trees, such as *Hagenia abyssinica.*

Outstanding or Distinctive Biodiversity Features

Much has already been published on the flora and fauna of these forests (Langdale-Brown et al. 1964; Lind and Morrison 1974; White 1983; Newmark 1991a; Meadows and Linder 1993; Wass 1995; Chapman and Chapman 1996; Young 1996; Davenport et al. 1996a, 1996b; Knox and Palmer 1998; Pócs 1998; Ngoile et al. 2001). These studies show that the ecoregion contains moderate levels of species richness but has low rates of endemism when compared with the other tropical forest ecoregions in eastern Africa. This is probably because the forests are not especially old, most being formed on volcanoes that are 1–2 million years old.

Only a few plant endemics are found in the forests, with higher rates of plant endemism in the grassland and rocky habitats (WWF and IUCN 1994). These mountains support a diverse avifauna, including eight endemic bird species (Stattersfield et al. 1998). Some of the endemic species, such as the Aberdare cisticola (*Cisticola aberdare*), Abbott's starling (*Cinnyricinclus femoralis,* VU), and Kenrick's starling (*Poeoptera kenricki*) occur on only two or three mountain ranges in the ecoregion. Others, such as Hunter's cisticola (*Cisticola hunteri*), Jackson's francolin (*Pternistis jacksoni*), and Sharpe's longclaw (*Macronyx sharpei*), occur in most (but not all) of the major mountains. Some of these species are typical of the montane forest, whereas others are found only in montane grasslands (Stattersfield et al. 1998; Bennun and Njoroge 1999; Byaruhanga et al. 2001; Baker and Baker 2002).

The mammal fauna includes eight strictly endemic species, which are either shrews (*Crocidura gracilipes,* CR; *Crocidura raineyi,* CR; *Crocidura ultima,* CR; *Surdisorex norae,* VU; and *Surdisorex polulus,* VU) or rodents (*Grammomys gigas,* EN; *Tachyoryctes annectens,* EN; and *Tachyoryctes audax,* VU). Near-endemic mammals include Jackson's mongoose (*Bdeogale jacksoni,* VU), Abbot's duiker (*Cephalophus spadix,* VU), sun squirrel (*Heliosciurus undulatus*), and the eastern tree hyrax (*Dendrohyrax validus,* VU) (Hilton-Taylor 2000). There are no endemic large mammals. A population of the bongo (*Tragelaphus eurycerus*) occurs in the Abadere Mountains, which is the most eastern distribution of this rainforest species in Africa. Other Central African elements in the mammal fauna are African golden cat (*Profelis aurata,* VU) and the giant forest hog (*Hylochoerus meinertzhageni*). The threatened black rhinoceros (*Diceros bicornis,* CR) and elephant (*Loxodonta africana,* EN) also occur.

The herpetofauna contains a number of strictly endemic species, particularly chameleons such as Tilbury's chameleon (*Chamaeleo marsabitensis*), Müller's leaf chameleon (*Bradypodion uthmoelleri*), Mount Kenya hornless chameleon (*Bradypodion excubitor*), and Ashe's bush viper (*Atheris desaixi*). The amphibian fauna, although important, has only three strictly endemic species (*Hyperolius montanus, H. cystocandicans,* and *Phrynobatrachus kinangopensis*), far less than the thirty found in the adjacent Eastern Arc Forests [19].

The lower rates of endemism in these mountains, when compared with other eastern African forest ecoregions, presumably result from the shorter length of time available for speciation or relictualization. Most of the forests of this ecoregion have developed on geologically recent volcanoes, and some of these are still active or have erupted in the past 10,000 years. Pócs (1998) has shown that the Eastern Arc Forests [19] contain many endemic mosses and bryophytes but also contain forty-five species that are also found on Madagascar. In contrast, the East African Montane Forests [18] have few endemics and do not contain species occurring on Madagascar, presumably reflecting their younger age. Nevertheless, some speciation has already occurred, as on the dormant volcano of Kilimanjaro, which is approximately 1 million years old and erupted less than 20,000 years ago (Newmark 1991a; Ngoile et al. 2001).

Status and Threats

Current Status

Historically, this ecoregion was a mosaic of forest, bamboo, and grasslands throughout its middle elevations, grading into extensive areas of savanna, woodlands, and other habitat types at lower elevations and moorland habitats at higher elevations. Although the habitats of the ecoregion were always naturally fragmented on different isolated mountains, over time and through human activity, the habitat has become more fragmented.

National parks in Uganda (Elgon), Kenya (Marsabit, Kulal, Mount Kenya, and Abadares), and Tanzania (Kilimanjaro and Meru), and the Ngorongoro Conservation Area contain well-protected areas of habitat. Other larger habitat patches are found in forest reserves, administered by forestry departments. Outside of the national parks and forest reserves, almost all forest and montane grassland habitats have been converted to agricultural or other human use, as around Mount Kenya National Park (Gathaara 1999) and Kilimanjaro (Ngoile et al. 2001; UNEP et al. 2002).

Types and Severity of Threat

At lower elevations, areas of forest and forest mosaic have been converted to tea and coffee estates, conifer plantations, and subsistence agriculture. The major loss of habitat from plantations occurred during the British colonial period in the early and mid-twentieth century. Since that time, increasing human populations have further converted montane habitats to farmland. In many places, land conversion has occurred up to the boundaries of the protected areas, including Kilimanjaro (Newmark 1991a), Kakamega and Mau forests (Wass 1995), and Mount Kenya (Gathaara 1999). Human-induced fire in the moorland zones is also believed to have depressed the upper limit of forest and replaced it with a fire-maintained border, as on Mount Kilimanjaro (Newmark 1991a; UNEP et al. 2002). Another problem is the continued hunting of large mammal populations both outside and inside protected areas.

Justification for Ecoregion Delineation

This ecoregion follows the "undifferentiated montane vegetation" unit of White (1983) but has been separated from similar montane habitats further south because of their significantly different biogeographic histories. The mountains further east and south (the Eastern Arc Forests [19]) are nonvolcanic, are geologically ancient, and support endemic-rich forests. By comparison, the East African Montane Forests [18] are mainly volcanic, are geologically younger, and support endemic-poor forests. The only modification to White's linework has been a refinement of the high-elevation areas to follow the 1,500-m elevation contour at the border between Sudan and Uganda.

Ecoregion Number: **19**

Ecoregion Name: **Eastern Arc Forests**

Bioregion: **Eastern and Southern Africa**

Biome: **Tropical and Subtropical Moist Broadleaf Forests**

Ecoregion Size: **23,700 km²**

Political Units: **Kenya, Tanzania**

Biological Distinctiveness: **Globally Outstanding**

Conservation Status: **Critical**

Conservation Assessment: **I**

Authors: *Jan Schipper, Neil Burgess*

Reviewed by: *Jon Lovett, Nike Doggart, Alan Rodgers*

Location and General Description

The Eastern Arc Forests [19] comprise a discontinuous chain of forested mountains in southeastern Kenya and eastern Tanzania (Lovett and Wasser 1993). From north to south, the main blocks are the Taita Hills in Kenya, the North and South Pare, the East and West Usambara, the North and South Nguru, the Ukaguru, the Uluguru, the Rubeho, and the Udzungwa in Tanzania. There are also smaller isolated outliers at Mahenge, Malundwe Hill, and the Uvidundwa Mountains of Tanzania.

Geologically these mountains are composed of metamorphosed Precambrian basement and rise dramatically from the subdued lowland plain. The main mountain blocks have been uplifted along ancient faults, with uplift events occurring periodically, probably at least since the Miocene (about 30 million years ago) (Griffiths 1993). Individual mountains rise to a maximum of 2,635 m altitude (Kimhandu peak in the Ulugurus), although maximum altitudes of 2,200–2,500 m are more typical. The tops of the Uluguru, Udzungwa, and East Usambaras are plateau-like. The soils are not rich but are often better than those of the surrounding lowlands.

The climate of the Eastern Arc Mountains is believed to have been stable over millions of years, as indicated by biological affinities to West Africa, Madagascar, and Asia (Hamilton 1982; Lovett and Friis 1996; Fjeldså and Lovett 1997; Fjeldså et al. 1997). The forests apparently survived the driest and coldest periods of the last ice ages, because the Indian Ocean did not cool appreciably and rainfall patterns may not have been greatly disrupted (Prell et al. 1980; Lovett and Wasser 1993). Currently these mountains are much wetter than the surrounding lowlands. The Ulugurus have up to 3,000 mm per year on the eastern slopes, and annual rainfall on other mountains exceeds 2,000 mm. There is some evidence that the climate has become drier and more seasonal in recent decades, with a lower likelihood of the forests being enveloped in mist (Hamilton and Bensted-Smith 1989).

Human population densities are substantially higher than those of the surrounding lowlands, with densities of up to 400 people/km² in the West Usambaras. In most areas populations are increasing, not only because of high birth rates (around 2.8 percent per annum) but also because of migration to the mountains to take advantage of the better agricultural potential.

The forest formations of the Eastern Arc have been divided into upper montane (2,635–1,800 m), montane (1,250–1,800 m), submontane (800–1,250 m), and lowland forest (Pócs 1976). At higher altitudes the canopy height decreases, and around 2,400 m altitude forest grades into Afromontane heathland plant communities with temperate affinities (Lovett 1993b). The upper altitudinal limit of forest vegetation is determined by the regular occurrence of frost.

The montane forest is characterized by large trees such as *Ocotea usambarensis, Allanblackia ulugurensis, Khaya anthotheca, Ochna holstii, Podocarpus latifolius, P. falcatus, Ilex mitis, Cola greenwayi, Cornus volkensii, Newtonia buchannii, Pachystela msolo,* and *Trichilia dregeana.* At lower altitudes the timber tree species *Milicia excelsa* also becomes important. Rubiaceae and Acanthaceae dominate the shrub layer. The mountain forests grade at lower altitudes into that of the lowland coastal forests of the Northern Zanzibar–Inhambane Coastal Forest Mosaic [20] (Lovett et al. 2000a). Further details on the vegetation composition are found in Lovett and Pócs (1993), Lovett and Wasser (1993), Lovett (1996, 1998a, 1998b), and Lind and Morrison (1974).

Outstanding or Distinctive Biodiversity Features

The Eastern Arc Mountains contain high species richness and many endemic species (Rodgers and Homewood 1982; Beentje 1988; Hamilton and Bensted-Smith 1989; Kingdon 1989; Lovett and Wasser 1993; Bennun et al. 1995; Burgess et al. 1998b; Stattersfield et al. 1998; Newmark 2002). Current estimates are that there are more than 2,000 plant species in 800 genera in the Eastern Arc (Lovett and Wasser 1993), of which at least 800 are believed to be endemic (Lovett 1998c). These forests are the center of endemism for the African violet (*Saintpaulia* spp.), which has been widely cultivated for house plants in Europe and America. When the Eastern Arc Mountains are combined with the Northern Zanzibar–Inhambane Coastal Mosaic [20], the density of endemic plant species is among the highest in the world (Myers et al. 2000). There are also high rates of endemism in the nonvascular bryophytes, including thirty-two known strict endemics (Pócs 1998). Endemic plants are found not only in the forests but also in the montane grasslands, in wetland areas, and on rocky outcrops.

There are ten strict endemic bird species in these mountains. Some near-endemic species have disjunct distribution patterns between the Eastern Arc and much further south in Malawi and Zimbabwe (Burgess et al. 1998a; Stattersfield et al. 1998). Other species have extremely limited distributions. The Taita thrush (*Turdus helleri,* CR) and Taita apalis (*Apalis fuscigularis,* CR) occur only in a few square kilometers of forest in the Taita Hills. The Udzungwa partridge (*Xenoperdix udzungwensis*, VU) is known only from a couple localities in the Udzungwa Mountains and one in the Rubeho Mountains (Dinesen et al. 1994; Baker and Baker 2002). The Uluguru bush shrike (*Malaconotus alius*, EN) is confined to one forest reserve on the Uluguru Mountains of less than 100 km^2 total forest area (Burgess et al. 2001). Other endemic bird species occur on several mountains, including the globally rare Usambara eagle-owl (*Bubo vosseleri*, VU), banded sunbird (*Anthreptes rubritorques*, VU), and Mrs. Moreau's warbler (*Bathmocercus winifredae*, VU) (Baker and Baker 2002).

Mammal endemism in the Eastern Arc is also quite high. There are two endemic diurnal monkeys (*Cercocebus galeritus sangei*, EN; *Procolobus gordonorum*, EN) and six strictly endemic small mammals, comprising five species of shrews (*Crocidura tansaniana*, VU; *Crocidura telfordi*, CR; *Crocidura usambarae*, VU; *Myosorex geata*, EN; and *Sylvisorex howelli*, VU) and one species of galago (*Galagoides orinus*). Other threatened mammals that occur in these forests include Abbot's duiker (*Cephalophus spadix*, VU), the eastern tree hyrax (*Dendrohyrax validus*, VU), and the black and rufous elephant-shrew (*Rhynchocyon petersi*, EN).

Amphibians and reptiles also exhibit high levels of species endemism. Notable among the thirty species of strictly endemic amphibians are reed frogs (*Hyperolius*, five endemic species), forest tree frogs (*Leptopelis*, two endemic species), tree toads (*Nectophrynoides*, five endemic species), and species in the Microhylidae family (four endemic species) and the Caeciliidae family (five endemic species). New species also continue to be discovered, such as the Kihansi spray toad (*Nectophrynoides asperginis*) (Poynton et al. 1998). High rates of endemism are also found in reptiles, including ten strictly endemic species of chameleons (seven *Chamaeleo* and three *Rhampholeon*), three species of worm snakes (*Typhlops*), and six species of colubrid snakes in four genera.

The invertebrates of the Eastern Arc also contain very high rates of endemism. Available information (see Lovett and Wasser 1993 and Burgess et al. 1998b) illustrate that up to 80 percent of certain invertebrate taxa can be strictly endemic to a single Eastern Arc mountain. The next mountain in the range will contain a similar high rate of strictly endemic species (Scharff 1992; Hoffman 1993; Brühl 1997).

This ecoregion is also notable from an evolutionary perspective, as both relict species and newly evolved species occur (Roy 1997; Roy et al. 1997). In the bryophytes, forty-five species are shared with Madagascar, indicating a very ancient link in the floras of these areas (Pócs 1998). In the plants, there are a number of affinities at the generic level with the forests of West Africa (Lovett 1998c; Burgess and Clarke 2000). Among the birds there are also species with affinities to those of West Africa and even with Asia (Udzungwa forest-partridge, *Xenoperdix udzungwensis*, and the African tailorbird, *Orthotomus metopias*).

Status and Threats

Current Status

The area of forest remaining on the Eastern Arc Mountain blocks has been estimated as follows: Taita Hills (6 km^2), Pare Mountains (484 km^2), West Usambaras (328 km^2), East Usambaras (413 km^2), Nguru (647 km^2, including Nguu), Ukaguru (184 km^2), Uluguru (527 km^2), Rubeho (499 km^2), Udzungwa (1,960 km^2), and Mahenge (291 km^2) (Newmark 1998). Many of these estimates are too high. For example, recent work in the Uluguru Mountains shows that the forest area is closer to 230 km^2 (Burgess et al. 2002a), and that on Mahenge is closer to 15 km^2. The mountains also support areas of natural grassland (Meadows and Linder 1993) and bamboo.

Forests of the Eastern Arc are protected in the 1,900-km^2 Udzungwa National Park (much of which is not montane forest vegetation) and in a larger number of forest reserves that are established for water catchment purposes. The Amani nature reserve has been recently established in the East Usambara forests.

Montane habitats are naturally scattered across the different mountain blocks. However, within each block these habitats have become more fragmented by conversion to agriculture. Large areas of forest have been lost from the lower slopes of all blocks, and forest has also been lost at higher elevations in some blocks (e.g., for tea plantations in the Usambaras and the Udzungwas). Outside the reserves, forest loss to agriculture continues (Burgess et al. 2002a).

Types and Severity of Threats

In the Udzungwa National Park the level of exploitation is low and protection is high. In catchment forest reserves the protection level is more variable and depends on access, presence of valuable timber, population pressure, and the capacity and interest of forestry officials to protect the forest. In some cases, there is significant commercial logging (using pitsawing techniques), and in other areas there is encroachment for farm plots and for the collection of wood products for firewood and poles. Outside catchment forest reserves the forests have been largely cleared on agricultural lands, except for small locally protected forest patches that are used for burial grounds and for traditional ceremonial purposes. More detailed summaries of the threats to these mountain forests are found in Lovett and Pócs (1993), Rodgers (1993), Burgess et al. (1998b), and Newmark (2002).

Justification for Ecoregion Delineation

This ecoregion covers the mountains in eastern Africa consisting of Precambrian rocks, under the direct climatic influence of the Indian Ocean. The Eastern Arc Forests [19] ecoregion is separated climatically from the Southern Rift Montane Forest-Grassland Mosaic [74] to the south (by the Makambako Gap between the Udzungwa Mountains and Mount Rungwe) and geologically from the East African Montane Forests [18] to the northwest. The mapped boundaries of the Eastern Arc Forests [19] are approximate to those of the "undifferentiated montane vegetation" unit of White (1983). Boundaries of the Udzungwa and Rubeho Mountains follow this boundary exactly, and experts refined the East and West Usambara, Nguru, Nguu, Mahenge, and Uluguru mountains.

Ecoregion Number:	**20**
Ecoregion Name:	**Northern Zanzibar–Inhambane Coastal Forest Mosaic**
Bioregion:	**Eastern and Southern Africa**
Biome:	**Tropical and Subtropical Moist Broadleaf Forests**
Political Units:	**Kenya, Tanzania, Somalia**
Ecoregion Size:	**112,600 km²**
Biological Distinctiveness:	**Globally Outstanding**
Conservation Status:	**Vulnerable**
Conservation Assessment:	**I**
Authors:	*Jan Schipper, Neil Burgess*
Reviewed by:	*Alan Rodgers, Jon Lovett, Nike Doggart*

Location and General Description

The Northern Zanzibar–Inhambane Coastal Forest Mosaic [20] ranges from coastal Somalia to southern Tanzania. The northernmost portion is an isolated forest outlier along the Jubba Valley of Somalia (Madgwick 1988). The ecoregion resumes in southern Somalia, extends into northern Kenya and inland along the Tana River, occupies a narrow coastal strip in central and southern Kenya to the border with Tanzania, bulges around the base of the East Usambara Mountains, and then extends south along the coast of Tanzania until grading into the Southern Zanzibar–Inhambane Coastal Forest Mosaic [21] around the town of Lindi. Outliers are found inland at the base of the Uluguru, Nguru, and Udzungwa Eastern Arc Mountains and on the larger offshore islands of Pemba, Zanzibar, and Mafia.

The climate of this ecoregion is tropical, with average temperatures above 25°C, little variation in day length, and generally high humidity. Mean annual rainfall varies from more than 2,000 mm on Pemba Island, to around 1,200–1,500 mm in southern Kenya and northern Tanzania and at the base of the Eastern Arc Mountains, to less than 1,000 mm in northern Kenya and lowland southern Tanzania. Most rainfall occurs in distinctive rainy seasons, although showers can occur at other times. In most of the ecoregion there are two rainy seasons, with a longer season from April to June and a shorter season

from November to December (Nicholson 1994; Burgess and Clarke 2000). To the south there tends to be one rainy and one dry season. The climatic pattern is estimated to have remained stable for millions of years (Axelrod and Raven 1978), even during the generally drier periods of the last ice age (Prell et al. 1980).

Tectonic activity has formed low ridges and swells (generally 100–300 m high) across this ecoregion during the past few tens of millions of years and has also caused the shoreline to move in and out from its present position. Most rocks are less than 30 million years old, but ancient Precambrian basement rocks occasionally reach the surface, particularly in the foothills of the Eastern Arc Mountains. The soils are complex, and most of them are fairly infertile (Hathout 1972).

The human population is highest around large coastal towns (Mombasa, Tanga, and Dar es Salaam). From southern Kenya to northern Tanzania there are around 100 people/km² in the rural areas, but population density falls rapidly in the north and south (to 10 people/km²), in line with declining rainfall and increasing seasonality.

This ecoregion forms a part of White's (1983) Zanzibar-Inhambane regional mosaic. Because of the exceptional level of plant endemism in the northern part of this regional mosaic, the Zanzibar-Inhambane regional mosaic has been reclassified as the Swahilian Regional Center of Endemism and the Swahilian-Maputaland regional transition zone (Clarke 1998). The Northern Zanzibar–Inhambane Coastal Forest Mosaic [20] falls entirely in the Swahilian Regional Center of Endemism. The vegetation is a mosaic of lowland forest patches, savanna woodlands, bushlands and thickets, and farmlands. Some of the more abundant trees in the remnant forests are *Afzelia quanzensis*, *Scorodophloeus fischeri*, *Dialium holtzii*, *Hymenaea verrucosa*, *Berlinia orientalis*, *Cynometra* spp., and *Xylia africana*. Lianas are also common, as are shrubs, herbs, grasses, sedges, ferns, and epiphytes.

Outstanding or Distinctive Biodiversity Features

There are numerous endemic species in the flora and fauna of this ecoregion, particularly in the forest habitats (Hawthorne 1993a; Burgess et al. 1998a; Stattersfield et al. 1998; Burgess and Clarke 2000). There is high degree of local turnover in the species between adjacent forest fragments in the landscape mosaic and a high incidence of rare species exhibiting disjunct distribution patterns.

More than 4,500 plant species and 1,050 genera occur in this vegetation mosaic. Of these, 3,000 species in 750 genera are found in the forest habitats, with great variation between forest patches (Robertson and Luke 1993; Burgess and Clarke 2000). At least 400 plant species are strictly endemic to the forest patches (which occupy around 1 percent of the area of the ecoregion), and at least 500 more species are strictly endemic to the nonforest vegetation (99 percent of the ecoregion area).

There are close generic relationships with lowland forests in West Africa. This implies an ancient forest history retaining evidence of connections across tropical Africa (Brenan 1978; Axelrod and Raven 1978; Lovett and Wasser 1993). Together with the Eastern Arc Forests [19], the Northern Zanzibar–Inhambane Coastal Forest Mosaic [20] forms a global center of botanical endemism (Lovett 1998b; Mittermeier et al. 1999; Lovett et al. 2000a; Myers et al. 2000).

Of the ten strictly endemic bird species, four are restricted to the island of Pemba (*Treron pembaensis*, *Nectarinia pembae*, *Zosterops vaughani*, and *Otus pembaensis*), one in the lower Tana River (*Cisticola restrictus*, DD), and the rest mainly in the mainland coastal forest remnants (*Erythrocercus holochlorus*; *Anthus sokokensis*, EN; *Ploceus golandi*, EN; and *Campethera mombassica*). The remaining strict endemic is found in coastal grasslands in Kenya, *Anthus melindae*. Some of the near-endemic species found in these forests include Sokoke scops owl (*Otus ireneae*, EN), Fischer's turaco (*Tauraco fischeri*), and Amani sunbird (*Anthreptes pallidigaster*, EN).

The mammals of this ecoregion include forest, savanna, and wetland species. Strict endemic mammals include Aders's duiker (*Cephalophus adersi*, EN), Pemba flying fox (*Pteropus voeltzkowi*, CR), Kenyan wattled bat (*Chalinolobus kenyacola*, DD), Dar-es-Salaam pipistrelle (*Pipistrellus permixtus*, DD), golden-rumped elephant shrew (*Rhynchocyon chrysopygus*, EN), Tana River Red colobus (*Procolobus rufomitratus*, CR), Tana River mangabey (*Cercocebus galeritus galeritus*, CR), Zanzibar red colobus (*Procolobus kirkii*, EN), and the rodent *Grammomys caniceps*. There are also a number of near-endemic threatened species, including the eastern tree hyrax (*Dendrohyrax validus*, VU) and the black and rufous elephant shrew (*Rhynchocyon petersi*, EN).

Of the reptile species occurring in the ecoregion, thirty-four are strictly endemic. Endemic reptiles are found in the geckos (Gekkonidae), chameleons (Chamaeleonidae), skinks (Scincidae), lacertid lizards (Lacertidae), worm-snakes (Typhlopidae), and true snakes (Atractaspididae, Elapidae, and Colubridae). The amphibians are also diverse and exhibit a moderate rate of endemism. Poynton (in Burgess and Clark 2000) lists fourteen amphibian species as largely confined to coastal forests, with two species being strictly endemic to this ecoregion (*Afrixalus sylvaticus* and *Stephopaedes usambarensis*).

Millipedes, mollusks, and butterflies also exhibit high diversity and rates of endemism. There are 1,200 mollusk species in the region, 86 of which are confined to forests. Butterflies are represented by 400 forest species, of which 75 are endemic (Burgess and Clarke 2000).

Status and Threats

Current Status

The lowland forest is the most biologically interesting component of the mosaic, but much has been destroyed. Approximately 1,000 km^2 of forest remains in the ecoregion, fragmented into more than 200 separate patches. Subfossil resin (gum copal) of the coastal forest tree species *Hymenea verrucosa* was mined widely along the coastal strip in the early 1900s, including in areas that now support fire-maintained savanna-woodland vegetation or farmland. This indicates a considerable loss of forest over perhaps thousands of years.

The remaining natural habitats are becoming more fragmented as agriculture and other human activities spread with increasing populations. The largest remaining block of habitat is the 370-km^2 Arabuko-Sokoke Forest near Malindi in Kenya (Wass 1995; Bennun et al. 1995). The next largest forest is in the Shimba Hills National Reserve in southern Kenya (around 68 km^2). In Tanzania forested areas are not larger than 40 km^2 and typically are much smaller (Burgess et al. 1996; Burgess and Clarke 2000). Small fragments of forest, often in burial groves, are found throughout the ecoregion.

Protected areas containing forest include the Tana River Primate Reserve, Shimba Hills National Reserve, Arabuko-Sokoke National Park, Sadaani Game Reserve (now including Mkwaja Ranch, soon to be a national park), and Mafia Island Marine Park. In total these areas cover less than 1,200 km^2 of land, of which more than 50 percent is savanna woodland. Most of the remaining forest areas are protected in central and local government-controlled forest reserves. These were originally established for watershed protection or controlled resource exploitation. Although many of the forest reserves are poorly managed because the forestry department lacks resources, local populations typically respect the boundaries, and forest vegetation remains at most sites. Local people protect some important sites as sacred forests.

Types and Severity of Threats

In the past, extensive areas of forest were converted to plantations of trees and sisal, especially at the base of the East Usambara Mountains in Tanzania. These large losses are no longer occurring, but the forests remain highly threatened. The most severe threat to the coastal forest patches comes from agricultural expansion. Other threats include extraction of woody materials (e.g., poles, timber, firewood, charcoal, and ropes) (Burgess and Clarke 2000). These occur in almost every forest that is not in a national park.

The nonforest habitats of the ecoregion are also threatened. The extensive areas of bushland and savanna are used extensively to provide wood for charcoal burning. This activity is most intense close to the large urban centers, especially Dar es Salaam in Tanzania. The nonforest habitats are also converted to farmland, especially along the southern Kenyan coast and close to Dar es Salaam in Tanzania. Some habitat has also been lost to mining of limestone for the production of concrete, especially near Mombasa and Dar es Salaam.

Justification for the Ecoregion Delineation

This ecoregion largely follows the Swahilian regional center of endemism of Clarke (1998), part of the Zanzibar-Inhambane regional mosaic of White (1983). The northern boundary includes the Jubba riverine forests in Somalia and the Tana riverine forests in northern Kenya. The western boundary largely follows that of White (1983). The eastern boundary follows the coast but includes the offshore islands of Pemba, Zanzibar, and Mafia (and smaller islands). The southern boundary is mapped further north than Clarke's (1998) southern boundary of his Swahilian regional center of endemism, in the vicinity of the Lukuledi River in southern Tanzania. This northern shift follows the division between coastal areas that have two rainy seasons per annum (this ecoregion) and those where there is one long rainy season (Southern Zanzibar–Inhambane Coastal Forest Mosaic [21]).

Ecoregion Number:	**21**
Ecoregion Name:	**Southern Zanzibar–Inhambane Coastal Forest Mosaic**
Bioregion:	**Eastern and Southern Africa**
Biome:	**Tropical and Subtropical Moist Broadleaf Forests**
Political Units:	**Mozambique, Malawi, Tanzania**
Ecoregion Size:	**147,000 km²**
Biological Distinctiveness:	**Bioregionally Outstanding**
Conservation Status:	**Critical**
Conservation Assessment:	**IV**
Authors:	*Jan Schipper, Neil Burgess*
Reviewed by:	*Alan Rodgers, John Burlison, Judy Oglethorpe*

Location and General Description

The Southern Zanzibar–Inhambane Coastal Forest Mosaic [21] runs approximately 2,200 km along the eastern coast of Africa, from the Lukuledi River in southern Tanzania to the Changane River just south of Xai-Xai in Mozambique. The ecoregion mostly extends less than 50 km inland of the Indian Ocean, although there are a few tiny outliers approximately 200 km inland in the foothills of mountains in easternmost Malawi and Zimbabwe (Burgess and Clarke 2000). The ecoregion also includes small offshore islands in Mozambique.

The topography is gently rolling, with some isolated higher plateaus and inselbergs, especially in the northern part. To the south, sand dunes are important features that support forest vegetation. Inland from Lindi, remnant areas of Miocene uplift

have been eroded into plateau areas that rise up to 800 m in altitude, especially at the Rondo Plateau. Further south, the Makonde Plateau in southern Tanzania and northernmost Mozambique rises up to 1,000 m altitude. The lowland forest outliers in Malawi and Zimbabwe are found at the base of Mount Mulanje in Malawi and the Chimanimani area of Zimbabwe (Chapman and White 1970; Dowsett-Lemaire 1990; Burgess and Clarke 2000).

The ecoregion has a tropical climate in the northern portion and borders the subtropical zone in the southern portion. In the Lindi area there is one prolonged dry season and one wet season, although showers are frequent. Rainfall, controlled by monsoon winds, is around 800–1,000 mm per year, although it is higher on some of the plateaus. In coastal Mozambique the climate follows the same general trends, but with lower rainfall in the north (around 800 mm per annum). Mean maximum temperatures are between 27°C and 30°C in the north and 24°C in the south of the ecoregion.

The portion of the ecoregion found in Tanzania consists mainly of Tertiary marine sediments that have been uplifted and then eroded. In northern Mozambique, Precambrian basement rocks form inselbergs that have been deformed and eroded over hundreds of millions of years. Along the coast there are recent dunes, and around river mouths, riverborne deposits are found (Buckle 1978).

The distribution of human populations varies across the ecoregion. In southern Tanzania there is a high population density on the Makonde Plateau, which extends just over the border into the similar upland habitats of northern Mozambique. In the lowlands of northern Mozambique, however, there are few people, and some areas are almost empty. Further south in Mozambique the coastal area is heavily settled, especially along the main road north from Maputo to Beira.

In terms of the phytogeographic classification of Moll and White (1978) and White (1983), this ecoregion falls in the Zanzibar-Inhambane regional mosaic and reaches the northern limits of the Tongaland-Pondoland regional transition zone. A recent reclassification of the phytogeographic framework of the region has divided the Zanzibar-Inhambane regional mosaic into two regions, with the majority of the Southern Zanzibar–Inhambane Coastal Forest Mosaic [21] ecoregion occupying the Swahili-Maputaland regional transition zone (Clarke 1998). The vegetation consists of a mosaic of savanna woodland, forest patches, thickets, swamps, and littoral vegetation types with a diverse species composition. At the coastline the vegetation grades into mangrove vegetation in sheltered bays and along tidal rivers, with some sand dunes and coastal lagoon habitats.

Outstanding or Distinctive Biodiversity Features

The species richness in this ecoregion is low for obligate forest species but is greatly elevated by the species found in the diverse mosaic of nonforest habitat types. A high number of endemic species occurs in the northern portion of the ecoregion (southern Tanzania), followed by an almost complete lack of data in the central portion (northern and central Mozambique) and high numbers of endemics again in the south (in Mozambique). The outliers of the ecoregion in Malawi and Zimbabwe also contain endemic species (Burgess and Clarke 2000).

Although data are poor, more than 150 strictly endemic plants occur in this ecoregion, with 100 species confined to southern Tanzania (Burgess et al. 1998; Burgess and Clarke 2000). Northern Mozambique is also expected to be rich in endemic plant species but has never been studied.

Current information on the avifauna is also limited, especially in northern and central Mozambique. The only strict endemic bird in this ecoregion is Reichenow's batis (*Batis reichenowi*), which many taxonomists do not recognize as a valid species (e.g., Sibley and Monroe 1990). Near-endemic species include the east coast akalat (*Sheppardia gunningi*, VU) and the spotted ground thrush (*Zoothera guttata*, EN), which have a scattered distribution in eastern Africa. A number of forest patches in southern Tanzania are regarded as important bird areas (Baker and Baker 2001).

The mammals in this ecoregion are somewhat poorly known, particularly in northern Mozambique. Larger mammal species shared with the neighboring Eastern Miombo Woodlands [52] include elephant (*Loxodonta africana*, EN), buffalo (*Syncerus caffer*), sable antelope (*Hippotragus niger*), roan antelope (*H. equinus*), and Lichtenstein's hartebeest (*Sigmoceros lichtensteinii*). The near-endemic primate *Galagoides rondoensis* occurs in southern Tanzania.

Among the reptiles, strictly endemic species include *Ancylocranium* and *Chirindia* (worm lizards), *Chamaeleo* and *Rhampholeon* (chameleons), *Scelotes* (burrowing skink), and *Typhlops* (burrowing snakes). Most of these endemics are small, ground-living or burrowing forms, and given the lack of basic biological research it is expected that more endemic species occur, especially in northern Mozambique. Only one near-endemic amphibian, the Mahenge toad (*Stephopaedes loveridgei*), is known from this ecoregion.

Important subcenters of endemism occur. The most significant is in the Lindi area of southern Tanzania, associated with a number of raised plateaus (WWF and IUCN 1994; Burgess et al. 1998). Rondo plateau has one of the most important concentrations of endemic plants in eastern Africa (more than sixty endemic species in less than 50 km^2 of forest). It also contains three endemic reptiles, *Melanoseps rondoensis*, *Typhlops rondoensis*, and *Chirindia rondoensis* (Burgess and Clarke 2000), the near-endemic primate *Galagoides rondoensis*, and one endemic bird subspecies, *Stactolaema olivacea hylophona*. Further to the south there is another area of endemism on the Bazarruto Archipelago. Three endemic reptile species are found here: *Lygosoma lanceolatum*, *Scelotes insularis*, and *Scelotes duttoni* (Broadley 1990, 1992).

Status and Threats

Current Status

The forests of this ecoregion probably have been reduced in extent over thousands of years (Burgess and Clarke 2000). Currently, around 4,000 km^2 of forest is believed to exist, mainly in Mozambique (Younge et al. 2002). The ecoregion also supports extensive areas of savanna woodland, bushland and thicket, and wetlands of various types, especially in Mozambique.

In southern Tanzania the remaining blocks of forest habitat are small (none more than 20 km^2), and most are found in government forest reserves. The most important forest in southern Tanzania is found on the Rondo Plateau. In the 1940s this plateau supported a diverse moist evergreen forest. This was logged in the 1950s, and most was replanted with pine (*Pinus caribea*) and the native hardwood mvule (*Milicia excelsa*). Logging and planting stopped in the early 1990s because of infrastructure decline. As a result, the natural forest is regenerating in some areas.

In northern and central Mozambique extensive areas of forest and thicket remain (Younge et al. 2002) but are poorly known. In southern Mozambique there are also a number of forests, but these are typically small (Burgess and Clarke 2000). Three protected areas cover this ecoregion in Mozambique (Quirimbas, Bazaruto, and Pomene 2002). The newly declared Quirimbas National Park in northern Mozambique covers around 5,000 km^2 of terrestrial habitats, much of it coastal vegetation types including large areas of coastal forest.

Types and Severity of Threats

Threats to this ecoregion vary from commercial logging in the newly accessible forests of northern Mozambique and pitsawing of valuable timber in Tanzania to forest clearing for agriculture and future tourism development. In many cases the severity of these threats is poorly known because the areas are remote and there is little available information on the sites.

One of the effects of the Mozambique war has been a severe reduction of the population density of larger mammals caused by hunting for food, although smaller species such as grey duiker (*Syvicapra grimmia*) and suni (*Neotragus moschatus*) are still numerous.

Justification for Ecoregion Delineation

The Southern Zanzibar–Inhambane Coastal Forest Mosaic lies between the Lukuledi River in southern Tanzania and the Changane River in southern Mozambique. These boundaries are similar to those presented by Clarke (1998), who divided the Zanzibar-Inhambane regional mosaic of White (1983) into a northern Swahilian regional center of endemism and a southern Swahilian-Maputaland regional transition zone, based on endemic plant species. We follow Clarke's southern boundaries; however, we have shifted the northern boundary to southern Tanzania to follow changes in the rainfall pattern.

Ecoregion Number:	**22**
Ecoregion Name:	**Maputaland Coastal Forest Mosaic**
Bioregion:	**Eastern and Southern Africa**
Biome:	**Tropical and Subtropical Moist Broadleaf Forests**
Political Units:	**Mozambique, South Africa, Swaziland**
Ecoregion Size:	**30,200 km²**
Biological Distinctiveness:	**Bioregionally Outstanding**
Conservation Status:	**Critical**
Conservation Assessment:	**IV**
Author:	*Don Kirkwood*
Reviewed by:	*Jeremy J. Midgley,*
	Anthony B. Cunningham,
	John Burlison, Judy Oglethorpe

Location and General Description

The Maputaland Coastal Forest Mosaic [22] extends from the Changane River in southern Mozambique close to the town of Xai Xai (25°S) to the Umfolosi River just north of Cape Saint Lucia in South Africa (28°S). The area consists largely of a flat to gently undulating, low-lying coastal plain with a maximum elevation of about 200 m. On the western margin, the narrow Lebombo Mountain Range rises to about 600 m (Maud 1980).

Most of the Maputaland coastal plain is covered with infertile, wind-distributed sands, forming a series of north-to-south dune ridges parallel to the current coastline (Goodman 1990). The oldest dunes are immediately adjacent to the Lebombo Mountains and appear to be of Plio-Pleistocene age (Davies 1976). Along the coast the dunes are young and may still be forming in places. Sometimes almost 200 m in height, these are among the tallest vegetated dunes in the world (van Wyk 1994a). In South Africa, the soils of western Maputaland are fertile, especially along the west bank of the Pongolo River, because they are derived from alluvium, river terraces, and Cretaceous sediments (Maud 1980).

The climate is moist subtropical along the coast, where rainfall is more than 1,000 mm per annum, becoming dry subtropical a short distance inland, with less than 600 mm rain per annum (Maud 1980; Watkeys et al. 1993). Summers are hot and humid, with the highest monthly precipitation between September and April. Winters are cool and dry. Mean annual temperature varies from 21°C along the Lebombo Mountains to 23°C in the center of the coastal plain, moderating slightly along the coast to 22°C.

The Maputaland Coastal Forest Mosaic contains extensive areas of wetland. Lake St. Lucia covers approximately 350 km² and is the largest estuarine system in Africa. Lake Sibayi (60 km²) is the largest freshwater lake in southern Africa, separated from the ocean by a narrow dune cordon. The Kosi Lake System (37 km²) (van Wyk 1994b) and several similar-sized water bodies such as Lake Piti in the Maputo Elephant Reserve also contain wetland habitat (Davies and Day 1998).

The vegetation of the Maputaland Coastal Forest Mosaic is complex and exceptionally diverse. Moll (1978, 1980) described at least fifteen major vegetation types in the South African part of the ecoregion. The vegetation of southern Mozambique has also been described in great detail (Myre 1964).

Forest grows on top of the Lebombo Range, especially in deeper valleys and moister southeastern slopes. Although species composition is varied, canopy species such as *Chrysophyllum viridifolium, Homalium dentatum, Combretum kraussii,* and various *Ficus, Celtis,* and *Strychnos* spp. are most common. Understory trees and tall shrubs beneath the fairly open canopy include *Buxus natalensis, Englerophytum natalense,* and *Rothmannia globosa.*

The dry Sand Forest occurs primarily on inland paleo-dune sands and is species rich, with a high number of woody endemics (Moll 1980; van Wyk 1994b; Kirkwood and Midgley 1999). *Cleistanthus schlechteri* and *Newtonia hildebrandtii* consistently dominate the canopy, although *C. schlechteri, Hymenocardia ulmoides, Psydrax fragrantissima,* and the understory trees *Croton pseudopulchellus* and *Drypetes arguta* are also characteristic.

The coastal dunes also support dense forests. *Mimusops caffra, Euclea natalensis,* and *Diospyros rotundifolia* form short forests or thickets on the seaward side of the dunes. With some protection from the salt wind, more diverse forests develop with canopies as tall as 30 m. Common species include *Diospyros inhacaensis, Apodytes dimidiata, Celtis africana, Drypetes gerrardii, Ziziphus mucronata, Strychnos* spp., and *Ficus* spp.

In some parts of the ecoregion there are extensive areas of palm veld. These consist of continuous grass cover of *Themeda, Eragrostis, Aristida,* and *Perotis,* with scattered *Hyphaene coriacea* palms. Another palm, *Phoenix reclinata,* is also common, as are the short trees *Dichrostachys cinerea* and *Strychnos madagascariensis.* Dwarf woody plants, especially *Parinaria curatellifolia,* become common closer to the coast.

Outstanding or Distinctive Biodiversity Features

The Maputaland Coastal Forest Mosaic corresponds to one of three clear foci of high floristic endemism in White's (1983) Tongaland-Pondoland Regional Mosaic (van Wyk 1994a). The region is also a transition zone at the southern end of the distribution of many tropical plant and animal species (Poynton 1961; Moll 1980; van Wyk 1994a). Most of the flora and fauna are of Afrotropical origin, and endemic taxa appear to have evolved only recently. Evidence for this includes the large number of infraspecific endemics (van Wyk 1994a) and the young age of most of the sandy coastal plain, perhaps less than 1 million years (Davies 1976).

The number of vascular plant species is at least 2,500 and may be as high as 3,000. Of these, at least 225 species or infraspecific taxa are endemic or near endemic (MacDevette et al. 1989; van Wyk 1994b). Among the plants, Scott-Shaw (1999) lists twenty-six threatened plants (IUCN status "vulnerable" and higher) in the Maputaland region (South African portion). Many of these are endemic or near endemic and also often rare or with restricted ranges.

Faunal diversity and endemism are also high (van Wyk 1994a). More than 100 mammal species are recorded, and although there are no strict or near-endemic mammal species there may be several endemic subspecies (van Wyk 1994a). Elephants (*Loxodonta africana*, EN) occur, as do the large predators such as leopard (*Panthera pardus*), lion (*Panthera leo*, VU), and cheetah (*Acinonyx jubatus*, EN), the last two being reintroduced.

In the birds, the endemic Neergaard's sunbird (*Nectarinia neergaardi*) is restricted almost entirely to the dry semi-deciduous sand forest of Maputaland (Harrison et al. 1997). Rudd's apalis (*Apalis ruddi*) is also near endemic, together with the pink-throated twinspot (*Hypargos margaritatus*) and the lemon-breasted seedeater (*Serinus citrinipectus*) (Stattersfield et al. 1998).

Amphibians and reptiles are also diverse and support some endemic species (van Wyk 1994a). Of the approximately 112 reptile species or subspecies found here, around 20 are near endemic and four are strictly endemic. Among the amphibians, forty-seven frog species or subspecies are found in this ecoregion, with one near-endemic species. This is also an area of endemism for freshwater fishes, with eight endemic or near-endemic species.

Status and Threats

Current Status

Much of the vegetation is still in excellent condition, particularly compared with that of areas further south (van Wyk 1994b). The presence of landmines slowed the return of people to the Mozambican areas, and human population densities remain low.

Almost 14 percent of the ecoregion is protected in reserves. The ecoregion is considered close to adequately protected in South Africa, with only areas of the Lebombos, riverine vegetation, and the coastal grasslands against the coastal dune cordon underrepresented in the present reserve system. The contiguous reserves making up the Greater St. Lucia Wetland Park cover 2,134 km². The biological richness of the region and the potential for ecotourism have resulted in the establishment of a number of private nature and game reserves, the most important of which is the 170 km² Phinda Resource Reserve adjacent to Mkuzi Game Reserve.

Protected areas in Mozambique include the Maputo Game Reserve (900 km²) and Ilhas de Inhaca e dos Portugeses Faunal Reserve (20 km²). Although almost all large game species have been exterminated in the Maputo Elephant Reserve (except for a sizable elephant population), it still contains examples of a wide range of important ecosystems including coastal grassland, sand forest, wetlands, and mangroves.

A proposed transfrontier conservation area would link Tembe Elephant Park and Ndumo Reserve in South Africa with the Futi Corridor, Maputo Special Reserve, other areas of southern Mozambique, and parts of eastern Swaziland.

Types and Severity of Threats

The biggest threat facing the natural vegetation in Maputaland is the spread of invasive exotic plants. *Chromolaena odorata* is the most aggressive invader, and another is guava (*Psidium guajava*). All vegetation types appear to be vulnerable to invasion, although riverine systems tend to be most severely affected.

Outside reserves in South Africa, major impacts on the environment have come from afforestation with exotic *Pinus* and *Eucalyptus* species in the vicinity of Lake St. Lucia and on its eastern shores (Moll 1980). Large-scale afforestation schemes have also been initiated in southern Mozambique, and more are planned.

Clearing for slash-and-burn subsistence farming is widespread and increasing. Swamp forest is highly favored for its fertile and moist soils and has already been heavily degraded by commercial banana and local cash crop farming, even in the gazetted reserve. Several plant species are highly threatened by overharvesting, such as the threatened *Warburgia salutaris*, a highly sought-after medicinal plant now almost extinct in the wild (Hilton-Taylor 1996; Scott-Shaw 1999). Canopy tree species such as *Erythrophleum lasianthum* and *Balanites maughamii* in Sand Forest and *Cassipourea gerrardii* and *Cassine papillosa* in Dune Forest are girdled on a large scale to supply the demand for herbal medicines (Cunningham 1988, 1991). Woodcutting for firewood and tourist carvings (mainly large bowls) is also a booming industry, using *Cleistanthus schlechteri* and *Balanites maughamii*, the two Sand Forest dominants.

Almost all game animals outside reserves have been eliminated by hunting, and poaching in reserves is common. International trade also threatens a range of species, from birds such as the brown-headed parrot (*Poicephalus cryptoxanthus*) to plants such as the rare and localized Retief cycad (*Encephalartos lebomboensis*), which has been decimated by cycad collectors (Scott-Shaw 1999).

Justification for Ecoregion Delineation

This ecoregion encompasses the northern section of the Tongaland-Pondoland regional center of endemism (White 1983). The ecoregion boundaries largely follow the Maputaland

Center of van Wyk (1994a, 1994b). The ecoregion is bordered by the Changane River in southern Mozambique, which is considered a significant biogeographic boundary, and the Umfolosi River just north of Cape St. Lucia in South Africa.

Ecoregion Number:	**23**
Ecoregion Name:	**KwaZulu-Cape Coastal Forest Mosaic**
Bioregion:	**Eastern and Southern Africa**
Biome:	**Tropical and Subtropical Moist Broadleaf Forests**
Political Units:	**South Africa**
Ecoregion Size:	**17,800 km²**
Biological Distinctiveness:	**Bioregionally Outstanding**
Conservation Status:	**Critical**
Conservation Assessment:	**IV**
Author:	*Charlotte Heijnis*
Reviewed by:	*Jeremy J. Midgley,* *Anthony B. Cunningham*

Location and General Description

The KwaZulu-Cape Coastal Forest Mosaic [23] is distributed in a narrow band along the eastern South African coastline and contains the coastal tropical and subtropical forest of South Africa. It extends from Cape St. Lucia (about 32°E), south along the eastern narrow coastal plain to Cape St. Francis (26°S 33°E). Its inland boundary lies at the foothills of the Drakensberg Escarpment.

Topographically the ecoregion is around 450 m in altitude and more steeply rolling in the north, falling to 350 m altitude and more gently rolling to the south. The underlying rocks are predominantly silt and mudstone sediments of the Karoo sequence interspersed by islands of Natal Group sandstones and Precambrian basement rocks (granites, gneisses, and schists). At the coast, calcareous Quaternary dunes overlie the rocks (Tinley 1985). Soils are generally alkaline and range from medium- to coarse-grained.

The coastal plain of KwaZulu-Natal has a subtropical climate and lacks frosts. Further south, frosts occur, but the proximity to the sea moderates both winter and summer temperature regimes. Mean maximum temperatures range between 15°C and 24°C, and mean minimum temperatures between 10°C and 15°C. Rainfall in the north ranges between 900 and 1,500 m per annum and falls in the summer, with lower rainfall in the south falling during the winter. Eighteen large rivers bisect the ecoregion, draining the Drakensberg Escarpment and the Grootwinterhoek and Baviaanskloof mountains. These rivers extend from the Umfolozi in the extreme north to the

Gamtoos in the south. River number and size decrease to the south.

The vegetation of this ecoregion consists of the narrower (minimum 8 km), southern part of Moll and White's (1978) Tongaland-Pondoland regional mosaic. This regional mosaic is part of a greater Indian Ocean Coastal Belt extending down from the extreme southeastern corner of Somalia. This ecoregion includes three of Acocks's (1953) five subdivisions of the Indian Ocean forest and thornveld: the Typical Coastal-Belt Forest, Transitional Coastal Forest, and Dune Forest. The more xeric southwest extension of this ecoregion across the Keiskamma River was described by Acocks (1953) as Alexandria Forest. These different forest types correspond roughly to the coastal lowlands forest, sand forest, and dune forest of Everard et al. (1994).

Common trees in Acocks's (1953) Typical Coastal-Belt Forest include *Millettia grandis* and *Protorhus longifolia,* which are indicator species for this forest type, as well as *Vepris undulata, Combretum kraussii,* and *Rhus chirindensis.* Widespread shrubs and climbers are *Uvaria caffra, Dalbergia obovata,* and *Tricalysia lanceolata.* Species of Typical Coastal-Belt Forest do not usually occur south of the Great Kei River, where species more typical of Transitional Coastal Forest such as *Ptaeroxylon obliquum, Schotia brachypetala, Cassine* spp., and *Euphorbia grandidens* replace them.

Acocks's (1953) Dune Forest occupies a narrow belt on high dunes running down the coast. Principal trees include *Mimusops caffra, Euclea natalensis,* and *Psydrax obovata.* Lianas are less important in this forest, with common shrubs and climbers including *Scutia myrtina, Allophylus natalensis,* and *Dracaena hookeriana.* In rocky coastal areas, the forest extends down to the high tide level. *Mimusops caffra* and *Allophylus natalensis* are diagnostic of Dune Forest.

The southern Alexandria Forest is a short (10 m), very dense forest with species such as *Ochna arborea* var. *arborea, Apodytes dimidiata dimidiata,* and *Cassine aethiopica.* Less common trees include *Euclea natalensis, Pittosporum viridiflorum,* and *Rapanea melanophloes.* In this short forest, shrubs are particularly important, and large proportions are scramblers. *Scutia myrtina, Azima tetracantha,* and *Grewia occidentalis* are all found here, and undergrowth species include various *Acanthaceae* spp., *Panicum deustum,* and *Sansevieria* spp.

Outstanding or Distinctive Biodiversity Features

The coastal Tongaland-Pondoland forests are a regional center of floral endemism. According to White (1978) this coastal belt supports a rich flora of about 3,000 species, with approximately 40 percent of larger woody species considered endemic to the region. A large number of plants are endemic or near endemic to the sandstone outcrops, and these taxa comprise a significant proportion of the Tongaland-Pondoland Regional Mosaic endemics (Van Wyk 1990). Moll and White (1978) attributed the Tongaland-Pondoland Regional Mosaic with twenty-three endemic genera, many of which are paleoendemics. Species of monotypic woody taxa are found in at least six genera (*Dahlgrenodendron, Eriosemopsis, Jubaeopsis, Pseudosalacia, Pseudoscolopia,* and *Rhynchocalyx*) in one family (Rhynchocalycaceae).

Some other plant genera that are widespread in Africa (e.g., *Bersama, Diospyros, Euclea,* and *Rhoicissus*) have their center of variation in Tongaland-Pondoland Regional Mosaic (Geldenhuys 1992). Other genera, such as *Cassine* and *Eugenia,* are more widespread in Africa and elsewhere and have markedly high concentrations of species in the area (Moll and White 1978). Many of the endemic and near-endemic species of the sandstone islands of KwaZulu-Cape Coastal Forest Mosaic are taxonomically isolated. Specific examples of these plant species include *Rhynchocalyx lawsonioides, Dahlgrenodendron natalensis,* and *Rinorea dematiosa.*

Because this ecoregion is a mosaic of forest and thornveld, the ecoregion is rich in faunal species that live in both elements. Around 100 mammal species occur, including the endemic Sclater's tiny mouse shrew (*Myosorex sclateri,* VU), the near-endemic Duthie's golden mole (*Chlorotalpa duthieae,* VU), and the giant golden mole (*Chrysospalax trevelyani,* EN).

Rare bird species are confined primarily to the forest components of the ecoregion, but there are no strictly endemic species. Birds that are found only in forests and bushlands of the southern African coasts include the brown scrub-robin (*Cercotrichas signata*), Knysna turaco (*Tauraco corythaix*), Knysna woodpecker (*Campethera notata*), Knysna scrub-warbler (*Bradypterus sylvaticus,* VU), Chorister robin-chat (*Cossypha dichroa*), and forest canary (*Serinus scotops*). Other notable bird species include the southern bald ibis (*Geronticus calvus,* VU), southern banded snake eagle (*Circaetus fasciolatus*), and wattled crane (*Grus carunculatus,* VU).

The Natal diving frog (*Natalobatrachus bonebergi*) is one of six near-endemic amphibians. In the reptiles the Transkei dwarf chameleon (*Bradypodion caffrum*), Günther's burrowing skink (*Scelotes guentheri*), the skink *Acontias poecilus,* and the gecko *Cryptactites peringueyi* are all regarded as strictly endemic to this ecoregion.

Status and Threats

Current Status

The KwaZulu-Cape Coastal Forest Mosaic [23] is highly fragmented. A few large forest complexes are found in large areas of nonforest, but the majority of remaining vegetation is found in small, isolated patches amounting to approximately 9,468 km^2 (Lubke and McKenzie 1996). This is little more than half of its previous extent.

Today an estimated 9.5 percent of KwaZulu-Cape Coastal Forest Mosaic [23] is conserved (Low and Rebelo 1996), mostly in a large number of small reserves that are not interconnected. Important reserves include the Amatikulu Nature Reserve, East

London Coast, Great Fish River, Geelkrans, Sunshine Coast, Vernon Crookes, Woody Cape, and Hluleka Wildlife Reserve. Other areas of natural habitat are protected in private and tribal areas and in conservancies and natural heritage sites (Geldenhuys and MacDevette 1989). Although this conservation network protects elements of the KwaZulu-Cape Coastal Forest Mosaic, it still does not adequately conserve large intact blocks of forest and interspersed thornveld as single units. Many of the forest patches and corridors outside conservation areas are not protected.

Types and Severity of Threats

The main threat to the ecoregion is the direct and indirect use of forest as a resource. Traditional uses of forests by local populations include building materials, traditional medicines (such as bark from *Protorhus longifolia*), food, water, and grazing. With burgeoning populations and decreased employment, demands for these resources have increased. The natural fragmentation of the forests has been aggravated by land-use practices such as clearing for agriculture, forestry, subsistence use, and burning practices for grazing and improved water runoff in catchments (Feely 1980; von Maltitz and Fleming 2000). Much of the northern, inland distribution of KwaZulu-Cape coastal forest in Natal, particularly the thicket component, has been replaced by sugar cane. Informal housing is also an ever-increasing threat to the integrity of the ecoregion.

Coastal dunes are especially sensitive to human interference. Disturbance to dunes includes mining and holiday resort expansion, alien plant invasion, recreational activities including off-road access, inappropriate planning and zoning, and ineffective administrative control (Tinley 1985). Where dune forest has been disturbed by urban development, *Brachylaena discolor, Strelitzia nicolai,* and *Chrysanthemoides monilifera* tend to dominate the secondary vegetation.

Fast-growing alien plant species also invade disturbed habitats, such as *Casuarina equisetifolia* (Geldenhuys et al. 1986). There are few true indigenous pioneer forest trees (e.g., *Trema orientalis, Apodytes dimidiata,* and *Acacia karoo*). The initial pure stand of alien pioneers is gradually colonized and enriched by the native shade-tolerant species (Huntley 1965; Knight et al. 1987).

Justification for Ecoregion Delineation

This ecoregion follows the southern part of Moll and White's (1978) Tongaland–Pondoland regional mosaic. Bordered to the north by the more tropical Maputaland Coastal Forest Mosaic [22] (White 1983) and to the south by subtropical evergreen and semi-evergreen bushland and thicket, it has affinities to both tropical and subtropical regions but also supports a high number of narrowly endemic species, including relict species.

Ecoregion Number:	**24**
Ecoregion Name:	**Knysna-Amatole Montane Forests**
Bioregion:	**Eastern and Southern Africa**
Biome:	**Tropical and Subtropical Moist Broadleaf Forests**
Political Units:	**South Africa**
Ecoregion Size:	**3,100 km²**
Biological Distinctiveness:	**Bioregionally Outstanding**
Conservation Status:	**Endangered**
Conservation Assessment:	**IV**
Author:	*Colleen Seymour*
Reviewed by:	*Jeremy J. Midgley*

Location and General Description

The Knysna-Amatole Montane Forests [24] ecoregion contains the southernmost Afromontane forest in Africa. It is divided into two distinct portions: the Knysna forest along the coast (568 km²) (Midgley et al. 1997) and the Amatole forests further inland (405 km²).

Much of the Knysna forest occurs on gentle to moderate slopes, ranging from 5 to 1,220 m above sea level, with a mean of 240 m. The forests of the Amatole Mountains are situated at higher altitudes, between 700 and 1,250 m (Geldenhuys 1989). The Knysna forest has a predominant geology of quartzite, shale, schist, conglomerate, and dune sand, whereas the Amatole Mountains have shale, sandstone, mudstone, and dolerite (Geldenhuys 1989). The soils of these forests are generally acidic and nutrient-poor (Van der Merwe 1998) but richer than adjacent fynbos areas.

Rain falls throughout the year, with maxima in early and late summer (Geldenhuys 1989). Mean daily maximum and minimum temperatures in the Knysna forest are 23.8°C in February (summer) and 18.2°C in August (winter), and in the Amatole forests the maxima and minima are 19.7°C and 8.9°C (Geldenhuys 1989). Annual rainfall varies between the sites as well, ranging from 525 to 1,220 mm in the Knysna forest and from 750 mm to 1,500 mm in the Amatole forests (Geldenhuys 1989). Rainfall appears to be the primary environmentally limiting factor of forest extent because forest is unable to persist in areas with rainfall of less than 500 mm (Rutherford and Westfall 1986; Geldenhuys 1989).

Although climatic factors appear to be responsible for large-scale forest distribution, small-scale patterns are determined primarily by fire (Midgley et al. 1997). Forest species readily invade neighboring fynbos when fire is excluded (Luger and Moll 1993). Isolated patches of fynbos in the Knysna forest are thought to have resulted from postglacial forest expansion (Geldenhuys 1989; Midgley and Bond 1990). Charcoal is often found in these forests (Scholtz 1983), suggesting that they do burn, although little is known of their fire regimes (Midgley et al. 1997).

Most trees here are of tropical origin, although members of an older local nontropical floral kingdom, such as the white (*Platylophus trifoliatus*) and red (*Cunonia capensis*) alders, are also successful (van der Merwe 1998). Among the more common trees in these forests are ironwood (*Olea capensis*), stinkwood (*Ocotea bullata*), Outeniqua yellowwood (*Podocarpus falcatus*), real yellowwood (*Podocarpus latifolius*), Cape holly (*Ilex mitis*), white pear (*Apodytes dimidiata*), Cape beech (*Rapanea melanophloeos*), bastard saffron (*Cassine peragua*), Cape plane (*Ochna arborea*), assegai tree (*Curtisia dentata*), and kamassi (*Gonioma kamassi*). Plants found in the understory include wild pomegranate (*Burchellia bubalina*), black witch-hazel (*Trichocladus clinitus*), seven weeks fern (*Rumohra adiantiformis*), and Cape primrose (*Streptocarpus* spp.).

Outstanding or Distinctive Biodiversity Features

The Knysna-Amatole Montane Forests [24] are part of the archipelago-like Afromontane regional center of plant endemism (White 1983). However, in this ecoregion tree endemism is low, and few plant species are rare (Midgley et al. 1997).

The most famous inhabitants of the Knysna forest are the remnants of the southernmost population of African elephant (*Loxodonta africana*, EN). Until recently, it was believed that only one elephant, an elderly cow, remained. The eradication of elephants has undoubtedly altered the natural processes in this ecosystem because they once caused many treefall gaps by destroying trees (Von Gadow 1973).

The ecoregion contains one strict endemic mammal species, the long-tailed forest shrew (*Myosorex longicaudatus*, VU), and one near-endemic mammal species, Duthie's golden mole (*Chlorotalpa duthieae*, VU) (Hilton-Taylor 2000). The leopard (*Panthera pardus*) is the largest predator of the Knysna forest, but intensive surveys have failed to find evidence of leopards, indicating that they may be extinct here.

Among the birds, the Knysna turaco (*Tauraco corythaix*), Knysna scrub-warbler (*Bradypterus sylvaticus*, VU), Knysna woodpecker (*Campethera notata*), chorister robin-chat (*Cossypha dichroa*), and forest canary (*Serinus scotops*) are all near endemic to this ecoregion (Barnes et al. 2001).

These forests are also home to the strictly endemic Knysna dwarf chameleon (*Bradypodion damarnum*) (Van der Merwe 1998). Amphibian species restricted mainly to the Knysna forest include the tree frog *Afrixalus knysnae*, the southern (Koyal) ghost frog (*Heleophryne regis*), and the plain rain frog (*Breviceps fuscus*) (Passmore and Carruthers 1995). The Amatole toad (*Bufo amatolicus*) and hogsback frog (*Anhydrophryne rattrayi*) are found mainly in the Amatole forests.

Status and Threats

Current Status

These forests have a long history of human inhabitation and use. Major exploitation of the Knysna forest began in the 1700s (Phillips 1931, 1963) and in 1891 in the Amatole forests. The settlers not only harvested timber but also cleared forest for crops and grazing (Geldenhuys 1989). In response to the continuing destruction, the forests were closed to exploitation in 1939 but were reopened in 1965 under controlled harvesting by forestry scientists (Von Breitenbach 1974).

Both publicly owned and privately owned forests are now in an advanced state of recovery from past timber exploitation. In the Knysna forest, timber and other economically important forest products are used conservatively and are collected from small, ecologically suitable areas of state forest (Milton 1987a, 1987b; Geldenhuys and Van der Merwe 1988). An innovative timber harvesting system, better known as a yield regulation system, involves selective harvesting of overmature trees in proportion to natural turnover rates (Seydack et al. 1995).

Public conservation and local authorities actively manage more than 70 percent of the total indigenous forest area in Knysna, and nearly 20 percent is conserved in proclaimed nature reserves and national parks (Van Dijk 1987; Geldenhuys 1989). In the Amatole region, it is estimated that about 90 percent of the forest area is under protection (Geldenhuys and MacDevette 1989).

Types and Severity of Threats

Although the forest patches are small and fragmented, their value to people is disproportionate to their size. Direct uses include timber for furniture and building, fuelwood, food, traditional medicines, home craft and decorative materials, hunting, recreation, tourism, and burial sites. Indirect uses include protection of water supply and soils in catchments and development of pharmaceutical products (Geldenhuys and MacDevette 1989). The principal large-scale disturbances today include exploitation for timber, clearing for agriculture, and fire (Geldenhuys 1989). Rural communities around the Amatole forests also collect fuelwood, medicinal plants, and construction materials. Alien plants, invasive animals such as ants, and forestry activities are also threats (Passmore and Carruthers 1995). For example, the spread of the aggressive Argentine ant (*Iridomyrmex humilis*) poses a serious threat to the swift moth (*Phalaena venus*).

Justification for Ecoregion Delineation

The Knysna-Amatole Montane Forests [24] form the southernmost extent of the Afromontane archipelago-like regional center of endemism (White 1983). This is the largest area of Afromontane forest in the region (Low and Rebelo 1996) and has several local endemic species.

Ecoregion Number: **25**

Ecoregion Name:	**Ethiopian Lower Montane Forests, Woodlands, and Bushlands**
Bioregion:	**Horn of Africa**
Biome:	**Tropical and Subtropical Moist Broadleaf Forests**
Ecoregion Size:	**248,800 km²**
Political Units:	**Ethiopia, Eritrea, Sudan, Djibouti, Somalia**
Biological Distinctiveness:	**Regionally Outstanding**
Conservation Status:	**Endangered**
Conservation Assessment:	**II**
Authors:	*Chris Magin, Miranda Mockrin*
Reviewed by:	*Ib Friis, Derek Yalden*

Location and General Description

The Ethiopian Lower Montane Forests, Woodlands, and Bushlands [25] ecoregion surrounds the highlands of Ethiopia and Eritrea, extending to the outlying massifs of Jebel Elba and Jebel Hadai Aweb, Jebel Ower near Port Sudan, and the Goda and Mabla massifs in Djibouti. Some authorities (e.g., Friis 1992) would also include the montane forest patches in the mountains of northern Somalia (currently in the Somali Montane Xeric Woodlands [102]).

The altitudinal limits of the ecoregion vary from one locality to another depending on annual precipitation but are generally between 1,100 and 1,800 m. From May to October, winds blow from the southwest and bring rainfall to the Ethiopian portion of the ecoregion. During the rest of the year, onshore winds from the Red Sea bring moisture to the Eritrean side of the mountains. Rainfall varies from 600 mm per annum in the driest sites to 2,300 mm in wetter areas. Humidity often is higher than would be expected, mainly because of cloud precipitation. Temperatures vary according to the season and elevation, but mean maxima lie between 18°C and 24°C, and mean minima are between 12°C and 15°C. Unlike in the moist equatorial mountains, the effects of cold in these dry highlands descend to lower altitudes.

Ancient Precambrian basement rocks form the substrate of the forests in southwestern Ethiopia and Eritrea and the woodlands and bushlands in deep river valleys. At higher altitudes Tertiary volcanic rocks are dominant. The topography is generally rugged, and soils are generally infertile. During the

Pliocene and Pleistocene there were major climatic fluctuations. During ice age periods glaciers formed on the peaks of the Ethiopian Highlands, with vegetation similar to Eurasian tundra at present lower altitudes.

Phytogeographically, the ecoregion is part of the Afromontane archipelago-like regional center of endemism (White 1983). The area supports East African evergreen and semi-evergreen forests, woodlands, and bushlands. At lower elevations, on the eastern slopes of the plateaus, and in the large river valleys where there is a rainshadow, woodland (*kolla*) habitats are dominated by *Terminalia, Commiphora, Boswellia,* and *Acacia* species. However, moister sites in southwest forest patches are dominated by tall trees, chiefly *Aningeria* and other Sapotaceae, species of Moraceae, and species of *Olea*. Transitional forests occur between 500 and 1,500 m in Welga, Ilubabor, and Kefa and have rainfall close to 2,000 mm per annum. These transitional forests change to Afromontane forests at approximately 1,500 m altitude in the southwest, where the rainfall is between 700 and 1,500 mm (Friis 1992).

The Afromontane forests include parts of the Harenna forest south of the Bale Mountains National Park, which is a type of forest that once covered a large part of Ethiopia and possibly Yemen. *Coffea arabica* is the dominant understory shrub, and wild coffee is still harvested. At the lower levels of the Harenna forest, the tall, closed canopy consists of *Warburgia ugandensis, Croton macrostachyus,* and *Syzygium guineense,* with emergent *Podocarpus falcatus*. At higher elevations, moist dense forest contains trees typical of eastern Africa, with *Aningeria* and *Olea* being dominant. In drier habitats and in higher locations the climatic zone is called *weyna dega* and is increasingly dominated by *Juniperus procera* (Friis 1992).

Outstanding or Distinctive Biodiversity Features

The main Ethiopian and Eritrean parts of this ecoregion support a variety of forest types with associated bushland and woodland habitats and consequently have high species richness and endemic species (Friis 1992; Lovett and Friis 1996; Friis et al. 2001). Plant and animal endemics are also found along the drier northeastern margins of the Ethiopian Highlands, which link to the mountains of northern Eritrea and Somalia as well as the Day forest in the Goda Massif in Djibouti. The small outlier in Djibouti is an important forest island in a sea of semi-desert, with at least four known endemic plant species (Magin 1999).

Four strict endemic birds occur: the Djibouti francolin (*Francolinus ochropectus,* CR), Harwood's francolin (*Pternistis harwoodi,* VU), Prince Ruspoli's turaco (*Tauraco ruspoli,* VU), and the yellow-throated seedeater (*Serinus flavigula,* VU) (Magin 2001a). Nonforest species found here are the Sidamo lark (*Heteromirafra sidamoensis,* EN), white-tailed swallow (*Hirundo megaensis,* VU), and Ethiopian bush-crow (*Zavattariornis strese-*

manni, VU). Other birds considered near endemic to the ecoregion include the dark-headed oriole (*Oriolus monacha*), Abyssinian catbird (*Parophasma galinieri*), and yellow-fronted parrot (*Poicephalus flavifrons*).

Mammals restricted to Ethiopia that occur in this ecoregion include the shrew *Crocidura harenna* (CR), the bat *Myotis scotti,* the narrow-footed woodland mouse (*Grammomys minnae*), and Menelik's bushbuck (*Tragelaphus scriptus meneliki*) (Yalden and Largen 1992). Other threatened mammals are lion (*Panthera leo,* VU) and Swayne's hartebeest (*Alcelaphus buselaphus swaynei,* EN). In the early 1900s, elephant (*Loxodonta africana,* EN), black rhinoceros (*Diceros bicornis,* CR), buffalo (*Syncerus caffer*), and oryx (*Oryx gazella*) were found in the Nechisar area, but have all been extirpated (Yalden et al. 1996).

Although accurately ascribing species of amphibian and reptile to this complex ecoregion has proven problematic, there are believed to be a number of strict and near-endemic species. Endemic reptiles include two strictly endemic chameleons, *Chamaeleo balebicornutus* and *Chamaeleo harennae*. There are also endemic amphibians, including tree frogs (*Afrixalus clarkeorum, A. enseticola, Leptopelis susanae, L. ragazzii, L. vannutelli*), ranid frogs (*Phrynobatrachus bottegi, Ptychadena harenna*), and caecilian (*Sylvacaecilia grandisonae*) (Largen 2001).

The extent to which current flora and fauna of this ecoregion survived in refugia during ice ages remains unknown. Studies by Friis et al. (2001) indicate that there are more plant endemics in open country habitats than in forests, which suggests that the forest fauna suffered more during the ice age periods. Yalden et al. (1996) draw attention to the poverty of forest mammal fauna in southwestern Ethiopia (about fifty species found further south in East Africa are apparently absent), and Dowsett (1986) shows that the forests have fewer endemic birds than the open habitats.

Status and Threats

Current Status

Because humans have intensively occupied the highlands of Ethiopia for thousands of years, it is difficult to assess how much of the ecoregion was formerly forested or supported woodland, wooded grassland, thicket, or forest mosaics. Remnant ancient trees in enclosed church compounds and cemeteries provide evidence that forest was previously much more widespread. A large portion of the ecoregion is now covered by farmland or secondary vegetation derived from agricultural or wood-harvesting activities (Friis 1992; Nievergelt et al. 1998). For example, 88 percent of the Day forest in Djibouti has been lost in the last two centuries, and more than 20 percent of the loss has occurred in the last 50 years (CNE 1991).

The largest areas of natural forest are found in the southwest, where rainfall is highest. Smaller areas of drier forest are also found to the north on the scarp slopes facing the Red Sea and Gulf of Aden. Nonforest habitats are found in areas of very high population density. Little habitat remains in its natural state, except in rocky ravines and other inaccessible areas.

The ecoregion is poorly protected, although some small areas are included in Ethiopian protected areas that primarily encompass other ecoregions. Patches of this ecoregion are contained in the Babile Elephant Sanctuary and the Awash, Omo, and Nechisar national parks. Many of these protected areas offer little protection to native flora and fauna (Yalden et al. 1996).

Types and Severity of Threats

All natural habitats in the ecoregion are highly threatened because they have been reduced to small patches, are severely fragmented, and are poorly protected. Agriculture is the main threat, coupled with exploitation of trees for fuelwood and timber.

In many areas poor agricultural methods and overgrazing have resulted in intense soil erosion. Cultivation, grazing, and firewood removal are all serious concerns in protected areas. Nechisar National Park is threatened by intensive natural resource use, fueled by the rapid growth in the nearby town of Arba Minch (Tilahun et al. 1996). The Harenna forest was previously a uniquely undisturbed habitat in Ethiopia, stretching without major human settlements from *Terminalia* woodland and low-altitude montane forest though high-altitude forest to Afroalpine moorland, crossing three ecoregions. It was traditionally used for gathering honey, coffee, and other forest products and for cattle grazing. An expanding urban population in this region now threatens this forest.

Justification for Ecoregion Delineation

This ecoregion is based on the "East African evergreen and semi-evergreen bushland" and "thicket and cultivation and secondary grassland replacing upland and montane forest" vegetation units mapped by White (1983). The ecoregion lies between 1,100 and 1,800 m in elevation. It contains endemic species, and its separation from other montane forests justifies its separation as an ecoregion. The tiny Day Forest in the Goda Massif in Djibouti has strong affinities with this ecoregion.

Ecoregion Number:	**26**
Ecoregion Name:	**Comoros Forests**
Bioregion:	**Madagascar–Indian Ocean**
Biome:	**Tropical and Subtropical Moist Broadleaf Forests**
Political Units:	**Federal Islamic Republic of the Comoros**
Ecoregion Size:	**2,100 km²**
Biological Distinctiveness:	**Globally Outstanding**
Conservation Status:	**Critical**
Conservation Assessment:	**I**
Author:	*Jan Schipper*
Reviewed by:	*Roger Safford, Elise Granek*

Location and General Description

The Comoros Forests [26] cover the Comoros Islands, which are located in the northern part of the Mozambique Channel about 300 km from northern Madagascar and about 300 km from the mainland of East Africa. The islands of Ngazidja (formerly Grande Comore, 1,146 km²), Mwali (formerly Mohéli, 211 km²), and Ndzuani (formerly Anjouan, 424 km²) make up the Federal Islamic Republic of the Comoros, and Mayotte (374 km²) is a French dependency.

The Comoros Islands are volcanic in origin, with Ngazidja formed during the Quaternary and Mwali and Ndzuani during the late Tertiary. Mount Karthala (2,361 m), an active volcano that erupts every 10 to 20 years, dominates Ngazidja. There is no surface water on this island because all rainfall percolates into the porous rock. The other islands are highly eroded, with numerous valleys and steep ridges.

The Comoros Islands have a maritime tropical climate. The rainy season is from November to April, when the predominant northerly winds off the Indian Ocean bring moist, warm air to the region. The average temperature during the wet season is 25°C, with temperatures reaching above 29°C in March, the hottest month. From May to September southerly winds dominate the region, bringing cooler (approximately 18°C) and drier air. The central, higher-elevation areas of an island are cooler and wetter than the coastal regions, with the driest coastal areas receiving around 1,000 mm rain per annum, whereas more than 5,000 mm per annum is recorded on some uplands. This climatic variation results in distinct microhabitats with correspondingly distinct flora and fauna. The islands are struck by cyclones about once every 10 years.

The human population of these islands is believed to be around 500,000 people, with up to 950 people/km² in some cultivable areas. Population growth rates are also high. At least 80 percent of the natural vegetation has already been destroyed

(Safford 2001a), and this clearance is still proceeding. The soil on these islands consists of laterite, which is rich in minerals but very poor in humus material, and is subject to extreme erosion.

In the past most of these islands were covered by forest. Semi-deciduous forests once covered the lowlands but are now largely cleared. Midelevation and montane evergreen moist forests occur above about 800 m altitude and range up to approximately 1,800 m in elevation on Mount Karthala. Some of the dominant species include *Ocotea comoriensis, Khaya comorensis,* and *Chrysophyllum boivinianum* (White 1983). Sparse herbaceous vegetation, including species such as *Nuxia pseydodentata, Breonia* sp., and *Winmannia* sp., grow on lava flows and cinder fields at the base of this active volcano. Major plant families include Sapotaceae, Ebenaceae, Rubiaceae, Myrtaceae, Clusiaceae, Lauraceae, Burseraceae, Euphorbiaceae, Sterculiaceae, Pittosporaceae, and Celastraceae (White 1983).

Outstanding or Distinctive Biodiversity Features

Of the approximately 1,000 native plant species, 30 percent are endemic to the Comoros (Adjanahoun et al. 1982; WCMC 1993; WWF and IUCN 1994; Moulaert 1998). Most of the flora of these islands has affinities with those of Africa and Madagascar; however, a small percentage is more closely related to that of Asia. There are also many introduced plant species.

The species richness of the fauna is low, although it is higher than that of most other Indian Ocean islands because of its proximity to both Madagascar and continental Africa. A high proportion of the fauna on the Comoros Islands is endemic.

Among the avifauna, sixteen species and twenty-seven subspecies are strictly endemic to the ecoregion (Louette 1988; Thibault and Guyot 1988; Adler 1994; Sinclair and Langrand 1998; Stattersfield et al. 1998; Safford 2001a, 2001c). Almost all of the endemic species are found in the forest areas or the montane heathlands. Of particular importance is Mount Karthala on Ngazidja, which has four strict endemic bird species: Comoro scops-owl (*Otus pauliani,* CR), Grande Comore flycatcher (*Humblotia flavirostris,* EN), Grande Comore drongo (*Dicrurus fuscipennis,* EN), and Mount Karthala white-eye (*Zosterops mouroniensis,* VU) (Safford 2001a). Three other bird species are threatened with extinction: the Mayotte drongo (*Dicrurus waldenii,* EN), Anjouan scops-owl (*Otus capnodes,* CR), and Mohéli scops-owl (*Otus moheliensis,* CR) (Hilton-Taylor 2000).

Of the native mammals present on these islands, two fruit bat species are strictly endemic (*Pteropus livingstonii,* CR, and *Rousettus obliviosus*), and there is one near-endemic bat subspecies (*Pteropus seychellensis comorensis*). Other mammals include the mongoose lemur (*Eulemur mongoz,* VU), introduced from Madagascar. Endemic reptiles include species of day gecko (*Phelsuma* spp.), shining-skink (*Cryptoblepharus* spp.), and chameleon (*Furcifer* sp.) (Henkel and Schmidt 2000). There are no endemic amphibian species. Among the eighty-three butterflies,

there are seventeen endemic species and fifteen endemic subspecies (Lewis et al. 1998), including the swallowtail butterflies *Papilio aristophontes* and *Graphium levassori* (Desegaulx de Nolet 1984; Turlin 1995).

Status and Threats

Current Status

The conservation status of forests in this ecoregion is poor. Little intact forest remains on Ndzuani and Mayotte, and much of the remaining forest on Mwali and Ngazidja is badly degraded except at higher elevations, where terrain is rugged or otherwise unproductive. The largest remaining block of forest is on the slopes of Mount Karthala on Ngazidja. On Ndzuani there are scattered fragments of native forest approximately 10 km² in extent. This provides the only remaining habitat for the surviving population of the Ndzuani scops owl (*Otus capnodes*) and the majority of habitat for Livingstone's fruit bat (*Pteropus livingstonii*). On Mayotte, forests still remain on mounts Sapéré, Bénara, and Choungi (Safford 2001c).

Although progress has been made in creating conservation laws on the Comoros, none of the remaining forest areas are effectively protected. There are just three protected areas in this ecoregion: the Saziley National Park on Mayotte, Lake Dziani Boudouni, a Ramsar wetlands site, and the Mwali Marine park, including the Mwali islets, both on Mwali. Creation of a Coelacanth Marine Park is in progress on the southern end of Ngazidja.

Types and Severity of Threats

The human population of the islands is high, with more than 500,000 residents on the four islands combined (with a population density greater than 330 people/km²). Population growth is around 3 percent per year and is putting increasing pressure on the forests. Little good agricultural land remains because of the rugged topography and porous volcanic soils. The majority of the human population of Grande Comoro and Mayotte is concentrated in the coastal lowland areas. However, on Ndzuani, and to a lesser extent Mwali, there are significant populations in mountain villages.

All mature forest habitats on the Comoros Islands are highly threatened by agricultural expansion. Deforestation or conversion of land along rivers further changes the microclimatic conditions in the forest ecosystem and may severely affect temperature-sensitive species including bats and reptiles. The regenerating forests on the recent lava flows of Mount Karthala are not being farmed because these lands are marginal for agriculture. Populations of other endemic species (e.g., several bird species) are in some cases critically small and extremely endangered (Stattersfield et al. 1998).

Current forest conservation measures are inadequate, and

there is a high risk that several species will go extinct without effective conservation of key forest patches. Wildlife exploitation includes poaching of green sea turtles for local consumption. Collection of day geckoes for the pet trade is a potential threat to these species. Currently, hunting of fruit bats is not a problem, but as food resources become scarcer with the growing population, the local taboos against eating bats may disappear.

The islands also suffer from introduced plant and animal species. Introduced carnivores include the small Indian mongoose (*Herpestes auropunctatus*), the Indian civet (*Viverricula indica*), feral domestic cats, and rats (*Rattus rattus*), which may have severe effects on bird populations. Introduced exotic plants, such as *Lantana camara,* outcompete or severely affect native plant species.

Cyclones and volcanic activity are major ecological processes that affect biodiversity. Cyclones bring strong winds, which can affect fruit bat populations, topple trees, and cause landslides. These effects are exacerbated by deforestation and logging that opens forest areas, increasing their exposure to wind. Volcanic activity affects biodiversity by covering intact habitat.

Justification for Ecoregion Delineation

The Comoros Islands were included with the African mainland in White's (1983) "coastal mosaic" vegetation unit. However, we regard these islands as a distinct ecoregion because they have been separated from both Madagascar and the African mainland for a long time, and possess a high level of endemism across all taxa.

Ecoregion Number:	**27**
Ecoregion Name:	**Granitic Seychelles Forest**
Bioregion:	**Madagascar–Indian Ocean**
Biome:	**Tropical and Subtropical Moist Broadleaf Forests**
Political Units:	**Republic of Seychelles**
Ecoregion Size:	**300 km²**
Biological Distinctiveness:	**Globally Outstanding**
Conservation Status:	**Critical**
Conservation Assessment:	**I**
Author:	*Winnie Roberts*
Reviewed by:	*Ross Wanless*

Location and General Description

The Seychelles Islands are the only midoceanic granitic islands in the world, located in the Indian Ocean, about 5° south of the equator and 930 km northeast of Madagascar. The 115 islands in the group can be divided into three types: ancient granitic islands, old volcanic islands, and young limestone islands. The forty an-

cient granitic islands (made of c750-million-year-old rocks) make up the core of this ecoregion and represent peaks rising from a largely submarine granite plateau, which separated from other parts of the Gondwana supercontinent around 65 million years ago. In addition to these granitic islands are others made up of old (65 million years) volcanic rocks (Silhouette and North Islands) and young raised coral reefs (e.g., Bird and Denis Islands).

Mahé, the largest and tallest island in the Seychelles (145 km², 914 m elevation), is typical of the granitic islands. A mountain ridge runs the length of the island. The lower regions have been extensively developed for residential and agricultural use, whereas the upper regions are still largely forested with a mixture of invasive species and patches of natural forest. The granitic islands generally have steep sides and impressive peaks, shaped by weathering and erosion. The erosion of the steepest inclines has produced large rocky outcrops, or glacis.

The islands experience a humid tropical climate with little seasonal variation in temperature. Heavy monsoon rains occur from November to February, and in the cooler and somewhat drier months (March to January) the trade winds blow steadily from the southeast. Mean annual rainfall varies with elevation and on the granitic islands ranges from 2,300 to 5,000 mm (Stoddart 1984).

The abundant rainfall and warm temperatures, along with soil enriched by guano, allowed lush palm forests to develop on portions of the islands. At elevations below 610 m, palms, pandans, and hardwoods characterize the remaining natural forests. Forest composition varies somewhat from island to island within the Seychelles, but common tree species include *Phoenicophorium borsigianum, Paraserianthes falcataria, Adenanthera pavonina, Morinda citrifolia, Phyllanthus casticum,* and coconut palms. In river valleys and marshes throughout the islands, various species of palms and screwpine (*Pandanus* spp.) were formerly abundant. Above 600 m are remnants of cloud forest, rich with tree ferns and mosses. Many of the specialist cloud forest plants are closely related to southeast Asian species, reflecting ancient biogeographic connections.

The first permanent human presence on these islands arrived in 1770. The human population of the granitic Seychelles is now around 80,000 people, of whom around 90 percent live on Mahé. The presence of people and introduced plants and animals on these islands has caused a dramatic loss of habitat and change to the natural ecology of most areas. Only tiny remnants of the natural vegetation remain on the Seychelles.

Outstanding or Distinctive Biodiversity Features

Because of their great age, extreme isolation, and varied habitats, the Seychelles support many endemic species in a variety of taxa. There are 1 endemic plant family, twelve endemic genera, and around 80 endemic species from a flora of about 233 native plants (White 1983; Procter 1984; Robertson 1989). The palms are a particularly unique group, with six endemics classified into six monotypic genera, including the imperiled, monotypic palm coco-de-mer (*Lodoicea maldivica*), which is restricted to the islands of Praslin and Curieuse (and a small introduced population on Silhouette). The pandans are also unusually diverse, with the granitic islands hosting eight species, of which five are endemic.

Many of the unique plants of the Seychelles have small populations and restricted distributions. For example, the jellyfish tree (*Medusagyne oppositifolia*) has a total population of fewer than thirty plants scattered over three hilltops on Mahé. It is the sole representative of the endemic family, Medusagynaceae, and is one of the rarest plant species in the world. Many other Seychelles endemic plants are also threatened with extinction.

The flora shows affinities with that of nearby islands, Madagascar and the Mascarenes, and with mainland Africa and Asia. The Seychellois species of *Impatiens, Pseuderanthemum,* and *Rothmannia* are more closely related to species found on the African continent than to species on Madagascar or the Mascarenes (WWF and IUCN 1994). Although Asia is twice as far away as mainland Africa, several species are found only on the Seychelles and in Indo-Malaysia and Polynesia. For example, *Amaracarpus pubescens* grows only in the Seychelles and in Java. These disjunct distribution patterns are thought to be relictual, reflecting the islands' geological history, when they were linked with Asia (Friedman 1994).

Unlike many other midoceanic island groups, the Seychelles support a fairly diverse herpetofauna. The most famous native species were the giant land tortoises, and although both original species are now extinct (*Dipsochelys arnoldi* and *D. hololissa*), tortoises from Aldabra atoll have been brought to the granitic Seychelles and are commonly kept as pets (Bourn et al. 1999). Giant tortoises were once abundant on both the granitic and coralline islands throughout the Seychelles, but the populations on the granitic Seychelles were decimated by human occupation and settlement. Some of the thirteen strictly endemic reptiles are *Calumma tigris, Lamprophis geometricus, Ailuronyx seychellensis,* and *Phelsuma astriata*. In the amphibians, there are seven species of caecilian and four species of frog (*Nesomantis thomasseti, Sooglossus gardineri, Sooglossus pipilodryas,* and *Sooglossus sechellensis*), all in the endemic family Sooglossidae (Rocamora and Skerrett 2001).

One endemic bat species, the Seychelles sheath-tailed bat (*Coleura seychellensis*, CR), and one endemic bat subspecies, the Seychelles giant fruit bat (*Pteropus seychellensis seychellensis*), occur on the Seychelles (Hilton-Taylor 2000).

The bird fauna of the Seychelles includes 48 species of resident landbirds and waterbirds, 18 breeding seabirds, 24 annual migrants, and 127 occasional visitors (Rocamora and Skerrett 2001). Two species, the Seychelles scops-owl (*Otus insularis*, CR) and the Seychelles paradise-flycatcher (*Terpsiphone corvine,* CR), are confined to single islands. Other endemics found on more than one island are the Seychelles kestrel (*Falco araea*, VU), Seychelles blue-pigeon (*Alectroenas pulcherrima*), Seychelles swiftlet

(*Collocalia elaphra,* VU), Seychelles bulbul (*Hypsipetes crassirostris*), Seychelles magpie-robin (*Copsychus sechellarum,* CR), Seychelles warbler (*Acrocephalus sechellensis,* VU), Seychelles white-eye (*Zosterops modestus,* CR), and Seychelles fody (*Foudia sechellarum,* VU). Two endemic subspecies of note are the black parrot (*Coracopsis nigra barklyi*) and Seychelles turtle-dove (*Streptopelia picturata rostrata*), although the latter probably has been made extinct through hybridization with the Madagascar turtle-dove (*Streptopelia picturata picturata*).

In the past, there were more endemic bird species on the Seychelles, but these have gone extinct (one probable species and three subspecies). The extreme rarity of some other taxa (Gerlach 1997; Stattersfield et al. 1998; Hilton-Taylor 2000) indicates that further extinctions are still possible (at least in the wild). Intensive species recovery programs, including reintroductions, have greatly assisted the survival of some birds. For example, the Seychelles warbler has increased from 26–29 birds on one island to more than 2,000 birds spread over three islands (Komdeur 1994). Similarly, the Seychelles magpie-robin increased from fewer than 20 birds on one island in the late 1980s to more than 100 individuals on three islands in 2002. In addition to endemic breeding birds, the Seychelles also support globally significant seabird breeding colonies of nine species (Rocamora and Skerrett 2001). In line with other groups, the invertebrate faunas are also rich in endemic species. However, little is known about most species' status or distribution.

Status and Threats

Current Status

Two centuries of human presence on the islands has led to widespread habitat loss, the introduction of exotic species, and habitat fragmentation. Replacement of original forests by plantations and exotic species is especially severe in the lowlands (up to 300 m). As forests were cut for plantations in the 1800s and early 1900s at middle elevations, erosion became a problem, although this was mitigated by control measures, including the planting of the invasive shrub *Chrysobalanus icaco*. Coconut, vanilla, and cinnamon plantations now occupy most of the coastal plateaus (Sauer 1967; Rocamora and Skerrett 2001). At higher altitudes some native forest remains. The Vallée de Mai on Praslin Island provides the best example of intact native palm forest and has been declared a World Heritage Site. National parks and preserves cover about 42 percent of the Seychelles Islands. The Morne Seychellois National Park is the largest terrestrial park (35 km^2) and contains important cloud forest. Other important reserves are the Aride Special Reserve (0.7 km^2), Cousin Special Reserve (0.3 km^2), La Digue Veuve Special Reserve (0.1 km^2), and Curieuse National Park (15 km^2). Although extremely small, the reserves protect critically endangered species and their habitats.

Types and Severity of Threats

Despite efforts to protect the flora and fauna of the Seychelles, there are still a number of threats to the native biota. Anthropogenic disturbance of native habitats is still a problem, and human-facilitated introduction of exotic species is a continuing threat. Alien species make up at least 57 percent of the total flora of the Seychelles (Procter 1984).

Introduced goats, pigs, and cattle inhibit regeneration of native forest, and introduced cats, dogs, common mynah (*Acridotheres tristis*), and tenrecs prey on native species, particularly birds, lizards, caecilians, and invertebrates. Introduced plants also outcompete the native vegetation and provide unsuitable habitat for the endemic animals. Besides habitat loss and alteration and introduced species, the main conservation concern in the Seychelles is the vulnerability of small populations with restricted ranges. Although many native species probably have always had small populations, the majority of them were spread over several islands. After human settlement, many species have been reduced to one or two relict populations. The endemic landbirds all occupy a mere fraction of their historic range, although translocations and reintroductions have helped improve the conservation status of several restricted populations.

Justification for Ecoregion Delineation

The millions of years of isolation of the granitic islands of the Seychelles and the exceptionally high levels of endemism in both flora and fauna warrant their classification as a unique ecoregion. Endemism at higher taxonomic levels is particularly notable, including the families Medusagynaceae and Sooglossidae.

Ecoregion Number:	**28**
Ecoregion Name:	**Mascarene Forests**
Bioregion:	**Madagascar–Indian Ocean**
Biome:	**Tropical and Subtropical Moist Broadleaf Forests**
Political Units:	**Republic of Mauritius, Réunion (France)**
Ecoregion Size:	**4,900 km^2**
Biological Distinctiveness:	**Globally Outstanding**
Conservation Status:	**Endangered**
Conservation Assessment:	**I**
Author:	*Jan Schipper*
Reviewed by:	*Anthony Cheke, John Mauremootoo*

Location and General Description

The Mascarene Islands of Réunion, Mauritius, and Rodrigues are a volcanic archipelago located 640 to 1,450 km east of Madagascar in the western Indian Ocean (Strahm 1996a). The largest islands are the French overseas *département* of Réunion (2,500 km^2) and the island of Mauritius (1,865 km^2), which together with Rodrigues (109 km^2), the Cargados Carajos Shoals or St. Brandon (3 km^2), and Agalega (21 km^2) form the Republic of Mauritius. The nearest large landmass is Madagascar, 680 km northwest of Réunion.

The rocks, soils, and topography of the three islands are strongly influenced by their volcanic histories and length of weathering since the last volcanic event (Montaggioni and Nativel 1988). Rodrigues is the oldest volcano, with low relief, rising to 390 m. Mauritius ceased to be active around 20,000 years ago, and its highest point, Piton de la Rivière Noire, reaches 828 m. Réunion is much younger, and its volcano, Piton de la Fournaise (2,631 m), is active several times each year. The topography is steep and rugged; the highest mountain, the Piton des Neiges, reaches 3,069 m. The island soils are predominantly weathered latosols in various stages of formation depending on the age of the volcanic parent material (Arlidge and Wong you Cheong 1975).

Temperatures along the coasts are warm and seasonal, averaging 27°C in the warm season (December–April) and around 22°C in the cold season (May–November). In the Mauritian uplands the temperature averages about 5°C cooler, and above 1,500 m in Réunion frosts are regular during the southern winter. In the lowlands of Mauritius, the average rainfall varies from 750 mm on the leeward side of the island to 2,400 mm on the southeast coast. In the uplands, the rainfall varies from 2,400 to 4,500 mm per annum. Réunion has higher rainfall, generally of 4,000–6,000 mm per annum on the eastern mountains and up to 10,000 mm in some places (Padya 1989; Soler 1997). The Mascarenes are also subject to intense cyclones.

People colonized these islands from 1638 onward. In 2000 the human population of Mauritius was estimated at 1.14 million (density 613 people/km^2) and Rodrigues at 35,800 (328 people/km^2). High densities also occur on Réunion, which had a population of 706,000 in 1999. Population growth rates are low.

The vegetation of these islands was originally diverse, ranging from mangroves (Mauritius only) and wetlands, through palm-rich woodland, lowland dry forest, and rainforest, to montane evergreen forests, and finally (on Réunion) to heathland vegetation types on the highest mountains (Vaughan and Wiehé 1937; Cadet 1977; Lorence 1978; Strahm 1996b). More than 90 percent of the original vegetation has been destroyed, and only tiny fragmented remnants remain. Major plant families include Sapotaceae, Ebenaceae, Rubiaceae, Myrtaceae, Clusiaceae, Lauraceae, Burseraceae, Euphorbiaceae, Sterculiaceae, Pittosporaceae, Sapindaceae, Celastraceae, Pandanaceae, and Arecaceae, with notable radiation of some families, especially on Mauritius.

Outstanding or Distinctive Biodiversity Features

These islands once hosted one of the richest flora and faunas of any oceanic archipelago. The flora contains approximately

955 species, of which approximately 695 are endemic (as are 38 genera) (WWF and IUCN 1994). Palms are particularly diverse, with five endemic genera, including the monotypic *Tectiphiala* and *Hyophorbe,* whose relatives are in South America. The flora's closest affinities are with Madagascar and to a lesser extent Africa, and then Asia.

The Mascarenes are well known for their exceptional ornithological importance (Barré 1988; Adler 1994; Safford 1997a; Safford and Jones 1998). Of the sixteen endemic birds on the islands, seven are confined to Mauritius, four to Réunion, and two to Rodrigues (Stattersfield et al. 1998; Safford 2001b). Many of the endemic species are also threatened with extinction: Rodrigues warbler (*Acrocephalus rodericanus,* EN), Rodrigues fody (*Foudia flavicans,* VU), Réunion cuckooshrike (*Coracina newtoni,* EN), Mauritius cuckooshrike (*Coracina typica,* VU), Mauritius kestrel (*Falco punctatus,* VU), Mauritius fody (*Foudia rubra,* CR), Mauritius black bulbul (*Hypsipetes olivaceus,* VU), Mauritius parakeet (*Psittacula eques,* CR), Mauritius olive white-eye (*Zosterops chloronothus,* EN), and Mauritius pink pigeon (*Columba mayeri,* EN). The endemic species are confined principally to the remaining forest patches, although some have started to colonize plantations (Barré 1988; Nichols et al. 2002). Elevation is important in explaining variations in endemism on these islands (Adler 1994; Sinclair and Langrand 1998).

Important colonies of breeding seabirds are also found. Mauritius has colonies of Trindade petrel (*Pterodroma arminjoniana,* VU), masked booby (*Sula dactylatra*), wedge-tailed shearwater (*Puffinus pacificus*), red-tailed tropicbird (*Phaethon rubricauda*), sooty tern (*Sterna fuscata*), white tern (*Gygis alba*), brown noddy (*Anous stolidus*), and lesser noddy (*A. tenuirostris*) (Safford 2001b). Barau's petrel (*Pterodroma baraui,* EN) and the endemic Mascarene black petrel (*Pterodroma aterrima,* CR) breed on the mainland of Réunion, as does Audubon's shearwater (*Puffinus lherminieri*) (Le Corre and Safford 2001).

Bats are the only native mammals and include the endemic greater Mascarene flying fox (*Pteropus niger,* VU) and the Rodrigues flying fox (*Pteropus rodricensis,* CR). Another similar fruit bat, the lesser Mascarene flying fox (*Pteropus subniger,* EX), is extinct (Cheke 1987).

Thirteen strictly endemic reptiles occur on these islands, but five species of giant tortoises, several geckos, and a giant skink have become extinct since 1600 (Cheke 1987). Several of the survivors are from the day-gecko genus, *Phelsuma* (Henkel and Schmidt 2000). Round Island, a small island off the northern tip of Mauritius, is particularly important for reptiles (Bullock 1986). Species include the Round Island boa (*Casarea dussumieri*), the recently extinct burrowing boa (*Bolyeria multocarinata*) in the endemic family Bolyeridae, the Round Island skink (*Leiolopisma telfairii*), and the Round Island day gecko (*Phelsuma guentheri*). There are no endemic amphibians. Although the invertebrates are less well known, there are numerous endemic species. Many landsnails are endemic to Mauritius; 30 percent of these have gone extinct, and another 30 percent are severely threatened (Griffiths 1996). Many of the endemic ants have also lost out to introduced competitors.

These islands have seen some of the highest rates of human-caused extinction in the world (Cheke 1987; Tonge 1989; Griffiths 1996; Henkel and Schmidt 2000). Extinctions include more than 50 percent of the bird fauna, including an entire family of birds: the Raphidae comprising the dodo (*Raphus cucullatus*) on Mauritius and the Rodrigues solitaire (*Pezophaps solitaria*) (Cheke 1987; Stattersfield et al. 1998; Mourer-Chauviré et al. 1999). Other birds have been rescued from extinction by intensive conservation work (Jones and Hartley 1995; Jones et al. 1995). Around half of the nonbird vertebrates (around twenty species) and around 30 percent of the endemic mollusks have also gone extinct on these islands (Safford 2001b). Furthermore, there have been extinctions of endemic plants, possibly totaling as many as 100 species (WWF and IUCN 1994). Plant species on the verge of extinction include *Drypetes caustica, Tetrataxis salicifolia, Hyophorbe amaricaulis,* and *Ramosmania rodriguesii* (Strahm 1989; Hilton-Taylor 2000).

Status and Threats

Current Status

There has been a huge loss of the original forest habitat on the Mascarene Islands (Cheke 1987; Safford 1997b). On Réunion, it is estimated that less than 25 percent of the island is covered with natural vegetation (around 650 km^2), much of this found at high elevations above the treeline. On Mauritius, only about 5 percent of the island contains any vestiges of natural vegetation, and less than 2 percent is close to natural (and even then invaded by alien plants). The largest patches of remaining habitat are found in the southwest around the Black River Gorge, in the southeast on the Bamboo Mountains, and in a few remnants in the northern mountain ranges. On Rodrigues, the severely degraded natural vegetation covers at most 1 percent of the total land area. Remaining habitat is restricted to small patches on the tops of hills and in (mostly dry) river valleys.

By virtue of lack of settlement and fewer introductions of invasive alien species, small offshore islets contain some of the best habitats. Round Island contains the last remnants of the palm-rich forest that once clothed much of northern Mauritius. It is also the only islet that has escaped invasion by woody species. Ile aux Aigrettes, a 26-ha islet less than 1 km from the southeast coast of Mauritius, contains the best remaining remnant of coastal ebony forest that used to surround much of the main island.

There are several small protected areas on Réunion. Two nature reserves, Mare Longue (68 ha) and the recently established

Roche Ecrite (3,600 ha), are well protected, and the latter protects the entire known population of the endemic *Coracina newtoni* (Le Corre and Safford 2001). Two protected biotopes have recently been established (2001), the islet of Petite Ile (2 ha) and mountainous region of Piton des Neiges and Grand Bénard (1,818 ha). Offering significantly less protection is the State Biological Reserves system, with six sites in total (131 km²) including Mazerin, Bébour, and Forêt des Hauts de St. Phillipe. Planning is in progress for the creation of a national park in Réunion. On Mauritius, there are also several protected areas. The largest is the Black River Gorges National Park (66 km²), which protects about 40 percent of the remaining native vegetation on the island. On the Rodrigues mainland there are only two protected areas (nature reserves comprising only about 20 ha), which include remnants of the natural vegetation. Some offshore islets around Mauritius and Rodrigues are protected as nature reserves.

Types and Severity of Threats

The habitats and species endemic to the Mascarene Islands are all under some degree of threat. Some species of plant are reduced to a single individual (WWF and IUCN 1994). On Mauritius, introduced herbivores, deer, pigs, and crab-eating macaques (*Macaca fascicularis*) and (also on Réunion) giant African land snails (*Achatina* spp.) lead to the destruction of habitat and endemic plant species. Introduced rats, cats, shrews, tenrecs, and mongooses (the last not on Réunion) prey on adult and young endemic animals, and introduced birds compete with endemic birds. Eighteen plant species have been identified as aggressive invaders in Mauritius (Strahm 1999). The tiny patches of lowland habitat are also destroyed for plantation forestry or (rarely) agricultural conversion. Endemic fruit bats are hunted for food. The endemic mollusks are threatened by introduced rats, house shrews (*Suncus murinus*), toads (*Bufo gutturalis*), and carnivorous snails (*Euglandina rosea*). Finally, the small populations of many endemic species make them vulnerable to cyclones and other catastrophic events.

Justification for Ecoregion Delineation

The Mascarene Islands are composed of three main islands: Réunion, Mauritius, and Rodrigues, and some islets. Although each island contains distinct flora and fauna, they were included in a single ecoregion because they share some endemic species, have a similar volcanic history, and are extremely isolated from other land masses.

Ecoregion Number:	**29**
Ecoregion Name:	**Madagascar Humid Forests**
Bioregion:	**Madagascar–Indian Ocean**
Biome:	**Tropical and Subtropical Moist Broadleaf Forests**
Political Units:	**Madagascar**
Ecoregion Size:	**112,400 km²**
Biological Distinctiveness:	**Globally Outstanding**
Conservation Status:	**Endangered**
Conservation Assessment:	**I**
Author:	*Helen Crowley*
Reviewed by:	*Steve Goodman, Achille Raselimanana, Frank Hawkins*

Location and General Description

The Madagascar Humid Forests [29] occupy the eastern side of Madagascar, including the littoral forests of the narrow coastal plain and up to the crest of the eastern escarpment around 1,200–1,600 m (punctuated by isolated massifs up to 2,000 m). The eastern escarpment traps humid air from the sea, forming a continuous cloud layer at 900–1,200 m. This moisture supports a band of forest from Andravory in the north to Andohalela in the south, with a narrow section along the Angavo Scarp extending just southwest of Lake Alaotra to the northeast of Antananarivo. At the northern edge of the ecoregion, around Vohémar, the humid forest changes to a transitional dry forest. At the ecoregion's southern limit, in the rainshadow of the Anosyennes Mountains, the humid forest changes over a short distance from a dry transitional forest to spiny forest.

Climate plays an important role in distinguishing the humid forests from the subhumid forests of the central highlands. The two most important factors are the distribution of rainfall and the length of the dry season. Rainfall generally exceeds 2,000 mm per year in the humid forests, and in some areas (such as the Masoala Peninsula) it can be as high as 6,000 mm per year. The dry season is less than 2 months long. Winter mists between May and September also characterize the humid forests. Cyclones occur in some years between December and March and can destroy habitat.

The geology of the eastern side of Madagascar is roughly homogeneous. Most of this ecoregion is underlain by metamorphic and igneous Precambrian basement rock, with patches of marble and quartz north of the Antongil Bay. There is also an eastern lowland belt of lava and a narrow coastal plain of unconsolidated sands (Du Puy and Moat 1996).

Plants from the following genera and families typify the humid forest: *Diospyros* (Ebenaceae), *Ocotea* (Lauraceae), *Symphonia* (Guttiferae), *Tambourissa* (Monimiaceae), *Dalbergia* (Leguminosae), *Leptolaena* (Sarcolaenaceae), *Weinmannia* (Cunoniaceae), and *Schefflera* (Araliaceae). The shrub and herb layers of the midelevation forests include Compositae, Rubiaceae, and Myrsinaceae. One distinctive feature of this forest is the diversity of *Pandanus* species (Pandanaceae), bamboos (Graminaceae), and epiphytic plants (Nicoll and Langrand 1989; Lowry et al. 1997). Amber Mountain (Montagne D'Ambre) contains a significant area of humid forest in the higher elevations, where there is a humid microclimate above 1,000 m. With increasing elevation the forests have decreasing stature, fewer straight unbranched and boled trees, less stratification, more epiphytes, more bryophytes and lichens, a better-developed and more diverse herb layer, and floristic changes (Lewis et al. 1996).

Outstanding or Distinctive Biodiversity Features

The forests of eastern Madagascar (including the Madagascar Humid Forests [29] and Madagascar Subhumid Forests [30]) are the most diverse habitats on this island and contain exceptionally high levels of endemism. As with many other areas of Madagascar, there is little detailed information on many species, and the number and distribution of species are continually changing as new research is completed. Amber Mountain also has both high floral and faunal diversity, with eight primate species and nearly eighty bird species, including the endemic Amber Mountain rock thrush (*Monticola erythronotus*).

Plant diversity and endemism are extremely high, with 82 percent specific endemism (Perrier de la Bathie 1936). For example, 97 percent of the 171 species of Malagasy palms (e.g., *Dypsis, Neophloga*) are endemic to the island, and the majority of these are found in the eastern forests (Dransfield and Beentje 1995). Orchids (Orchidaceae) are also diverse, with many species confined to these eastern forests. One of the most famous is the *Angreaceum sesquipedale*, with a 35-cm spur that has evolved in tandem with its pollinator, a species of sphinx moth with a 30-cm-long tongue (Jolly et al. 1984).

The forest mammal fauna includes all five families of Malagasy primates, seven endemic genera of Rodentia, six endemic genera of Carnivora, and several species of Chiroptera. Lemurs found here include the recently rediscovered hairy-eared dwarf lemur (*Allocebus trichotis*, EN), two subspecies of ruffed lemurs (*Varecia variegata variegata*, EN; *V. v. rubra*, EN), indri (*Indri indri*, EN), eastern woolly lemur (*Avahi laniger*), diademed sifaka (*Propithecus diadema diadema*, EN), Milne-Edwards's sifaka (*P. d. edwardsi*, EN), silky sifaka (*P. d. candidus*, EN), golden bamboo lemur (*Hapalemur aureus*, CR), greater bamboo lemur (*H. simus*, CR), white-collared lemur (*Eulemur fulvus albocollaris*), and collared lemur (*E. f. collaris*).

The region harbors the greatest diversity of birds in Madagascar. Of the 165 breeding species recorded in the eastern forests, 42 are endemic (Langrand 1990). Many of the species belong to the endemic families or subfamilies Mesitornithidae, Brachypteraciidae, Philepittinae, Vangidae, and Couinae. Some of the birds are extremely rare, and the Madagascar serpent ea-

gle (*Eutriorchis astur*, CR) and the Madagascar red owl (*Tyto soumagnei*, EN) have only recently been rediscovered (BirdLife International 2000). The snail-eating coua (*Coua delalandei*, extinct [EX]), a humid forest species from the endemic subfamily Couinae, was last recorded in 1834 and is almost certainly extinct. All specimens of this species with precise locality information come from the offshore island of Isle Sainte Marie.

The humid rainforest is home to numerous amphibians and reptiles species. About fifty reptile species and twenty-nine amphibian species are strictly endemic to the humid forest ecoregion (A. Raselimanana, pers. comm., 2001). These numbers will increase as new taxa are discovered and described. As an illustration, the following chameleon and dwarf chameleon forms are restricted to this ecoregion: *Calumma gallus, C. cucullata, C. furcifer balteatus, Furcifer bifidus, Brookesia superciliaris*, and *B. therezieni*. The ecoregion holds an assortment of locally endemic geckos: *Paroedura masobe, Ebenavia inunguis, Matoatoa spannringi, Phelsuma antanosy, P. serraticauda, P. masohoala, P. flavigularis, Microscalabotes bivittis, Uroplatus lineatus, U. malama*, and *Homopholis antongilensis*. A large number of skinks are found only in the ecoregion, such as *Amphiglosssus astrolabi* and *A. frontoparietalis*. Moreover, an assortment of snakes occurs in this ecoregion that are unknown from elsewhere on the island: *Pseudoxyrhopus sokosoko, P. tritaneatus, P. microps, P. heterurus, Micropisthodon ochraceus, Ithycyphus goudoti, I. pereneti, I. oursi*, and *Brygoophis coulagesi*. For amphibians, several species are confined to the humid forest: *Heterixalus alboguttatus, Mantella aurantianca, M. pulchra, M. laevigata, Mantidactylus argenteus, M. flavobruneus, M. klemmeri, M. microtympanum, M. microtis*, and *M. redimitus*.

Although animal and plant species change significantly with elevation, the transition often is gradual across the complete latitudinal range of eastern Madagascar (Goodman 1996, 1999, 2000). This is illustrated by the distribution among the subspecies of diademed sifaka (*Propithecus diadema*) in the humid forests, where the important delineators are the major rivers flowing from the central highlands to the east, such as the Mangoro and Mananara (Harcourt 1990; Mittermeier et al. 1994).

Status and Threats

Current Status

The forested habitats of this ecoregion have been heavily degraded and fragmented, and much has already disappeared. Little remains of the lowland forest between sea level and about 800 m, the zone that was separated as the Sambirano Domain by Perrier de la Bâthie (1936) and Humbert (1955). In fact, the littoral forests are among Madagascar's most degraded vegetation types, with little remaining undisturbed forest and patches that are too small to support viable populations of large lemurs (Ganzhorn et al. 2001). Deforestation has been going on for centuries, and there is intense pressure on the remaining forests.

Recent estimates of remaining forest cover vary with differences in methods (COEFOR/CI 1993; Nelson and Horning 1993; Du Puy and Moat 1996). Du Puy and Moat (1996) estimated that 33,000 km² of lowland forest and 32,500 km² of midelevation forest remains.

In only a few locations (e.g., Masoala Peninsula) do the forests extend continuously across a significant elevational gradient, starting at sea level. Most large blocks of forest are found in protected areas, such as Masoala National Park, Mananara Biosphere Reserve (including Verezanantsoro National Park), Ambatovaky Special Reserve, and Zahamena Integral Nature Reserve and National Park. Lowland forest outside of protected areas includes the forests near Rantabe in the northeast and the eastern slopes of the northern Vohimena Mountains in the southeast.

Types and Severity of Threats

The major threat to the Madagascar lowland forests is the shifting cultivation practice of *tavy*. The forest is cut and burned and the land used to grow crops such as manioc (cassava) and hill rice. After 2 or 3 years, the land is abandoned and regenerates into bushland and thicket. Ideally, after more than 10 years the land can be used again for agricultural crops. Because of increasing population pressure, fallow periods have become shorter. In some areas, the land has become degraded to the point where crops cannot be planted and then becomes secondary grassland or is invaded by species such as bracken (*Pteridium*). A reduction of the farming fallow period has led to lower crop yields and further pressure on the forests because they are the only source of new farmlands.

Secondary threats to the forest include unintentional burning from wildfires and legal and illegal commercial logging. In some areas, overexploitation of selected forest species for decorative purposes, such as palms and *Cyathea* tree ferns, can critically undermine the forest's integrity (Lowry et al. 1997). Given the established agricultural practices and increasing human population, damage to the natural resources of the remaining eastern forests will continue. The extinction of numerous narrowly endemic species over the next few decades cannot be discounted in this ecoregion.

Justification for Ecoregion Delineation

This ecoregion is based on a simplified derivation (Schatz 2000) of Cornet's (1974) humid bioclimatic division. This humid zone runs along the length of Madagascar's eastern coast, with the inland boundary delineated at the junction where the length of the dry season is between 2 and 3 months. This corresponds roughly to the crest of the eastern escarpment between 1,200 and 1,600 m. This ecoregion includes a narrow strip (1–5 km wide) of littoral forest that runs along portions of the coast and contains a high number of endemic species.

Ecoregion Number:	**30**
Ecoregion Name:	**Madagascar Subhumid Forests**
Bioregion:	**Madagascar–Indian Ocean**
Biome:	**Tropical and Subtropical Moist Broadleaf Forests**
Political Units:	**Madagascar**
Ecoregion Size:	**199,300 km²**
Biological Distinctiveness:	**Globally Outstanding**
Conservation Status:	**Critical**
Conservation Assessment:	**I**
Author:	*Helen Crowley*
Reviewed by:	*Steve Goodman, Achille Raselimanana, Frank Hawkins*

Location and General Description

The Madagascar Subhumid Forests [30] are located in the highlands of Madagascar, where natural habitat is now reduced to fragmented patches in a sea of anthropogenic grasslands and agriculture. Included in the ecoregion are some wetlands and lakes (e.g., Lake Alaotra), sclerophyllous forest, and tapia forest. To the east, the ecoregion grades into the Madagascar Humid Forests [29] at the crest of the eastern escarpment, which extends up to 1,600 m in elevation. The western boundary gently merges into the Madagascar Dry Deciduous Forests [33] around 600 m elevation. At higher elevations (generally above 1,800 m) the Madagascar Subhumid Forests ecoregion is replaced by ericoid thickets of the Madagascar Ericoid Thickets [84].

Although it is in the tropics, the ecoregion's altitude means that the climate is closer to temperate. Temperatures at higher elevations are between 15°C and 25°C. The cool, dry season occurs between July and September, and a warmer wet season occurs in the rest of the year. Rainfall is approximately 1,500 mm per year, although it may be as much as 2,000 mm in the Sambirano area in the northwest and as little as 600 mm in the southwest. The underlying geology is mainly of highly deformed Precambrian basement rocks, although there are a few areas with more recent lava flows and some alluvial deposits associated with wetlands (Du Puy and Moat 1996).

The largest remaining areas of subhumid forest habitat are found in the Sambirano region in the northwest, portions of Amber Mountain (Montagne d'Ambre), and areas of the northern highlands (sensu Carleton and Goodman 1998). A few remnant regions of humid forest, isolated and highly fragmented, also remain scattered across the central highlands (e.g., Ambohitantely, Ambohijanahary). The Sambirano Domain (sensu Humbert 1955) includes the Sambirano River valley, up to an elevation of 800 m at the foothills of Tsaratanana and Manongarivo Massifs, a portion of the northwestern coast between Baie d'Ambaro and Baie d'Ampasindava, and nearby offshore islands. This area is a transition zone between the western and the eastern regions of Madagascar (Gautier and Goodman 2002).

Southwest of the ecoregion and along the plateau lie several isolated blocks of forest, including Isalo, Analavelona, Makay, Ambohitantely, Ankaratra, and Ivohibe. They are depauperate compared with the eastern forests but do include some locally endemic species in sclerophyllous and fire-resistant vegetation types such as tapia. The largest intact areas of this habitat are found in the Isalo and Itremo massifs between 800 and 1,800 m altitude on sandstone and quartzite. Members of the families Sarcolaenaceae and Euphorbiaceae, including the fire-resistant *Uapaca bojeri* and the genus *Sarcolaena*, characterize these forests. There are also some endemic *Kalanchoe* and *Aloe* species. The ecoregion also contains a number of wetland areas, with very different species composition to the forests or grassland areas.

There is some debate regarding the degree to which the central highlands were formerly forested and the degrees to which humans have affected the fauna and flora (Burney 1997; Lowry et al. 1997). However, it is certainly true that vast grasslands, composed of alien or pantropical grass species, now cover most of the central highlands (Jolly et al. 1984; Du Puy and Moat 1996). Only three or four grass species cover huge areas, resulting in a landscape with extremely low species diversity and endemism. Several species of *Eucalyptus* and *Acacia* have also been introduced and are now the most common trees in the highlands. A few native fire-resistant trees persist, including the endemic palms *Bismarckia nobilis* and *Ravenala madagascariensis* and the tapia tree (*Uapaca bojeri*).

Outstanding or Distinctive Biological Features

The central plateau was once home to a remarkable array of endemic species. These included several species of elephant birds (Aepyornithidae), including the world's largest bird species (*Aepyornis maximus*), a giant tortoise, and several lemur species, most of which were large-bodied species, some larger than female gorillas today. All of these species have become extinct since the arrival of humans on the island around 2,000 years ago. Today most natural vegetation cover has been destroyed, and species are restricted to tiny pockets of habitat.

Numerous extant mammals are endemic to this ecoregion. These include the Alaotran gentle lemur (*Hapalemur griseus alaotrensis*), two other species of bamboo lemurs (*Hapalemur aureus,* CR; and *H. simus,* CR) (Harcourt 1990; Mittermeier et al. 1994), *Microgale nasoloi,* and a number of shrews, tenrecs, and rodents, such as the Malagasy mountain mouse (*Monticolomys koopmani*) (Garbutt 1999).

Two endemic bird species are found in the wetlands of this ecoregion, and others are confined to the subhumid forests or shared with other Madagascar ecoregions. In the wetlands, both the Alaotra little grebe (*Tachybaptus rufolavatus,* CR) and the Madagascar pochard (*Aythya innotata,* CR), have not been ob-

served recently and may be extinct (Hilton-Taylor 2000). In the forests a new genus and species was named only a few years ago, the cryptic warbler (*Cryptosylvicola randrianasoloi*). Other endemics include yellow-browed oxylabes (*Crossleyia xanthophrys*), the brown emu-tail (*Dromaeocercus brunneus*), and Appert's greenbul (*Phyllastrephus apperti,* VU). Bird species in this ecoregion that are limited to marshland habitats on Madagascar include the slender-billed flufftail (*Sarothrura watersi,* EN), Madagascar snipe (*Gallinago macrodactyla*), and Madagascar rail (*Rallus madagascariensis*) (Langrand 1990).

Many species of chameleon and dwarf chameleon occur only in this ecoregion, including *Calumma oshaughnessyi ambreensis, C. tsaratananensis, Furcifer petteri, Brookesia ambreensis, B. antakarana, B. lineata,* and *B. lolontany* in the northern and northwestern portion and *C. fallax, Furcifer campani,* and *F. minor* in the central and southern portions. Other lizard species endemic to the ecoregion include the skinks *Mabuya madagascariensis, M. nancycoutouae, Amphiglossus meva,* and *Androngo crenni,* the geckos *Lygodactylus blanci* and *Phelsuma klemmeri,* and the plated lizard *Zonosaurus ornatus.* There are also a few endemic snake species, including *Pseudoxyrhopus ankafinensis* and *Liopholidophis sexlineatus.* A few groups of amphibians also contain more than one endemic species, such as *Scaphiophryne goettliebi, Mantella cowani, Mantidactylus domerguei, Boophis laurenti,* and *B. microtympanum.*

Status and Threats

Current Status

The remaining natural habitats of the central highlands are extremely fragmented except for the zone spanning its eastern edge and the upper portion of the eastern escarpment. Some habitat is protected in the forests of Ambohijanahary and Ambohitantely and the eastern slopes of Andringitra. However, the degree to which the protected areas can maintain and manage the integrity of these habitats varies. Lack of resources, inadequate training, and limited personnel, in addition to the absence of clear management plans, all contribute to the difficulty of preventing habitat destruction in the reserves.

Types and Severity of Threats

Humans have modified nearly all of the ecoregion, either directly or indirectly (Lowry et al. 1997). Remaining patches of forest and woodlands of the central highlands face continuous and intensive pressure from encroaching agriculture, increasing exploitation by growing human populations, and fire. Introduced plants and animals are affecting the integrity of habitats.

The wetland areas on the central highlands are threatened by conversion to rice farming, siltation, and pollution. Most marshlands and wetlands of Madagascar have already been degraded or converted to rice cultivation. A few undisturbed areas remain, including the marshlands and associated forest of the Torotorofotsy area near Andasibe and other sites scattered throughout the ecoregion. Many of these marshlands are still polluted with runoff from surrounding agricultural lands. Extinctions of freshwater fish have occurred in these wetlands in the last few years (P. Loiselle, pers. comm., 2000). Lake Alaotra, in particular, is an important wetland, perhaps still supporting the endemic Alaotra little grebe and Madagascar pochard. However, these species may be extinct. The lake is also home to an endemic primate subspecies, the Alaotra gentle lemur (*Hapalemur griseus alaotrensis*), and several species of endemic fish. Lake Itasy used to be a significant habitat for waterbirds but has been degraded by intensive rice cultivation.

Justification for Ecoregion Delineation

This ecoregion occupies central Madagascar and is based on a simplification (Schatz 2000) of Cornet's subhumid bioclimatic division (Cornet 1974). It is distinguished from the humid forests to the east by the greater seasonal and daily variation in temperature and humidity and a longer dry season of 3–7 months. The eastern border corresponds roughly to the crest of the eastern escarpment between 1,200 and 1,600 m. Although the western boundary differs from Humbert's vegetation map (Humbert 1959), the 600-m contour better reflects climatic patterns that distinguish moist evergreen forest from dry deciduous forest. Seasonal mists in the central high plateau attenuate the length of the dry season to 3 or 4 months, compared with 7 months further west where there are no mists. The ecoregion also includes disjunct subhumid areas such as Mount Amber in the north and the Analavelona and Isalo massifs in the southwest.

Ecoregion Number:	**31**
Ecoregion Name:	**Cape Verde Islands Dry Forests**
Bioregion:	**African Palearctic**
Biome:	**Tropical and Subtropical Dry Broadleaf Forests**
Political Units:	**Republic of Cape Verde**
Ecoregion Size:	**4,600 km²**
Biological Distinctiveness:	**Regionally Outstanding**
Conservation Status:	**Critical**
Conservation Assessment:	**II**
Author:	*Jan Schipper*
Reviewed by:	*Cornelis J. Hazevoet*

Location and General Description

The Cape Verde Islands are an archipelago of ten islands and five islets located in the eastern Atlantic Ocean approximately

500 km from the coast of Senegal, West Africa (16°00′N, 24°00′W). These islands occur in two groups: the Barlavento, or windward islands in the north, and Sotavento, or leeward islands in the south. Size varies dramatically between islands, with Santiago (São Tiago, 991 km²) the largest and Raso (7 km²) among the smallest. The total land area for the archipelago is 4,033 km², scattered over 58,000 km² of ocean.

The archipelago is volcanic in origin and located in the southwestern portion of the Senegalese continental shelf. The landscape is rugged on the younger islands (Fogo, Santo Antão, Santiago, and São Nicolau), with peaks reaching more than 2,000 m (the highest mountain is the active volcano Mount Fogo, 2,829 m) but comparatively flat on the older islands (Maio, Sal, and Boa Vista). The major rocks are basalt and limestone, and there are deposits of salt and kaolin.

Cape Verde has a tropical climate with two seasons: a dry season from December to July and a warm and wet season between August and November. The mountainous islands receive significantly more rainfall than the lower, flatter islands. Some places have an annual rainfall around 1,200 mm, whereas others have basically no rainfall for years at a time. Temperatures range between 20°C and 40°C, and average between 25°C and 29°C. The volcanic soils are fertile, but most parts of the islands are too arid for agriculture. Periodically the islands suffer from prolonged droughts and serious water shortages.

In their virgin state these islands supported savanna or steppe vegetation on the lower parts of the islands, with an arid shrubland at higher altitudes. Much of the vegetation is drought-resistant. In only 500 years, the native biota of the Cape Verde Islands has been severely affected by human habitation. Native vegetation is fragmented and is confined largely to mountain peaks, steep slopes, and other inaccessible areas.

Outstanding or Distinctive Biodiversity Features

Out of a total flora of 621 species, 85 are endemic and a further 88 species have a wider distribution but are believed to be naturally part of the flora (Hazevoet 2001). Introduced species make up the bulk of the remainder, with some species that could be either native or introduced. Intense habitat loss and invasive alien plant species have resulted in many of the plant species being threatened with extinction (WWF and IUCN 1994).

Out of 171 bird species (Hazevoet et al. 1999), 4 species of landbird are endemic to these islands (Stattersfield et al. 1998), and there are a number of endemic forms that may be valid species (Hazevoet 1995). Two of the endemic bird species, the Iago sparrow (*Passer iagoensis*) and Alexander's swift (*Apus alexandri*), are widely distributed in these islands and occur on at least nine of the ten major islands. The remaining two species have a narrower distribution: the Raso lark (*Alauda razae*, CR) on Raso Island and the Cape Verde warbler (*Acrocephalus brevipennis*, EN) on Santiago Island and São Nicolau (Hazevoet et al. 1999). Historically the Cape Verde warbler also occurred on the island of

Brava, but there have been no records there since 1969, and its present status on that island remains uncertain.

The islands are also important for rare breeding seabirds. There is also a breeding population of the near-endemic Fea's (or Cape Verde) petrel (*Pterodroma feae*) (BirdLife International 2000). Other important breeding seabird populations are the magnificent frigatebird (*Fregata magnificens*), red-billed tropicbird (*Phaethon aethereus*), Cape Verde shearwater (*Calonectris edwardsii*), Cape Verde little shearwater (*Puffinus boydi*), and white-faced storm petrel (*Pelagodroma marina*).

There are no endemic mammals on the islands, although five species of bats occur: naked-rumped tomb bat (*Taphozous nudiventris*), Kuhl's pipistrelle (*Pipistrellus kuhlii*), Savi's pipistrelle (*Pipistrellus savii*), gray big-eared bat (*Plecotus austriacus*), and Schreibers's long-fingered bat (*Miniopterus schreibersi*) (Hazevoet 1995). The amphibian fauna is also extremely depauperate. Fifteen lizard species occur on Cape Verde, of which thirteen are endemic. These include a giant skink on Raso Island, *Macroscincus coctei,* and a giant gecko, *Tarentola gigas,* found on both Raso and Branco. The other endemics are five Mabuya skinks, three *Hemidactylus* lizards, and three *Tarentola* geckos.

Status and Threats

Current Status

In the 500 years since humans first colonized the islands, the loss of natural habitats has been severe. Remaining areas of natural habitat are confined to steep rocky areas and ravines in the mountainous islands and to patches in the flatter islands. None of these areas are protected.

Breeding seabirds have been greatly reduced in numbers and restricted to small islands by the combined effects of habitat loss and predation from introduced feral animals (e.g., cats, rats, and green monkeys). Human exploitation of wildlife resources has also had an effect; in particular, the eggs and nestlings of seabirds are a traditional source of food for the islanders. Recently some of the important habitats for breeding seabirds, typically small islets offshore of the coast of the main islands, have been declared protected areas, but there has been little action to protect the breeding seabirds at these sites, and killing seabirds for food continues unabated. Law 79/III/90 gave Raso Island the status of nature reserve, but the requirements of the law are ignored.

Types and Severity of Threats

The remaining habitats and their notable flora and fauna are all threatened by the activities of humans and the presence of introduced species. Threats include overgrazing by livestock, overfishing, improper land use that often results in extensive soil erosion, and the demand for wood, which has resulted in deforestation and desertification.

The introduction of exotic animals such as rats, sheep, goats, green monkeys, and cattle has devastated the native flora and fauna. Rats and other introduced mammals can ravage nesting areas of seabirds and wipe out entire colonies over time. Livestock is responsible for denuding soil, which results in extensive erosion, water loss, and compaction, which hinders native plant regeneration.

Justification for Ecoregion Delineation

The Cape Verde Islands lie 500 km off the western coast of Africa and are sufficiently distant and distinct to warrant their own ecoregion. Biogeographically they are part of Macronesia, together with the Canary Islands ecoregion (with which they share some affinities). The islands are home to a number of endemic plant and vertebrate species, particularly birds and reptiles.

Ecoregion Number:	**32**
Ecoregion Name:	**Zambezian *Cryptosepalum* Dry Forests**
Bioregion:	**Eastern and Southern Africa**
Biome:	**Tropical and Subtropical Dry Broadleaf Forests**
Political Units:	**Zambia, Angola**
Ecoregion Size:	**38,200 km²**
Biological Distinctiveness:	**Globally Outstanding**
Conservation Status:	**Relatively Stable**
Conservation Assessment:	**III**
Author:	Suzanne Vetter
Reviewed by:	Jonathan Timberlake

Location and General Description

The Zambezian *Cryptosepalum* Dry Forests [32] consist almost entirely of dense evergreen forest dominated by *Cryptosepalum exfoliatum* ssp. *pseudotaxus,* known locally as *mavunda.* It is found at 1,100–1,200 m elevation in the higher-rainfall areas of the Kalahari sands of northern Barotseland in western Zambia and marginally extends into Angola. The two main blocks of *Cryptosepalum* forest are found to the north and south of the Kabompo River, which forms a part of the Upper Zambezi River catchment and drains the ecoregion. To the west this ecoregion adjoins *Loudetia simplex*–dominated grassland, where seasonal waterlogging suppresses tree growth, whereas to the southwest it is bordered by the Barotse floodplain. These grasslands comprise the Western Zambezian Grassland [56] ecoregion. To the north, east, and west the ecoregion grades into miombo woodlands [49, 50].

The soils of the *Cryptosepalum* Dry Forests are derived from Kalahari sands and are nutrient poor and free draining. Soil depth, moisture storage, and the presence of calcium and hard-pans determine the composition of the vegetation cover. The lack of surface water and nutrient-poor soils has resulted in a low human population density (fewer than five people per square kilometer). The ecoregion has a tropical climate, with a maximum temperature of 30°C and a minimum of 7°C (Schulze and McGee 1978). Mean annual precipitation ranges from 800 to 1,200 mm. Three climatic seasons can be distinguished: a hot dry season from August to October, a hot wet season from November to April, and a cool dry season from May to July.

The Zambezian *Cryptosepalum* Dry Forests [32] form part of a larger complex of caesalpinoid woodland ecoregions that support wet and dry miombo, mopane, thicket, dry forests, *Baikiaea* woodland, and flooded grassland habitats, among others. The dominance of caesalpinoid trees is a defining feature of this bioregion (i.e., a complex of biogeographically related ecoregions), which has been called the Zambezian regional center of endemism by White (1983).

This *Cryptosepalum*-dominated dry forest is similar to Zambezian *Baikiaea* Woodland [51] in structure but floristically shares more affinities with the dry forests in the Zambezi Valley, which are characterized by *Combretum* and *Commiphora,* and also has affinities to the Itigi-Sumbu Thicket [48]. Although a few trees exceed 25 m, the height and structure of *Cryptosepalum* are its main distinguishing biological feature. There is a continuous canopy at 15–18 m, with a lower canopy of different species at 7–14 m. Beneath this lies a discontinuous shrub layer comprising a tangle of shrubs and lianas (Cottrell and Loveridge 1966; Fanshawe 1960). Other tall canopy trees include *Brachystegia spiciformis, Syzygium guineense* ssp. *afromontanum, Bersama abyssinica,* and *Erythrophleum africanum,* although the composition of these species and structure of the forest can vary by location (Cottrell and Loveridge 1966; Werger and Coetzee 1978). Common lianas include *Landolphia camptoloba, Combretum gossweileri, Uvaria angolensis,* and *Artabotrys monteiroae. Diospyros undabunda* is conspicuous where there is a dense thicket understory and epiphytic lichens are common. Grass is sparse on the forest floor, which is covered predominantly in mosses. After clearing, nearly impenetrable regeneration stages with numerous lianas, shrubs, and small trees develop (Cottrell and Loveridge 1966; Werger and Coetzee 1978). The sparse herbaceous undergrowth makes it difficult for fires to penetrate, and fire is not an important disturbance factor in mature forest stands. However, most plant species in the *Cryptosepalum* dry forest are killed by fire if they are burnt.

Outstanding or Distinctive Biodiversity Features

The ecoregion does not contain endemic plant species and is distinguished more by its structure than its floral composition. *Cryptosepalum* dry evergreen forests contain moderate vertebrate species richness and low, if any, endemism. Ansell (1960) states that none of the known mammals are endemic to the *Crypto-*

sepalum habitat, which shares most of its fauna with the surrounding *Brachystegia* woodland or the Guineo-Congolian forest. The only near-endemic mammal is Rosevear's striped grass mouse (*Lemniscomys roseveari*), known from two localities in the Zambezi District and Solwezi (van der Straeten 1980). The former is in this ecoregion, and the latter falls in the Central Zambezian Miombo Woodlands [50] ecoregion.

The *Cryptosepalum* forests provide habitat for larger mammals such as yellow-backed duiker (*Cephalophus silvicultor*), blue duiker (*C. monticola*), red river hog (*Potamochoerus porcus*), and greater kudu (*Tragelaphus strepsiceros*). *Cryptosepalum* forest is also used as cover by elephant (*Loxodonta africana,* EN) and buffalo (*Syncerus caffer*) from the Kabompo-Mwinilunga areas.

Cryptosepalum forests have a distinct and rich avifauna comprising moist evergreen forest species, *Brachystegia* woodland species, and widespread species. The highest species richness is found where local habitat disturbance (such as shifting cultivation) results in a mosaic of tree savanna, thicket, savanna woodland, and forest (Oatley 1969). Benson and Irwin (1965) regarded fifteen bird species as representative of *Cryptosepalum* dry evergreen forest (bird names have been updated according to Sibley and Monroe 1993). These are olive long-tailed cuckoo (*Cercococcyx olivinus*), Ross's turaco (*Musophaga rossae*), yellow-rumped tinkerbird (*Pogoniulus bilineatus*), Cabanis's greenbul (*Phyllastrephus cabanisi*), Boulton's batis (*Batis margaritae*), African crested-flycatcher (*Trochocercus cyanomelas*), square-tailed drongo (*Dicrurus ludwigii*), black-fronted bushshrike (*Telophorus nigrifrons*), Perrin's bushshrike (*T. viridis*), olive sunbird (*Nectarinia olivacea*), forest weaver (*Ploceus bicolor*), and black-tailed waxbill (*Estrilda perreini*). Many of these species are found in similar dry evergreen forest fragments elsewhere in Africa (Winterbottom 1978). Benson and Irvin (1965) and Oatley (1969) note that several birds coexist with closely related species in *Cryptosepalum* forests. Examples include chinspot batis (*Batis molitor*), Boulton's batis, and golden-tailed woodpecker (*Campethera abingoni*). This coexistence suggests that *Cryptosepalum* forest fragments were isolated during drier climate cycles and have more recently combined.

Few reptiles inhabit evergreen forest, and those that do are most likely to be encountered at the forest edge in clearings where the sun penetrates (Poynton and Broadley 1978). The reptile and amphibian fauna of the *Cryptosepalum* dry forests fall into a broad transition zone between the tropical fauna of much of Africa and the cape fauna of southwestern South Africa. Butterflies have also been listed, but only one species, *Charaxes manica,* is restricted to the *Cryptosepalum* forest (Cottrell and Loveridge 1966).

Status and Threats

Current Status

Although the ecoregion is small and fragmented by edaphic and climatic factors, much of the habitat is still in a natural, undisturbed state. This is largely because the absence of surface water in the *Cryptosepalum* forests prevents human settlement. In addition, *Cryptosepalum exfoliatum* is not an economically important timber species, the soils are nutrient poor and not suitable for agriculture, and the native vegetation is difficult to clear. The southern parts of the forests are used for fuelwood and timber by the inhabitants of the nearly treeless Barotse floodplain, but the impacts seem limited to a few localized parts of the forest margins (Collar and Stuart 1985).

Only one protected area, the West Liuwa National Park, falls within this ecoregion. It is surrounded in the east, north, and west by game management areas, which provide protection to the larger mammals in the form of hunting restrictions and also maintain the habitat.

Types and Severity of Threats

This ecoregion does not appear to be seriously threatened, although a lack of knowledge prevents a detailed assessment. Habitat fragmentation and destruction have not yet occurred on a large scale and are unlikely to happen in the short- to medium-term given the sparse (if growing) human population, lack of water, and poor agricultural potential.

Hunting is a general threat to wildlife in southwestern Zambia, even in protected areas. Lack of management, infrastructure, and funds in protected areas make poaching difficult to control, and although no data are available, West Liuwa is unlikely to be an exception. However, because of their remoteness and impenetrability, the *Cryptosepalum* forests probably are less heavily hunted than other, more open reserves in more populated areas.

Justification for Ecoregion Delineation

The boundaries for this ecoregion follow the "Zambezian dry evergreen forest" vegetation unit of White (1983). The underlying Kalahari sands determine the extent of *Cryptosepalum* forest. Further west and southwest the vegetation changes to the Western Zambezian Grasslands [56], where seasonal waterlogging suppresses tree growth. Elsewhere the *Cryptosepalum* vegetation grades into deciduous miombo and *Baikiaea* forests where there is a transition to different soils.

Ecoregion Number:	**33**
Ecoregion Name:	**Madagascar Dry Deciduous Forests**
Bioregion:	**Madagascar–Indian Ocean**
Biome:	**Tropical and Subtropical Dry Broadleaf Forests**
Political Units:	**Madagascar**
Ecoregion Size:	**152,100 km^2**
Biological Distinctiveness:	**Globally Outstanding**
Conservation Status:	**Endangered**
Conservation Assessment:	**I**
Author:	*Helen Crowley*
Reviewed by:	*Steve Goodman, Achille Raselimanana, Frank Hawkins*

Location and General Description

The Madagascar Dry Deciduous Forests [33] of western Madagascar are one of the world's richest and most distinctive tropical dry forests. The ecoregion occurs in two separate geographic regions. The first is on the western side of Madagascar from the Ampasindava Peninsula in the north to Belo-sur-Tsiribihina and Maromandia in the south. The second is in the northern part of the island, excluding Mount Amber (Montagne d'Ambre) above 1,000 m. This ecoregion is contiguous with the Madagascar Succulent Woodlands [114] in the southwest and the Madagascar Subhumid Forests [30] to the north and east; the latter limit largely coincides with the western edge of the central highlands around 600 m.

The elevation rises gradually from sea level at the western edge to around 600 m at the eastern edge, where it meets the margin of the central highlands. There is significant topographic variation, such as the limestone massifs of Ankarana, Namoroka, and Bemaraha and the volcanic cone of Mount Amber. The geology of the ecoregion is varied, being rather complex in some zones, and includes ancient Precambrian basement rocks, unconsolidated sands, and Tertiary and Mesozoic limestone (Du Puy and Moat 1996).

The climate of the ecoregion is tropical, with a mean maximum temperature between 30°C and 33°C and a mean minimum between 8°C and 21°C. The ecoregion occupies the rainshadow on the western side of the central highlands of Madagascar and has a long, pronounced dry season. Most of the rain falls from October to April, with an annual average of around 1,500 mm in the north to around 1,000 m in the south.

The habitats of the ecoregion are a mosaic of dry deciduous forest, degraded secondary forests, and grasslands. Before human colonization of Madagascar around 2,000 years ago, most of the area was covered with dry deciduous forest. The secondary grasslands are a result of frequent burning and are similar to those of the central highlands, with very low faunal and floral diversity, dominated by alien plant species. In contrast, the largely undisturbed dry deciduous forests of western Madagascar have a high diversity of endemic plant and animal species. The forest is essentially deciduous, with most trees losing their leaves during the dry season (May–October). The most luxuriant forests with the highest canopies (10–15 m) are found on the richer soils. Forests on sandy soils have shorter canopies (10–12 m), and those on calcareous rocks and soils are stunted (Nicoll and Langrand 1989).

Trees include the flamboyant tree (*Delonix regia*, family Leguminosae), several species of baobabs (*Adansonia* spp.), and *Pachypodium* spp. on the drier, calcareous soils of the west. The *tsingy* massifs support *Dalbergia* and *Cassia* spp. (Leguminosae), *Ficus* spp. (Moraceae), and *Adansonia madagascariensis*. In the scrub layer, there are diverse liana species of the Asclepiadaceae family and shrubs of the Leguminosae and Rubiaceae families. The herb layer is sparse, although some forests have a carpet of *Lissochilus* orchids. The *tsingy* plateau includes excellent habitat for drought-adapted succulents. Species found on Ankarana include *Pachypodium decaryi*, *Adenia neohumbertii* (family Passifloraceae), and several species of *Euphorbia*: *E. ankarensis*, *E. pachypodiodes*, and *E. neohumbertii* var. *neohumbertii* (Preston-Mafham 1991).

Outstanding or Distinctive Biodiversity Features

The Madagascar Dry Deciduous Forests [33] form a major part of the western center of endemism in Madagascar (White 1983). Although plant species diversity is not as high as in the moist eastern forests, levels of endemism are higher. White (1983) estimated generic and specific plant endemism at 20 and 70 percent, respectively. Succulence is rarely seen in the leaves but is more common in the main tissues. Bottle trees and bottle lianas are common, including those of the genus *Adenia* and the thorny long-necked bottles of *Pachypodium*. Pachycaul tree species include four species of *Adansonia*, several *Moringa*, and *Delonix hildebrandtii* (Guillaumet 1984; Rauh 1995). Some of the plants are threatened by extinction, including two of the six Malagasy endemic baobab species (*Adansonia grandidieri*, EN, and *A. suarezensis*, EN).

The fauna of the Madagascar Dry Deciduous Forests [33] has some overlap with that of the Madagascar Succulent Woodlands [114], but it is mostly distinct, endemic, and diverse.

Endemic mammal species include the golden-crowned sifaka (*Propithecus tattersalli*, CR), mongoose lemur (*Eulemur mongoz*, VU), western forest rat (*Nesomys lambertoni*), golden-brown mouse lemur (*Microcebus ravelobensis*, EN), northern rufous mouse lemur (*M. tavaratra*), western rufous mouse lemur (*M. myoxinus*, EN), Perrier's sifaka (*Propithecus diadema perrieri*, EN), Milne-Edwards's sportive lemur (*Lepilemur edwardsi*), and a species of forest mouse, *Macrotarsomys ingens* (CR). There is high turnover of lemur species, with five subspecies of *Propithecus*, three species of *Lepilemur*, and five species of *Microcebus*. The dry deciduous forests are one of the primary habitats for the island's largest predator, the fossa (*Cryptoprocta ferox*, EN), and some of the smaller endemic Carnivora.

The ecoregion contains important habitats for 131 of the 186 resident terrestrial bird species listed for Madagascar by Langrand (1990). Several of these species are associated with lakes and rivers of the region, such as the Manambolo, Betsiboka, Mahajamba, and their satellite lakes. These species include Bernier's teal (*Anas bernieri*, EN), Madagascar fish-eagle (*Haliaeetus vociferoides*, CR), Humblot's heron (*Ardea humblotii*), and Sakalava rail (*Amaurornis olivieri*, CR) (Stattersfield et al. 1998). Some of these species also use the fringes of the mangroves on the western coast of Madagascar. Endemic birds of the dry forests include Van Dam's vanga (*Xenopirostris damii*, EN).

A number of chameleon species are endemic to this ecoregion, including *Brookesia bonsi*, *B. decaryi*, *Furcifer tuzetae*, *F. rhinoceratus*, and *F. angeli*. The dwarf chameleons *Brookesia exarmata* and *B. perarmata* are endemic to the Tsingy de Bemaraha World Heritage Site. Several geckos are endemic, including *Paroedura maingoka*, *P. vazimba*, *P. tanjaka*, *Uroplatus guentheri*, and *Lygodactylus klemmeri*, the latter known only from Tsingy de Bemaraha. The region also holds several endemic skink species, including *Mabuya tandrefana*, *Pygomeles braconnieri*, and *Androngo elongatus*. New species of plated lizard have also been recently described from the ecoregion: *Zonosaurus bemaraha* in the southern portion and *Z. tsingy* in the northern portion (Raselimanana et al. 2000). The rivers and lakes are important habitats for the endemic Madagascar sideneck turtle (*Erymnochelys madagascariensis*, EN), and the scrubland and bamboo forests support the endemic ploughshare tortoise (*Geochelone yniphora*, EN).

Status and Threats

Current Status

A majority of the western forests of Madagascar have already been destroyed (Jenkins 1987). The dry forest habitat, as mapped by White (1983), shows that the main areas of intact forest are located near the western coast, and the areas closest to the central highlands are already composed of secondary grassland. More recent surveys indicate that the remaining forest is smaller than originally thought and ranges from 12,000 to 20,000 km^2 (Du Puy and Moat 1996). Most of this forest is in small fragments of 35 km^2 or less (Morris and Hawkins 1998). Many of the isolated blocks of remaining primary forest are found in a number of protected areas, but there are some important areas of dry forest habitat that are not protected. Protected areas include Ankarafantsika National Park (605 km^2) and Tsingy de Bemaraha World Heritage Site (1,520 km^2). Other special reserves include Tsingy de Namoroka (217 km²), Ankarana (182 km²), Analamerana (347 km²), Bemarivo (116 km²), Maningozo (79 km²), Ambohijanahary (600 km²), Manongarivo (352 km²), and Bora (48 km²).

Types and Severity of Threats

The major threat to the dry, deciduous forests is destruction and fragmentation through intentional burning to clear land for grazing and agricultural lands and through wildfires sparked by burning adjacent secondary grasslands. With an expanding rural population and increasing degradation of existing arable lands, the pressure on the remaining forest is extremely high. Selective logging and the removal of large trees pose additional threats of forest habitat degradation. It is likely that much of the remaining forest is already selectively logged and has lost the largest of its trees. These degraded forests do not support viable populations of at least seven of the eight lemur species found in more intact forests (Ganzhorn et al. 1999). Several species of diurnal lemurs are hunted for food, and this may be adversely affecting the regeneration of the forests (Ganzhorn et al. 1999).

River, wetland, and lake systems embedded in this ecoregion are threatened with siltation resulting from deforestation of adjacent forests and soil erosion and runoff from the central highlands. Lakes and wetland habitats are also being destroyed by rice paddy cultivation, overfishing, and invasive species (e.g., water hyacinth, *Eichhornia crassipes*).

Justification for Ecoregion Delineation

This ecoregion follows the dry bioclimatic zone of Cornet (1974) that extends inland to the 600-m contour. This is believed to provide a better reflection of original vegetation than the 950-m contour that Humbert and Cours Darne (1965) used to define their Western Domain phytochoria. This ecoregion has its southern limit in the Belo-sur-Tsiribihina and Maromandia region and extends to the northern portion of the island (not including Mount Amber). There is an eastern extension of this region running from the Ampasindava Peninsula to the Vohémar area. The southern border of this ecoregion is still under discussion. The division between Madagascar Succulent Woodlands [114] and Madagascar Dry Deciduous Forests [33] is regarded by some as reflecting "fundamental biogeographic patterns"(Lowry et al. 1997). However, other schemes have treated these two ecoregions as a single, "western" domain (Humbert and Cours Darne 1965; see Lowry et al. 1997).

Ecoregion Number:	**34**
Ecoregion Name:	**Mediterranean Conifer and Mixed Forests**
Bioregion:	**African Palearctic**
Biome:	**Temperate Coniferous Forests**
Political Units:	**Morocco, Algeria, Tunisia**
Ecoregion Size:	**23,100 km² (in Africa)**
Biological Distinctiveness:	**Globally Outstanding**
Conservation Status:	**Critical**
Conservation Assessment:	**I**
Authors:	*Nora Berrahmouni, Pedro Regato*
Reviewed by:	*Abdelmalek Benabid, Hans Peter Müller*

Location and General Description

The Mediterranean Conifer and Mixed Forests [34] comprise relict stands of fir and pine forest in the humid, medium to high elevations of major mountain massifs in North Africa (Ozenda 1975; Dallman 1998) and in the southernmost mountains of Cádiz and Málaga in Spain. In Morocco, this forest ecoregion can be found on the Rif, with its highest elevation at 2,448 m on Mount Tidirhin, and in the Middle Atlas, with its highest elevation at 3,340 m on Mount Bou Naceur. In northern Algeria, examples of this ecoregion occur in the Tellien Atlas and the Saharan Atlas Mountain Ranges at about 2,000 m. In northwestern Tunisia, the ecoregion occurs in the Kroumerie and Mogod Mountain Ranges at around 1,150 m.

The North African Mountains have a complex lithological composition, including sandstone, dolomite, limestone and marl, quartzite and schist, and igneous granites and volcanic formations. The landforms are steep and rugged mountain summits and high plateaus ranging from about 500 to 1,200 m. This particular landform pattern, together with the distance of the Middle Atlas and Saharan Atlas from sea influence, creates an intense continental gradient. Sharp changes in microclimates occur between the northern humid slopes and the cold, dry southern slopes (Charco 1999). The ecoregion receives an average annual rainfall of 1,000 mm, but at high elevations average annual rainfall can range from 1,600 to 2,200 mm. Snow falls frequently in winter, when average minimum temperatures drop below 0°C (Djellouli 1990; INRAT 1967).

Two forest zones are found; the conifer zone includes the higher elevations from 1,200 to 2,500 m and a mixed broadleaf zone, which includes the lowland and medium elevation from sea level to about 1,500 m (Quézel 1983). The dominant canopy tree species of the montane conifer forests is the endemic Atlas cedar (*Cedrus atlantica*), which normally constitutes mixed stands with the evergreen holm oak (*Quercus ilex ballota*) and less frequently with deciduous oak species (*Quercus faginea, Q. canariensis*). The Atlas cedar forests extend over 150 km² in the

Rif Mountains and 1,000 km² in the Middle Atlas (Benabid and Fennane 1994). In the Tellien Atlas, they form scattered populations over 300 km² of land. Fir forests (*Abies numidica* in the Algerian mountains Djebel Babor and Tababort and *Abies marocana* in the Moroccan Rif range) and pine forests (*Pinus nigra mauretanica* in the Moroccan Rif and the Algerian Djurdjura and *Pinus pinaster hamiltonii* var. *Maghrebiana* in the Rif and Middle Atlas) cover only few thousand hectares. Another fir species, *Abies pinsapo,* closely related to the Moroccan fir, characterizes three relict forest areas in South Spain (the Sierras of Grazalema, Sierra de las Nieves, and Bermeja), which are similar to the North African fir forests (Barbéro and Quézel 1975; Franco 1986; Charco 1999). Above the timberline, at 2,700 m in Morocco and 2,100 m in Algeria, the vegetation is placed in the Mediterranean High Atlas Juniper Steppe [69].

Broadleaf mixed oak forests dominate moist positions at medium and low elevations of the Rif, Tellien Atlas, and Kroumerie-Mogod mountain ranges. These forests are dominated by different oak tree species: *Quercus canariensis* occupies large areas of the Tellien Atlas and Kroumerie Mountains, *Quercus faginea* and *Q. pyrenaica* forests occupy about 250 km² in a few areas of the Rif Mountains, and *Quercus afares* and *Quercus lusitanica* occupy small areas. *Pinus pinaster hamiltonii* var. *maghrebiana* characterizes diverse mixed oak and pine forest stands, which spread in the low elevations of the Algerian coastal Kabilye in a humid and warm climate (Charco 1999; Mediouni 2000; WWF MedPO 2001).

Outstanding or Distinctive Biodiversity Features

Flowering plant endemism is more than 20 percent on the main mountain ranges: the Rif, the Middle Atlas, and the Tellien Atlas. The Rif Massif has at least 190 endemic plants, the Middle Atlas Range has 237 endemics, and the Tellien Atlas contains 91 endemics. The high rate of endemism is partly the result of the long isolation of Holarctic taxa in the high elevations of these North African mountain ranges.

All pine, cedar, and fir species that characterize the North African conifer forests are endemic to the ecoregion. There are also endemic oak species, such as *Quercus afares* in the Kabylie Mountains in Algeria and the Krumerie Mountains in Tunisia. Numerous endemic plants are found in the forest understory, including *Paeonia maroccana, P. corallina, Doronicum atlanticum, Calamintha baborensis, Geranium malviflorum, Rubia laevis, Scilla hispanica* ssp. *algeriensis, Crataegus laciniata, Cotoneaster atlanticus, Cytisus megalanthus,* and *Argyrocytisus battandieri* (Mediouni and Boutemine 1988; Mediouni and Yahi 1989; Charco 1999). There is a high number of threatened plants in this ecoregion; for example, Morocco has 186 species, Algeria has 141 species, and Tunisia has 24 species (Walter and Gillett 1998; Hilton-Taylor 2000). *Abies maroccana,* with a total range of 40 km² in the Rif, is an endangered tree species that should also be included on the IUCN Red List. The Atlas cedar forests also con-

tain Palearctic tree and shrub species at their southernmost limits, including *Taxus baccata, Sorbus aria, S. torminalis, Acer opalus granatensis, A. campestre, A. monspessulanum, Prunus mahaleb, P. insititia, Ilex aquifolium, Betula pendula fontqueri, Populus tremula, Lonicera etrusca,* and *L. arborea.*

The faunal diversity is among the most significant of the Palearctic realm. This was the last refuge of the Atlas lion (*Panthera leo leo*), which survived in the wild until 1930. Rare and endemic mammals still present include the Barbary macaque (*Macaca sylvanus,* VU) and the last individuals of the Barbary leopard (*Panthera pardus panthera,* CR) (Hilton-Taylor 2000). The *Quercus canariensis* and *Quercus suber* forests of the Tellien Atlas and Kroumerie Mountains host the last existing populations of the only African endemic deer species, *Cervus elaphus barbarus.* These oak forests are also the last refuge for the serval (*Leptailurus serval*), which has been almost extirpated in the Mediterranean region. Other species include Palearctic and North African taxa such as the Maghrebian wild cat (*Felis silvestris libyca*), red fox (*Vulpes vulpes*), golden jackal (*Canis aureus*), wild boar (*Sus scrofa*), crested porcupine (*Hystrix cristata*), common genet (*Genetta genetta*), North African long-tailed wood mouse (*Apodemus sylvaticus hayi*), and European polecat (*Mustela putorius*) (Nowell et al. 1996; Blondel and Aronson 1999; Charco 1999).

The moist conifer and broadleaf mixed forest ecosystems contain one endemic bird, the Algerian nuthatch (*Sitta ledanti,* EN), which is found in Djebel Babor and in a few other Algerian and Tunisian forest areas (Stattersfield et al. 1998). According to some bird taxonomists, Levaillant's woodpecker (*Picus vaillantii*) is also an endemic species, although others regard it as a subspecies of green woodpecker (*Picus viridis levaillanti*). Some Palearctic birds have their southernmost distribution in these forests, including the lesser spotted woodpecker (*Dendrocopos minor*) and the great spotted woodpecker (*Dendrocopos major*). The Rif and Algeciras mountains on both sides of the Gibraltar Strait and the Middle Atlas Mountain lakes are crucial for large-scale bird migrations to and from northern Europe to Africa (Fishpool and Evans 2001).

Endemic reptiles also occur, including Lataste's lizard (*Lacerta pater*), Koelliker's glass lizard (*Ophisaurus koellikeri*), and mountain viper (*Vipera monticola*). Palearctic and North African species of reptile and amphibian also occur, such as the viperine snake (*Natrix maura*), Algerian sand lizard (*Psammodromus algirus*), eyed lizard (*Lacerta lepida*), European fire salamander (*Salamandra salamandra*), olive midwife toad (*Alytes obstetricans*), painted frog (*Discoglossus pictus*), and Mediterranean tree frog (*Hyla meridionalis*) (Blondel and Aronson 1999).

Status and Threats

Current Status

Large forest stands occurred in Roman times (reported by Strabo, Herodotus, and Pliny) and in medieval times (reported by Leo

Africanus). The medieval forests provided enough timber for Fez to serve as a major lumber center until the twelfth century. Pollen evidence indicates clearing and widespread deforestation from about 1600 to 1900, presumably because of the advent of livestock raising and settlement in the mountains. Statistics indicate rapid deforestation in the twentieth century. A quarter of Morocco's forests (10,000 km²) vanished between 1940 and 1982. Algeria suffered huge deforestation during its colonial period. Forest cover in the Tunisian Mountains shrank by one-third between 1919 and 1970 (Brandt and Thornes 1996). These countries also lost 75 percent of their cedar forests between 1956 and 1971: 3,300 km² in Morocco and about 1,000 km² in Algeria.

Protected areas include El Feija National Park in Tunisia, Chrea National Park, Djurdjura National Park and biosphere reserve in Algeria, and Ifrane Forest Reserve and Tazekka National Park in Morocco. All three countries intend to increase their protected area networks. In Morocco 154 natural sites have been identified and designated as candidate new protected areas (Ministère de l'Agriculture et de la Mise en Valeur Agricole 1992). The Algerian government also has plans to enlarge its protected area network and to classify a number of key areas, such as Djebel Babor, as nature reserves. In addition, a biodiversity strategy and action plan have been prepared, with attention to the design and management of protected areas in the country for the coming years.

Types and Severity of Threats

Human impact on the habitats of this ecoregion has been severe and remains intense on the remaining forest patches. The collapse of the seminomadic Berber pastoral system has transformed summer camps in the high mountain grasslands into permanent human settlements. A large amount of firewood is collected, in many cases illegally. This has led to intense pressure on cedar trees, in particular. Conversion to agriculture is another important human impact in the broadleaf forests (Thirgood 1981; Médail and Quézel 1997). The need for livestock fodder in winter gives rise to extensive overgrazing in the forests that also causes soil degradation.

Threats in the forested Djebel Babor Nature Reserve (Atlas Tellien, Algeria) and Talassemtane Reserve (Rif, Morocco) consist mostly of fires, overgrazing, and illegal logging. In Djebel Babor, the number of *Abies numidica* trees has decreased by half since the 1950s, and in the Tazaot area in the Rif Mountains a 1977 forest fire destroyed almost all of the *Abies maroccana* var. *tazzaota* forest stand. The threat from fire is still significant, and plant regeneration is still seriously affected by continuing pressure from domestic livestock grazing (WWF MedPO 2001). Human impacts have reduced the forest's resilience to natural disturbances (Gómez-Campo 1985). During periods of intense drought, cedar forest stands can become very dry and prone to fires.

The rural population in this ecoregion is high and still growing in and around protected areas. Tunisian national parks are fenced areas where human presence is completely forbidden, and people's rights of use are not clearly established, resulting in many illegal activities such as logging and overgrazing. Political instability on the countries of the ecoregion complicates conservation activities.

Justification for Ecoregion Delineation

This ecoregion follows the vegetation unit that White (1983) defines as "Mediterranean montane forest and altimontane shrubland," although the altimontane area (above 3,200 m) was extracted as a separate ecoregion [70]. Lower-elevation areas from Tunisia are included because their mixed deciduous and evergreen oak forests were once representative of this ecoregion. The high levels of endemics and relict taxa, particularly among plants, make this ecoregion distinct.

Ecoregion Number:	**35**
Ecoregion Name:	**Sahelian *Acacia* Savanna**
Bioregion:	**Western Africa and Sahel**
Biome:	**Tropical and Subtropical Grasslands, Savannas, Shrublands, and Woodlands**
Political Units:	**Senegal, Mauritania, Mali, Niger, Nigeria, Cameroon, Chad, Central African Republic, Sudan, Eritrea, Ethiopia, Burkina Faso**
Ecoregion Size:	**3,053,600 km²**
Biological Distinctiveness:	**Bioregionally Outstanding**
Conservation Status:	**Relatively Stable**
Conservation Assessment:	**V**
Author:	*Chris Magin*
Reviewed by:	*Chris Duvall*

Location and General Description

The Sahelian *Acacia* Savanna [35] stretches across Africa from northern Senegal and Mauritania on the Atlantic coast to Sudan and Eritrea on the Red Sea, varying in width from several hundred to more than a thousand kilometers. Commonly called the Sahel (a word meaning "shore" in Arabic, signifying its geographic relationship to the "sea" of the Sahara), this ecoregion comprises a grassland-dominated transition zone between savanna woodlands to the south and the Sahara Desert to the north.

Most of the Sahel ranges between 200 and 400 m in altitude, generally rising from the lowlands near the Atlantic coast to the edges of the Ethiopian Highlands, although there are also a few massifs more than 3,000 m high. The upper elevations of these massis have been assigned to the West Saharan Montane Xeric Woodlands [96].

The Sahelian climate is tropical, hot, and strongly seasonal. The monthly mean maximum temperatures vary from 33°C to 36°C, and monthly mean minimum temperatures are between 18°C and 21°C (Leroux 2001). Large average variations in precipitation are characteristic, with frequent droughts (Ellis and Galvin 1994). Average annual rainfall is around 600 mm in the south but declines to around 200 mm in the north, with most rain falling between June and October (Leroux 2001). During the 7- to 9-month dry season, most woody species lose their leaves, and herbaceous vegetation dries up (Breman and Kessler 1995). This pattern leads to the long-distance, north-south movements of livestock and, formerly, the wildlife that once characterized the Sahel (Ellis and Galvin 1994; Raynaut 1997). In January and February, the Harmattan wind brings dry, hot air laden with dust from the Sahara.

The Sahel overlies a complex mosaic of various Precambrian to Quaternary deposits (Gritzner 1988). Lake deposits are also found, indicating the positions of lakes in the wetter parts of the late Pleistocene and early Holocene (Petit-Maire and Riser 1983), especially around Lake Chad [61] and the Inner Niger Delta [62] (Lézine 1989). The soils of the ecoregion are mainly entisols and aridisols, and most are sandy and highly permeable, so that permanent surface water is rare. Soil fertility is low, limiting vegetation development in most areas (van Keulen and Breman 1990). Human population density is also generally low but rather patchy, ranging from fewer than 5 people/km^2 in the north to 50–100 people/km^2 around larger cities in the south and around some water sources (Raynaut 1997).

The Sahelian *Acacia* Savanna [35] falls mainly in White's (1983) Sahelian regional transition zone, although part of its southern margin lies in the Sudanian regional center of endemism. Wooded grassland is widespread on sandy soils in the southern Sahel, with many thorny shrubs and small trees, including many *Acacia* and several *Ziziphus* species, *Commiphora africana*, *Balanites aegyptiaca*, and *Boscia senegalensis*. Grass cover is continuous but often dominated by short annual species such as *Aristida mutabilis*, *Chloris prieurii*, and *Cenchrus biflorus*. In the northern Sahel, short grasslands grow on deep, sandy soils, with widely dispersed shrubs of *Acacia*, *Boscia senegalensis*, and *Calotropis procera*. Although most grasslands are dominated by annual species, some perennial species may be present, including *Panicum turgidum*, *Cymbopogon schoenanthus*, and *Aristida sieberana*.

Outstanding or Distinctive Biodiversity Features

The Sahelian *Acacia* Savanna [35] is not especially species rich; most characteristic taxa are also found in neighboring ecoregions, reflecting the Sahel's transitional character.

Most plant species in the Sahel are widespread and fairly common (Geerling 1985). However, there are a number of endemic plants, most widespread and common, such as *Indigofera*

senegalensis and *Panicum laetum,* but others are highly localized and rare, such as *Andrachne gruveli* and *Barleria schmittii* (Walter and Gillett 1998).

This ecoregion hosts several endemic animals, mainly small rodents adapted to arid conditions. There is a center of endemism for the genus *Gerbillus* in the highlands of western Sudan, where four endemic gerbils occur (*Gerbillus bottai, G. muriculus, G. nancillus,* and *G. stigmonyx*). Endemic mammals also include one bat species, *Eptesicus floweri,* found in Mali and Sudan, a zebra mouse species, *Lemniscomys hoogstraali* (data deficient [DD]), found in Sudan, and two other gerbils from the genus *Taterillus: T. petteri* and *T. pygargus*. Three bird species are considered near endemic: the rusty lark (*Mirafra rufa*), the masked shrike (*Lanius nubicus*), and the sennar penduline-tit (*Anthoscopus punctifrons*) (Borrow and Demey 2001). For reptiles, endemism is more pronounced, with ten species regarded as strictly endemic, of which the only species of conservation concern is the African spurred tortoise (*Geochelone sulcata,* vulnerable [VU]).

Before the twentieth century vast herds of ungulates and other large animals including elephant, giraffe, and ostrich occurred in numbers rivaling or surpassing those found today in eastern and southern Africa. Most of the large animal populations have been reduced to scattered remnants through a century of unregulated hunting with modern firearms and vehicles. The scimitar-horned oryx (*Oryx dammah,* extinct in the wild [EW]) is presumed to be extinct in the wild (East 1999). Dama gazelle (*Gazella dama,* endangered [EN]), dorcas gazelle (*Gazella dorcas,* VU), red-fronted gazelle (*Gazella rufifrons,* VU), western giraffe (*Giraffa camelopardus peralta,* EN), and ostrich (*Struthio camelus*) are now found in a handful of protected areas, mainly in the central Sahel (East 1999; Hilton-Taylor 2000). Additionally, elephants (*Loxodonta africana,* EN) are now found only in Mali's Gourma-Rharous Partial Faunal Reserve and adjacent areas in Burkina Faso and Eritrea's Gash-Setit Wildlife Reserve (Said et al. 1995). Predators used to include African wild dog (*Lycaon pictus,* EN), cheetah (*Acinonyx jubatus,* VU), striped hyena (*Hyaena hyaena*), lion (*Panthera leo,* VU), and Nile crocodile (*Crocodylus niloticus*), which have now been extirpated over most of the ecoregion (Hilton-Taylor 2000).

The pronounced dry season signals a significant migration of fauna within the ecoregion. This includes annual passage of large numbers of migrant birds on the Afrotropical-Palearctic flyway and intra-African migration of birds and bats (Borrow and Demey 2001; Fishpool and Evans 2001). In the past there were major migrations of large mammals, but these have largely ceased mainly because the wildlife populations have been decimated by hunting (East 1999).

Status and Threats

Current Status

Vegetation cover has varied significantly because of climate change over the past several thousand years (Lézine 1989;

Tucker et al. 1991), and recently there has been much concern about the role humans may play in altering the Sahel (Gritzner 1988; Adams et al. 1996; Swift 1996). Although substantial natural changes in vegetation structure may result from short-term climatic variation, most recent studies have found little evidence of desertification (Geerling 1985; Tucker et al. 1991; Lawesson 1995). Agricultural land covers less than 30 percent of the total area of the ecoregion, and more than 80 percent of agricultural land is permanent pasture (Wood et al. 2000b). Away from permanent lakes, rivers, and wetlands, the dominant form of land use is seminomadic pastoral livestock production, with zebu cattle as the most common type of livestock (Raynaut 1997). However, the role of humans in decimating wildlife populations in the Sahel is clear: in the twentieth century, the introduction of modern firearms and vehicles led to highly destructive overhunting and the near extirpation of most large wildlife species.

Recent conservation efforts have strongly emphasized protection of the Sahelian fauna. The total area of protected land is around 224,825 km², or about 5 percent of the ecoregion. There are no Sahelian protected areas in Sudan, despite the high level of endemicity. Important protected areas include Aïr and Ténéré National Nature Reserve, Niger; Chad Basin National Park, Nigeria; Sahel Partial Faunal Reserve, Burkina Faso; Gash-Setit Wildlife Reserve, Eritrea; and Ansongo-Menaka Partial Faunal Reserve, Mali. Many important sites for migratory birds are also found, including Djoudj National Park and four Ramsar sites, the largest of which is Lac Fitri in Chad (Fishpool and Evans 2001). However, many of the protected areas shown on maps are essentially nonexistent on the ground, and wildlife may survive in these reserves simply because of their remoteness from areas of human activity.

Types and Severity of Threats

The shift in biomass in the Sahel from native wildlife species to livestock probably has not profoundly degraded vegetation, except around waterholes and cities, where activity may be intensive. However, agriculture probably has had a limited impact on vegetation in the Sahel because areas suitable for agriculture are limited and fires do not spread well or burn fiercely in the sparse, short grasslands (Breman and Kessler 1995; Raynaut 1997). Firewood collection has led to a reduction in woody cover in limited, densely populated areas but has not led to an overall loss of woody vegetation across the Sahel (Benjaminsen 1993). Some limited areas of particularly dense or valuable vegetation may be threatened by high local demand, such as woodlands dominated by *Acacia senegal* in northern Senegal or *Albizia aylmeri* (DD) in Sudan. The southern portion of the ecoregion faces a higher threat of habitat destruction, particularly in southern Niger and northern Nigeria, in Senegal, and along the Nile River in Sudan, where human population density is higher and agriculture is more widely practiced.

Nearly all large animal species have been removed from this ecoregion, facilitated by modern hunting methods (automatic rifles and off-road vehicles) and exacerbated by civil disturbance, poor law enforcement, and competition for grazing and water access with domestic livestock (Newby 1988). Hunting has slowed in some areas, such as Mauritania, but this is mainly because there is little left to hunt. Civil disturbance, especially in Chad, Central African Republic, and Sudan, has increased hunting in protected areas in the past decade, and poaching is still rife in most protected areas in other Sahelian countries.

Justification for Ecoregion Delineation

The Sahelian *Acacia* Savanna [35] ecoregion follows two of White's vegetation units: the "Sahel *Acacia* wooded grassland and deciduous bushland" and the "Northern Sahel semi-desert grassland and shrubland" (White 1983). Two small modifications to White's mapping unit included moving the Adrar des Iforhas (Mali) from this ecoregion to the West Saharan Montane Xeric Woodlands [96] ecoregion and extending the eastern boundary of the ecoregion to the foothills of the Ethiopian Highlands.

Ecoregion Number:	**36**
Ecoregion Name:	**West Sudanian Savanna**
Bioregion:	**Western Africa and Sahel**
Biome:	**Tropical and Subtropical Grasslands, Savannas, Shrublands, and Woodlands**
Political Units:	**Senegal, the Gambia, Guinea, Mali, Côte d'Ivoire, Burkina Faso, Ghana, Togo, Benin, Niger, Nigeria, Mauritania**
Ecoregion Size:	**1,638,400 km²**
Biological Distinctiveness:	**Regionally Outstanding**
Conservation Status:	**Vulnerable**
Conservation Assessment:	**II**
Author:	*Chris Magin*
Reviewed by:	*Chris Duvall, James Shambaugh*

Location and General Description

The West Sudanian Savanna [36] stretches across West Africa from Senegal and the Gambia to the eastern border of Nigeria. It lies between the Guinean Forest-Savanna Mosaic [38] to the south and the Sahelian *Acacia* Savanna [35] to the north.

The climate is tropical and strongly seasonal. The highest average daily temperatures (35°C–40°C) occur in March, April, and May, and the lowest average daily low temperatures (15°C–

20°C) are in November and December. Mean annual precipitation ranges up to 1,600 mm to the south but declines to 600 mm per annum on the northern border with the Sahelian *Acacia* Savanna [35]. Rainfall is highly seasonal: the dry season in the north and can last up to 9 months, whereas it lasts about 6–7 months in the south (Leroux 2001). Spatial and temporal variation in rainfall abundance is high both within and between years.

This ecoregion overlies a mixture of Precambrian basement rocks and a number of post-Jurassic sedimentary basins (Adams et al. 1996). The soils are mainly ultisols and alfisols to the south, with aridisols found to the north (d'Hoore 1964). Soil fertility is low, with heavily weathered lateritic soils, including ancient outcrops of ferricrete (soils cemented by accumulated iron oxides) (Goudie 1973). Although there are few prominent topographic features greater than 500 m in elevation, several sandstone massifs occur and contribute significantly to local biodiversity by creating a wide range of microhabitats. The Niger River drains most of the ecoregion, and the Senegal, Gambia, and Volta rivers also drain large areas.

Average human population density is 50–100 people/km^2. However, this population is patchy, and density figures range from fewer than 10 people/km^2 in remote areas to more than 1,000 people/km^2 in major cities. Agricultural land covers around 30 percent of the ecoregion, and about 80 percent of this is permanent pasture (Wood et al. 2000b). Agricultural land is concentrated on the northern and southern margins. Anthropogenic woodland types characterize many agricultural areas, where useful species such as baobab (*Adansonia digitata*) and shea (*Vitellaria paradoxa*) predominate (Boffa 1999).

This ecoregion roughly follows the western half of White's (1983) Sudanian regional center of endemism, whereas the East Sudanian Savanna [37] is congruent with the eastern half. The vegetation consists of woodland with an understory of long grasses, shrubs, and herbs. Combretaceae and Fabaceae are the dominant plant families, with *Combretum, Terminalia,* and *Acacia* the most abundant genera. The southern portion of this ecoregion consists mainly of woodlands, where tree canopies cover at least 40 percent of the ground surface (Lawesson 1995). Here, grasses may include various *Hyparrhenia* species that may grow to 4 m high. The northern portion hosts mainly grasslands dominated by numerous species of short grasses, where woody plants cover less than 10 percent of the area (Lawesson 1995). Shrubland is scattered in patches throughout the ecoregion and usually consists of thorny shrubs, especially various *Acacia* and *Ziziphus* species, dispersed among various short grass species. Riparian forest occurs along many waterways and may include plant and animal species characteristic of the more humid Guinean Forest-Savanna Mosaic [38]. Relict plant and animal communities survive in the sandstone massifs, which are important for maintaining biodiversity (White 1983; Lawesson 1995; Duvall 2001). Small areas of edaphic vegetation, such as grassy floodplains, or fadamas, also occur throughout the area.

Outstanding or Distinctive Biodiversity Features

Although there are several hundred endemic plants in the West Sudanian Savanna [36], most of the approximately 900 plants endemic to White's Sudanian region are widespread (White 1983), and few are rare, threatened, or endangered (Geerling 1985). The few species with limited ranges occur mainly in sandstone massifs, including the Bandiagara Cliffs (central Mali) and the Manding Mountains (western Mali, eastern Senegal, and northern Guinea). Woody plants found only in the West Sudanian Savanna ecoregion include the shrubs *Euphorbia sudanica, Gardenia sokotensis,* and *Tephrosia mossiensis* (Arbonnier 2001). The rarest woody plants are the ecoregion endemics *Gilletiodendron glandulosum* (VU) and *Kleinia cliffordiana,* and the nonendemics *Encephalartos barteri, Pteleopsis habeensis* (EN), and *Vepris heterophylla* (EN) (Walter and Gillett 1998). The status of herbaceous species, mosses, and fungi in the ecoregion is poorly known, although there are several highly localized species known from the sandstone massifs. Compared with the miombo woodlands of East Africa, the flora of the West Sudanian Savanna [36] is impoverished (White 1983).

The ecoregion supports a rich fauna. Common large animals are bushbuck (*Tragelaphus scriptus*), warthog (*Phacochoerus africanus*), vervet monkey (*Chlorocebus aethiops*), baboon (*Papio hamadryas papio* and *P. h. anubis*), and savanna monitor lizard (*Varanus exanthematicus*). Most large mammals have been heavily hunted. Predators, such as lion (*Panthera leo,* VU), leopard (*Panthera pardus*), cheetah (*Acinonyx jubatus,* VU), spotted hyena (*Crocuta crocuta*), and several smaller species survive sparsely, mainly in protected areas. Large herbivores have been nearly extirpated and survive predominantly in protected areas. These include elephant (*Loxodonta africana,* EN), hippopotamus (*Hippopotamus amphibius*), roan antelope (*Hippotragus equinus*), western buffalo (*Syncerus caffer brachyceros*), kanki hartebeest (*Alcelaphus buselaphus major*), and waterbuck (*Kobus ellipsiprymnus defassa*).

Three endemic small mammals occur: Senegal one-striped grass mouse (*Lemniscomys linulus,* DD), the shrew *Crocidura longipes* (EN), which is restricted to wetlands in northwestern Nigeria, and Felou's gundi (*Felovia vae,* VU), known only from western Mali (Kingdon 1997). Additionally, several subspecies of larger mammals are endemic, including the near-endemic korrigum (*Damaliscus lunatus korrigum,* VU) and the strictly endemic western giant eland (*Taurotragus derbianus derbianus,* EN), which survives only in Senegal's Niokolo-Koba National Park (East 1999). Additional threatened mammals are western chimpanzee (*Pan troglodytes verus,* EN), African wild dog (*Lycaon pictus,* EN), lion, cheetah, and West African manatee (*Trichechus senegalensis,* VU) (Hilton-Taylor 2000). Most exist at very low populations in a few protected areas.

Several bird species, particularly warblers and finches, are near endemic to the ecoregion, including two estrildid finches: *Lagonosticta virata* and *Lagonostica sanguinodorsalis* (Borrow and

Demey 2001). The distribution of other vertebrates is poorly known, but some strictly endemic species are found, including the brown running frog (*Kassina fusca*), Mali screeching frog (*Arthroleptis milletihorsini*), Wagler's blind snake (*Leptotyphlops albiventer*), Thierry's cylindrical skink (*Chalcides thierryi*), and Mocquard's cylindrical skink (*Chalcides pulchellus*) (Poynton 1999). Two threatened reptiles occur: the slender-snouted crocodile (*Crocodylus cataphractus,* DD) and the dwarf crocodile (*Osteolaemus tetraspis,* VU).

The pronounced dry season signals a significant migration of fauna in the ecoregion. This includes an annual passage by migrant birds on the Afrotropical-Palearctic flyway and intra-African migration associated with seasonal weather changes (Borrow and Demey 2001; Fishpool and Evans 2001). Many bats also migrate seasonally. In the past there were major migrations of large mammals, but these have largely ceased because of habitat alteration and intense hunting (East 1999). However, seasonal movement continues for several hundred elephant from northern Burkina Faso and very small numbers of giraffe (*Giraffa camelopardus peralta,* EN) from Niger into to the Sahelian *Acacia* Savanna [35] ecoregion in Mali (Said et al. 1995; East 1999).

Status and Threats

Current Status

There has been significant loss and fragmentation of habitat, especially in areas of high human population density. Remaining blocks of habitat suitable for wildlife are found in protected areas or agriculturally marginal zones (Happold 1995).

Almost all countries in the ecoregion have established protected areas. The total area of protected lands (not including forest reserves) is about 76,000 km^2, about 5.6 percent of this ecoregion (IUCN 1998). Protected areas include Boucle du Baoulé National Park Complex (Mali), Comoé National Park (Côte d'Ivoire), Kainji Lake National Park (Nigeria), Kéran National Park (Togo), Mole National Park (Ghana), and River Gambia National Park (the Gambia), as well as the transboundary areas Niokolo-Koba and Badiar national parks (Senegal and Guinea), Pendjari and Arli national parks (Benin and Burkina Baso), and W national parks (Niger, Burkina Faso, and Benin). All protected areas have been subject to some degree of poaching and agricultural encroachment, although several hold internationally significant populations of rare, threatened, or endangered species. Many of the protected areas shown on maps are poorly protected, and wildlife may survive simply because of its remoteness from areas of human activity.

Types and Severity of Threats

For the larger mammals, hunting for food and for sport has removed species over wide areas (Happold 1995; Said et al. 1995;

East 1997). Urban population growth proceeds at a high rate throughout West Africa, and meat from wild animals is a popular and valuable source of protein. Thus, poaching is a potentially profitable occupation for many people (Caspary 1999). Until 1997, small numbers of hunting permits were issued for the last surviving population of western Derby's eland, found in the binational Niokolo-Koba National Park (East 1999, 2000). The expansion of commercial agriculture has disrupted traditional forms of agriculture, which were generally not destructive of wildlife habitat. Finally, many development projects, especially dams, have directly caused habitat destruction and have led to the expansion of commercial exploitation of forest products in areas that did not previously experience such pressure (Laurance 1999; Wood et al. 2000b). In northern Nigeria, for instance, some agricultural schemes have seriously damaged wetlands, with negative consequences for wildlife and the people whose livelihoods depend on fishing or farming in the wetlands (Hollis et al. 1993). Similarly, dam construction has damaged wildlife habitat in Senegal, Mauritania, Mali, and Ghana.

Justification for Ecoregion Delineation

This ecoregion is largely congruent with the western half of White's (1983) Sudanian regional center of endemism, whereas the East Sudanian Savanna [37] follows the eastern half. Although these two ecoregions essentially share a single fauna, they have been separated at the Mandara Plateau, which forms part of the chain of high-elevation areas separating West and Central Africa. This division reflects a boundary for several plant taxa and a number of small mammals.

Ecoregion Number:	**37**
Ecoregion Name:	**East Sudanian Savanna**
Bioregion:	**Western African and Sahel**
Biome:	**Tropical and Subtropical Grasslands, Savannas, Shrublands, and Woodlands**
Political Units:	**Nigeria, Cameroon, Chad, Central African Republic, Sudan, Uganda, Democratic Republic of the Congo, Ethiopia, Eritrea**
Ecoregion Size:	**917,600 km^2**
Biological Distinctiveness:	**Bioregionally Outstanding**
Conservation Status:	**Relatively Stable**
Conservation Assessment:	**V**
Author:	*Chris Magin*
Reviewed by:	*Derek Pomeroy, Marc Languy*

Location and General Description

The East Sudanian Savanna [37] lies south of the Sahel in central and eastern Africa and is divided into a western block and an eastern block by the Sudd Flooded Grasslands [59]. The western block stretches from the Nigeria-Cameroon border through Chad and the Central African Republic to western Sudan. The eastern block is found in eastern Sudan, Eritrea, and the low-lying parts of western Ethiopia and also extends south through southern Sudan, into northwestern Uganda, and marginally into the Democratic Republic of the Congo around Lake Albert.

The ecoregion is flat, mainly lying between 200 and 1,000 m in altitude, although elevation rises slightly in western Ethiopia and around Lake Albert. The climate is tropical and strongly seasonal. The annual rainfall is as high as 1,000 mm in the south but declines to the north, with only 600 mm found on the border with the Sahelian *Acacia* Savanna [35]. Almost all rainfall occurs in a single rainy season from April to October, during which time large areas of southern Chad and northern parts of the Central African Republic become inundated and inaccessible. During the dry season the ground dries out, most of the trees lose their leaves, and the grasses dry up and may burn.

The ecoregion overlies a mixture of Precambrian basement rocks and a number of post-Jurassic sedimentary basins. The soils are mainly ultisols and alfisols in the south, with entisols in the north. Some oxisols and vertisols are also found in the east. The ecoregion is sparsely populated, with typical population densities ranging between one and five people per square kilometer, although there may be as many as thirty people per square kilometer in some places.

The ecoregion falls in the Sudanian regional center of endemism of White (1983). The vegetation is undifferentiated woodland with trees that are mainly deciduous in the dry season, with an understory of grasses, shrubs, and herbs. Typical trees in the western block of the ecoregion include *Anogeissus leiocarpus*, *Kigelia aethiopica*, *Acacia seyal*, and species of *Combretum* and *Terminalia*. In the eastern block *Combretum* and *Terminalia* species, *Anogeissus leiocarpus*, *Boswellia papyrifera*, *Lannea schimperi*, and *Stereospermum kunthianum* dominate the woody vegetation. The solid-stemmed bamboo *Oxytenanthera abyssinica* is prominent in the western river valleys of Ethiopia. Dominant grasses include tall species of *Hyparrhenia*, *Cymbopogon*, *Echinochloa*, *Sorghum*, and *Pennisetum* (Tilahun et al. 1996).

Outstanding or Distinctive Biodiversity Features

Both blocks of the East Sudanian Savanna [37] closely resemble the West Sudanian Savanna [36] in habitat and species composition. The two ecoregions differ somewhat in terms of their species assemblages and the degree to which the habitat and mammal assemblages are intact. The East Sudanian Savanna [37] has low rates of faunal endemism, with only one strictly endemic mammal (a mouse, *Mus goundae*, VU), and two strictly endemic reptiles (*Rhamphiophis maradiensis* and *Panaspis wilsoni*). Five bird species are considered endemic, including two strict endemics, Reichenow's firefinch (*Lagonosticta umbrinodorsalis*) and Fox's weaver (*Ploceus spekeoides*). The near-endemic Karamoja apalis (*Apalis karamojae*, VU) is found elsewhere in East Africa, and two other near-endemic species, the white-crowned robin-chat (*Cossypha albicapilla*) and Dorst's cisticola (*Cisticola dorsti*, DD), are shared with the West Sudanian Savanna [36]. However, the situation is different for plants because the ecoregion is largely congruent with part of the Sudanian regional center of endemism and is thus part of an important area for endemic plants. There are approximately 2,750 species of higher plants in the entire Sudanian regional center of endemism, and roughly one-third probably are endemic to this ecoregion (White 1983). However, because the ecoregion is vast, plant endemism per unit area is low.

Threatened mammal species include large herds of elephant (*Loxodonta africana*, EN) in Chad and Central African Republic (DPNRF 1997), wild dog (*Lycaon pictus*, EN), cheetah (*Acinonyx jubatus*, VU), and lion (*Panthera leo*, VU). Black rhinoceros (*Diceros bicornis*, critical [CR]) and northern white rhinoceros (*Ceratotherium simum cottoni*, CR) have been extirpated from the ecoregion, although occasional unconfirmed reports of the former (e.g., from southern Chad) are sometimes received (DPNRF 1997). The eastern giant eland (*Taurotragus derbianus gigas*) still survives in good numbers in the Central African Republic, especially in the western regions of the country, out of reach from Sudanese poachers (East 1999). Giant eland are less susceptible to poachers than more sedentary and less wary antelope species but have been almost completely eliminated from Sudan. The roan antelope's (*Hippotragus equinus*) cautious behavior has also allowed it to withstand poaching pressure to some degree. However, uncontrolled poaching in Chad and Sudan has resulted in decreasing roan antelope populations in the rest of this ecoregion.

Status and Threats

Current Status

The original wooded savanna habitat has been significantly reduced, although to a lesser extent than in the West Sudanian Savanna [36], primarily because of the lower human population density. There are a good number of protected areas, and outside formal protection habitats remain in reasonable condition in many regions.

The total area of protected lands is more than 136,000 km^2. This is approximately 18 percent of the ecoregion. However, many of these protected areas are not adequately enforced or policed. They include Dinder, Radom, and Boma National Parks in Sudan, Zakouma National Park in Chad, Manovo-Gounda-Saint Floris and Bamingui-Bangoran national parks in Central African Republic, Gambella National Park in Ethiopia, and Mount Kei in Uganda.

Types and Severity of Threats

The habitats of the ecoregion are threatened principally by seasonal shifting cultivation, overgrazing by livestock, cutting of trees and bushes for wood, burning of woody material for charcoal, and uncontrolled wild fires. Climatic change is a further threat, exacerbating the impacts of human activities because the ability of the ecosystem to recover from overuse is reduced when there is little rainfall. The main threats to the species in the ecoregion come from overgrazing and, in the case of large animals, poaching or overhunting for meat. Poaching is particularly pronounced in politically unstable areas such as southern Sudan.

Justification for Ecoregion Delineation

This ecoregion, along with the West Sudanian Savanna [36], forms part of the Sudanian regional center of endemism. The West [36] and East Sudanian Savannas [37] are similar in terms of their broader species assemblages, but they were split into two separate ecoregions near the Mandara Plateau because of a change in species distributions around this point. Modifications to White's original vegetation boundaries include placing an extension of the Sahelian *Acacia* Savanna [35] below Lake Chad and moving the northern boundary of the East Sudanian Savanna [37] further south. Edaphic grassland and communities of *Acacia* and broadleaved trees identified by White have also been largely excluded. The eastern portion of this ecoregion follows White's "Ethiopian undifferentiated woodland" and "Ethiopian transition from undifferentiated woodland to *Acacia* deciduous bushland and wooded grassland" vegetation units. This eastern part was extended to include the Sudanian undifferentiated woodland vegetation unit south toward Lake Albert and Mount Elgon.

Ecoregion Number:	**38**
Ecoregion Name:	**Guinean Forest-Savanna Mosaic**
Bioregion:	**Western Africa and Sahel**
Biome:	**Tropical and Subtropical Grasslands, Savannas, Shrublands, and Woodlands**
Political Units:	**Senegal, the Gambia, Guinea-Bissau, Guinea, Sierra Leone, Côte d'Ivoire, Ghana, Togo, Benin, Nigeria, Cameroon**
Ecoregion Size:	**673,600 km^2**
Biological Distinctiveness:	**Regionally Outstanding**
Conservation Status:	**Endangered**
Conservation Assessment:	**II**
Authors:	Illisa Kelman, Paul Burgess
Reviewed by:	Chris Duvall, James Shambaugh

Location and General Description

The Guinean Forest-Savanna Mosaic [38] runs east to west across West Africa, forming the transition between the Eastern and Western Guinean Lowland Forests [1, 3] and the West Sudanian Savanna [36]. The ecoregion reaches the coast at the Dahomey Gap, a narrow strip of semi-arid habitats in eastern Ghana and Benin that separate the Guinean rainforests from those of the Congolian region. The landscape is typified by gently rolling plains, averaging 200–500 m above sea level. Scattered inselbergs rise hundreds of meters above the plains, up to 1,500 m. The Niger River passes through the ecoregion, and the Senegal, Gambia, and Volta rivers also drain large areas.

This ecoregion lies in the humid tropical savanna zone (Bailey 1998), with mean annual high temperatures ranging between 30°C and 33°C and lows ranging between 14°C and 21°C. Annual rainfall averages 1,600–2,000 mm but declines in the Dahomey Gap, parts of which receive 1,000 mm or less. To the north, there is a single rainy season of 5 to 7 months, peaking in August. In the southern portion rainfall is bimodal, with two rainy seasons of 2 to 4 months each, separated by a short dry period in July or August (Adams et al. 1996; Leroux 2001). In winter, the Harmattan, a dry northeastern wind, brings dust clouds from the Sahara.

In Guinea-Bissau and the Volta Basin, Precambrian basement rocks dominate, whereas Cretaceous sediments occur in Nigeria, most notably in the Benue and lower Niger valleys. Alfisols and inceptisols underlie much of the central and eastern portions, and acidic and highly weathered ultisols are prominent in the west (d'Hoore 1964). Soil fertility is low in most areas. Barren, ancient outcrops of ferricrete (soils cemented by accumulated iron oxides) are scattered across the ecoregion (Adams et al. 1996). Average human population density is 60–140 people/km^2 in most areas. However, the human population is patchy, and density figures range from fewer than 10 people/km^2 in remote areas to 1,000 people/km^2 or more in major cities.

The Guinean Forest-Savanna Mosaic [38] forms the western half of White's (1983) Guineo-Congolian/Sudanian regional transition zone. The gallery and groundwater forests represent extensions of the drier types of Guineo-Congolian peripheral semi-evergreen forest (White 1983; Lawson 1986; Lawesson 1995). Tall grasses (up to 4 m high) characterize the savanna areas, including fire-tolerant species of the genera *Andropogon*, *Hyparrhenia*, and *Pennisetum* (Lawson 1986). Short grasses characterize the Dahomey Gap. Agricultural land covers around 40 percent of the total area of the ecoregion (Wood et al. 2000b). Anthropogenic woodland types characterize agricultural areas, where tree species useful to people predominate, such as oil palm (*Elaeis guineensis*), locust bean (*Parkia biglobosa*), and various figs (*Ficus* sp.) (Boffa 1999).

Outstanding or Distinctive Biodiversity Features

The history of vegetation change in this ecoregion is closely tied to that of the Guinean rainforest and is contentious because of uncertainty about the relationships between human activities, climate change, and vegetation. Paleoenvironment researchers have identified as many as twenty major fluctuations between arid and humid climate conditions in West Africa over the last 10 million years (de Menocal 1995). At the time of the last glacial maximum (about 18,000 years ago) rainforests covered about 10 percent of their present area, in patches along the coast of modern Liberia, Côte d'Ivoire, and Ghana (Maley 1996). Most of the current area of rainforest was covered by grass-dominated vegetation (Adams et al. 1996). Approximately 8,000 years ago, wetter conditions led to an expansion of the rainforest into areas today occupied by the Guinean Forest-Savanna Mosaic [38] and West Sudanian Savanna [36], including the Dahomey Gap. The climate dried again about 4,000 years ago, and the Dahomey Gap has existed continuously since then. At about this time, the first clear evidence appeared of the effects of human activities on forest vegetation in West Africa: the frequency of forest fires and the abundance of cultivated plants increased, and the abundance of some noncultivated plants declined (Maley 1996). Several smaller climate changes have occurred in the past 4,000 years (Fairhead and Leach 1996, 1998b).

Highland areas and inselbergs are especially important sites for biodiversity because complex topography creates a number of microhabitats. The uplands of Sierra Leone and Guinea support the ecoregion endemic trees *Bafodeya benna*, *Diospyros feliciana* (VU), and *Fleurydora felicis*. Isolated patches of semi-evergreen forest on inselbergs on eastern Ghana's coastal plain also support several endemics, such as *Commiphora dalzielii*, *Grewia megalocarpa*, *Talbotiella gentii* (EX/EN), and *Turraea ghanensis* (EN). These inselbergs also host populations of several species that are otherwise found only in distant parts of Africa. For instance, the trees *Crossandra nilotica* and *Ochna ovata* are also found in East Africa, and the shrubs *Capparis fascicularis* and *Grewia villosa* and the grasses *Aristida sieberana*, *Chloris prieurii*, and *Schoenefeldia gracilis* normally are found in the Sahel (White 1983). Several more widespread plants of concern occur, including the timber trees *Afzelia africana* (VU), *Albizia ferruginea* (VU), and *Khaya senegalensis* (VU) (Walter and Gillett 1998).

The wide variety of habitat types in the Guinean Forest-Savanna Mosaic [38] support a high diversity of animals. Forest patches contain characteristic rainforest species, such as the African palm civet (*Nandinia binotata*), lesser spot-nosed monkey (*Cercopithecus petaurista petaurista*), and Maxwell's duiker (*Cephalophus maxwelli*), along with species of conservation concern such as the western giant forest hog (*Hylochoerus meinertzhageni ivoriensis*, VU), white-thighed black and white colobus (*Colobus vellerosus*, VU), and bongo (*Tragelaphus eurycerus*), which is endangered in West Africa (East 1999). Woodland and grassland areas provide habitat for species more characteristic of the West Sudanian Savanna [36]. Common species include baboon (*Papio hamadryas papio* and *P. h. anubis*), common duiker (*Sylvicapra grimmia*), helmeted Guinea fowl (*Numida meleagris*), and side-striped jackal (*Canis adustus*), and rarer species include the dwarf crocodile (*Osteolaemus tetraspis*, VU), lion (*Panthera leo*, VU), and red-flanked duiker (*Cephalophus rufilatus*). Several rare species are found in both forest and woodland or grassland habitats, including western chimpanzee (*Pan troglodytes verus*, EN) and elephant (*Loxodonta africana*, EN) (Hilton-Taylor 2000). Rare wetland or riverine species include African manatee (*Trichechus senegalensis*, VU), spotted-neck otter (*Lutra maculicollis*, VU), and sitatunga (*Tragelaphus spekii*), which is endangered in West Africa (East 1999).

Two rodent species are the only strictly endemic mammals in the Guinean Forest-Savanna Mosaic: *Mus mattheyi* (DD), found in Ghana, and *Steatomys jacksoni* (VU), found only in Ghana and Nigeria. The eleven species of strictly endemic reptiles in the ecoregion include the ugly worm lizard (*Cynisca feae*), known only from Guinea-Bissau, and the Gambia blind snake (*Leptotyphlops natatrix*), found in the Gambia. No amphibians are believed to be strictly endemic, and only one, *Phrynobatrachus francisci*, is near endemic. The same is true of the birds, of which there are no strict endemics and just two near-endemic species (Borrow and Demey 2001).

Status and Threats

Current Status

Most habitat types are highly fragmented and degraded across this ecoregion, especially in areas of high human population density. Remaining large habitat blocks are found mainly in protected areas or agriculturally marginal zones. Protected areas cover about 32,000 km², or 5.9 percent of the ecoregion. Some of the larger protected areas in this ecoregion are Old Oyo National Park (Nigeria), Bui and Digya national parks (Ghana), Badiar National Park (Guinea), and Marahoué National Park (Côte d'Ivoire).

Sacred groves (forest patches that are protected by social or cultural use restrictions) are an important indigenous conservation measure in this ecoregion (Decher 1997). In sacred groves, a wide variety of plant and animal species survive, and humans may actively seek to improve the species richness of groves by transplanting species between different groves (Fairhead and Leach 1996).

Types and Severity of Threats

Hunting and loss of habitat to agriculture or logging are the main threats to biodiversity. Hunting has extirpated most larger animals, and those that survive tend to be wide-ranging habitat generalists, common throughout Africa (Happold 1995). Whereas predatory species have been eliminated out of fear to protect people and property, other species are hunted to provide meat. Because few tourists visit the protected areas of the ecoregion, they generate little income to fund their management. Instead, impoverished residents exploit wildlife to supply the bushmeat trade, governments permit logging concessions, and farmers clear forest habitats for farmland.

Fire is a threat to forest components of the ecoregion. Humans use fire both to flush game for hunting and to clear land for agriculture. Frequent fires may also alter fallow succession and the species composition of vegetation types by killing forest species that are less fire-resistant (Nyerges 1989). The loss or degradation of all types of vegetation is likely to continue into the future because of population growth and the expansion of markets (Wood et al. 2000a).

Justification for Ecoregion Delineation

The Guinean Forest-Savanna Mosaic ecoregion extends from the Atlantic Ocean to the West Sudanian Savanna [36], with its boundaries following White's (1983) "mosaic of lowland rainforest and secondary grassland" vegetation unit. It is separated from the Northern Congolian Forest-Savanna Mosaic [40] by the Cameroon Highlands Forests [10], which act as a distribution range limit for several taxa characteristic of Central African forest-savanna mosaics. The southern boundary is defined by the transition to more continuous forest cover.

Ecoregion Number: **39**

Ecoregion Name: **Mandara Plateau Mosaic**

Bioregion: **Western Africa and Sahel**

Biome: **Tropical and Subtropical Grasslands, Savannas, Shrublands, and Woodlands**

Political Units: **Cameroon, Nigeria**

Ecoregion Size: **7,500 km²**

Biological Distinctiveness: **Locally Important**

Conservation Status: **Critical**

Conservation Assessment: **IV**

Author: *Colleen Seymour*

Reviewed by: *Steve Gartlan*

Location and General Description

The Mandara Plateau Mosaic [39] is located along the border between northeastern Nigeria and northwestern Cameroon (11°N, 14°E). Although within the semi-arid Sudano-Sahelian belt of Africa, it is wetter and more vegetated than surrounding areas. The ecoregion can be divided into three altitudinally distinct zones: mountains, plateau, and plains (FAO 2001a). Most of the botanical value is found on the plateau areas above 1,200 m.

The ecoregion receives 800–1,000 mm of rain during a 6-month wet season, from May to October, with the rest of the year being dry (FAO 2001a). Mean temperature ranges from 15°C to 30 °C and is moderated by altitude. Unlike the volcanic rocks of the Cameroon Mountains further to the southwest, the Mandara range is composed of ancient granites, which weather to produce an infertile soil (Morton 1986). However, there are also areas of more fertile soil, resulting in a long history of human settlement (Riddell and Campbell 1986). Agricultural activities are undertaken across most available land.

This area falls in White's (1983) Sudanian regional center of endemism. It is thought that *Isoberlinia doka* woodlands were once the dominant vegetation on Mount Mandara, which now supports only extremely degraded forms of this vegetation (MacKinnon and MacKinnon 1986; Bellefontaine et al. 1997). The trees here reach heights of 12–18 m, and woody cover averages 50 percent or more. Grasses (e.g., *Andropogon* spp. and *Beckeropsis* spp.) largely dominate the herbaceous layer (Bellefontaine et al. 1997).

On the whole, the ecoregion is heavily grazed and burnt. There are a few montane species and a number of submontane species in its flora. The highest points, from 1,200 to 1,494 m, hold a mix of Sudanian and Afromontane species such as the large succulent *Euphorbia desmondi*, *Olea hochstetteri*, and *Pittosporum viridiflorum*.

Outstanding or Distinctive Biodiversity Features

The Mandara Mountains are known to be home to a number of rare and endemic plants (Stuart et al. 1990) with East African montane affinities, but they have not been well studied. Their ecological needs and major threats are thus poorly understood, and the conservation needs of this area are not fully appreciated. The current state of degradation of this ecosystem will certainly affect its prospects for future conservation action.

The dry forests of Mozogo-Gokoro National Park and the Mayo Louti Forest Reserves (Stuart et al. 1990) are only marginally associated with the Mandara Plateau Mosaic [39] proper because they are found at lower altitudes between two parts of the range. The savanna woodland of this area is dominated by *Acacia albida*, with *A. senegal* and *A. nilotica* also present. Other species include *Balanites aegyptiaca, Ziziphus* spp., *Crateva adansonii, Celtis integrifolia, Ficus* spp., and *Khaya senegalensis*. Recorded fauna includes vervet monkey (*Chlorocebus aethiops*), patas monkey (*Erythrocebus patas*), warthog (*Phacochoerus aethiopicus*), bushbuck (*Tragelaphus scriptus*), bush duiker, (*Sylvicapra grimmia*), and python (*Python sebae*).

The mountains themselves may still harbor a population of the western subspecies of mountain reedbuck (*Redunca fulvorufula adamauae,* EN) (Stuart et al. 1990; Hilton-Taylor 2000). Bird species include Rüppell's griffon (*Gyps rupelli*) and Egyptian vulture (*Neophron percnopterus*) (Scholte 1998). The region also contains three near-endemic reptile species: Mount Lefo chameleon (*Chamaeleo wiedersheimi*), *Mabuya langheldi,* and African wall gecko (*Tarentola ephippiata*).

Status and Threats

Current Status

The distinctive plant communities associated with the Mandara Plateau receive no protection from the current network of protected areas in the region (MacKinnon and MacKinnon 1986). It is believed that habitats are highly degraded in most areas.

Most data for this ecoregion relate mainly to Cameroon's 14 km^2 Mozogo-Gokoro National Park, which is situated in the lower-lying regions and is not fully representative of the ecoregion at large. In 1998, Mozogo-Gokoro National Park had not been burnt for more than 40 years (Culverwell 1998). As a result of this exclusion of fire, much of it has become far more densely wooded, with the establishment of dry thickets and the almost total elimination of grass cover. Although the park has been treated effectively as a strict nature reserve since its inception in 1932, there are plans to open it for tourism.

Types and Severity of Threats

The Mandara Mountains are intensively used for cropping and livestock grazing (FAO 2001a). Subsistence farmers also collect large amounts of fuelwood. Heavy grazing probably is detrimental to biodiversity in the area, although the farmers in the mountain region tend to stall feed their cattle during the growing season. Not only does this protect crops, but also the manure from the stalls is collected and used to fertilize areas under cultivation (FAO 2001a). Firewood collection has added to the severe degradation of the indigenous woodlands.

Throughout Cameroon the major threat to biodiversity is posed by clearance of both lowland and montane forests (Stuart et al. 1990). This appears to have occurred on the Mandara Plateau, with little natural forest vegetation remaining. Grazing and burning for clearing of new land are also extensive (Stuart et al. 1990).

Justification for Ecoregion Delineation

The extent of this ecoregion is taken directly from the "Mandara plateau mosaic" vegetation unit (White 1983). This plateau was considered a distinct ecoregion because of its isolation from other highland areas and the unusual biogeographic affinities with montane areas in eastern Africa.

Ecoregion Number:	**40**
Ecoregion Name:	**Northern Congolian Forest-Savanna Mosaic**
Bioregion:	**Central Africa**
Biome:	**Tropical and Subtropical Grasslands, Savannas, Shrublands, and Woodlands**
Political Units:	**Cameroon, Democratic Republic of the Congo, Central African Republic, Uganda, Sudan**
Ecoregion Size:	**708,100 km^2**
Biological Distinctiveness:	**Regionally Outstanding**
Conservation Status:	**Relatively Stable**
Conservation Assessment:	**III**
Author:	*Illisa Kelman*
Reviewed by:	*Steve Gartlan*

Location and General Description

The Northern Congolian Forest-Savanna Mosaic [40] extends from the Cameroon Highlands [10] east through the Central African Republic, through northeastern Democratic Republic of the Congo (DRC) and southwestern Sudan, and into a sliver of northwestern Uganda. The Ubangi and Uele rivers demarcate the central and eastern borders with the Northeastern Congolian Lowland Forests [16], and the Bar al Ghazall, part of the

Upper Nile drainage, delineates the transition to the Sudd Flooded Grasslands [59] to the east.

The ecoregion lies in the tropical savanna climate zone (Faniran and Jeje 1983) and is climatically transitional between the Sudanian and Guineo-Congolian regions (White 1983). A single wet season and a single dry season characterize the region, with mean annual precipitation ranging from 1,200 to 1,600 mm per year. The precipitation declines to the north as the ecoregion grades into the East Sudanian Savanna [37]. Temperatures range from 34°C in the rainy season to 13°C in the dry season (White 1983).

Most of the ecoregion sits on a dissected plateau 500 m in elevation and rises to 700 m toward the Cameroon Highlands. Precambrian basement rocks underlie the area, outcropping as inselbergs. Heavily weathered oxisols are found in the central and eastern portions, whereas the western CAR has unweathered entisols (Brady and Weil 1999). Changes in soils partially explain the abrupt shift from rainforest to open grassland (Cole 1992; Hopkins 1992).

Forest, woodland, and secondary grassland intergrade in patterns controlled by annual precipitation, duration of water stress, and the severity of dry season fires and human activity (Cole 1992; Longman and Jenik 1992; Maley 1996). Gallery forests grow along watercourses and elsewhere with sufficient groundwater (Mayaux et al. 1999). Widespread gallery species include *Berlinia grandiflora, Cola laurifolia, Cynometra vogelii, Diospyros elliotii, Parinari congensis,* and *Pterocarpus santalinoides.* Remnant peripheral semi-evergreen rainforest occurs in the south, supporting species such as *Afzelia africana, Aningeria altissima, Chrysophyllum perpulchrum, Cola gigantea, Morus mesozygia,* and *Khaya grandifolia.* The ecoregion also contains expansive areas of moist wooded grasslands (Mayaux et al. 1999). Grasses include *Andropogon* spp., *Hyparrhenia* spp. and *Loudetia* spp. Trees common throughout these grasslands include *Annona senegalensis, Burkea africana, Combretum collinum, Hymenocardia acida, Pariniari curatelifolia, Stereospermum kunthianum, Strychnos* spp., and *Vitex* spp. Where human population remains sparse in central Cameroon and the Central African Republic, patches of dense dry forest remain, dominated by *Isoberlinia doka* with *Afzelia africana, Burkea africana, Anogeissus leiocarpus, Terminalia* spp., and *Borassus aethiopum* (White 1983).

Outstanding or Distinctive Biodiversity Features

Over the past million years Central Africa has experienced repeated climatic fluctuations that have caused rainforest and savanna expansion and contraction. Plants and animals adapted, migrated, or became extinct with each climatic oscillation. During dry periods, savanna communities invaded far into the Congo Basin. The moist riparian forests became isolated from one another and formed forest island communities in the savanna matrix (White 1983; Kingdon 1989; Smith et al. 1997).

During wetter periods this area was covered by moist Congolian rainforest.

This ecoregion supports moderate levels of faunal diversity, including many species with broad distributions in tropical Africa (Millington et al. 1992). Mammals include a mixture of savanna and forest-adapted species such as red-flanked duiker (*Cephalophus rufilatus*), black rhinoceros (*Diceros bicornis,* CR) (limited to a few individuals remaining in Cameroon), giant eland (*Taurotragus derbianus*), bongo (*Tragelaphus eurycerus*), and northern savanna giraffe (*Giraffa camelopardalis congoensis*). The top predator is the lion (*Panthera leo,* VU). The ecoregion also supports both the forest elephant (*Loxodonta africana cyclotis,* EN) in the forest patches and savanna elephant (*Loxodonta africana africana,* EN) in the savanna woodlands (Belsky and Amundson 1992; Hopkins 1992).

The ochre mole rat (*Cryptomys ochraceocinereus*) and Pousargue's mongoose (*Dologale dybowskii*) are near-endemic species. There is one near-endemic bird species, Oberlaender's ground-thrush (*Zoothera oberlaenderi*). There are a number of endemic amphibians and reptiles, including the Mauda River frog (*Phrynobatrachus albomarginatus*), Buta River frog (*P. scapularis*), Bamileke Plateau frog (*Rana (Amnirana) longipes*), eastern dwarf clawed frog (*Hymenochirus boulengeri*), and Inger's grassland frog (*Ptychadena ingeri*). Reptiles include the strictly endemic Sudan beaked snake (*Rhinotyphlops sudanensis*), *Ichnotropis chapini,* and *Helophis smaragdina.*

Status and Threats

Current Status

The ratio of forest to savanna in this area in this ecoregion has fluctuated over time (Kingdon 1989). Humans have also modified the ecosystem through their use of fire, agriculture, and livestock over at least 3,000 years. People have become part of the "natural" disturbance cycle, maintaining this vast mosaic.

Garamba National Park in DRC and Mbem-Djerem National Park in Cameroon are the only two protected areas in this ecoregion. However, there are believed to remain large areas of habitat in regions of low human population density.

Types and Severity of Threats

Increasing human population, poverty, the ongoing civil wars in Sudan and DRC, strife between government and rebel groups in the Central African Republic, and incursions by well-armed poaching gangs from the Sudan mean that this ecoregion faces many threats. The distribution of large mammals has been drastically reduced in recent times. Hunting camps are found far into the bush, with ivory and other valuable animal products targeted to pay for weapons and food. Gallery forests are logged for timber. Fuelwood is collected in all areas, and charcoal is produced where there is an urban demand. Ongoing economic, political, and social instability has drained the already limited conservation budgets.

Justification for Ecoregion Delineation

This ecoregion is based on the "mosaic of lowland rainforest and secondary grassland" vegetation unit of White (1983). It is separated from the Guinean Forest-Savanna Mosaic [38] ecoregion by the Cameroon Highlands Forests [10], which act a range limit for some forest-savanna mosaic species. The southern boundary is defined by the transition to more continuous forest cover. The northern boundary was verified with 1 km classified land-cover data derived from advanced very-high-resolution radiometer satellite imagery (Loveland et al. 2000).

Ecoregion Number:	**41**
Ecoregion Name:	**Victoria Basin Forest-Savanna Mosaic**
Bioregion:	**Eastern and Southern Africa**
Biome:	**Tropical and Subtropical Grasslands, Savannas, Shrublands, and Woodlands**
Political Units:	**Uganda, Rwanda, Tanzania, Democratic Republic of the Congo, Burundi, Sudan, Ethiopia, Kenya**
Ecoregion Size:	**165,800 km²**
Biological Distinctiveness:	**Globally Outstanding**
Conservation Status:	**Critical**
Conservation Assessment:	**I**
Author:	*Illisa Kelman*
Reviewed by:	*Derek Pomeroy, Marc Languy, Alan Rodgers*

Location and General Description

Centered on Lake Victoria, the Victoria Basin Forest-Savanna Mosaic [41] encompasses most of south-central Uganda and the eastern half of Rwanda and extends marginally into Tanzania, Burundi, Democratic Republic of the Congo (DRC) and Kenya. A small outlier sits on the Sudan-Ethiopia border. Historically the ecoregion supported large areas of tropical forest mixed with savanna woodlands (Hamilton 1984; Howard 1991; Hamilton et al. 2001).

Annual mean maximum temperatures range from 24°C to 27°C, and mean minimum temperatures range from 15°C to 18°C. Most of the ecoregion receives 1,000–1,400 mm rainfall, but rainfall exceeds 2,000 mm on the Sesse Islands in Lake Victoria and in the Sudanese outlier. Rain generally falls in two seasons, from March to May and August to November. Some

areas, such as Lake Mburo National Park and the western Rift Valley, which contains Queen Elizabeth National Park, receive as little as 700 mm of rain per year (Government of Uganda 1967).

Precambrian bedrock underlies the ecoregion, producing old and leached soils. Middle Pleistocene tectonic events shaped the western branch of the Great Rift Valley, and uplift reversed the flow of the Kagera River, which now demarcates parts of the national boundaries of Tanzania with Rwanda and Uganda (White 1983; Cahen et al. 1984). The altered hydrology created the shallow (maximum 75 m deep) and young (about 0.75 m.y.o.) Lake Victoria. Lake Victoria sits high (1,134 m) on the Equatorial Lake Plateau, which averages 800–1,500 m above sea level (Government of Uganda 1967; White 1983). Human densities are high close to Lake Victoria, throughout Rwanda, and in the larger urban centers.

This ecoregion is largely congruent with the Lake Victoria regional mosaic of White (1983), one of the major phytogeographic divisions in Africa. This region contains a mix of floristic associations because it is surrounded by six other African phytochoria. During the colder and drier periods associated with the northern ice ages, rainfall was lower than today, and forest disappeared from most of what is now Uganda (Hamilton 1974; Hamilton et al. 2001), except perhaps for a small refuge area now occupied by the Sango Bay forests. At the end of the last ice age, about 12,500 years ago, the rainfall increased and the forests began to return (Dale 1954; Beuning et al. 1997; Jolly et al. 1997). However, much has been lost to deforestation, which started around 3,000 years ago (Dale 1954; Hamilton 1974, 1984; Plumptre 1996; Jolly et al. 1997; Hamilton et al. 2001). The species composition of these forests is similar to that of the Congo Basin but is impoverished.

The unique Sango Bay (Uganda) to Minziro (Tanzania) alluvial forests grow on a floodplain where the Kagera River enters Lake Victoria. These forests support both Guineo-Congolian lowland species, dominated by *Baikiaea insignis,* and Afromontane species (Langdale-Brown et al. 1964; White 1983; Howard 1991; Katende and Pomeroy 1997). This forest may represent a relict area from a time when the climate was cooler and montane forest habitats occurred at lower altitudes (Jolly et al. 1997; Hamilton et al. 2001). It survives in this area now because of the permanent soil moisture.

A great variety of savanna types occupy areas where rainfall is too low for forest (Langdale-Brown et al. 1964; White 1983). Papyrus (*Cyperus papyrus*) swamp occurs widely in the ecoregion. Outside protected areas, years of overexploitation have left only remnant patches of these savanna habitats, in a matrix of secondary habitat and farmland.

Outstanding or Distinctive Biodiversity Features

There is a diverse flora in this ecoregion, but endemism rates are low. The Sango Bay-Minziro forest supports Afromontane elements in the flora and some endemic plant species. This forest type may represent a remnant from periods when the climate was cooler and drier and survives here because of the constant soil moisture.

The avifauna is species rich but supports only two strict endemic species: the papyrus canary (*Serinus koliensis*) and Fox's weaver (*Ploceus spekeoides*), both found in *Papyrus* swamp vegetation (Pomeroy 1993; Byaruhanga et al. 2001). Two near-endemic species also occur, Nahan's francolin (*Pternistis nahani,* EN) which is confined to a few forests, and the red-faced barbet (*Lybius rubrifacies*), which is found in drier habitats. The Kibale ground-thrush (named *Turdus kibalensis*), known only from Kibale Forest, is no longer considered a separate species (Urban et al. 1997). The large swamps in this ecoregion support a diverse assemblage of waterbirds, including the shoebill stork (*Balaeniceps rex*), which occurs at low densities.

A high diversity of mammals occurs, but there are only three strictly endemic species: the dark shrew (*Crocidura maurisca*), moon shrew (*Crocidura selina,* EN), and Issel's groove-toothed swamp rat (*Pelomys isseli,* VU). Near-endemic mammals include Glen's wattled bat (*Chalinolobus gleni*), Uganda large-tooth shrew (*Crocidura mutesae*), eastern needle-clawed galago (*Galago matschiei*), Aellen's pipistrelle (*Pipistrellus inexspectatus*), Ankole mole rat (*Tachyoryctes ankoliae,* VU), and dwarf multimammate mouse (*Mastomys pernanus,* DD) (Hilton-Taylor 2000). The chimpanzee (*Pan troglodytes,* EN) is also found in many of the forested areas of the western parts of the ecoregion. Kibale forest is one of the most primate-rich forests in Africa, supporting twelve species (NBDB 1995; Struhsaker 1997).

In the drier parts of the ecoregion where savanna dominates over forest-savanna mosaic, there are significant but diminished populations of African elephant (*Loxodonta africana,* EN). The Queen Elizabeth/Virunga parks lack some typical eastern African savanna species; for example, there are no zebra (*Equus grevyi*), impala (*Aepyceros melampus*), eland (*Taurotragus oryx*), or oribi (*Ourebia ourebi*). Predators include lion (*Panthera leo,* VU).

The amphibians and reptiles are diverse, but again there are few endemic species. The banded toad (*Bufo vittatus*) is a strict endemic, and Roux's river frog (*Phrynobatrachus rouxi*) is a near endemic. Similarly, there are three strict endemic reptile species: the snake *Atheris acuminata,* the agamid lizard *Agama mwanzae,* and *Gastropholis rwanda.*

Status and Threats

Current Status

The vegetation in this ecoregion reflects its long history of human use and has been significantly modified. Pollen records show that anthropogenic deforestation started about 2000 years B.P. (Hamilton et al. 2001). The current vegetation dynamics, characterized by shifting forest and savanna patches, com-

plexly intertwined with human settlement patterns, have become the "natural" disturbance cycle (Jolly et al. 1997).

Most of the important sites are protected as national parks, game reserves, and forest reserves. In the Albertine Rift Valley are Queen Elizabeth National Park (QENP) and the Kigezi and Kyambura wildlife reserves. Along its eastern flanks are a series of forest reserves stretching from Kalinzu in the south to Budongo in the north (Howard and Davenport 1996). The DRC's Parc de Virunga, to the west of Lake Edward, adjoins the QENP. Outside this concentration, protected areas are fewer and further apart. They include Lake Mburo and Murchison Falls national parks in Uganda, Saiwa Swamp and Kakamega Forest National Reserve in Kenya, Ibanda and Rumanyika game reserves in Tanzania, Lac Rwihinda Nature Reserve in Burundi, and Akagera National Park in Rwanda.

Most of the forest in the ecoregion is found in forest reserves managed by forestry departments (Howard 1991; Wass 1995). These reserves were originally established to control timber extraction and for the residual benefits of watershed protection (Howard 1991; Struhsaker 1997). The Uganda Government agreed to set aside 20 percent of the former production forests for nature protection, and a research program defined the areas of highest biodiversity value, which have been turned into nature reserves (Howard and Davenport 1996; Howard et al. 1998; 2000).

Types and Severity of Threats

Before the mid-1970s, Uganda had an effective forest and wildlife conservation program through systems of national parks, game reserves, and forest reserves. In 1979, the war in Uganda decimated the elephant population and extirpated populations of white and black rhinoceros. Uganda's conservation efforts were reduced for 15 years, but relative internal stability through the 1990s enhanced conservation management once again, and large mammal populations began to increase (Arinaitwe et al. 2000). Furthermore, people who had encroached into forest reserves were resettled elsewhere.

Civil war affected Burundi, Rwanda, DRC, and Sudan in the 1980s and 1990s. Rebel armies poached bushmeat, hardwood, and elephant ivory for survival and to sell for arms. Refugees from these conflicts also devastated animal populations for bushmeat and forests for fuelwood. By the end of 1998, Uganda supported some 200,000 refugees, mostly from Sudan. In the mid-1990s, the Tutsi ethnic group that had fled Rwanda to Uganda returned to Rwanda, bringing at least 650,000 cattle to graze in Akagera National Park. The Minister of Agriculture retracted protected area status from 60 percent of the park as a way to help returning refugees.

Aquatic and lakeshore habitats are also threatened. By 1990, Burundi's Lac Rhiwindi Nature Reserve suffered serious human impact, and Lake Mburo National Park has been legally halved in area because of encroachment. Many other wetlands throughout the ecoregion are also imperiled.

Justification for Ecoregion Delineation

This ecoregion largely follows White's (1983) Lake Victoria Regional Mosaic, based on the following vegetation types: "mosaic of lowland rain forest and secondary grasslands," "mosaic of East Africa evergreen bushland and secondary *Acacia* wooded grassland," and pockets of "drier type Guineo-Congolian rainforest." The northeastern boundary has been modified from White's units to extend further north into Sudan.

Ecoregion Number: **42**

Ecoregion Name: **Southern Congolian Forest-Savanna Mosaic**

Bioregion: **Central Africa**

Biome: **Tropical and Subtropical Grasslands, Savannas, Shrublands, and Woodlands**

Political Units: **Democratic Republic of the Congo, Angola**

Ecoregion Size: **569,700 km²**

Biological Distinctiveness: **Bioregionally Outstanding**

Conservation Status: **Vulnerable**

Conservation Assessment: **V**

Author: *Illisa Kelman*

Reviewed by: *Jo Thompson, John Hart, Terese Hart*

Location and General Description

The Southern Congolian Forest-Savanna Mosaic [42] covers much of south central Democratic Republic of the Congo (DRC) and the northeast corner of Angola. Unlike the abrupt transition from forest to savanna that occurs north of the equator, the boundaries between this ecoregion and its neighbors are notably diffuse. To the north, this ecoregion grades into the peripheral semi-evergreen broadleaf moist forests of the Central Conglian Lowland Forests [15]. To the south and west, miombo floral elements blur the border with adjacent miombo ecoregions [49, 50] (Mayaux et al. 1999).

The annual mean maximum temperature is between 27°C and 30°C, and the annual mean minimum temperature is between 18°C and 21°C. Rainfall averages 1,400 mm per annum, higher than that of the Western Congolian forest-savanna mosaic. In the drier southeast, annual precipitation falls to 1,200 mm, whereas at the northern margins and in the southern highlands, annual precipitation rises to 1,600 mm, supporting forest.

This ecoregion gradually rises from the southern part of the Congo Basin at 300–400 m onto the Central African Plateau at about 1,000 m. This elevational change was caused by pre- and mid-Cretaceous uplift and down warping (Cahen et al. 1984). Deep river valleys provide much of the topographic relief and expose pre-Cretaceous "Karoo" sedimentary rock, which underlies much of the region (White 1983). Pockets of Precambrian basement are also exposed in some places. Quaternary Kalahari sands stretch across the southern portion of the ecoregion. The variations in the parent materials are reflected by the soils, with entisols on the Kalahari sands, oxisols on post-Jurassic rocks, and ultisols on the Precambrian basement (Brady and Weil 1999).

The Southern Congolian Forest-Savanna Mosaic [42] occupies the central portion of White's (1983) Guineo-Congolian/Zambezian regional transition zone, a mosaic of dry semi-evergreen Guineo-Conglian rainforest and grassland. The degree to which these vegetation complexes are interwoven is influenced by the severity and duration of the dry season, rainfall, groundwater availability, land-use and fire history, soils, past climate, topography, underlying geology, geomorphology, and extent of elephant herbivory (Hopkins 1992). Characteristic of the ecoregion are long, ribbon-like forests running north to south along rivers fanning from the Congo Basin into the savanna woodlands to the south. Larger tree species characteristic of the drier peripheral rainforest include *Albizia zygia, Lovoa trichilioides,* and *Parkia filicoidea.* The southern part of the ecoregion is characterized by secondary grassland and wooded grassland. Principal grasses include *Andropogon schirensis, Hyparrhenia confinis,* and *Pennisetum unisetum.* Although some species, such as *Albizia adianthifolia,* have Guineo-Congolian linkages, most are Zambezian species.

Outstanding or Distinctive Biodiversity Features

Over the last 10 million years, Central Africa may have experienced more than twenty climatic fluctuations, causing major vegetation shifts. At times forests were restricted to river margins, and at other times humid tropical forests extended into land currently under savanna. Flora and fauna shifted location, adapted, or disappeared. This ecoregion is in the zone of dynamic change between forest and savanna. There are few endemics and moderate levels of species richness in all groups.

Widespread forest mammals occurring here include bongo (*Tragelaphus eurycerus*), blue duiker (*Cephalophus monticola*), and yellow-backed duiker (*Cephalophus silvicultor*). More typical savanna animals found across the mosaic include waterbuck (*Kobus ellipsiprymnus*), southern reedbuck (*Redunca arundinum*), elephant (*Loxodonta africana,* EN), roan antelope (*Hippotragus equinus*), buffalo (*Syncerus caffer*), and hippopotamus (*Hippopotamus amphibius*). The top predator is the lion (*Panthera leo,* VU) (Kingdon 1989; IUCN 1992b).

Although rates of endemism are generally low, there are a few endemic small mammals (*Chaerephon gallagheri*, CR; *Congosorex polli*, CR; *Malacomys lukolelae*; *Mus kasaicus*, CR; and *Praomys minor*, VU). Some northern parts of this ecoregion also support populations of bonobo (*Pan paniscus*, EN) and Congo peacock (*Afropavo congensis*, VU). In the Lukuru Wildlife Reserve bonobo move into the savannas on a seasonal basis (J. Hart, pers. comm., 2002).

No bird species are strictly endemic to the ecoregion (Demy and Louette 2001). Low rates of endemism are also found in amphibians and reptiles. Herpetologic endemics include a strictly endemic reed frog (*Hyperolius polli*) and two other amphibians, known from only one other ecoregion (*Hyperolius obscurus* and *Arthroleptis phrynoides*). Three endemic reptile species are the African bighead snake (*Hypoptophis wilsoni*), *Monopeltis kabindae*, and *Feylinia macrolepis*.

Status and Threats

Current Status

Fossil evidence suggests that forest-savanna mosaic habitats have persisted since the last glacial maximum around 20,000 years ago but that more recently people have become important factors in shaping the forest and savanna (Furley et al. 1992). Significant areas of forest remain close to the various rivers that feed into the Congo River. Most of the forest is found in the Kasai Occidental province of DRC. Much of the rest of the ecoregion has savanna woodlands, often fragmented by agriculture.

Only small areas of the ecoregion are protected. Outliers of savanna are protected in the southern portion of Salonga National Park (J. Hart, pers. comm., 2002). The Shaba Elephant Reserve and the Lukuru Wildlife Reserve are in the ecoregion. Managed areas include Bushimaie, Luama, and Mangai. Since the start of the armed conflict in the 1990s most protected areas of the DRC have received little investment from the government or from international conservation organizations (Hart and Mwinyihali 2001). Some protection has continued in the Lukuru Reserve, despite the war.

Types and Severity of Threats

Humans are believed to have accelerated the spread of grasslands across this region, especially in the drier south, by increasing the frequency and intensity of the fire cycle and clearing land for agriculture. The most extensive areas of burning occur in the southern Bandundu, Kasai, and Shaba provinces of the DRC, where the boundary between the dense humid forest and savanna is clearly demarcated. Elephants have also contributed to the spread of grasslands (Hopkins 1992).

Since 1994, the DRC has suffered serious ethnic strife and civil war. Although the impacts of armed conflict have been felt across the country, it remains unclear how severely biodiversity in this ecoregion has been affected. Available evidence and high human population densities in some areas (Bandundu, Kikwit, Kasai) suggest that fauna has been severely depleted (J. Hart, pers. comm., 2002).

Justification for Ecoregion Delineation

This forest-savanna mosaic ecoregion lies between lowland rainforest to the north and miombo woodland ecoregions to the south. It is separated from the Western Congolian Forest-Savanna Mosaic [43] to the west of the Kwango River by higher rainfall and some differences in species composition in the flora and fauna. The majority of the ecoregion consists of "mosaic of lowland rain forest and secondary grassland" but also incorporates small areas of "riverine forest" and "Kalahari Sand grasslands" of White (1983).

Ecoregion Number:	**43**
Ecoregion Name:	**Western Congolian Forest-Savanna Mosaic**
Bioregion:	**Central Africa**
Biome:	**Tropical and Subtropical Grasslands, Savannas, Shrublands, and Woodlands**
Political Units:	**Angola, Democratic Republic of the Congo, Republic of the Congo, Gabon**
Ecoregion Size:	**413,400 km²**
Biological Distinctiveness:	**Bioregionally Outstanding**
Conservation Status:	**Relatively Stable**
Conservation Assessment:	**V**
Author:	*Illisa Kelman*
Reviewed by:	*Lee White*

Location and General Description

The Western Congolian Forest-Savanna Mosaic [43] covers the dissected plateaus bordering the lower Congo River as it flows through the Democratic Republic of the Congo (DRC), Republic of the Congo, and northern Angola. There are also distinctive outliers in the humid rainforest of Gabon and along the Atlantic coast from the Republic of the Congo through to the Democratic Republic of the Congo (DRC). The Congo River bisects the ecoregion as it passes between Kinshasa and Brazzaville.

Close to the Congo River, as far inland as Kinshasa, the ecoregion is fairly flat and reaches an altitude of 200 m. North of the Congo River the ecoregion rises onto plateaus that average 650 m in elevation, interspersed with spectacular canyons as deep as 300 m in Lefini Park. To the south in Angola the ecoregion may reach 900 m altitude. Much of the ecoregion has Precambrian basement rocks (Cahen et al. 1984), with post-Jurassic sediments and Tertiary-aged Kalahari sands in the east. Heavily leached oxisols cover the post-Jurassic and Precambrian rocks, with entisols over the Kalahari sands.

The current climate is tropical, with limited seasonality. Mean maximum temperatures range from 30°C in the lowlands and 21°C on the high plateaus. Mean minimum temperatures vary similarly, from 21°C to 15°C respectively. Where this ecoregion borders the Guineo-Congolian forest region, mean annual rainfall is more than 1,400 mm, but across most of the region the mean is around 1,200 mm.

The climatic history of this ecoregion is key to understanding its current vegetation (Weber et al. 2001). During the climatic fluctuations associated with ice ages over the last 2.5 million years, forest vegetation has expanded and contracted (Maley 1996, 2001). The last glacial maximum was around 18,000 years ago, when savanna dominated this area. Forest also contracted 3,000–2,500 years ago when there was a drier climatic period and agricultural expansion (Schwartz 1992b). The forests have been expanding again over the past 2,000 years, and there is evidence that the current savanna areas are being colonized by young successional stages of forest. Shifting cultivation and even fire along the forest margin have not prevented this expansion (Maley 2001).

White (1983) classifies most of this ecoregion in the Guineo-Congolian/Zambezian regional transition zone, a mosaic of wooded grasslands with patches of forest. Guineo-Congolian semi-evergreen forests (White 1983) extend many kilometers into the savanna habitats along the broad valleys of the Congo River tributaries. These gallery forests have a similar species composition to those of the interior Congolian rainforests. Stands of dense dry evergreen forest occupy the Bateke Plateau, which stretches across southern Republic of the Congo (White 1983; Mayaux et al. 1999). Mabwati, a dense dry forest community growing on Kalahari sands, includes canopy species found only in the Southern and Western Congolian Forest-Savanna Mosaic ecoregions [42, 43]: *Marquesia macroua, M. acuminata, Daniella alsteeniana,* and *Berlinia giorgii.* The remainder of the ecoregion consists of savanna grassland and savanna-woodland habitats supporting widespread species of the Zambezian regional center of endemism (White 1983; White and Abernethy 1997). Drier miombo woodland, characterized by *Brachystegia* spp. and *Julbernadia* spp., demarcates the western border of the ecoregion. Wet miombo, characterized by the addition of *Isoberlinia* spp., forms the eastern boundary (White 1983).

Outstanding or Distinctive Biodiversity Features

This ecoregion supports moderate species richness in all taxonomic groups, principally because of the interdigitation of habitats, the presence of several large rivers, and associated gallery forests. The mosaic landscape creates a high ratio of ecotone to interior habitat and may be a region of tropical differentiation and speciation (Smith et al. 1997). Although there are few assessments of plant diversity, the plants of the Lopé Reserve have been described in detail (White et al. 2000).

One frugivorous bat (*Rhinolophus adami,* DD) and a mouse (*Dendroprionomys rousseloti*) are the only mammals strictly endemic to this ecoregion. One other bat is regarded as a near endemic, *Rhinolophus silvestris.* There are also several threatened mammal species: elephant (*Loxodonta africana,* EN), gorilla (*Gorilla gorilla,* EN), chimpanzee (*Pan troglodytes,* EN), sun-tailed guenon (*Cercopithecus solatus,* VU), and Bouvier's red colobus (*Procolobus pennanti bouvieri,* CR). The latter may now be restricted to Lefini Reserve (Hilton-Taylor 2000). Widespread mammals include buffalo (*Syncerus caffer*), waterbuck (*Kobus ellipsiprymnus*), bushbuck (*Tragelaphus scriptus*), reedbuck (*Redunca arundinum*), and yellow-backed and common duikers (*Cephalophus silvicultor, Sylvicapra grimmia*) (Millington et al. 1992). The lion (*Panthera leo,* VU) is the top predator in the area.

Two strictly endemic birds are found in the ecoregion: the white-headed robin-chat (*Cossypha heinrichi,* VU), which is protected only in the DRC's Bombo Lueme Hunting Reserve, and the orange-breasted bush-shrike (*Laniarius brauni,* EN). Both species are threatened by continued forest clearing. Five other species are regarded as near endemics: red-backed mousebird (*Colius castanotus*), pale-olive greenbul (*Phyllastrephus fulviventris*), black-chinned weaver (*Ploceus nigrimentum*), African river-martin (*Pseudochelidon eurystomina*), and red-crested turaco (*Tauraco erythrolophus*).

Herpetologic strict endemics include the western Congo worm lizard (*Monopeltis guentheri*), the lizard *Mabuya bocagii,* and the Cambondo screeching frog (*Arthroleptis carquejai*), known from only one site in Angola (Frost 1999). In general this is not an important area for herpetofaunal endemism.

In the northern part of the ecoregion there is a gradual encroachment of forest into the savanna woodland habitats (White and Abernethy 1997). In these places the flora and fauna of the savanna portion of the ecoregion are being replaced by species more typical of the Congo Basin forests (of the Northwestern and Atlantic Equatorial forest ecoregions).

Status and Threats

Current Status

The current distribution of forest and savanna habitats is a consequence of long-term climatic fluctuations, causing forest expansion and retreat, and more recent human influences. Although large tracts of wooded savanna remain, savanna animals in this ecoregion are severely threatened by intense hunting pressures.

A number of protected areas exist. In the Republic of the Congo these include Lefini, Nyanga Nord, Tsoulou, and Mount

Fouari Faunal Reserve and Mount Mavoumbou and Nyanga Sud Hunting Reserve. In Gabon there are the Moukalaba-Doudou, Plateau Batéké, and Lopé national parks. A number of new protected areas have recently been declared in Gabon. The Forestière de Luki Biosphere Reserve is found in DRC. There are no protected areas in Angola.

Types and Severity of Threats

Two processes are running in parallel in this ecoregion, which may have different outcomes for future biodiversity values. First, a natural expansion of the rainforest component of the ecoregion has been observed in several regions over periods of decades and may have been continuing for 2,000 years (Maley 2001; White 2001). This process is likely to eventually destroy the Lopé savanna outlier (forest is currently kept in check by controlled burning). This process may reduce the habitat available for savanna-woodland species. Alternatively, human-caused deforestation in this ecoregion may push back the natural forest expansion and degrade the savanna woodlands. The savanna woodland habitats are under threat from subsistence farming, and close to dense urban centers (e.g., Kinshasa) they are being exploited to supply charcoal to the urban market. Timber is also harvested for local use and export, with a significant timber industry operating in the forest patches in this ecoregion. Mammal populations have been dramatically reduced in many areas by hunting; porcupines, duikers, and primates are the most heavily hunted species.

The Gabonese portion of this ecoregion has been politically stable and peaceful since independence. Civil warfare has affected other parts of this ecoregion in Angola, Republic of the Congo, and DRC. In Angola alone hundreds of thousands of people have been left homeless and 1.5 million killed during 25 years of periodic civil war.

Justification for Ecoregion Delineation

This ecoregion represents the western portion of White's (1983) Guineo-Congolian/Zambezian regional transition zone and incorporates the "mosaic of lowland rain forest and secondary grassland" with occasional pockets of drier rainforest. It is separated from the Southern Congolian Forest-Savanna Mosaic [42] to the east of the Kwango River because of the presence of some narrowly endemic mammals and birds on either side of this divide. The Lopé savannas, which are surrounded by forest habitats, are somewhat different from the other savannas in this ecoregion but are clearly distinct from the surrounding forests.

Ecoregion Number:	**44**
Ecoregion Name:	**Somali *Acacia-Commiphora* Bushlands and Thickets**
Bioregion:	**Horn of Africa**
Biome:	**Tropical and Subtropical Grasslands, Savannas, Shrublands, and Woodlands**
Political Units:	**Sudan, Eritrea, Ethiopia, Somalia, Kenya**
Ecoregion Size:	**1,053,900 km²**
Biological Distinctiveness:	**Globally Outstanding**
Conservation Status:	**Relatively Stable**
Conservation Assessment:	**III**
Authors:	*Chris Magin, Miranda Mockrin*
Reviewed by:	*Ib Friis, Mats Thulin*

Location and General Description

The Somali *Acacia-Commiphora* Bushlands and Thickets [44] occupy the majority of the Horn of Africa to the east of the Ethiopian Highlands, including the Ogaden Desert and northeast Kenyan semi-deserts. A narrow corridor of the ecoregion occupies the floor of the Ethiopian section of the Rift Valley, separating the northwestern and southeastern Ethiopian Highlands. A solid finger to the Sudan border encircles the southwestern lowlands of Ethiopia, and a slim finger extends north to the Eritrea-Sudan border. The ecoregion is mainly flat and low-lying (more than half lies below 500 m), rising toward the west and north. However, it is defined more by rainfall and vegetation type than by altitude and thus extends from sea level on the coast of Somalia to more than 1,500 m in the Rift Valley and Sidamo region of southern Ethiopia.

The mean maximum temperatures are around 30°C, and the mean minimum temperatures are 15° to 18°C. Annual rainfall varies from below 100 mm in the Ogaden Desert to around 600 mm in areas bordering the Ethiopian Highlands. There are only three permanent rivers: the Awash, Wabi Shebele, and Jubba, all of which originate in the Ethiopian Highlands. Most of the area is underlain by post-Cretaceous rocks that are mainly marine in origin, over which soils indicative of high aridity (xerosols and yermosols) have developed. Deep, infertile sands characterize the Somali hinterland, or Haud. Precambrian granites outcrop as inselbergs, or *burs,* in southern Somalia.

The ecoregion is sparsely populated, typically with fewer than twenty people per square kilometer. In the heart of the Ogaden Desert and other dry parts of the ecoregion there are no permanent inhabitants, whereas in the Ethiopian Rift Valley there is smallholding agriculture. In the lower basins of the three major rivers, the Awash, Wabi Shebele, and Jubba, there are extensive areas of irrigated farming.

Phytogeographically, the ecoregion lies in the Somali-Masai regional center of endemism, dominated by deciduous bushland and thicket (White 1983). The most common tree species are deciduous and belong to the genera *Acacia* and *Commiphora*. The understory consists of shrubby herbs less than 1 m high, such as *Acalypha*, *Barleria*, and *Aerva*. At lower elevations where rainfall is less consistent, vegetation becomes semi-desert scrubland. Here *Acacia* and *Commiphora* are joined by *Euphorbia*, *Aloe*, and grass species such as *Dactyloctenium aegyptium* and *Panicum turgidum* (Tilahun et al. 1996). *Crotalaria* and *Indigofera* are also found. Forest vegetation once surrounded the bases of some inselbergs in southern Somalia and lined permanent watercourses but has largely been destroyed by human activity (Friis 1992). Important evergreen genera include *Boscia*, *Dobera*, *Salvadora*, *Grewia*, and *Cadaba*.

Outstanding or Distinctive Biodiversity Features

The Horn of Africa is a well-recognized center of endemism (Kingdon 1989). Around 2,500 plant species are recorded from the Somalia-Masai phytochorion (White 1983), of which around half are believed to be endemic. Up to fifty plant genera are also endemic to the ecoregion (White 1983). Several subcenters of plant endemism are found in northeastern Somalia, in the Ogaden Desert, in the region of the *burs* of southern Somalia, in the southern part of the former Bale Region of southeastern Ethiopia, and in the Borana region of southern Ethiopia (Thulin 1994; Friis et al. 2001). These areas often coincide with exposed limestone (Friis et al. 2001). Examples of endemic trees are *Boswellia rivae* (a frankincense tree), *Commiphora guidottii* (the source of scented myrrh), *Steganotaenia commiphoroides* (a member of the largely herbaceous family Apiaceae that grows to 10 m), and *Hildegardia gillettii* (with a flask-shaped trunk).

The Horn of Africa is also a notable center of endemism for mammals, particularly for arid-adapted antelopes such as the dibatag (*Ammodorcas clarkei*, VU), beira (*Dorcatragus megalotis*, VU), hirola (*Damaliscus hunteri*, CR), and Speke's gazelle (*Gazella spekei*, VU). Between 400 and 1,000 hirola live in the vicinity of Bura east of the Tana River in Kenya. The ecoregion also supports several strict endemic small mammals, including four *Gerbillus* species, one *Microdillus* species, one white-toothed shrew (*Crocidura greenwoodi*, VU), and the walo (*Ammodillus imbellis*, VU), known only from Somalia. The Somali warthog (*Phacochoerus aethiopicus delamerei*, VU) is also near-endemic to this ecoregion.

The widely distributed but threatened ungulate species dorcas gazelle (*Gazella dorcas*, VU) and Sömmerring's gazelle (*Gazella soemmerringii*, VU) are also found. In northern Ethiopia a small population of African wild ass (*Equus africanus somaliensis*, CR) inhabits the Yangudi Rassa National Park, and a herd

of Grevy's zebra (*Equus grevyi*, EN) occurs in the Alledeghi Game Reserve. Small numbers of the formerly widespread but now endangered subspecies Swayne's hartebeest (*Alcelaphus buselaphus swaynei*, EN) inhabit the Senkelle Wildlife Sanctuary and Nechisar National Park in Ethiopia's Rift Valley. Elephants (*Loxodonta africana*, EN) and buffalo (*Syncerus caffer*) were previously widespread in the wetter portions of this ecoregion (Barnes et al. 1999). Elephant populations are decreasing, with limited numbers found in protected areas. The Babile Elephant Reserve in Ethiopia was established to protect the only known population of the isolated, ecologically distinct subspecies *Loxodonta africana orleansi* (Barnes et al. 1999). This subspecies may also still occur in the Somali Alifuuto (Arbowerow) Nature Reserve (WCMC 1993). Lion (*Panthera leo*, VU), leopard (*Pathera pardus*), cheetah (*Acinonyx jubatus*, VU), and striped and spotted hyenas (*Hyaena hyaena* and *Crocuta crocuta*) are the main large carnivores, although wild dog (*Lycaon pictus*, EN) is also found in Mago and Omo national parks in Ethiopia (Woodroffe et al. 1997).

The Abyssinian yellow-rumped seedeater (*Serinus xanthopygius*), the short-billed crombec (*Sylvietta philippae*, DD), and Sidamo bushlark (*Heteromirafra sidamoensis*, VU) are restricted to this ecoregion, and the sombre chat (*Cercomela dubia*, DD), white-winged collared-dove (*Streptopelia reichenowi*), Salvadori's weaver (*Ploceus dicrocephalus*), and scaly babbler (*Turdoides squamulatus*) are near endemic. Most endemic species are associated with dry habitats. However, the riverine habitats along the Jubba and Wabi Shebele support two strictly endemic birds, the Degodi lark (*Mirafra degodiensis*, VU) and the Bulo Burti bush-shrike (*Laniarius liberatus*, CR) (Stattersfield et al. 1998).

The ecoregion is of very high importance for endemic reptiles, including the strictly endemic Somali agama (*Agama bottegi*), Lanza's racerunner (*Eremias ercolinii*), and Somali writhing skink (*Lygosoma somalicum*).

Status and Threats

Current Status

The habitats of the ecoregion are mainly dry woodlands and scrub, with a gradation to grasslands and deserts in the driest places. Most of these areas remain unfragmented and intact because the human population is low and agriculture is concentrated along watercourses. However, the ecoregion has been severely affected by political instability and war over the past few decades. Large mammal populations have been depleted, especially in Somalia, where there has been conflict in the past decade. Stable government must return to Somalia before large-scale conservation work can occur in the Somalian portion of this ecoregion. The Ethiopian portion is currently stable, but conflicts between protected areas and people continue.

There are several protected areas in this ecoregion. Protected areas in Ethiopia include Yangudi Rassa, Nechisar, Awash, Omo, and Mago national parks, Chew Bahr Wildlife Reserve, and Babile Elephant Sanctuary. Most of these parks are not well protected. In Kenya, the Malka Mari National Park falls in this ecoregion, and in Somalia, the Alifuuto (Arbowerow) Nature Reserve occurs, although there is no recent information about the status of this site.

Types and Severity of Threats

Habitats are partially degraded through grazing by livestock and fuelwood collection, particularly close to villages and towns. Overuse is more intense in areas where large-scale farming and irrigation schemes have been launched (e.g., around Lake Ziway in Ethiopia). Agricultural schemes have also denuded the landscape in Ethiopia in the Gode Plain along the Wabi Shebele River and in the Awash Valley in the Afar region. Similar projects have existed for many years along the Wabi Shebele and Jubba rivers in Somalia. Natural gas fields are being developed in the region of Ethiopia closest to the Somali border (Tilahun et al. 1996). A severe problem in northern Somalia is the uncontrolled exploitation of the *Acacia* woodlands (particularly *A. bussei*) for charcoal production. Riverine vegetation often is extremely degraded, mostly by the cutting of fuelwood and poles. Some economically important species, such as the yeheb nut (*Cordeauxia edulis*), may be declining because of overgrazing.

Protected areas are also degraded by human use. Recent reports indicate that more people are moving into Mago, Omo, and Nechisar national parks, with serious impacts on the vegetation and wildlife.

Justification for Ecoregion Delineation

This ecoregion is a major center of endemism for plants, reptiles, mammals, and birds in the dry Horn of Africa. The ecoregion is mapped to cover the northeastern part of the "Somali-Masai *Acacia-Commiphora* bushland and thicket" and the "Somalia-Masai semi-desert grassland and shrubland" vegetation units of White (1983). It is bound on the west by the Omo River, to the south by the Guiba and Tana rivers, and inland by areas of sandy and rocky plateau. These boundaries reflect the southern extent of the dibatag and the northern extent of the common eland (*Taurotragus oryx*), Bohor reedbuck (*Redunca redunca*), and Burchell's zebra (*Equus burchelli*).

Ecoregion Number: **45**

Ecoregion Name: **Northern *Acacia-Commiphora* Bushlands and Thickets**

Bioregion: **Eastern and Southern Africa**

Biome: **Tropical and Subtropical Grasslands, Savannas, Shrublands, and Woodlands**

Political Units: **Kenya, Sudan, Uganda, Ethiopia, Tanzania**

Ecoregion Size: **326,000 km²**

Biological Distinctiveness: **Regionally Outstanding**

Conservation Status: **Relatively Stable**

Conservation Assessment: **III**

Authors: *Mary Rowen, Colleen Seymour*

Reviewed by: *Derek Pomeroy, Marc Languy, Alan Rodgers*

Location and General Description

The Northern *Acacia-Commiphora* Bushlands and Thickets [45] extend from the southeast corner of Sudan and northeast Uganda through much of lowland Kenya, reaching as far as the border with the Northern Zanzibar-Inhambane Coastal Forest Mosaic [20]. To the north it is replaced by drier savanna and semi-desert vegetation, and to the south it grades into the Southern *Acacia-Commiphora* Bushlands and Thickets [46] around the Kenya-Tanzania border.

The ecoregion falls in the seasonal tropics, with seasonality controlled by movement of the Inter Tropical Convergence Zone. Mean maximum temperatures are 30°C in the lowlands, falling to around 24°C in the higher elevations. Mean minimum temperature ranges from 18°C to 21°C. Annual rainfall ranges from 200 mm in the drier areas near Lake Turkana to about 600 mm closer to the Kenyan Coast. Most precipitation falls in the long rains, typically from March to June, with less falling during the short rains of October–December. The timing and amounts of rainfall vary greatly from year to year, and it is not uncommon that one or both rainy seasons fail. In northern Kenya and southern Sudan, there is typically one rainy period per year. During drier periods the desiccated vegetation becomes highly flammable, and large parts of the ecoregion burn every year.

The underlying rocks are Precambrian basement, Tertiary volcanic lavas, and Quaternary basin and dune formations. These sediments outcrop in many places because of shallow soils, but elsewhere the ecoregion is level or gently undulating. The lowest elevation is in the east, north, and northwest (elevations of 200–400 m), increasing toward the south and southwest, where elevations rise up to 1,000 m. There are also a number of mountains, such as Moroto in Uganda, which exceeds

3,000 m, and several others over 2,000 m (e.g., Lenoghi, Matthews, and Kulal). The soils indicate aridity and are mainly aridisols, with entisols around Lake Turkana Basin. Along the moister western margins, vertisols can also be found.

In terms of the phytogeographic classification of African vegetation (White 1983), this ecoregion is part of the Somali-Masai regional center of endemism. The vegetation is predominantly *Acacia-Commiphora* bushland and thicket. Common plant genera include *Acacia, Commiphora, Boswellia, Aristida, Stipa,* and *Chloris* grasses.

Outstanding or Distinctive Biodiversity Features

This ecoregion is essentially a transition zone between the drier Somali *Acacia-Commiphora* Bushlands and Thickets [44] to the north and the wetter Southern *Acacia-Commiphora* Bushlands and Thickets [46] to the south. Much of the flora and fauna overlaps from these adjacent ecoregions, resulting in a mixture of drought-adapted and tropical savanna species. Although the dry climate means that there is an insufficient vegetation base to sustain the vast migratory herds of large mammals found in the south, it is too wet to be inhabited solely by the more arid-adapted species found in the north.

Although there is no assessment of the plant richness or endemism across this ecoregion, biodiversity studies in southern Karamoja (Pomeroy and Tushabe 1996) illustrate high species richness. Mammalian species diversity in this ecoregion is high, but there are few endemics, all rodents: Cosens's gerbil (*Gerbillus cosensi,* CR), diminutive gerbil (*G. diminutus*), Percival's gerbil (*G. percivali*), and Loring's rat (*Thallomys loringi*). The highly threatened antelope hirola (*Damaliscus hunteri,* CR) occurs as a small introduced population in Tsavo National Park (south of the species' natural range) (East 1999). The ecoregion also supports several species of arid-adapted ungulates: Grevy's zebra (*Equus grevyi,* EN), beisa oryx (*Oryx gazella beisa*), gerenuk (*Litocranius walleri*), and lesser kudu (*Tragelaphus imberbis*).

More than 10,000 elephants (*Loxodonta africana,* EN) are found in the ecoregion, but populations have declined greatly over the past 30 years. Between 1975 and 1980, elephant numbers in Tsavo National Park declined from nearly 35,000 to about 14,000 after a period of drought followed by heavy poaching (Ottichilo 1987). The population is currently around 7,500 animals. Once numerous in the central and southern portions of this ecoregion, the black rhinoceros (*Diceros bicornis,* CR) has been extirpated in most places. Of an estimated 65,000 individuals present in East Africa in the 1960s, only about 420 remain in Kenya (Emslie and Brooks 1999). Most of the surviving animals are found in heavily guarded areas of national parks and fenced sanctuaries, many in this ecoregion. Wild dogs (*Lycaon pictus,* EN) also occur but have declined greatly. In this ecoregion, they are now extinct in Amboseli and Nairobi national parks and Buffalo Springs National Reserve and are scarce elsewhere (Maggi 1995; Woodroffe et al. 1997). Other large carni-

vores include lion (*Panthera leo,* VU) and cheetah (*Acinonyx jubatus,* VU). Approximately 80 percent of the remaining 4,500 Grevy's zebras in the wild are found in the Laikipia-Samburu areas of this ecoregion, and there are important populations of Masai and reticulated giraffe (*Giraffa camelopardalis tippelskirchi* and *reticulata*).

The ecoregion has a species-rich avifauna, but endemism is low. Strict endemics include Friedmann's lark (*Mirafra pulpa,* DD), Williams's lark (*Mirafra williamsi,* DD), and Hinde's pied-babbler (*Turdoides hindei,* VU) (Bennun and Njoroge 2001).

Reptile species richness is also high, although the number of endemic species is low. Strictly endemic reptiles include Scheffler's dwarf gecko (*Lygodactylus scheffleri*), side spotted dwarf gecko (*Lygodactylus laterimaculatus*), and Teitana purple-glossed snake (*Amblyodipsas teitana*). One amphibian is believed to be strictly endemic to the area, the Aruba dam reed frog (*Hyperolius sheldricki*). One threatened reptile is the pancake tortoise (*Malacochersus tornieri,* VU), which is overexploited for the pet trade.

Status and Threats

Current Status

Historically, human use of the area was limited to pastoralist and hunter-gatherer societies because little of the land is suitable for intensive agriculture, with less than 600 mm rainfall annually. The herds of pastoralists' cattle, sheep, goats, and camels are a significant component of the system (Burney 1996). Large areas of habitat still remain.

The ecoregion contains a number of well-functioning national parks and other reserves. Protected areas in Kenya include South Turkana and Samburu national reserves and a number of national parks: Meru, Kora, Longonot, Ol Donyo Sabuk, Nairobi, Amboseli, Chyulu, Tsavo East and West, and Maralai. In Tanzania, Umba and Mkomazi game reserves are in this ecoregion, and in Uganda, the Bokora corridor connects Matheniko and Pian Upe wildlife reserves. In Kenya, private ranches and sanctuaries also support important wildlife populations.

Types and Severity of Threats

The habitats and species of this ecoregion are increasingly threatened by unsustainable water use, frequent grassland burning, tree cutting, and farmland expansion. Illegal hunting for skins, ivory, and rhino horn have also severely reduced populations of large animals, particularly elephants and rhinoceros (Barnes et al. 1999; Emslie et al. 1999). Although much of this illegal hunting has been curtailed, an increasing level of bush-meat hunting threatens many wildlife species (TRAFFIC 2000). Tourists are also problematic in some of the protected areas, harassing species of conservation concern such as Grevy's zebra and cheetah (IUCN 1987). Throughout the ecoregion, cutting of trees for firewood and charcoal production is a major threat to the maintenance of the *Acacia-Commiphora* bushland and thicket. Some plant species, such as the African blackwood (*Dalbergia melanoxylon*), are also threatened by overharvesting. The wood is used mainly to make carvings for sale to tourists, and large trees have been removed from most areas.

The human dominance of standing waterholes and springs and the diversion of rivers poses serious threats to wildlife. Upstream extraction from the Ewaso Nyiro River for food and flower production is particularly problematic because it has prevented this river from flowing year-round into the Buffalo Springs, Samburu, and Shaba reserves.

Justification for Ecoregion Delineation

This ecoregion forms a part of the "Somali-Masai *Acacia-Commiphora* bushland and thicket" vegetation unit of White (1983). This larger unit was separated into three ecoregions based on different bioclimatic and associated floral and faunal patterns. This ecoregion is a transition zone between drier and moister habitats. It contains the southern extent of species such as the gerenuk, which are more common in the Somali *Acacia-Commiphora* Bushlands and Thickets [44] to the north. It also contains species more commonly found further south in moister habitats, such as the Masai giraffe.

Ecoregion Number:	**46**
Ecoregion Name:	**Southern *Acacia-Commiphora* Bushlands and Thickets**
Bioregion:	**Eastern and Southern Africa**
Biome:	**Tropical and Subtropical Grasslands, Savannas, Shrublands, and Woodlands**
Political Units:	**Tanzania, Kenya**
Ecoregion Size:	**227,800 km²**
Biological Distinctiveness:	**Globally Outstanding**
Conservation Status:	**Vulnerable**
Conservation Assessment:	**I**
Authors:	*Colleen Seymour, Mary Rowen*
Reviewed by:	*Alan Rodgers, Richard Estes*

Location and General Description

The Southern *Acacia-Commiphora* Bushlands and Thickets [46] ecoregion is located in north central Tanzania and extends into southwestern Kenya on the eastern margins of Lake Victoria. The ecoregion grades into miombo woodland to the south and into the Northern *Acacia-Commiphora* Bushlands and Thickets [45] to the north. The boundaries and transitions to other ecoregions are somewhat indistinct.

The climate is tropical, with a bimodal rainfall pattern. The long rains occur from March to May and the short rains from November to December, but the short rains often fail, and the long rains sometimes do as well. Mean rainfall is 600–800 mm annually through most of the ecoregion, with extremes of 500 mm in the dry southeastern plains and 1,200 mm in the northwestern portion in Kenya. Mean maximum temperatures are around 30°C at lower elevations and as low as 24°C in the highest part of the ecoregion. Mean minimum temperatures are between 9°C and 18°C, and normally between 13°C and 16°C.

Precambrian basement rocks (up to 2.5 billion years old) underlie most of the ecoregion and often outcrop as inselbergs (locally called kopjes) or small mountains. The ecoregion is generally between 900 and 1,200 m altitude, becoming higher from east to west. Some areas are rugged, and in the Masai Steppe region of Tanzania mountains are found. For example, Lolkisale (2,132 m) and Lossogonoi (2,124 m) both support montane forests on their peaks (Baker and Baker 2002).

This ecoregion forms the southernmost part of the Somali-Masai phytochorion of White (1983). There is appreciable variation in the floristic composition, but *Acacia, Commiphora,* Capparidaceae, and *Grewia* are nearly always present (White 1983; McNaughton and Banyikwa 1995). In southwest Kenya, *Acacia* and *Combretum* (without *Commiphora*) dominate the vegetation. During the long dry season (August–October) the ecoregion becomes extremely desiccated: most trees lose their leaves, and the grasslands dry out and often burn. Some fires occur naturally, but most are started by pastoralists to promote new vegetative growth for their livestock, by hunters to drive game animals, or by agriculturists to clear their fields.

The human population ranges between ten and fifty people per square kilometer. The highest populations occur close to Lake Victoria and in the foothills of mountains, such as the Pare and Usambaras, in Tanzania. Masai pastoralists occur widely throughout the ecoregion, and cultivation by the Masai and other tribes is increasing.

Outstanding or Distinctive Biodiversity Features

This ecoregion supports some of the largest aggregations and migrations of large mammals in the world, including large numbers of wildebeest, zebra, Thompson's gazelle, buffalo, impala, eland, elephant, and lesser kudu. Part of this ecoregion contains the route of the annual Serengeti-Mara migration of approximately 1.3 million blue wildebeest (*Connochaetes taurinus*), 200,000 Burchell's zebra (*Equus burchelli*), and 400,000 Thomson's gazelle (*Gazella thomsonii*). The Ruaha-Rungwa ecosystem at the southern end of the ecoregion has the second largest population of elephants in Tanzania (approximately 40,000) (Barnes et al. 1998). The Masai Steppe of Tanzania provides wet season habitat for the elephants from Tarangire National Park, which return to that site in the dry season. The Masai Steppe is also an important habitat for lesser kudu (*Tragelaphus imberbis*), which ranges as far south as Ruaha National Park. Lion (*Panthera leo*, VU), leopard (*Panthera pardus*), wild dog (*Lycaon pictus*, EN), cheetah (*Acinonyx jubatus*, VU), and spotted hyena (*Crocuta crocuta*) also occur. Although there are large populations of many large mammal species, endemism is low and no strictly endemic species occur.

The species richness of birds is high, with Tarangire and Serengeti national parks each containing 350–400 bird species. The Serengeti Plains is also an endemic bird area (Stattersfield et al. 1998), supporting the restricted-range rufous-tailed weaver (*Histurgops ruficauda*) (monotypic genus), grey-crested helmet-shrike (*Prionops poliolophus*), Fischer's lovebird (*Agapornis fischeri*), and Karamoja apalis (*Apalis karamojae*, VU). Most of these species are found primarily in *Acacia-Commiphora* habitats.

Diverse assemblages of reptiles also occur, with three strict endemics: Mpwapwa purple-glossed snake (*Amblyodipsas dimidiata*), Mpwapwa worm lizard (*Chirindia mpwapwaensis*), and Mpwapwa wedge-snouted worm lizard (*Geocalamus modestus*). One species of special concern is the pancake tortoise (*Malacochersus tornieri*, VU), which is threatened by collection for the international pet trade (Hilton-Taylor 2000).

Status and Threats

Current Status

Protected areas cover around 20 percent of the ecoregion, including Masai Mara National Reserve and Ruma and Ndere Island national parks in Kenya, part of Serengeti National Park, Tarangire National Park, and Ruaha National Park, and Rungwa, Mkungunero, Swagaswaga, Maswa, Grumeti, Ikorongo, and Mkomazi game reserves in Tanzania. Ngorongoro Conservation Area also falls mainly in this ecoregion. There are also game controlled areas (e.g., Loliondo) and a few forest reserves in Tanzania. Habitats are well conserved in the protected areas but are subject to conversion outside these areas. In the past, severe elephant poaching has occurred across this region, including in the protected areas. This has been largely stopped by the Convention on International Trade in Endangered Species (CITES) restrictions on the ivory trade and the efforts of the Tanzanian government. Poaching of black rhinoceros (*Diceros bicornis*, CR) has been even more severe, and the species has essentially been extirpated from most of the ecoregion, except for a few heavily protected sites. Hunting of animals for meat has also expanded greatly in this ecoregion, including from within protected areas (TRAFFIC 2000).

Types and Severity of Threats

Major threats to the long-term viability of the ecoregion's flora and fauna include the conversion of habitat to farmland, the loss of viable corridors between protected areas, unsustainable charcoal production or logging, and the unsustainable killing of wildlife. Outside protected areas, habitat conversion continues throughout the ecoregion, with much of the land suitable for agriculture already converted. Commercial farms have transformed habitats in some parts of the Masai Steppe (Baker and Baker 2002), and smallholder farming is expanding. Increasingly, the Masai are starting to undertake farming in this ecoregion during the long rains, a change from their traditional pastoralism. Parks and reserves in the ecoregion are also becoming isolated habitat islands, and the number of habitat corridors allowing seasonal and drought-related movements and the natural migrations of species is declining. For example, the number of available corridors between Tarangire National Park and nearby protected areas has decreased from thirty to four since the early 1970s (Kahurananga and Silkiluwasha 1997).

Local populations and the urban centers of Arusha and Moshi in Tanzania are increasing demands for natural resources throughout this ecoregion. Charcoal is the main cooking fuel in towns and surrounding areas, and it comes mainly from unprotected woodlands in this ecoregion. Legal and illegal hunting for bushmeat is widespread, both for subsistence and for trade (Campbell and Hofer 1995; TRAFFIC 2000). The extirpation of the black rhinoceros outside heavily protected reserves (a few heavily guarded animals remain in Mkomazi [Coe et al. 1999], Ngorongoro, and Serengeti) and the CITES ban on ivory export have lowered rates of poaching. However, uncontrolled trophy hunting in some of the Tanzanian hunting concessions, especially in Loliondo, Grumeti, and Ikoronogo Game Reserves, has also caused wildlife losses and led to much debate (Baker and Baker 2001, 2002). In recent years there has been a great expansion of mining, particularly for gold and tanzanite, in parts of the ecoregion.

Justification for Ecoregion Delineation

This ecoregion occupies the southern section of the "Somali-Masai *Acacia-Commiphora* bushland and thicket" vegetation unit of White (1983). The southern boundary of this ecoregion directly follows that of White. The northern boundary is less defined in that there is a gradation between the northern and the southern *Acacia-Commiphora* bushland ecoregions. In the Southern *Acacia-Commiphora* Bushlands and Thickets [46] there is a gradual loss of species typical of the Northern *Acacia-Commiphora* Bushlands and Thickets [45], such as the gerenuk (*Litocranius walleri*). The ecoregion boundary is somewhat arbitrarily placed close to the Tanzania-Kenya national boundary.

Ecoregion Number: **47**

Ecoregion Name: **Serengeti Volcanic Grasslands**

Bioregion: **Eastern and Southern Africa**

Biome: **Tropical and Subtropical Grasslands, Savannas, Shrublands, and Woodlands**

Political Units: **Tanzania**

Ecoregion Size: **18,000 km²**

Biological Distinctiveness: **Globally Outstanding**

Conservation Status: **Relatively Stable**

Conservation Assessment: **III**

Authors: *Colleen Seymour, Mary Rowen*

Reviewed by: *Alan Rodgers, Richard Estes, Markus Borner*

Location and General Description

The Serengeti Volcanic Grasslands [47] ecoregion is found just south of the Tanzania-Kenya border, close to the equator (between 2°S and 4°S). The ecoregion falls in the seasonal tropics, with mean maximum temperatures between 24°C and 27°C and mean minimum temperatures between 15°C and 21°C. Mean annual rainfall in the Serengeti varies from 1,050 mm in the northwest to 550 mm in the southeast (Sinclair et al. 2000). This rainfall is strongly seasonal, with peaks from March to May and November to December (Schaller 1972; Norton-Griffiths et al. 1975; Sinclair 1979). Rainfall is the main determinant of vegetation growth and hence ungulate food supply (Sinclair 1977; McNaughton 1979, 1983).

The Serengeti Plains consist of volcanic ash derived from local volcanoes such as the extinct caldera of Ngorongoro and the dormant volcanoes of Kerimasi and Oldonyo Lengai (last eruption in 1966). Most of the topography is flat to slightly undulating, interrupted by scattered rocky outcrops (kopjes), which are part of the Precambrian basement protruding through ash layers.

The Serengeti Volcanic Grasslands [47] are classified as part of the Somali-Masai regional center of plant endemism (White 1983) and cover the short grassland portion of the Greater Serengeti ecosystem (Sinclair et al. 2000). Different plant species predominate depending on depth, stability, and age of the underlying ash and disturbance (Dublin 1991; Sinclair et al. 2000). Among the dominant species on the dunes and short and intermediate grasslands are a variety of *Sporobolus* spp., *Pennisetum mezianum, Eragrostis tenuifolia, Andropogon greenwayi, Panicum coloratum, Cynodon dactylon, Chloris gayana, Dactyloctenium* sp., *Digitaria macroblephara,* and sedges of the genus *Kyllinga* (White 1983). In periods of severe drought, the grasslands become largely denuded of standing vegetation.

Outstanding or Distinctive Biodiversity Features

The primary importance of the Serengeti Volcanic Grasslands [47] ecoregion is its migrating large mammal community. Although populations fluctuate, there are an estimated 1.3 million blue wildebeest (*Connochaetes taurinus*), 200,000 Burchell's zebra (*Equus burchelli*), and 400,000 Thomson's gazelle (*Gazella thomsonii*) migrating between this ecoregion and the Southern *Acacia-Commiphora* Bushlands and Thickets [46] each year (Campbell and Borner 1995). A large number of associated mammalian predators are also involved in these movements. By the onset of the dry season (late May), the grasses on these plains have either dried out or been grazed, and water is scarce (Bell 1971; Belsky 1986; Schaller 1972). This triggers the migration of wildebeest and zebra, followed by Thomson's gazelle, from the plains to the Southern *Acacia-Commiphora* Bushlands and Thickets [46]. At the beginning of the wet season, these animals return to the plains (Schaller 1972). Several mammal species occurring here are of international importance because of their abundance, including eland (*Tragelaphus oryx*), waterbuck (*Kobus ellipsiprymnus*), blue wildebeest, hartebeest (*Alcelaphus buselaphus*), impala (*Aepyceros melampus*), Grant's (*Gazella granti*) and Thompson's gazelles, zebra, and buffalo (*Syncerus caffer*) (Stuart et al. 1990; East 1999).

Floral and faunal endemism is low. The Serengeti covers part of the Serengeti Plains endemic bird area (Stattersfield et al. 1998). Although five restricted-range species occur, they are found mainly in the adjacent Southern *Acacia-Commiphora* Bushlands and Thickets [46]. Notable bird species include the rufous-tailed weaver (*Histurgops ruficauda* [monotypic genus]), Usambaro barbet (*Trachyphonus usambiro*), grey-crested helmet shrike (*Prionops poliolophus*), grey-breasted spurfowl (*Francolinus rufopictus*), and Fischer's lovebird (*Agapornis fischeri*) (Baker and Baker 2001).

The black rhinoceros (*Diceros bicornis,* CR) still remains in the ecoregion but has been nearly extirpated by poaching for its horn. Remnant populations are found on the floor of the Ngorongoro crater, just outside the ecoregion border, at Moru, and one in the Mara Game Reserve. Wild dog (*Lycaon pictus,* EN) disappeared from Serengeti National Park in 1991. Although a rabies epidemic killed three of the packs, the full cause of the disappearance remains contentious (Morell 1995; Dye 1996; East and Hofer 1996). In the last 2 years, two or three packs of wild dogs have re-emerged in the Ngorongoro Conservation Area, in the Gol Mountains (M. Borner, pers. comm., 2003). Despite the loss of wild dog from the ecoregion, the area still has a wide array of mammalian predators including cheetah (*Acinonyx jubatus,* VU), lion (*Panthera leo,* VU), leopard (*P. pardus*), spotted and striped hyena (*Crocuta crocuta, Hyaena hyaena*), sidestriped (*Canis adustus*), common (*C. aureus*), and black-backed (*C. mesomelas*) jackal, honey badger (*Mellivora capensis*), caracal (*Caracal caracal*), serval (*Leptailurus serval*), wild cat (*Felis syl-*

vestris), bat-eared fox (*Otocyon megalotis*), and genet and mongoose species. Avian predators are also plentiful, with Serengeti National Park having thirty-four raptor species and six vulture species (Schmidl 1982; Mundy et al. 1992).

Status and Threats

Current Status

Much of the ecoregion's habitat occurs in protected areas, most of which are joined in a continuous habitat block. The protected area network includes parts of Serengeti National Park and Ngorongoro Conservation Area. Other protected areas containing parts of the ecoregion are the Grumeti, Maswa, and Ikorongo game reserves in Tanzania. There has been little loss of habitat in the protected areas, except for small areas used for tourist hotels.

Types and Severity of Threats

No people live in the Serengeti National Park or the adjoining game reserves. Therefore, direct threats to the habitat of this ecoregion are low. However, human populations outside the reserves, especially to the north and west, are growing at up to 4 percent a year (Packer 1996), with some of these people attracted to the area by the wildlife resources and tourism opportunities (Campbell and Hofer 1995; Leader-Williams et al. 1996). At present, large numbers of animals are poached annually for their meat (Mduma et al. 1998). However, it is hoped that schemes to give local communities legal rights to manage the wildlife around their villages will reduce this problem. There are also plans to channel more money earned from tourist activities in the park back into the community because the contribution from tourism to the local economy has been low (Leader-Williams et al. 1996). To the north in Kenya conversion of the habitat to wheat farming is a particular problem, especially the Isiria Escarpment, a vital drought refuge for wildebeest and zebra.

   Another increasing problem is the infection of wild animals with diseases from surrounding domesticated animals. The wild dog was nearly extirpated in the area. Possible explanations include stress-related diseases as a result of handling, infections acquired from local domestic dogs, competition from lions and hyenas, demographic stochasticity, food shortage, and emigration (Dye 1996; East and Hofer 1996). An outbreak of canine distemper between January and October 1994 is estimated to have killed more than a third of all lions in Serengeti National Park and neighboring Masai Mara. Hyenas and bat-eared foxes were also affected. The outbreak is believed to have originated among the roughly 30,000 domestic dogs that live in the area, most of which are not vaccinated (Morell 1995; Roelke-Parker et al. 1996). A program to vaccinate domestic dogs on the western boundaries of the park was initiated in 1996 (Bristow 1996).

Justification for Ecoregion Delineation

This ecoregion was taken directly from the vegetation unit "edaphic grassland on volcanic soils" mapped by White (1983). It reflects the discrete boundaries provided by deposits of volcanic ash, which in turn affect the composition of the grasslands and help explain the high concentrations of mammals at certain times of the year.

Ecoregion Number:	**48**
Ecoregion Name:	**Itigi-Sumbu Thicket**
Bioregion:	**Eastern and Southern Africa**
Biome:	**Tropical and Subtropical Grasslands, Savannas, Shrublands, and Woodlands**
Political Units:	**Tanzania, Democratic Republic of the Congo, Zambia**
Ecoregion Size:	**7,800 km²**
Biological Distinctiveness:	**Bioregionally Outstanding**
Conservation Status:	**Relatively Stable**
Conservation Assessment:	**V**
Authors:	*Lyndon Estes*
Reviewed by:	*Jonathan Timberlake, Alan Rodgers, Mike Bingham*

Location and General Description

The Itigi-Sumbu Thicket [48] is a unique but poorly understood ecoregion found mainly in central Tanzania and northeastern Zambia. Stands occur in Tanzania close to its namesake town of Itigi near Dodoma and also in Zambia between lakes Mweru Wantipa and Tanganyika (Wild and Barbosa 1967). Elements of these thickets are also found elsewhere, such as in Zimbabwe around Abercorn (Burtt 1942; Wild and Barbosa 1967; White 1983), in the Lower Shire (S. Malawi), and in the mid-Zambezi valley (northern Zimbabwe, southeastern Zambia). In all cases these thicket units are discrete and clearly demarcated from the surrounding mopane, miombo, or *Acacia* woodlands.

   The Itigi deciduous thicket of White (1983) falls in the Zambezian regional center of endemism and is characterized by dense, primarily deciduous vegetation (Wild and Barbosa 1967; White 1983). The rainfall is low; for example, at Manyoni in the Tanzanian Itigi it is around 700 mm per annum. Precambrian basement rocks underlie the ecoregion. Above the bedrock is a characteristic soil structure, consisting of seasonally well-aerated and well-watered sandy soils of 0.6–3 m in depth that desiccate and harden during the dry season, with an impermeable duricrust of cement-like consistency beneath (Burtt 1942;

White 1983). The soils typically are acidic (pH 4.0–4.5) (Milne 1937).

The distribution of the Itigi-Sumbu Thicket follows that of the duricrust soils. In regions where there is no duricrust, then miombo, mopane, or *Acacia* savanna woodlands dominate. The vegetation is generally deciduous during the 4-month dry period (White 1983), although in the lower canopy some of the shrubs are evergreen (Edmonds 1976). During the wetter seasons when leaves are developed, little sunlight penetrates to the ground, and there is a poorly developed herbaceous layer (White 1983). Fire is also unable to easily penetrate Itigi-Sumbu Thicket, although it is a common feature of the surrounding woodland (Burtt 1942; White 1983).

At least 100 woody species are found in this ecoregion (Fanshawe 1969), which are mostly spineless, coppicing shrubs of 3–5 m in height. Characteristic species are *Baphia burttii, B. massaiensis, Burttia prunoides, Combretum celastroides* ssp. *orientale, Grewia burttii, Pseudoprosopsis fischeri,* and *Tapiphyllum floribundum* (Wild and Barbosa 1967; White 1983). *Albizia petersiana, Craibia brevicaudata* ssp. *burtii,* and *Bussea massaiensis* occasionally emerge from the lower canopy (White 1983) to form an open, upper canopy of 6–12 m height (Edmonds 1976; White 1983).

Outstanding or Distinctive Biodiversity Features

The acidic duricrust soils of the Itigi-Sumbu Thicket [48] allow otherwise rare plant species to dominate and outcompete species that are widespread in the surrounding woodlands. A few plants are endemic in the Itigi thicket of Tanzania and Zambia, although precise numbers are not known. There are no endemic plants in the patches of this habitat in Malawi and Zimbabwe.

Compared with the unusual plant composition in this ecoregion, vertebrates are neither distinct nor abundant. Few large vertebrates can navigate through the dense thicket, although elephants (*Loxodonta africana,* EN) pass with ease. The thickets were once important refuges for elephants during the dry season, where they could feed on the seeds of *Grewia burttii* and *Grewia platyclada* during the day (Burtt 1942). Most of the elephants have been hunted out of the ecoregion. The Itigi-Sumbu Thicket [48] also supported black rhinoceros (*Diceros bicornis,* CR) before its eradication by poaching. Frugivorous birds and mammals also disperse the seed of *Grewia burttii,* although saplings are never seen growing outside the thicket (Burtt 1942).

Some subspecific endemism has evolved in the invertebrates of this ecoregion, particularly in the butterfly family Papilionoideae. The Itigi-Sumbu Thickets [48] also support an abundance of termites (Burtt 1942). Their large termitaria support plants that are otherwise foreign to the community, such as the large candelabra euphorbia (*Euphorbia bilocularis*) (White 1983). The termitaria may also be responsible for the clumped vegetation structure seen in aerial photographs (Fanshawe 1969).

Status and Threats

Current Status

The largest blocks of intact Itigi-Sumbu Thicket [48] are found on the northern shores of Lake Mweru Wantipa. Parts of the Zambian Itigi-Sumbu Thicket are found in two protected areas on the shores of Lake Mweru Wantipa, in Mweru Wantipa National Park and Tabwa Reserve on the northern and eastern shores, respectively (Almond 2000). Another portion falls in Nsumbu National Park. Unfortunately, protected status has not prevented the removal of significant amounts of Itigi thicket in Zambia (Almond 2000). The part of the ecoregion occurring in Tanzania is completely unprotected (Stuart et al. 1990; Kideghesho 2001) and believed to be heavily degraded.

Types and Severity of Threats

The greatest threat to this ecoregion comes from its cultivation by an increasing human population (Moyo et al. 1993). Once cultivated, the duricrust in the soil is destroyed, so the special soil conditions that are needed by Itigi thicket cannot return, even if cultivation stops. Woodland vegetation and not Itigi thicket regenerates on abandoned farms. Thicket deforestation patterns are correlated with proximity to human settlements, which have grown on average by 15 percent between 1984 and 1999 in Zambia (Almond 2000). In both Tanzania and Zambia the specialized habitats of this ecoregion could be removed within the next 50 years.

Justification for Ecoregion Delineation

Itigi-Sumbu Thicket [48] vegetation occurs on specialized, acidic soils with a duricrust. The distribution of these soils controls the distribution of the ecoregion, which is otherwise replaced by various kinds of savanna woodland. Although thicket occurs throughout Caesalpinoid woodlands, only the two larger areas mapped by White (1983) as "Itigi deciduous thicket" have been mapped.

<table>
<tr><td>Ecoregion Number:</td><td>49</td></tr>
<tr><td>Ecoregion Name:</td><td>Angolan Miombo Woodlands</td></tr>
<tr><td>Bioregion:</td><td>Eastern and Southern Africa</td></tr>
<tr><td>Biome:</td><td>Tropical and Subtropical Grasslands, Savannas, Shrublands, and Woodlands</td></tr>
<tr><td>Political Units:</td><td>Angola, Democratic Republic of the Congo, Zambia</td></tr>
<tr><td>Ecoregion Size:</td><td>660,100 km^2</td></tr>
<tr><td>Biological Distinctiveness:</td><td>Regionally Outstanding</td></tr>
<tr><td>Conservation Status:</td><td>Relatively Intact</td></tr>
<tr><td>Conservation Assessment:</td><td>III</td></tr>
<tr><td>Author:</td><td>Suzanne Vetter</td></tr>
<tr><td>Reviewed by:</td><td>Brian Huntley</td></tr>
</table>

Location and General Description

The Angolan Miombo Woodlands [49] cover all of central Angola and extend into the Democratic Republic of the Congo (DRC). Most of this ecoregion is found at elevations between 1,000 and 1,500 m above sea level and includes the highlands of Huíla, Huambo, and Bié (Huntley 1974a). The majority of the ecoregion drains east into the Zambezi River.

The area experiences a tropical climate, with rainfall strongly concentrated in the summer. Mean annual rainfall ranges from less than 800 mm in the south to about 1,400 mm in the north and west (Huntley 1974a). Mean maximum temperatures are around 30°C in the south, falling to 24°C at the higher elevations. Minimum temperatures are between 15°C and 18°C in the low-lying areas and 9°C at higher-elevation areas, where frosts are possible.

The geology of the area comprises a mixture of Karoo sandstones, Kalahari sands, and metamorphosed Precambrian basement (Huntley 1974a). Soils developed over these rocks are highly leached, well drained, and nutrient poor, tend to be acidic, and have low organic matter (Frost 1996). In some areas, drainage is restricted and there is seasonal waterlogging.

This ecoregion is a part of the Zambezian regional center of endemism (White 1983). Miombo woodland is distinguished from other African savanna, woodland, and forest formations by the dominance of tree species in the family Fabaceae, subfamily Caesalpinioideae, particularly in the genera *Brachystegia, Julbernardia,* and *Isoberlinia* (Campbell et al. 1996). Large trees widespread in the Angolan miombo are *Brachystegia spiciformis, Julbernardia paniculata,* and *Copaifera baumiana,* and *Brachystegia floribunda, B. boehmii, B. gossweilerii, B. wangermeeana, B. longifolia, B. bakerana, Guibourtea coleosperma,* and *Isoberlinia angolensis* are locally dominant (Huntley 1974a; Werger and Coetzee 1978; Dean 2000). The grass layer is up to 2 m tall, and several species of *Loudetia, Hyparrhenia, Tristachya,* and *Monocymbium ceresi-*

iforme predominate. Most miombo woody species shed their leaves in the late dry season. A few weeks before the onset of the rains, the trees flush again, and many species also flower at that time (Werger and Coetzee 1978). Small areas of *Marquesia acuminata, Guibourtia coleosperma,* and *Cryptosepalum exfoliatum* evergreen dry woodland occur in the northeast of the ecoregion, and transitions to *Colophospermum mopane* and *Baikiaea plurijuga* communities occur in the drier south. On seasonally waterlogged soils along drainage lines, especially on Kalahari sands, the woodland gives way to grasslands dominated by *Loudetia, Andropogon, Trachypogon,* and *Tristachya* species (Huntley 1974a). Open woodlands with scattered *Uapaca, Piliostigma, Annona, Entadopsis,* and *Erythrina* species often develop on the ecotone between the woodland edge and drainage lines (Dean 2000).

Human populations generally are low because of the nutrient-poor soils that limit agricultural potential and the presence of tsetse fly (*Glossina* spp.), the vector of trypanosomiasis that affects humans and domestic livestock. In most of Angola, human population density is less than five people per square kilometer (Huntley and Matos 1994). Population density increases in the higher-elevation areas in the southwest and is lowest in the southeast, where large areas are almost uninhabited.

Outstanding or Distinctive Biodiversity Features

Overall floral richness is high, although the diversity of canopy tree species is low. The number of endemic plants is unknown. Miombo is notable among dry tropical woodlands for the dominance of tree species with ectomycorrhizal rather than vesicular-arbuscular mycorrhizal associations (Frost 1996) that enable the trees to grow on porous, infertile soils.

Faunal richness is moderate. More than 170 mammal species occur, including four near-endemic small mammal species and one strict endemic: Vernay's climbing mouse (*Dendromus vernayi,* CR). The giant sable antelope (*Hippotragus niger variani,* CR), is an endemic subspecies of large mammal. The nutrient-poor soils, together with the harsh dry season and long droughts, limit the density of herbivores and bias their composition toward larger-bodied species such as elephants (*Loxodonta africana,* EN) and buffaloes (*Syncerus caffer*). Larger antelopes with more specialized feeding habits are sable antelope (*Hippotragus niger*), roan antelope (*H. equinus*), and Lichtenstein's hartebeest (*Sigmoceros lichtensteinii*) (Frost 1996). Patches of better grazing are scattered through the drier miombo, typically associated with rivers and wetlands. These areas support a wider variety of species, such as giraffe (*Giraffa camelopardalis*), bushbuck (*Tragelaphus scriptus*), blue duiker (*Cephalophus monticola*), and yellow-backed duiker (*C. silvicultor*), with sitatunga (*Tragelaphus spekii*), waterbuck (*Kobus ellipsiprymnus*), and tsessebe (*Damaliscus lunatus*) in the wettest areas. Threatened species include lion (*Panthera leo,* VU), cheetah (*Acinonyx jubatus,* VU), and African wild dog (*Lycaon pictus,* EN).

Although the avifauna is rich, endemism levels are low, with only the slender-tailed cisticola (*Cisticola melanurus*) considered near endemic. The *Brachystegia*-restricted avifauna in the ecoregion is similar to that of western Zambia and southern DRC (Benson and Irwin 1966). Typical miombo species include miombo tit (*Parus griseiventris*), miombo rock-thrush (*Monticola angolensis*), red-capped crombec (*Sylvietta ruficapilla*), and Böhm's flycatcher (*Muscicapa boehmi*) (Dean 2000). The wattled crane (*Grus carunculatus*, VU) occurs in some wetlands.

The herpetofauna of the ecoregion is moderately rich, with two strictly endemic frog species, the Angola ornate frog (*Hildbrandtia ornatissima*) and the Huila forest tree frog (*Leptopeltis anchietae*), and one other near-endemic frog, the Luita River reed frog (*Hyperolius vilenai*). Among the reptiles are strictly endemic species, including Bocage's horned adder (*Bitis heraldica*). According to Poynton and Broadley (1978), the upland areas of Angola (such as the Bié Plateau), which form the heart of the ecoregion, do not appear to be notable centers of reptile or amphibian endemism.

Fire is an important ecological factor in miombo woodland. The strong seasonality in precipitation leaves the vegetation dry for several months of the year, and thunderstorms at the start of the rainy season can easily set the vegetation alight (Werger and Coetzee 1978). Invertebrates (termites and caterpillars in particular) are also important ecological agents and probably remove more biomass than large mammals. Termites produce enormous mounds that are richer in nutrients and organic matter than the surrounding nutrient-poor landscape (Malaisse 1978; Frost 1996).

Status and Threats

Current Status

From 1974 to 2002, Angola experienced an almost continuous, intense civil war that affected every town and inhabitant of the country (Huntley and Matos 1994). Many rural people moved to cities, leaving large stretches of habitat unaffected by human settlement and activities. However, large areas of the country are inaccessible because of landmines.

A number of protected areas fall in the ecoregion, all of them in Angola. Luando National Park and the nearby Kangandala Integral Nature Reserve were both formed to protect the giant sable antelope and are the only known areas where populations of this species survive. Proposals have been made to join these two reserves (Stuart et al. 1990). Kameia National Park is an enormous, seasonally inundated grassy plain drained by three permanent rivers. Bikuar and Mupa National Parks fall into the transition zone between *Baikiaea* woodlands of the southwest arid biome and the miombo woodland belt (Huntley 1974a). The total area officially protected amounts to roughly 34,700 km², or 6.3 percent of the ecoregion. However, the status of these reserves is poor because of decades of warfare, and populations of animals in the parks are presumed to be low.

Types and Severity of Threats

The civil war led to poor security, mass displacement of people, and a depressed economy. As a result, conservation is a low priority of government, nongovernment organizations, and the general public. Most protected areas have been abandoned for periods of time, as park wardens were forced to leave for economic or security reasons. The three most important protected areas in the ecoregion, Kangandala, Kameia, and Luando, all have substantial human settlements (Huntley 1974a; Huntley and Matos 1994). The impact of the war on the fauna has been catastrophic (Huntley and Matos 1992). Most of the ecoregion is sparsely settled at present. Thus, habitat fragmentation and modification through settlement and agriculture, woodcutting, and livestock impacts are minimal. However, charcoal manufacture is extensive close to cities (Dean 2000).

Justification for Ecoregion Delineation

The Angolan Miombo Woodlands [49] form the western portion of "wetter Zambezian miombo woodland" of White (1983) and include the "mosaic of *Brachystegia bakerana* thicket and edaphic grassland" vegetation. Bordered on the east by the Zambezi River, the ecoregion is a part of the Zambezian regional center of endemism (White 1993) but has some floral and faunal differences from other parts of the miombo, most notably the presence of the giant sable antelope.

Ecoregion Number:	**50**
Ecoregion Name:	**Central Zambezian Miombo Woodlands**
Bioregion:	**Eastern and Southern Africa**
Biome:	**Tropical and Subtropical Grasslands, Savannas, Shrublands, and Woodlands**
Political Units:	**Angola, Burundi, Democratic Republic of the Congo, Malawi, Tanzania, Zambia**
Ecoregion Size:	**1,184,200 km²**
Biological Distinctiveness:	**Globally Outstanding**
Conservation Status:	**Relatively Stable**
Conservation Assessment:	**III**
Author:	*Karen Goldberg*
Reviewed by:	*Alan Rodgers, Jonathan Timberlake*

Location and General Description

The Central Zambezian Miombo Woodlands [50] ecoregion covers about 70 percent of central and northern Zambia, the southeastern third of DRC, western Malawi, much of western Tanzania, and parts of Burundi and northeastern Angola. It occupies the Central African Plateau at altitudes between 1,000 and 1,600 m, with a few localized areas of higher relief, such as Mount Mulumbe in southern DRC. Valleys of major rivers such as the Limpopo, Zambezi, Shire, Luangwa, Rufiji, and Rovuma and the upper Congo River drainage dissect the plateau.

Precambrian basement rocks form the underlying geology. Over tens of millions of years they have been eroded and weathered into a low-relief peneplain, with inselbergs projecting from the surface. Seasonal wetlands, or *dambos,* occupy shallow depressions in the surface of the plateau, and larger examples, such as Bangweulu and Busanga, are separated into the Zambezian Flooded Grasslands [63]. The miombo soils typically are highly leached, nutrient poor, and well drained and tend to be acidic, with a low proportion of organic matter (Frost 1996; Frost et al. 2002). Termites are abundant in this woody biomass rich habitat, and by concentrating organic matter and nutrients in their mounds, termites produce nutrient-rich patches in an otherwise nutrient-poor landscape (Frost et al. 2002).

Rainfall is unimodal, concentrated during the hottest months of November–April (Frost 1996). A pronounced drought occupies the cooler season, lasting up to 7 months. Rainfall typically is between 1,000 and 1,200 mm annually, with up to 1,400 mm falling at higher elevations in the southwest. Mean maximum temperatures range from 24°C to 27°C, depending on altitude. Mean minimum temperatures range from 9°C to 18°C, and frosts are possible. The human density is below twenty people per square kilometer except in areas that have access to permanent water, such as along the margins of Lake Tanganyika and Lake Malawi, as well as the Copperbelt in northern Zambia and southern DRC (Campbell 1996).

This ecoregion is a part of the Zambezian regional center of endemism (White 1983). Trees of the legume subfamily Caesalpinioideae, particularly from the genera *Brachystegia, Julbernardia,* and *Isoberlinia,* dominate (Campbell 1996; Campbell et al. 1996). This is the center of diversity for miombo tree species, and the following are frequent canopy dominants: *Brachystegia floribunda, B. glaberrima, B. taxifolia, B. wangermeeana, Julbernardia globiflora, J. paniculata,* and *Isoberlina angolensis.*

Outstanding or Distinctive Biological Features

The Central Zambezian Miombo Woodlands [50] have the highest floral richness of the African miombo ecoregions, with the peak of miombo plant species richness in Zambia. The ecoregion may support more than 3,000 plant species, with perhaps several hundred endemics. It is the center of distribution and endemism for the genera *Crotalaria, Indigofera,* and *Brachystegia,* the latter with seventeen of its thirty species located in Zambia (Rodgers et al. 1996).

This ecoregion supports a low density of large mammals, attributed to the harsh dry season, long droughts, and poor soils,

which results in vegetation of low nutrient content. These conditions favor large-bodied generalist animals, such as elephant (*Loxodonta africana,* EN), African buffalo (*Syncerus caffer*), and in the past black rhino (*Diceros bicornis,* CR). Other species make more use of non-miombo habitats, including sable antelope (*Hippotragus niger*), roan antelope (*H. equinus*), Lichtenstein's hartebeest (*Sigmoceros lichtensteini*), and southern reedbuck (*Redunca arundinum*). Rates of mammal endemism are low, with three strict endemics: Rosevear's striped grass mouse (*Lemniscomys roseveari*) and two white-toothed shrews, *Crocidura ansellorum* (CR) and *C. zimmeri* (VU). Threatened species include lion (*Panthera leo,* VU), cheetah (*Acinonyx jubatus,* VU), and African wild dog (*Lycaon pictus,* EN).

The avifauna is rich and contains distinctive miombo species (Benson and Irwin 1966). However, rates of endemism in this ecoregion are low, with the only strict endemics being Ruwet's masked weaver (*Ploceus ruweti,* DD) and black-faced waxbill (*Estrilda nigriloris,* DD), both from DRC. Other range-restricted species include grey-crested helmetshrike (*Prionops poliolophus*), an uncommon species endemic to the woodlands of Kenya and northern Tanzania (Collar and Stuart 1985). The slender-tailed cisticola (*Cisticola melanurus,* DD) is confined to grassy places in well-developed miombo (Urban et al. 1997).

Species richness in reptiles and amphibians is high, and there are endemic species. Fifteen reptiles and thirteen amphibians are considered strictly endemic to the ecoregion. Several of these species are confined to the area in and around the Upemba National Park in the Shaba district of DRC, which is a local center of importance for endemics in the ecoregion as a whole.

Status and Threats

Current Status

Throughout the ecoregion stands of apparently pristine miombo have been cleared or heavily modified in the past and then regenerated (Kjekshus 1977; Misana et al. 1996). The long interaction between people and the habitats and species of the miombo are important in interpreting its current status (Frost et al. 2002). There is little remaining miombo in Burundi and Malawi, but approximately 40 percent of Zambia and Tanzania are still covered by miombo. Large areas also remain in southern DRC.

The ecoregion has an extensive protected areas network, with fourteen national parks, thirteen game and wildlife reserves, and game controlled areas, forest reserves, and communally managed conservation areas. Of the national parks, two stand out as being of particular importance. Kafue in Zambia consists of a mosaic of extensive grasslands, miombo, mopane, and riverine woodland (Stuart and Stuart 1992). Upemba National Park in southern DRC is an area of endemism for plants (WWF and IUCN 1994), the threatened black-faced waxbill (*Estrilda nigriloris*), and endemic amphibians, reptiles, and small mammals.

Types and Severity of Threats

Although much of the ecoregion is sparsely populated, few areas have not been affected by anthropogenic activities in some way. High population densities in Burundi and Malawi have already resulted in severe loss of miombo. More than 80 percent of people living in miombo depend on fuelwood and charcoal for cooking, heat, and light (Misana et al. 1996). Cutting woody vegetation for charcoal production, especially close to major roads and large urban centers, is having a marked impact. In Zambia, it is estimated that between 1937 and 1983, 51 percent of the Copperbelt region had been deforested for industrial and household fuelwood (Chidumayo 1987). The levels of charcoal production and deforestation are increasing constantly.

Large-scale cultivation is uncommon, but subsistence agriculture is practiced by as much as 75 percent of the population. Growing staple and cash crops such as maize, cassava, sorghum, millet, and tobacco has converted significant areas. Growing tobacco for export has led to large losses of miombo in Tanzania and Malawi for both cultivation and fuelwood (Moyo et al. 1993; Misana et al. 1996). In Zambia, *citimene,* a traditional form of ash-fertilizing agriculture, is also practiced (Van Wilgen et al. 1997), and increased human populations mean that this is no longer sustainable (Frost et al. 2002).

The high incidence of fires in the area poses further threats to the ecoregion. Although fire is an integral part of miombo ecology, human setting of fires is believed to have increased the frequency of fire far above the natural level. Most of the deliberate burning and the uncontrolled fires occur at the end of the dry season, just before the onset of the summer rains. The fires burn with greater intensity as quantities of dry fuel accumulate. These hotter fires are destructive even to fire-tolerant trees and can also have negative impacts because this time coincides with miombo trees breaking their dormancy (Chidumayo and Frost 1996). Repeated late-season fires in many areas have decreased forest regeneration and seed germination, and seedling survival and growth can be severely disturbed (Chidumayo et al. 1996). In addition, fire removes species that are less fire tolerant from the miombo, thereby reducing species diversity.

Poaching and illegal hunting for bushmeat have a significant impact on the wildlife throughout the ecoregion (TRAFFIC 2000). Elephant and rhino poaching, in particular, has been extremely severe. Most areas outside parks and reserves have little wildlife left, except in the remote interior parts of Tanzania. The live animal trade in Tanzania is one of the biggest in Africa, especially in birds and tortoises, and can severely deplete local populations. Some of the species involved are captured in the miombo habitats.

Justification for Ecoregion Delineation

Comprising primarily "wetter Zambezian miombo woodland," this ecoregion contains the highest levels of richness and endemism in White's (1983) Zambezian center of endemism. Patches of "mosaic of Zambezian dry evergreen forest and wetter miombo woodland" and small areas of "grasslands on Kalahari Sands" are also incorporated. Lake Malawi forms the eastern boundary of this ecoregion. Areas of miombo woodland to the east of Lake Malawi form the Eastern Miombo Woodlands [52].

Ecoregion Number:	**51**
Ecoregion Name:	**Zambezian *Baikiaea* Woodlands**
Bioregion:	**Eastern and Southern Africa**
Biome:	**Tropical and Subtropical Grasslands, Savannas, Shrublands, and Woodlands**
Political Units:	**Angola, Namibia, Botswana, Zambia, Zimbabwe**
Ecoregion Size:	**264,400 km²**
Biological Distinctiveness:	**Bioregionally Outstanding**
Conservation Status:	**Relatively Intact**
Conservation Assessment:	**V**
Author:	Suzanne Vetter
Reviewed by:	Jonathan Timberlake

Location and General Description

The Zambezian *Baikiaea* Woodlands [51] ecoregion is a mosaic of *Baikiaea plurijuga*–dominated forest, woodland, thicket, and secondary grassland. It is equivalent to the Zambezian deciduous forest and scrub forest of White (1983). To the north, the vegetation changes gradually into evergreen Zambezian *Cryptosepalum* Dry Forests [32] and wet miombo woodland [49, 50]. On the Barotse floodplain, seasonal waterlogging or flooding suppresses tree growth, and *Baikiaea* woodlands give way to grasslands [56, 63].

The ecoregion lies on an extensive plain of 800–1,000 m in elevation and is drained by the Okavango, Cuando, and Upper Zambezi rivers and their numerous tributaries. Fossil dunes deposited in the Pleistocene are characteristic features, as are extensive *dambos* (shallow, seasonally inundated pans, or *vleis*) that have formed in river valleys and dune troughs. The low clay content of the Kalahari Sand soils means nutrients exist only when there is organic matter. Exposure of the soil surface to the sun through clearing and burning of the vegetation destroys much of the organic matter, and such areas tend to remain bare (Bingham 1995).

The ecoregion experiences a hot, semi-arid climate. Mean annual rainfall ranges from less than 400 mm in the drier southwest to 600 mm in the eastern parts and approximately 800 mm in the northernmost portion in Angola and Zambia. Rainfall is strongly concentrated from November to April. The rainfall is absorbed and retained deep in the Kalahari sands. Consequently, forest and woodland vegetation with deep-rooted trees is able to grow into, or even through, the long dry season. The mean maximum temperature is between 27°C and 30°C, and the mean minimum temperature ranges from 9°C to 12°C. Frost occurs in the southern part of the ecoregion, which is occasionally severe (below −5°C), killing young tree growth. However, many plant species in the woodlands are adapted to frost, fire, and herbivory and can coppice readily.

The Zambezian *Baikiaea* Woodlands [51] ecoregion represents a transition from moist southern savanna woodlands to drier southwestern deserts. These woodlands are nutrient deficient and dominated by slow-growing plants that have secondary compounds to protect against browsing. Based on the distribution of woody species, White (1965) recognized the Barotse center—comprising the *Baikiaea* and allied woodlands and grasslands—as a subcenter of endemism within the Zambezian regional center of endemism. It is defined by species confined to Kalahari sand and by the dominant tree species, *Baikiaea plurijuga* (Zambezian teak or mukusi), which is near endemic. The following species are also endemic or near endemic: *Baphia massaiensis* ssp. *obovata*, *Dialium englerianum*, *Paropsia brazzeana*, *Bauhinia petersiana* ssp. *macrantha*, *Copaifera baumiana*, and *Guibourtia coleosperma*. In well-developed *Baikiaea* communities, species of *Brachystegia* and *Julbernardia* are uncommon, and *Colophospermum mopane* is absent (Werger and Coetzee 1978).

Outstanding or Distinctive Biodiversity Features

Much of the ecoregion falls in the Barotse subcenter of plant endemism (White 1965). However, the fauna of the area has low levels of endemism, representing a merging of elements from the southern savannas, arid southwest, and miombo woodlands. Many animal species use the ecoregion only seasonally, or are dependent on the grasslands and wetlands within the woodlands.

Moderate species richness is found in most taxonomic groups. Mammal species include many widespread savanna species, such as Burchell's zebra (*Equus burchelli*), roan antelope (*Hippotragus equinus*), sable antelope (*H. niger*), bushbuck (*Tragelaphus scriptus*), greater kudu (*T. strepsiceros*), impala (*Aepyceros melampus* ssp. *melampus*), black-fronted duiker (*Cephalophus nigrifrons*), oribi (*Ourebia ourebi*), steenbok (*Raphicerus campestris*), common eland (*Taurotragus oryx*), blue wildebeest (*Connochaetes taurinus*), buffalo (*Syncerus caffer*), giraffe (*Giraffa camelopardalis*), tsessebe (*Damaliscus lunatus*), and waterbuck (*Kobus ellipsiprymnus*). Threatened large mammals include black rhinoceros (*Diceros bicornis*, CR), lion (*Panthera leo*, VU), African wild dog (*Lycaon pictus*, EN), cheetah (*Acinonyx jubatus*, VU), and elephant (*Loxodonta africana*, EN). Transborder movements of elephant still occur (Barnes et al. 1999).

The ecoregion's avifauna is characterized by moderately high species richness but low endemism. *Baikiaea* woodlands are the preferred habitat of the near-endemic Bradfield's hornbill (*Tockus bradfieldi*) and slaty egret (*Egretta vinaceigula*, VU). The ecoregion also has a rich variety of raptor species including secretary bird (*Sagittarius serpentarius*), white-backed vulture (*Gyps africanus*), lappet-faced vulture (*Torgos tracheliotus*), white-headed vulture (*Trigonoceps occipitalis*), hooded vulture (*Necrosyrtes monachus*), lesser kestrel (*Falco naumanni*, VU), Dickinson's kestrel (*F. dickinsoni*), bateleur (*Terathopius ecaudatus*), martial eagle (*Polemaetus bellicosus*), and African hawk eagle (*Hieraaetus spilogaster*). Riparian vegetation supports Pel's fishing owl (*Scotopelia peli*) (Barnes 1998).

The ecoregion also has three near-endemic reptiles and one near-endemic amphibian species. The near-endemic reptiles are the Annabon lidless skink (*Panaspis annobonensis*), Caprivi rough-scaled lizard (*Ichnotropis grandiceps*), and *Dalophia ellenbergeri*. The single near-endemic amphibian is the Khwai River toad (*Bufo kavangensis*).

Status and Threats

Current Status

Generally, the ecoregion is sparsely settled with fewer than five people per square kilometer. This, combined with the arid environment and nutrient-poor sandy soils, means much of the habitat has not been modified or fragmented. However, especially in Zambia, Angola, and Zimbabwe, timber logging together with frequent wildfires has significantly reduced the area of mature *Baikiaea* woodland and forest.

Ten protected areas cover 8 percent of the ecoregion across the five countries. There are three protected areas in Angola (Bikuar and Mupa national parks and Luiana Partial Reserve), four in Namibia (Mudumu Nature Reserve, Mahango and Caprivi game reserves, and Popa Game Park), Kazuma Pan and Hwange national parks in Zimbabwe, and Simoa Ngwezi National Park in Zambia. In addition, the West Zambezi Game Management Area in the southwestern corner of Zambia is in the ecoregion. A large area of *Baikiaea* woodland in Zimbabwe is protected as Forest Land, which is used mostly for hunting rather than logging. In Angola, more than 25 years of civil war have devastated wildlife (Huntley and Matos 1994). In contrast, Hwange National Park, the largest in the ecoregion, is well managed and has a developed tourist infrastructure and active research programs.

Types and Severity of Threats

The *Baikiaea* forests and woodlands are easily penetrated by fire, especially in the late dry season. After frequent fires, a dense shrub layer develops, which is dominated either by shrubs and climbers or by grasses and herbs. When fire damage is severe or after cultivation, *Baikiaea plurijuga* can disappear completely because it does not regenerate easily in frequently burned sites.

Poaching is widespread, even in some protected areas. Commercial poaching is a significant problem in Sioma Ngwezi and the surrounding game management area. Cross-border smuggling of wildlife products in remote areas is also a major concern for wildlife management and protection.

Annual migration routes for large mammals often are blocked. None of the parks in the ecoregion cover the entire migratory ranges of animals such as wildebeests and elephants. Protected areas also fail to extend to rivers, where animals move in search of drinking water. Cattle fences can also cause increased rates of mortality when animals are cut off from grazing and water resources (e.g., those erected in Botswana between the Caprivi Strip and the Okavango Delta in 1995 to control the spread of cattle lung disease).

Timber logging is a threat to the *Baikiaea* woodland and forest habitats. Annual production of mukusi timber peaked at 100,000 m^3 in the 1930s and again in 1964 with the construction of railway lines (Bingham 1995). Since the mid-1970s, logging has declined to around 20,000 m^3 per year, largely because of a decline of harvestable timber (van Gils 1988). In Zambia, a recent inventory found no exploitable reserves in the prime teak forest areas of Sesheke District (Bingham 1995). Recently, the conservation status of *Baikiaea plurijuga* in Zambia was elevated to vulnerable as a result of exploitation and habitat destruction (Bingham et al. 2000). The status of *Baikiaea* in the other countries—especially in Angola and Namibia, where the most extensive stands occur—is not certain.

Justification for Ecoregion Delineation

The Zambezian *Baikiaea* Woodlands [51] ecoregion follows White's (1983) "Zambezian dry deciduous forest and secondary grassland." It is dominated by Zambezian teak or mukusi (*Baikiaea plurijuga*) vegetation associated with deep Kalahari sands, which delineate the eastern and western extents of the ecoregion. The southern boundary is determined by frost. The ecoregion covers most of the Barotse subcenter of endemism in the Zambezian regional center of endemism for plants (White 1983).

Ecoregion Number:	**52**
Ecoregion Name:	**Eastern Miombo Woodlands**
Bioregion:	**Eastern and Southern Africa**
Biome:	**Tropical and Subtropical Grasslands, Savannas, Shrublands, and Woodlands**
Political Units:	**Malawi, Mozambique, Tanzania**
Ecoregion Size:	**483,900 km²**
Biological Distinctiveness:	**Globally Outstanding**
Conservation Status:	**Relatively Stable**
Conservation Assessment:	**III**
Author:	*Karen Goldberg*
Reviewed by:	*Alan Rodgers, Jonathan Timberlake, John Burlison, Judy Oglethorpe*

Location and General Description

The Eastern Miombo Woodlands [52] cover southeastern Tanzania and the northern half of Mozambique, with a few patches extending into southeastern Malawi. Unlike other miombo woodland ecoregions (Angolan [49], Central Zambezian [50], and Southern [53]), which are found above 1,000 m altitude on the Central African Plateau, this ecoregion is confined mostly to lower elevations between 200 m toward the coast and 900 m further inland (Bridges 1990). Undulating ridges mixed with shallow, flat-bottomed valleys, or *dambos*, that are often seasonally waterlogged characterize the landscape. Inselbergs are common, especially in northern Mozambique.

The underlying geology of the Eastern Miombo Woodlands [52] is mainly Precambrian basement, although in Tanzania there are also Karoo sandstone and younger marine sediments (Bridges 1990). Tens of millions of years of weathering have produced highly leached, nutrient-poor, well-drained, and acidic soils with low organic matter content (Frost 1996). These soils change in a regular sequence across the undulating landscape of the miombo, with poorest soils on ridgetops and the best in valley bottoms (Rodgers 1996).

The ecoregion experiences a seasonal tropical climate, with a unimodal rainfall pattern where most rain falls in the hotter months from November through March. Little rain falls in the cooler dry season, which can last up to 6 months (Werger and Coetzee 1978). Mean annual rainfall ranges from 700 to 1,000 mm. Mean maximum temperatures range between 21°C and 30°C depending on elevation. The ecoregion's mean minimum temperatures are between 15°C and 21°C, and the area is almost frost-free. Human populations in this ecoregion are low (Moyo et al. 1993). This is largely because of the nutrient-poor soils, which limit agricultural potential, and the widespread presence of tsetse fly (*Glossina* spp.), the vector of trypanosomiasis that affects humans and their domestic livestock (Matzke 1996).

This ecoregion is a part of the Zambezian center of endemism (White 1983). Trees of the subfamily Caesalpinioideae, particularly species of *Brachystegia and Julbernardia,* dominate (Campbell et al. 1996). These species are ectomycorrhizal, with fungal symbionts helping them obtain essential nutrients in a low–soil-nutrient environment. In southern Tanzania and northern Mozambique *Brachystegia spiciformis, B. boehmii,* and *B. bussei* are dominant, and further to the west species such as *B. utilis* and *B. taxifolia* are also found. Trees associated with these dominants include members of the genera *Afzelia, Burkea, Erthrophleum, Ficus, Monotes, Pterocarpus, Swartzia, Uapaca,* and *Xeroderris* (Rodgers 1996).

Outstanding or Distinctive Biodiversity Features

Most plant species in this ecoregion are widespread, and rates of endemism are low (White 1983; Rodgers 1996; Rodgers et al. 1996). An endemic cycad (*Encephalartos turneri*) is found on some of the inselbergs in northern Mozambique (IUCN 1987; Goode 1989), and the inselbergs rising from miombo woodland may support additional endemics.

The species diversity of mammals in the ecoregion is high, but there are no endemic species (Rodgers 1996; Rodgers et al. 1996). Characteristic miombo mammals include the sable antelope (*Hippotragus niger*), Lichtenstein's hartebeest (*Sigmoceros lichtensteinii*), and common duiker (*Sylvicapra grimmia*). More widespread mammal species are African buffalo (*Syncerus caffer*), greater kudu (*Tragelaphus strepsiceros*), elephant (*Loxodonta africana,* EN), eland (*Taurotragus oryx*), black rhinoceros (*Diceros bicornis,* CR), and Burchell's zebra (*Equus burchelli*). Large carnivores include lion (*Panthera leo,* VU), leopard (*P. pardus*), spotted hyena (*Crocuta crocuta*), and African wild dog (*Lycaon pictus,* EN). Smaller predators include the bushy-tailed mongoose (*Bdeogale crassicauda*), a species restricted to eastern Africa. Because of annual droughts and frequent fires, many species are at least seasonally dependent on non-miombo vegetation in or adjacent to the ecoregion for food, water, or shelter. These refuges provide a greater variety of habitats, resulting in higher richness in ecotonal areas, such as near inselbergs or rivers, than in areas of uniform miombo woodland (Rodgers 1996; Rodgers et al. 1996).

Although species richness for birds is high in this ecoregion, with 450 bird species recorded in the Selous Game Reserve alone, only Stierling's woodpecker (*Dendropicos stierlingi*) is near endemic (Baker and Baker 2001). Two globally threatened species occur in the area: corncrake (*Crex crex,* VU) and lesser kestrel (*Falco naumanni,* VU) (BirdLife International 2000; Baker and Baker 2001).

Near-endemic reptiles include Pitman's shovelsnout snake (*Prosymna pitmani*) and spotted flat lizard (*Platysaurus maculatus*). The Nile crocodile (*Crocodylus niloticus*) is widely distributed in the ecoregion but is found in higher densities in protected areas. The Masiliwa shovelsnout frog (*Hemisus brachydactylus*) is the only near-endemic amphibian of the ecoregion.

Status and Threats

Current Status

Poor soil and the presence of tsetse fly have resulted in low human population densities across much of the ecoregion (Rodgers 1996). In the 1970s and 1980s the armed conflict in Mozambique caused the mass exodus of people from rural areas in northern Mozambique to coastal or urban areas, or along major transport routes, resulting in some areas being entirely depopulated. Extensive areas of miombo habitat still remain in both southern Tanzania and northern Mozambique.

Six protected areas are found in the ecoregion. In Tanzania, the Selous Game Reserve covers roughly 48,000 km² and is Africa's largest uninhabited protected area. It supports the largest population of elephants in Africa, with more than 50,000 individuals recorded in 1994 (Barnes et al. 1999). There are also 120,000 buffalo, 1,300 wild dog, 3,000–4,000 lion, more than 20,000 Lichtenstein's hartebeest, and more than 5,000 sable antelope (East 1999). Black rhinoceros also still occur. The 3,230-km² Mikumi National Park in Tanzania is contiguous with the Selous, and both join the Kilombero Valley Game Controlled Area to the west. Three small game reserves are found to the north of the Ruvuma River.

Two protected areas are also found in Mozambique. The largest is Niassa Game Reserve and its associated buffer zones, covering 42,000 km², where as many as 8,700 elephants occur (Barnes et al. 1999). Gile Game Reserve, the smallest of the ecoregion's protected areas (2,100 km²), consists mainly of miombo woodland and associated *dambos* and supports a similar flora and fauna to those of Niassa. Parts of the new Quirimbas National Park in Mozambique also contain these habitats.

Types and Severity of Threats

The low human population densities across most of this ecoregion have ensured that the vegetation has remained nearly intact. Furthermore, abandonment of large areas of the ecoregion because of the civil war in Mozambique and the Ujamaa villagization program in Tanzania resulted in the regeneration of miombo. However, there are still a range of threats, including deforestation, unsustainable hunting or poaching, and mining, such as for emeralds in Tunduru District, Tanzania.

Fire is a natural ecological factor in miombo woodland, where thunderstorms at the start of the rainy season would have caused miombo fires (Werger and Coetzee 1978). However, over the past few thousand years people have used fire to clear land for cultivation, to maintain pastures for livestock, to drive game animals to positions where they can be easily hunted, to harvest honey, or for a variety of other reasons (religious, cultural, accidental). The greater frequency of fire and change in its seasonality mean that fire is now perceived as a significant threat to miombo (Rodgers 1996).

Miombo contains the valuable timber species muninga (*Pterocarpus angolensis*) and African blackwood (*Dalbergia melanoxylon*), which is used to make musical instruments, such as clarinets and piano keys, as well as carvings. Both of these species are heavily exploited, with logging concessions in Mozambique. Miombo woodlands are also heavily used to supply charcoal to urban centers, particularly to Dar-es-Salaam in Tanzania (Moyo et al. 1993).

In both Tanzania and Mozambique trophy hunting is allowed and provides a major source of revenue to the government. However, illegal hunting and poaching for bushmeat pose a threat to the large mammals of the ecoregion (TRAFFIC 2000). In the Selous Game Reserve, poaching reduced elephant populations from 100,000 in 1981 to roughly 50,000 in 1994 (Barnes et al. 1999). Black rhinoceros has also suffered serious declines.

Justification for Ecoregion Delineation

This ecoregion largely follows the boundaries of the floristically impoverished "drier Zambezian miombo woodland" vegetation unit of White (1983). This ecoregion is mainly below 1,000 m, whereas other miombo ecoregions typically are found above this altitude, and the rainfall is between 700 and 1,000 mm per annum, making this ecoregion drier than other miombo ecoregions.

Ecoregion Number:	**53**
Ecoregion Name:	**Southern Miombo Woodlands**
Bioregion:	**Eastern and Southern Africa**
Biome:	**Tropical and Subtropical Grasslands, Savannas, Shrublands, and Woodlands**
Political Units:	**Malawi, Mozambique, Zambia, Zimbabwe**
Ecoregion Size:	**408,300 km²**
Biological Distinctiveness:	**Bioregionally Outstanding**
Conservation Status:	**Vulnerable**
Conservation Assessment:	**V**
Author:	*Karen Goldberg*
Reviewed by:	*David Cumming, John Burlison, Judy Oglethorpe*

Location and General Description

The Southern Miombo Woodlands [53] extend across the Central African Plateau over central Zimbabwe, Mozambique, southern Zambia, and Malawi. Most of the ecoregion is found between 1,000 and 1,500 m, although it has variable topography.

Parts are flat or undulating plains, but much of the Zambian portion covers rugged country, and in eastern Zimbabwe and western Mozambique intrusive granites and gneisses often rise above the woodland as rounded hills (also known as *dwalas*) or inselbergs. Sections of the ecoregion in northwestern Zimbabwe and Mozambique are found at much lower elevations, from 200 to 800 m, and are mostly on sedimentary Karoo sandstones (Barnes 1998). Numerous grassy wetlands are interspersed along drainage lines in *vleis* or *dambos* (Barnes 1998). Highly weathered, acidic, and nutrient-poor soils, mainly alfisols, are typical, with oxisols in wetter locations. Many of the roots of trees in these woodlands have symbiotic relationships with fungi, which helps them grow in nutrient-poor conditions (Högberg and Piearce 1986).

The ecoregion has a tropical savanna climate with three distinct seasons: a hot dry season from mid-August through October, a hot wet season from November through March, and a warm dry season from April through early August. Mean maximum temperatures range between 18°C and 27°C but are typically around 24°C. Mean minimum temperatures are between 9°C and 15°C, and frost occurs most years, except in the lowland sections. Rainfall is 600 to 800 mm in the main part of the ecoregion in Zimbabwe and increases to about 1,000 mm in the lower-elevation portion in Mozambique. It is highly seasonal, with a prolonged dry season usually lasting 4–7 months (Cole 1986). This ecoregion includes the cities of Harare and Lusaka (the capitals of Zimbabwe and Zambia, respectively). Population densities are between 50 and 200 people/km² for most of the area, falling to fewer than 20 people/km² in Mozambique (Byers 2001).

This ecoregion forms part of a belt of miombo woodland from Angola in the west to Tanzania in the east, occupying much of the Zambezian regional center of endemism (White 1983). Miombo trees are dominated by the subfamily Caesalpiniodeae and characterized by *Brachystegia* and *Julbernardia* species, especially *B. spiciformis* and *J. globiflora*. Other common tree species include *Uapaca kirkiana*, *B. boehmii*, *Monotes glaber*, *Faurea saligna*, *F. speciosa*, *Combretum molle*, *Albizia antunesiana*, *Strychnos spinosa*, *S. cocculoides*, *Flacourtia indica*, and *Vangueria infausta*. Grass cover usually is sparse. Where drainage is poor, *Acacia* savannas or grassland may become locally dominant (Werger 1978). Other associated vegetation includes dry deciduous forest and thicket and deciduous riparian vegetation and patches of mopane woodland (White 1983).

Outstanding or Distinctive Biodiversity Features

This ecoregion, mapped by White (1983) as drier Zambezian miombo, is floristically impoverished, although areas of serpentine soils in Zimbabwe provide localized sites of speciation and endemism (Frost 1996).

The overall faunal diversity of this ecoregion is high, although most of the species are widespread and endemism is very low. Many species are only seasonally dependent on non-miombo habitats to provide food, water, or shelter (Rodgers et al. 1996). The low soil nitrogen and phosphorus produces foliage with low nutritional quality, affecting the animal communities. In general, animal densities are quite low. Typical dry miombo species are common duiker (*Syvicapra grimmia*), greater kudu (*Tragelaphus strepsiceros*), Sharpe's grysbok (*Raphicerus sharpei*), and sable antelope (*Hippotragus niger*). Elephants (*Loxodonta africana*) and buffalo (*Syncerus caffer*) are wide ranging but are found at low densities. Several threatened animals occur, including the black rhinoceros (*Diceros bicornis*, CR), elephant (EN), and the predators lion (*Panthera leo*, VU), cheetah (*Acinonyx jubatus*, VU), and African wild dog (*Lycaon pictus*, EN) (Frost et al. 2002).

Of the nearly 500 bird species occurring, none are strictly endemic. However, a few species are largely confined to the ecoregion or have extremely small distribution ranges. Lilian's lovebird (*Agapornis lilianae*) inhabits mostly mopane woodland in the Zambezi Valley but seasonally wanders into more mixed woodland on alluvial terraces (Harrison et al. 1997). The boulder chat (*Pinarornis plumosus*) is found in well-wooded terrain with large boulders. Stierling's woodpecker (*Dendropicos stierlingi*) is found in two small areas of southern Malawi, with the remainder of the population restricted to the Eastern Miombo Woodlands [52]. Chaplin's barbet (*Lybius chaplini*), endemic to south central Zambia, is a locally common resident of miombo woodland. Two globally threatened bird species are also found: the Cape griffon (*Gyps coprotheres*, VU) and the lesser kestrel (*Falco naumanni*, VU) (Barnes 1998; BirdLife International 2000).

There are several near-endemic reptiles in this ecoregion but few strict endemics, such as the ocellated flat lizard (*Platysaurus ocellatus*) and Pungwe flat lizard (*Platysaurus pungweensis*).

Status and Threats

Current Status

Historically, miombo vegetation was sparsely populated, partially because of poor soils, which made it largely unsuitable for cultivation (Chenje and Johnson 1994). The great rinderpest epidemic of the late nineteenth century further contributed to the depopulation of both people and livestock in the area. Many miombo areas were uninhabitable, and some were turned into game parks at the beginning of the twentieth century.

In Zimbabwe protected areas include Chizarira National Park, Chirisa Safari Area (includes mopane woodland), Matusadona National Park, and Mavuradonha Wilderness Area. Other smaller parks are the Mazowe Botanical Reserve, Sebakwe, Lake Chivero, Lake Kyle, and Ngezi recreational parks, and Mushandike Sanctuary. Both Lake Chivero and Kyle have populations of white rhinoceros (*Ceratotherium simum*) and rea-

sonable antelope populations. Zimbabwe has also spearheaded community-based conservation in southern Africa through the Communal Areas Management Programme for Indigenous Resources (CAMPFIRE). Recent political developments in Zimbabwe are resulting in widespread poaching, a drastic decline in tourism, and failure of some of the community-based natural resource management schemes.

Three Zambian protected areas fall in the ecoregion: North Luangwa National Park, Lukusuzi National Park, and Lower Zambezi National Park. Game management areas (GMAs) cover most of the remaining area in Zambia. Mozambique suffered serious upheavals through years of civil war, resulting in the hunting of many large mammals for meat and to finance the war. The wildlife sector is being rehabilitated, and management of Gorongosa National Park recommenced in 1995. Although wildlife numbers in the park were severely reduced from 1970s levels, all species are still present (with the exception of black rhinoceros, which was already scarce or absent before the war) (National Directorate of Forestry and Wildlife 1997). In Malawi, Kasungu National Park and Majete Game Reserve fall in this ecoregion.

Types and Severity of Threats

Habitat is generally fairly well conserved in many protected areas. The main threats in the Zimbabwean protected areas come from high elephant population densities. In the Zambezi Escarpment in Zimbabwe, elephants have heavily degraded more than 85 percent of the miombo woodland (Cumming et al. 1997). Elephant populations in Zimbabwe have grown from about 4,000 to 88,000 in the last 100 years.

Outside protected areas in Zimbabwe and Malawi much of the habitat has been converted to farmland or is degraded through cutting and fire (Chenje and Johnson 1994). More than 80 percent of the people living in miombo depend on firewood for cooking, heat, and light. Charcoal making is common in miombo along main transport routes to supply villages and urban areas and is having serious impacts in some places (e.g., Mozambique). Hunting for bushmeat was once conducted primarily for subsistence and cultural traditions. Now, the trade is becoming commercialized and urbanized, often catering to the urban market (TRAFFIC 2000). The current political and economic situation in Zimbabwe poses a threat to the extensive system of protected areas, and land invasions threaten the communal and private wildlife conservancies.

Large areas of habitat exist in Mozambique, but postwar development is occurring rapidly and land is being allocated on a concession basis for agriculture, logging, plantation forestry, tourism, hunting, mining, and other activities. Logging is occurring in many areas, taking advantage of the timber growth that occurred during the war, when many parts of the miombo were inaccessible. Commercial agriculture is expanding, and plantation forestry is being planned.

Justification for Ecoregion Delineation

This ecoregion comprises White's (1983) "drier Zambezian miombo woodland" to the southwest of Lake Malawi and includes several other areas east of the Zimbabwe Highlands, west of Lake Malawi, and southeast of Lake Kariba because of their similar vegetation compositions. It is separated from drier Zambezian miombo in the northeast woodlands (Eastern Miombo Woodlands [52]) because of differences in fauna and flora.

Ecoregion Number:	**54**
Ecoregion Name:	**Zambezian and Mopane Woodlands**
Bioregion:	**Eastern and Southern Africa**
Biome:	**Tropical and Subtropical Grasslands, Savannas, Shrublands, and Woodlands**
Political Units:	**Botswana, Malawi, Mozambique, Namibia, South Africa, Swaziland, Zambia, Zimbabwe**
Ecoregion Size:	**473,300 km²**
Biological Distinctiveness:	**Regionally Outstanding**
Conservation Status:	**Relatively Stable**
Conservation Assessment:	**III**
Authors:	*Lyndon Estes, Leticia Greyling*
Reviewed by:	*Jonathan Timberlake, John Burlison, Judy Oglethorpe*

Location and General Description

The Zambezian and Mopane Woodlands [54] are widespread in lower-lying areas of eastern southern Africa. The largest section extends from Swaziland through northeastern South Africa, north through Mozambique, and into Zimbabwe along the Zambezi Valley. A smaller portion runs through northern Botswana and the eastern extremity of Namibia's Caprivi Strip up through Zambia, including the Luangwa Valley. A smaller portion occurs in southern Malawi in the Shire River Valley.

The ecoregion falls largely in the tropical summer rainfall zone, with precipitation largely confined to November–April (White 1983). Annual average rainfall generally varies between 450 mm and 710 mm (Wild and Barbosa 1967; Farrell 1968; White 1983; Low and Rebelo 1996; Smith 1998), and mean annual temperature varies from 18°C to 24°C (White 1983).

The basement rocks are mainly Precambrian granites and gneisses, with basalt and Permian sedimentary rocks in some areas. The terrain is generally flat or gently undulating along the floors of the major river valleys, with average elevations ranging from 200 m to 600 m (White 1983; Smith 1998). The southeastern portion of the ecoregion occurs east of the Drakensberg Escarpment at elevations generally from 170 to 800 m (Low and Rebelo 1996), although heights of 1,525 m are attained in places (White 1983). Soil conditions may vary, but mopane soils generally possess an impervious B-horizon, or zone of accumulation (White 1983; Smith 1998).

The ecoregion falls in the Zambezian regional center of endemism (White 1983) and covers parts of three vegetation mapping units: *Colophospermum mopane* woodlands and scrub woodlands, north Zambezian undifferentiated woodland and wooded grassland, and south Zambezian undifferentiated woodland and scrub woodland. The tree *Colophospermum mopane* (Caesalpinioideae) is characteristic (Wild and Barbosa 1967; White 1983). Mopane can form pure stands but is generally associated with several other trees and shrubs, such as *Kirkia acuminata, Dalbergia melanoxylon, Adansonia digitata, Combretum apiculatum, C. imberbe, Acacia nigrescens, Cissus cornifolia,* and *Commiphora* spp. (Wild and Barbosa 1967; Farrell 1968; White 1983; Timberlake 1995, 1999; Low and Rebelo 1996; Smith 1998).

The herbaceous component of mopane woodland differs according to soil conditions and vegetation structure: dense swards are found beneath gaps in the mopane canopy on favorable soils, whereas grasses are almost completely absent in shrubby mopane communities on heavy, impermeable alkaline clays (White 1983; Smith 1998; Low and Rebelo 1996). Typical grasses include *Aristida* spp., *Eragrotis* spp., *Digitaria eriantha, Brachiaria deflexa, Echinochloa colona, Cenchrus ciliaris, Enneapogon cenchroides, Pogonarthria squarrosa, Schmidtia pappophoroides, Stipagrostis uniplumis,* and *Urochloa* spp. (Wild and Barbosa 1967; Timberlake 1995, 1999; Low and Rebelo 1996; Smith 1998).

The eastern and northeastern areas of the ecoregion support White's (1983) north Zambezian undifferentiated woodlands and wooded grasslands. Vegetation grows on fertile, well-drained, and slightly basic soils at elevations intermediate to the river valleys and uplands (White 1983). Characteristic woody plants are *Acacia* spp., *Albizia* spp., *Combretum* spp., *Adansonia digitata,* and *Xeroderris stuhlmannii.* The south Zambezian undifferentiated woodlands and scrub woodlands form the southeastern part of the ecoregion and are transitional between the Zambezian center of endemism and phytochoria to the south (White 1983). Important woody constituents are *Acacia gerrardii, A. nigrescens, A. nilotica, Combretum apiculatum, C. collinum, Dichrostachys cinerea, Kirkia acuminata, Peltophorum africanum, Piliostigma thonningii, Sclerocarya birrea,* and *Terminalia sericea* (White 1983; Low and Rebelo 1996).

Elephant browsing activity and fire are the two major natural factors shaping the vegetation and associated fauna in the Zambezian and Mopane Woodlands [54] (White 1983; Trollope et al. 1998). Elephants crop the larger trees in savannas and woodlands, creating openings exploited by grasses, thus increasing fire frequency and intensity (White 1983; Smith 1998; Trollope et al. 1998). This interplay between fire and elephants normally results in an open, two-tiered savanna consisting of large trees interspersed with shrubs at varying stages of growth (Smith 1998; Trollope et al. 1998).

Outstanding or Distinctive Biodiversity Features

The flora of the Zambezian and Mopane Woodlands [54] is not characterized by high species diversity, although White's (1983) two Zambezian woodland types mapped in this ecoregion are considered floristically rich. Mopane trees are characteristic of much of the low-lying parts of the area and in many places dominate to the exclusion of other tree species (Wild and Barbosa 1967; White 1983).

This ecoregion is one of the most important areas for large mammal diversity and biomass in southern Africa (Turpie and Crowe 1994). Vegetation here is more nutritive than that in surrounding ecoregions with higher rainfall, and as a result the area is well known for supporting large concentrations of ungulates (Huntley 1978; Mills and Hes 1997). This includes some of the most significant remaining populations of black rhinoceros (*Diceros bicornis*, CR), elephant (*Loxodonta africana*, EN) (Hilton-Taylor 2000), white rhinoceros (*Ceratotherium simum*), hippopotamus (*Hippopotamus amphibius*), buffalo (*Syncerus caffer*), blue wildebeest (*Connochaetes taurinus*), giraffe (*Giraffa camelopardalis*), greater kudu (*Tragelaphus strepsiceros*), and nyala (*Tragelaphus angasii*) (Stuart et al. 1990). Predators are also abundant, and lion (*Panthera leo*, VU), cheetah (*Acinonyx jubatus*, VU), African wild dog (*Lycaon pictus*, EN), spotted hyena (*Crocuta crocuta*), and leopard (*Panthera pardus*) are found in a number of the ecoregion's large protected areas. Sharpe's grysbok (*Raphicerus sharpei*) is also well represented in this ecoregion (East 1999), where it favors dense mopane vegetation (Mills and Hes 1997).

Although the ecoregion is rich in vertebrate species, it has few strict endemics (Pinhey 1978). Near-endemic birds include Lilian's lovebird (*Agapornis lilianae*), black-cheeked lovebird (*Agapornis nigrigenis*, VU), and Chaplin's barbet (*Lybius chaplini*). One mammal species, Juliana's golden mole (*Amblysomus julianae*, CR), is near endemic. Two endemic ungulate subspecies, Cookson's wildebeest (*Connochaetes taurinus cooksoni*) and Thornicroft's giraffe (*Giraffa camelopardalis thornicrofti*), are confined to the Luangwa Valley (Stuart et al. 1990; East 1999). The ecoregion contains a number of near-endemic reptiles, such as the two-toed burrowing skink (*Scelotes bidigittatus*), bluetail scrub lizard (*Nucras caesicaudata*), and Sabi quill-snouted snake (*Xenocalamus sabiensis*).

Status and Threats

Current Status

The majority of Zambezian and Mopane Woodlands [54] occur in areas of low human population density (Murphree 1990 cited in Els and Bothma 2000), so much of the habitat is in good condition. Agriculture and cattle farming are found mainly in the southeastern portion of the ecoregion in South Africa and Swaziland and the mid-Zambezi valley. In these areas, as much as 43 percent of the land has been changed by agriculture and settlement, and livestock grazing and resource use activities degrade the bulk of the remaining natural habitat (Peel and Stalmans 1999).

The low agricultural potential of this ecoregion and the large attendant mammal populations have prompted the establishment of an extensive protected area network (Huntley 1978). Government conservation areas protect more than 40 percent of the ecoregion, with private game ranches, nature reserves, and conservancies in South Africa, Zimbabwe, and Botswana adding another 5 to 10 percent. The most significant national parks are Kruger in South Africa, Gonarezhou and Hwange in Zimbabwe, Banhine, Gorongosa, Zinave, and Limpopo in Mozambique, Luangwa North and South in eastern Zambia, and the whole complex of protected areas along the middle Zambezi in Zambia and Zimbabwe.

South Africa, Zimbabwe, and Mozambique have recently declared the Greater Limpopo Transfrontier Park. The park includes Kruger, Gonarezhou, and Limpopo national parks and land in Zimbabwe connecting Kruger and Gonarezhou. A much larger Greater Limpopo Transfrontier Conservation Area (TFCA) is under discussion in this region, which will include the transfrontier park and hunting areas, private ranches, community areas, and Zinave and Banhine National Parks (GLTPJMB 2002). The Dongola/Limpopo Valley TFCA is another transborder national park that will span 4,900 km^2 of South Africa, Botswana, and Zimbabwe.

Types and Severity of Threats

The most widespread threat to the ecoregion is over-exploitation of wildlife. Black rhinos are still threatened by poaching for their horn, and farmers kill wild dogs as pests (Stuart et al. 1990). Poaching is common in poorly funded parks, particularly in Zambia and Mozambique, and wildlife populations in many areas in Mozambique were devastated during the war. However, wildlife management is improving, and populations are slowly recovering after the end of the war.

The most immediate threat to the ecoregion is the present land redistribution in Zimbabwe, where many large private farms have been broken into much smaller units since 2000. The large Save Valley and Chiredzi Conservancies in southeastern Zimbabwe near Gonarezhou National Park have been badly affected by increased poaching and habitat loss from slash-and-burn farming methods in areas unsuitable for agriculture. Community conservation initiatives, such as the CAMPFIRE program, are also reported to have collapsed in the areas where land invasions have occurred.

Justification for Ecoregion Delineation

Lower elevation, lower rainfall, and the presence of *Colophospermum mopane* separate the Zambezian and Mopane Woodlands [54] from neighboring ecoregions. Stretching from the Okavango to Lake Malawi and south to Swaziland, it comprises parts of three of White's (1983) vegetation mapping units: "*Colophospermum mopane* woodlands and scrub woodlands," "north Zambezian undifferentiated woodland and wooded grassland," and "south Zambezian undifferentiated woodland and scrub woodland."

Ecoregion Number: **55**

Ecoregion Name: **Angolan Mopane Woodlands**

Bioregion: **Eastern and Southern Africa**

Biome: **Tropical and Subtropical Grasslands, Savannas, Shrublands, and Woodlands**

Political Units: **Angola, Namibia**

Ecoregion Size: **133,500 km²**

Biological Distinctiveness: **Bioregionally Outstanding**

Conservation Status: **Relatively Intact**

Conservation Assessment: **V**

Author: *Suzanne Vetter*

Reviewed by: *Phoebe Barnard, Antje Burke, Jonathan Timberlake*

Location and General Description

The Angolan Mopane Woodlands [55] ecoregion lies along the Owambo Basin, which stretches between Namibia and Angola. It is found inland of the Namib Escarpment Woodlands [109] and surrounds the Etosha Pan Halophytics [67].

This ecoregion lies on the western edge of the Central African Plateau, at around 1,000 m in elevation along the Owambo Basin. The area is mostly flat but gains elevation toward the Waterberg Mountains (1,857 m), which lie near its southeastern border (Barnard 1998). The landscape consists of sand, silt, clay, low-grade coal, and tillate derived from the Karoo Sequence dating from 300 to 130 million years ago (Mendelsohn et al. 2000). Millions of years of deposition and erosion from wind and water have produced a complex variety of soils

and vegetation. Soils include halomorphic soils around the Etosha Pan, weakly developed shallow soils of arid origin to the south, and soloetzic and planosolic soils, as well as arenosols to the north (le Roux 1980). Two main river systems, both originating in the highlands in Angola, drain much of this ecoregion. The Kunene River is the only perennial river flowing through this ecoregion. The Cuvelai Drainage Basin is a 7,000-km² area of ephemeral, shallow, parallel channels or *oshanas* that fill Lake Oponono when flooded, about twice in 3 years (Barnard 1998). This lake is the source of the Ekuma River, which flows into the Etosha Pan when the Cuvelai Basin is flooded. Two other ephemeral rivers, the Oshigambo and Omuramba Owambo, also flow into the pan (Berry et al. 1973).

Mean annual rainfall is 350–500 mm, increasing inland of the coastal desert areas (Barnard 1998). The rain normally falls in the summer, between August and April, with most falling in late summer. February usually has the mean maximum rainfall, about 110 mm. Annual rainfall is unpredictable. In the Etosha National Park, for example, the total annual rainfall in 1946 was 90 mm, but in 1950 it was 975 mm (Berry 1972).

The variations in soils and climate contribute to a wide local variety of vegetation types (Mendelsohn et al. 2000). The main vegetation units include mopane shrublands, western Kalahari woodlands, western part of the karstveld, and Cuvelei. The vegetation types form intricate mosaics. Mopane (*Colophospermum mopane*) occurs as a shrub or a tree depending on local conditions (Giess 1971). In Angola, the mopane grows over vast areas in a low, thorny bushveld. It is associated with *Acacia kirkii*, *A. nilotica* ssp. *subalata*, *A. erubescens*, *Combretum apiculatum*, *Commiphora* spp., *Dichanthium papillosum*, *Dichrostachys cinerea*, *Grewia villosa*, *Indigofera schimperi*, *Jatropha campestris*, *Peltophorum africanum*, *Rhigozum brevispinosum*, *R. virgatum*, *Securinega virosa*, *Spirostachys africana*, *Terminalia sericea*, *Ximenia americana*, and *X. caffra* (Werger 1978; White 1983). On alluvial soils *Acacia kirkii* becomes abundant. During the rainy season species-rich aquatic communities associated with ephemeral pans and drainage lines (i.e., *oshanas*) develop.

Outstanding or Distinctive Biodiversity Features

Levels of endemism are lower in the Angolan Mopane Woodlands [55] than in the neighboring Namib Escarpment Woodlands [109] and Kaokoveld Desert [106] ecoregions to the west (Simmons et al. 1998b).

The once-abundant mammalian fauna has declined in most of this ecoregion (Mendelsohn et al. 2000). The largest remaining populations of large mammals occur in and around Etosha National Park, especially during the dry winter. Damara zebra (*Equus burchelli antiquorum*), blue wildebeest (*Connochaetus taurinus*), springbok (*Antidorcas marsupialis*), and giraffe (*Giraffa camelopardalis*) are the most numerous mammals. Threatened species also occur, such as elephant (*Loxodonta africana*, EN), black rhinoceros (*Diceros bicornis*, CR), lion (*Panthera leo*,

VU), cheetah (*Acinonyx jubatus,* VU), and the near-endemic black-faced impala (*Aepyceros melampus petersi,* VU) (Joubert and Mostert 1975; Lindeque and Lindeque 1997; Hilton-Taylor 2000). In the wetter summer months, wildlife migrates away from the waterholes, primarily to the west (Balfour and Balfour 1992; Lindeque and Lindeque 1997). Near-endemic mammals include Thomas's rock rat (*Aethomys thomasi*) and two white-toothed shrews (*Crocidura erica,* VU, and *C. nigricans*).

This ecoregion has a diverse avian fauna, higher than those of the arid ecoregions to the west and south or the Zambezian *Baikiaea* Woodlands [51] to the east (Robertson et al. 1998). At least 340 bird species have been recorded in the Etosha National Park (Hall 1994; Barnes 1998). Etosha supports the only breeding population of the threatened blue crane (*Grus paradisea*) outside South Africa (Simmons et al. 1998a). Ten bird species are near endemic, most of which are shared with the Namib Escarpment Woodlands [109] ecoregion. These include the Damara rock jumper (*Achaetops pycnopygius*), grey kestrel (*Falco ardosiaceus*), Carp's tit (*Parus carpi*), southern violet wood-hoopoe (*Phoeniculus damarensis*), Bradfield's hornbill (*Tockus bradfieldi*), Monteiro's hornbill (*Tockus monteiri*), bare-cheeked babbler (*Turdoides gymnogenys*), and black-lored babbler (*Turdoides melanops*). The rufous-tailed palm thrush (*Cichladusa ruficauda*) and Cinderella waxbill (*Estrilda thomensis*) are not savanna species but are restricted to *Hyphaene* palms and the vegetation along the Kunene River (Sinclair and Hockey 1996; Simmons et al. 1998a).

The ecoregion has a large number of reptile species, including the endemic *Afrogecko ansorgii* and the skaapsteker (*Psammophylax ocellatus*), a venomous snake that reaches up to 1 m in length. The ecoregion also has the second highest spider richness in Namibia, with 115 species native to the ecoregion, second only to the adjacent Kalahari *Acacia* Woodlands [58] (Griffin 1998b).

Status and Threats

Current Status

Two national parks occur on the Angolan side of the ecoregion. These are Bikuar National Park (7,900 km²) and Mupa National Park (6,600 km²). Although these two parks cover a representative area of the Angolan Mopane Woodlands, they do not offer adequate protection, a result of the 30-year civil war in the country (Dean 2000).

Game Reserve No. 2 originally covered about 80,000 km² of northern Namibia and was the largest nature reserve in the world. It included much of the Kaokoveld Desert [106], the northern portion of the Namib Escarpment Woodlands [109], and the central area of the Angolan Mopane Woodlands [55]. The reserve stretched from the Kunene River border south more than 200 km to the Hoarusib River. In 1968 this reserve was reduced by 72 percent to become the present Etosha Na-

tional Park, an area of 22,912 km² (du Plessis 1992). The excised land was reallocated to communal homelands (Owambo, Kaokoland, and Damaraland). Today about 50 percent of the land formerly occupied by Game Reserve No. 2 will be protected, mainly by large communal conservancies (Barnard et al. 1998). Private nature reserves and game farms are also found in this ecoregion, and game farming is increasing in popularity (Barnard et al. 1998).

Types and Severity of Threats

In the first half of the twentieth century, severe overhunting of mammals was a major threat to wildlife outside Namibia's protected areas. However, in 1967 legislation shifted the ownership of wildlife on freehold lands from the state to the individual landowner. This significantly reduced the hunting threat to wildlife as landowners began to commercialize wildlife (Griffin 1998a). After Game Reserve No. 2 was dissolved in 1968, wildlife in this area was nearly decimated by poaching gangs, colonial officials, contractors, and local people armed from Namibia's war of liberation. The poaching problem has continued to represent a threat to the wildlife of the ecoregion, particularly to the black rhino population (Lindeque and Lindeque 1997; Barnard 1998). Rhinos were also dehorned as a last resort to deter poachers (Berger and Cunningham 1994). The Nature Conservation Amendment Act of 1996 extends wildlife ownership rights to people living in communal areas with the hope that rural dwellers will accrue benefits from wildlife and then manage it sustainably (through conservancies).

The situation on the Angolan side of the ecoregion is less encouraging. The 30-year civil war in Angola has devastated conservation efforts in the area (Dean 2000). Protected areas are open to poachers, timber harvesting, human settlement, and agriculture. Few, if any, viable populations of larger mammals have survived, and populations of lion, black rhino, and giraffe have become locally extinct. The situation is similar outside the protected areas, and overexploitation of wildlife and other natural resources is common. On the positive side, the Angolan government has recently established a State Secretariat for the Environment and has begun training demobilized soldiers as park wardens. Angola is also a signatory to the Convention on Biological Diversity (Dean 2000). The serious landmine situation in Angola, though devastating for humans and other large mammals, has afforded some protection to other species by rendering some areas off-limits to settlement or large-scale exploitation.

The Etosha National Park faces several management challenges. Anthrax has become a significant disease in the park and together with rinderpest is one of the most dramatic diseases affecting wild animals in Africa (Ebedes 1976). Another disease, feline immunodeficiency virus, is also prevalent, particularly among cheetah (Simmons et al. 1998a). Fences erected around the park between 1962 and 1973 have disturbed the natural herbivore migration patterns in the area (Mendelsohn et al. 2000). This disturbance is a particular threat during drought years when the ungulates cannot move away from drought-stricken areas (Simmons et al. 1998a).

Justification for Ecoregion Delineation

The linework for this ecoregion is based on the western portion of White's (1983) "*Colophospermum mopane* woodland and scrub woodland." This is distinguished from the Zambezian and Mopane Woodlands [54] by different floral and faunal compositions.

Ecoregion Number:	**56**
Ecoregion Name:	**Western Zambezian Grasslands**
Bioregion:	**Eastern and Southern Africa**
Biome:	**Tropical and Subtropical Grasslands, Savannas, Shrublands, and Woodlands**
Political Units:	**Angola, Zambia**
Ecoregion Size:	**34,000 km²**
Biological Distinctiveness:	**Locally Important**
Conservation Status:	**Relatively Intact**
Conservation Assessment:	**V**
Author:	*Suzanne Vetter*
Reviewed by:	*Jonathan Timberlake*

Location and General Description

The Western Zambezian Grasslands [56] are located in western Zambia, in two main sections; the northern section comprises mostly the Liuwa plains, and the southern section covers the Silowana plains and Mulonga plain. The ecoregion is defined by seasonal waterlogging of the soils, which prevents tree growth.

Most of this ecoregion lies around 1,000 m in elevation, although the Liuwa plains are somewhat higher. The ecoregion experiences a tropical savanna climate with three seasons: a hot dry season (August–October), a hot wet season (November–April), and a cool dry season (May–July). Mean annual rainfall ranges from 800 to 1,000 mm. The mean maximum temperature is around 27°C, and the mean minimum temperature is between 12°C and 15°C.

Much of this ecoregion falls in the traditional kingdom of Barotseland. The human population density is generally less than five people per square kilometer (Turpie et al. 1999). The grasslands in Barotseland are part of a transhumant farming system in which people and their livestock move between the floodplain in the dry season and the more wooded uplands in the rainy season.

The ecoregion falls in the Zambezian center of endemism (White 1983). The vegetation consists of short, sparse, wiry grassland dominated by *Loudetia simplex*, which is used as a thatching grass, and *Monocymbium ceresiiforme*. These species often are associated with other wiry grasses including species of *Andropogon, Eragrostis, Aristida, Elionurus, Melinis,* and *Tristachya*. Different sedge species of the family Cyperaceae are common where the soil contains more humus (Werger and Coetzee 1978; White 1983). Trees are largely absent and are replaced by rhizomatous geoxylic suffrutices, or woody plants with most of their modified stems underground. They form "underground forests" (White 1976) less than 0.6 m tall. Most of these species are closely related to forest or woodland trees or lianas. They flower precociously before the end of the dry season while grasses are still dormant. This is probably an evolutionary adaptation to being burned almost annually, which suggests that fires have been a part of the ecology of these grasslands for a long time (Lock 1998). The early flowering may also result from poor drainage and limited rooting depth caused by anaerobic conditions (White 1976; Timberlake et al. 2000).

Outstanding or Distinctive Biodiversity Features

Species richness of plants is low, and there are no known endemic species.

The mammalian fauna is representative of the southern African savannas, and some 140 species are known to occur in the ecoregion, including large numbers of ungulates. The Liuwa plains are home to about 30,000 migratory blue wildebeest (*Connochaetes taurinus*), the largest herd in Zambia. The wildebeest herds migrate into Angola in June and return to the southern part of Liuwa plains in October. In total, the migration takes 5 months and covers more than 200 km. Similarly, approximately 3,000 lechwe (*Kobus leche*) move east from the Liuwa plains to the Zambezi floodplain in the dry season (Muleta et al. 1996). Other ungulate species found in the area include tsessebe (*Damaliscus lunatus*), oribi (*Ourebia ourebi*), reedbuck (*Redunca arundinum*), Burchell's zebra (*Equus burchelli*), roan antelope (*Hippotragus equinus*), sable antelope (*H. niger*), greater kudu (*Tragelaphus strepsiceros*), and Lichtenstein's hartebeest (*Sigmoceros lichtensteinii*). Large carnivores include lion (*Panthera leo,* VU), leopard (*P. pardus*), African wild dog (*Lycaon pictus,* EN), cheetah (*Acinonyx jubatus,* VU), and spotted hyena (*Crocuta crocuta*). Lions are reported to be largely extinct in Liuwa plain (Muleta et al. 1996).

Bird species richness is moderate, but there are no endemic species. Notable species include wattled crane (*Grus carunculatus,* VU), white-bellied bustard (*Eupodotis senegalensis*), and martial eagle (*Polemaetus bellicosus*).

The reptile and amphibian fauna is part of a broad transition zone between the tropical fauna of Africa and the cape fauna of southwestern South Africa (Poynton and Broadley 1978). Elements from the adjoining ecoregions, as well as more widespread species, are represented here, such as the Angola green snake (*Philothamnus angolensis*) and Cape rough-scaled lizard (*Ichnotropis capensis*).

Status and Threats

Current Status

This ecoregion has a long history of human habitation, and parts of it have been settled for centuries (Turpie et al. 1999). The vegetation has adapted to some human disturbance, most notably frequent fires. In fact, fires and other disturbances aid the expansion of grasslands at the expense of woody vegetation. A little fragmentation has occurred as a result of cropping and settlement, but overall the habitat is largely intact and continuous.

The Liuwa Plain National Park is the only protected area that contains significant areas of Western Zambezian Grasslands [56]. The large Western Zambezi Game Management Area (GMA) surrounds this national park. Although the GMA is mostly *Baikiaea* and miombo woodland, it contains several areas of grassland. Liuwa Plain National Park supports wildlife representative of the ecoregion, although population numbers have declined and are not known in many cases (Muleta et al. 1996; Turpie et al. 1999). The GMA is now largely devoid of wildlife, except for wildlife migrating through it, because of poaching and lack of management (Muleta et al. 1996).

Types and Severity of Threats

The GMA and national park experience subsistence hunting by local residents and commercial poaching, mostly by outsiders. The poaching problem has been exacerbated by the availability of firearms acquired from the liberation struggles in Angola and Namibia (Muleta et al. 1996). Migrating animals outside the parks are not protected at all, and hunting pressure is intense.

Cattle numbers are increasing in the grassland plains, although they are still below the carrying capacity of the system (van Gils 1988; Simwinji 1997) because of diseases. Burning is common because the wiry grasses, particularly *Loudetia simplex*, are otherwise unpalatable. Although the vegetation is adapted to burning, the increased frequency is thought to create ecological problems for nesting birds (Turpie et al. 1999). Uncontrolled late burning is considered to be a major cause of rangeland degradation (Simwinji 1997).

In some parts of the ecoregion, including inside Liuwa Plain National Park, settlements and cultivation have modified and fragmented the grassland vegetation. The extent of the damage to this ecoregion has not been assessed, nor has vegetation recovered on abandoned fields.

Although land-use practices such as agriculture have not changed greatly over the long time that the area has been settled, they are intensifying because of population pressures. In addition, people living in the Barotse floodplain area have reported declines in useful plants, fish, and wildlife since the 1960s (Turpie et al. 1999). Hunting and poaching have increased since independence, a change that goes against a cultural heritage through the Barotse Royal Establishment, which has placed great importance on sustainable resource use (Simwinji 1997).

Justification for Ecoregion Delineation

The Western Zambezian Grasslands ecoregion roughly follows White's (1983) "edaphic and secondary grassland on Kalahari Sand." It is distinguished from the neighboring Zambezian *Cryptosepalum* Dry Forests [32] by seasonal waterlogging that prevents tree growth in this ecoregion.

Ecoregion Number: **57**

Ecoregion Name:	**Southern Africa Bushveld**
Bioregion:	**Eastern and Southern Africa**
Biome:	**Tropical and Subtropical Grasslands, Savannas, Shrublands, and Woodlands**
Political Units:	**Botswana, Mozambique, South Africa, Zimbabwe**
Ecoregion Size:	**223,100 km²**
Biological Distinctiveness:	**Bioregionally Outstanding**
Conservation Status:	**Endangered**
Conservation Assessment:	**IV**
Author:	*Amy Spriggs*
Reviewed by:	*Brian Huntley*

Location and General Description

The Southern Africa Bushveld [57] covers the southeast corner of Botswana, southern Zimbabwe, and northern South Africa. It has a well-defined southern boundary, the Highveld Grasslands [77], which is a cool, high-elevation (1,500–2,000 m) grassland that is exposed to frequent, severe frosts in winter. To the east, the ecoregion is bounded by the mountain ranges of the Drakensberg, Strydpoortberg, and Soutpansberg.

The ecoregion occurs on an extensive, undulating interior plateau, which lies at an elevation between 700 and 1,100 m. The soils of this plateau are mostly coarse, sandy, and shallow, overlying granite, quartzite, sandstone, or shale (Low and Rebelo 1996). The most distinctive topographic feature is the rugged and rocky Waterberg Mountains, which rise up from the plateau to an elevation between 1,200 and 1,500 m. A flat plain known as the Springbok Flats extends into the southwest of the ecoregion. An unusual soil type characterizes this plain: black or red vertic clay, derived from basalt (Low and Rebelo 1996). The area is drained by the Limpopo and Olifants rivers. The

Limpopo Basin covers most of the ecoregion, from Botswana, Zimbabwe, and northern South Africa, whereas the Olifants River Basin is smaller and drains the eastern portion. The two rivers converge before flowing into the Indian Ocean, just north of Maputo in Mozambique.

The climate is tropical and seasonal, with hot, wet summers and cool, dry winters. It has an average annual rainfall between 350 and 750 mm (Nix 1983), with a slightly higher average (650–900 mm) in the Waterberg Mountains (Low and Rebelo 1996). The highveld plateau forms a ridge running east to west to the south and provides shelter against cooler air masses (Van der Meulen 1979). As a result, the temperatures in the bushveld are higher than on the more elevated highveld and range from −3°C to 40°C, with an average of 21°C. The bushveld also experiences mild frosts a few times per year, whereas the highveld is exposed to frequent, severe frosts in winter.

The Southern Africa Bushveld [57] is located in White's (1983) Zambezian phytogeographic region, which is floristically impoverished at this southern end (Goldblatt 1978). Cowling et al. (1997b) divide the vegetation of this ecoregion into mixed savanna and mopane savanna. Mopane savanna extends from southeastern Botswana into the main plateau of Zimbabwe and down into northern South Africa, known as the Tuli Block. The ecoregion is dominated by often monospecific stands of the winter-deciduous mopane tree (*Colophospermum mopane*) in the subfamily Caesalpinoideae. This tree ranges from a dense shrubland to a tall, open woodland, reflecting variability in growth conditions and a history of disturbance, especially by elephants (Cowling et al. 1997). Toward the Tuli Block in northern South Africa, *C. mopane* is intermixed with other tree species, such as *Acacia tortilis* and *A. mellifera*. In the northern part of the ecoregion (around Bulawayo in Zimbabwe) the vegetation grades into woodland savanna. *Hyparrhenia filipendula* and *H. dissolute* are the most common grass species. The silver clusterleaf (*Terminalia sericea*) is the dominant tree, intermixed with varying proportions of *Burkea africana*. To the north of this, *Acacia* species, such as *Acacia nilotica*, *A. karoo*, and *A. rehmanniana*, become dominant.

Clay thorn bushveld occurs on the Springbok Flats in the southern part of the ecoregion (Low and Rebelo 1996). This veld type is associated with the unusual basalt-derived clays of the flats. It is an open savanna dominated by many *Acacia* species, such as *Acacia tortilis*, *A. nilotica*, *A. nigrescens*, *A. gerrardii*, and *A. karoo*. The grasses are dense, tall, and coarsely tufted. Turf grass (*Ischaemum afrum*), deck grass (*Sehima galpinii*), and canary millet (*Setaria incrassata*) are the dominant species. The savanna surrounding the Waterberg Mountains is characterized by African beachwood (*Faurea saligna*), common hookthorn (*Acacia caffra*), red seringa (*Burkea africana*), *Terminalia sericea*, and *Peltophorum africanum*. The rugged, rocky slopes of these mountains are dominated by white seringa (*Kirkia acuminata*), *Combretum apiculatum*, *C. molle*, and common sugarbush (*Protea caffra*).

Outstanding or Distinctive Biodiversity Features

This ecoregion supports low plant species richness and has few endemics. Many of the elements characteristic of the region are also lacking, such as the genera *Tamarindus* (Fabaceae) and *Monotes* (Dipterocarpaceae). It also contains only one species of the widespread *Brachystegia* genus (Fabaceae) and one species of *Julbernardia* (Fabaceae) (White 1983). The Waterberg Mountains have the highest levels of plant species richness and endemism in the ecoregion.

The ecoregion contains a high faunal species richness but a low per unit area richness and low numbers of endemic species. There is only one near-endemic mammal, Juliana's golden mole (*Amblysomus julianae*, CR). The ecoregion supports many of the charismatic large mammals associated with African savannas. Although these species are not endemic, some are threatened, including black rhinoceros (*Diceros bicornis*, CR), cheetah (*Acinonyx jubatus*, VU), and elephant (*Loxodonta africana*, EN) (Hilton-Taylor 2000).

African savannas are well known for their rich, diverse, and colorful bird life. However, only one of these is near endemic to the ecoregion, the boulder chat (*Pinarornis plumosus*). The ecoregion also includes important colonies of Cape griffon (*Gyps coprotheres*, VU) (Fishpool and Evans 2001). There is a rich reptile fauna, with higher levels of endemism than in other taxa. Many of the strict endemics, such as the Waterberg girdled lizard (*Cordylus breyeri*), the Waterberg dwarf gecko (*Lygodactylus waterbergensis*), and the Waterberg flat lizard (*Platysaurus minor*), are endemic to the Waterberg Mountains. Other species strictly endemic to this ecoregion include Loboatse hinged tortoise (*Kinixys lobatsiana*), Muller's velvet gecko (*Homopholis mulleri*), and dwarf flat lizard (*Platysaurus guttatus*).

Status and Threats

Current Status

Many areas of natural habitat are conserved in the form of provincial nature reserves, conservation areas, and private game farms. Pilanesberg National Park is one of the most important reserves. This 553-km^2 national park encompasses an isolated ring complex of volcanic hills (one of three such ring complexes in the world). The Hans Strydom and Doorndraai Dam Nature Reserves, situated in the Waterberg Mountain Range, support a number of endemic plants and animals, such as *Euphorbia waterbergensis*, *Hibiscus waterbergensis*, and *Aloe petrophila*. The most threatened veld type is found on the fertile clays of the Springbok Flats. Low and Rebelo (1996) estimate that 60 percent of this veld type has been transformed to agriculture, and only 1 percent is conserved, in the Nylsvlei Nature Reserve.

In the Limpopo Province of South Africa there has been a recent trend to restock privately owned savanna areas with indigenous herbivores. These private game farms contribute significantly to conservation for species such as reintroduced white rhinoceros (*Ceratotherium simum*). Although the percentage of habitat protected by the game farms is high, the habitat is somewhat fragmented because these small conservation areas often are interspersed with cattle ranches. This is the case for most of the Northern Province but not for the Waterberg Mountains area, where unprotected areas are largely undisturbed, mainly because of the rocky nature of the area (Low and Rebelo 1996). In Botswana and Zimbabwe human population density is low and is restricted to small settlements, so large areas of continuous habitat remain. No protected areas are found in this ecoregion in Botswana. In Zimbabwe, the most important protected area is Matopos National Park (425 km^2).

Types and Severity of Threats

The major land-use practices in the Limpopo Province of South Africa are game and cattle farming. Whereas game farming better preserves the natural habitat, cattle farming can lead to overgrazing, contributing to erosion, bush encroachment, and altered fire frequency. In addition, the predatory and scavenging fauna of the bushveld are perceived as pests by farmers and routinely exterminated. Black-backed jackal (*Canis mesomelas*), caracal (*Caracal caracal*), and Cape griffon are common target species, although nontarget species such as bat-eared fox (*Otocyon megalotis*), aardwolf (*Proteles cristatus*), and aardvark (*Orycteropus afer*) often are killed.

Direct habitat loss is prevalent on the Springbok Flats, the most threatened habitat in the ecoregion. The fertile clay soils of these flats are ideal for crops such as wheat, maize, and sunflowers. In addition to direct habitat loss, the agriculture in this area has had a negative impact on many bird species through the use of organochloride insecticides and DDT (Barnes 1998). Direct habitat loss is also threatening the south of the ecoregion through the expansion of the Pretoria-Witwatersrand-Vereeniging complex. This "Vaal Triangle" is one of South Africa's most densely populated industrial areas (Cowling et al. 1997).

There are fewer threats to the north of the ecoregion in Botswana and Zimbabwe, where low-intensity goat and cattle farming cause some impacts. The removal of dead wood for firewood may also negatively affect tree-hole nesting birds and small mammals (du Plessis 1995). In large areas of Botswana and Zimbabwe, wildlife contributes significantly to the local economy. Wildlife use was originally mostly licensed trophy hunting but is now increasingly oriented toward nonconsumptive recreation and tourism.

Four exotic plant species have invaded the Southern Africa Bushveld [57]. These are *Jacaranda mimosifolia*, *Lantana camara*, syringa (*Melia azedarach*), and white mulberry (*Morus alba*).

However, the impact of these invasive species is not as serious as elsewhere in southern Africa (such as the Lowland Fynbos and Renosterveld [89] ecoregion).

Justification for Ecoregion Delineation

The Southern Africa Bushveld [57], stretching from Bulawayo in the north to Pretoria in the south, combines White's (1983) "*Colophospermum* mopane woodland and scrub woodland" and "South Zambezian undifferentiated woodland" and includes portions of "Kalahari deciduous *Acacia* wooded grassland and deciduous bushland." Lying on a plateau, it has a higher elevation gradient than surrounding ecoregions to the west, north, and east. The Highveld forms a distinct southern boundary, with elevations between 1,500 and 2,000 m.

Ecoregion Number:	**58**
Ecoregion Name:	**Kalahari *Acacia* Woodlands**
Bioregion:	**Eastern and Southern Africa**
Biome:	**Tropical and Subtropical Grasslands, Savannas, Shrublands, and Woodlands**
Political Units:	**Botswana, Namibia, Zimbabwe, South Africa**
Ecoregion Size:	**335,500 km^2**
Biological Distinctiveness:	**Locally Important**
Conservation Status:	**Relatively Intact**
Conservation Assessment:	**V**
Author:	*Amy Spriggs*
Reviewed by:	*Jonathan Timberlake*

Location and General Description

The Kalahari *Acacia* Woodlands [58] occupy a broad band of land that runs from northern Namibia and diagonally across Botswana toward South Africa. In Botswana, a smaller section extends north between the Okavango Delta and the Makgadigkadi Pans toward the border of the Chobe Nature Reserve and then extends east, just crossing the border into Zimbabwe, and into the Tuli Block to the south.

The ecoregion lies in the center of the Great African Plateau at an elevation of about 1,000 m (Werger 1978; Moyo et al. 1993; Penry 1994). Granitic hills are scattered across the central region (the Tsau and Khwebe Hills) and in the northwest (the Tsodilo, Aha, and Gcwihaba Hills). There are also outcrops of Precambrian rocks on the hill massifs around the towns of Mahalapye and Palapye, such as the Mokgware, Tswapong, and Shoshong hills. These hills rise to just above 1,300 m and are isolated topographic features in the otherwise flat landscape. Arenosols dominate the parts of the ecoregion underlain by Kalahari sands. These soils are poorly developed, but the sands are deep and have a high moisture storage capacity. To the southeast, the sandveld grades into an undulating plain, called the hardveld. The soils of the hardveld are highly leached and ferruginous but are richer than those on sand.

The climate of the ecoregion is semi-arid, with droughts occurring on an approximate 7-year cycle (Moyo et al. 1993; Penry 1994). Rainfall occurs mostly in the summer, from October through March. During the winter, from May through August, there is little or no rain. The annual rainfall ranges from about 300 mm in the southwest to 600 mm in the north, varying from year to year. Surface water is scarce (Moyo et al. 1993; Penry 1994). In the hardveld, a number of ephemeral rivers drain into the Limpopo, itself a seasonal river. The Mhalatswe and Lotsane are the largest of these rivers. Surface drainage over the sandveld areas is limited mainly to pans and dry valleys, and surface water is present only after rain. Deception, Quoxo, and Groot Laagte are three major fossil valleys found in the sandveld. One ephemeral river, the Boteti River, flows out of the Okavango Delta and across this ecoregion to the Makgadigkadi Pans.

Temperatures are typical of a continental climate, with high diurnal and seasonal ranges. In June and July, temperatures can drop below freezing, but in the summer temperatures may exceed 40°C. The mean maximum temperature is between 27°C and 30°C, and the mean minimum temperature is between 9°C and 12°C.

The ecoregion spans the southern extent of the Zambezian regional center of endemism and the northern portion of the Kalahari-Highveld Regional transition zone. *Acacia* woodlands are typical, transitional between the caesalpinoid woodlands to the north and west (especially mopane and *Baikiaea*) and the true *Acacia* woodlands or microphyllous savannas to the south (Wild and Barbosa 1967; Werger 1978; White 1983; Campbell 1990). The main tree species in the core of the ecoregion are *Lonchocarpus nelsii, Terminalia sericea, Burkea africana, Combretum* spp., *Acacia erioloba,* and *A. luederitzii.*

In the sandveld, taller trees are mainly confined to low sand ridges, which are dominated by *Terminalia sericea, Burkea africana, Peltophorum africanum, Croton gratissimus, Acacia erioloba, A. fleckii, A. luederitzii,* and *Combretum zeyheri.* A shrub savanna occurs on the gently rolling plains between the sand ridges and is composed mainly of *Dichrostachys cinerea, Grewia flava, G. flavescens, Acacia mellifera, Bauhinia macrantha,* and *Commiphora pyracanthoides.* The grass cover includes *Stipagrostis uniplumis, Aristida meridionalis, A. congesta, Eragrostis pallens, E. superba, E. lehmanniana, Heteropogon contortus,* and *Digitaria eriantha.*

On the hardveld, vegetation patterns are complex as a result of changing soil types, linked with complex changes in the underlying geology. In the area south of the Makgadigkadi Pans,

Colophospermum mopane begins to dominate the tree savanna. The dominant grasses are *Cenchrus ciliaris, Digitaria milanjiana, Eragrostis* spp., and *Dichanthium papillosum.* Around the town of Palapye, the vegetation changes and resembles the *Terminalia sericea*–dominated tree savanna of the sandveld. To the west, in an area surrounding the town of Mahalapye, the vegetation becomes a more open tree savanna of *Acacia nigrescens,* sometimes found with *Sclerocarya birrea, Peltophorum africanum, Acacia erioloba, Combretum apiculatum,* and *Combretum imberbe.* The grass cover is mixed but includes *Panicum maximum, Digitaria eriantha,* and *Eragrostis lehmanniana.* South of the town of Mahalapye, *Peltophorum africanum* dominates the tree savanna, with a dense shrub cover of *Dichrostachys cinerea, Acacia tortilis,* and *Grewia flava* present under the canopy.

Outstanding or Distinctive Biodiversity Features

The ecoregion has no plant endemics and only 500–700 vascular plant species (Barnard 1998; Van der Walt and Le Riche 1999; van Rooyen 2001).

Faunal species richness and endemism are also low. However, many of the charismatic large mammals associated with African savannas occur, such as blue wildebeest (*Connochaetes taurinus*), hartebeest (*Alcelaphus buselaphus*), greater kudu (*Tragelaphus strepsiceros*), and eland (*Taurotragus oryx*). Although no species is endemic, several are listed as threatened, including elephant (*Loxodonta africana,* EN), cheetah (*Acinonyx jubatus,* VU), lion (*Panthera leo,* VU), and African wild dog (*Lycaon pictus,* EN) (Hilton-Taylor 2000). More arid-loving animals such as the gemsbok (*Oryx gazella*) and springbok (*Antidorcas marsupialis*) are also found. Many of the large herbivores undertake seasonal migrations, especially during droughts. Blue wildebeest, eland, Burchell's zebra (*Equus burchelli*), buffalo (*Syncerus caffer*), and hartebeest all migrate in this ecoregion. Increasingly, veterinary foot-and-mouth disease control fences are restricting migration routes and preventing these migratory movements.

The ecoregion has a rich avifauna (Penry 1994; Maclean 1984). The black-lored babbler (*Turdoides melanops*) is near endemic, found in the area west of the Okavango Delta and extending into Namibia. The threatened lappet-faced vulture (*Torgos tracheliotus,* VU) (BirdLife International 2000) is found throughout the ecoregion. Of the amphibian and reptile species, the Tsodilo gecko (*Pachydactylus tsodiloensis*) is strictly endemic, found only on the Tsodilo hills in the northwest of the ecoregion.

Status and Threats

Current Status

Much of the ecoregion's natural vegetation remains intact and is conserved in several protected areas: the Central Kalahari Game Reserve in Botswana, the Khaudom Game Reserve in

Namibia, and the Nxai pan complex north of Makgadikgadi (Stuart et al. 1990; Stuart and Stuart 1992). The Central Kalahari Game Reserve covers 51,800 km² of sandveld and is still inhabited by groups of bushmen. Khaudom Game Reserve covers 3,840 km² of flat sandveld in the northeast. Other protected areas that extend into this ecoregion include Hwange in Zimbabwe and Chobe in Botswana. There is only one reserve in the hardveld, the Stevensford Private Game Reserve, which is a small reserve located on the north bank of the Limpopo, southeast of the town of Palapye.

In the mid-1970s the Botswana government proposed a network of wildlife management areas (WMAs). The WMAs are mostly in areas of land adjacent to reserves and aim to provide areas where a wildlife industry can be developed (Williamson 1994). Though designated, none of the WMAs has been gazetted (Dikobe 1995). The WMAs increase the area of sandveld under protection but do not include the hardveld.

Types and Severity of Threats

The most important threat to this ecoregion is the expansion of livestock agriculture. This has been made possible by the wide-scale sinking of boreholes to tap into groundwater sources and the elimination of competing wildlife species. Veterinary cordon fences have been erected to control the spread of foot-and-mouth disease to cattle but pose a major threat to wildlife populations (Owens and Owens 1980; Campbell 1990; Moyo et al. 1993; Williamson 1994). In the past 20 years, the networks of fences that cross much of Botswana have disrupted the migratory patterns of a number of large mammals, causing a decline in populations of wildebeest, hartebeest, zebra, and buffalo (Thomas and Shaw 1991; Mothoagae 1995; Hanks 2002).

Overgrazing also changes the composition of natural vegetation, particularly by the replacement of sweet grasses such as *Chloris gayana, Cynodon dactylon,* and *Eragrostis pilosa* by less palatable species such as *Aristida congesta* (Moyo et al. 1993; Williamson 1994). The poisoning of predatory mammals has caused a large decline in lappet-faced vultures (BirdLife International 2000).

Justification for Ecoregion Delineation

This ecoregion, stretching from northeastern Namibia to the South African Highveld, follows White's (1983) "Zambezian transition from undifferentiated woodland to *Acacia* deciduous bushland and wooded grassland." The ecoregion's low levels of endemism highlight its transitional nature.

Ecoregion Number:	**59**
Ecoregion Name:	**Sudd Flooded Grasslands**
Bioregion:	**Western Africa and Sahel**
Biome:	**Flooded Grasslands and Savannas**
Political Units:	**Sudan, marginally in Ethiopia**
Ecoregion Size:	**179,700 km²**
Biological Distinctiveness:	**Globally Outstanding**
Conservation Status:	**Relatively Stable**
Conservation Assessment:	**III**
Author:	*Colleen Seymour*
Reviewed by:	*Philip Winter*

Location and General Description

The Sudd Flooded Grasslands [59] ecoregion is located in southern Sudan and comprises one of the largest swamps in Africa, about 600 km long and similarly wide. The White Nile (known in various sections as the Bahr-el-Abiad, Bahr-el-Jebel, Albert Nile, and Victoria Nile) rises in the headwaters of Lake Victoria in a region of year-round rainfall. After running through Uganda, it overflows in southern Sudan into a shallow depression, creating the Sudd swamps at 380–450 m above sea level (Beadle 1981). Since 1961, inflow to the Sudd has increased substantially, presumably because of increased rainfall in the headwaters around Lake Victoria. The inflow before 1960 was 26,831 billion m³/year of water, but from 1960 to 1980 it averaged 50,324 billion m³/year (Hughes and Hughes 1992). The wetland area consequently increased dramatically until 1980, although trends in recent years are not known.

The southern portion of the floodplain has more rainfall than the northern, receiving on average about 800 mm/year, compared with the north's 600 mm/year. These rains fall between April and September (Moss 1998), during the hot season, when temperatures average 30–33°C, dropping to an average of 18°C in the cooler season. Vertisols have developed in the waterlogged conditions over these nutrient-poor sediments, although fluvisols and patches of luvisols can be found along the river courses.

The Dinka, Nuer, and Shilluk tribes used to coexist in the Sudd with tens of thousands of large herbivores. These people depend on the annual floods and rain to regenerate floodplain grasses that feed their herds of cattle (Denny 1991). The human population density is less than twenty people per square kilometer and is concentrated along major rivers, lakes, and floodplains.

The floodplain ecosystem supports a variety of plant species ranging from those adapted to mesic environments to those adapted to more xeric environments. Moving from the interior of the swamps, the ecological zones grade from the open water and submerged vegetation of a river-lake, to floating fringe vegetation, to seasonally flooded grassland, to rain-fed wetlands, and finally to floodplain woodlands (Hickley and Bailey 1987). *Cyperus papyrus* is dominant at riversides and in the wettest swamps. *Phragmites* and *Typha* swamps are extensive behind the papyrus stands, and there is an abundance of submerged macrophytes in the open waterbodies. Seasonal floodplains, up to 25 km wide, are found on both sides of the main swamps. Wild rice (*Oryza longistaminata*) and *Echinochloa pyramidalis* grasslands dominate the seasonally inundated floodplains. Beyond the floodplain, *Hyparrhenia rufa* grasslands cover the rain-fed wetlands. Mixed woodlands of *Acacia seyal, Ziziphus mauritiana, Combretum fragrans,* and *Balanites aegyptica* border the floodplain ecosystem (Denny 1991).

Outstanding or Distinctive Biodiversity Features

The swamps and floodplains of the Sudd are among the most important wetlands in Africa, supporting a rich biota, which includes more than 400 bird species and more than 90 mammal species (Hughes and Hughes 1992). Major migrations of birds and mammals occur or used to occur in this ecoregion.

In the 1980s, southern Sudan had one of the highest antelope populations in Africa (East 1999). The endemic Nile lechwe (*Kobus megaceros*) occurs, numbering between 30,000 and 40,000 individuals (Estes 1991). The most abundant antelopes are the kob (*Kobus kob*), tiang (*Damaliscus lunatus tiang*), and Mongalla gazelle (*Gazella thomsonii albonotata*). These three antelopes undertake (or used to undertake) large-scale migrations over the largely undisturbed habitat to the east of the Sudd. For example, a million individuals of white-eared kob used to migrate more than 1,500 km near the Ethiopian border, following the availability of floodplain grasses (Denny 1991). Moreover, the tiang, which migrated from the Zeraf area to the South West in the 1980s, were thought to number half a million. The current status of these migrations is unknown but their numbers probably are much lower. More than 800,000 individual antelopes were estimated to inhabit Boma National Park in 1982–1983, with population densities up to 1,000/km^2 near food sources during the dry season (Fryxell and Sinclair 1988). Surveys in 2001 indicated a population of around 176,120 white-eared kob and no more than 100,000 for all other antelopes combined (Wildlife Survey Team 2001).

The floodplains provide important habitat for wetland birds, with more than 2.5 million using the floodplains of the Sudd annually, mainly migratory species moving between Europe and Africa (Robertson 2001c). These wetlands also support the largest population of shoebill (*Balaeniceps rex*) in the world, estimated at around 6,400 individuals (Robertson 2001c). The area is also a stronghold for the great white pelican (*Pelecanus onocrotalus*), ferruginous duck (*Aythya nyroca*) (Robertson 2001c), and black-crowned crane (*Balearica pavonina*) (Newton et al. 1996; Shmueli et al. 2000). Werner's garter snake (*Elapsoidea laticincta*) is a near-endemic snake, but other amphibians and reptiles are unremarkable.

Annual floods, capable of inundating more than 15,000 km^2 of land, are crucial to the maintenance of biological diversity in the Sudd. Welcomme (1979) estimated that only 11 percent of the total flooded area was permanent water, although this proportion may have increased in recent wetter years.

Status and Threats

Current Status

The Sudd swamp covers at least 30,000 km^2, and the peripheral effects of the swamp are believed to extend over the entire ecoregion. Much of the area remains as a vast near-wilderness area.

There are three designated protected areas in the ecoregion: Shambe National Park (1,000 km^2), Zeraf Island Game Reserve (6,750 km^2), and Fanyikang Game Reserve (45 km^2) (Hughes and Hughes 1992; Stuart et al. 1990). Boma and Badingilo national parks and Mongalla Game Reserve protect adjacent ecosystems. White-eared kob, which migrate south from Boma during the wet season, are believed to share a wet season grazing area with the migratory tiang in the short grass plains east of Badingilo. A recent assessment regarded the protection and management in Boma and Badingilo national parks as effectively nil (East 1999).

A large extension of Shambe Game Reserve has been proposed to ensure the protection of the ecoregion's biodiversity (Stuart et al. 1990). However, Zeraf Island might be a more feasible proposition because it has probably suffered less from warfare than Shambe because of the protection offered by the two rivers that define it.

Types and Severity of Threats

The civil war in southern Sudan poses the largest threat to the large mammals of the region, and in many regions they have been shot out (P. Winter, pers. comm., 2002). Lack of effective management and protection in the parks means that poaching is uncontrolled, and it is believed that populations of elephant (*Loxodonta africana*, EN) and other large ungulates have been decimated. Furthermore, it appears as if many large carnivores have been extirpated over much of Sudan, although lack of data makes it difficult to ascertain their status in the Sudd (Ginsberg 2001).

Plans have existed since 1902 to divert the waters of the White Nile around the Sudd swamps via the Jonglei Canal. Work on the canal began in 1978 but was stopped in 1984 because of the outbreak of the second civil war (Hughes and Hughes 1992). Diversion would cause the Sudd swamps and associated floodplains to shrink dramatically, possibly threatening the fauna and flora that depend on them for their survival. The empty canal is detrimental to wildlife in the area, particularly to some populations of large mammals. The canal blocks the annual movement of the tiang southwest to their wet season grazing area, and many thousands have been shot as they try to find crossing points (P. Winter, pers. comm., 2002).

Justification for Ecoregion Delineation

This ecoregion is a simplification of White's (1983) complex vegetation mosaic for the area that includes "edaphic grassland mosaics with semi-aquatic vegetation" and "edaphic grassland in the Upper Nile basin." The boundaries reflect the maximum extent of the influence of the Sudd, based primarily on faunal distribution patterns.

Ecoregion Number:	**60**
Ecoregion Name:	**Nile Delta Flooded Savanna**
Bioregion:	**African Palearctic**
Biome:	**Flooded Grasslands and Savannas**
Political Units:	**Egypt**
Ecoregion Size:	**51,000 km²**
Biological Distinctiveness:	**Locally Important**
Conservation Status:	**Critical**
Conservation Assessment:	**IV**
Authors:	*Miranda Mockrin, Michelle Thieme*
Reviewed by:	*Mahmoud A. Zahran*

Location and General Description

The Nile Delta Flooded Savanna [60] extends along the Nile River from the Aswan High Dam 1,100 km downstream to the mouth of the Nile as it enters the Mediterranean Sea. It also includes the delta, which is about 175 km long and 260 km wide (Hughes and Hughes 1992). The delta experiences a Mediterranean climate, with summer temperatures at a maximum in July and August (mean 30°C, maximum 48°C) and winter temperatures of 5–10°C. Only 100–200 mm of rain falls each year and is concentrated in the winter. The rainfall has an insignificant influence on the habitats of the ecoregion, which are tied to the floodplain of the Nile River.

The Nile is the longest river in the world, extending 6,695 km from the mountains on the eastern side of Lake Tanganyika to the Mediterranean Sea, with an estimated basin area of 3,026,000 km². The main tributaries are the Bahr el Jebel, the Bahr el Ghazal, and the Sobat River, which combine and become the White Nile, as well as the Blue Nile and the Atbara River. Approximately 84 percent of the water reaching Aswan comes from the Ethiopian Highlands and 16 percent from Equatorial East Africa.

The Nile River previously braided into many channels as it flowed through the delta, depositing and moving unconsolidated, alluvial sediments brought from upper reaches of the river. Construction of the Aswan High Dam greatly reduced flow levels and stopped the flooding of the riverine floodplains and delta. The remaining marshland is associated with lakes and lagoons along the seaward face of the delta. The main lakes are El-Manzalah, El-Burullus, and Edko. The outer margins of the delta are eroding, and salinity levels of the lakes and coastal lagoons are rising as their connection to the sea increases.

The Nile Delta was once known for large papyrus (*Cyperus papyrus*) swamps. This species is now largely absent from the delta, although stands do remain downstream around the Damietta Branch of the Nile River (Serag 2000; Zahran and Willis 2002) and in islands along the Nile River around Cairo. The current vegetation of the delta consists mainly of *Phragmites australis*, *Typha domingensis*, *Juncus* spp. (*J. acutus*, *J. subulatus*, *J. rigidus*), and other wetland plants. The large Manzala coastal lagoon supports beds of *Ceratophyllum demersum*, *Potamogeton crispus*, and *P. pectinatus*, with *Najas pectinata* and *Eichhornia crassipes* along lakeshores (Hughes and Hughes 1992). The salt-tolerant *Halocnemum* spp. and *Nitraria retusa* grow in marshes along the Mediterranean coast.

Outstanding or Distinctive Biodiversity Features

The ecoregion supports a mixture of Palearctic (European) and African species. The Nile was connected to the Niger and Chad water systems at various times in the late Pleistocene, through a series of shallow lakes in the Sahara Desert. Therefore, the three river systems share a similar flora and fauna, and endemism in the Nile is low (Kingdon 1989). The Nile River in Egypt has at least 553 plant species associated with it, of which at least 8 species are endemic. Two additional endemics live in the oases close to the Nile (El Hadidi and Hosni 1994).

The Nile Delta is part of one of the world's most important migration routes for birds. Every year, millions of birds pass between Europe and Africa along the eastern African flyway, and the wetland areas of Egypt are especially important as stopover sites (Denny 1991). Species of large birds that pass through the ecoregion in huge numbers include white stork (*Ciconia ciconia*), black stork (*Ciconia nigra*), common crane (*Grus grus*), and great white pelican (*Pelecanus onocrotalus*), as well as birds of prey such as short-toed snake-eagle (*Circaetus gallicus*), booted eagle (*Hieraaetus pennatus*), steppe eagle (*Aquila nipalensis*), lesser spotted eagle (*Aquila pomarina*), common buzzard (*Buteo buteo*), and European honey-buzzard (*Pernis apivorus*). Millions of smaller birds also pass through during the spring and autumn migration seasons.

Several hundred thousand waterbirds winter in the delta, including the world's largest concentrations of little gull (*Larus minutus*) and whiskered tern (*Chlidonias hybridus*) in Lake Manzala (Baha El Din 1999). Other waterbirds include northern shoveler (*Anas clypeata*), common teal (*A. crecca*), Eurasian wigeon (*A. penelope*), garganey (*A. querquedula*), grey heron (*Ardea cinerea*), common pochard (*Aythya ferina*), ferruginous duck (*A.*

nyroca), Kentish plover (*Charadrius alexandrinus*), and great cormorant (*Phalacrocorax carbo*) (Hughes and Hughes 1992). This ecoregion also contains the largest breeding population of slender-billed gull (*Larus genei*) in the Mediterranean Sea.

Among mammals, European species include the common otter (*Lutra lutra*) and the red fox (*Vulpes vulpes*). Flower's shrew (*Crocidura floweri*) is endemic to the ecoregion. Healthy populations of swamp cat (*Felis chaus*) are found around Lake Manzala. The ecoregion also supports the endemic Damietta toad (*Bufo kassasii*). Aquatic reptiles include Nile monitor (*Varanus niloticus*), Nile crocodile (*Crocodylus niloticus*), and two marine turtles that breed at Lake Bardawil in the delta, the loggerhead (*Caretta caretta*, EN) and the green turtle (*Chelonia mydas,* EN). The African softshell turtle (*Trionyx triunguis*) was once found in the delta but has been eradicated from Egypt. The remaining Mediterranean population is considered to be critically endangered (Schleich et al. 1996). The Egyptian tortoise, *Testudo kleinmanni* (EN), lives in the dunes and islets of this ecoregion.

Status and Threats

Current Status

Practically no areas of delta habitat remain undisturbed. The completion of the first Aswan Dam (between 1912 and 1934) dampened the annual flood pulse in the Nile Delta. The completion of the second Aswan (High) Dam totally stopped flooding, and most of the former seasonally or permanently flooded habitats have been converted to settled agriculture. Only fragments of the former wetlands remain, with the best-protected examples being in lakes El Mannah, El Qatta, Faraontya, Sinnéra, and Sanel Hagar and the coastal lagoons of Manzala and Miheishar.

The ecoregion is largely unprotected. Ashtoun el Gamil-Tanee Island Natural Area and the Lake Burullus Ramsar site are the only two protected areas in the delta and cover a total area of less than 500 km². Lake Burullus is threatened by fishing and pollution, although it remains the most unspoiled of the delta wetlands. Ashtoun El Gamil Protected Area was created largely to protect gravid fish and fry as they journey in and out of nearby Manzala Lake and is not large enough to contain suitable waterfowl habitat. There are plans to enlarge this protected area, which may give it a greater significance in the conservation of Egypt's resident and transient avifauna (Baha El Din 1999).

Types and Severity of Threats

Almost all of the 63 million people of Egypt live in the Nile Delta Flooded Savanna [60]. Population densities average 1,000 people/km², with much higher densities occurring in major towns (e.g., Cairo). Population pressure on natural resources is immense, and humans have altered all of the natural vegetation outside small reserves. People arrived in the area around 250,000 years ago and have been farming intensively here for more than 5,000 years.

The delta ecosystem no longer receives a yearly input of sediments and nutrients from upstream. Consequently, large amounts of fertilizers are applied to the land each year. Runoff of fertilizers and dumping of wastewater and sewage sludge is leading to the accumulation of trace elements in the sediments of the delta (Elsokkary 1996). At least one species, the catfish *Clarias lazera,* has been shown to be accumulating metals, including mercury, iron, and copper in its muscle and liver (Adham et al 1999). Fertilizers, along with saltwater intrusion, have also caused the upper delta to become more saline.

There are three major threats to the remaining habitats and species. Salinity may continue to increase in the delta from infiltration by seawaters as the delta face erodes and as erosion opens the existing lagoons to the sea. Wetlands and other migrating birds will increasingly be hunted and trapped to provide a food source for local populations and for sale to other countries (e.g., quail trapping, which occurs along the coast). Finally, inappropriate siting of windmills for electricity generation could cause considerable mortality in migrating birds.

Other concerns include rising sea levels caused by changing global climatic conditions. For example, El-Raey et al. (1997) estimate that with only a half-meter rise in sea level, 26 percent of the city of Rosetta and the estuary of the Nile River would be inundated. Ongoing efforts to halt erosion along the Nile banks by lining them with rocks probably will make these areas less attractive to waterfowl (Baha El Din 1999). In addition, political conflicts between upstream and downstream countries could intensify. Egypt is totally dependent on the Nile River and uses its regional power to prevent upstream countries from developing water use schemes that threaten its water supply (Singh et al. 1999).

Justification for Ecoregion Delineation

The boundary of the ecoregion along the Nile River follows the border between the desert and the floodplain, and in the delta it includes the former flooded area of the Nile River. It is distinguished as an important stopover for migrating birds between the Palearctic and Afrotropical realms.

Ecoregion Number:	**61**
Ecoregion Name:	**Lake Chad Flooded Savanna**
Bioregion:	**Western Africa and Sahel**
Biome:	**Flooded Grasslands and Savannas**
Political Units:	**Cameroon, Chad, Niger, Nigeria**
Ecoregion Size:	**18,800 km²**
Biological Distinctiveness:	**Bioregionally Outstanding**
Conservation Status:	**Relatively Stable**
Conservation Assessment:	**V**
Authors:	*Miranda Mockrin, Michelle Thieme*
Reviewed by:	*Philip Hall, David Thomas*

Location and General Description

The Lake Chad Flooded Savanna [61] includes Lake Chad and the nearby Hadejia-Nguru wetlands. Lake Chad straddles the borders between four African countries: Cameroon, Chad, Niger, and Nigeria. It currently covers 2,500 km², one-tenth of its area in the 1960s. During wetter periods of the Pleistocene the lake expanded to cover 2.5 million km² (Monod 1963; Hughes and Hughes 1992). An outlier of the ecoregion is found at the Hadejia-Nguru floodplain in northern Nigeria, covering about 6,000 km². This area is connected to Lake Chad through the Yobe River (Hollis et al. 1993; Acreman and Hollis 1996).

The climate in this ecoregion is dry, with an annual average of 320 mm of rain. Rainfall occurs from June through October. Most water input comes from rivers, with the majority of input originating as precipitation on the Adama Plateau, brought to Lake Chad via the Chari and Logone rivers. The Logone-Chari system is estimated to contribute 95 percent of the total riverine inflow, whereas the Yobe River system carries less than 2.5 percent (Hughes and Hughes 1992). Evaporation is extremely high, reaching rates of 2,300 mm per year (Hammer 1986). Despite the high rates of evaporation, Lake Chad has low levels of salinity because the more saline waters sink and exit the lake through subterranean conduits in the north (Beadle 1981; Hughes and Hughes 1992).

Vast expanses of dark, cracking Pleistocene clays line the southern shore of the lake (Dumont 1992). Areas of open water between 1.5 m and 5 m depth remain, mostly near the Chari River inflow, with swamps to the west (Sarch and Birkett 2000). Vegetation in the south basin consists of *Cyperus papyrus, Phragmites mauritianus, Vossia cuspidata,* and other wetland plants. *Phragmites australis* and *Typha australis* grow in the more saline north basin (Beadle 1981). Occasionally, the floating plant Nile lettuce (*Pistia stratiotes*) covers large areas of open water (Denny 1991).

Seasonal *yaére* grasslands grow on the southern lakeshore, where annual flooding is prolonged and water depth reaches 2 m. Vegetation consists of *Echinochloa pyramidalis, Vetiveria nigri-*

tana, Oryza longistaminata, and *Hyparrhenia rufa*. In areas with less prolonged flooding, *karal* or *firki* woodland vegetation is present. *Acacia seyal* is the dominant species but is replaced by *A. nilotica nilotica* in depressions. Below the trees, a layer of tall herbs and grasses grow to 2–3 m in height, including *Caperonia palustris, Echinochloa colona, Hibiscus asper, Hygrophila auriculata,* and *Schoenfeldia gracilis* (White 1983).

Outstanding or Distinctive Biodiversity Features

This ecoregion is important for large numbers of migrant birds, especially ducks and waders that spend the Palearctic winter period in Africa (Polet 2000; Roux and Jarry 1984; Scott and Rose 1996). Seventeen waterfowl species and forty-nine other wetland bird species are recorded, with abundance varying from year to year. More than 1 million ruff (*Philomachus pugnax*) have been recorded at Lake Chad (Keith and Plowes 1997). In the Hadejia-Nguru wetlands the most common waterbirds are white-faced whistling duck (*Dendrocygna viduata*), garganey (*Anas querquedula*), northern pintail (*Anas acuta*), and ruff (Garba-Boyi et al. 1993; Scott and Rose 1996; Dodman et al. 1999; Polet 2000).

Lake Chad supports two near-endemic bird species, the river prinia (*Prinia fluviatilis*) and the somewhat more widespread rusty lark (*Mirafra rufa*). Other birds of note include the marbled teal (*Marmaronetta angustirostris,* VU), which is occasionally seen on Lake Chad, and the near-threatened ferruginous duck (*Aythya nyroca*), which has also been found to occur at the lake in large numbers. The lakeshore is also important for large concentrations of Palearctic birds of prey, such as the steppe eagle (*Aquila nipalensis*) and booted eagle (*Hieraaetus pennatus*). The West African subspecies of black-crowned crane (*Balearica pavonina pavonina*) also occurs.

Two near-endemic rodent species are found: *Mastomys verheyeni* and the Lake Chad gerbil (*Taterillus lacustris*). The wetlands of Lake Chad and the Hadejia-Nguru wetlands formerly supported herds of large mammals. Savanna species included redfronted gazelle (*Gazella rufifrons,* VU), dama gazelle (*G. dama,* EN), dorcas gazelle (*G. dorcas,* VU), patas monkey (*Erythrocebus patas*), striped hyena (*Hyaena hyaena*), cheetah (*Acinonyx jubatus,* VU), and caracal (*Caracal caracal*) (Happold 1987). Species more adapted to the wetland habitats include African elephant (*Loxodonta africana,* EN), two otter species (*Lutra maculicollis, Aonyx capensis*), hippopotamus (*Hippopotamus amphibius*), sitatunga (*Tragelaphus spekii*), and kob (*Kobus kob*). Most of the large mammals have been hunted out and replaced by large numbers of cattle. Nile crocodiles (*Crocodylus niloticus*) are also now extremely rare and may have been wiped out.

Large numbers of fish, corresponding with seasonal inflows, migrate to the rich floodplains to feed and breed (Beadle 1981). Many of these species also occur in the Nile, Niger, and Congo rivers (Sarch and Birkett 2000; WCMC 1993; Beadle 1981).

Status and Threats

Current Status

Local communities occupy most of the land for settlements, farms, and cattle grazing, and as bases for fishing. Part of the area is found in the Chad Basin National Park. In the Hadejia-Nguru wetlands former forest reserves have been upgraded in status to a disjointed national park. In July 2000, the Lake Chad Basin Commission (LCBC) met and declared all of Lake Chad a transboundary Ramsar site of international importance.

Disagreements over water use can become volatile, sparking conflicts between neighboring provinces and countries (Hollis et al. 1993). The LCBC was formed in 1964 to regulate and plan uses of the water and natural resources in the Lake Chad Basin (Jauro 1998).

Types and Severity of Threats

Severe drought through the 1970s devastated the Sahelian region and led to decreased water inflows to Lake Chad. During the drought, increased human pressure on the dwindling lake and surrounding habitats put intense pressure on the entire area (Jauro 1998). In recent years the rains have been better, and the lake has increased in area.

Large water management schemes are also a threat. For example, a major project promoted by the LCBC aims to replenish the Lake Chad Basin by augmenting it with water from the Congo River Basin. Still in the conceptual stage, this plan proposes moving 100 billion m^3 of water annually from the Congo River in a navigable canal 2,400 km long (Jarou 1998). Dams and irrigation projects also threaten the Hadejia-Nguru wetlands (Barbier et al. 1996). The extent of local irrigation also increased after the introduction of small gasoline-powered pumps (FAO 1997).

The fish populations in the lake have suffered declines recently from drought, overfishing, diversion or blockage of instream flows, and increased juvenile catch caused by the use of smaller mesh (World Bank 2000). The most important fish in Lake Chad are the characin (*Alestes baremoze*) and the Nile perch (*Lates niloticus*). Characin populations have decreased drastically, whereas Nile perch catch sizes have decreased substantially so that they seldom exceed 8 kg (Keith and Plowes 1997).

Justification for Ecoregion Delineation

Like the Inner Niger Delta Flooded Savanna [62], the Lake Chad Flooded Savanna [61] is distinguished by the ecological role it plays for migrating birds. The ecoregion boundaries around Lake Chad follow the "herbaceous swamp and aquatic vegetation" unit of White (1983). An outlier in Nigeria covers the Hadejia-Nguru wetlands. These wetlands are connected to Lake Chad by rivers and provide habitat to species that still move between the two areas (especially birds).

Ecoregion Number:	**62**
Ecoregion Name:	**Inner Niger Delta Flooded Savanna**
Bioregion:	**Western Africa and Sahel**
Biome:	**Flooded Grasslands and Savannas**
Political Units:	**Mali**
Ecoregion Size:	**46,000 km^2**
Biological Distinctiveness:	**Locally Important**
Conservation Status:	**Endangered**
Conservation Assessment:	**IV**
Authors:	Miranda Mockrin, Michele Thieme
Reviewed by:	Eddy Wymenga

Location and General Description

The Inner Niger Delta Flooded Savanna [62] contains the inland delta of the Niger River in central Mali between Djenné in the south and Tombouctou in the north. Containing a mixture of channels, swamps, and lakes, the delta expands to cover 30,000 km^2 during the flood season and contracts to 3,900 km^2 or less during the dry season (Welcomme 1986). The floodplain is remarkably level, dropping only 8 m over its course (Hughes and Hughes 1992). As waters flow through the delta, they pass over Pleistocene and recent alluvium overlying Paleozoic sandstone (Hughes and Hughes 1992).

The local rainfall over the delta varies from 750 to 250 mm but has a negligible impact on the flooding. The floodwaters come primarily from the Niger River; its main tributary, the Bani River; and much smaller and temporal streams that flow down from the Dogonland Plateau. Rains fall on the Niger's headwaters in the Fouta Djalon Highlands of Guinea from May through September, creating a flood surge that reaches the inland delta in October (Wymenga et al. 2002). This surge dissipates as it continues through the delta (John et al. 1993; Quensière 1994; FAO 1997).

A diverse mix of vegetation occupies the wetland (Hiernaux 1982; John et al. 1993). Flooded forests dominated by *Acacia kirkii* with some *Ziziphus mauritiana* form a characteristic but dwindling type of vegetation in the area. Trees such as *Diospyros* sp. and *Kigelia africana* grow on higher levees, whereas palms such as *Hyphaene thebaica* and *Borassus aethiopum* occur near villages (Hiernaux 1982; Hughes and Hughes 1992). The southern half of the delta is low-lying floodplain with grasses such as *Acroceras amplectens*, *Echinochloa pyramidalis*, and bourgou (*E. stagnina*). Other typical species of these flooded pastures are *Vetivera nigritiana* and *Vossia cuspidata*. Along the heavily grazed outer fringes, *Andropogon gayanus*, *Cynodon dactylon*, and *Hyparrhenia dissoluta* dominate.

Outstanding or Distinctive Biodiversity Features

The Inner Niger Delta provides essential habitat for huge numbers of wetland birds, including Afrotropical resident species and

Palearctic migrants (Roux and Jarry 1984; Dodman et al. 1999; Wymenga et al. 2002). More than 500,000 garganey (*Anas querquedula*) and up to 200,000 northern pintail (*Anas acuta*) occur during the northern winter, along with large numbers of ferruginous duck (*Aythya nyroca*), white-winged tern (*Chlidonias leucopterus*), ruff (*Philomachus pugnax*), black-tailed godwit (*Limosa limosa*), and other waterbirds (Frazier 1999; van der Kamp and Diallo 1999; van der Kamp et al. 2001; Wymenga et al. 2002). Total bird numbers can reach more than 1 million in suitably wet years. In addition, more than 1 million wetland-related passerines, particularly sand martin (*Riparia riparia*) and yellow wagtail (*Motacilla flava*), pass through the delta during their migration. The Inner Niger Delta also contains large waterfowl breeding colonies, with a total of 80,000 breeding pairs of fifteen species of cormorant, heron, spoonbill, and ibis reported in the mid-1980s (Skinner et al. 1987). Of the seven mixed breeding colonies present in 1985–86 (Skinner et al. 1987) only two remained in 1998–2002 (Wymenga et al. 2002). These contain an estimated breeding population of 50,000–60,000 cattle egrets (*Bubulcus ibis*), 18,000–20,000 African cormorant (*Phalacrocorax africanus*), and 250–300 pairs of African darters (*Anhinga rufa*). Breeding occurs during the high flood season (September–November), and the colonies are situated in flooded forests of (mainly) *Acacia kirkii*. Two nonwetland bird species are near endemic to this ecoregion: the Mali firefinch (*Lagonosticta virata*) (Wheatley 1996) and the river prinia (*Prinia fluviatilis*).

Mammals include a mixture of Sahelian species and those of more mesic environments. Buffon's kob (*Kobus kob kob*) was once numerous in the Inner Niger Delta but is no longer present. This also seems the case for roan antelope (*Hippotragus equinus*), dorcas gazelle (*Gazella dorcas,* VU), and dama gazelle (*Gazella dama,* EN). Small populations of red-fronted gazelle (*Gazella rufifrons,* VU) are believed to be still present, although little information exists (Wymenga et al. 2002). Other species include warthog (*Phacochoerus africanus*), Saharan striped polecat (*Ictonyx libyca*), side-striped jackal (*Canis adustus*), patas monkey (*Erythrocebus patas*), pale fox (*Vulpes pallida*), and African wild cat (*Felis silvestris*) (Boitani et al. 1999; Happold 1987). A small population of elephants (*Loxodonta africana,* EN) lives east of the delta and migrates between Burkina Faso and southern Mali (Shumway 1999). The delta also harbors an important but dwindling population of African manatees (*Trichechus senegalensis,* VU). Hippos (*Hippopotamus amphibius*) are present, but the population is reduced to only 40–60 individuals (Wymenga et al. 2002).

The vast floodplains also provide habitat for Nile crocodile (*Crocodylus niloticus*), Nile monitor (*Varanus nilotica*), and African rock python (*Python sebae*). Nile crocodile is believed to be on the edge of extinction, and the Nile monitor and rock python are facing heavy human pressure (Wymenga et al. 2002). There is also a diverse fish fauna, including endemic species (Lowe-McConnell 1985).

Status and Threats

Current Status

Three Ramsar sites were declared in 1987: Lac Horo, Lac Débo, and the Séri floodplain complex, together comprising 1,620 km^2. The Ramsar sites are owned by the state and are used by local residents and nomadic pastoralists for drinking water, fishing, seasonal agriculture, and livestock rearing. Lac Horo is separated from the Niger by a dam and a sluice gate. The rest of the area is unprotected and heavily used for different forms of human activity (Wymenga et al. 2002).

Types and Severity of Threats

Mali is among the poorest countries in the world, with 65 percent of its land area desert or semi desert. Economic activity is confined largely to the riverine area irrigated by the Niger, and fish makes up 60 percent of the total animal protein consumed by Malians (FAO 1998; Quensière 1994).

People have intensively exploited the Niger Delta over the last 1,000 years. In the past resource use was dictated by a complex traditional system of management (Gallais 1967). After independence in 1960 the traditional systems of management were discontinued, which caused many environmental problems. With decentralization in the 1990s, local populations regained responsibility of the management and conservation of natural resources. Today, approximately 600,000 people, 2 million cattle, and 3 million sheep inhabit the delta (Hassane et al. 2000; Heringa 1990).

Farming varies from basic subsistence level to larger, irrigated projects such as the FAO special program for food security pilot project in the Mopti region (Coche 1998). Most of the flooded forests—essential as breeding habitat for colonial breeding waterbirds and potentially very important as fish fry areas—have been removed by a combination of drought and deforestation (Wymenga et al. 2002). Many fish species disappeared after serious droughts in the 1980s, but water management projects and overexploitation have compounded the decline (Quensière 1994; Ticheler 2000). The Selingue Dam and other smaller dams (Markala) have a negative impact on fisheries, cutting off fish migrations, lowering fish production, and altering the timing and extent of the annual delta flood.

Justification for Ecoregion Delineation

The Inner Niger Delta Flooded Savanna [62] is distinguished by the important habitat it provides for wetland birds, including both Palearctic migrants and Afrotropical residents. The linework for this ecoregion follows the "mosaic of edaphic grassland and aquatic vegetation" unit of White (1983).

Ecoregion Number:	**63**
Ecoregion Name:	**Zambezian Flooded Grasslands**
Bioregion:	**Eastern and Southern Africa**
Biome:	**Flooded Grasslands and Savannas**
Political Units:	**Angola, Botswana, Democratic Republic of the Congo, Malawi, Mozambique, Tanzania, Zambia**
Ecoregion Size:	**153,500 km²**
Biological Distinctiveness:	**Globally Outstanding**
Conservation Status:	**Relatively Stable**
Conservation Assessment:	**III**
Author:	*Karen Goldberg*
Reviewed by:	*Jonathan Timberlake*

Location and General Description

Embedded predominantly in miombo and mopane woodlands, the Zambezian Flooded Grasslands [63] occur as isolated patches from northern Botswana in the south to Tanzania in the north. The ecoregion includes flooded grasslands in the Kilombero Valley, the Moyowosi-Malagarasi system, and the Ugalla River in Tanzania; the Okavango Delta in Botswana; Lake Chilwa in Malawi; the Barotse Floodplain, the Kafue Flats, Busanga and Lukanga swamps, Lake Mweru, Mweru Marsh, and the Bangweulu-Luapala-Chambezi system in Zambia; and a number of smaller floodplains and wetlands. These wetlands are scattered across the Central African Plateau between 1,000 and 1,200 m above sea level. The underlying geology consists largely of Precambrian granite, volcanics, serpentine, and sandstone, although components in western Zambia and the Okavango are found on deep Kalahari sands. The substrate of the wetlands is generally alluvial (Moyo et al. 1993) and often comprises a seasonally waterlogged gley soil. The pH of these fertile soils ranges from weakly alkaline to weakly acidic, although several wetlands have a peaty organic horizon with a pH as low as 3.5.

The climate is seasonal and tropical, with rainfall occurring during the hot summer. Most rain in the south falls between November and March (Frost 1996), whereas in the north the rainy season goes on through the end of May. Droughts lasting up to 7 months typify the cooler season. The variation in rainfall across the ecoregion is fairly high, with Lake Bangweulu experiencing annual rainfall greater than 1,400 mm, whereas the Okavango of Botswana and Ugalla River and Moyowosi system in Tanzania can receive as little as 450 mm (Moyo et al. 1993). However, inundation of these wetlands is contingent not only on localized rainfall but also on water supply from rivers that arise in wetter catchment areas. Mean maximum temperatures can range between 18°C and 27°C and mean minimum temperatures between 9°C and 18°C, depending on altitude.

The ecoregion falls in the Zambezian phytochorion (White

1983). Although it is characterized mostly by a mosaic of edaphic grassland and semi-aquatic vegetation, local variation in flora between wetlands is fairly high because of the wide distribution of the flooded areas that constitute this ecoregion (Timberlake 2000). Grasses belonging to the genera *Acrocera, Echinochloa, Leersia, Oryza, Phragmites, Typha,* and *Vossia,* together with *Cyperus papyrus,* dominate most of these wetlands. Herbaceous freshwater swamps and aquatic vegetation are also common. Some wetlands, such as Lake Chilwa, support halophytic vegetation. Elevated areas in these wetlands support different types of vegetation. For instance, termite mounds support distinctive patches of grassland and thicket, Zanzibar-Inhambane lowland forest in the Kilombero Valley, and woodland vegetation in Ugalla. Miombo or mopane woodlands mostly surround the wetlands, although dry forest, secondary grasslands, and Itigi thicket also occur locally.

Outstanding or Distinctive Biodiversity Features

There are few endemic species in this ecoregion, but species richness is high. The impressive mammal fauna includes huge herds of large mammals that undertake seasonal migrations. Lechwe (*Kobus leche*) populations are known to exceed 20,000 in Moremi Game Reserve, and more than 35,000 individuals of the endemic subspecies Kafue lechwe (*Kobus leche kafuensis*) have been recorded in Lochinvar National Park. Moreover, the largest remaining population of puku (*Kobus vardonii*) is found in the Kilombero Valley, and more than 20,000 buffalo (*Syncerus caffer*) occur in the Moyowosi delta of northwestern Tanzania (IUCN 1987; East 1999). There are also significant populations of waterbuck (*Kobus ellipsiprymnus*), southern reedbuck (*Redunca arundinum*), tsessebe (*Damaliscus lunatus*), wildebeest (*Connochaetes taurinus*), oribi (*Ourebia ourebi*), and sitatunga (*Tragelaphus spekii*). Hippopotamus (*Hippotragus amphibius*) is common.

There is also a high diversity of birds. The only strict endemic bird is the Kilombero weaver (*Ploceus burnieri,* VU). Near-endemic birds include the Tanzania masked weaver (*Ploceus reichardi*) (Clements 1991) and the Katanga masked weaver (*Ploceus katangae*) (Stattersfield et al. 1998). In addition, this ecoregion supports populations of slaty egret (*Egretta vinaceigula,* VU), an uncommon resident of the Okavango, Chobe, Caprivi Strip, and Zambezi Valley north to the Bangweulu swamps (Brown et al. 1982). Other globally threatened species are wattled crane (*Grus carunculatus,* VU) and corncrake (*Crex crex,* VU) (Hilton-Taylor 2000).

Amphibian and reptile species richness and endemicity are not particularly high, with only two strict endemic species. The Merera toad (*Bufo reesi*) is known only from the Kilombero Floodplain (Frost 1999), and the Barotse water snake (*Crotaphopeltis barotseensis*) occurs only on the Kalabo Floodplain (Uetz 2001). The Nile crocodile (*Crocodylus niloticus*) is also common in many parts of the ecoregion (Broadley 1971).

Information on plant and invertebrate diversity of these wetlands is sparse. However, it is known that sixteen butterfly species are found only in the Zambezi River Basin (Pennington 1978).

Status and Threats

Current Status

The largest wetland in this ecoregion is the Okavango Delta (68,640 km^2), which is protected mainly in the Moremi Game Reserve and two adjacent wildlife management areas. It has been designated a Ramsar site because of its bird richness, with more than 650 species recorded (Ramsar Convention Bureau 2002). Zambia contains several protected areas in this ecoregion; Lochinvar and Blue Lagoon national parks and the Bangweulu-Kafinda Game Management Areas are considered key locations for threatened antelopes in sub-Saharan Africa (East 1999). Bangweulu is also designated as a Ramsar site, with more than 400 bird species.

The ecoregion supports three important wetlands in Tanzania: the Kilombero (3,000 km^2), Moyowosi (1,000 km^2), and Ugalla systems (Baker 1996). All three are under some form of protection as game controlled areas or game reserves. Both Kilombero and Moyowosi are also Ramsar sites.

Types and Severity of Threats

The combination of historically low human population densities, largely caused by waterborne diseases and the presence of tsetse flies (Chabwela and Mumba 1998), and the more recent establishment of parks and reserves has ensured that many natural habitats remain (East 1999). Wetland organisms and communities have evolved and adapted to thrive in an environment that is in constant flux, which also explains the low number of endemic species.

Although several of the wetlands such as the Barotse Floodplain and the Kafue Flats have been occupied for centuries, large changes are becoming evident in many of the areas as human activities and land use intensify, including conversion of areas to farmland (Timberlake 1998; Turpie et al. 1999). As many as 250,000 head of cattle are said to graze in the Kafue Flats alone (Chabwela and Mumba 1998), and large mammals are now confined almost exclusively to two national parks (Blue Lagoon and Lochinvar national parks, a combined area of only 860 km^2). In addition, overfishing is also becoming an issue of increasing concern. For example, more than 50 percent of the fish production for Zambia comes from the Bangweulu Basin and Kafue Flats (Chabwela 1992).

Water diversion for irrigation and hydroelectric dams have already affected some floodplain systems. For example, the Itezhitezhi Dam on the Kafue River has changed the flood regime such that unseasonal flooding occurs on the flats, threat-ening the breeding sites of the wattled (*Grus carunculatus*) and grey-crowned (*Balearica regulorum*) cranes (Dodman et al. 1996). The area of land available for wildlife and human uses such as fishing and recession farming has been reduced by dam operations (Chabwela 1992; Kalapula 1992). The population of Kafue lechwe, which numbered about 100,000 in 1971 before hydroelectric dams were constructed, dropped to nearly half that number in 1987 after dam construction altered the flood regime. A similar fate could await the Okavango Delta. Botswana, Namibia, Angola, and Zambia would like to extract large quantities of water from the Okavango and Kwando rivers and their tributaries for irrigation and urban water use.

Justification for Ecoregion Delineation

All the seasonally flooded areas in the Zambezian phytochorion of White (1983) were mapped in this ecoregion. The borders of each area roughly follow the "herbaceous swamp and aquatic vegetation" unit of White (1983).

Ecoregion Number:	**64**
Ecoregion Name:	**Zambezian Coastal Flooded Savanna**
Bioregion:	**Eastern and Southern Africa**
Biome:	**Flooded Grasslands and Savannas**
Political Units:	**Mozambique**
Ecoregion Size:	**19,500 km^2**
Biological Distinctiveness:	**Bioregionally Outstanding**
Conservation Status:	**Critical**
Conservation Assessment:	**IV**
Author:	*Karen Goldberg*
Reviewed by:	*Jonathan Timberlake, John Burlison, Judy Oglethorpe*

Location and General Description

The Zambezian Coastal Flooded Savanna [64] ecoregion is restricted to the coastal regions of Mozambique and comprises the floodplain deltas of the Zambezi, Pungwe, Buzi, and Save rivers. The Zambezi Delta forms the most extensive portion of the ecoregion, covering roughly 200 km along the coastline and penetrating up to 120 km inland (Timberlake 1998). A number of small swamps at the mouth of the Save also fall into this ecoregion. These wetlands typically are below 50 m in altitude. The substrate is dominated mainly by recently deposited alluvial and floodplain fluvisols, overlaying sediments that were deposited by the rivers during the Quaternary and Cretaceous periods.

This ecoregion is located in the subtropical climatic belt of

Africa. Annual rainfall is 800–1,400 mm, and most falls from October to March. Mean maximum temperatures range between 27°C and 30°C, with mean minimum temperatures averaging 18°C. Until the 1960s, much of the ecoregion was subject to seasonal flooding that spread out onto the floodplains at the onset of the rainy season. It was only with the construction of large dams such as the Cabora Bassa and the Kariba on the Zambezi River that these important inundation events were severely reduced.

Occurring in White's (1983) Zanzibar-Inhambane regional mosaic, the vegetation of this wetland ecoregion contains both open grassland–dominated communities and mixed freshwater swamp forests. Much of the area is a lightly wooded savanna, typified by either the palms *Borassus aethiopum* and *Hyphaene coriacea* or by acacias such as *A. sieberiana* and *A. polyacantha*. Secondary grassland occurs on former sugar plantations, the main species being *Pennisetum polystachion, Hyperthelia dissoluta,* and *Hyparrhenia* spp. Closer to the coast, papyrus (*Cyperus papyrus*), *Phragmites, Vossia cuspidata,* and other emergent and floating-leaved aquatics dominate large areas. Small patches of swamp forest are found, including *Barringtonia racemosa, Phoenix reclinata,* and *Syzygium* spp. Dune grasslands and shrubland occur on the coast, whereas dense mangrove forests flank the larger channels (Timberlake 2000).

Outstanding or Distinctive Biodiversity Features

The Zambezi Delta has a high diversity of habitats from forest through grassland to papyrus swamp, mangrove, and dunes. This is reflected in a high species richness of mammals and birds. The lengthy civil war and instability in Mozambique decimated many of the formerly abundant large mammal populations. Species such as buffalo (*Synerus caffer*), waterbuck (*Kobus ellipsiprymnus*), southern reedbuck (*Redunca arundium*), Burchell's zebra (*Equus buchellii*), and hippopotamus (*Hippopotamus amphibius*) occurred in high numbers at the onset of the civil war in the late 1970s but had declined by as much as 90 percent by the early 1990s (Goodman 1992). Other herbivores found in lower numbers include eland (*Taurotragus oryx*), Lichtenstein's hartebeest (*Sigmoceros lichtensteinii*), sable (*Hippotragus niger*), oribi (*Ourebia ourebi*), bushbuck (*Tragelaphus scriptus*), suni (*Neotragus moschatus*), and blue duiker (*Cephalophus monticola*). The number of elephants (*Loxodonta africana,* EN) is low, although it is now increasing. Black rhinoceros (*Diceros bicornis,* CR), once abundant in the area, was hunted out during the war.

The delta floodplains provide habitat for many waterbird species. About 20 percent (or 2,750 individuals) of the world population of the wattled crane (*Grus carunculatus,* VU) is thought to breed in the Zambezi Delta (Beilfuss and Allan 1996; Goodman 1992; Singini 1996). Waterbirds at the Marromeu Complex Game Management Area in the Zambezi Delta include open-billed storks (*Anastomus lamelligerus*), saddle-billed storks (*Ephippiorhynchus senegalensis*), great white pelicans (*Pelecanus* *onocrotalus*), great snipe (*Gallinago media*), African skimmer (*Rynchops flavirostris*), and lesser flamingo (*Phoenicopterus minor*) (Timberlake 1998). The rare eastern African coastal forest bird species, East Coast akalat (*Sheppardia gunningi,* VU), is also recorded in forest habitat of the Marromeu Reserve.

The only strictly endemic vertebrate is the Pungwe worm snake (*Leptotyphlops pungwensis*). The Nile crocodile (*Crocodylus niloticus*), Nile monitor (*Varanus niloticus*), and African rock python (*Python sebae*) are among the more common larger reptiles. There is also one near-endemic amphibian, *Afrixalus delicates,* which occurs in the Zambezi Delta area and then in the coastal area south of Maputo to Durban.

Status and Threats

Current Status

Two decades of armed conflict had a severe impact on the ecoregion's biota. The military occupation of most of the protected areas resulted in the extermination of many herd animals. For example, herds of up to 55,000 buffalo used to roam the floodplains of Marromeu and the Zambezi Delta (Tello and Dutton 1979), but in 1995 fewer than 1,000 were counted in a survey of the Marromeu area (Singini 1996). Waterbuck were also seriously affected, declining from around 48,000 in 1978 (Tello and Dutton 1979) to 143 in 1994 (Cumming et al. 1994). Hippopotamus fell from a population of roughly 2,820 in 1977 to 260 by 1990 (Anderson et al. 1990). Since the end of the war in the early 1990s, populations are beginning to recover.

Although the animal populations of the ecoregion suffered severely as a result of the civil war, much of the habitat is still intact. At present, the Marromeu Complex Game Management Area in the Zambezi Delta is the only officially protected area in this ecoregion. Calls have been made to designate the Zambezi Delta as a wetland of international significance under the Ramsar convention (Anderson et al. 1990; Singini 1996).

Types and Severity of Threat

The Zambezi Delta has already experienced much of the damage caused by the construction of Kariba and Cabora Bassa Dams. Formerly, the Zambezi used to flood extensively and occasionally crossed over into the Pungwe Basin through Gorongosa National Park along the low-lying rift valley that cuts across the two catchments (Tinley 1977). With the reduction in flooding, productivity of the alluvial plain is now determined largely by rainfall. The Marromeu grasslands are clearly drying out, and the changing hydrology of this area creates concern for the future of the wattled crane (Timberlake 1998). Two new dams are proposed at Batoka Gorge below Victoria Falls and at Mepanda Uncua below Cabora Bassa. The Buzi, Pungwe, and Save may face similar regulation of their flow regimes as more dams are constructed along their courses.

Poaching poses a continued threat to many larger mammals. One consequence of drastically reduced hippo numbers has been the clogging of water channels by vegetative growth, often by alien species such as *Eichhornia crassipes*. Coupled with the greatly reduced grazing pressure, this has dramatically altered the topography of the floodplain in some areas. If the wetlands continue to dry out, there is little doubt that subsistence farmers will increasingly encroach on the floodplains in search of land for pastures or cultivation.

Justification for Ecoregion Delineation

The Zambezian Coastal Flooded Savanna [64] is included in White's (1983) Zanzibar-Inhambane regional mosaic [ecoregion 21]. However, such a large wetland area is not typical of this ecoregion and thus has been separated here. The largest area of flooded savanna is delineated at the mouth of the Zambezi River, with smaller patches of flooded savanna at the Pungue River, Buzi River, and Save River in the south. In each case, the boundaries follow the maximum range of flooding from these rivers.

Ecoregion Number:	**65**
Ecoregion Name:	**Saharan Halophytics**
Bioregion:	**African Palearctic**
Biome:	**Flooded Grasslands and Savannas**
Political Units:	**Western Sahara, Mauritania, Morocco, Algeria, Tunisia, Libya, Egypt**
Ecoregion Size:	**53,900 km^2**
Biological Distinctiveness:	**Locally Important**
Conservation Status:	**Relatively Stable**
Conservation Assessment:	**V**
Authors:	*Nora Berrahmouni, Pedro Regato*
Reviewed by:	*Salima Benhouhou, Hans Peter Müller*

Location and General Description

The Saharan Halophytics [65] cover a number of saline depressions scattered across northern Africa. The most extensive areas from west to east are Iriki and Oued Draa in Morocco; Rhir Valley, Sebkhat Tidikelet, Sebkha of Timimoun, Chott Hodna, and Chott Melghir in Algeria; and the Chott el Gharsa (600 km^2), Chott Djerid (4,600 km^2), and Chott Fedjedj (800 km^2) in Tunisia.

Perhaps the most important saline depressions are the Qattara Depression and the Siwa Depression in Egypt. The Qattara Depression is 285 km long and 135 km wide and covers approximately 19,500 km^2 of land below sea level (Hughes and

Hughes 1992). Salt marshes occupy approximately 300 km^2, with some areas having encroaching wind-blown sands. Extensive playas cover about a quarter of the total area. The Siwa Depression extends east to west over 82 km and is up to 28 km wide. This depression has eighteen lakes covering approximately 7 km^2, with a maximum depth of 25 m below sea level. Each lake is surrounded by brackish marshlands (Zahran and Willis 1992). All of the lakes are saline but are supplied with freshwater by eighteen underground springs. Although there are no permanent settlements in the Qattara Depression, the freshwater springs enabled the Siwa Depression to host a low but permanent population of a few hundred residents.

Many of the halophytic areas were once larger lakes that have dried up over the past 7,000 years as the climate has become drier (van Zinderen Bakker 1979; Nicholson and Flohn 1980; Petit-Maire et al. 1980). Currently the ecoregion is in a hyper-arid phase, with high summer temperatures reaching more than 50°C in the summer and low winter temperatures dropping below 0°C at night. Rainfall is irregular, varying between 10 and 100 mm per year in years with rain. Here, evaporation from surface water and groundwater exceeds rainfall, leading to the deposition of soluble salts to form saline soils, or solonchaks. These saline depressions, also known as *chotts* (shallow, irregularly flooded depressions) and *sebkhas* (irregularly flooded depressions), vary in size and geomorphologic origin.

The vegetation of the ecoregion belongs to the "azonal halophytic vegetation" mapped by White (1983) and includes halophytic (salt-loving) species such as *Halocnemum strobilaceum, Atriplex littoralis, Salicornia fruticosa, Salsola tetrandra, Atriplex halimus,* and *Zygophyllum album* (Guinochet 1951; Ozenda 1983; Benhouhou et al. 2001). The halophytes belong mainly to the Frankeniaceae, Chenopodiaceae, Tamaricaceae, and Zygophyllaceae families (Zahran and Willis 1992).

Outstanding or Distinctive Biodiversity Features

These halophytic habitats contain low species richness, with most species showing Palearctic and Afrotropical affinities (Happold 1984). Among the small mammals, gerbils (e.g., *Gerbillus campestris* and *G. nanus*) are the most abundant species, followed by sand rats (*Psammomys obesus* and *Psammomys vexillaris*), jerboas (*Jaculus jaculus* and *J. orientalis*), and jirds (*Massoutiera mzabi* and *Meriones crassus*) (Lay 1991; Le Houérou 1991). These rodents are particularly well adapted to feed on the succulent leaves and stems of dominant species of the Chenopodiaceae family found in those saline habitats. Predators in the ecoregion include Rüppel's fox (*Vulpes rueppelli*) and Libyan striped weasel (*Ictonyx libyca*). In the past addax (*Addax nasomaculatus*, CR) probably would have occurred in this ecoregion, but it is likely to have been eradicated. Small numbers of scimitar-horned oryx (*Oryx dammah*, EX) may also have occurred in the past, but this species is believed extinct in the wild. Other desert antelopes may still be found in small numbers, such as the slender-horned gazelle (*Gazella leptoceros*, EN), dama gazelle (*Gazella dama*, EN), dorcas gazelle (*Gazella dorcas*, VU), and red-fronted gazelle (*Gazella rufifrons*, VU).

A number of desert-adapted birds also occur, such as the thick-billed lark (*Ramphocoris clotbey*), desert wheatear (*Oenanthe deserti*), red-rumped wheatear (*Oenanthe moesta*), spotted sandgrouse (*Pterocles senegallus*), and pin-tailed sandgrouse (*Pterocles alchata*). A greater diversity of birds can be found in the wetland areas, particularly for wintering and migrant Palearctic waders, waterfowl, and birds of prey, especially when they are flooded. In the Siwa Depression some of the species include the lesser flamingo (*Phoenicopterus minor*) and greater flamingo (*Phoenicopterus ruber roseus*), which nest. It is possible that the globally threatened slender-billed curlew (*Numenius tenuirostris*, CR) uses some of the ephemeral wetlands in this ecoregion because this species is also found during the winter in a few North African coastal wetlands.

Reptile diversity is high in and around this ecoregion and includes the Egyptian fringe-fingered lizard (*Acanthodactylus pardalis*), eyed skink (*Chalcides ocellatus*), and desert monitor (*Varanus griseus*). Amphibians include the Berber toad (*Bufo mauritanicus*) and the European green toad (*Bufo viridis*).

Status and Threats

Current Status

There are no protected areas in the ecoregion. Furthermore, little is known about the conservation status of many of the saline depressions. An exception would be the Siwa Oasis, which is well known and frequently visited around the edge of the lakes. In Tunisia, the Chott el Djeride and Chott Fedjej are also well known and visited in the rainy season. In general the drier sites, with rarely inundated saline depressions, are largely untouched by humans. However, in areas of the ecoregion that support permanent water sources, such as the Siwa Depression, larger mammals have been hunted out.

Types and Severity of Threats

The main threats to these saline habitats are not well known. However, for some areas it appears that overgrazing by camels during the dry season reduces the overall cover of halophytic communities. A major threat with regard to the Qattara Depression is the proposal to flood the depression with seawater to make an inland sea. This would have a devastating effect on the whole area and destroy all existing natural habitats. An additional threat is the continued process of natural drying of the area (perhaps now exacerbated by global warming), which might result in the complete loss of wetland habitats and their replacement with salt flats and sand areas similar to those observed elsewhere in the Sahara Desert.

Justification for Ecoregion Delineation

The boundaries for this ecoregion are taken directly from the "azonal halophytic vegetation" unit of White (1983). Although White does not separate the northern African halophytic areas from the southern ones, experts felt that the geographic distance between them and the significant differences in the floristic and faunistic composition were sufficient to justify the separation into distinct ecoregions.

1 Ecoregion Number: **66**

Ecoregion Name:	**East African Halophytics**
Bioregion:	**Eastern and Southern Africa**
Biome:	**Flooded Grasslands and Savannas**
Political Units:	**Tanzania**
Ecoregion Size:	**2,600 km²**
Biological Distinctiveness:	**Locally Important**
Conservation Status:	**Endangered**
Conservation Assessment:	**IV**
Author:	*Amy Spriggs*
Reviewed by:	*Alan Rodgers*

Location and General Description

The East African Halophytics [66] encompass two saline (soda) lakes, Lake Natron and Lake Eyasi, both situated in the eastern arm of the Rift Valley in Tanzania. Other similar saline lakes in the Rift Valley that support flamingo populations include Lake Magadi, Lake Elmenteita, Lake Nakuru, and Lake Bogoria in Kenya and Lake Manyara in Tanzania (Hughes and Hughes 1992). Lake Natron is the larger of the two lakes in this ecoregion, measuring a maximum of 58 km long and 15 km wide, but this varies widely from year to year. Natron is situated in the Arusha District of northern Tanzania (south from the Kenyan border). Lake Eyasi is about 80 km long and 14 km wide and is situated in Mbulu District of northern Tanzania, abutting the boundary of the Ngorongoro Conservation Area.

Volcanic lava and ash form the substrate for most of the ecoregion (Greenway and Vesey-Fitzgerald 1969). This was deposited in the past, but in recent times rapid weathering continues to deposit the material in depressions, resulting in deep sodium-rich soils.

Both Natron and Eyasi are situated in a semi-arid region, which receives erratic rainfall of around 600 mm per year. Most of the rain falls between December and May, followed by a long dry season. The rains fail completely in some years. Daily temperatures often are above 40°C. The hot, dry conditions expose these shallow lakes to high evaporation rates. Lake Natron is fed principally by the Ewaso Ngiro River, which has its catchment in the central Kenyan Highlands. The lake is also fed by hot, mineral-laden springs. Lake Eyasi is fed from the Sibiti River, which flows from Lake Kitangire. Water loss from both Natron and Eyasi lakes occurs solely through evaporation because neither lake has an outlet. The low rainfall, low inflow rates, high temperatures, volcanic substrates, and input of hot, mineral-rich waters result in lake waters that contain a saturated salt solution of pH 9–10, with water temperatures that reach 41°C near the mineral springs. These conditions are not constant and can change abruptly during heavy rains.

These lakes are devoid of macrophytic vegetation but are productive in terms of blue-green algae (Cyanophyta) such as *Spirulina* spp. (Finlayson and Moser 1991). The mudflats surrounding the permanent water are inhospitable to most other forms of life. A few halophytic (salt-loving) plant species are able to grow on the saline soils fringing the lakes, with the dominant species including *Cyperus laevigatus, Dactyloctenium* spp., *Juncus maritimus, Salvadora persica, Sporobolus spicatus, Sporobolus robustus,* and *Suaeda monoica.* The slightly less alkaline plains surrounding the lakes are dominated by grasses and by *Sesbania sesban,* with scattered *Acacia xanthophloea* trees (White 1983).

Outstanding or Distinctive Biodiversity Features

The species richness in the ecoregion is very low as a result of the extreme and highly variable environments. However, the population sizes of the few species adapted to these environments are large.

Lake Natron, Lake Eyasi, and similar lakes in the Rift Valley have large populations of wetland birds, especially greater and lesser flamingos (*Phoenicopterus ruber* and *P. minor*) (Bennun and Njoroge 1999; Baker and Baker 2002). Lake Natron is the only significant and regular breeding ground for the majority of the world's lesser flamingos (Simmons 1996; Baker and Baker 2002). The Makgadikgadi Pan in Botswana is the main breeding site for the greater flamingo in Africa (see the Makgadikgadi Halophytics [68]), but Lake Natron is also important. The major feeding sites are lakes Nakuru and Bogoria in Kenya, which are slightly less saline and have a greater abundance of small crustaceans (for greater flamingo) and blue-green algae (for lesser flamingo). Africa's flamingo populations are not isolated, and many migrate between the soda lakes of East Africa and the Etosha and Makgadikgadi Pans in southern Africa (Berry 1972; Simmons 1996).

The nonbird fauna of the lake margins has not been well studied, although it is evident that neither the water of these lakes nor their surrounding soda-encrusted mudflats provide suitable habitat for most species. The habitats surrounding the lakes support large mammals such as African elephant (*Africana loxodonta,* EN), eland (*Taurotragus oryx*), and caracal (*Caracal caracal*).

A single species of fish known as the white-lipped or alkaline tilapia (*Oreochromis alcalica*) is found in abundance in the waters of these lakes. This small fish lives on the edges of the hot spring inlets, where the water is between 36°C and 40°C. It is endemic to the saline lakes of the Rift Valley (Lott 1992) and is adapted to withstand the high temperatures and salinity of the lakes and the dramatic changes in conditions brought about by rain.

Status and Threats

Current Status

The environment of these saline lakes is inhospitable, and the lakes are largely inaccessible, surrounded by expansive mudflats. As a result, the ecoregion is uninhabited and has not been fragmented by human activities.

Lake Natron has recently been declared a Ramsar site in recognition of its large importance for wetland birds (especially lesser flamingo) (Ramsar Convention Bureau 2002). The area surrounding Lake Natron falls in a large game controlled area. This offers some protection to wild animals in the area but no protection to the ecosystem as a whole. Lake Eyasi falls outside the protected area network.

Types and Severity of Threats

The major threat to Lake Natron is the hydroelectric power scheme proposed by the Kenyan government for the Ewaso Ngiro River (Johnson and Bennun 1994), which is the main source of water into Lake Natron. Altering its flow could change the hydrology and ecology of the lake, threatening the world's largest and most secure breeding site of the lesser flamingo.

Lake Natron has also been proposed as a site for a sodium bicarbonate (baking soda) extraction plant since 1951 (Guest and Stevens 1951). There is already a similar plant on Lake Magadi, 25 km further north in Kenya. If built, such a plant could pose a significant threat to the habitats of Lake Natron. There are also plans to log forests in Lake Natron's Mau catchment, which could result in increased siltation of the lake (Ramsar Convention Bureau 2002).

Justification for Ecoregion Delineation

This ecoregion follows White's (1983) "halophytic vegetation" unit around Lake Eyasi and Lake Natron. If mapping were undertaken at a smaller scale, then other saline lakes in the Rift Valley that support flamingo populations should be included in the ecoregion, such as Lake Magadi, Lake Elmenteita, Lake Nakuru, and Lake Bogoria in Kenya and Lake Manyara in Tanzania.

Ecoregion Number:	**67**
Ecoregion Name:	**Etosha Pan Halophytics**
Bioregion:	**Eastern and Southern Africa**
Biome:	**Flooded Grasslands and Savannas**
Political Units:	**Namibia**
Ecoregion Size:	**7,200 km²**
Biological Distinctiveness:	**Locally Important**
Conservation Status:	**Relatively Intact**
Conservation Assessment:	**V**
Author:	*Amy Spriggs*
Reviewed by:	*Phoebe Barnard*

Location and General Description

The Etosha Pan Halophytics [67] ecoregion is the remnant of a large, inland Pliocene lake. The pan is located on the interior southern plain of the Owambo Basin, in the Oshikoto, Oshana, and eastern Omusati Regions of Namibia. Elevation ranges from 1,071 to 1,086 m. It is the largest pan system in Namibia and consists of a flat, saline depression roughly 4,850 km². Numerous smaller salt and clay pans, such as Fisher's and Beiseb pans to the east, the Naktukanaoka, Otjivalunda, and Adamax pans to the west, and Ondangwa Pan to the north, surround the main pan (Mendelsohn et al. 2000).

The mean annual rainfall of the Etosha National Park is about 430 mm, varying from 419 mm at Okaukuejo on the western edge of the pan to 440 mm at Namutoni Fort on the eastern edge (Berry 1972). Most of the rain falls in late summer, January–April, with February having a mean maximum rainfall of 110 mm. There are large rainfall fluctuations from year to year. Climatically there are normally three distinct seasons: hot and wet (January–April), cool and dry (May–August), and hot and dry (September–December). Temperatures can be extreme, ranging from below zero in winter to more than 45°C in summer (Simmons et al. 1998a).

In Pliocene times, the Kunene River flowed into a large inland lake, of which Etosha Pan is a present-day remnant. Continental uplift changed the course of the river toward Ruacana Falls in the west, cutting off the lake's water supply. During the drying process, the soil of the pans in and to the north of Etosha became mineral rich. Wind erosion deepened the depression over millions of years (Wellington 1938; Mendelsohn et al. 2000).

Today, the pan is subject to periodic partial flooding during the rainy season. Direct rainfall accounts for only a small proportion of the pan's water; three rivers supply the majority: the Ekuma, Oshigambo, and Omuramba Owambo. The Ekuma River flows seasonally from the southern shores of Lake Oponono, situated about 70 km north of the pan. This lake receives input from numerous perennial watercourses and *oshanas* (linearly linked, shallow, parallel ponds or lakes) that flow into the Cuvelai inland delta from Angola (Berry et al. 1973; Mendel-

sohn et al. 2000). The Oshigambo River also draws its water from southern Angola. The Ekuma and Oshigambo rivers form deltas in the northwestern corner of the pan, about 13 km apart. The Omuramba Owambo receives its water from a catchment to the northeast of Etosha, and it flows into Etosha Pan through Fisher's Pan, a small eastern extension. All three rivers flow erratically during the rainy season and, depending on their levels, flood the pan to varying degrees. In unusually dry years, the rivers may not flow at all. Once in about 7–10 years, during exceptional rains, the oshanas and rivers fill with rainwater and sometimes with floodwater from the Cuvelai River in Angola. However, most of the time the pan is a dry, saline desert, and water is found only in numerous waterholes surrounding the pan. These waterholes are fed by aquifers, which occur at various depths throughout the region (Mendelsohn et al. 2000).

The Etosha Pan is almost devoid of vegetation and is classified by Giess (1971) as a saline desert. The dominant plants are blue-green algae that cover the surface of the pan during the rainy season (Berry 1991). One of the few vascular plants is the annual grass *Sporobolus salsus* (Lindeque and Lindeque 1997). Halophytic vegetation lines the edge of the pan, consisting principally of *Sporobolus spicatus, S. ioclados, S. tenellus, Odyssea paucinervis*, and the small shrubs *Suaeda articulata* and *Salsola tuberculata* (Mendelsohn et al. 2000). *Atriplex vestita* and *Sporobolus tenellus* are also present, as are the occasional patches of annuals such as *Chloris virgata, Diplachne fusca, Dactyloctenium aegyptium*, and *Eragrostis porosa* (White 1983).

Outstanding or Distinctive Biodiversity Features

The harsh climate of the pan makes it an unsuitable habitat for most animals. Vegetation is scarce, and floodwater, when present, is extremely salty. Only highly specialized, salt-tolerant fauna adapted to withstand long, dry periods and respond rapidly to rainfall inhabit the pan (Kok 1987; Lindeque and Lindeque 1997). These are mainly crustacea, which dominate this environment because of their short life cycles and desiccation-tolerant eggs (Curtis et al. 1998). Ground squirrels and the endemic Etosha agama (*Agama etoshae*) are found on the fringes (Branch 1998).

In contrast to the desolate pan, the waterholes at the fringes of the pan (particularly in the south) are the sites of spectacular congregations of large ungulates, such as Burchell's zebra (*Equus burchelli*), blue wildebeest (*Connocheatus taurinus*), and springbok (*Antidorcas marsupialis*). Other species include elephant (*Loxodonta africana*, EN), giraffe (*Giraffa camelopardalis*), black rhinoceros (*Diceros bicornis*, CR), gemsbok (*Oryx gazella*), eland (*Taurotragus oryx*), greater kudu (*Tragelaphus strepsiceros*), steenbok (*Raphicerus campestris*), Damara dik dik (*Madoqua kirkii*), and black-faced impala (*Aepyceros melampus petersi*, VU) (Joubert and Mostert 1975; Lindeque and Lindeque 1997; Hilton-Taylor 2000; Mendelsohn et al. 2000). Predators such as lion (*Panthera leo*, VU), leopard (*Panthera pardus*), cheetah (*Aci-*

nonyx jubatus, VU), spotted hyena (*Crocuta crocuta*), and brown hyena (*Hyaena brunnea*) also occur. During the wet summer, the wildlife moves away from the waterholes to use the lush grazing and temporary pools, mainly to the west of the park. Springbok, zebra, and wildebeest undergo some of the largest movements, numbering in the thousands.

The Etosha National Park has a rich bird fauna (Hamunyela et al. 1998; Simmons et al. 1998a). Etosha, along with Sua Pan in Botswana, is especially important as a breeding site for the greater and lesser flamingo (*Phoenicopterus ruber* and *P. minor*). In flood years, up to 1.1 million flamingos have been recorded. However, the pan does not hold a viable population of flamingos because of their low level of natural recruitment (Simmons 1996, 2000). When it is flooded, great white pelicans (*Pelecanus onocrotalus*) also breed in large numbers (Berry et al. 1973; Simmons et al. 1998a). Etosha is also home to the only breeding population of blue crane (*Grus paradisea*, VU) outside South Africa (Simmons et al. 1998a).

Status and Threats

Current Status

The Etosha Pan is completely protected in Etosha National Park, although some nearby pans fall outside the park boundary (du Plessis 1992). In addition, it is one of four wetlands in Namibia designated as a Ramsar site of international importance (Curtis et al. 1998; Ramsar Convention Bureau 2002).

Types and Severity of Threats

Although the pan itself is well protected in Etosha National Park, the Cuvelai drainage system that extends to the north of the ecoregion is essential to the ecology of the pan. However, this system falls outside the protected area network. Any alteration of water flow in this system could devastate the ecology of the pan. For example, a major diversion of water flow would prevent the pan from flooding, and the huge flocks of flamingos and pelicans would lose their breeding ground (Hails 1997).

Another major threat to the ecology of the Etosha Pan is the introduction of pesticides and insecticides into the system. National campaigns to combat human and stock disease vectors and agricultural pests have had a negative impact on the aquatic invertebrate fauna of the Cuvelai system that feeds the pan (Curtis et al. 1998), as well as birds such as the lesser flamingo and white pelican (Berry 1972; Berry et al. 1973). Large mammals surrounding the pan have also declined because of predation and diseases such as anthrax (Mendelsohn et al. 2000).

Justification for Ecoregion Delineation

The boundaries for this ecoregion are based on the extent of halophytic vegetation in the area mapped by White (1983), al-

though nearby smaller pans have not been mapped at this scale. Although it supports plant communities similar to those of other southern African pans, such as Makgadikgadi, its remoteness with respect to conservation planning was considered a factor in making Etosha its own ecoregion.

Ecoregion Number: **68**

Ecoregion Name: **Makgadikgadi Halophytics**

Bioregion: **Eastern and Southern Africa**

Biome: **Flooded Grasslands and Savannas**

Political Units: **Botswana, Mozambique**

Ecoregion Size: **30,400 km^2**

Biological Distinctiveness: **Locally Important**

Conservation Status: **Relatively Intact**

Conservation Assessment: **V**

Author: *Amy Spriggs*

Reviewed by: *Jonathan Timberlake*

Location and General Description

The Makgadikgadi Halophytics [68] ecoregion covers the Makgadikgadi Pan complex in northeastern Botswana (White 1983), some surrounding grasslands, and an outlier of saline grasslands in the Changane Valley in southern Mozambique. The Makgadikgadi Pan complex covers more than 12,000 km^2, one of the largest in the world (Hughes and Hughes 1992). It contains two major pans, Ntwetwe Pan (106 by 96 km) and Sua Pan (112 by 72 km), surrounded by a number of smaller pans. Halophytic communities are also widespread in the Changane Valley in southern Mozambique.

The Makgadikgadi Pans complex is situated in the Kalahari, a flat, semi-desert plain extending across much of central southern Africa. The climate of the Kalahari is semi-arid, with droughts occurring in approximately 7-year cycles (Moyo et al. 1993; Penry 1994). The rainfall is 450 and 500 mm per annum, with 40 to 50 percent variability (Penry 1994). Most rain falls as thunderstorms during the summer, from October to March. There is little rain in winter (May–August). Summer temperatures average around 35°C and reach a maximum of 44°C, and during the winter midday temperatures reach 17°C.

The Makgadikgadi Pans occupy the center of a depression (the Kalahari Basin) that once held an enormous lake that spanned most of northern Botswana. The Zambezi, Okavango, and Chobe rivers once fed this ancient lake (Shaw et al. 1997). The formation of various geological faults on the southern extremity of the East African Rift Valley diverted the flow of these rivers away from the basin, causing the lake to slowly dry up. This drying process concentrated sodium carbonate in the lakebed, eventually leaving flat, soda-saturated clay pans (the

Makgadikgadi Pans complex). Two rivers currently supply water to these plans: the Boteti and the Nata. In wetter years the Boteti River flows out of the Okavango Delta and empties into the southern portion of the Ntwetwe Pan (Penry 1994). The Nata River, which originates in Zimbabwe, flows more reliably and empties into the northeastern part of the Sua Pan at the Nata Delta, an area that rarely dries out completely.

The pans of the Makgadikgadi complex have solonetz soils that are almost devoid of macrophytic vegetation. The dominant plant life is a thin layer of blue-green algae, which covers surfaces during the rainy season. On the saline fringes two macrophytes dominate: *Sporobolus spicatus* and the spiny grass *Odyssea paucinervis* (White 1983). Salt marshes are also found scattered around the wetter fringes of the pans, supporting species such as *Portulaca oleracea* and *Sporobolus tenellus* (Hughes and Hughes 1992). Surrounding the pans on less brackish soil are grasslands dominated by *Odyssea paucinervis* and *Cynodon dactylon*, with *Cenchrus ciliaris* and *Eriochloa meyeriana* dominating the crests of calcrete escarpments. These grasslands have few trees, except to the west, where *Hyphaene* palms fringe the drainage lines, extending north to Nxai Pan. To the north and northwest of Ntwetwe Pan, baobab (*Adansonia digitata*), *Acacia kirkii,* and *Acacia nigrescens* trees are found scattered throughout the grassland (Comley and Meyer 1994).

In the moderately saline areas of the Changane Valley grasslands predominate, scattered with islands of *Acacia nilotica*. The more frequently flooded areas have a higher salinity and are dominated by grasslands interspersed with extensive bare patches. The dominant grasses in these saline flooded grasslands are *Eriochloa meyeriana*, *Sporobolus nitens*, and *Aristida adscensionis*. Near the Changane River, salinity is even higher, and species of *Sarcocornia, Salicornia, Atriplex,* and *Suaeda* predominate (White 1983).

Outstanding or Distinctive Biodiversity Features

The harsh climate of the Makgadikgadi Pans is unsuitable for most animals. Only highly specialized invertebrates permanently inhabit the pans (Kok 1987; Curtis et al. 1998). Ostriches (*Struthio camelus*) nest on the pans to avoid scavengers, such as the black-backed jackal (*Canis mesomelas*). Animals confined to the grasslands surrounding the pans include hartebeest (*Alcelaphus buselaphus*), gemsbok (*Oryx gazella*), springbok (*Antidorcas marsupialis*), steenbok (*Raphicerus campestris*), greater kudu (*Tragelaphus strepsiceros*), giraffe (*Giraffa camelopardus*), Burchell's zebra (*Equus burchelli*), blue wildebeest (*Connocheatus taurinus*), brown hyena (*Hyaena brunnea*), spotted hyaena (*Crocuta crocuta*), lion (*Panthera leo,* VU), cheetah (*Acinonyx jubatus,* VU), and elephant (*Loxodonta africana,* EN) along the Boteti River (Penry 1994). The Nxai Pan has a large springbok population and is one of the few places where springbok and impala (*Aepyceros melampus*) cohabit (Comley and Meyer 1994; Williamson 1994).

For most of the year the pans are devoid of birds, except for ostriches and species such as the chestnut-banded and Kittlitz's plovers (*Charadrius pallidus, C. pecuarius*) (Penry 1994). The grasslands surrounding the pans support important populations of kori bustard (*Ardeotis kori*), ostrich, secretary bird (*Sagittarius serpentarius*), Burchell's and yellow-throated sandgrouse (*Pterocles burchelli* and *P. gutturalis*), and francolins, among others (Penry 1994; Tyler and Bishop 2001). After good rains the pans attract thousands of waterbirds, most of which come to breed. The most spectacular concentrations include the greater and lesser flamingos (*Phoenicopterus ruber, P. minor*), which flock in the thousands. Sua Pan is the most important breeding area in Africa for the greater flamingo (Simmons 1996). Other birds that congregate around the pans include wattled crane (*Grus carunculatus,* VU), black-necked grebe (*Podiceps nigricollis*), great white and pink-backed pelican (*Pelecanus onocrotalus, P. rufescens*), black-winged stilt (*Himantopus himantopus*), marsh sandpiper (*Tringa stagnatilis*), black-winged pratincole (*Glareola nordmanni*), and many Anatidae (Finlayson and Moser 1991; Penry 1994; Tyler and Bishop 2001).

Status and Threats

Current Status

The Makgadikgadi Halophytics [68] ecoregion is largely undisturbed, human populations are very low, and large blocks of habitat remain intact. The Makgadikgadi and Nxai Pan National Park covers more than 5,500 km² of this ecoregion. Although this park is extensive, it does not include Sua Pan, which is the only viable breeding site of the greater and lesser flamingo in southern Africa. The Nata community have established their own informally protected area, the Nata Sanctuary, which includes the northern area of Sua Pan and the Nata Delta. The seasonally flooded saline wetlands of Changane area in southern Mozambique are unprotected.

Types and Severity of Threats.

The major threat to wildlife throughout Botswana has been the erection of veterinary cordon fences (Owens and Owens 1980). Wildebeest and hartebeest, which spend the wetter months along the boundary between Ghanzi and Kgalagadi (in the Kalahari Xeric Savanna [105] ecoregion), move northeast toward the Central Kalahari Game Reserve during the dry season. Here they encounter the Kuke and Makalamakedi veterinary fences and are channeled into the Lake Xau and Boteti River areas. This has caused major overgrazing and competition with local wildlife. In the drought of the 1980s, 90 percent of the wildebeest population and 83 percent of the hartebeest population died (Owens and Owens 1980; Campbell 1990). Fences to the north of the Makgadikgadi Pans have similarly prevented the southward migration of buffalo and zebra from the Okavango delta during the wet season, leading to a large decline in populations since 1987. A veterinary fence also protrudes into the eastern area of Sua Pan. This fence reduces the flamingo population because significant numbers of adults die every year in collisions with the fence wires (Williamson 1994). Uncontrolled tourism, particularly motorbike tours, is another major threat to the fauna of the Makgadikgadi Pans (Simmons 1996).

A soda ash extraction plant is situated at the northeastern edge of Sua Pan (Steyn 1990). Although the plant is small and is not expected to have much direct impact on the pan, the associated infrastructure of roads and railways extends beyond the immediate area. There is also a small salt mine on the shores of Ntwetwe Pan (Steyn 1990).

Another threat to the Makgadikgadi Pans complex is the potential diversion of water for irrigation schemes elsewhere in the region. For example, a major diversion of water from the Nata River, would reduce flooding in Sua Pan, resulting in flamingos, pelicans, and other water birds losing their breeding grounds (Hails 1997). Water diversion could also result in the drying up of the Nata Delta and Boteti River, essential areas for waterbird and Palearctic migrants during the dry winter months.

Justification of Ecoregion Delineation

This ecoregion based on White's (1983) "halophytic vegetation" unit in southern Africa and encompasses the Makgadikgadi Pans and associated fringing vegetation in Botswana, as well as the halophytic vegetation in the Changane Valley in Mozambique.

Ecoregion Number: **69**

Ecoregion Name: **Mediterranean High Atlas Juniper Steppe**

Bioregion: **African Palearctic**

Biome: **Montane Grasslands and Shrublands**

Political Units: **Morocco**

Ecoregion Size: **6,300 km²**

Biological Distinctiveness: **Globally Outstanding**

Conservation Status: **Endangered**

Conservation Assessment: **I**

Authors: *Nora Berrahmouni, Pedro Regato*

Reviewed by: *Abdelmalek Benabid, Hans Peter Müller*

Location and General Description

The Mediterranean High Atlas Juniper Steppe [69] is found in the Moroccan High Atlas, the highest mountain massif in North Africa (Toubkal summit at 4,167 m). Additional juniper steppe may be found in Algeria, on summits of the Saharan At-

las. Both the High Atlas and the Saharan Atlas constitute a barrier between the Mediterranean biogeographic region and the Saharo-Sindic desert region. These mountain massifs exhibit a sharp precipitation gradient ranging from annual rainfall greater than 1,000 mm to average annual rainfall between 200 and 600 mm. A similar gradient exists between cold north-facing slopes, where snow remains for more than 7 months above 3,500 m, and the cold semi-arid south-facing slopes, subject to dry Saharan winds.

These mountain massifs, like the rest of the North African mountains, contain a diversity of rock formations such as sandstone, dolomite, limestone and marl, quartzite, schist, granite, and volcanic materials. Precipitous slopes and deep canyons characterize the landforms. Between the High Atlas and the Saharan Atlas are extensive eastern Moroccan and Algerian high plateaus.

The wide elevational range of this ecoregion encompasses two distinct forest and woodland zones: the xeric conifer zone and the evergreen broadleaf zone. The latter also contains cedar forests. Although this ecoregion is delineated at altitudes exceeding 2,700 m, the xeric conifer zone that characterizes the driest and cold, high elevations has an elevational range of 1,500–3,150 m. The evergreen broadleaf zone covers the humid and cold north-facing slopes, ranging from the lowest hills up to elevations of 2,800 m, often overlapping with the Mediterranean Conifer and Mixed Forests [34] ecoregion. At the highest elevations, above the treeline (roughly 3,200 m), there is a mosaic of alpine meadow vegetation interspersed with rock outcrops (Quézel 1981; WWF MedPO 2001).

The dominant canopy tree species of the xeric conifer woodlands is *Juniperus thurifera,* a hardy endemic species of the western Mediterranean (Quézel 1971; Benabid and Fennane 1994). It often constitutes mixed stands with the evergreen holm oak (*Quercus ilex ballota*). Juniper woodlands cover 310 km² in Morocco, mainly in the High Atlas and the Middle Atlas, with several hundred individuals in the Anti-Atlas Range. There are also a few square kilometers in the Saharan Atlas, on Aures Massif in eastern Algeria. The plant composition of anthropogenically altered juniper woodlands includes a large number of cushion and thorny shrubs, including *Cytisus balansae, Erinacea anthyllis, Prunus prostrata,* and *Astragalus armatus.* Well-preserved juniper and holm oak mixed stands are characterized by the presence of more shade-loving species, including *Fraxinus dimorpha, Lonicera arborea, Crataegus laciniata, Buxus sempervirens,* and *Berberis vulgaris australis,* as well as other high shrubs such as *Juniperus oxycedrus* and *Ephedra nebrodensis.*

Holm oak (*Quercus ilex*) may be the most widespread tree species in the Mediterranean region. This oak is able to adapt to many different environmental conditions, from a warm and dry coastal climate to cold and humid high mountain areas. The high mountain holm oak (*Quercus ilex ballota*) once grew extensively over the north-facing slopes of the High Atlas. Today, well-preserved coppice woodlands are found in the central and western part of this massif, but anthropogenic pressure has substantially degraded the holm oak woodlands to the east. The forest understory includes many shrub species from the mountain juniper woodlands, as well as a herbaceous layer with *Cephalantera rubra, Helleborine latifolia, Festuca triflora,* and *F. rubra.*

High meadow vegetation is characterized by open grasslands comprising a few plant species adapted to the strong wind conditions of the mountain summits. Species include *Avena montana, Festuca mairei, F. alpina,* and *Ranunculus geraniifolius.*

Outstanding or Distinctive Biodiversity Features

About 30 percent of all flowering plants in the High Atlas are endemic. This is the result of the extended isolation of high-elevation Holarctic taxa in the North African mountain ranges (Dallman 1998). The relict *Cupressus atlantica* is endemic to the ecoregion, whereas *Juniperus thurifera* and *Quercus ilex ballota* are endemic to the western Mediterranean. Other endemic plant species associated with these juniper and holm oak mountain woodlands include *Cirsium chrysacanthum, Euphrasia minima, Genista florida maroccana, Retama dasycarpa, Cytisus balansae,* and *Artemisia atlantica.* Many of the plant species have a restricted distribution range and are threatened (Walter and Gillett 1998). *Cupressus atlantica* has been reduced to several hundred individuals scattered in the western High Atlas Mountains (Benabid and Fennane 1994). Other threatened species are *Alyssum flahaultianum, Epilobium psilotum,* and *Erodium atlanticum.* The holm oak forests from the north-facing slopes of the central part of the High Atlas (Jbel Tazerkunt) contain species of high biogeographic significance, such as the Macronesian laurel (*Laurus azorica*), which provides a link to the offshore islands of the Azores (Benabid 1985).

The faunal diversity of this ecoregion is similar to that of the rest of the North African mountain ranges. Together with the Middle Atlas, the High Atlas represents the last refuge for the Barbary leopard (*Panthera pardus panthera,* CR) and for a wild population of Barbary sheep (*Ammotragus lervia,* VU) (Hilton-Taylor 2000). Other species represent a mixture of Palearctic, Afrotropical, and more restricted North African taxa. Some examples are red fox (*Vulpes vulpes*), common jackal (*Canis aureus*), Barbary ground squirrel (*Atlantoxerus getulus*), wild boar (*Sus scrofa*), common otter (*Lutra lutra*), Egyptian mongoose (*Herpestes ichneumon*), crested porcupine (*Hystrix cristata*), and European polecat (*Mustela putorius*) (Blondel and Aronson 1999).

The mountain juniper and holm oak woodlands support a notable mountain bird community, with alpine accentor (*Prunella collaris*) and rufous-tailed rock thrush (*Monticola saxatilis*). Locally rare raptors include golden eagle (*Aquila chrysaetos*) and lammergeier (*Gypaetus barbatus*); however, these have disappeared in the Haut Atlas Oriental National Park (Magin 2001b). Reptile endemics distributed in the High Atlas woodlands include Atlas day gecko (*Quedenfeldtia trachyblepharus*), High Atlas lizard (*Lacerta andreanskyi*), mountain skink (*Chal-*

cides montanus), and mountain viper (*Vipera monticola*) (Blondel and Aronson 1999; Charco 1999).

Status and Threats

Current Status

The original vegetation has been dramatically reduced to degraded open woodlands as a result of intense land clearance for agriculture. Crop terraces cover large areas on steep mountain slopes. Grazing, timber production, and firewood collection are also problematic (Thirgood 1981). Morocco may have already lost 90 percent of its mountain juniper woodlands, from 3,270 km² of original cover to 310 km² at present.

The degree of degradation seems to be higher in the eastern portion of the High Atlas, where junipers and holm oaks cannot stand the combination of human pressure and increasing aridification. The southernmost cedar stands in the eastern High Atlas are under threat of complete removal. Moreover, junipers do not readily regenerate in degraded environments, and "old-growth tree cemeteries" cover many high mountain areas (Médail and Quézel 1997; WWF MedPO 2001).

Three protected areas are located in this ecoregion: Toubkal and Haut Atlas Oriental National Parks in Morocco and Belezma National Park in Algeria. Management plans for the Haut Atlas Oriental and Belezma have been written, but implementation is weak, and the areas still lack appropriate legal status. There is no management plan for Toubkal National Park. In general, resources (human, financial, and equipment) and knowledge (of species and habitat distribution and ecology) are insufficient to help implement adequate conservation and sustainable use programs. Furthermore, key decision-making agencies do not adequately focus on habitats and wildlife populations to secure their long-term survival and the maintenance of the ecological processes related to them. The rural human population is high in this ecoregion and is still growing in and around protected areas. People's needs are normally in strong conflict with protected areas, and rights of use are not clearly established.

Types and Severity of Threats

Anthropogenic impact remains high in this ecoregion and is exacerbated by the socioeconomic instability of the Maghreb countries. The collapse of the seminomadic Berber pastoral system has transformed summer camps in the high mountain grasslands into permanent human settlements. Local people collect large amounts of firewood, in many cases illegally. The intensive collection of juniper and holm oak branches often kills the trees. Furthermore, livestock's need for fodder during winter, when the grass is covered by snow for several months, gives rise to extensive overgrazing and soil degradation in the forest understory.

According to some projections, North African montane coniferous forests are extremely threatened by climate change.

The intensification of the summer drought season and the increase of the average annual temperature could exceed physical thresholds within which these forests and their species can persist. Additionally, human impact, mainly through soil degradation, has reduced the forest's resilience to survive natural disturbances (Brandt and Thornes 1996; Dallman 1998).

The growing trend of high mountain tourism (skiing, off-road car and motorbike driving, and hiking) constitutes a new challenge to the conservation of the Mediterranean High Atlas Juniper Steppe [69] because these activities hasten soil erosion and increase firewood demand during winter. From a positive perspective, certain environment-friendly mountain tourism activities, well adapted to the High Atlas environmental constraints, are a potential source of income to local communities to help reduce the practices of unsustainable grazing and firewood collection.

Justification for Ecoregion Delineation

White (1983) defines only one vegetation unit for the high elevation areas in Mediterranean region: "Mediterranean montane forest and altimontane shrubland." In keeping with the method applied to the rest of the continent and because of the high number of endemic plants found in this region, together with some animal endemics, the altimontane areas were separated from the lower-elevation conifer forests. Experts recommended the separation of the altimontane shrublands along the 2,700 m elevation contour, which delineates a portion of the High Atlas Mountains in Morocco.

Ecoregion Number:	**70**
Ecoregion Name:	**Ethiopian Upper Montane Forests, Woodlands, Bushlands, and Grasslands**
Bioregion:	**Horn of Africa**
Biome:	**Montane Grasslands and Shrublands**
Political Units:	**Ethiopia, Eritrea, Sudan**
Ecoregion Size:	**245,400 km²**
Biological Distinctiveness:	**Globally Outstanding**
Conservation Status:	**Critical**
Conservation Assessment:	**I**
Authors:	*Chris Magin, Miranda Mockrin*
Reviewed by:	*Ib Friis, Derek Yalden*

Location and General Description

The Ethiopian Upper Montane Forests, Woodlands, Bushlands, and Grasslands [70] is a biologically rich and severely threat-ened ecoregion that covers the majority of two Ethiopian Mountain massifs (northwestern and southeastern), separated by a part of the Rift Valley. This ecoregion ranges from 1,800 to 3,000 m in elevation, with montane forest at lower altitudes and Afroalpine habitat at higher altitudes. The climate of these highlands is greatly affected by their topography and by the movement of the Inter Tropical Convergence Zone (ITCZ). As the ITCZ moves north between May and October, warm moist air from the Indian Ocean brings rain to the southern slopes of the Ethiopian Highlands. In the remainder of the year, winds from the Red Sea typically contain less moisture, which falls mainly on the northern side of the highlands. Overall, the highest annual rainfall (up to 2,500 mm) occurs on the southwestern faces of the highlands, which support montane or transitional forests, whereas average annual rainfall is closer to 1,600 mm over most of the ecoregion.

The Ethiopian Highlands first began to rise in the Eocene 55 million years ago. The main dome was split into two halves, the Northern and Southern Highlands, by the development of the Rift Valley during the Miocene, about 13 million years ago (Yalden and Largen 1992). Geologically the area consists of thick Tertiary basaltic lava flows in most places. However, Precambrian rocks outcrop in southwestern Ethiopia and Eritrea, and Mesozoic rocks form the surface outcrops of the southeastern highlands of Ethiopia. During the last ice age higher areas of the Ethiopian plateau were glaciated. Glaciation peaked 14,000–18,000 years ago and the ice started to melt from around 11,000 years ago.

Phytogeographically, the ecoregion is part of the Afromontane archipelago-like regional center of endemism (White 1983). The natural vegetation probably was a mixture of closed forest in areas with higher rainfall (mainly to the southwest of the two main massifs and on some higher mountains), grassland, bushland, and thicket in other lower-rainfall areas (Friis 1992; Friis and Demissew 2001). Forest structure and composition vary with locality and elevation. There is a cloud forest belt at 2,000–2,500 m in the south, whereas in drier locations (particularly on steep hillsides) the forest is dominated by *Podocarpus falcatus* and *Juniperus procera*, often with *Hagenia abyssinica*. In the north, between 2,300 and 2,700 m in the Simien Mountains, there is evergreen broadleaf montane forest dominated by *Syzygium guineense, Juniperus procera,* and *Olea africana* (Nievergelt et al. 1998). On the moist upper slopes of the Harenna Forest, a shrubby zone of *Hagenia* and *Schefflera* grows along with giant lobelias (*Lobelia gibberroa, L. rhynchopetalum*). Below 2,400 m, small pockets of dense, moist forest occur. The Harenna Forest receives more than 1,000 mm per year because the southwest-facing orientation of the steep scarp face attracts orographic rainfall. Dominant trees include *Aningeria* and *Olea*, often draped with lianas and epiphytes. Woodland and shrubland dominated by *Acacia* species probably was the natural vegetation over the majority of the lower plateau.

Outstanding or Distinctive Biodiversity Features

The ecoregion contains a high number of endemic plants, with most endemics occupying open habitats (Tilahun et al. 1996; Friis et al. 2001). The highest number of plant species that are strictly endemic to Ethiopia and Eritrea occur at altitudes between 2,000 and 2,500 m, peaking at 1,500–1,800 m. There are several local centers of endemism for plants throughout the highlands (Friis et al. 2001). Because of the complexity of this ecoregion and the different regional vegetation classification schemes that exist, it is not possible to give a precise figure for the total number of endemic plants in the ecoregion. (For consistency with other parts of Africa we have followed White 1983, which differs significantly from Friis 1992.)

The high-altitude regions of Ethiopia and Eritrea have a high diversity of mammals, with ten near-endemic species, many shared with the Ethiopian Montane Moorlands [71] ecoregion, such as the Walia ibex (*Capra walie*, CR) (Nievergelt et al. 1998), mountain nyala (*Tragelaphus buxtoni*, EN), and gelada baboon (*Theropithecus gelada*). An antelope commonly found at lower elevations is Menelik's bushbuck (*Tragelaphus scriptus meneliki*), a subspecies endemic to Ethiopia. Two different subspecies of *Chlorocebus aethiops* are found in this ecoregion: the Djam-djam or Bale monkey (*Chlorocebus aethiops djamdjamensis*, DD), which is restricted to the Southern Highlands, and the black-faced vervet (*Chlorocebus aethiops aethiops*).

The ecoregion covers the majority of the South and Central Ethiopian Highlands endemic bird areas (Stattersfield et al. 1998). Species include the strictly endemic lineated pytilia (*Pytilia lineata*) and the near-endemic Ruppell's chat (*Myrmecocichla melaena*) and Ankober serin (*Serinus ankoberensis*, EN). The white-winged flufftail (*Sarothrura ayresi*, EN) is a threatened species that occurs in this ecoregion.

At least ten amphibians are endemic or near endemic to the ecoregion, including the strictly endemic grassland forest tree frog (*Leptopelis yaldeni*), together with five species of near-endemic reptiles shared with other ecoregions in the Ethiopian Highlands [25, 71] (Largen 2001).

Status and Threats

Current Status

By the early twentieth century, only 5 percent of the Ethiopian Highlands was forested, although it is believed that at one time forest covered most of the area (Friis 1992). For example, *Podocarpus* and *Juniperus species* once covered 176,000 km^2 in the central, eastern, and northern regions. Less than 1 percent of these forests remains because *Podocarpus* and *Juniperus* provide the most commonly used timber in Ethiopia. Remaining grassland and thorn scrub patches are generally confined to rocky and steep areas. Besides these inaccessible areas, the only remaining forested areas are church graveyards. Even these

forests are not totally protected. A survey of *Syzygium* forest in graveyards in the Simien region found that trees were harvested for local use and cattle grazed in the graveyards, impeding forest regeneration (Nievergelt et al. 1998). Parts of the ecoregion are officially protected in the Bale Mountains National Park, but about 2,500 people currently live in the park along with 10,500 head of livestock. Controlled hunting areas and wildlife reserves offer little or no protection for native flora and fauna (Yalden et al. 1996). The forests, woodlands, and grasslands remaining to the southwest of the Nechisar National Park are not protected. The proposed Termaber-Wufwasha-Ankober conservation area in the western highlands would protect much of the ecoregion's biodiversity, as would the proposed areas for protecting the forests further to the southwest.

Several species face global extinction. The most severely threatened is the Walia ibex, which numbers fewer than 400 individuals (Nievergelt et al. 1998). It is threatened by habitat loss and hybridization with free-ranging domestic goats.

Types and Severity of Threats

The human population density is very high, ranging from 100 to 400 people per square kilometer. Most of the population practices subsistence farming, and the demand for natural products and land for farming is huge. This has been the case for many hundreds of years and is the predominant reason for the widespread loss of natural vegetation (Tilahun et al. 1996). The Bale Mountains National Park is one of the few places in this ecoregion where examples of intact vegetation types can be found, including the large and nearly unexplored Harenna Forest. However, even these areas are increasingly threatened by human activities.

Justification for Ecoregion Delineation

White (1983) maps most of this ecoregion as "undifferentiated montane vegetation," which includes parts of the Ethiopian Lower Montane Forests, Woodlands, and Bushlands [25]. The ecoregion boundaries follow the 1,800-m contour for the lower elevation and 3,000-m contour for the upper elevation. Although the south and central highlands are recognized as two areas of bird endemism (Stattersfield et al. 1998), they are not dissimilar in terms of their flora and fauna and are included here in a single ecoregion.

Ecoregion Number: **71**

Ecoregion Name: **Ethiopian Montane Moorlands**

Bioregion: **Horn of Africa**

Biome: **Montane Grasslands and Shrublands**

Political Units: **Ethiopia**

Ecoregion Size: **25,200 km²**

Biological Distinctiveness: **Globally Outstanding**

Conservation Status: **Critical**

Conservation Assessment: **I**

Authors: *Chris Magin, Miranda Mockrin*

Reviewed by: *Ib Friis, Derek Yalden*

Location and General Description

The Ethiopian Montane Moorlands [71] ecoregion covers the higher parts of the Ethiopian Highlands Massif, from 3,000 m to over 4,500 m. Below 3,000 m the ecoregion would have formerly graded into montane forests and grasslands, as it still does at the Harenna Forest in the Bale Mountains, southeastern Ethiopia. However, most of these areas are now farmed or used for grazing.

The climate of the ecoregion is presumed to be complex, but data are largely lacking. Annual rainfall is highest in the southwest (perhaps as high as 2,500 mm), where the dry season may only last for 2 months, and declines to as little as 1,000 mm in the north, where the dry season may last as long as 10 months. Estimates of the mean maximum temperature on the higher peaks are between 6°C and 12°C, and mean minimum temperature is below 0°C (Miehe and Miehe 1994). Frosts are common throughout the year, especially in the winter (November–March) (Rundel 1994). In the Simien Mountains National Park, recorded temperature ranges from −2.5°C to 4°C minimum and 11°C to 18°C maximum. On the Sanetti Plateau of the Bale Mountains diurnal ranges of more than 40°C (−15°C to +26°C) have been recorded (Hedberg 1997).

The ecoregion lies on Tertiary volcanic deposits, which are extremely thick in the Simien Mountains. The soils developed over these rocks are principally nitosols and in some areas lithosols. The volcanic highlands were divided into two parts by the development of the Rift Valley around 13 million years ago, and volcanic activity ended 4 to 5 million years ago. During the last ice age, thick glaciers covered this ecoregion. As the climate warmed around 10,000 years ago, the ice melted, and alpine vegetation moved to ever-higher altitudes. Ice remained on the peaks of the highlands until a few thousand years ago. Some of the highest areas remain barren, such as the central peaks in the Bale Mountains.

The Ethiopian Highlands are extremely rugged, with the highest point at 4,620 m on Mount Ras Dashan. Other high peaks are found in Tigray (Amba Alaghe), Amhara (Welo: Kollo, Amba Ferit), Gojjam (Choké Mountains), and the southern mountains: Mount Zuquala, Guraghe Mountains, Galama Mountains (Chilalo, Badda, Mount Damota), and the Gughé Highlands. The highest point of the Simien Mountains National Park is 4,430 m, and the highest point of the Bale Mountains National Park is 4,377 m. The Bale Mountains contain the largest area above 3,000 m in Africa. Numerous alpine lakes are found in this region, some of which persist year-round. These higher elevations are sparsely populated, although agricultural activities continue wherever it is still possible to grow sufficient food to survive.

The ecoregion is a part of the Afroalpine archipelago-like region of extreme floristic impoverishment (White 1983) and is mapped as "altimontane vegetation in tropical Africa." Trees are absent at such high elevations, although some bushes and shrubs such as *Hypericum revolutum* occur. Much of the montane vegetation is a heathland scrub around 0.5–1.0 m high, dominated by the tree heathers *Erica trimera* (formerly *Philippia*) and *Erica arborea* (Miehe and Miehe 1994). Between the shrubs, the soil is bare, with some small plant species in the genera *Helichrysum, Alchemilla, Cerastium,* and the grasses *Koeleria* and *Aira.* Steep rocky slopes and cliffs in the high-elevation regions support very little vegetation, and the sedge *Carex monostachya* dominates flat swampy areas. A distinctive feature of the vegetation is the giant *Lobelia rynchopetalum,* which reaches a height of 6 m when flowering. All plant species found in the Afroalpine zone show xeromorphic characteristics designed to reduce transpiration. As examples, *Lobelias* have thick leathery leaves, Ericaceae and *Helichrysum* have highly revolute leaf margins, and most of these species have a small leaf surface area (Hedberg and Hedberg 1979).

Outstanding or Distinctive Biodiversity Features

The flora and fauna of these highlands contain species that indicate links to the Palearctic realm to the north and to the surrounding Afrotropical realm. Although the ecoregion possesses considerable endemism, this is much less than in the high mountains of Kenya and Tanzania or along the Albertine Rift. For example, there are no *Dendrosenecio* species and very few shrubby *Alchemilla* species. Moreover, species of *Rosa* and *Primula* are found, which are typical of Palearctic montane areas (Hedberg and Hedberg 1979; Vuilleumier and Monasterio 1986).

WWF and IUCN (1994) estimate that there are about 950 vascular plant species with at least 15 site endemics in the Bale Mountains. However, the number confined to the highest montane areas is not known. WWF and IUCN (1994) estimate five to ten site endemics in the Simien Mountains. This level of endemism probably is repeated in other high mountains of the ecoregion.

Strictly endemic mammals include the Ethiopian wolf (*Canis simensis,* CR), giant climbing mouse or Nikolaus's mouse

(*Megadendromus nikolausi,* VU), big-headed mole-rat (*Tachyoryctes macrocephalus*), Ethiopian narrow-headed rat (*Stenocephalemys albocaudata*), gray-tailed narrow-headed rat (*Stenocephalemys griseicauda*), and black-clawed brush-furred rat (*Lophuromys melanonyx*) (Sillero-Zubiri et al. 1995, 1997; Yalden and Largen 1992). The northern population of Ethiopian wolf is a separate subspecies, *C. s. simensis,* distinct from the population in Bale Mountains National Park and surrounding areas, *C. s. citernii* (Stuart and Stuart 1996). Near-endemic mammals include the Walia ibex (*Capra walie,* CR), the majority of which are found in the Simien Mountains National Park; the mountain nyala (*Tragelaphus buxtoni,* EN), the gelada baboon (*Theropithecus gelada*), and a number of other rodents and shrews. Klipspringer (*Oreotragus oreotragus*) and rock hyrax (*Procavia capensis*) can also both be found in rocky habitats.

Endemic species often display a number of unique behavioral, morphologic, and physiologic adaptations to the high-altitude environment. Gigantism is seen in several plant species (giant lobelias, tree heather, and giant St. John's wort), and the perennial plants have evolved dry, paper-like flowers to withstand harsh winds. An endemic species of small toad, Malcolm's Ethiopia toad (*Altiphrynoides malcolmi*), has internal fertilization, and eggs develop in moist soil (Sillero-Zubiri et al. 1997).

This ecoregion also contains important bird habitat, including alpine lakes and streams. Significant populations of Palearctic birds winter here, with several thousand Eurasian wigeon (*Anas penelope*) and northern shoveler (*Anas clypeata*) observed in the Bale Mountains, along with waders such as ruff (*Philomachus pugnax*) and common greenshank (*Tringa nebularia*). Globally threatened species found in the Bale Mountains include the greater spotted eagle (*Aquila clanga,* VU), imperial eagle (*Aquila heliaca,* VU), lesser kestrel (*Falco naumanni,* VU), and wattled crane (*Grus carunculatus,* VU). The Abyssinian longclaw (*Macronyx flavicollis*) is considered near endemic, as are the moorland chat (*Cercomela sordida*), Abyssinian waxbill (*Estrilda ochrogaster*), moorland francolin (*Scleroptila psilolaemus*), Rueppell's chat (*Myrmecocichla melaena*), ankober serin (*Serinus ankoberensis*), and spot-breasted lapwing (*Vanellus melanocephalus*). Three Palearctic species, golden eagle (*Aquila chrysaetos*), red-billed chough (*Pyrrhocorax pyrrhocorax*), and ruddy shelduck (*Tadorna ferruginea*), breed in the Bale Mountains.

Status and Threats

Current Status

Substantial blocks of habitat are protected in the Bale Mountains and Simien Mountains national parks. People and their livestock reside in both national parks, although there has been some forced relocation. Although the Bale Mountains National Park has been under management since 1970, it has never been formally gazetted. The Arsi-controlled hunting area is also located in this ecoregion. Although controlled hunting areas in Ethiopia tend to offer little protection (Kingdon 1989), sport hunting is currently banned in Ethiopia. The ecoregion is naturally fragmented because it occurs only in the highest portions of the Ethiopian Highlands. Outside the protected areas, habitat blocks are relatively intact when they are too high to be used by people for agriculture or grazing. Agriculture decreases above 3,200 m, where barley is the only crop that can be cultivated (Demissew 1996).

A survey of Simien Mountains National Park in 1996 found that human activities, especially overgrazing, are having a severe effect on the park (Nievergelt et al. 1998). The majority of the Geech Plateau was considered overgrazed or heavily grazed. One area of *Erica arborea–Hypericum revolutum* forest had deteriorated such that there was no undergrowth and no regeneration. In grassland areas, the number of *Lobelia* had increased because the species germinates readily on bare soil patches. However, even *Lobelia* decrease as overgrazing intensifies. The Simien Mountains National Park is currently designated a World Heritage in Danger site by United Nations Educational, Scientific and Cultural Organization (UNESCO).

Types and Severity of Threats

Ethiopia's highlands are among the most densely populated agricultural areas in Africa. An increasing human population has removed large areas of habitat through high-altitude agriculture, heather fires, and overgrazing by livestock. Delicate Afroalpine vegetation is not resilient; any tampering, especially fire, may have irreversible consequences (Demissew 1996).

With the expansion of human habitation, many wildlife populations are restricted to national parks. Even here the pressures are heavy. About 2,500 people currently live in the Bale Mountains National Park along with 10,500 head of livestock, and about 1,500 people inhabit the Simien Mountains National Park.

Several endemics face global extinction. The Walia ibex numbers fewer than 400 individuals (Nievergelt et al. 1998) and is threatened by habitat loss and hybridization with free-ranging domestic goats. The Bale Mountains National Park contains the largest population of wolves, but that population has steadily decreased throughout the 1990s so that it now numbers only approximately 200 individuals (Sillero-Zubiri et al. 1997). They face a wide range of threats, including being killed by vehicles on the Sanetti Plateau road, persecution by humans, and competition, diseases, and hybridization with domestic dogs. The mountain nyala is threatened both by habitat loss and by shooting for meat.

Management strategies for Bale Mountains and Simien Mountains national parks and surrounding areas must be implemented, based on community participation. Such strategies must incorporate earlier management plans written for the parks in the 1980s (Hillman 1986, 1990; Hurni 1986).

Justification for Ecoregion Delineation

The lower boundaries of this ecoregion follow the altimontane vegetation unit, roughly along the 3,000 m contour (White 1983). The Ethiopian Montane Moorlands [71] are separated from similar higher-altitude moorlands further south in East Africa and on the Rwenzori Mountains by differences in the flora and fauna. For example, the Ethiopian moorlands lack typical Afroalpine species such as *Dendrosenecio* but possess genera typical of the Palearctic realm, such as *Rosa* and *Primula* (Hedberg and Hedberg 1979; Vuilleumier and Monasterio 1986). Numerous species of endemic plants and animals are also confined to the Ethiopian montane moorlands ecoregion.

Ecoregion Number: **72**

Ecoregion Name: **East African Montane Moorlands**

Bioregion: **Eastern and Southern Africa**

Biome: **Montane Grasslands and Shrublands**

Political Units: **Kenya, Tanzania, Uganda**

Ecoregion Size: **3,300 km²**

Biological Distinctiveness: **Globally Outstanding**

Conservation Status: **Relatively Stable**

Conservation Assessment: **III**

Author: *Jan Schipper*

Reviewed by: *Derek Pomeroy, Marc Languy, Alan Rodgers*

Location and General Description

The East African Montane Moorlands [72] occupy the peaks of large mountains close to the equator in Kenya, northernmost Tanzania, and the border area between Kenya and Uganda. The lower altitudinal limit for the moorland ecoregion is at the transition to the East African Montane Forests [18] ecoregion, around 3,000 m (depending on location). This boundary also corresponds with the lower limit of the Afroalpine belt (Hedberg 1951).

This ecoregion has extreme relief and high altitudes. The highest peaks are Mount Kilimanjaro (5,899 m) and Mount Meru (4,200 m) in Tanzania, Mount Kenya (5,195 m) and the Aberdare Mountains (3,800 m) in Kenya, and Mount Elgon (4,324 m) on the border between Kenya and Uganda. There are also small areas of ericaceous vegetation on some other mountains in the region, such as Mount Kinyeti (3,187 m) in the Imatong Mountains (Killick 1979), Mount Kadam (3,068 m), Mount Moroto and Mount Napak in Uganda (Langdale-Brown et al. 1964), and Mount Hanang in Tanzania. These mountains are all volcanic in origin. The oldest, Mount Elgon, is 25 million

years old, whereas Mount Kilimanjaro, Mount Kenya, the Aberdares, and Mount Meru are all much younger (2–5 million years old). The soils are not well developed at these high altitudes, and bare volcanic rock is common, particularly in the driest areas. Glaciers and snowfields are present on Kilimanjaro and Mount Kenya, although they are retreating (Nilsson 1940; Hastenrath 1992; Osmaston and Kaser 2001).

Rainfall (or snowfall at higher altitudes) has two peaks, one in April and the other from November to December. On Mount Kenya, about 2,500 mm of precipitation falls from 1,400 to 2,200 m, and about 850 mm falls at the summit, mainly as snow. Mount Kilimanjaro has a peak of 2,000 mm of rainfall at 2,200 m but only around 200 mm of precipitation above 4,200 m. Precipitation is lower on the northern slopes because of rainshadow effects. Occult precipitation from clouds and mist becomes important at higher altitudes, mainly on the windward slopes of the mountains.

These mountains have a severe climate with extreme and rapid fluctuations, best expressed as "summer every day and winter every night" (Hedberg 1964). Daily nocturnal frosts occur above 4,000 m altitude. The mean minimum temperatures over 30 days on Mount Kenya are 1.7°C at 3,048 m and −3.9°C at 4,770 m. Absolute minimum temperatures at these elevations were −1.5°C and −8.3°C, respectively. On Mount Kilimanjaro, the mean minimum was 1.9°C at 4,160 m (Vuilleumier and Monasterio 1986; Rundel 1994). At 5,000 m, the mean annual temperature is −1°C.

At these high altitudes, the flora and fauna must withstand extreme cold, heat, and desiccation in a single day. At the highest altitudes (around 5,000 m), vegetation is small and cushion-like (*Agrostis sclerophylla* and *Sagina afroalpina*), ideal adaptations to these stresses. At altitudes between 3,500 and 4,500 m there are many larger plants (up to 8 m tall) of the genera *Senecio* (subgenus *Dendrosenecio*) and *Lobelia*. Many plant species have developed characteristic protective measures against the harsh conditions (Hedberg and Hedberg 1979; Mabberley 1986; Smith and Young 1987), including giant rosettes (*Lobelia*, *Dendrosenecio*, and *Cardus*), massive tussocks, and gigantism, as among heather (*Erica* and *Phillipia*) and St. John's wort (*Hypericum*).

This ecoregion is part of the Afroalpine archipelago-like region of extreme floristic impoverishment (White 1983). Some authors have suggested that the flora of this and other high montane ecoregions in Africa should be placed in a separate Afroalpine phytogeographic region (Hedberg 1961, 1986, 1994). Seven vegetation formations are recognized: ericaceous woodland and wooded grassland, *Dendrosenecio* woodland and wooded grassland, tussock grassland, *Helichrysum* scrub, and swamp or mire vegetation (Hedberg 1951; Lind and Morrison 1974; Davenport et al. 1996a, 1996c). *Dendrosenecios* are scattered through grasslands from 3,500 m to the vegetation limit at 5,000 m. Tussock grasslands are found in fire-prone or dry areas. *Helichrysum* scrub is found mainly on dry and rocky locations. Acidic mires are found at lower altitudes in places where water can accumulate. Above 4,350 m on Mount Kilimanjaro, the rainfall is only 150 mm per annum, and a montane desert with few plant species is found.

Outstanding or Distinctive Biodiversity Features

The Afroalpine habitats of the mountains in East Africa contain high levels of endemism, notably among the plants (Hedberg 1957, 1997) and invertebrates (Salt 1954). When combined, the East African Montane Moorlands [72] and the Rwenzori-Virunga Montane Moorlands [73] contain 81 percent endemism in their 208 plant species (WWF and IUCN 1994). Numbers of endemic plant species vary between the different mountain blocks, as follows: Mount Elgon (sixteen species), Aberdare Mountains (six species), Mount Kenya (nine species), Mount Kilimanjaro (six species), and Mount Meru (two species) (WWF and IUCN 1994). Most of the species are recently evolved on these mountains and may have arrived through long-distance transport processes and then diversified independently on the different mountains.

The richness and endemism in the plants and invertebrates are not mirrored in vertebrates. The number of species is low, and there are few endemics. Among the avifauna, there are no strict endemic species. However, there are species that exhibit biogeographic distributions similar to those of islands (Dowsett 1986). For example, Hunter's cisticola (*Cisticola hunteri*) is found between 1,500 and 4,500 m on all mountaintops in this ecoregion (Stattersfield et al. 1998). Two other species, Jackson's francolin (*Pternistis jacksoni*) and Sharpe's pipit (*Macronyx sharpei*, EN), occur on the northern three mountains but not on Mount Meru and Mount Kilimanjaro to the south. Finally, the Aberdare cisticola (*Cisticola aberdare*, EN) is found only in the Aberdare and in the neighboring Mau ranges (East African Montane Forests [18] ecoregion).

Most of the mammals are opportunist visitors. However, there are also three endemic small mammals: the King mole rat (*Tachyoryctes rex*, EN), Emdi mole rat (*T. spalacinus*), and highland shrew (*Crocidura allex*, VU).

The reptiles and amphibians are adapted to the cold and unpredictable climate. There are two strictly endemic amphibians (*Phrynobatrachus keniensis* and *P. kinangopensis*) and near-endemics such as the high montane specialists *Xenopus borealis* and *Hyperolius montanus*. There are also a number of strict and near-endemic reptiles, such as *Adolfus alleni*, *Montatheris hindii*, *Chamaeleo schubotzi*, and *C. sternfeldi*. The alpine meadow mabuya (*Mabuya irregularis*), in particular, is common in the alpine grasslands.

Status and Threats

Current Status

This ecoregion is naturally scattered across the few suitable high mountaintops in East Africa. Much of the ecoregion is found in national parks, such as Mount Kenya National Park, Mount Elgon National Park, and Aberdare National Park in Kenya and Arusha National Park (which includes Mount Meru) and Kilimanjaro National Park in Tanzania. The Ugandan side of Mount Elgon is also a national park, as is Mount Kadam Forest Reserve. Because the ecoregion is found at very high altitudes that are not suitable for farming and is located mainly in protected areas, little habitat has been lost. However, fire has become a serious problem, and the frequent burning of the heathlands is changing the vegetation structure, causing the loss of woody elements and probably changing species composition. Fire control is regarded as one of the most important management challenges on Kilimanjaro (Ngoile et al. 2001).

Types and Severity of Threats

This ecoregion is largely protected and is not threatened by habitat conversion to agriculture. Fire has become a huge problem, with large areas of the heathland burning each year. The incidence of fire has been linked to dramatic increases in the numbers of tourists walking in these mountains, as have other problems such as littering, erosion, and the cutting of woody vegetation to make fires (Ngoile et al. 2001). On Mount Kenya rodent numbers have increased considerably along climbing routes and huts. Because climbing is a seasonal activity, during the nontourist season the rodents switch from eating tourist litter to giant groundsel (*Dendrosenecio*), which is beginning to degrade these endemic plant species. On the Kenyan side of Mount Elgon, outside the national park, human habitation occurs into the ericaceous belt; residents and cattle regularly traverse the alpine belt, and fires in the caldera are not uncommon (Hedberg 1994).

Justification for Ecoregion Delineation

This ecoregion follows the "altimontane vegetation unit" mapped by White (1983). The boundary approximates to the 3,000 m contour, which is also the boundary for Afroalpine vegetation. The ecoregion is separated from similar ones in the Albertine Rift and in Ethiopia by differing floral composition, climate, and endemic plant and animal species composition.

Ecoregion Number: **73**

Ecoregion Name: **Rwenzori-Virunga Montane Moorlands**

Bioregion: **Eastern and Southern Africa**

Biome: **Montane Grasslands and Shrublands**

Political Units: **Democratic Republic of the Congo, Uganda, Rwanda**

Ecoregion Size: **2,700 km^2**

Biological Distinctiveness: **Globally Outstanding**

Conservation Status: **Endangered**

Conservation Assessment: **I**

Author: *Jan Schipper*

Reviewed by: *Derek Pomeroy, Marc Languy, Tim Davenport*

Location and General Description

The Rwenzori-Virunga Montane Moorlands [73] occupy the high-elevation (above 3,000 m) portions of the Rwenzori and Virunga mountains. The Rwenzori range is located along the borders of southwestern Uganda and the Democratic Republic of the Congo (DRC), and the Virungas are at the borders of Rwanda, DRC, and Uganda. Both areas grade at lower elevations into the montane forests of the Albertine Rift Montane Forests [17] ecoregion.

Topographically, this is an extremely rugged area. The Virunga Mountains contain active volcanoes with pronounced topographic relief. These volcanoes rise up to 4,300 m at their highest point. The Rwenzori Mountains differ from the other high mountains in East Africa because they are made up of Precambrian basement rocks uplifted in the middle Tertiary and are not volcanic. The Rwenzori range—the fabled Mountains of the Moon—extends up to 5,108 m at the Margherita summit, which is Africa's third tallest peak, after Mount Kilimanjaro and Mount Kenya (Yeoman 1989). The highest parts of the Rwenzoris contain snowfields and glaciers, although these are retreating (Osmaston and Kaser 2001; Osmaston et al. 1998).

The extreme climate of this ecoregion is governed by its close proximity to the equator and the high altitude. Most days the nighttime temperatures dip below freezing and then rise above freezing during the daytime (Hedberg 1964; Rundel 1994). However, the temperature fluctuations are not as severe here as in the other altimontane ecoregions in eastern Africa, largely because of the more frequent cloud cover (Hedberg 1997). The seasonality of the precipitation is similar to that in the Albertine Rift Montane Forests [17], and although the quantity of precipitation declines at higher altitudes, it is still wetter here than in the other East African high mountain ecoregions (Hedberg 1997). The moist climate is indicated by the presence of moss layers up to 30 cm thick on branches of ericaceous vegetation between 2,900 and 3,800 m on the Rwenzori Mountains (Killick 1979). Lichens also festoon the vegetation on the Virunga Mountains, again indicating high rainfall and cloudborne moisture.

This ecoregion forms part of the Afroalpine archipelago-like region of extreme floristic impoverishment (White 1983). Others regard the flora of high peaks of tropical East Africa (Rwenzori, Virunga, Mount Elgon, the Aberdare Mountains, Mount Kenya, Mount Kilimanjaro, and Mount Meru) as sufficiently distinct to justify their recognition as a separate Afroalpine phytogeographic region (e.g., Hedberg 1961, 1994; Knox and Palmer 1998).

The vegetation belts of these mountains and others in eastern Africa have been mapped in detail (Hedberg 1951). Although not as high as Mount Kilimanjaro or Mount Kenya, the Rwenzori Mountains support a significantly larger altimontane area. Seven main vegetation formations are recognized: ericaceous woodland and wooded grassland with *Philippia* and *Erica arborea*, *Dendrosenecio* woodland and wooded grassland, tussock grassland, *Helichrysum* scrub, and swamp or mire vegetation. *Hagenia-Hypericum* woodland occurs on the more humid slopes in the south and west of Volcanoes National Park (Rwanda) between 2,600 and 3,600 m elevation. Lind and Morrison (1974) mapped the complex distribution of the vegetation belts on the Virunga and Rwenzori Mountains.

Outstanding or Distinctive Biodiversity Features

This ecoregion supports high numbers of endemic plant species, predominantly at the highest altitudes, and several endemic and near-endemic animal species, concentrated at lower elevations.

Of the 278 woody plant taxa so far recorded from the altimontane zone of Africa, 81 percent are endemic to East Africa (Hedberg 1961; Vuilleumier and Monasterio 1986). Most impressive are the giant heathers, groundsels, ericas, and lobelias (Butynski and Kalina 1993), which are adapted to harsh climatic conditions (Hedberg and Hedberg 1979; Mabberley 1986; Smith and Young 1987). Fourteen plant species are strictly endemic to the Rwenzoris and five to the Virungas in the Afromontane and altimontane zones.

This ecoregion supports endemic and near-endemic bird species (Dowsett 1986; Bober et al. 2001; Dehn and Christiansen 2001). One species, Stuhlmann's double-collared sunbird (*Nectarinia stuhlmanni*), is confined to high (2,600–3,500 m) elevations in the Rwenzori Mountains (Bober et al. 2001). The majority of the near-endemic species use several mountains along the Albertine Rift chain and range over a variety of elevations, most spending at least part of their time in the lower elevation Albertine Rift Montane Forests [17]. These birds include Rockefeller's sunbird (*Nectarinia rockefelleri*, VU), Archer's robin-chat (*Cossypha archeri*), handsome francolin (*Pternistis nobilis*), and stripe-breasted tit (*Parus fasciiventer*). Other restricted-range montane birds include the Dusky crimson-wing (*Cryptospiza jacksoni*), Shelley's crimson-wing (*C. shelleyi,* VU), and strange weaver (*Ploceus alienus*).

Larger mammals tend to be visitors and are often opportunists or generalists. The exception to this is the Ruwenzori black-fronted duiker (*Cephalophus rubidus*), whose distribution is confined to the Rwenzori Mountains and is shared between this ecoregion and the lower elevation Albertine Rift Montane Forests [17]. The mountain gorilla (*Gorilla beringei beringei,* CR) occasionally uses the border habitat with the moorlands and the montane forests. Larger mammals that regularly move between lowland or montane forests and this region include elephant (*Loxodonta africana,* EN), buffalo (*Syncerus caffer*) (Young 1996), leopard (*Panthera pardus*), African golden cat (*Profelis aurata,* VU), and side-striped jackal (*Canis adustus*) (Davenport 1996). The golden monkey (*Cercopithecus mitis kandti*), which is possibly a separate restricted-range species, is sometimes found above 3,000 m (D. Twinomugisha, pers. comm., 2002). The crescent shrew (*Sylvisorex lunaris*) is also near endemic.

The reptiles and amphibians in this high-altitude environment are far less diverse than in lower-elevation ecoregions. There is one endemic amphibian, the Karissimbi Forest tree frog (*Leptopelis karissimbensis*) (Schiøtz 1999). Near-endemic reptiles include Rwanda five-toed skink (*Leptosiaphos graueri*) and the coarse chameleon (*Chamaeleo rudis*).

Status and Threats

Current Status

The ecoregion is naturally disjunct because of its position on two high-altitude portions of the Albertine Rift Mountains. However, because the habitats are at extremely high altitudes there has been little fragmentation or habitat degradation from human use.

The entire ecoregion is protected in several national parks: Mgahinga Gorilla (33.7 km²) and Rwenzori Mountains National Parks (996 km²) in Uganda, Volcanoes National Park (160 km²) in Rwanda, and Virunga National Park in DRC (240 km²). These areas also hold additional status as either World Heritage sites or biosphere reserves. Tourism remains at fairly low levels, partly because of the past insecurity of the area. However, since reopening in 2001, the Rwenzori National Park is again attracting tourists.

Type and Severity of Threats

Few people live in this ecoregion because it falls in national parks, although population densities are increasing outside the parks. Poachers, honey collectors, and nontimber forest product gatherers use the high mountain forests, and tourists visit the areas to hike and to see the mountain gorillas.

Although currently largely intact, the habitats of the ecoregion are potentially threatened along border zones that are periodically occupied by various armed rebel and bandit groups. This makes the management of the protected areas and the gorilla populations problematic and often dangerous. The large mammal fauna, including the mountain gorilla, is also threatened by hunting for meat and for use as fetishes by the local population. Current problems in Volcanoes National Park include encroachment, illegal wood and bamboo cutting, feral dogs, the absence of a buffer zone, and the lack of technical and administrative staff. Mgahinga National Park is threatened by increased agricultural and pastoral activities on its borders, illegal hunting with snares, uncontrolled fire, reduced occult precipitation, and the invasion of exotic plant species such as black wattle (*Acacia mearnsii*) (Davenport 1996). The vegetation of the Rwenzori area could also be threatened by the development of tourism.

Justification for Ecoregion Delineation

The delineation of the two parts of this ecoregion follows the 3,000-m contour and the boundaries of the "altimontane vegetation unit" of White (1983). Although small in size, the distinctiveness of the vegetation justified the distinction of this area from the surrounding montane vegetation. The lower daily temperature range, higher moisture levels, more frequent cloud cover, and different species composition including strict endemics allowed the separation of this ecoregion from the other altimontane areas in eastern Africa.

Ecoregion Number:	**74**
Ecoregion Name:	**Southern Rift Montane Forest-Grassland Mosaic**
Bioregion:	**Eastern and Southern Africa**
Biome:	**Montane Grasslands and Shrublands**
Political Units:	**Malawi, Mozambique, Tanzania, Zambia**
Ecoregion Size:	**33,500 km²**
Biological Distinctiveness:	**Globally Outstanding**
Conservation Status:	**Endangered**
Conservation Assessment:	**I**
Author:	*Lyndon D. Estes*
Reviewed by:	*Tim Davenport*

Location and General Description

The Southern Rift Montane Forest-Grassland Mosaic [74] consists of several discontinuous mountain chains and plateaus around the western and northern shores of Lake Malawi/Nyasa and extends to the eastern ranges of Lake Tanganyika. The northwestern extent is the Ufipa Plateau in Tanzania, between lakes Tanganyika and Rukwa, and the Tanzanian Southern Highlands are made up of several different ranges, including the Mbeya and Kipengere ranges, the Umalila Highlands, the Poroto and Livingstone mountains, and the Kitulo Plateau. In Malawi the ecoregion extends from the Misuku Hills in the north to the Kirk Range, straddling the border between Malawi and Mozambique. In between are the extensive Nyika and North and South Viphya plateaus and a number of smaller mountains. The Nyika Plateau extends into Zambia.

The underlying geology of the ecoregion is Precambrian granite, gneiss, and schist (Cribb and Leedal 1982; Kerfoot 1964a; Chapman and White 1970). Younger sedimentary rocks overlay these ancient rocks in some places, and Mount Rungwe is an ancient volcano. The ecoregion's soils are primarily well-drained andosols and ferrisols (McKone and Walzem 1994; Chapman and White 1970), and its topography is characterized by large plateaus surrounding high peaks and ridges, bounded on all sides by escarpments or deeply dissected hill country (Chapman and White 1970; Cribb and Leedal 1982). Plateau altitudes range from 1,400 to 2,400 m, with the two highest peaks in Tanzania (Rungwe and Mtorwi) attaining 2,961 m.

The climate pattern is dictated largely by Lake Nyasa, which provides moisture to the highlands (Lovett 1993a; Chapman and White 1970). Average annual rainfall ranges from 823 mm on the Ufipa Plateau to 2,850 mm in the Livingstone and Poroto Mountains (Cribb and Leedal 1982), with the mean rainfall around 1,500 mm (Dowsett-Lemaire 1989a; Chapman and White 1970). Precipitation is largely confined to the wet season, November through April (Cribb and Leedal 1982; Dowsett-Lemaire 1989a), although light rains or mist may occur at the higher altitudes during the dry season between May and August (Kerfoot 1964a; Dowsett-Lemaire 1989a). Mean annual temperatures are between 13°C and 19°C (Dowsett-Lemaire 1989a), with an average maximum of 22°C and an average minimum of 9.8°C (McKone and Walzem 1994). At the highest altitudes, temperatures as low as −7°C have been recorded, and frosts are common (Cribb and Leedal 1982).

This ecoregion is composed of several distinct vegetation communities. The most dominant is grassland (White 1983), commonly attributed to the high frequency and extent of fire (White 1983; Dowsett-Lemaire 1989a). Dominant grass species are *Loudetia simplex, Exotheca abyssinica, Monocymbium ceresiiforme, Themeda triandra, Andropogon* spp., *Pennisetum* spp., and *Setaria* spp. (White 1983; Chapman and White 1970; Cribb and Leedal 1982). A number of herbs, sedges, and geophytes also occur in the grassland community (Kerfoot 1964a), as does the occasional fire-resistant shrub, usually of the genus *Protea* (White 1983). In areas of impeded drainage, permanent and seasonal bogs known as *dambos* may be found. These habitats, dominated by grasses and sedges, have rich orchid floras (Cribb and Leedal 1982; Kerfoot 1964a).

Several other vegetation types are set in the grassland matrix, the most prominent of which is Afromontane forest, although this constitutes less than 5 percent of the landscape and is confined to fire-sheltered pockets, moist escarpments, valleys, and watercourses (Chapman and White 1970; Kerfoot 1964a; Dowsett-Lemaire 1989a). Dominant tree and shrub species are *Apodytes dimidiata, Bersama abyssinica, Chrysophyllum gorungosanum, Entandophragma excelsum, Ficalhoa laurifolia, Garcinia* spp., *Ilex mitis, Kiggelaria africana, Myrianthus holstii, Ocotea usambarensis, Parinari excelsa, Podocarpus latifolius, Polyscias fulva, Rapanea melanophloeos,* and *Syzygium guineense* (Chapman and White 1970; Dowsett-Lemaire 1989a; Kerfoot 1964a; White 1983). The most common constituents of the herbaceous layer are species of the family Acanthaceae and of the genera *Impatiens, Begonia, Streptocarpus, Plectranthus,* and *Peperomia* (White 1983; Dowsett-Lemaire 1989a; Kerfoot 1964a). An ericaceous zone also occurs on some of these mountains (Hedberg 1951), generally above the forest zone, although examples may be found at lower elevations in areas of shallow soil and frequent mists (White 1983).

At lower altitudes, the vegetation grades into miombo woodland dominated by *Brachystegia, Julbernardia,* and *Isoberlinia.* These ascend as high as 2,050 m on the xeric western escarpment of the Nyika Plateau (Dowsett-Lemaire 1989a) and up to 2,100 m on the western slopes of the Livingstone Mountains.

Outstanding or Distinctive Biodiversity Features

Levels of endemism in the flora and fauna of the Southern Rift are much lower than those in the Eastern Arc Mountains to the northeast, probably because the latter has been much more climatically stable over millions of years and derives its rainfall from the Indian Ocean (Lovett 1993a). However, the Southern Rift is particularly rich in orchids and contains six endemic *Protea* species (Beard 1992). The Nyika Plateau is home to south-central Africa's richest orchid flora, totalling 214 species (Kurzweil 2000). Four orchid species and two subspecies are endemic to this area, with four near-endemics also found on nearby highlands (Kurzweil 2000). The Nyika Plateau also holds thirteen other endemic plant species and seven endemic subspecies (Willis et al. 2000). The Kitulo Plateau in the Southern Highlands of Tanzania supports 350 plant species, 3 of which are strictly endemic, and 15 are restricted to the Kitulo Plateau and other Southern Highlands mountain ranges (Lovett and Prins 1994). The Ufipa Plateau also hosts a rich assemblage of plants (Carter 1987; T. Davenport, pers. comm., 2002). Many northern plant taxa reach their southernmost distributions in this ecoregion, and many southern plants reach their northernmost limit at Mount Mulanje just to the south (Williamson 1979; Chapman and White 1970; Kerfoot 1964b).

The fauna contains a number of strict and near-endemic species. Near-endemic mammals include Abbott's duiker (*Cephalophus spadix*, VU), the desperate shrew (*Crocidura desperata*, CR), and three small rodents: greater long-tailed pouched rat (*Beamys major*), black and red bush squirrel (*Paraxerus lucifer*) (two endemic races), and Swynnerton's squirrel (*Paraxerus vexillarius*, VU).

Near-endemic birds include the churring cisticola (*Cisticola njombe*), buff-shouldered widowbird (*Euplectes psammocromius*), Chapin's apalis (*Apalis chapini*), black-lored cisticola (*Cisticola nigriloris*), Uhehe fiscal (*Lanius marwitzi*), and Iringa akalat (*Sheppardia lowei*, VU) (Fishpool and Evans 2001; Stattersfield et al. 1998). The declining wattled crane (*Grus carunculatus*, VU) has an important breeding site in the Nyika National Park and is also found in the Southern Highlands of Tanzania and on the Ufipa Plateau. The Nyika National Park also supports the world's largest breeding population of blue swallow (*Hirundo atrocaerulea*, VU) and the locally rare Stanley bustard (*Neotis denhami*) (Dowsett-Lemaire et al. 2001). Blue swallows also breed on Kitulo and Mbeya range (T. Davenport, pers. comm., 2002).

Among the herpetofauna are a number of strict and near-endemic amphibians: Stewart's river frog (*Phrynobatrachus stewartae*), Rungwe river frog (*Phrynobatrachus rungwensis*), and Usambara big-fingered frog (*Probreviceps macrodactylus*). Strict and near-endemic reptiles include Ngosi volcano chameleon (*Chamaeleo fuelleborni*), Poroto Mountain chameleon (*C. incornutus*), *Cordylus nyikae*, Ukinga spinytail lizard (*C. ukingensis*), *Eumecia johnstoni*, Braun's mabuya (*Mabuya brauni*), and the pitless pygmy chameleon (*Rhampholeon nchisiensis*).

Examples of endemic invertebrates are the dragonfly *Teinobasis malawiensis*, which is known only from montane streams in northern Malawi (Stuart et al. 1990), the Kitulo-endemic satyrid butterfly, *Neocoenyra petersi* (Kielland 1990), *Papilio ufipa* (Ufipa endemic), *Neocoenyra mittoni* (Mbeya range endemic), *Harpendyreus boma*, and *Alaena bicolora* (Tanzania Southern Highlands endemics). Other endemic invertebrates include the millipede *Iringius rungwe* (Rungwe) and arachnids *Bursellia paghi* (Tanzania Southern Highlands endemic) and *Callitrichia pileata* (Rungwe).

Status and Threats

Current Status

The archipelago-like nature of the Southern Rift Highlands means that its component habitat blocks are naturally isolated from one another by topography. However, human intervention has caused further fragmentation and degradation in the habitat islands. Cultivation is widespread and is increasing rapidly (Cribb and Leedal 1982; McKone 1995). Most of the Mbeya range below 2,200 m has been subjected to shifting cultivation for decades (Kerfoot 1964a). In the Rungwe District of the Mbeya Region, remaining natural vegetation is confined largely to government-proclaimed and traditional forest reserves, but even these conserved areas are subject to anthropogenic disturbances (McKone and Walzem 1994; McKone 1995). An analysis of satellite images spanning 20 years up to 1989 has shown that at least

24 percent of the Kitulo Plateau had been transformed by cultivation and pasture (Lovett and Prins 1994). The majority of grassland on Malawi's second largest plateau, the South Viphya, has been planted with exotic *Pinus* spp. (Dowsett-Lemaire 1989a), and other areas of the ecoregion have been similarly afforested (Dowsett-Lemaire 1989a; McKone and Walzem 1994). In Malawi this deforestation is particularly pronounced: all that remains of the once extensive midaltitude montane forests are small relictual groves used as graveyards by local people (Dowsett-Lemaire 1989a).

The protected area network throughout most of the ecoregion is inadequate, with the exception of the Nyika Plateau area, the majority of which is protected inside Zambia and Malawi's contiguous Nyika National Parks. Part of Chipata Mountain is protected in Malawi's Nkhotakhota Game Reserve (Carter 1987), and Chirobwe Mountain in the Dedza-Chirobwe Highlands has a forest reserve, although this is under pressure from wood collectors (Dowsett-Lemaire 1989a). The Government of Tanzania announced in February 2002 that 13,500 ha of Kitulo Plateau will be gazetted as the country's next national park, increasing protection for the rare orchid flora (Davenport 2002b). Mbeya Region of Tanzania's Southern Highlands contains seventeen forest reserves that fall inside the ecoregion's boundaries. A number of other forest reserves lie in the administrative regions of Rukwa and Iringa. Most of these have low levels of management and are often subject to illegal pitsawing, fuelwood collection, grazing, agricultural encroachment, hunting, and uncontrolled burning. Besides these official forest reserves, there are numerous smaller traditional forest reserves in the Southern Highlands, established by local communities for a variety of cultural reasons (McKone 1995). At least ninety-four are known from the Rungwe district (McKone 1995).

Types and Severity of Threats

Each year, large fires of primarily anthropogenic origin sweep the Southern Rift Highlands (Chapman and White 1970; Cribb and Leedal 1982; Lovett and Prins 1994; Kerfoot 1964a). This burning regime is believed to have been the main cause of the replacement of Afromontane forests with grassland and scrub grassland (Dowsett-Lemaire 1989a; Kerfoot 1964a; Chapman and White 1970).

Land is increasingly being converted to crops such as tea, coffee, banana, finger millet, maize, beans, potatoes, and pyrethrum (Chapman and White 1970; Lovett and Prins 1994). In areas of the Kitulo Plateau where the soil has been disturbed, grassland tends to be replaced by a "shrubby sward" (Lovett and Prins 1994).

The trade in orchid tubers for consumption in Zambia is seriously threatening as many as eighty-five terrestrial orchid species across the Southern Highlands (Davenport 2002a; Davenport and Ndangaklasi 2003). More than 2 million plants from three genera (*Disa, Habenaria,* and *Satyrium*) are being harvested in Tanzania each year.

Exotic timber trees of *Pinus* and *Eucalyptus* are invading the grasslands of this ecoregion (McKone and Walzem 1994). A bramble *Rubus* sp. has spread extensively throughout the Nyika National Park.

Justification for Ecoregion Delineation

The boundaries of this ecoregion follow the "undifferentiated montane vegetation" unit in the Afromontane archipelago-like regional center of endemism (White 1983), with one refinement to include the Ufipa Plateau to the northwest. This ecoregion derives its climate from Lake Malawi/Nyasa and lacks the exceptional endemism of the Eastern Arc Mountains to the northeast. It also differs significantly from the forest-grassland mosaics of the Mount Mulanje area to the south, where species of the southern flora have their northern limits.

Ecoregion Number:	**75**
Ecoregion Name:	**South Malawi Montane Forest-Grassland Mosaic**
Bioregion:	**Eastern and Southern Africa**
Biome:	**Montane Grasslands and Shrublands**
Political Units:	**Malawi**
Ecoregion Size:	**10,200 km²**
Biological Distinctiveness:	**Globally Outstanding**
Conservation Status:	**Critical**
Conservation Assessment:	**I**
Author:	*Amy Spriggs*
Reviewed by:	*Françoise Dowsett-Lemaire, Robert Dowsett*

Location and General Description

The South Malawi Montane Forest-Grassland Mosaic [75] is situated at the southern end of Malawi about 100 km south of Lake Malawi. The ecoregion is made up of Mount Mulanje and some lower-altitude mountains to the west and northwest. Mount Mulanje rises sharply from the surrounding Phalombe Plain (Dowsett-Lemaire 1988), covering an area of 650 km², with high plateaus about 2,000 m above sea level, incised by several deep ravines. The plateaus are surmounted by twenty rocky peaks, which generally reach about 2,500 m in altitude. One of these, Sapitwa Peak (3,002 m), is the highest point in South-Central Africa. To the west and northwest of Mount Mulanje, across the Tuchila Plain, lie the Shire Highlands, dominated by Thyolo Mountain (1,462 m). Michuru and Chiradzulu Mountains lie

further north in the vicinity of Blantyre and Limbe. At the northern end of the Shire Highlands lies the extensive Zomba Plateau. It contains many peaks, such as Chagwa, Nawimbe, and Mulunguzi, ranging between 1,761 and 2,018 m in elevation. The northern half of Zomba Plateau is often called Malosa Mountain (2,077 m). The Liwonde Hills (including Chikala, 1,626 m) comprise a chain of four round hills immediately north of the Shire Highlands.

Mount Mulanje consists of a cluster of plutonic intrusions of syenite, quartz-syenite, and granite, which are uplifted and faulted (Chapman 1994). The soils of Mount Mulanje are classified as humic ferrisols (Chapman and White 1970). Information on the geology and soils of other smaller mountains is scarce.

The ecoregion experiences a single, austral summer rainy season, extending from November to April (Chapman 1994). The remainder of the year is drier, but maritime air from the Mozambique Channel brings occasional mist and rain to southeast-facing slopes (called Chiperone weather). The quantity of rainfall varies throughout the ecoregion, particularly on Mount Mulanje, largely because of rainshadow effects. The average annual rainfall ranges from 1,600 mm at the foot of Mount Mulanje to 2,800 mm at higher elevations. Although rainfall is much lower on the western side of the mountain, mists are prevalent at higher elevations year-round. At moderate altitudes average maximum temperatures are 24°C in summer and 12°C in winter, and minimum temperatures are 15°C and 9°C in summer and winter, respectively. At higher altitudes temperatures drop to −3°C in winter, and frosts are common (Chapman and White 1970).

The vegetation of Mount Mulanje has been studied in greater detail than the other mountains (Brenan 1953–54; Chapman 1962; Dowsett-Lemaire 1988, 1990). Five indigenous vegetation types occur on Mulanje, namely miombo woodland, lowland and midaltitude forest, Afromontane-*Widdringtonia* (endemic Malawi cedar) forest, plateau grassland, and high-altitude vegetation (Chapman 1962).

Brachystegia or transition woodland occurs in a band along the foothills of Mount Mulanje. Until recently, the southern and southeastern slopes of Mulanje were vegetated with lowland rainforest (~600 to 950 m), dominated by *Newtonia buchananii* and *Khaya anthotheca* (Chapman 1994). Remnants of this lowland rainforest occur along streams on the lower slopes and on the tea estates at the foot of the mountain.

Midelevation forest is found between 950 and 1,500 m. Dominant canopy trees are *Newtonia buchananii,* associated with *Albizia adianthifolia, Funtumia africana,* and *Chrysophyllum gorungosanum* (Dowsett-Lemaire 1988). Most of this forest has been destroyed, and the largest intact patch is in the Great Ruo Gorge. Other important midaltitude forests are found on Soche Mountain and Chikala Hill (Dowsett-Lemaire et al. 2001) and on Thyolo Mountain.

At higher altitudes on Mount Mulanje (over 1,500 m)

Afromontane-*Widdringtonia* forests are found in gorges and ravines and on some of the lower plateaus (particularly Lichenya), where they are sheltered from dry winds and fire. *Olea capensis* and the Mulanje cedar (*Widdringtonia whytei*) are the most common emergent trees. Mount Mulanje also represents the northern limit of *Widdringtonia nodiflora*, a species widely distributed throughout southern Africa (White et al. 2001). Currently, the cedar forests occur as small fragments on the mountain, covering about 15 km² (Pauw 1998). Other important canopy trees present at high elevations (1,800–2,000 m) include *Podocarpus latifolius, Ekebergia capensis, Cassipourea malosana,* and *Rapanea melanophloeos*.

The plateaus of Mulanje are dominated by tussock grasslands. Small groups of trees include *Ilex mitis, Philippia benguelensis,* and *Syzygium cordatum*. At the highest altitudes the vegetation becomes a heathland similar to that on the Cape Mountains of South Africa. Species include *Phylica tropica, Aloe arborescens,* and the endemic *Erica milanjiana*. Grasses include large tussocks of *Festuca costata* and *Danthonia davyi* interspersed with cushions of *Eragrostis volkensii* and the grass *Alloeochaete oreogena,* a Mulanje endemic that grows up to 3 m tall with a tree trunk–like structure.

Outstanding or Distinctive Biodiversity Features

The ecoregion forms part of the Afromontane archipelago-like regional center of endemism of White (1983). It has a low level of specific endemism when compared with other Afromontane forest islands in the region, but it has a high level of species richness (WWF and IUCN 1994). However, a few species of trees and large shrubs appear to be endemic to Mount Mulanje, such as *Widdringtonia whytei, Rawsonia burtt-davyi, Ficus modesta,* and *Encephalartos gratus* (White et al. 2001).

The highest rate of faunal endemism is found in the reptiles and amphibians. Strict endemics include Mulanje mountain chameleon (*Bradypodion mlanjense*), Malawi stumptail chameleon (*Rhampholeon platyceps*), king dwarf gecko (*Lygodactylus rex*), and Mitchell's flat lizard (*Platysaurus mitchelli*). Among amphibians, Tshiromo frog (*Afrana johnstoni*), Ruo River screeching frog (*Arthroleptis francei*), and Broadley's grassland frog (*Ptychadena broadleyi*) are strictly endemic. Tshiromo frog and the Ruo River screeching frog are restricted to Mount Mulanje (there is a possible record of *A. francei* from Zomba), and Broadley's grassland frog is found on Mulanje and on the Zomba Plateau (van Strien 1989; Broadley 2001).

All of these forested mountains are important for birds (Dowsett-Lemaire 1989b; Dowsett-Lemaire et al. 2001; Stattersfield et al. 1998; Collar and Stuart 1988). The endangered Thyolo alethe (*Alethe choloensis*, EN) is near endemic to the forests, mainly at medium altitudes. Other threatened forest birds are the spotted ground thrush (*Zoothera guttata*, EN) and white-winged apalis (*Apalis chariessa*, VU) (Dowsett-Lemaire et al. 2001). Additional notable species are the olive-flanked robin-

chat (*Cossypha anomala anomala*), bar-throated apalis (*Apalis thoracica flavigularis*), green-headed oriole (*Oriolus chlorocephalus*), and moustached green tinker bird (*Pogoniulus leucomystax*).

Herds of large mammals, such as eland (*Taurotragus oryx*) and sable (*Hippotragus niger*), once roamed the foothills but are long gone. The only antelopes remaining are bushbuck (*Tragelaphus scriptus*), red duiker (*Cephalophus natalensis*), and blue duiker (*Cephalophus monticola*), which live in dense vegetation. Klipspringer (*Oreotragus oreotragus*) has also survived because it occupies inaccessible high, rocky slopes. Yellow baboon (*Papio hamadryas cynocephalus*) is common in the woodlands, and the blue monkey (*Cercopithecus mitis*) and vervet monkey (*Chlorocebus aethiops*) occur along forest edge (Ansell and Dowsett 1988).

Status and Threats

Current Status

Much of the ecoregion is protected in government forest reserves, but the current conservation status is regarded as poor (Sakai 1989; Chapman 1990; Chapman 1994; Dowsett-Lemaire et al. 2001). Some of these reserves have been converted to pine and eucalyptus plantations, with fragments of indigenous vegetation remaining in more inaccessible areas. Thyolo Mountain Forest Reserve used to support midaltitude forest dominated by fig (*Ficus* spp.) trees (Dowsett-Lemaire 1989a). Illegal deforestation by agriculturalists accelerated in the late 1990s, destroying more than 80 percent of the forest (Dowsett-Lemaire et al. 2001). The lower-lying land around Thyolo Mountain was almost entirely cleared for tea in the early twentieth century, but several estates conserved important patches of lowland rainforest (~1,050 m), including *Albizia gummifera, Ficus vallis-choudae, Khaya anthotheca, Macaranga capensis,* and *Trilepisium madagascariense* (Dowsett-Lemaire 1990). The midaltitude forest in the Lisau Saddle at Chiradzulu (an important site for the spotted ground thrush and white-winged apalis) was almost totally destroyed in the 1990s by the Forestry Department, which replaced the forest with *Eucalyptus* plantations. Zomba Mountain has also been degraded by *Pinus* plantations, although small patches of midaltitude and montane forest remain. The adjacent Malosa Mountain is in much better condition because the reserve was never altered by plantations.

At Mulanje, the situation is critical. All of the southeastern slopes of the mountain, once covered with the most extensive midaltitude forest in the country (~40 km²), have been deforested in recent decades, mainly by tea estate workers short of land. The drier western and northern slopes of the mountain have been less affected, largely because of the absence of tea plantations. Plants such as bamboo, thatching grass, and *Raphia* palms are also harvested from the lower slopes. The high plateaus suffer from uncontrolled fires and illegal extraction of timber (especially Mulanje cedar).

Types and Severity of Threats

The most serious threat is the acute shortage of land caused by large tea plantations in the foothills between Mulanje and Thyolo Mountains. Tea plantation workers are obliged to grow their annual maize crops in forest reserves, thus deforesting the lower slopes of Mulanje and Thyolo Mountain Forest Reserve. Forestry is another major threat, with large areas converted to pine and eucalyptus plantations. Nearly all of the indigenous vegetation on the Zomba Plateau has been supplanted by Mexican pine (*Pinus patula*) (Verboom 1992). The Mulanje cedar (*Widdringtonia whytei*) is under serious threat from overexploitation (Chapman 1994). Other serious threats include the uncontrolled invasion by the exotic Himalayan raspberry (*Rubus ellipticus*) and Mexican pine (Verboom 1992), the hunting of large mammals, and uncontrolled fires. Another potential threat is the exploitation of bauxite (Chapman 1994).

Justification for Ecoregion Delineation

The ecoregion delineation follows the vegetation map of White (1983) that outlines Mulanje, Thyolo, Shire, and Chiradzulu mountains in the larger "undifferentiated montane vegetation" and "altimontane vegetation" units. The ecoregion ranges from about 600 m to more than 3,000 m altitude to capture the transitions between lower- and higher-elevation fauna. The ecoregion is centered on Mount Mulanje, with a number of endemic species shared with other mountains in this ecoregion. It is distinguished as a center of plant diversity (WWF and IUCN 1994) and as part of a larger endemic bird area (Stattersfield et al. 1998).

Ecoregion Number:	**76**
Ecoregion Name:	**Eastern Zimbabwe Montane Forest-Grassland Mosaic**
Bioregion:	**Eastern and Southern Africa**
Biome:	**Montane Grasslands and Shrublands**
Political Units:	**Zimbabwe, Mozambique**
Ecoregion Size:	**7,800 km²**
Biological Distinctiveness:	**Globally Outstanding**
Conservation status:	**Endangered**
Conservation Assessment:	**I**
Author:	*Amy Spriggs*
Reviewed by:	*Jonathan Timberlake*

Location and General Description

The Eastern Zimbabwe Montane Forest-Grassland Mosaic [76] covers the mountains that extend along the border between Zimbabwe and Mozambique. From north to south, these are the Rukotso Plateau, Catandica Mountains, Nyanga Mountains, Bvumba Highlands, Himalaya Mountains, Mount Tsetserra, Chimanimani Mountains, and Chipinge Highlands, the Chirinda Forest on Mount Selinda, and Mount Gorongosa in Mozambique.

The northern highland areas have deeply cut valleys, heavily affected by east-west faulting, with flat-topped or rounded hills and steep slopes along the eastern escarpment. The Chimanimani Mountains further south are more rugged, with jagged peaks and deep gorges (Childes and Mundy 2001). The entire ecoregion is above 1,000 m in elevation, with a maximum elevation of 2,592 m on the Nyangani Massif in the north and 2,400 m in the Chimimimani Mountains. The isolated Gorongosa Mountain in Mozambique rises to 1,863 m. Most of the underlying geology is mid-Precambrian basement rocks of quartz-mica schists, quartzites, granite, and dolerite.

Habitats comprise montane grassland and ericaceous shrubland, with patches of montane, submontane, medium-altitude, and lowland moist evergreen forests. Differences in vegetation result from variation in altitude, water availability (particularly during the dry season), anthropogenic disturbance, and to a lesser degree soil, aspect, and topography (Müller 1999). The predominant vegetation type is montane grassland. Evergreen rainforests cover a smaller area. The forests are more adapted to seasonally drier conditions than similar forests in the equatorial belt. Annual rainfall in the ecoregion varies from 741 to 3,000 mm per year, largely related to differences in aspect (Müller 1999). Most of the rain falls in the austral summer (November–April), whereas the austral winter (May–July) is drier. Windward slopes of the highlands extract moisture from the air, which condenses into rain, low clouds, or mist. The mists are especially important in allowing forest to persist during the dry season (Childes and Mundy 2001). Annual mean temperatures range from a minimum between 9°C and 12°C to a maximum between 25°C and 28°C (Phipps and Goodier 1962). Night frosts are common on nonsloping terrain during the dry and cool season.

At higher elevations a short, tufted grassland occurs, dominated by *Loudetia simplex*, *Trachypogon spicatus*, *Exotheca abyssinica*, and *Monocymbium ceresiiforme*. At lower elevations *Themeda triandra* is dominant on the more fertile red soils, and *Loudetia simplex* is dominant on the poorer white sands. Fire plays an important ecological role in these grasslands. Above about 1,200 m on the Chimanimani, Nyanga, and Gorongosa Mountains, heathlands are found on poor, acidic soils (Phipps and Goodier 1962; Killick 1979). In the Chimanimani Mountains the dominant species include *Phylica ericoides*, *Passerina montana*, *Erica eylesii*, *E. pleiotricha*, *E. johnstoniana*, and *Protea gazensiss*.

The forests of this ecoregion form part of White's (1983) Afromontane archipelago-like regional center of endemism. Moist evergreen forest occurs on south- and east-facing mountain slopes and in the deep valleys and ravines below 2,100 m elevation (Timberlake and Shaw 1994). Small patches occur on the eastern slopes of the Nyanga Mountains and also on the Chimanamani, Bvumba, Himalaya, and Gorongosa mountains. The southernmost area of forest is Chirinda Forest, on Mount Selinda. Both Chirinda and Gorongosa differ from the other forest areas in that they are of medium altitude. Elevation can be used to distinguish four forest zones: montane (above 1,650 m), submontane (1,350–1,650 m), medium altitude (850–1,350), and lowland (350–850 m) (Müller 1999). Six forest types make up the montane forest zone, occurring primarily on the Nyangani, Nyanga, and Bvumba mountains. Dominant species include *Syzygium masukuense, Afrocrania volkensii, Ilex mitis, Schefflera umbellifera, Maesa lanceolata,* and *Syzygium guineense.* One forest type with limited distribution is *Widdringtonia nodiflora* forest, which is confined to streams in the Nyanga and Chimanimani Mountains in rainshadow areas where soils are well drained. The submontane zone includes species of montane and medium altitudes in four forest types. Dominant species include *Cassipourea malosana, Craibia brevicaudata, Albizia gummifera,* and *A. schimperiana.* The medium-altitude forest zone reflects the gradual transition from wetter to drier vegetation types. Dominant canopy species are *Chrysophyllum gorungosanum, Craibia brevicaudata,* and *Trichilia dregeana.* Little rainforest remains in the lowland forest zone, although small fragments remain in the Pungwe and Rusitu Valleys, outside this ecoregion.

The western slopes of the highlands contain drier forest types, which grade into miombo woodlands. Occurring on well-drained slopes up to an elevation of 1,800 m, these woodlands vary from closed to open and are dominated by *Brachystegia spiciformis, B. utilis, B. tamarindoides,* and *Uapaca kirkiana.* The ground flora consists of grasses such as *Digitaria diagonalis, Loudetia simplex,* and *Themeda triandra,* dicotyledonous herbs, ferns (particularly *Pteridium aquilinum*), and creepers such as *Smilax kraussiana* (Phipps and Goodier 1962).

Outstanding or Distinctive Biodiversity Features

This ecoregion supports high plant species richness as a result of the complex mosaic of different vegetation types (Timberlake and Müller 1994). The highest levels of endemism and richness occur in the montane grasslands and heathlands, where many species have restricted or scattered distributions. For example, the quartzite soils of the Chimanimani contain many endemic species, five of which are endemic *Aloe* species (Mapaura 2002). Restricted-range grassland species in the Nyanga area include *Aloe inyangensis, Moraea inyangani, Erica simii, Euphorbia citrina,* and *Protea inyangensis* (Childes and Mundy 2001). By contrast, the forests have higher species richness but lower levels of endemism because many of their plants are fairly widespread in montane forest areas throughout eastern Africa.

There are at least two endemic mammals in this ecoregion: Arend's golden mole (*Chlorotalpa arendsi*) and the Silinda rock

rat (*Aethomys silindensis*). Characteristic forest species include African palm civet (*Nandinia binotata*), mutable sun squirrel (*Heliosciurus mutabilis*), greater galago (*Otolemur crassicaudatus*), and dark-footed forest shrew (*Myosorex cafer*). Leopards (*Panthera pardus*) are found throughout this ecoregion, moving between the different vegetation types. The more open grasslands and heathlands contain fewer mammal species, with the klipspringer (*Oreotragus oreotragus*), sable antelope (*Hippotragus niger*), eland (*Taurotragus oryx*), and rock hyrax (*Procavia capensis*) being the most conspicuous.

A rich and varied bird life is found, including the strictly endemic Chirinda apalis (*Apalis chirindensis*) and Roberts's prinia (*Prinia robertsi*) and the globally threatened Swynnerton's robin (*Swynnertonia swynnertoni,* VU) (Childes and Mundy 2001). The Chirinda apalis is restricted to the deep forest above 1,500 m, whereas Roberts's prinia is found on the forest margins above 1,350 m (Sinclair and Hockey 1996; Stattersfield et al. 1998). Other notable species include the green-headed oriole (*Oriolus chlorocephalus*) and plain-backed sunbird (*Anthreptes reichenowi*) in forests and the Taita falcon (*Falco fasciinucha*) and blue swallow (*Hirundo atrocaerulea,* VU) in grasslands. In the heathlands of the Chimanimani Mountains, fynbos species occur, such as Gurney's sugarbird (*Promerops gurneyi*) and the malachite sunbird (*Nectarinia famosa*).

This ecoregion contains a number of endemic reptiles and amphibians. Strict endemic amphibians include *Probreviceps rhodesianus, Afrana johnstoni, Arthroleptis troglodytes,* and *Stephopaedes anotis* (Childes and Mundy 2001). Endemic reptiles include the ferocious round-headed worm lizard (*Zygaspis ferox*), Zimbabwe girdled lizard (*Cordylus rhodesianus*), FitzSimons's dwarf gecko (*Lygodactylus bernardi*), and Arnold's skink (*Proscelotes arnoldi*) in montane grasslands and Marshall's dwarf chameleon (*Rhampholeon marshalli*) in montane forests (Branch 1998). The grasslands and heathlands also support the Berg adder (*Bitis atropos*) (Broadley 1994).

Butterflies include at least five endemics confined to the Nyanga grasslands. Restricted-range species in the Chirinda Forest include *Anthene sheppardi* and *Pentila swynnertoni* and the endemic *Mimacrea neokoton* (Childes and Mundy 2001). Other large and attractive species common in the Chirinda Forest are the forest king charaxes (*Charaxes xiphares*), golden-banded forester (*Euphaedra neophron*), and forest green (*Euryphura achlys*) (Gardiner 1994).

Status and Threats

Current Status

The higher-altitude vegetation types are reasonably well protected. The Chimanimani National Park (171 km^2) includes representative portions of all vegetation types in the ecoregion (Stuart and Stuart 1992). In particular, it conserves significant areas of high-elevation heathland vegetation. The Nyanga National

Park (471 km^2) covers habitats between 1,500 and 2,100 m, including both montane grassland and forest. The Mozambique side of the Chimanimani Mountains is less protected, and Mount Gorongosa is unprotected.

Types and Severity of Threats

Threats vary within the ecoregion, depending mainly on altitude and habitat type. The most threatened areas are the lower- and medium-elevation moist evergreen rainforests. These forests grow on fertile, well-watered slopes that are ideal for farming. Much of the former forest vegetation has already been cleared for tea and coffee plantations and for dairy farming (Timberlake et al. 1994). There are still patches of unprotected forest that could also be lost to agriculture.

Forests at higher elevations are mostly moderately or well protected by steep slopes. However, on the disturbed margins of the forest at medium altitudes, exotics such as *Lantana camara*, Mauritius thorn (*Caesalpinia decapetala*), wattle (*Acacia mearnsii*), and eucalyptus species are invading (Timberlake and Musokonyi 1994). Threats to Chirinda Forest include poaching (particularly for blue duiker and guineafowl), the collection of firewood, and the gathering of traditional herbal medicines.

The high-altitude montane forests, grasslands, woodlands, and heathlands face fewer threats. They are restricted to areas that are not suitable for farming and have a low human population density. However, invasion by exotic trees, such as wattle and pine, is an increasing problem (Childes and Mundy 2001). This is particularly serious in the Nyanga grasslands. Pine and wattle plantations are also affecting forest regeneration in the Bvumba Highlands.

Justification for Ecoregion Delineation

The boundaries for this ecoregion are taken from the "undifferentiated montane vegetation unit" of White (1983) and include the Nyanga Mountains, Bvumba Highlands, Chimanimani Mountains, Chipinge Highlands, and isolated Mount Gorongosa to the east. The area is separated as an ecoregion because of its high endemism, and has been previously recognized as both an endemic bird area (Stattersfield et al. 1998) and center of plant diversity (WWF and IUCN 1994).

Ecoregion Number:	**77**
Ecoregion Name:	**Highveld Grasslands**
Bioregion:	**Eastern and Southern Africa**
Biome:	**Montane Grasslands and Shrublands**
Political Units:	**South Africa, Lesotho**
Ecoregion Size:	**186,200 km^2**
Biological Distinctiveness:	**Bioregionally Outstanding**
Conservation Status:	**Critical**
Conservation Assessment:	**IV**
Authors:	*Rauri Bowie, Aliette Frank*
Reviewed by:	*Brian Huntley*

Location and General Description

The Highveld Grasslands [77] draws its name from the high interior plateau of South Africa, known as the Highveld. Once covered by species-rich grassland habitats, the ecoregion is now largely converted to agriculture.

The Highveld is flat, with elevations ranging from 1,400 m to 1,800 m above sea level. The landscape is traversed by many meandering rivers, with the grassland community historically playing an important role in natural water purification of the westward-flowing rivers that originate on the Drakensberg Escarpment (Davies and Day 1998). This ecosystem process has been disrupted in many areas by water transfer projects that have been built to provide greater Johannesburg with its water supply (Davies and Day 1998).

High rainfall maintains the grasslands during the summer, with the mean annual rainfall between 400 and 900 mm. Frequent fires, frost, and heavy grazing (formerly by wild animals and now by cattle and sheep) suppress the presence of shrubs and trees (Low and Rebelo 1996). The mean maximum temperature ranges from 21°C to 24°C, and the mean minimum ranges from 3°C to 6°C, with temperatures sometimes reaching 38°C in the summer and −11°C in the winter. Summer rainfall is not evenly distributed throughout the region, resulting in several different habitat types. Differences in habitat types are further accentuated by the variable soil characteristics of the region. Over most of the area sandstones and shales of the Karoo sequence are dominant. Deep red sand-loam soils dominate the cooler and wetter northeast and transition to shallower lithosols in the extreme northeast (Low and Rebelo 1996).

The vegetation of this ecoregion can be divided into three main types: Kalahari-Karoo highveld transition zone, sweet grasslands, and sour grasslands (see also Harrison et al. 1997; Huntley 1984). Several authors (White 1983; Acocks 1988; Low and Rebelo 1996) provide a more detailed subdivision. Dominant grass species include *Panicum coloratum, Themeda trianda, Eragrostis curvula, E. lehmanniana,* and *Brachiaria serrata.* Nongrassy forb species include *Helichrysum rugulosum, Crabbea acaulis,* and *Rhynchosia totta* (Bredenkamp et al. 1989; Coetzee et al. 1993; Eckhardt et al. 1993; Fuls et al. 1993; Cowling et al. 1997a).

Outstanding or Distinctive Biodiversity Features

Although highly fragmented, the Highveld contains the greatest expanse of grassland remaining in southern Africa. Analyses of pollen spores from the Winterberg Escarpment suggest that grasses have dominated the floral community since at least the early Holocene (Meadows and Meadows 1988; Meadows and Linder 1993). Despite the severely degraded nature of this ecoregion, it provides the last remaining stronghold of several grassland species that have suffered major reductions in abundance in the grassland biome, such as blue crane (*Grus paradisea,* VU) (Allan 1992).

Bird species richness is fairly high in this ecoregion (Harrison et al. 1997). However, Botha's lark (*Spizocorys fringillaris,* EN) is the only bird species strictly endemic to the ecoregion, where it inhabits heavily grazed grassland. Near-endemic species include Rudd's lark (*Heteromirafra ruddi,* CR), buff-streaked chat (*Oenanthe bifasciatai*), and yellow-breasted pipit (*Anthus chloris,* VU) (Harrison et al. 1997).

This ecoregion supports a high number of mammal species, although only the orange mouse (*Mus orangiae*) is endemic. Populations of several large mammal species are found, some of which are rare in southern Africa (Stuart and Stuart 1995). Among these are the brown hyena (*Hyaena brunnea*), ground pangolin (*Manis temminckii*), African striped weasel (*Poecilogale albinucha*), and aardwolf (*Proteles cristatus*). Herds of large mammals, including black wildebeest (*Connochaetes gnou*), blesbok (*Damaliscus pygargus*), springbok (*Antidorcas marsupialis*), and white rhinoceros (*Ceratotherium simum*), used to occur here but were extirpated by early settlers. The current populations of these species have been reintroduced.

Few reptile species occur in the ecoregion, although there is one strictly endemic species: the giant spinytail lizard (*Cordylus giganteus*) (Branch 1998). Several additional reptile species are near endemic, including Drakensberg rock gecko (*Afroendura niravia*) and Breyer's whiptail (*Tetrodactylus breyeri*) (Branch 1998). There are no endemic amphibians.

Status and Threats

Current Status

Most of the near-pristine grassland that remains in this ecoregion is found in nature reserves. The main protected areas are the Valei, Nooitgedacht Dam, Bronkhortspruitdam, Vaal Dam, Willem Pretorius, Rustfontein Dam, and Koppies Dam nature reserves and the Ermelo Game Park. Together with a number of smaller reserves, these currently conserve only 0.5 percent of the ecoregion. Outside the reserves, even the areas of grassland that have remained in a near-natural state are declining steadily in area and quality. The present state of habitat fragmentation, together with anthropogenic changes predicted for the coming years, may lead to the extinction or near-extinction of some larger animal species, such as the blue crane (Allan 1992).

Types and Severity of Threats

These grasslands have already suffered extensive transformation. Because this is one of the best farming areas in South Africa, large tracts of land were converted to agriculture decades ago, mainly for maize production. Urban expansion, fire, and overgrazing have led to increased fragmentation, as has coal mining and afforestation with exotic trees, particularly *Eucalyptus* species (Low and Rebelo 1996; Cowling et al. 1997a). Planted black wattle (*Acacia mearnsii*) has become invasive and spreads rapidly along rivers.

Justification for Ecoregion Delineation

This ecoregion, distinguished from surrounding ecoregions by its higher elevations on the Highveld Plateau, follows Low and Rebelo's (1996) highveld grasslands. These include "rocky highveld grassland," "moist clay highveld grassland," "dry clay highveld grassland," "moist sandy highveld grassland," "moist cool highveld grassland," and "moist cold highveld grassland." The lumping of these finer units corresponds closely to the "Highveld grassland" vegetation unit of White (1983).

Ecoregion Number:	**78**
Ecoregion Name:	**Drakensberg Montane Grasslands, Woodlands, and Forests**
Bioregion:	**Eastern and Southern Africa**
Biome:	**Montane Grasslands and Shrublands**
Political Units:	**South Africa, Swaziland, Lesotho**
Ecoregion Size:	**202,200 km²**
Biological Distinctiveness:	**Globally Outstanding**
Conservation Status:	**Endangered**
Conservation Assessment:	**I**
Authors:	*Raurie Bowie, Aliette Frank*
Reviewed by:	*Anthony B. Cunningham, Rob Scott-Shaw*

Location and General Description

The Drakensberg Montane Grasslands, Woodlands, and Forests [78] are found in southeastern South Africa, primarily in the

KwaZulu-Natal province. The ecoregion is composed of several habitat types, including Afromontane and upland grasslands around the Drakensberg Escarpment, lowveld bushveld near Pongola, and arid mountain bushveld of the Soutpansberg. The elevational range of this ecoregion is between 150 and 2,500 m, with the Drakensberg Alti-Montane Grasslands and Woodlands [79] occurring on the higher slopes of the Drakensberg. Sandstone and shale from the Karoo sequence, together with dolerite intrusions, characterize the geology (Deall et al. 1989; WWF and IUCN 1994; Low and Rebelo 1996).

Rainfall ranges from 450 mm per year in the southwest to more than 1,100 mm in the northeast, with the highest altitudes receiving 1,900 mm annually. Because of the high precipitation, soils (mostly lithosols) typically are leached and are rocky and shallow in most places. Cold and wet conditions pervade most of the area, excluding the Lesotho Plateau, where drought can occur in the rainshadow (Low and Rebelo 1996). Temperatures vary between 40°C and −13°C, averaging 15°C (White 1983; Low and Rebelo 1996). The large climatic variation results in a diverse vegetative community, with montane grassland prevalent on the wet exposed slopes and forest patches lining the valleys (for details see White 1983; Lubke et al. 1986; Matthews et al. 1993; WWF and IUCN 1994). As the elevation decreases toward the southeast, the montane habitat grades into thicket and forest in wetter areas and grassland and bushveld savannas in drier areas (White 1983; Low and Rebelo 1996).

In the forest habitat, *Podocarpus* is the most common tree genus, and *Widdringtonia* is also present (WWF and IUCN 1994). Dominant canopy trees include *Maytenus peduncularis, Podocarpus latifolius, Pterocelastrus rostratus, Olinia emarginata,* and *Scolopia mundii.* In the grasslands, common grass species include *Monocymbium ceresiiforme, Diheteropogon filifolius, Sporobolus centrifugus, Harpochloa falx, Cymbopogon dieterlenii,* and *Eulalia villosa.* These occur on leached, rocky, and shallow soils that are characteristic of moist, cool, and steep mountain slopes. Herbs include *Helichrysum cerastioides, H. spiralepis, Rhus discolor, Selago galpinii, Clutia monticola,* and *Sebaea sedoides.* Grasses on the lower slopes of the Drakensberg include *Poa annua, Hyparrhenia hirta, Aristida diffusa,* and *Trachypogon spicatus.* Herbs include *Rhus dentata* and *Leucosidea sericea* (White 1983; Lubke et al. 1986; Matthews et al. 1993; WWF and IUCN 1994; Cowling et al. 1997a).

Outstanding or Distinctive Biodiversity Features

The lower Maloti-Drakensberg forms the center of southern Africa's Afromontane region. Unlike tropical African mountains, where Afromontane communities are found only above 2,000 m, here latitude compensates for altitude, allowing Afromontane communities to occur around sea level (Dowsett 1986; Cowling et al. 1997a).

An estimated 30 percent of plant species are endemic to both the Drakensberg Montane [78] and Drakensberg Alti-Montane Grassland [79] ecoregions (Huntley 1994; Matthews et al. 1993).

This ecoregion has a number of local hotspots of plant endemism that are usually associated with geology and isolated mountainous areas. The Barberton Center, Wolkberg Center, and Soutpansberg Center have recently been described along with fifteen others in southern Africa (van Wyk and Smith 2001). Plant endemics include *Disa intermedia, Gladiolus serpenticola, Hemizygia thorncroftii, Protea curvata,* and *Syncolostemon comptonii* in the Barberton Center; *Crocosmia mathewsiana, Dierama adelphicum, Encephalartos nubimontanus, Erica revoluta, Encephalartos dolomiticus,* and *Gladiolus rufomarginatus* in the Wolkberg Center; and *Cineraria cyanomontana, Encephalartos hirsutus, Zoutpansbergia caerulea,* and *Tylophora coddii* in the Soutpansberg Center.

Animal endemism is not high in most groups but is pronounced among reptiles and Lycaenid butterflies. Strictly endemic reptiles include the Drakensberg dwarf chameleon (*Bradypodion dracomontanum*), woodbush legless skink (*Acontophiops lineatus*), Transvaal flat lizard (*Platysaurus relictus*), and prickly girdled lizard (*Pseudocordylus spinosus*). Many of the endemic reptile species have been discovered in the last 10 years (Branch 1998). Recent surveys found an additional six species of dwarf chameleons (*Bradypodion* spp.) that are still awaiting description (B. Branch, pers. comm., 2000). The isolated grasslands occurring in the Soutpansberg and Blouberg mountain ranges are unique and contribute significantly to the high degree of endemism observed. These mountains were once the home of Eastwood's whip lizard (*Tetradactylus eastwoodae*), which has not been rediscovered since its description in 1913. Its likely extinction probably resulted from the destruction of its original habitat for pine plantations (Branch 1998).

Bird species richness is fairly high. Near-endemics include the bush blackcap (*Lioptilus nigricapillus*), Drakensberg siskin (*Serinus symonsi*), forest canary (*Serinus scotopus*), yellow-breasted pipit (*Anthus chloris*, VU), orange-breasted rockjumper (*Chaetops aurantius*), buff-streaked chat (*Oenanthe bifasciata*), Rudd's lark (*Heteromirafra ruddi*, CR), and Botha's lark (*Spizocorys fringillaris*) (Hilton-Taylor 2000; Barnes et al. 2001). The ecoregion also contains the last stronghold of the bearded vulture or lammergeier (*Gypaetus barbatus*) in southern Africa and the threatened blue swallow (*Hirundo atrocaerulea*, VU).

Some of the largest ungulate populations in southern Africa are found in the ecoregion, including eland (*Taurotragus oryx*), southern reedbuck (*Redunca arundinum*), mountain reedbuck (*Redunca fulvorufula*), grey rhebok (*Pelea capreolus*), black wildebeest (*Connochaetes gnou*), white rhinoceros (*Ceratotherium simum*), and oribi (*Ourebia ourebi*) (Stuart and Stuart 1995). The Hluhluwe-Umfolozi Park (in KwaZulu-Natal province) contains important populations of black (*Diceros bicornis*, CR) and white rhinoceros, elephant (*Loxodonta africana,* EN), lion (*Panthera leo,* VU), cheetah (*Acinonyx jubatus,* VU), wild dog (*Lycaon pictus,* EN), and other large mammals. Strictly endemic mammals include the thin mouse shrew (*Myosorex tenuis*) and Gunning's golden mole (*Amblysomus gunningi*), and the Natal red rockhare (*Pronolagus crassicaudatus*) is near endemic.

Only one frog is strictly endemic (*Breviceps sylvestris*), and another five (*Leptopelis xenodactylus, Breviceps maculatus, Breviceps verrucosus, Cacosternum poyntoni,* and *Rana dracomontana*) are near endemic to this ecoregion (Passmore and Carruthers 1995).

Status and Threats

Current Status

The grassland biome is one of southern Africa's most endangered habitats (Cowling et al. 1997a), primarily as a result of extensive fragmentation caused by agriculture and afforestation with *Eucalyptus* and *Pinus* species. Low and Rebelo (1996) estimate that 32–45 percent of this ecoregion has been converted.

Two of southern Africa's largest and most well-developed reserves occur here. Higher-elevation parts of the important Hluhluwe-Umfolozi Park are found in this ecoregion. The second largest park is uKhahlamba-Drakensberg Park, which includes Giant's Castle, Cathedral Peak, and Cobham Nature Reserves. It contains southern Africa's largest breeding population of bearded vultures (Barnes et al. 2001) and has high plant species richness and endemism, almost equaling that of the Drakensberg Alti-Montane Grasslands and Woodlands [79]. Other large protected areas in or extending into this ecoregion include the recently created Tse'hlanyane National Park in central Lesotho. Golden Gate and Qwa Qwa National Park also contribute significantly to the preservation of pristine natural habitat in the higher altitudes of this ecoregion. Double Drift, Opathe, Songimvelo, and Itala game reserves are lower-altitude reserves with vegetation transitional to savanna.

These reserves fulfill important conservation functions, yet they protect only 1 to 2 percent of the area. The central portion of these grasslands, moist upland grassland (Low and Rebelo 1996) and moist midlands mistbelt (Scott-Shaw 1999), is poorly conserved, with the major conservation areas being the Blinkwater and Karkloof reserves in KwaZulu-Natal. In the southwestern portion, there are few conservation areas, although the vegetation is represented in the Mountain Zebra National Park. The northeastern portion is also poorly protected, but there are areas of the natural vegetation at Blyde River Canyon, Gustaf Klingiel, Mount Sheba, and Pilgrim's Rest Nature Reserves.

Types and Severity of Threats

Vast areas of the ecoregion have been transformed to agriculture and forestry plantations. The humid grasslands have suffered most, with some (e.g., Moist Midlands Mistbelt, also known as Camp BioResource Group 5) more than 90 percent transformed (Scott-Shaw 1999). In areas that have been heavily grazed, the native grassland has been invaded by dense and less palatable grass species, such as where *Themeda triandra* has been replaced by *Aristida junciformis* (Low and Rebelo 1996). In addition, fires (both natural, caused by lightning strikes, and

controlled burns for tick removal) and heavy grazing by domestic animals have extensively altered the habitat. These changes have contributed to the loss of indigenous plant species over large areas. It is believed that the Afromontane forest was previously widespread throughout the ecoregion. The remaining mosaic of isolated forest patches is prone to fire and species-specific exploitation, such as bark-stripping of medicinal trees such as *Rapanea melanophloeos*. Most are found in sheltered locations in river valleys. Remaining grassland areas are a focus for the commercial collection of medicinal plants, including the destructive harvest of bulbs, tubers, aromatic stems, and rhizomes (e.g., *Gunnera perpensa, Dianthus, Gladiolus, Scilla natalensis, Eucomis autumnalis, Alepidea amatymbica,* and *Helichrysum odoratissimum*). Overuse also threatens the pepperbark tree (*Warburgia salutaris*); the largest remaining population in South Africa is in the Blouberg Mountains.

Justification for Ecoregion Delineation

The boundaries of this ecoregion were adopted from the "northeastern mountain grassland," "southeastern mountain grassland," "moist upland grassland," "short mistbelt grassland," "west cold highveld grassland," "coast-hinterland bushveld," "Natal central bushveld," "subarid thorn bushveld," "eastern thorn bushveld," and "Natal lowveld bushveld" units of Low and Rebelo (1996). This corresponds with White's (1983) "undifferentiated montane vegetation" unit. The disjunct Soutpansberg montane area, delineated in Low and Rebelo as "Soutpansberg arid mountain bushveld," was included because of its similar biological patterns.

Ecoregion Number:	**79**
Ecoregion Name:	**Drakensberg Alti-Montane Grasslands and Woodlands**
Bioregion:	**Eastern and Southern Africa**
Biome:	**Montane Grasslands and Shrublands**
Political Units:	**South Africa, Lesotho**
Ecoregion Size:	**11,900 km²**
Biological Distinctiveness:	**Globally Outstanding**
Conservation Status:	**Critical**
Conservation Assessment:	**I**
Authors:	*Raurie Bowie, Aliette Frank*
Reviewed by:	*Rob Scott-Shaw*

Location and General Description

The Drakensberg Alti-Montane Grasslands and Woodlands [79] span the steep and treeless upper slopes of the Drakensberg Escarpment that form a semicircular border between Lesotho and KwaZulu-Natal in South Africa. The ecoregion covers the land above 2,500 m altitude, including Thabana-Ntlenyana (3,482 m), the highest point in southern Africa (Barnes 2001). This area is also known as uKhahlamba and the Maloti-Drakensberg.

The Drakensberg forms the southernmost extent of a discontinuous mountain chain extending from eastern Africa, and although it is not as high in elevation as some of the mountains at the equator, climatic conditions are similar because the higher latitude compensates for the lower elevation (Moreau 1966; White 1983; Dowsett 1986). Precipitation averages about 1,000 mm annually, and temperatures vary from −8° to 32°C, with temperatures dropping to −20°C on the summit plateau. The Drakensberg is an important water catchment area and the source of both the Tugela and Senqu (Orange) rivers. The geology comprises basalts from the Stormberg Group that weather to produce shallow, acidic lithosols. Topographic variation creates a variety of habitats with many different plant communities (Cowling et al. 1997b).

The Drakensberg forms the southernmost extent of the Afromontane regional center of endemism (White 1983; Dowsett 1986; Linder 1998). The ecoregion also coincides with the Drakensberg alpine center, one of eighteen centers of plant endemism in southern Africa (van Wyk and Smith 2001). The basalt rocks form plateaus and steep slopes that support a treeless alpine vegetation consisting mostly of tussock grasses, creeping or mat-forming plants, and ericoid dwarf shrubs (White 1993). In the ericaceous and Afroalpine belts, characteristic grass species are from the genera *Agrostis, Deschampsia, Festuca, Koeleria, Pentaschistis,* and *Poa.* On the plateau, dominant and diagnostic grass species include *Merxmuellera disticha, M. drakensbergensis, Festuca caprina, Eragrostis caesia, Poa binata,* and *Pentaschistis galpinii.* Other common plant species include *Carex clavata, Scirpus falsus, Helichrysum flanaganii, H. trilineatum, H. witbergense,* and *Erica frigida* (Low and Rebelo 1996; Cowling et al. 1997b).

Outstanding or Distinctive Biodiversity Features

The Drakensberg hosts a remarkable number of endemic plant species and an unusually high diversity of plant communities, largely as a result of the topographic heterogeneity of the area. Of particular note are the wetlands and heaths on the flatter summit. Important plant communities include a wide range of wetlands (including tarns, mires, and bogs), heaths, fynbos, cliff and cave communities, and grasslands, each harboring many endemic plants. Genera largely confined to the region include *Merxmuellera, Strobilopsis, Heteromma, Saniella, Schoenoxiphium, Rhodohypoxis,* and *Glumicalyx.* Although the number of plant species solely restricted to the high Drakensberg is unknown, in the KwaZulu-Natal region of the Drakensberg 1,750 vascular plant species have been recorded. Of these, 394 species are endemic to the southern Drakensberg (22.5 percent) (Hilliard and Burtt 1987). The highest levels of endemism occur on the highest peaks, including *Helichrysum palustre, H. bellum, H. qahlambanum, Hy-*

poxis ludwigii, Kniphofia caulescens, Strobilopsis wrightii, Aponogeton ranunculiflorus, Gladiolus microcarpus, Erica thodei, and *Disa sankeyi* (van Wyk and Smith 2001). Threatened species include *Aponogeton ranunculiflorus* (VU), *Crocosmia pearsei* (VU), and *Brachystelma alpina* (VU). *Protea nubigena* (CR) and *Encephalartos ghellinckii* (VU) are also found in adjacent lower-altitude ecoregions (Scott-Shaw 1999). The spiral aloe (*Aloe polyphylla,* EN), a spectacular endemic plant of Lesotho, is found only above 2,000 m (Huntley 1994). The altimontane habitats of the Drakensberg have only about twenty vascular plant species in common with the Afroalpine flora of east and northeast Africa (Hedberg 1986). The region lacks the conspicuous alpine giant lobelias (*Lobelia* spp.) and giant senecios (*Dendrosenecio* spp.) of the high mountains of central and eastern Africa, which evolved recently from an East African ancestor (Knox and Palmer 1995). Instead, many of the endemic Drakensberg plants have a Cape center of origin.

The ecoregion supports an interesting avifauna with three near-endemic bird species: orange-breasted rockjumper (*Chaetops aurantius*), Drakensberg siskin (*Serinus symonsi*), and mountain pipit (*Anthus hoeschi*) (Barnes 2001; Barnes et al. 2001). In addition, the ecoregion supports rare bird species such as the yellow-breasted pipit (*Anthus chloris,* VU), blue crane (*Grus paradisea,* VU), southern bald ibis (*Geronticus calvus,* VU), Cape vulture (*Gyps coprotheres,* VU), and Hottentot buttonquail (*Turnix hottentotta*). There is also an isolated race of lammergeier (*Gypaetus barbatus meridionalis*).

The mammal fauna is not particularly important but includes mountain species such as klipspringer (*Oreotragus oreotragus*) and mountain reedbuck (*Redunca fulvorufula*) (Skinner and Smithers 1990; Stuart and Stuart 1995). The threatened mouse *Mystromis albicaudatus* (VU) also occurs.

Three river frogs (*Rana dracomontana, R. vertebralis,* and *Strongylopus hymenopus*) are endemic to fast-flowing streams of the altimontane grassland (Passmore and Carruthers 1995). Grass banks next to altimontane rivers are also home to the cream-spotted mountain snake (*Montaspis gilvomaculata*) (Bourquin 1991), the only member of its genus. The high alpine habitats are also home to three endemic lizard species, *Pseudocordylus langi, Tropidosaura cottrelli,* and *T. essexi.* The ecoregion also supports important populations of a threatened lizard, Breyer's whip lizard (*Tetradactylus breyeri,* VU), and the five-toed whip lizard (*Tetradactylus seps*), Drakensberg crag lizard (*Pseudocordylus melanotus*), and near-endemic Drakensberg rock gecko (*Afroedura nivaria*) (Branch 1998). The Drakensberg minnow (*Pseudobarbus quathlambe*) and two species of fairy shrimp have very limited distribution ranges and are endemic to this ecoregion.

Status and Threats

Current Status

At the highest elevations of the ecoregion, habitats are relatively intact. Elsewhere on the high plateau severe livestock trampling, frequent fire, and heavy selective grazing have degraded the gently sloping grasslands, heaths, and wetlands.

Until recently, only around 200 km² of the ecoregion was conserved in protected areas. However, 13,000 km² of the northeastern border between Lesotho and South Africa is being developed as the Maloti-Drakensberg Transfrontier Conservation and Development Project. This initiative is operational through an understanding between the governments of Lesotho and South Africa to mutually manage Sehlabathebe National Park in Lesotho and uKhahlamba-Drakensberg Park in South Africa. The uKhahlamba-Drakensberg Park was recently declared a World Heritage Site. Two new high-altitude nature reserves, Tse'Hlanyane National Park and Bokong Nature Reserve, have also been declared recently.

Types and Severity of Threats

The greatest threats to the habitats and species of this ecoregion stem from overgrazing by livestock and a high frequency of human-induced fires, especially in the dry winter. This has particularly affected the yellow-breasted pipit and a number of other threatened bird species (Barnes 2001). Other concerns include the expansion of invasive plants such as *Chrysocoma ciliata* and *Helichrysum trilineatum,* soil erosion, the indiscriminate harvesting of medicinal plants, and clearance for cultivation.

Justification for Ecoregion Delineation

The boundaries of this ecoregion follow the altimontane vegetation in southern Africa, mapped by White (1983), and the "alti mountain grassland" unit of Low and Rebelo (1998), following the 2,500-m elevation contour. The ecoregion extent also coincides with the Drakensberg Alpine Center, one of eighteen centers of plant endemism in southern Africa (van Wyk and Smith 2001). The ecoregion supports numerous endemic plants at the highest elevations but lacks species such as giant *Lobelia* and *Dendrosenecio* typical of the altimontane areas in eastern Africa.

Ecoregion Number:	**80**
Ecoregion Name:	**Maputaland-Pondoland Bushland and Thickets**
Bioregion:	**Eastern and Southern Africa**
Biome:	**Montane Grasslands and Shrublands**
Political Unit:	**South Africa**
Ecoregion Size:	**19,500 km²**
Biological Distinctiveness:	**Regionally Outstanding**
Conservation Status:	**Endangered**
Conservation Assessment:	**II**
Authors:	*Karen Goldberg, Alliette Frank*
Reviewed by:	*Anthony B. Cunningham, Rob Scott-Shaw*

Location and General Description

The Maputaland-Pondoland Bushland and Thickets [80] are found alongside the numerous rivers of the Eastern Cape and KwaZulu-Natal Provinces of South Africa. A narrow band of KwaZulu-Cape Coastal Forest Mosaic [23] separates this ecoregion from the subtropical waters of the Indian Ocean to the east, and it is mostly surrounded by the Drakensberg Montane Grasslands, Woodlands, and Forests [78].

The climate is seasonal and dry, with most areas experiencing less than 800 mm rainfall per annum, falling to 450 mm in southern areas. Up to 75 percent of the annual precipitation falls in the warm summer, between October and March (Moll 1976). Temperature ranges from 12°C to 26°C, and the ecoregion is frost-free because of its close proximity to the sea.

The rifting and breakup of Gondwana and subsequent cycles of uplift and erosion have shaped the landscape. These processes formed the Great Escarpment, which separates the elevated interior of southern Africa from the coastal margins. Geologically the ecoregion covers basement granites, schists, gneisses, lavas, and sedimentary strata from the Cretaceous and Cenozoic periods (Low and Rebelo 1996). The soils are deep and well drained (Cowling 1984) and are moderately to highly leached (Moll 1976).

This ecoregion is distinguished from other African thicket types by a predominance of evergreen sclerophyllous plants and is differentiated from the Albany Thickets [91] ecoregion to the south by its paucity of succulent trees and shrubs (Everard 1987). It generally consists of a closed canopy formation up to 6 m in height and often forms an impenetrable tangle of spinescent shrubs, low trees, and vines. It is generally not divided into strata and usually does not have a pronounced herbaceous or grass layer (Low and Rebelo 1996). Characteristic arborescent plant species include *Diospyros dichrophylla, Euphorbia triangularis, E. tetragona, Bauhinia natalensis, Encephalartos princeps,* and *Aloe pluridens.* More widespread species include *Putterlickia pyracan-*

tha, Rhoicissus tridentata, Grewia accidentalis, Phyllanthus verrucosus, and the grass *Panicum maximum.*

Outstanding or Distinctive Biodiversity Features

The ecoregion broadly falls into White's (1983) Tongoland-Pondoland evergreen and semi-evergreen thicket and is commonly called subtropical transitional thicket (Cowling 1984; Everard 1987) because it traverses a number of formal biomes (Low and Rebelo 1996). The ecoregion is situated in one of the most diverse areas in Africa and shows significant floristic overlap with Afromontane forest, coastal forest, broadleaved Zambezian woodland, and Karoo shrubland (White 1983).

Cowling (1983b) has shown that the thickets of this ecoregion have plant diversity similar to that of other vegetation types in the southeastern Cape and are as species-rich as the fynbos formations. Between 6,000 and 7,000 plant species occur in the ecoregion, with more than forty-nine species per 100 square meters recorded in the mesic Kaffrarian thicket areas (Everard 1987). Species endemism is low, with most found in succulent genera such as *Euphorbia, Crassula, Delosperma,* and *Aloe* (van Wyk and Smith 2001). Cycads show high diversity, with the Kei cycad (*Encephalartos princeps*), Bushman's river cycad (*E. trispinosus*), Alexandria cycad (*E. arenarius*), cerinus cycad (*E. cerinus*) and the Albany cycad (*E. latifrons*) being endemic or near endemic (Goode 1989). The low levels of endemism may result from climatic instability that has favored generalist species (Gibbs Russel and Robinson 1981).

The overall faunal diversity is moderate to poor, and there are few endemic species. Only two near-endemic mammal species are found in this ecoregion: the giant golden mole (*Chrysospalax trevelyani*, EN) and the Natal red rockhare (*Pronolagus crassicaudatus*) (Kingdon 1997). Populations of large mammals have declined markedly, and some larger carnivores have become locally extinct (Smithers 1983). The black-backed jackal (*Canis mesomelas*) was historically found throughout the area, but control measures have resulted in an enormous decline (Smithers 1983). Bush pigs (*Potamochoerus larvatus*) remain common, whereas blue duiker (*Cephalophus monticola*) is mostly confined to forest and dense thicket (Smithers 1983). Other antelopes include bushbuck (*Tragelaphus scriptus*), greater kudu (*T. strepsiceros*), common duiker (*Sylvicapra grimmia*), and mountain reedbuck (*Redunca fulvorufula*).

Bird richness is high, but all species are shared with surrounding ecoregions. Two near-endemics occur marginally in this ecoregion: the chorister robin-chat (*Cossypha dichroa*) and the forest canary (*Serinus scotops*). The Cape griffon (*Gyps coprotheres*, VU) also occurs, with important breeding colonies at Collywobbles vulture colony, found along the cliffs of the convoluted gorge formed by the mBashe River (Barnes 1998). Another globally threatened bird is the spotted ground-thrush (*Zoothera guttata*, EN).

Amphibian and reptile diversity and endemism are fairly

high. Two of the reptiles are near endemic to the ecoregion (*Bradypodion thamnobates* and *Kinixys natalensis*) and are regarded as locally rare (Hilton-Taylor 2000), whereas two of the amphibians (*Hyperolius pickersgilli* and *Leptopelis xenodactylus*) are vulnerable, and *Cacosternum poyntoni* is assessed as data deficient (Hilton-Taylor 2000).

Status and Threats

Current Status

This ecoregion is naturally fragmented because it occupies only the narrow river valleys running through the Drakensberg Mountain foothills. In addition, the Eastern Cape and KwaZulu-Natal have been subject to a long history of human occupation. High human densities in these areas have resulted in significant landscape degradation.

Because of the large human population throughout the ecoregion, many wildlife populations have been severely depleted. However, only about 50 percent of the habitat has been transformed (Low and Rebelo 1996). This is partly because bushland thicket largely occupies the steep slopes of river valleys, which are generally unsuitable for cultivation and often less accessible to livestock. In addition, thicket is known to encroach into grasslands in the absence of large browsers or reduced fire regimes.

Roughly 7.5 percent of the Maputaland-Pondoland center of plant diversity is conserved in twelve protected areas (WWF and IUCN 1994). The Oribi Gorge Nature Reserve protects part of the Mzimkulwana River in KwaZulu-Natal. In the Eastern Cape the Thomas Baines Nature Reserve covers hilly country with a mixture of bushland thicket, mixed grassland, and fynbos. The endemic plant *Oldenburgia arbuscula* occurs here (Greyling and Huntley 1984). The Andries Vosloo Kudu Reserve has been expanded to include the newly established Sam Knott Nature Reserve and the Double Drift Nature Reserve. Together these three protected areas cover 450 km^2 (Stuart and Stuart 1992). Endemic plant species include *Pachypodium bispinosum, P. succulentum,* and *Encephalartos trispinosus* (Greyling and Huntley 1984).

Types and Severity of Threats

Whereas much of the southern portion of the ecoregion is found on steep slopes that are unsuitable for cultivation, the northern areas are found on more even ground, where extensive cultivation of staple and cash crops occurs (WWF and IUCN 1994). In addition, rapidly expanding populations are clearing marginal or unsuitable land, threatening the remaining vegetation.

Parts of the area suffer from overgrazing, mostly from goats, sheep, and cattle (Skead 1987). Cattle farming has limited impact because these animals prefer grasslands, and their grazing and trampling can encourage thicket growth by reducing grass cover. High livestock densities also threaten wildlife because high numbers of domesticated animals generally displace game from suitable habitat. Furthermore, wild predators and scavengers such as the black-backed jackal, caracal, leopard, and Cape vulture have been eradicated by livestock farmers who see these animals as a threat to their livelihoods. Poisoned carcasses often are used for this purpose.

Unsustainable medicinal plant harvesting poses significant threat to several plant species that are collected in great numbers using methods that kill the plant (Ellis 1986; Hutchings et al. 1996). A further threat is the invasion of natural vegetation by alien plants. Some of the more aggressive invading species are *Chromoleana odorata, Lantana camara, Psidium guajava, Rubus* spp., *Solanum mauritianum, Acacia cyclops,* and *A. mearnsii*. The last has been commercially planted on a vast scale, and *A. dealbata* has also invaded along watercourses. Several large urban centers are also found: Durban, Umtata, King William's Town, East London, and Grahamstown. The expansion of these towns will lead to further loss of natural habitat.

Justification for Ecoregion Delineation

This ecoregion forms part of the "South African evergreen and semi-evergreen bushland" vegetation unit of White (1983), although it more closely follows the "valley thicket" vegetation type of Low and Rebelo (1996). It contains transitional Maputaland-Pondoland and Afromontane affinities and forms part of the Maputaland-Pondoland center of plant diversity and Endemism (WWF and IUCN 1994; van Wyk and Smith 2001), recognized as one of the most floristically diverse areas in Africa.

Ecoregion Number:	**81**
Ecoregion Name:	**Angolan Scarp Savanna and Woodlands**
Bioregion:	**Eastern and Southern Africa**
Biome:	**Montane Grasslands and Shrublands**
Political Units:	**Angola**
Ecoregion Size:	**74,400 km²**
Biological Distinctiveness:	**Globally Outstanding**
Conservation Status:	**Relatively Stable**
Conservation Assessment:	**III**
Author:	*Suzanne Vetter*
Reviewed by:	*Brian Huntley*

Location and General Description

The Angolan Scarp Savanna and Woodlands [81] comprises a long narrow strip of land running from about 6°S to 14°S lati-

tude between the Atlantic Ocean, the southwest arid biome of Angola, and the top of the scarp face of the Central African Plateau.

The ecoregion covers an elevational range from sea level to about 1,000 m. It includes two main geomorphologic regions: the Coastal Belt and the Transition Zone (Huntley 1974b). The Coastal Belt varies in width from 200 km in the north to less than 40 km further south and does not exceed 300 m in elevation. This area is made up of marine sediments (marls and limestones) and recent sands. The Transition Zone is a discontinuous escarpment belt formed through erosion of the ancient massif, which runs roughly parallel to the coast. It rises sharply in the south, whereas the increase is more gradual farther north, forming a series of steps (Texeira 1968; Huntley 1974b). This area consists of gneisses and granites of the Precambrian basement complex. A great variety of soil types are found. These include heavy black cotton soils, red or reddish sandy soils, calcareous patches on the coastal plain and lower escarpment, and fertile paraferralitic and ferralitic soils in the moist escarpment forests (Texeira 1968; White and Werger 1978).

The ecoregion has a tropical climate with summer rain. Along the coast, the cold Benguela Current influences the climate so that humidity is high year-round but annual rainfall is low, ranging from 400 to 800 mm. Offshore, the Benguela Current meets warm equatorial waters and produces mists that are precipitated by the escarpment. A narrow belt on the escarpment combines high summer rainfall of the inland areas with the year-round humidity of the coastal plains (Hall 1960b). Total precipitation in this zone is thought to exceed 1,600 mm (Huntley 1974b). Temperatures vary according to elevation and latitude, with the highest mean annual temperatures along the inner margin of the Coastal Belt north of the Cueve River, where they exceed 25°C (Huntley 1974b). The lowest mean annual temperatures occur along the border with the highlands of the Angolan Montane Forest-Grassland Mosaic [82] and are approximately 20°C.

Four African phytochoria are found in this ecoregion: the Afromontane archipelago-like regional center of endemism at the highest elevations, and at lower altitudes the Guineo-Congolian and the Zambezian centers of endemism and the Guineo-Congolia–Zambezia regional transition zone. Two other phytochoria border on the ecoregion: the Kalahari-Highveld regional transition zone and the Karoo-Namib regional center of endemism in the arid southwest. The vegetation is consequently highly varied and ranges from dry woodland and wooded grassland to humid mist forest.

North of the Cuanza River, the vegetation is a mosaic of tall, tropical gallery forest and tall grassland, interdigitated by mangrove and swamp communities along the major rivers and their mouths. The forest patches are dominated by tree species such as *Piptadeniastrum africanum, Milicia excelsa, Ceiba pentandra,* and *Musanga cecropioides*. The grasslands contain scattered fire-tolerant woody plants, including *Hymenocardia acida, Erythrina abyssinica, Piliostigma thonningii,* and *Cussonia angolensis* (Huntley and Matos 1994).

South of the Cuanza River, the Zambezian component of the ecoregion comprises a mosaic of closed woodlands, grasslands, and palm savanna (Huntley and Matos 1994), whose distribution is controlled by soil type (Werger and Coetzee 1978). These woodlands are floristically rich but lack mopane and miombo dominant species (White and Werger 1978). Typical species include *Sterculia setigera, Euphorbia conspicua, Strychnos* spp., *Acacia welwitschii,* and baobab (*Adansonia digitata*). Extensive sand plateaus near the coast are occupied by palm savanna, with the grass layer dominated by *Eragrostis superba, Schizachyrium semiberbe,* and *Digitaria milanjiana.* To the north and south of Luanda, grassland occupies large areas of smoothly rounded hills on marine clays, dominated by *Setaria welwitschii* (Texeira 1968; Barbosa 1970; Huntley and Matos 1994).

On the upper slopes of the escarpment, where rain and mist provide year-round moisture and there are fertile soils, mist (or cloud) forest occurs in a discontinuous band 1–15 km wide (White and Werger 1978). These forest patches total between 1,300–2,000 km^2 and are most extensive in the Gabela and Amboim areas (Hawkins 1993). The vegetation is of Guineo-Congolian affinity, and dominant tree species include *Celtis prantlii, Morus mezozygia, Albizia glaberrima, A. gummifera, Ficus mucuso,* and *F. exasperata* (Hall 1960b; Texeira 1968; Huntley 1974a; Hawkins 1993; Huntley and Matos 1994). Two wild coffee species, *Coffea canephora* and *C. welwitschii,* are among the understory species.

Outstanding or Distinctive Biodiversity Features

The importance of the Angola scarp was recognized by Hall (1960b) in her study of Angolan birds and supported with mammal evidence by Cabral (1966, in Huntley 1974b). First, the escarpment allows subspecies to develop in the drier southwest arid and *Brachystegia* biomes by forming a barrier between them. Second, the escarpment zone with its great range of elevations and high humidity has provided refugia for forest species in periods of climatic desiccation. Despite being of great biological interest, the flora and fauna have been poorly studied. For example, no plant lists exist for any of the ecoregion's many vegetation communities (B. J. Huntley, pers. comm., 2001). Of the ecoregion's fauna, only birds have been studied in any detail.

Thirteen bird species are either strictly endemic or near endemic to this ecoregion, of which seven are threatened (Hilton-Taylor 2000). These species are grey-striped francolin (*Francolinus griseostriatus,* VU), red-crested turaco (*Tauraco erythrolophus*), Angola helmetshrike (*Prionops gabela,* EN), white-fronted wattle-eye (*Platysteira albifrons*), Angola slaty-flycatcher (*Dioptrornis brunneus*), Gabela akalat (*Sheppardia gabela,* EN), Angola cave-chat (*Xenocopsychus ansorgei*), Pulitzer's longbill

(*Macrosphenus pulitzeri,* EN), golden-backed bishop (*Euplectes aureus*), orange-breasted bush shrike (*Laniarius brauni,* EN), Gabela bushshrike (*Laniarius amboimensis,* EN), and Monteiro's bushshrike (*Malaconotus monteiri,* DD). There are also a number of endemic bird subspecies: the brown-chested alethe (*Alethe poliocephala hallae*), yellow-necked greenbul (*Chlorocichla falkensteini falkensteini*), Hartert's camaroptera (*Camaroptera brachyura harteri*), and two subspecies of Lühder's bushshrike, *Laniarius luehderi brauni* and *L. l. amboimensis.* Some taxonomists consider the latter two full species (Dean 2000).

In the northernmost part of the ecoregion, between the Congo and Cuanza rivers, are forests and grasslands of the Guineo-Congolia–Zambezia regional transition zone. Forest mammals include Beecroft's scaly-tailed squirrel (*Anomalurus beecrofti*), forest giant squirrel (*Protoxerus stangeri*), forest elephant (*Loxodonta africana cyclotis,* EN), potto (*Perodicticus potto*), bay duiker (*Cephalophus dorsalis*), and water chevrotain (*Hyemoschus aquaticus*). The large mammal fauna of the grasslands south of the Cuanza River used to include roan antelope (*Hippotragus equinus*), red buffalo (*Syncerus caffer nanus*), elephant (*Loxodonta africana,* EN), southern reedbuck (*Redunca arundinum*), bushbuck (*Tragelaphus scriptus*), and eland (*Taurotragus oryx*) (Huntley 1974b). Among the mammals in the montane escarpment forest are yellow-backed duiker (*Cephalophus silvicultor*), black-fronted duiker (*C. nigrifrons*), blue duiker (*C. monticola*), and tree pangolin (*Manis tricuspis*) (Huntley 1974b). Grazing antelopes are largely absent. Heavy hunting during the war probably eliminated many of the larger mammals.

There are two strictly endemic reptile species, Barbosa's leaf-toed gecko (*Hemidactylus bayonii*) and *Monopeltis luandae,* and four strictly endemic amphibian species, Cuanza reed frog (*Hyperolius punctulatus*), Congulu forest tree frog (*Leptopelis jordani*), Quissange forest tree frog (*Leptopelis marginatus*), and Congolo frog (*Hylarana parkeriana*). Most of the strict endemics are found in the Angola scarp forests, although some are also restricted to the drier areas in the lowlands.

Status and Threats

Current Status

With the movement of tens of thousands of refugees from the interior to the coastal areas, substantial transformation of formerly sparsely settled areas has occurred since 1974, despite the low agricultural potential of this arid zone. Around the larger urban centers, particularly Luanda, human settlement and activities such as woodcutting and livestock grazing have had considerable but mostly localized impacts on the vegetation and soils.

The escarpment forests were used almost entirely for coffee production, which peaked between 1950 and 1970, when an estimated 95 percent of the forest area was underplanted by shade coffee (Hawkins 1993). Coffee berry disease in the 1950s, a drop in coffee prices in the mid-1970s, and the upheaval of the civil war since 1974 have resulted in the abandonment of many of the coffee plantations, allowing forest to recover (Hawkins 1993; Huntley and Matos 1994). However, subsistence cultivation on the fertile forest soils is increasing, and Hawkins (1993) estimated that some 30 percent of the forest areas were affected, especially around Gabela.

There is one protected area in the ecoregion, with three further areas proposed for protection (Huntley 1974a; Huntley and Matos 1994). The Kisama National Park borders the Atlantic coast and the banks of the Cuanza and Longa rivers. The proposed Gabela and Chingoroi strict nature reserves would cover patches of biologically important escarpment forest. The Pungo Andongo Natural Monument comprises a series of large rocky outcrops between Gabela and the coast (Huntley 1974a).

The almost continuous civil war in Angola since 1974 led to great instability, poor security, economic depression, displacement of the rural population, and a lack of infrastructure and basic services. Its effects on conservation, particularly on large mammals, were devastating. Most of Angola's protected areas were abandoned as their wardens were forced to leave for economic or security reasons, opening the areas to poachers and settlers (Huntley and Matos 1994). However, since the war ended, there is new hope for conservation in Angola.

Type and Severity of Threats

The most immediate and important threat to the ecoregion's biodiversity is the encroachment of subsistence agriculture in the fertile escarpment forest areas (Hawkins 1993). With the end of the war deforestation might now accelerate and coffee plantations may be re-established in the escarpment forests.

Hunting is uncontrolled in most of Angola, including the protected areas. Huntley and Matos (1992) estimated that twenty-one species of larger mammals are close to extinction in Angola, including lion (*Panthera leo*, VU), cheetah (*Acinonyx jubatus*, VU), and forest elephant (*Loxodonta africana cyclotis*, EN). There are no data on the extent and impact of subsistence hunting on populations of smaller mammals and birds, but these species may be an important source of protein in the more populated rural areas (Hawkins 1993).

The human population density in the ecoregion is highly variable, being densest in and around Luanda, the country's capital city, with a population of more than 1.5 million. This large population has had a serious impact on the natural resources of the adjacent regions.

Justification for Ecoregion Delineation

This ecoregion is based primarily on BirdLife's western Angola endemic bird area (EBA) (Stattersfield et al. 1998). It encompasses portions of White's (1983) "North Zambezian undifferentiated woodland" and "mosaic of lowland forest and secondary grassland," extending inland 1–15 km along the Angola escarpment, which forms its eastern border. Although included in the western Angola EBA, the Bailundu Highlands have been separated into the Angolan Montane Forest-Grassland Mosaic [82] ecoregion because the characteristic elements of the ecoregion's fauna and flora are more closely related to those of other Afromontane areas than to those of the surrounding Angolan biomes (Huntley 1974b; Dean 2000).

Ecoregion Number:	**82**
Ecoregion Name:	**Angolan Montane Forest-Grassland Mosaic**
Bioregion:	**Eastern and Southern Africa**
Biome:	**Montane Grasslands and Shrublands**
Political Units:	**Angola**
Ecoregion Size:	**25,500 km^2**
Biological Distinctiveness:	**Regionally Outstanding**
Conservation Status:	**Endangered**
Conservation Assessment:	**II**
Author:	*Suzanne Vetter*
Reviewed by:	*Brian Huntley*

Location and General Description

The Angolan Montane Forest-Grassland Mosaic [82] comprises a number of small montane forest patches surrounded by grasslands and *Protea* savanna in the west-central highlands of Angola. The ecoregion lies on the Marginal Mountain Chain of Angola, which is restricted to a narrow band running along the inland margin of the escarpment from 11°S to 16°S. The forest patches are restricted to the deep ravines or remote valleys of the highest mountains in the Huambo and Cuanza Sul provinces and an area of Afromontane forest mosaic further south, on the Serra da Chela in the Huíla province. Residual land surfaces that possibly date back to the Gondwanan age (King 1963) form the highest points of the ecoregion, reaching 2,620 m on Mount Môco, 2,582 m on Mount Mepo, and 2,554 m on Mount Lubangue (Huntley 1974b). The mountain chain is part of the Precambrian basement complex that comprises gneisses and granites (Barbosa 1970). The soils are deep and highly weathered.

Mean annual rainfall is between 1,200 and 1,600 mm, increasing with elevation. Rainfall is concentrated in the summer, although precipitation from mists, which rise as the cold Benguela Current meets warmer tropical waters offshore, occurs through most of the year. The coolest months are July and August, when subzero temperatures are recorded often in the mountains (Huntley 1974b). Mean annual temperatures range from 17°C to 20°C (Texeira 1968).

The ecoregion represents a small fragment of White's (1983) Afromontane archipelago-like center of endemism, which consists of widely scattered "islands" of forest on mountain systems in southern, eastern, and western Africa (Werger 1978). The characteristic elements of the ecoregion's fauna and flora are more closely related to those of other such Afromontane areas than to those of the surrounding Angolan ecoregions (Huntley 1974b; Dean 2000).

Pockets of forest survive mainly in deep, humid ravines and on isolated peaks higher than 1,800 m (Huntley and Matos 1994). The forests patches range from 1 to 20 ha in size and attain a canopy height of 8–15 m. The dominant forest tree species is the yellowwood *Podocarpus latifolius*. Other common tree species include *Polyscias fulva*, *Apodytes dimidiata*, *Pittosporum viridiflorum*, *Syzygium guineense afromontanum*, *Halleria lucida*, *Olea* spp., and *Ilex mitis*. Hardly any grass grows in these shady forests, and they are less heavily overgrown with epiphytes than similar forests elsewhere in Africa (Huntley and Matos 1994). The canopy tends to be irregular because of the steep and rocky slopes on which the forest patches are found.

Open grasslands with widely scattered trees and shrubs cover large areas of the highland plateau above 1,600 m and make up most of the ecoregion's area. In well-drained areas, this vegetation is fire prone and includes shrub species such as *Philippia benguelensis*, *Erica* spp., *Stoebe vulgaris*, and *Cliffortia* sp. and grasses such as *Themeda triandra*, *Tristachya inamoena*, *T. bequertii*, *Hyparrhenia andogensis*, *H. quarrei*, *Festuca* spp., and *Monocymbium ceresiiforme*. On waterlogged plateaus, representative plants include *Parinari capensis*, *Myrsine africana*, *Protea welwitschii*, *Dissotis canescens*, *Cyathea* spp., *Loudetia* spp., *Fimbristylis* spp., and *Xyris* spp. (Huntley and Matos 1994). The grasslands in the ecoregion are partly of edaphic origin and partly maintained by fire, much of anthropogenic origin. Although the forest vegetation is not very flammable, fires can intrude into the forest in hot, dry periods or when logging has thinned forests and grass has been able to grow. The abrupt boundaries of the forest fragments and their remaining distribution in ravines and moist south-facing slopes demonstrate that the extent of forests is determined largely by fire.

Outstanding or Distinctive Biodiversity Features

The montane forests of Angola are of great biogeographic interest because they are the sole surviving relics of a much larger moist forest biome that existed during more favorable past climatic conditions (Huntley 1974b). A few narrow endemic species and subspecies are found in the Angolan montane forests, and many elements of their fauna and flora are shared not with the surrounding ecoregions but with Afromontane vegetation occurring on mountain formations thousands of kilometers away. The ecoregion has been extremely poorly studied, with

the exception of the avifauna, which is described by Hall (1960a).

Few large mammal species occur in the ecoregion. Burchell's zebra (*Equus burchelli*), eland (*Taurotragus oryx*), southern reedbuck (*Redunca arundinum*), oribi (*Ourebia ourebi*), and roan antelope (*Hippotragus equinus*) formerly occurred in the montane grassland areas (Huntley 1974a), but it is unlikely that populations of these large mammals survive today. The mammal fauna of the forests and forest margins used to include blue duiker (*Cephalophus monticola*) and bushpig (*Potamochoerus larvatus*) but had already been severely reduced through hunting nearly 30 years ago (Huntley 1974a). The small mammal fauna is poorly known. Two shrew species, *Crocidura erica* (VU) and *Crocidura nigricans*, are considered near endemic to this ecoregion.

Of the approximately 360 bird species found in the ecoregion, five are endemic or near endemic. Boulton's batis (*Batis margaritae*) is known only from here and the border area between Zambia and southern Democratic Republic of the Congo. Swierstra's francolin (*Francolinus swierstrai*, VU) is known only from a few montane areas in Angola (Collar and Stuart 1985). The near-endemic Angola cave-chat (*Xenocopsychus ansorgei*) is limited to a few rocky hills and cliffs and surrounding forest habitat in four isolated areas, two of which fall in the ecoregion: eastern Namibe and Huíla in the southern outlier of the ecoregion (Dean 2000). The near-endemic grey-striped francolin (*Francolinus griseostriatus*, VU) and Angolan slaty-flycatcher (*Dioptrornis brunneus*) occur mainly on the escarpment of the Angolan Scarp Savanna and Woodlands [81] but have also been recorded in this ecoregion (Dean 2000). A number of endemic and near-endemic bird subspecies are separated by more than 2,000 km from their nearest relatives in the mountains of Fernando Po, Cameroon, Rwenzori, Tanzania, Ethiopia, and Malawi. They include long-billed pipit (*Anthus similis moco*), mountain wheatear (*Oenanthe monticola nigricauda*), mountain nightjar (*Caprimulgus poliocephalus koesteri*), western green tinkerbird (*Pogoniulus coryphaeus angolensis*), evergreen forest-warbler (*Bradypterus lopezi boultoni*), Abyssinian hill babbler (*Pseudoalcippe abyssinica ansorgei*), and thick-billed seedeater (*Serinus burtoni tanganjicae*). Some other birds have an Afromontane distribution, such as bar-tailed trogon (*Apaloderma vittatum*), scarce swift (*Schoutedenapus myoptilus*), orange ground-thrush (*Zoothera gurneyi*), and African black swift (*Apus barbatus sladeniae*) (Dean 2000). The Fernando Po swift (*Apus sladeniae*, DD) occurs on Mount Môco (BirdLife International 2000).

The herpetofauna of the ecoregion is poorly documented. The link-marked sand racer (*Psammophis ansorgii*) is considered strictly endemic, and Marx's rough-scaled lizard (*Ichnotropis microlepidota*) near endemic, but research is severely lacking, and data are likely to be inaccurate. The green night adder (*Causus angolensis*) has recently been reported to be endemic in this ecoregion (D. Broadley, pers. comm., 2000). Notable among the amphibians is the strictly endemic *Hyperolius erythromelanus*.

Status and Threats

Current Status

The forests of this ecoregion are highly fragmented as a result of fires, agriculture, and woodcutting. The remaining forest patches seldom exceed 20 ha in size, and their total area probably is less than 200 ha (Huntley 1974b). The most extensive forest areas, at Mount Namba, were exploited for timber during the colonial period and are devoid of pristine patches. Undisturbed patches remain at Mount Môco between 1,800 and 2,400 m elevation (Huntley and Matos 1994). However, because of the lack of data, it is not known how extensive the forest patches once were and at what rate their extent and quality have changed since the 1970s. No protected areas currently exist in this ecoregion. Unless drastic conservation efforts are implemented, it is possible that little or nothing will remain of the forest patches and their fauna (Huntley 1974b; Huntley and Matos 1994).

Types and Severity of Threats

The almost continuous civil war in Angola since 1974 led to great instability, poor security, economic depression, displacement of the rural population, and a lack of infrastructure and basic services. As a result, water, sanitation, health, energy, and food are the most important items on the environmental agenda, against which most other issues, including conservation, pale in significance (Moyo et al. 1993). The highest human population densities in Angola outside Luanda are encountered in the highlands of this ecoregion. Population densities exceed thirty people per square kilometer (Moyo et al. 1993) as a result of the high agricultural potential of the area. The high population pressure, instability caused by war, and the lack of protected areas make conservation of the forest fragments and surrounding grasslands a daunting task.

Probably the greatest threat to the forest areas is from logging and other harvesting of forest products because the forests are already confined to steep, inaccessible slopes and ravines that are unsuitable for farming. The moist, often waterlogged grassland areas are unsuitable for agriculture and not greatly affected by fires. Better-drained areas are subject to frequent fires, although the vegetation is fire adapted. Clearing for agriculture probably is the greatest threat to the Afromontane grasslands and savannas, given the high agricultural potential and the dense human population. Hunting probably has eliminated large mammals.

Justification for Ecoregion Delineation

Much of the ecoregion follows the four areas of Afromontane vegetation in the Huambo and Cuanza Sul provinces and an area of Afromontane forest mosaic further south, on the Serra da Chela in the Huíla province, mapped by White (1983). The ecoregion also corresponds to the Bailundu Highlands portion of the western Angola endemic bird area (Stattersfield et al. 1998), although it was modified to the 1,200-m elevation contour. The ecoregion is distinguished from surrounding areas because its fauna and flora have Afromontane affinities.

Ecoregion Number:	**83**
Ecoregion Name:	**Jos Plateau Forest-Grassland Mosaic**
Bioregion:	**Western Africa and Sahel**
Biome:	**Montane Grasslands and Shrublands**
Political Units:	**Nigeria**
Ecoregion Size:	**13,300 km²**
Biological Distinctiveness:	**Bioregionally Outstanding**
Conservation Status:	**Critical**
Conservation Assessment:	**IV**
Author:	*Jan Lodewijk R. Were*
Reviewed by:	*David Happold*

Location and General Description

The Jos Plateau Forest-Grassland Mosaic [83] rises above the surrounding forest-grassland mosaic and savanna ecoregions of northern Nigeria. The plateau is the largest landmass above 1,000 m in Nigeria and is approximately 250 km by 150 km in size. It comprises high plains about 1,300 m above sea level and a number of granite hill ranges that reach heights over 1,900 m. On the west and south sides, the plateau is demarcated by a scarp 500–700 m high, but on the plateau's northern side the transition to the lower plains is less abrupt.

The Jos Plateau consists mainly of granites that are part of the Precambrian Basement Complex and form part of the land surface of the Gondwana supercontinent (Buchanan and Pugh 1955). The granites are particularly resistant to erosion and generally form shallow and sandy soils. In some areas there are also basaltic rocks, which weather to form deep clay loams (Buchanan and Pugh 1955).

Temperatures on the Jos Plateau are lower than the surrounding areas, with a minimum between 15.5°C and 18.5°C and a maximum between 27.5°C and 30.5°C. Rainfall is around 2,000 mm in the southwestern part of the plateau and declines to around 1,500 mm in the northeast (Happold 1987). Average rainfall for the town of Jos is 1,411 mm per year (Payne 1998). The heavier rains in the south and west result from the moisture-bearing winds meeting the escarpment at this point. The watershed pattern on the Jos Plateau is unusual in that its streams drain into different larger river systems. Some streams flow northeast to Kano and Lake Chad, east to the Gongola River (which enters the Benue), south to the Benue, and west to the Kaduna River, which feeds into the Niger River (Payne 1998).

White (1983) identifies this ecoregion as an isolated Afromontane vegetation unit in the Guinea-Congolian–Sudanian regional transition zone. Dense savanna woodland is likely to have been the climax vegetation for this ecoregion, but human activities have resulted in extensive and severe degradation. Vegetative formations can be divided into various savanna mosaics: woodland, riparian forest, and scrub regions. Although much of the original vegetation has been converted to agriculture, some remaining fragments show affinities with the East African Highlands and upland areas in West Africa (White 1993). Today, only a few remnants of woodland remain, and these are restricted to the steep and less accessible margins of the plateau, with open grassland occupying the remainder of the plateau. Forests are limited to the southern and western escarpments, river edges, and the base of rocky outcrops (Payne 1998).

White (1983) lists the following plant species in bushland and scrub forest on the plateau: *Carissa edulis, Dalbergia hostilis, Diospyros abyssinica, D. ferrea, Dodonaea viscosa, Euphorbia desmondii, E. kamerunica, E. poissonii, Ficus glumosa, Kleinia cliffordiana, Rhus longipes, R. natalensis, Ochna schweinfurthiana, Olea capensis, Opilia celtidifolia,* and *Pachystela brevipes.* The most numerous tree species in the woodland areas are *Isoberlinea doka, Vitex doniana, Lannea schimperi,* and *Uapaca somon.* The domination by species common to the lowland savanna areas that surround the plateau may not be a natural occurrence. The destruction of the original woodland over at least several hundred years and the presumed increased incidence of fires may have allowed these elements to spread to the Jos Plateau.

Outstanding or Distinctive Biodiversity Features

The Jos Plateau Forest-Grassland Mosaic [83], though small, contains a large number of endemic species. Two small mammals are strictly endemic: the Nigerian mole rat (*Cryptomys foxi*) and Fox's shaggy rat (*Dasymys foxi*). Several other mammals occur in Nigeria only on the Jos Plateau but are also found elsewhere in Africa, including the bushveld horseshoe bat (*Rhinolophus simulator*), greater long-fingered bat (*Miniopterus inflatus*), gray climbing mouse (*Dendromus melanotis*), and West African subspecies of klipspringer (*Oreotragus oreotragus porteousi*, EN). The two bat species and the dark-eared climbing mouse are also found in the Cameroon Highlands Forests [10], suggesting that there may have been a faunal and climatic link between these two areas in the past (Happold 1987). The distribution of klipspringer is of special biogeographic interest because this is the only West African population. With the exception of some small, isolated populations in Central African Republic (East 1999), the Nigerian klipspringers are separated by about 3,000 km from the nearest East African population, and much of that intervening habitat is not suitable for them because they inhabit

rocky outcrops and inselbergs exclusively. Now one of the rarest antelopes in Nigeria, klipspringers were at one point found in Bauchi, Bornu, and Zaria provinces and were widespread on the Bauchi Plateau (Happold 1987).

Two endemic birds occupy the Jos Plateau: the rock firefinch (*Lagonosticta sanguinodorsalis*) and its brood-parasite, the Jos Plateau indigobird (*Vidua maryae*) (Payne 1998). The Adamawa turtle dove (*Streptopelia hypopyrrha*) also occurs, further indicating affinities with the Cameroon Highlands Forests [10] (Payne 1998). Other notable bird species include Gambaga flycatcher (*Muscicapa gambagae*), booted eagle (*Hieraaetus pennatus*), red-headed lovebird (*Agapornis pullarius*), double-toothed barbet (*Lybius bidentatus*), grey-winged robin-chat (*Cossypha polioptera*), and pale-winged indigobird (*Vidua wilsoni*) (Ezealor 2001).

No endemic reptiles or amphibians are known from this plateau; however, several uncommon reptiles are found. These include Rodenburg's mabuya (*Mabuya rodenburgi*), Gambia agama (*Agama weidholzi*), *Cynisca rouxae*, and *C. senegalensis*.

Status and Threats

Current Status

The human population of the area is high, with about 200–300 people/km^2. Jos was a mining center through the 1920s, and development workers and Christian missions have been located here for decades (Payne 1998). Much of the habitat has been converted for agriculture, and firewood collection has been another major cause of deforestation. In and around villages most trees are useful species such as the oil palm (*Elaeis guineensis*), canarium (*Canarium schweinfurthii*), and mango (*Mangifera indica*), many of which are planted (Netting 1968). Native trees are largely to wholly absent in most human-inhabited areas, except for an occasional sacred grove. There are no officially protected areas.

Types and Severity of Threats

Most of the original woodland vegetation has been cleared or is limited to inaccessible areas or river margins. Agricultural activities are intense, particularly in areas located over basalt rock, where the soil is more fertile and well suited for crops, such as potatoes, that cannot be grown in most other parts of Nigeria. Although the threats are lower in areas where the substrate is granite, these areas are also heavily degraded. Firewood collection is intense and has resulted in the removal of most trees.

Justification for Ecoregion Delineation

The Jos Plateau ecoregion follows the "Jos Plateau mosaic" delineated by White (1983). It includes a number of species with South or East African affinities, such as the West African population of klipspringer, and a number of endemic mammals and birds. Although it shares some biological affinities with the Cameroon Highlands, it is considered sufficiently geographically distinct to warrant a separate ecoregion. The boundary of the ecoregion is taken from White (1983) but has been modified using the 1,000-m elevation contour.

Ecoregion Number:	**84**
Ecoregion Name:	**Madagascar Ericoid Thickets**
Bioregion:	**Madagascar–Indian Ocean**
Biome:	**Montane Grasslands and Shrublands**
Political Units:	**Madagascar**
Ecoregion Size:	**1,300 km²**
Biological Distinctiveness:	**Globally Outstanding**
Conservation Status:	**Endangered**
Conservation Assessment:	**I**
Author:	*Helen Crowley*
Reviewed by:	*Steve Goodman, Achille Raselimanana, Frank Hawkins*

Location and General Description

The Madagascar Ericoid Thickets [84] include the ericoid thicket habitats found above approximately 1,800 m on the upper slopes of Madagascar's four major massifs (listed from north to south): Tsaratanana (2,876 m), Marojejy (2,133 m), Ankaratra (2,643 m), and Andringitra (2,658 m). The transition from montane sclerophyllous forest to ericoid thicket occurs at different elevations on these massifs. On Tsaratanana, montane sclerophyllous forest exists up to about 2,500 m and then changes to ericoid thicket, whereas on these other three massifs the transition to ericoid thicket commences just below 2,000 m. The Anjanaharibe-Sud Special Reserve in the north and Andohahela National Park in the extreme southeast contain small areas of ericoid thicket above 1,950 m.

There is wide daily and seasonal fluctuation in the temperature of these montane areas. Snow has been recorded on the Andringitra Massif during the cold season, and the temperature can fall as low as −11°C (Saboureau 1962; Paulian et al. 1971). The maximum daily temperatures can exceed 30°C (Langrand and Goodman 1997). Rainfall probably is more than 2,500 mm per year on the wetter eastward-facing slopes of these massifs but is far less in the rainshadow on the western slopes. The great temperature range and intense sunlight can lead to temporary arid conditions. The massifs are made of metamorphic and igneous Precambrian basement rocks. Thin, nutrient-poor soils overlay these rocks (Du Puy and Moat 1996).

The upper montane sclerophyllous forest is dominated by plant species from the families Podocarpaceae, Cunoniaceae, Er-

iaceae, and Pandanaceae, and the trees are shrouded with mosses, lichens, and epiphytes. At higher altitudes, this forest gives way to the ericoid thicket, which is dominated by the Asteraceae (*Psiadia, Helichrysum, Stoebe*), Ericaceae (*Erica, Agauria, Vaccinium*), Podocarpaceae (*Podocarpus*), Rhamnaceae (*Phylica*), and Rubiaceae plant families. A wide variety of lichens and bryophytes are represented. Small, damp, peat-filled depressions harbor specialized, endemic plants, whereas rock outcrops host a drought-tolerant flora including *Aloe, Kalanchoe,* and *Helichrysum.* Plants that grow on rock outcrops are less threatened than other ericoid thicket species because the rocky areas block the local passage of fire.

Outstanding or Distinctive Biodiversity Features

On the major mountain massifs of Madagascar, species richness of numerous groups, such as birds, reptiles, and amphibians, decreases with altitude, whereas other taxa such as rodents and insectivores (Lipotyphla) show distinct midelevation bulges. At the elevations of the ericoid thicket there is a pronounced decrease in species richness for all of these groups (Goodman 1996, 1998, 1999, 2000; Hawkins 1999).

The flora is notable for containing a number of endemic species and elements linked to other biogeographic regions. There are more than 150 vascular plant endemics on the Andringitra massif, including 25 orchid species (Preston-Mafham 1991). Although endemism on these mountains is very high at the species level, they often belong to widespread genera. For example, the true heathers in the genus *Erica* are also represented in the Mascarenes and mainland Africa but have undergone extensive speciation in Madagascar (Guillaumet 1984; Dorr and Oliver 1999). There are other elements in the flora that link to other high montane areas in Africa and Europe (Vuilleumier and Monasterio 1986). For example, the Andohariana Plateau on Andringitra harbors *Gunnera perpensa*, which is also found in the Ethiopian Highlands, as well as species from temperate genera (*Rubus, Ranunculus, Geranium,* and *Alchemilla*) that are also found in East Africa. In addition, *Sedum madagascariense* is the only Malagasy representative of this predominantly northern latitudinal genus. Southern African species represented here include *Kniphofia* spp. and *Stoebe* spp. (Guillaumet 1984).

Until recently, very little was known about the composition of the vertebrate fauna in these montane habitats. A detailed survey of Andringitra Massif was carried out in 1993 (Goodman 1996), of Anjanaharibe-Sud in 1994 (Goodman 1998), of Andohahela in 1995 (Goodman 1999), and of Marojejy in 1996 (Goodman 2000). Two mammals are considered endemic to this ecoregion or the ecotone between it and the upper limit of the Madagascar Subhumid Forests [30]. Both have been described as new genera: *Monticolomys koopmani*, known from the massifs of Ankaratra, Andringitra, and Andohahela, and *Voalavo gymnocaudus*, apparently endemic to the Marojejy-Anjanaharibe-Sud Massifs. The upper reaches of Tsaratanana have never been surveyed for small mammals. Other near-endemic mammal species living at the middle to upper reaches of eastern mountains include the four-toed rice tenrec (*Oryzorictes tetradactylus*), found only in the south-central highlands, highland streaked tenrec (*Hemicentetes nigriceps*), several shrew-tenrecs (e.g., *Microgale gracilis*, VU; *M. gymnorhyncha*; and *M. monticola*), and a species of tuft-tailed rat (*Eliurus majori*, EN).

The characteristic bird fauna of these high mountains includes several species narrowly limited to this habitat, such as cryptic warbler (*Cryptosylvicola randrianasoloi*) and yellow-bellied sunbird-asity (*Neodrepanis hypoxantha*, EN). Other species are more common in lower open areas, such as stonechat (*Saxicola torquata*) and Malagasy brush-warbler (*Nesillas typica*) (Hawkins 1999).

The ericoid thicket supports endemic reptiles including the geckos *Millotisaurus mirabilis* and *Lygodactylus arnoulti*. Andringitra also has a newly discovered endemic Gekkonidae, *Lygodactylus montanus* (Raxworthy and Nussbaum 1996). At least one amphibian (*Boophis williamsi*) is strictly endemic, and five other species are nearly endemic to the ecoregion. Two chameleon species are restricted to the high-elevation zone of the Marojejy National Park (*Calumma peyrierasi*) and Andohahela National Park (*C. capuroni*). A new day gecko subspecies (*Phelsuma lineata*), a new plated lizard species (*Zonosaurus*), and some frogs have been discovered recently in the upper reaches of the Tsaratanana Massif.

Status and Threats

Current Status

There has been notable degradation of the natural vegetation of this ecoregion over the past century. The major threat is conversion to highland cattle pasture, as has already occurred on the Plateau d'Andohariana on the Andringitra Massif. Associated with these pasturelands are regular burns to stimulate young grass growth. In Tsaratanana and Ankaratra, frequent burning has degraded significant areas of the montane habitat. However, to some extent fires from lightning strikes may have always been part of the natural cycle in this region (C. Raxworthy, pers. comm., 2000), as indicated by the presence of high densities of fire-dependent plants such as Asteriaceae and Ericaceae.

Three of the four most important montane areas represented in this ecoregion are included in protected areas: Tsaratanana, Andringitra, and Marojejy, which comprise 171 km². Furthermore, there are also small areas of montane habitat protected in the Anjanaharibe-Sud and Andohahela reserves. The Andringitra and Marojejy Massifs have the best-preserved montane habitats. Ankaratra has only very small areas of native forest and no formal protection. The Manjakatompo Forest Station in the Ankaratra Massif includes 6.5 km² of native forest, but this area has no management plan. A priority-setting workshop held in 1995 identified the protection of high-altitude ecosystems as essential to preserve unknown ecosystems and to protect water quality and also recognized the Ankaratra region as an area of high biodiversity importance (Ganzhorn et al. 1997).

The high-elevation patches that make up this ecoregion are naturally fragmented by the dispersed location of the mountain ranges throughout Madagascar. Recent studies of the vicariant distribution patterns of several montane amphibians and reptiles suggest that there was a period in Madagascar's recent geological history during cool and dry glacial periods when there was a continuous belt of montane habitats between the Andringitra and Ankaratra Massifs (e.g., Raxworthy and Nussbaum 1996).

Type and Severity of Threats

Fire is the biggest threat to the ericoid thicket habitats. Fires are lit to promote pasture for cattle grazing, particularly on the drier western slopes of the mountains. This ecoregion is buffered, to some extent, by the surrounding forests of lower altitudes. However, these forests are experiencing increasing pressure from expansion of domestic animal rangelands. Furthermore, these animals may disperse through their feces the seeds of introduced plants into ericoid thickets.

Justification for Ecoregion Delineation

The boundaries of this ecoregion follow Cornet's (1974) montane bioclimate boundaries. The ecoregion includes the summits of Marojejy, Tsaratanana, Ankaratra, and Andringitra massifs above the 1,800-m contour.

Ecoregion Number:	**85**
Ecoregion Name:	**Mediterranean Woodlands and Forests**
Bioregion:	**African Palearctic**
Biome:	**Mediterranean Forests, Woodlands, and Scrub**
Political Units:	**Algeria, Ceuta (Spain), Libya, Melilla (Spain), Morocco, Tunisia**
Ecoregion Size:	**358,300 km²**
Biological Distinctiveness:	**Globally Outstanding**
Conservation Status:	**Endangered**
Conservation Assessment:	**I**
Authors:	*Nora Berrahmouni, Pedro Regato*
Reviewed by:	*Abdelmalek Benabid, Hans Peter Müller*

Location and General Description

The Mediterranean Woodlands and Forests [85] include the low- and mid-elevation of the northern half of Morocco, Algeria, and

Tunisia and two Spanish sovereign areas, Ceuta and Melilla, located in Morocco. An additional, isolated portion of the ecoregion occurs in the Cyrenaic Peninsula of Libya (Jebel al Akhdar). Coastal plains characterize the northern half of the Atlantic coast of Morocco and the eastern coast of Tunisia. Hilly land, valleys, and plateaus alternate in the hinterland. Geologically, the ecoregion is extremely diverse and consists of a large variety of Mesozoic and Quaternary sedimentary rocks such as sand, sandstone, conglomerate, mudstone, limestone, dolomite, marl, and evaporite sediments, formed by desiccation in endorheic and coastal areas.

The ecoregion experiences hot and dry summers, with mild and humid winters. Mean annual temperatures range from 13°C to 19°C, and the mean minimum temperature ranges from 1°C to 10°C. The portion of this ecoregion located on the Atlantic coast of Morocco is influenced by cold offshore currents, which tend to moderate the temperatures. Annual rainfall ranges from 350 to 800 mm. A combination of different climates, geologies, and landforms results in five major forest types in this ecoregion. These are xeric pine forests, Berber thuya (*Tetraclinis articulata*) forests (Fennane 1989), cork oak (*Quercus suber*) forests, holm oak (*Quercus ilex ballota*) and holly oak (*Quercus ilex*) forests, and wild olive (*Olea europaea* and *O. maroccana*) and carob (*Ceratonia siliqua*) woodlands and maquis.

Xeric pine forests constitute mixed stands of Aleppo pine (*Pinus halepensis*) with evergreen holm oak (*Quercus ilex ballota*) and xeric juniper species (*Juniperus phoenicea, J. oxycedrus*) (Djebaili 1978, 1990; Benabid 1985; Kaabache 1993; Mediouni 2000). Berber thuya forests are distributed mainly in the dry and mild lowlands and hills of the northern half of the Atlantic and Mediterranean coasts of Morocco, the western half of the Algerian coast, and some mountain areas along the northeastern coast of Tunisia. These forests extend over almost 10,000 km² in North Africa: 7,500 km² in Morocco, 1,600 km² in Algeria, and about 220 km² in Tunisia, mainly on limestone substrates (Fennane 1989; Charco 1999; Mediouni 2000). Cork oak forests are widely distributed throughout the western Mediterranean along the coast from low and medium elevations on siliceous substrates. In this ecoregion, they are found in northern Morocco along the coastal plains between Casablanca and the Rif and in several hinterland areas around the Rif and Middle Atlas. In northern Algeria they occur along the Tellien Atlas and in northern Tunisia along the Kroumerie-Mogod mountain ranges. Cork oak forests grow from sea level up to 1,500 m in humid and warm climates (Benabid 1985; Mediouni 2000; WWF MedPO 2001). Holm oak and holly oak forests extend in North Africa over 20,000 km²: 14,320 km² in Morocco, 6,800 km² in Algeria, and 15 km² in Tunisia. They are widely distributed from the coast to the high altitudes (2,500–2,900 m) of the main mountain ranges. Holm oak can withstand large temperature and rainfall ranges and grows on a variety of substrates. Holly oak normally constitutes dense maquis and small forest stands in humid and warm climates on all types of substrates (Nabli 1995). Wild olive and carob woodlands and secondary dense shrub maquis were once widespread in the fertile soils of the dry coastal and inland plains. Now, much of the region has been transformed into agricultural land. Only a few remnants—small stands of trees kept in sacred areas called *marabout*—resemble the original forest structure (Sadki 1995; Mediouni 2000).

Outstanding or Distinctive Biological Features

Little quantitative data exist on the endemic vascular plant species of the ecoregion. Nevertheless, endemism rates are assumed to be high, as in other Mediterranean ecoregions. One of the dominant tree species of this ecoregion, the Berber thuya, is an endemic Tertiary relict whose only living relatives (*Callitris* spp.) are found in South Africa and Australia. Many plant taxa related to these forest ecosystems have a very restricted distribution range and are threatened with extinction (Walter and Gillett 1998; Oldfield et al. 1998).

The fauna is composed mainly of generalist species; nevertheless, species richness is among the highest in the Palearctic realm. Small mammals include a number of species restricted to northern Africa, such as white-toothed pygmy shrew (*Suncus etruscus*), Barbary striped grass mouse (*Lemniscomys barbarus*), North African hedgehog (*Atelerix algirus*), North African elephant shrew (*Elephantulus rozeti*), Barbary ground squirrel (*Atlantoxerus getulus*), and North African gerbil (*Gerbillus campestris*). Large mammals include species typical of the Palearctic, such as the red fox (*Vulpes vulpes*), wild boar (*Sus scrofa*), and European polecat (*Mustela putorius*) and those more typical of Africa, such as the striped hyena (*Hyaena hyaena*), small-spotted genet (*Genetta genetta*), and Egyptian mongoose (*Herpestes ichneumon*). This ecoregion may also harbor Barbary leopards (*Panthera pardus panthera*, CR), although these rare cats are confined mostly to remote montane and rugged foothill areas (Nowell et al. 1996). The near-endemic Barbary macaque (*Macaca sylvanus*, VU), Cuvier's gazelle (*Gazella cuvieri*, EN), and aoudad, or Barbary sheep (*Ammotragus lervia*, VU), may also occur here.

The avian community, with more than 120 species, includes an endemic subspecies of great spotted woodpecker, *Dendrocopos major numidus;* an endemic subspecies of grey shrike, *Lanius meridionalis algeriensis;* the threatened Algerian nuthatch (*Sitta ledanti*, EN); and the northwest African endemic Moussier's redstart (*Phoenicurus moussieri*). Other North African endemic subspecies present are the North African red crossbill (*Loxia curvirostra poliogyna*) and North African green woodpecker (*Picus viridis levaillanti*).

Reptiles are well represented and include a number of strict endemic species, such as the banded lizard-fingered gecko (*Saurodactylus fasciatus*), three cylindrical skinks (*Chalcides colosii, C. ebneri,* and *C. mauritanicus*), and two fringe-fingered lizards (*Acanthodactylus blanci* and *A. savignyi*). There are also a similar number of near-endemic reptiles that are shared only with other North African Mediterranean ecoregions, such as Koelliker's glass lizard (*Ophisaurus koellikeri*), the only anguid in North Africa. The threatened spurred tortoise (*Testudo graeca*, VU) also occurs. Widespread amphibian species include the Mediterranean tree frog (*Hyla meridionalis*), Spanish ribbed newt (*Pleurodeles waltl*), Moroccan spadefoot toad (*Pelobates varaldii*), and painted frog (*Discoglossus pictus*).

Status and Threats

Current Status

The original forest cover has been dramatically reduced and converted into agriculture and pastureland. Clearance has occurred at least since Roman times, when favorable weather conditions and fertile soils suited the growth of human settlements. This ecoregion is the most populous in all the Maghreb, containing 80–90 percent of the population.

The holm oak forests and the wild olive and carob forests once covered 50,000 km^2 each. Today in Algeria, only 1,000 km^2 of the original 10,000 km^2 of wild olive and carob forests remains, and only 6,800 km^2 of the original 18,000 km^{2}of holm oak forest remains. In Morocco, 5,000 km^2 of the estimated 36,240 km^2 original wild olive and carob forests remains, and 14,320 km^2 of the original 24,500 km^2 holm oak forest remains. Degraded shrublike communities represent the majority of this forest coverage, and the area of remaining good forest is much smaller.

The original extent of cork oak forests in North Africa is estimated to have been 30,000 km^2, but less than one-third of this exists today (3,500 km^2 in Morocco, 4,500 km^2 in Algeria, and 455 km^2 in Tunisia). Tannin production from the cork oak trunk, which occurred at the end of the nineteenth century and beginning of the twentieth century, contributed to the destruction of significant forest areas.

The southernmost cedar stands in the Eastern High Atlas are under imminent threat of extinction (Oldfield et al. 1998). Holly oak forests have never been inventoried because they are severely degraded and hold little value for forestry. Nonetheless, most of the original cover has been converted into agricultural land. Degraded shrublike communities dominate elsewhere as a result of intense overgrazing and fire management on pasture land, intense pruning and uprooting for firewood collection, and tannin production.

Although resources and funding are low, there are several national parks in this region. Al Hoceima National Park in Morocco includes a large marine area, but also contains some holm oak woodlands. Boukournine, Chaambi, Ichkeul, and the proposed Zaghouan national parks in Tunisia conserve holm oak and Aleppo pine forests, as do the Tlemcen and Gouraya national parks in Algeria.

Types and Severity of Threats

Human impact remains high, especially because this ecoregion contains most of the North African human settlements. High population growth results in rapid and intense land conversion, mainly for agriculture, urban development, industries, and quarries.

The northern Africa evergreen oak forest is susceptible to climate change. The intensified summer drought season and increasing average annual temperatures create stressful conditions for many species. In addition, human overexploitation of silvopastoral systems and soil degradation have reduced the forest's resilience to natural disturbances; during periods of intense drought, extensive tree stands suddenly die.

Justification for Ecoregion Delineation

The boundaries of this ecoregion are taken directly from the "Mediterranean sclerophyllous forest" vegetation unit of White (1983), with the addition of White's "Mediterranean anthropic landscape." The latter was added because the potential vegetation of this region before cultivation is likely to have been sclerophyllous forest. The ecoregion contains high numbers of endemic plants and some endemic animals.

Ecoregion Number:	**86**
Ecoregion Name:	**Mediterranean Dry Woodlands and Steppe**
Bioregion:	**African Palearctic**
Biome:	**Mediterranean Forests, Woodlands, and Scrub**
Ecoregion Size:	**292,200 km^2**
Biological Distinctiveness:	**Locally Important**
Conservation Status:	**Relatively Stable**
Conservation Assessment:	**V**
Authors:	*Nora Berrahmouni, Pedro Regato*
Reviewed by:	*Abdelmalek Benabid, Hans Peter Müller*

Location and General Description

The Mediterranean Dry Woodlands and Steppe [86] forms a wide band across North Africa, just inland of the moister Mediterranean Woodlands and Forests [85]. In the west, it extends from eastern Morocco, across northern Algeria, and into Tunisia, where it reaches the Mediterranean Sea. To the east are two disjunct regions located on opposite sides of the Gulf of Sirte in Libya. The first occurs to the north of the Jebel Nefussa (the Jefara Plain), and the second lies in the Jebel el Akhdar (Cyrenaica). Another disjunct area lies in northeast Egypt to the west of the Nile Delta. The ecoregion is particularly characterized by the presence of calcareous slabs called *dalles*.

The climate is quite arid, with an annual rainfall of 100–300 mm. Storms are part of the rainfall regime, occurring mostly during the winter. The temperature can fall close to 0°C in winter and rise over 40°C in the summer. The annual mean temperature is approximately 18°C.

In terms of the phytogeographic classification of White (1983), this ecoregion is considered part of the Mediterranean-Sahara regional transition zone. The main vegetation types are forests, matorrals, *steppes arborées*, chamaephytic and graminean steppes. A large area is also occupied by sand and halophile landscapes. On the south side of the two Atlas (Tellien and Saharien) mountain blocks, characteristic vegetation contains a mixture of *Pinus halepensis* and *Juniperus phoenicea* but is replaced by *steppe arborée* of *Juniperus phoenicea* and *Stipa tenacissima* further inland. Other species typical of the *steppe arborée* are *Globularia alypum*, *Salsola vermiculata*, *Thymus ciliatus*, *Helianthemum virgatum*, *Cistus libanotis*, *Rosmarinus tournefortii*, and *Asparagus stipularis*. On the higher plateau, vegetation types vary according to soil. *Stipa tenacissima* steppe is found on Atlas slopes and glacis with argilo-sandy soils, *Artemisia* "herba alba" steppe in silty glacis and depressions, and *Lygeum spartum* steppe in sandy accumulations. In sandy areas there is also a mosaic of vegetation types such as *Thymelaea microphylla*, *Aristida pungens*, *Retama retam*, and *Tamarix* sp. *Dayas* (depressions with good quality soil) vegetation may include betoum (*Pistacia atlantica*), *Ziziphus lotus*, *Anvillea radiata*, *Bubonium graveolens*, and *Malva aegyptiaca*.

Outstanding or Distinctive Biodiversity Features

The flora contains mainly widespread species, although there are some endemic species and some centers of localized endemism. For example, the area around the Jebal al Akhdar in the Libyan Arab Jamahiriya contains an enclave of nearly 100 endemic plant species, including *Arbutus pavarii*, *Crocus boulosii*, and *Cyclamen rohlfsianum* (White 1983).

The fauna is also composed mainly of widespread species, with no strict endemic species in any vertebrate group. Small mammals include *Gerbillus latastei*, *G. syrticus* (CR), *G. grobbeni* (CR), *Allataga tetradactyla* (EN), and *Microtus guentheri*. There are still populations of the Barbary sheep, or aoudad (*Ammotragus lervia*, VU), but they are severely threatened by habitat destruction and hunting (De Smet 1989). A few populations of gazelles still survive; there are about 400 individuals of Cuvier's gazelle (*Gazella cuvieri*, EN) in the Mergub Nature Reserve in Algeria, and in Tunisia there are more than 600 Cuvier's gazelles in the protected areas between the Chambi National Park and the Algerian border. The local subspecies of striped hyena (*Hyaena hyaena barbara*, DD) also persists (Hilton-Taylor 2000). Near-endemics and otherwise notable species are uncommon among reptiles, amphibians, and birds.

The ecoregion supports a number of mammal species typical of the Palearctic realm, such as wild boar (*Sus scrofa*), European otter (*Lutra lutra*), and red fox (*Vulpes vulpes*). Palearctic reptiles and amphibians are also found, including grass snake (*Natrix natrix*), viperine snake (*Natrix maura*), and the amphibian *Bufo viridis*.

Many of the species survive as highly reduced populations, and some species have been extirpated because of human activities (Le Houérou 1991). For example, the ostrich (*Struthio camelus*) was fairly common in the northern Sahara at the end of the nineteenth century but was extirpated from the area by the early twentieth century.

Status and Threats

Current Status

Debate continues on whether this ecoregion was originally forested. Patches of forest consisting principally of *Pinus halepensis*, *Juniperus phoenicea*, and *Quercus ilex* remain, especially in the mountains (White 1983). Elsewhere forest does not exist; instead the vegetation is dominated by species typical of drier areas. In both the Libyan Arab Jamahiriya and in Egypt, the vegetation has been severely degraded by grazing pressure and collection of woody plants for fuel. In the High Atlas, the vegetation has also been damaged by removal of the *Argania* scrub, which has allowed replacement in the west by species of *Euphorbia*. Currently much of the landscape is dominated by a mosaic of grassland consisting almost entirely of *Stipa tenacissima* or *Lygeum spartum*, alternating with patches of dwarf *Artemisia* "herba alba" shrubland.

There are a limited number of protected areas in this ecoregion, including the Karabolli National Park (150 km^2), Bier Ayyad Nature Reserve (20 km^2), and Nefhusa (200 km^2) in Libya. In Algeria, protected areas include Mergueb Nature Reserve (1,200 km^2), which has not yet attained legal recognition, and Djelfa Hunting Reserve (320 km^2). There are two national parks in Tunisia, Bou-Hedma (165 km^2) and Chaambi (67 km^2), and a permanent hunting reserve in Morocco, Bouarfa (2,200 km^2).

Types and Severity of Threats

The main threats to this ecoregion are the removal of remaining woody vegetation and overgrazing. Most ecoregion residents are pastoralists with herds of sheep and goats. During the dry season pastoralists move to the wetter Mediterranean Woodlands and Forests [85], where grazing opportunities are better; they move back to the Mediterranean Dry Woodlands and Steppe [86] during the rainy season. Permanently settled nomads often engage in a combination of agriculture and pastoralism, which is contributing to desertification.

Justification for Ecoregion Delineation

This ecoregion is based largely on the "sub-Mediterranean semi-desert grassland and shrubland" vegetation unit of White (1983). It contains a distinctive flora and fauna, with low rates of endemism. The extension west of 30°N and 8°W is included with the Mediterranean *Acacia-Argania* Dry Woodlands and Succulent Thickets [87] because of the area's distinctive flora, including *Argania spinosa* (White 1983).

Ecoregion Number:	**87**
Ecoregion Name:	**Mediterranean *Acacia-Argania* Dry Woodlands and Succulent Thickets**
Bioregion:	**African Palearctic**
Biome:	**Mediterranean Forests, Woodlands, and Scrub**
Political Units:	**Morocco, Western Sahara (Morocco), Canary Islands (Spain)**
Ecoregion Size:	**100,100 km²**
Biological Distinctiveness:	**Regionally Outstanding**
Conservation Status:	**Vulnerable**
Conservation Assessment:	**II**
Authors:	*Nora Berrahmouni, Pedro Regato*
Reviewed by:	*Abdelmalek Benabid, Hans Peter Müller*

Location and General Description

The Mediterranean *Acacia-Argania* Dry Woodlands and Succulent Thickets [87] fall mainly in Morocco, extending into the northwestern corner of Western Sahara and the eastern islands of the Canaries. On the African mainland, the ecoregion occupies the Atlantic coastal plain, the lowlands of Haouz-Tadla, the Souss and Draa valleys, and the western end of the High and Anti-Atlas Mountains. The mainland section of the ecoregion is flat and lies below 800 m. The geology is varied and includes Cretaceous- to Tertiary-aged calcareous and sandy deposits. Soils are poor.

The two easternmost Canary Islands of Fuerteventura and Lanzarote and numerous associated islets (e.g., Graciosa) are also included in the ecoregion because they are similar in climate and topography. The eastern islands of the Canaries are older and less rugged than the western islands. Although the Canary Islands are volcanic in origin, neither Fuerteventura nor Lanzarote, estimated to be 16–20 million years old, is still active; their highest point reaches only 807 m.

The ecoregion is subtropical, with mild, frost-free winters and cool summers due to the moderating influence of the sea

(especially on the Canaries). Generally, rainfall is between 100 and 500 mm per year, although it may fall as low as 50 mm in the Algerian part of the ecoregion. The mean annual temperature is constant, generally ranging between 18°C and 20°C. This stability is maintained by the moderating influence of the Atlantic Ocean. However, the maximum temperature can reach 50°C further inland, closer to the Sahara Desert.

Argania spinosa forest and *Euphorbia*-dominant succulent shrubland (along the coast represented by the species *Euphorbia regis-jubae* and farther inland by *Euphorbia balsamifera*) are the predominant vegetation types. The major species associated with the *Argania* forests are *Periploca laevigata, Launaea arborescens, Warionia saharae, Acacia gummifera, Rhus tripartitum, Withania frutescens, Euphorbia officinarum, Cytisus albidus, Ephedra altissima,* and *Tetraclinis articulata. Acacia gummifera* is a common companion species to the argan tree. *Balanites aegyptiaca* and *Maerua crassifolia* commonly grow among *Acacia-Argania* woodlands in the eastern part of the ecoregion.

Outstanding or Distinctive Biological Features

The islands of Fuerteventura and Lanzarote are important areas for plant endemism and in general contain more biological features of interest than the mainland. The islets between the lava-covered terrain in Timanfaya National Park support three plant species that are endemic to Lanzarote: *Echium pitardii, Odontospermum intermedium,* and *Polycarpa robusta.* On the coastal dunes of these two islands, local endemics, such as *Androcymbium psammophilum,* and North African plants, such as *Limonium tuberculatum* and *Traganum moquinii,* are found. The mainland portion of the ecoregion also possesses a number of endemic plant species that are related to the Macaronesian flora on the Canary Islands. For example, a subspecies of the Canary Islands *Dracaena drago* was recently discovered on the African mainland in the mountains northeast of Anezi between Tiznit and Tafraout. However, this area is less important for plant endemics than the Canary Islands.

The ecoregion contains a few strictly endemic mammals, including the Canary shrew (*Crocidura canariensis,* VU), Hoogstraal's gerbil (*Gerbillus hoogstraali,* CR), and Occidental gerbil (*Gerbillus occiduus,* CR). Other mainland species include common jackal (*Canis aureus*), striped hyena (*Hyaena hyaena*), honey badger (*Mellivora capensis*), wildcat (*Felis silvestris*), Egyptian mongoose (*Herpestes ichneumon*), crested porcupine (*Hystrix cristata*), Barbary ground squirrel (*Atlantoxerus getulus*), North African elephant shrew (*Elephantulus rozeti*), Barbary striped grass mouse (*Lemniscomys barbarus*), and wild boar (*Sus scrofa barbarus*). Two gazelle species, dorcas (*Gazella dorcas,* VU) and Cuvier's (*Gazella cuvieri,* EN), and the Barbary sheep (*Ammotragus lervia,* VU) are under threat of extinction.

The eastern Canary Islands are home to the endemic Canary Islands chat (*Saxicola dacotiae*) and used to support the now extinct Canary Island oystercatcher (*Haematopus meadewaldoi*, EX). There are also endemic subspecies of kestrel (*Falco tinnunculus dacotiae*), Houbara bustard (*Chlamydotis undulata fuertaventurae*), barn owl (*Tyto alba gracilirostris*), Eurasian thick-knee (*Burhinus oedicnemus insularum*), and cream-colored courser (*Cursorius cursor bannermani*). A number of other Canary Island endemics occur, which are also found on other islands in the group, including Berthelot's pipit (*Anthus berthelotii*) and subspecies of spectacled warbler (*Sylvia conspicillata orbitalis*), great grey shrike (*Lanius excubitor koenigi*), and lesser short-toed lark (*Calandrella rufescens polatzeki*).

The mainland fauna comprises a mixture of Palearctic, Afrotropical, and locally endemic species. Between the Cap Rhir in the north of Agadir and the Bas Draa estuary lie the estuaries of Tamri, Souss, and Massa, which are important nesting areas and wintering places for a large number of birds. The Souss-Massa National Park includes an important breeding population of the near-endemic northern bald ibis (*Geronticus eremita*, CR), which is also found in one or two other arid parts of the Mediterranean Basin and the Middle East.

Endemic amphibians and reptiles include the Brongersmai toad (*Bufo brongersmai*) and the near-endemic Morocco lizard-fingered gecko (*Saurodactylus mauritanicus*) and Mionecton cylindrical skink (*Chalcides mionecton*). The Draa Valley supports the endemic Böhme's gecko (*Tarentola boehmei*), Hoggar gecko (*Tarentola ephippiata hoggarensis*), and Quedenfeldtia gecko (*Quedenfeldtia moerens*). In addition, both the eastern Canary gecko (*Tarentola angustimentalis*) and the eastern Canary skink (*Chalcides polylepis occidentalis*) are endemic to the two larger islands and associated smaller islets in the eastern Canaries (Clarke and Collins 1996; Hilton-Taylor 2000).

Status and Threats

Current Status

In Morocco, the argan tree covers around 6,500 km^2 and makes up about 7 percent of the total forest cover of the country. Constituting a multiple-use silvopastoral system, argan woodlands are managed for a large number of products, such as oil, pasture, honey, charcoal, and construction wood. The most intact areas are found in protected areas. The Souss-Massa National Park includes the Oued Massa Biological Reserve (Peltier 1983). The new Bas Draa National Park (3,000 km^2) was established in September 2002. The Khnifiss–Puerto Cansado area is a Ramsar and Biosphere Reserve site. The Arganeraie Biosphere Reserve was established in 1998 and was the first of its kind in Morocco. It includes argan woodlands in addition to urban and agricultural areas.

Habitats on Fuerteventura, Lanzarote, and the smaller islets, including Graciosa, are much less fragmented than their western Canary counterparts. They have fewer inhabitants and

have a more arid climate that is generally unsuitable for agriculture. Unfortunately, they have also been badly affected by overgrazing. The system of protected areas in the eastern Canaries includes Timanfaya National Park (51 km^2) and the large Islotes y Famara (89 km^2), Pozo Negro (92 km^2), and Jandîa (143 km^2) nature parks. In addition, the United Nations Educational, Scientific and Cultural Organization (UNESCO) Man and Biosphere program has declared Lanzarote to be one of three Canary Island World Heritage Sites.

Types and Severity of Threats

In Morocco, overgrazing and overexploitation of argan trees are serious threats to natural vegetation. Despite legal protection, argan woodlands suffer from continuous degradation caused by the abandonment of traditional management practices and the intensification of their use. In the last 20 years, intensive farming activities have been increasing, mainly in the Souss region (M'Hirit et al. 1998).

Nonnative species are a serious problem in the Canary Islands, and the introduction of some species of fauna (squirrel, rabbit, mice, cats) has posed a serious threat to several of the endemic animals and to breeding seabirds. The use of off-road vehicles on the dunes of Fuerteventura and in the south of Lanzarote at Playa de los Papagayos has badly damaged the dunes and the vegetation of these areas.

Justification for Ecoregion Delineation

The boundaries of this ecoregion reflect a combination of White's (1983) vegetation units, including the southwestern portion of the "Northern Sahel semi-desert grassland and shrubland," the "transition from Mediterranean Argania scrubland to succulent semi-desert shrubland," and an anthropic landscape that probably once reflected these former vegetation types. Experts also included the eastern Canary Islands (including Lanzarote and Fuerteventura) because the lower-lying and more arid nature of these islands made them more similar to the mainland ecoregion than to the dry woodlands of the western Canary Islands.

Ecoregion Number: **88**

Ecoregion Name: **Canary Islands Dry Woodlands and Forests**

Bioregion: **African Palearctic**

Biome: **Mediterranean Forests, Woodlands, and Scrub**

Political Units: **Spain**

Ecoregion Size: **5,000 km^2**

Biological Distinctiveness: **Globally Outstanding**

Conservation Status: **Vulnerable**

Conservation Assessment: **I**

Authors: *Jaime A. de Urioste, Maria Jose Bethencourt Linares*

Reviewed by: *José María Fernández-Palacios, Tony Clarke*

Location and General Description

The Canary Islands lie a minimum of 96 km off the northwest coast of Africa between 27°20′N and 29°25′N and between 13°20′W and 18°10′W. The archipelago is divided into two ecological groups. The eastern islands form part of the Mediterranean *Acacia-Argania* Dry Woodlands and Succulent Thickets [87], and the western islands are in the Canary Islands Dry Woodlands and Forests [88]. The western islands forming this ecoregion are younger, wetter, and more topographically varied than those of the east and include Gran Canaria, Tenerife, La Gomera, La Palma, and El Hierro.

The Canary Islands have a recent geological past, with the oldest eastern islands being around 20 million years old (González et al. 1986). La Palma and El Hierro are the youngest in the archipelago, dating back 2–3 million years. The geology is entirely volcanic, and some islands have active volcanoes, with the most recent eruption on La Palma in 1971.

The climate of the Canary Islands is subtropical and varies according to elevation and slope, but it is especially influenced by the northeasterly trade winds, called *alisios*. The *alisios* deposit their rain on the northern slopes of the higher mountains in the Canary Islands (over 750 m). The southern parts of the islands remain drier and have proportionally higher temperatures and lower humidity levels. Sometimes, easterly dust-laden dry winds blow from the Sahara called *calima* or *calina* (Bacallado et al. 1984; González et al. 1986; Marzol 1998). Most rain falls during autumn and winter (November–March). In coastal zones, precipitation ranges between 100 and 350 mm per year. At elevations from 250 to 600 m rainfall is approximately 650 mm, and the annual precipitation at elevations above 600 m is around 1,000 mm (Bacallado et al. 1984; González et al. 1986; Marzol 1998). Tenerife has snow at the

highest elevations in the winter. Most rivers on the islands are dry during the summer, but El Teide flows year-round, fed by the snowmelt on Tenerife.

Vegetation can be described according to elevation zones. At the lowest elevations, subdesert scrub vegetation predominates, although groves of endemic palms (*Phoenix canariensis*) are also present. Coastal vegetation occurs from sea level to 400 m in the north and up to 1,000 m in the south. Along the transition zone from 400 to 600 m, between the sea-level coastal community and below the laurilsilva, there are thermophile woodlands. The species found here are common to both the lower and higher vegetation formations. This zone has been damaged for decades because of its good climate and growing medium for crops. Humid laurilsilva grows between 500 and 1,400 m in elevation only at the windward slope, with some species reaching more than 20 m in height. Laurisilva is formed by several taxa grouped in different families, but there are four typical species: *Ocotea foetens*, *Apollonias barbujana*, *Laurus azorica*, and *Persea indica*. Macronesian heaths, also known as *fayal-brezal*, grow from 500 to 1,700 m, in humid zones in young soils. There are three distinctive heath species: *Myrica faya*, *Erica arborea*, and *E. scoparia* (González et al. 1986). Canarian endemic pine forests (*Pinus canariensis*) range down to almost 500 m altitude in southern areas but in the northern parts of the islands are found between 1,200 and 2,000 m in elevation. On the highest tops of the mountains of La Palma and Tenerife there is scrub vegetation above 2,000 m.

Outstanding or Distinctive Biodiversity Features

This ecoregion contains high levels of animal and plant endemism. Every plant family has endemic representatives, often including endemic genera. More than one-fourth of the 1,992 vascular plants on these islands are endemic (Machado 1998). Endemicity in nonmigratory vertebrates is 17 percent, 44 percent in arthropod invertebrates and 29 percent in nonarthropod invertebrates (Machado 1998). All the terrestrial reptiles are endemic. These figures include only animals restricted to this ecoregion, but there are other species and subspecies with small distributions, endemic to all of the Canaries (including the eastern islands).

Each of the various plant vegetation zones contains endemic plant species. In the coastal lowlands endemic species are mainly from the Euphorbiaceae, such as *Euphorbia canariensis*. Other endemics include *Ceropegia fusca*, *Plocama pendula*, *Salvia canariensis*, *Argyranthemum frutescens*, *Rumex lunaria*, *Convolvulus floridus*, and *Messerschmidia fruticosa* (Bramwell and Bramwell 1983; González et al. 1986; Strasburger et al. 1986). In the medium-altitude woodlands, endemic plants include *Bosea yervamora*, *Echium strictum*, *Greenovia aurea*, *Aeonium* sp., *Monanthes laxiflora*, and *Dracaena draco*. Some of the endemic plants in laurisilva are *Ocotea foetens*, *Apollonias barbujana*, *Laurus azor-*

ica, Persea indica, Arbutus canariensis, Ilex canariensis, Visnea mocanera, Picconia excelsa, Heberdenia excelsa, Salix canariensis, and *Viburnum tinus.* Although Canarian endemic pine forests contain fewer plant species than other vegetation formations, they have a large number of endemics in all plant groups, including fungi and lichens. Some of these Canarian endemic plants are *Bystropogon plumosus, Aeonium spathulatum, Asparagus plocamoides, Tolpis laciniata,* and *Teline* sp. Some of the endemic plant species in the high mountain areas are *Spartocytisus supranubius, Erysimum scoparium, Nepeta teydea, Plantago webbii, Senecio palmensis, Juniperus cedrus, Polycarpaea tenuis,* and *Echium* sp. (Bramwell and Bramwell 1983; González et al. 1986; Marzol 1998).

The Madeira pipistrelle (*Pipistrellus maderensis,* VU) and Canary big-eared bat (*Plecotus teneriffae,* VU) are near endemic to this ecoregion (but endemic to the Canary Archipelago) (Bacallado et al. 1984; Trujillo 1991). Four birds are strictly endemic to the ecoregion: Bolle's pigeon (*Columba bollii*), laurel pigeon (*Columba junoniae,* VU), blue chaffinch (*Fringilla teydea*), and Canary Islands kinglet (*Regulus teneriffae*) (Bacallado et al. 1984; Heinzel et al. 1992; Moreno 1988). Berthelot's pipit (*Anthus berthelotii*), plain swift (*Apus unicolor*), and Atlantic canary (*Serinus canaria*) are near endemic to this ecoregion (Bacallado et al. 1984; Moreno 1988; Heinzel et al. 1992). The Canary Island oystercatcher (*Haematopus meadewaldoi*) is now considered extinct (Hilton-Taylor 2000). Bird subspecies restricted to the Canary Islands include a grey wagtail (*Motacilla cinerea canariensis*), a long-eared owl (*Asio otus canariensis*), three chaffinch subspecies (*Fringilla coelebs tintillon, F. c. ombriosa,* and *F. c. palmae*), three European blue tit subspecies (*Parus caeruleus teneriffae, P. c. ombriosus,* and *P. c. palmensis*), two great spotted woodpecker subspecies (*Dendrocopos major canariensis* and *D. m. thanneri*), and two kestrel subspecies (*Falco tinnunculus teneriffae,* and *F. t. dacotiae*) (Bacallado et al. 1984; Moreno 1988; Heinzel et al. 1992; Helbig et al. 1996). Marine birds use the archipelago as a nesting place, such as the Manx shearwater (*Puffinus puffinus*), which nest in gullies in laurisilva (Martín 1987).

Each island has its own endemic species or subspecies of lizard, skink, or gecko. In the recent past (decades in some cases) giant lizards (near 150 cm long) inhabited the Canaries, but these are now extinct. Smaller relatives of these extinct reptiles still live in cliffs and crevices of islands such as El Hierro, La Gomera, and Tenerife. Strictly endemic reptiles include Tenerife wall gecko (*Tarentola delalandii*), Gomero wall gecko (*Tarentola gomerensis*), six-lined cylindrical skink (*Chalcides sexlineatus*), and Canaryan cylindrical skink (*Chalcides viridanus*). Island-specific endemicity is even more spectacular in invertebrates, such as beetles and butterflies (Machado 1998). The Geometridae family (Lepidoptera) contains approximately 50 percent endemicity (García et al. 1992); *Orthoptera* and *Diptera* species are almost 45 percent and 40 percent endemic, respectively (García et al. 1992; Machado 1998).

Status and Threats

Current Status

A complex network of protected areas has been developed throughout the islands. These include the Frontera Rural Park, Roques de Salmor Integral Natural Reserve, and Tibataje Special Nature Reserve in El Hierro; La Caldera de Taburiente National Park and El Pinar de Garafía Integral Natural Reserve in La Palma; Garajonay National Park and Valle Gran Rey Rural Park in La Gomera; Teide National Park, Anaga Rural Park, Teno Rural Park, Corona Forestal Natural Park, and Las Palomas Natural Reserve in Tenerife; and Doramas Rural Park, Los Tilos de Moya Special Natural Reserve, and El Brezal Special Natural Reserve in Gran Canaria (Gobierno de Canarias 1995).

The Spanish-Canarian government and European Community authorities have made a significant effort to preserve and protect the natural habitats and biota in the archipelago. Accomplishments thus far include the reintroduction of the giant lizard of El Hierro (*Gallotia simonyi machadoi*), the conservation of Chiroptera and invertebrates in volcanic cavities, the conservation of five priority species of the Canarian subhumid montane layer (monteverde), and the conservation of endemic birds, such as the Gran Canaria blue chaffinch (*Fringilla teydea polatzeki*) and dark and white-tailed laurel pigeons (*Columba bolli, C. junoniae*).

Types and Severity of Threats

A variety of factors threaten the Canarian biota and habitats. Local and foreign enterprises have developed tourist facilities in different areas of the islands, causing enormous habitat destruction. The illegal construction of houses inside protected areas is also a large threat. Pollution and uncontrolled dumping sites are further concerns, despite some success by local authorities in regulating them.

The illegal capture and hunting of wild taxa also threaten some species. Bird species, such as canaries, goldfinches, and warblers, are trapped for the pet trade. Finally, one of the most dangerous factors is the introduction of alien species, which threaten local taxa with foreign diseases, hybridization risks, competition, and predation. The most significant introduced predators are feral cats and rats. The recent trend for exotic pets has led to the establishment of several alien species, such as *Psittacula krameri* and *Myiopsitta monachus*. Even snakes, caiman turtles, red swamp crayfish, and green iguanas have been found in the wild recently, some of them with breeding populations (Urioste 1999; Rodríguez and Urioste 2000).

Justification for Ecoregion Delineation

The Canary Islands are a volcanic archipelago rising up to 3,720 m from the Atlantic Ocean 96 km from mainland Africa. Their flora and fauna are highly distinct. The western islands form a distinct ecoregion because they include a number of distinctive vegetation types (including Macronesian laurel forests) containing many endemic species. The islands in this ecoregion are La Palma, El Hierro, Gomera, Tenerife, and Gran Canaria. The eastern islands of Fuerteventura and Lanzarote are much drier and lack forest habitats; they are placed in the African mainland Mediterranean *Acacia-Argania* Dry Woodlands and Succulent Thickets [87].

Ecoregion Number:	**89**
Ecoregion Name:	**Lowland Fynbos and Renosterveld**
Bioregion:	**Cape Floristic Region**
Biome:	**Mediterranean Forests, Woodlands, and Scrub**
Political Units:	**South Africa**
Ecoregion Size:	**32,800 km^2**
Biological Distinctiveness:	**Globally Outstanding**
Conservation Status:	**Critical**
Conservation Assessment:	**I**
Authors:	*Shirley M. Pierce, Richard M. Cowling*

Location and General Description

The Lowland Fynbos and Renosterveld [89] ecoregion is located at the southwestern tip of the African continent, where it forms part of the Cape Floristic Region (CFR), one of the most biologically diverse regions on Earth. This ecoregion is located on the coastal lowlands and interior valleys of the CFR, whereas the Montane Fynbos and Renosterveld [90] ecoregion comprises the uplands and mountains of the CFR. Both the fynbos and renosterveld ecoregions are easily distinguishable from neighboring ecoregions by their climate, soils, and resultant vegetation and flora. These fire-prone ecosystems, where small-leaved, evergreen shrubs are the most common growth form, are associated with predominantly winter rainfall between 250 and 2,000 mm annually, with mostly infertile soils (Cowling et al. 1997a).

This ecoregion encompasses most of the heavily transformed lowland portion of the CFR. Fynbos covers 19,227 km^2 (53.8 percent) of this ecoregion, and renosterveld covers 16,490 km^2 (46.2 percent). The lowland areas receive annual rainfall between 300 and 750 mm annually (Deacon et al. 1992). Temperatures are generally mild: frost is seldom recorded, and summer maximum temperatures seldom exceed 30°C, except in the interior valleys. Coastal areas are generally windy, especially in summer. The forelands along the West Coast are influenced by the cold Benguela Current and are prone to fog.

The CFR is roughly coincident geologically with the Cape Supergroup, a Devonian-Ordovician series of sedimentary strata consisting of alternating layers of quartzitic sandstones (the Table Mountain and Witteberg Groups) and fine-grained shales (Bokkeveld Group) (Deacon et al. 1992). The mountains of the cape are made of the resistant quartzitic sandstone, and softer shale forms the gentle valleys. Deposits of Tertiary limestone, Pleistocene sands, and Holocene dunes mantle the coastal margin.

There are five major perennial river systems in the CFR, all of which traverse this ecoregion. These rivers are important habitats for locally endemic freshwater fish (Skelton et al. 1995). They are also important migratory routes for fauna and flora and provide opportunities for exchange between the biotas of the coastal forelands and interior basins (Cowling et al. 1999c). The Olifants River forms the northern boundary of the West Coast Forelands, and the Berg River drains most of this region. The Breede River is the largest river on the south coast. The Olifant-Gourits-Groot River System, which bisects the Little Karoo and drains much of the South Coast Forelands, is a particularly important migratory corridor. Finally, the Groot-Baviaanskloof-Gamtoos system is the major river system in the eastern end of the ecoregion.

The predominant vegetation types are fynbos and renosterveld. Fynbos is a hard-leaved, evergreen, fire-prone shrubland characterized by four major plant types: restioids, ericoids, proteoids, and geophytes (Cowling and Richardson 1995). Restioids, members of the Gondwanan family Restionaceae, are evergreen rush or reedlike plants that are the uniquely diagnostic plant type of fynbos (Campbell 1985). The ericoids include many small-leafed shrubs, ranging from 0.5 to 2 m tall, which give fynbos its heathlike appearance. The proteoids are the tallest fynbos shrubs, 2–4 m in height, and comprise showy members of the Proteaceae, another Gondwanan family. Geophytes, or bulblike plants, usually are most conspicuous after fires. Many of these have been developed into valuable horticultural plants. Fynbos thrives in several locations in this ecoregion: on leached, aeolian sands of the coastal forelands and on the nutritionally imbalanced coastal dune and limestone sands of the coastal margin.

Renosterveld is Afrikaans for "rhinoceros veld," a possible reference to the historic habitation by the black rhinoceros (*Diceros bicornis*). Unlike fynbos, renosterveld lacks restioids, and proteoids are very rare (Cowling and Richardson 1995). This vegetation type comprises a low shrub layer 1–2 m tall, composed mainly of ericoids and usually dominated by the renosterbos (*Elytropappus rhinocerotis*) (Asteraceae), with a ground layer of grasses and seasonally active geophytes. Renosterveld always grows on fine-grained, shale-derived soils of the coastal plain and inland valleys where the annual rainfall is between 250 and 650 mm. At rainfalls higher and lower than this, it is replaced by fynbos and succulent karoo, respectively.

The vegetation of the ecoregion may be further subdivided into eight major types, or primary broad habitat units, including dune pioneer, fynbos-thicket mosaic, sand plain fynbos, limestone fynbos, grassy fynbos, fynbos-renosterveld mosaic, coast renosterveld, and inland renosterveld (Cowling and Heijnis 2001).

Outstanding or Distinctive Biodiversity Features

Both the lowland and montane forms of fynbos and renosterveld share similar biodiversity features. These two ecoregions comprise about 80 percent of the CFR, which is recognized as one of the world's six floral kingdoms. The CFR is home to about 9,000 vascular plant species, 69 percent of which are endemic (Goldblatt and Manning 2000) and 1,435 (16 percent) of which are listed in the South African Red Data Book (Hilton-Taylor 1996). The CFR includes 5 endemic plant families and 160 endemic genera. A spectacular feature of the flora is the diversification of some taxa: 13 genera have more than 100 species, and a very large genus, *Erica*, has 658 spp. In total, *Erica* and *Aspalathus* in the family Fabaceae and eleven other genera have more than 100 species (Goldblatt and Manning 2000).

Together, the Lowland and Montane Fynbos and Renosterveld [89, 90] ecoregions harbor approximately 7,000 of the CFR's 9,000 species (Cowling et al. 1996). Regional richness is among the highest in the world and certainly the highest outside some tropical rainforest areas (Cowling et al. 1992). The high regional plant richness is a consequence of the extremely rapid turnover of moderately rich communities along habitat (beta turnover) and geographic (gamma turnover) gradients (Cowling et al. 1992). Geophyte diversity is particularly high, with about 1,500 species occurring, mostly belonging to the petaloid monocot families, notably Iridaceae, Orchidaceae, Hyacinthaceae, and Amaryllidaceae. Comparable levels have been recorded only for islands such as Madagascar and New Zealand (Cowling and Hilton-Taylor 1994). About 80 percent of the plants in fynbos and renosterveld are endemic. A large number of these endemics are point endemics, restricted to areas of 100 km^2 or less (Cowling and McDonald 1999). Most endemic species grow in fynbos vegetation, although locally endemic geophytes are common in renosterveld vegetation. Goldblatt and Manning (2000) have recognized six centers of endemism for the CFR (both lowland and montane areas), all of which are dominated by fynbos and renosterveld ecosystems: the Northwestern Center, Southwestern Center, Agulhas Plain Center, Karoo Mountain Center, Langeberg Center, and Southeastern Center.

Regional-scale species endemism declines from the western to the eastern parts of the ecoregion (Cowling et al. 1992). This is largely a result of lower numbers of local endemics (habitat specialists and geographic vicariants) in eastern landscapes relative to western areas (Cowling and McDonald 1999). Because of the spectacularly high plant diversity and endemism of the fynbos and renosterveld, this ecoregion has been identified as a biodiversity hotspot by Conservation International (Myers et

al. 2000) and as a WWF-IUCN center of plant diversity (WWF and IUCN 1994).

The fynbos and renosterveld ecoregions are roughly coincident with the Cape Faunal Center (CFC), a distinct zoogeographic zone characterized by the phylogenetic antiquity of much of its invertebrate fauna (Stuckenberg 1962). Many of these ancient lineages are Gondwanan relicts. Examples include freshwater crustaceans (Phreatoicidea, Paramelitidae, and the unique, cave-dwelling *Spelaeogrypus lepidops*), harvestmen (the endemic Triaenonychidae), flies (*Pachybates, Trichantha*, and *Peringueyomina*), *Megaloptera, Dermaptera*, bugs of the tribe Cephalelini, caddis flies (*Trichoptera*), and various beetles, notably stagbeetles (Lucanidae) of the genus *Colophon*. Not only is the CFR characterized by the phylogenetic antiquity of its invertebrate fauna, but this center is also a region of endemic species richness for reptiles, amphibians, and freshwater fish.

The region is also a major zone of endemic species richness for freshwater fish, especially the drainage systems of the Olifants, Berg, Breede, and Gouritz rivers (Skelton et al. 1995). Barbine fish account for 81 percent of the fauna, with thirty species, and total endemicity is 50 percent, with 45 percent of endemics occurring in a single drainage system.

In stark contrast with the plants, invertebrates, and freshwater fishes, the terrestrial vertebrate fauna of the CFC is neither especially rich nor distinctive (Branch 1988; Crowe 1990). Bird diversity is not particularly high because of the structural uniformity of the vegetation and the shortage of food (McMahon and Fraser 1988). Approximately 290 species (excluding seabirds) have been recorded from the region, and just 7 of these are endemic or near endemic. Among the strict and near-endemic species, most are found in both the Lowland and Montane Fynbos and Renosterveld [89, 90] ecoregions, such as Victorin's scrub warbler (*Bradypterus victorini*), cape rock-jumper (*Chaetops frenatus*), orange-breasted sunbird (*Nectarina violacea*), Cape sugarbird (*Promerops cafer*), Cape siskin (*Serinus totta*), and Cape francolin (*Francolinus capensis*). Among the ecoregion's mammal species, 5 endemic or near-endemic species remain. Sadly, two others are now extinct: the blue buck antelope (*Hippotragus leucophaeus*), hunted to extinction by 1800, and the quagga (*Equus quagga*), hunted to extinction the 1850s. A flagship mammal is the strictly endemic bontebok (*Damaliscus pygargus*), which once grazed the renosterveld plains of the South Coastal Forelands but is now found mainly in protected sanctuaries. Other endemics include the Cape spiny mouse (*Acomys subspinosus*), the Cape dune mole rat (*Bathyergus suillis*), Duthie's golden mole (*Chlorotalpa duthiae*), and Verraux's mouse (*Myomys verreauxii*).

The CFC is home to 109 reptile species, 19 (17.4 percent) of which are endemic (Cowling and Pierce 1999a), including Kasner's burrowing skink (*Scelotes kasneri*) and common burrowing skink (*Scelotes bipes*). Tortoise diversity is impressive, and the CFC is a significant part of South African center of diversity for this group of terrestrial chelonians; notable species include the endemic geometric tortoise (*Psammobates geometricus,* EN). Although low in overall diversity, amphibians exhibit moderately high endemism. In all, there are thirty-eight amphibian species, nineteen of them endemic to the CFC, including *Breviceps rosei, Cacosternum capense,* and *Microbatrachella capensis.*

Status and Threats

Current Status

Lowland fynbos and renosterveld have been severely impacted by agriculture, invasive alien plants, and urbanization (Cowling et al. 1999c). About 41 percent of the original extent of fynbos and 75 percent of renosterveld have been transformed, principally by agriculture. Given the region's global significance as a biodiversity hotspot (Cowling and Pierce 1999a) and its long-standing recognition as a regional conservation priority (Rebelo 1997), the current conservation status is very poor. As of 1999, only 827.5 km^2, or 4.3 percent, of the original extent of fynbos was conserved in statutory reserves; corresponding data for renosterveld are 92.5 km^2, or 0.6 percent (Cowling et al. 1999c). Most of the protected areas in this ecoregion are small; exceptions are the West Coast National Park and De Hoop Nature Reserve, which are 260 and 400 km^2, respectively.

Of the twenty-nine broad habitat units (BHUs; Cowling and Heijnis 2001) recognized in lowland fynbos, only six have more than 10 percent of their original extent conserved, and none has achieved a reservation target proposed by Cowling et al. (1999c). Six of these BHUs require more than 75 percent of the extant habitat to achieve a reservation target. The situation is even worse for renosterveld. Only one of the seven BHUs has more than 1 percent conserved, and four require more than 80 percent of the extant habitat to achieve a reservation target.

Types and Severity of Threats

There are five major threats facing remnant lowland fynbos and renosterveld habitat. Invasive vegetation, such as alien trees and shrubs, is seriously threatening the native vegetation. Novel forms of agriculture can use otherwise marginal agricultural land. Some examples of these new forms include the cultivation of indigenous species for cut flowers, beverages, and medicinal purposes. Urbanization is a serious concern, especially in the two metropolitan centers (Cape Town in the west and Nelson Mandela in the east) and along the coastal margin. Habitat loss is compounded by fragmentation effects, which lead to biodiversity loss on small remnants of irreplaceable habitat. However, research has shown that plant species can persist in very small fragments, even in an agricultural matrix (Cowling and Bond 1991; Kemper et al. 1999). Finally, global climate change is likely to have a major negative influence on the biodiversity of fynbos, given the specialized habitat requirements of the numerous local and point plant endemics (Rutherford et al. 1999).

Justification for Ecoregion Delineation

This ecoregion was delimited by amalgamating all of the fynbos and renosterveld BHUs in the CFR that are associated with lowland habitats (below roughly 300 m elevation) (Cowling and Heijnis 2001). BHUs are surrogates for plant and animal biodiversity that were identified on the basis of concordant patterns of geology, topography, climate, and, in some cases, vegetation types (sensu Low and Rebelo 1996). The Lowland Fynbos and Renosterveld ecoregion comprises all of the dune pioneer, fynbos-thicket mosaic, sand plain fynbos, grassy fynbos, fynbos-renosterveld mosaic, coastal renosterveld, and two inland renosterveld (Waveren-Bokkeveld and Kannaland) BHUs.

Ecoregion Number:	**90**
Ecoregion Name:	**Montane Fynbos and Renosterveld**
Bioregion:	**Cape Floristic Region**
Biome:	**Mediterranean Forests, Woodlands, and Scrub**
Political Units:	**South Africa**
Ecoregion Size:	**45,800 km^2**
Biological Distinctiveness:	**Globally Outstanding**
Conservation Status:	**Relatively Stable**
Conservation Assessment:	**III**
Authors:	*Shirley M. Pierce, Richard M. Cowling*

Location and General Description

The Montane Fynbos and Renosterveld [90] ecoregion comprises the upland and less transformed portion of the world-renowned Cape Floristic Region (CFR), in southwestern South Africa. Fynbos makes up 29,475 km^2 (81.5 percent) of this ecoregion, and renosterveld makes up 6,684 km^2 (18.5 percent). The ecoregion is located in the mountains and uplands of the CFR, which are distributed throughout the region; some mountain ranges run inland, and others are parallel to the coast. There are four major physiographic regions: the Western Mountains, from the Cape Peninsula to the Bokkeveld Mountains; the South Coastal Mountains, from the Elgin Basin in the west to near Port Elizabeth; the Interior Mountains, located to the north of the Little Karoo Basin; and the Little Karoo Inselbergs, which comprise several isolated, fynbos-clad ranges in the Little Karoo.

Most of this ecoregion receives annual rainfall between 300 and 2,000 mm, although some sites in the southwest receive as much as 3,000 mm (Deacon et al. 1992). West of Cape Agulhas, rainfall is concentrated in the winter, and east of this zone rainfall distribution is less seasonal. Temperature lows are more extreme than on the adjacent lowlands. Frost is widespread on upper peaks, where snow may lie for several weeks in the winter. Summer temperatures seldom exceed 25°C, except in the interior valleys. The CFR overlies the Cape Supergroup, a Devonian-Ordovician series of sedimentary strata, where alternating quartzitic sandstones such as Table Mountain and the Witteberg Groups alternate with fine-grained shales, as seen in the Bokkeveld Group (Deacon et al. 1992). Faulting associated with the breakup of Gondwanaland violently folded the cape sediments, but since then the region has remained stable. Erosion has worn down the once high mountains into today's resistant quartzitic sandstone mountains, with the softer shales forming low valleys. The mountains that characterize this ecoregion do not reach exceptional heights, but the topography is nonetheless impressive because the mountains rise very steeply from the adjacent lowlands (Deacon et al. 1992).

Fynbos and renosterveld are the main vegetation communities found in this ecoregion. Fynbos is commonly identified as a hard-leaved, evergreen, fire-prone shrubland characterized by four major plant types: restioids, ericoids, proteoids, and geophytes (Cowling and Richardson 1995). Among these four plant types, restioids, evergreen reedlike plants, are uniquely diagnostic of fynbos (Campbell 1985). Ericoids are small-leafed shrubs from 0.5 to 2 m tall that give fynbos its heathlike appearance. The proteoids belong to another Gondwanan family, the Proteaceae, and are known for their showy blooms and their height. At 2–4 m, they are the tallest fynbos shrubs. Finally, geophytes, or bulblike plants, usually are most conspicuous after fires and also have attractive blooms. Geophytes are especially prized as horticultural plants. Fynbos thrives on the low nutrient soils of the rocky sandstone mountains. Trees are rare in fynbos, although the Clanwilliam cedar (*Widdringtonia nodiflora*) is restricted to this ecoregion, found only in the Cedarberg Mountains.

Montane fynbos encompasses a bewildering diversity of plants, especially in the winter rainfall west, where high turnover along habitat and geographic gradients results in extremely complex vegetation patterns (Cowling et al. 1992). The most recent region-wide treatment of the vegetation of the CFR (Cowling and Heijnis 2001) identified a number of mountain fynbos complexes, which are habitats that are characterized on the basis of homogeneous climates, topographies, and geologies and include a wide range of vegetation types. Although these complexes, called broad habitat units (BHUs; Cowling and Heijnis 2001), are surrogates for biodiversity, they also reflect the major biogeographic patterns of the montane flora. Thirty such mountain fynbos complexes have been recognized.

Outstanding or Distinctive Biodiversity Features

Both the lowland and montane forms of fynbos and renosterveld share similar biodiversity features. Genera tend to be exceptionally speciose: *Erica* and *Aspalathus* (Fabaceae) and eleven other genera have more than 100 species each (Gold-

blatt and Manning 2000). As a result, the species per genus ratio is 9:1, one of the highest in the world and more typical of an island biota than a continental area. About 80 percent of the species found in fynbos and renosterveld are endemic. Most endemic species grow in fynbos vegetation, although locally endemic geophytes are fairly common in renosterveld in the western, winter rainfall part of the lowland and montane fynbos ecoregions.

The Cape Fold Mountains are important habitats for fynbos vegetation. The linear sections of the mountains have promoted landscape-level diversity in plant communities rather than on a local community scale. Regional richness is more important than alpha diversity (number of plant species in a homogenous community). Because these mountains were of only moderate height after erosion, extreme cold has not been a limiting factor, even at the low temperatures during the Pleistocene. The maximum depression in mean annual temperature under glacial conditions was approximately 5°C lower than present temperatures. Terrain and relief diversity have combined with edaphic and climatic factors to create and preserve upland centers of species richness in the Cape Mountains (Deacon et al. 1992). In the lithosols of the mountains, podzolization is the main soil-forming process, and a wide diversity of variable and localized soil conditions have been selected for microhabitat plant specialists. This contrasts with the slightly richer soils found in the residual duplex and alluvial soils on the coastal platform of the Lowland Fynbos and Renosterveld [89].

The invertebrate fauna is also rich in species, with high rates of endemism. Many species come from ancient lineages, suggesting that they are Gondwanaland relicts (Stuckenberg 1962). Examples include freshwater crustaceans (Phreatoicidea, Paramelitidae, and the unique, cave-dwelling *Spelaeogrypus lepidops*), harvestmen (the endemic Triaenonychidae), flies (*Pachybates, Trichantha,* and *Peringueyomina*), Megaloptera, Dermaptera, bugs of the tribe Cephalelini, caddis flies (Trichoptera), and various beetles, notably stagbeetles (Lucanidae) of the genus *Colophon*.

The vertebrate fauna of the CFC is neither especially rich nor distinctive (Branch 1988; Crowe 1990). However, the fynbos fauna is unique, with some species specifically adapted to living on the low-nutrient vegetation available. Endemic mammals include the Cape mountain zebra (*Equus zebra zebra*, EN), Cape spiny mouse (*Acomys subspinosus*), and Verraux's mouse (*Myomys verreauxii*). The most common antelope species in the Cape Mountains are the Cape grysbok (*Raphicerus melanotis*), bush duiker (*Sylvicapra grimmia*), and klipspringer (*Oreotragus oreotragus*). Medium-sized mammals, such as honey badger (*Mellivora capensis*), aardvark (*Orycteropus afer*), and rock hyrax (*Procavia capensis*) are all found in this ecoregion.

Bird diversity in fynbos and renosterveld is not particularly high because the structurally uniform vegetation offers fewer niches and little high-quality food, such as fleshy fruits and large insects (McMahon and Fraser 1988). Near-endemic birds include

Victorin's scrub warbler (*Bradypterus victorini*), rufous rock-jumper (*Chaetops frenatus*), orange-breasted sunbird (*Nectarinia violacea*), Cape sugarbird (*Promerops cafer*), and Cape siskin (*Serinus totta*).

This ecoregion is a major zone of endemic species richness for freshwater fish, especially the drainage systems of the Olifants, Berg, Breede, and Gouritz Rivers (Skelton et al. 1995). The CFC is also home to 109 reptile species, 19 (17.4 percent) of which are strict endemics (Cowling and Pierce 1999a). Endemic reptiles include the graceful crag lizard (*Pseudocordylus capensis*) and Cape Mountain lizard (*Tropidosaura gularis*). Tortoise diversity is especially impressive, and South Africa has more terrestrial chelonians than any other country in the world. Although amphibians are low in overall diversity, they exhibit high endemism. In all, half of the species present in both fynbos and renosterveld are endemic, including montane marsh frog (*Poyntonia paludicola*), Table Mountain ghost frog (*Heleophryne rosei*, VU), and Swellendam Cape toad (*Capensibufo tradouwi*).

Status and Threats

Current Status

Because of its inaccessibility and poor agricultural potential, this ecoregion has undergone less transformation than its lowland counterpart (Cowling et al. 1999c). Only 9.1 percent of the original extent of montane fynbos and 11.7 percent of montane renosterveld have been transformed, principally by agriculture and forestry. The conservation status of this ecoregion is among the best in South Africa, a legacy of protection aimed at safeguarding mountain catchments for sustainable water production. By 1999, approximately 7,700 km², or 26.2 percent, of the original extent of fynbos was conserved in statutory reserves. The situation is not nearly as good for renosterveld, where corresponding data are 320 km², or 4.76 percent (Cowling et al. 1999c). Many of the protected areas in this ecoregion are large, especially when compared with those in the lowlands. However, none is large enough to sustain the full suite of ecological and evolutionary processes needed for the long-term conservation of the biota (Cowling et al. 1999c). Protected areas wholly or partially contained in this ecoregion include the Anysberg and Baviaanskloof nature reserves, Cederberg Wilderness Area, Groot Winterhoek Wilderness Area, Grootvadersbosch State Forest complex, Outeniqua Protected Areas complex, and Tsitsikamma National Park.

Of the thirty mountain fynbos BHUs, twenty have more than 10 percent of their original extent conserved, sixteen have more than 25 percent conserved, and fifteen have achieved reservation targets derived on the basis of biodiversity patterns and retention of habitat (Cowling et al. 1999c). Only in the case of nine BHUs do extant protected areas achieve less than 50 percent of the recommended reservation target. The situation for renosterveld is not nearly as encouraging. Only two of the six inland renosterveld BHUs have more than 5 percent conserved, and the reservation target has been achieved for only one of these.

Types and Severity of Threats

The three major threats facing montane fynbos and renosterveld habitat are: (1) invasive alien trees and shrubs; (2) novel forms of agriculture (e.g., the cultivation of indigenous species for cut flowers, cultivation and collection of plants for beverages and for their medicinal properties, such as rooibos and honeybush tea) that will transform otherwise marginal agricultural land; and (3) global climate change, which is likely to have a major negative influence on the biodiversity of fynbos, given the specialized habitat requirements of the numerous local and point plant endemics (Rutherford et al. 1999).

Justification for Ecoregion Delineation

This ecoregion was delimited by amalgamating all fynbos and renosterveld BHUs in the Cape Floristic Region that are associated with montane habitats (Cowling and Heijnis 2001). BHUs are surrogates for plant and animal biodiversity that were identified on the basis of concordant patterns of geology, topography, climate, and, in some cases, vegetation types (sensu Low and Rebelo 1996). Montane Fynbos and Renosterveld [90] comprises all of the Mountain Fynbos Complex and all but two of the inland renosterveld BHUs.

Ecoregion Number:	**91**
Ecoregion Name:	**Albany Thickets**
Bioregion:	**Eastern and Southern Africa**
Biome:	**Mediterranean Forests, Woodlands, and Scrub**
Political Units:	**South Africa (Eastern and Western Cape Provinces)**
Ecoregion Size:	**17,100 km²**
Biological Distinctiveness:	**Regionally Outstanding**
Conservation Status:	**Critical**
Conservation Assessment:	**II**
Author:	*Amy Spriggs*
Reviewed by:	*Jan Vlok, Anthony B. Cunningham*

Location and General Description

The Albany Thickets [91] ecoregion is a dense, spiny shrubland with a canopy up to 2.5 m in height and usually abundant succulents. It is located in South Africa along the Fish, Sundays, and Gamtoos river valleys in the Eastern Cape and is also found along the intermontane valleys of the inland Cape Fold Mountains (Kouga, Baviaanskloof, and Swartberg). Toward the east, it grades into Maputaland-Pondoland Bushland and Thickets [80], a sparser scrub forest up to 6 m tall, which is less spinescent and succulent, contains more grasses, and is dominated by subtropical woody evergreen species.

Geologically the quartzites of the Cape Fold Belt are replaced by the shales and sandstones of the Karoo Supergroup in the Eastern Cape (du Toit 1966). Erosion has greatly subdued the topography of the Cape Fold Belt, resulting in a fairly level basin, which is broken by the wide, deep valleys of the Fish, Sundays, and Gamtoos rivers. The soils of the valleys are well-drained, deep, lime-rich, sandy loams (Low and Rebelo 1996). Toward the coast, soils may include consolidated dune sands, with the densest thickets occurring on the deepest sandy loams. Soil characteristics may be one of the factors responsible for the confined distribution of the Albany Thicket (Cowling 1984).

The Fish, Sundays, and Gamtoos river valleys have high diurnal and annual temperature ranges and low, sporadic rainfall (Cowling 1983a). The three summer months (December to February) are the driest. Annual temperatures in the inland valleys are extreme, ranging from 0°C to more than 40°C. The coastal areas of the valleys have a more moderate climate, with a temperature range of 10–35°C and a slightly higher annual rainfall of 450–550 mm per year. Valley mists are common toward the coast, providing additional moisture (Low and Rebelo 1996).

The ecoregion supports a diversity of habitats in three regions: the dry, inland areas of the Fish, Sundays, and Gamtoos river valleys; the more moderate coastal areas of these river valleys; and the intermontane valleys to the north and west. The thickets of the Fish, Sundays, and Gamtoos river valleys are sparse, succulent, and spinescent. The vegetation contains a high proportion of both leaf- and stem-succulent shrubs such as Spekboom (*Portulacaria afra*), *Euphorbia bothae* (dominant along the Fish River Valley), *Euphorbia ledienii*, and Noorsdoring (*Euphorbia coerulescens*, dominant along the Sundays River Valley). Characteristic woody species include the Karoo crossberry (*Grewia robusta*), small bitterleaf (*Brachylaena ilicifolia*), *Maytenus capitata*, and *Lycium campanulatum* (Lubke et al. 1986; Everard 1987; Low and Rebelo 1996).

The thickets of the coastal areas of the Fish, Sundays, and Gamtoos river valleys are extremely dense and often impenetrable. Common woody species are white milkwood (*Sideroxylon inerme*), dune kokotree (*Maytenus procumbens*), and Septemberbush (*Polygala myrtifolia*) (Everard 1987; Low and Rebelo 1996). Characteristic succulent species include the Uitenhage aloe (*Aloe africana*), bitter aloe (*Aloe ferox*), *Euphorbia ledienii*, and *Euphorbia grandidens*. Dune thicket, which is less succulent, is confined to the coast and includes such species as *Cotyledon adscendens*, *Mimusops caffra*, *Brachylaena discolor*, and *Strelitzia nico-*

lai (Lubke et al. 1986; Low and Rebelo 1996; van Wyk and Smith 2001).

The thicket vegetation of the intermontane valleys to the north and west is a dense shrubland dominated by Spekboom (*Portulacaria afra*). Other species include kurky (*Crassula ovata*), large honeythorn (*Lycium austrinum*), jacketplum (*Pappea capensis*), gwarrie (*Euclea undulata*), *Grewia robusta*, *Aloe* spp., *Rhus* spp., and boerboon (*Schotia afra*). Many species of plakkies (*Crassula* spp.) and succulent herbs and grasses also occur. Toward the western limits Spekboom becomes overwhelmingly dominant and can form pure stands (Acocks 1953; Lubke et al. 1986; Low and Rebelo 1996).

Outstanding or Distinctive Biodiversity Features

Botanists have long recognized the Eastern Cape as a complex transition zone where four phytochoria converge (Goldblatt 1978; Cowling 1983a; White 1983; Lubke et al. 1986; Cowling and Hilton-Taylor 1994). There are sixty-one endemic plant species (of which thirty-one are threatened), with few endemics of cape origin. Most of these are found in the mesic thicket type, which has the highest endemism of the three thicket types (Lubke et al. 1986; WWF and IUCN 1994; van Wyk and Smith 2001). Croizat (1965) recognized the Albany area as a center of endemism for succulent *Euphorbia* species. Endemic succulents include sheep fig (*Delosperma* spp.), *Lampranthus productus*, *Euphorbia fimbriata*, *Euphorbia gorgonis*, *Gasteria armstrongii*, *Aloe africana*, and *Haworthia fasciata* (Cowling 1983). The Albany area is also a center of endemism for cycads (*Encephalartos* spp.) and geophytes including *Cyrtanthus*, *Albuca*, and *Ornithogallum* (Cowling 1983).

Although this ecoregion is an important center of floral endemism, rates of faunal endemism are low. Endemic reptiles include Tasman's girdled lizard (*Cordylus tasmani*), Essex's dwarf leaf-toed gecko (*Goggia essexi*), and Albany adder (*Bitis albanica*).

Large mammal species include elephant (*Loxodonta africana*, EN), black rhinoceros (*Diceros bicornis*, CR), leopard (*Panthera pardus*), bushbuck (*Tragelaphus scriptus*), common rhebok (*Pelea capreolus*), mountain reedbuck (*Redunca fulvorufula*), eland (*Taurotragus oryx*), greater kudu (*Tragelaphus strepsiceros*), hartebeest (*Alcelaphus buselaphus*), cape grysbok (*Raphicerus melanotis*), and common duiker (*Sylvicapra grimmia*). Bird species include the Knysna woodpecker (*Campethera notata*), Cape bulbul (*Pycnonotus capensis*), and martial eagle (*Polematetus bellicosus*).

Status and Threats

Current Status

Most of the ecoregion is underprotected, although the xeric thickets of the inland river valleys are found in the 513-km² Addo Elephant National Park (Greyling and Huntley 1984).

Other protected areas include the Groendal Wilderness Area, which contains mesic thicket vegetation; Kouga-Baviaanskloof Mountain Catchment Area, which contains a large area of dense Spekboomveld; Silaka Wildlife Reserve (4 km²); and the Kabeljousriver Nature Reserve (2 km²).

Types and Severity of Threats

In its natural state, Albany thicket forms impenetrable stands up to 3 m in height. Where browsing elephants and other ungulates once maintained ecological balance in the system by acting as patch disturbance agents, in many areas indigenous animals have been replaced by domestic herbivores, which have degraded much of the landscape (Kerley et al. 1995). Around 51 percent of the Albany Thicket has been converted to other land uses such as pastoralism, crop production, and urban expansion (Low and Rebelo 1996). Overgrazing, primarily by domestic goats, has severely degraded vegetation and led to soil erosion and loss of biodiversity. These effects are most noticeable in the more arid areas (xeric thickets and Spekboomveld) (Everard 1988).

Other threats include bush clearing for agriculture and development, resource exploitation, and invasive species. Land has been cleared along rivers for lucerne and other crops grown under irrigation and for orange orchards in the Addo region. Mesic thicket is under the greatest threat. Many of the endemic *Encephalartos* and *Euphorbia* species and species of the Aizoaceae family are also exploited by plant collectors.

In recent years there has been a shift toward game farming as an alternative to domestic pastoralism (Kerley et al. 1995; Kerley et al. 1999). This change has the potential to be both economically and ecologically sustainable over the long-term. Another area of opportunity is ecotourism, following the model developed by Addo Elephant National Park (Kerley et al. 1999).

Justification for Ecoregion Delineation

This ecoregion occurs most prominently along the Sundays, Gamtoos, and Fish river valleys in the Eastern Cape and mountain slopes in the eastern portion of the Western Cape. The ecoregion boundaries follow Low and Rebelo's (1996) mesic thicket, xeric succulent thicket, and Spekboom succulent thicket, with further refinement to accommodate the BHUs of the CFR that are associated with thicket vegetation (Cowling and Heijnis 2001).

Ecoregion Number:	**92**
Ecoregion Name:	**Atlantic Coastal Desert**
Bioregion:	**African Palearctic**
Biome:	**Deserts and Xeric Shrublands**
Political Units:	**Western Sahara (Morocco), Mauritania**
Ecoregion Size:	**40,000 km²**
Biological Distinctiveness:	**Locally Important**
Conservation Status:	**Relatively Intact**
Conservation Assessment:	**V**
Author:	*Chris Magin*
Reviewed by:	*John Newby*

Location and General Description

The Atlantic Coastal Desert [92] comprises the westernmost portion of the Sahara, the world's largest desert. This ecoregion covers most of Western Sahara's 1,110 km coastline, from La'ayoune southward, and roughly two-thirds of Mauritania's 754 km of coastline. It includes Mauritania's capital city of Nouakchott and the large port town of Nouâdhibou to the northwest. It merges into the Mediterranean *Acacia-Argania* Dry Woodlands and Succulent Thickets [87] to the north and is bordered to the east by the North Saharan Steppe [93]. An outlier of the Saharan Halophytics [65] is embedded in the southern portion of this ecoregion.

This ecoregion lies between sea level and a maximum of 200 m in elevation. Much of the coast is formed of cliffs 20–50 m high, and a sandy or gravelly hamada plateau stretches inland. The climate is extremely hot and arid, with only low amounts of episodic rainfall. However, mists from the Atlantic are common. Condensation of these mists permits the growth of lichens on shrubs and on bare ground between vascular plants. The human population density of the ecoregion is extremely low and traditionally practiced pastoral nomadism, raising camels, goats, and sheep. Working in mines and in the oil industry is becoming increasingly important.

In terms of the phytogeographic classification of White (1983), this ecoregion is classified as part of the Sahara regional transition zone. The area has much denser vegetation cover than other parts of the Sahara and is also richer in plant species. The *Euphorbia*-dominated succulent shrublands and the *Argania spinosa* scrub forests that characterize the Mediterranean *Acacia-Argania* Dry Woodlands and Succulent Thickets [87] to the north essentially disappear at Seguia el Hamra (White 1983). However, some of the species of succulent shrubs have a scattered distribution in the northern part of this ecoregion, most notably *Euphorbia regis-jubae, E. echinus,* and *Senecio anteuphorbium.* In the south (i.e., coastal Mauritania), Sahelo-Saharan–linked species such as *Acacia tortilis, Maerua crassifolia, Salvadora persica,* and *Balanites aegyptiaca* are well represented. Along the coast, halophytic genera such as *Suaeda, Atriplex,* and *Zygophyllum* are common, together with species such as *Salsola longifolia, Heliotropium undulatum,* and *Lycium intricatum.*

Outstanding or Distinctive Biodiversity Features

The Atlantic Coastal Desert is rich in endemic plants, but there are few endemic or near-endemic animals. Some of the more important biodiversity features are on the Atlantic Coast. The Mediterranean monk seal (*Monachus monachus,* CR) (Hilton-Taylor 2000) has its last stronghold in the coves along the Cap Blanc Peninsula near the Mauritanian town of Nouâdhibou on the border with Western Sahara.

The larger coastal bays, particularly the Baie d'Ad Dakhla and the Gulf of Cintra in Western Sahara and the Banc d'Arguin in Mauritania, are immensely important for more than 2 million wintering western Palearctic waders from fifteen different species (Shine et al. 2001). The most abundant are dunlin (*Calidris alpina*), bar-tailed godwit (*Limosa lapponica*), curlew sandpiper (*Calidris ferruginea*), and redshank (*Tringa totanus*), all with populations over 100,000 birds (Dodman et al. 1997). More than 100,000 other waterbirds also use the wetlands for breeding or on migration, including more than 30,000 greater flamingos (*Phoenicopterus ruber*).

Notable terrestrial fauna include the dorcas gazelle (*Gazella dorcas,* VU), golden jackal (*Canis aureus*), fennec (*Vulpes zerda*), Rüppell's fox (*Vulpes rueppelli*), sand cat (*Felis margarita*), honey badger (*Mellivora capensis*), and striped hyena (*Hyaena hyaena*). The globally threatened dama gazelle (*Gazella dama,* EN), addax (*Addax nasomaculatus,* CR), and cheetah (*Acinonyx jubatus,* VU) may have previously occurred here, but they are no longer found (East 1999). Similarly, the scimitar-horned oryx (*Oryx dammah*) may have once occurred but is now almost certainly extinct in the wild (Newby 1988; UNEP/CMS 1998, 1999; Newby, pers. comm., 2002). Reptiles include the Algerian whip snake (*Coluber algirus*).

Status and Threats

Current Status

The ecoregion has been severely degraded by prolonged droughts and overgrazing by livestock. However, little habitat has been totally lost because the arid climate prohibits farming, so extensive blocks of natural but degraded habitat remain throughout the ecoregion. There are only two protected areas: the 11,000 km² Banc d'Arguin National Park and the 300 km² Réserve Intégrale de Cap Blanc, both in Mauritania.

The coastal sector, called Aguerguer or Côte des Phoques, of the proposed 15,000–20,000 km² Parc National de Dakhla in Western Sahara would protect a significant part of this ecoregion.

Types and Severity of Threats

Overgrazing, cutting of trees for firewood and timber, and soil erosion aggravated by drought are contributing to desertification. The large mammal species have suffered from uncontrolled hunting, particularly by military personnel in Western Sahara. Invasion by brown rats (*Rattus norvegicus*) threatens seabird colonies along the coast. The population of monk seals is threatened by accidental capture and drowning in fishing gear, disturbance from tourists and fishers, and collapse of their breeding caves caused by coastal erosion. Wildlife in coastal areas is threatened by increasing pressure from the fishing industry and pollution from industrial developments at Nouâdhibou (Shine et al. 2001).

Justification for Ecoregion Delineation

The boundaries of this ecoregion, extending from the coast to 40 km inland, are taken directly from the "Atlantic coastal desert" vegetation unit of White (1983). Its uniqueness derives from the mist it receives from the Atlantic, making it rich in endemic plants, and its position along a major bird migration flyway means that the few coastal wetlands in the ecoregion support huge bird populations, especially during the Palearctic winter.

Ecoregion Number:	**93**
Ecoregion Name:	**North Saharan Steppe**
Bioregion:	**African Palearctic**
Biome:	**Deserts and Xeric Shrublands**
Political Units:	**Morocco, Algeria, Tunisia, Libya, Mauritania, Western Sahara, and Egypt**
Ecoregion Size:	**1,676,100 km²**
Biological Distinctiveness:	**Bioregionally Outstanding**
Conservation Status:	**Relatively Intact**
Conservation Assessment:	**V**
Authors:	*Nora Berrahmouni, Salima Benhouhou*
Reviewed by:	*John Newby, Hans Peter Müller*

Location and General Description

The North Saharan Steppe [93] covers a belt through northern Africa from Mauritania in the west to the Egyptian Red Sea in the east. In Morocco, Algeria, and Tunisia, the area forms a transition between the Mediterranean habitats to the north and true desert in the south. The Saharan Halophytics [65] ecoregion is scattered through this ecoregion in areas with saline conditions. Most of the region lies on Precambrian basement rocks overlain by Mesozoic marine sediments. The landforms comprise stony plateaus (hamadas and regs), dry river beds (wadis), dunes (ergs), mountains (djebels), and silty depressions (dayas). Major uplands, such as the Ougarta Mountains in the northwestern part of the Sahara (600 m), are uncommon in this ecoregion.

Hot, dry summers and cooler, wet winters characterize the climate. Rain comes from the Mediterranean Basin associated with weather depressions that track southward during winter and early spring (October–April). Annual rainfall varies from 50 mm in the south up to 100 mm in the north, but there are often years with no rainfall (especially in the southern parts of the ecoregion). The highest temperatures are recorded in July, with maximum temperatures often exceeding 45°C.

The ecoregion occurs primarily in White's (1983) Sahara regional transition zone and extends farther north into the Mediterranean-Sahara regional transition zone, particularly in Libya and Egypt. The flora belongs to the Saharo-Sindien region, which is derived from Mediterranean and Sudanian areas. This has led to some difficulty in its classification. Engler and Diels (1936) consider the flora to belong to the Paleotropical kingdom, whereas Eig (1931), Kassas (1967), Zohary (1973), and Ozenda (1983) relate it to the Holarctic realm. Other authors consider the region to be partially in both regions (Monod 1957; Quézel 1965; Wickens 1984). Vegetation varies according to the landforms. Typical species of the saline wadis are *Tamarix gallica*, *Salsola vermiculata*, *Salicornia fruticosa*, and *Suaeda fruticosa*, whereas the nonsaline wadis support *Acacia raddiana*, *Panicum turgidum*, *Pennisetum dichotomum*, and *Pituranthos chloranthus*. The dayas systems support *Anvillea radiata*, *Bubonium graveolens*, *Pistacia atlantica*, *Ziziphus lotus*; ergs contain *Calligonum arich*, *C. azel*, *C. comosum*, *Genista saharae*, *Ephedra alata*, and *Retama retam*; and the djebels are typified by *Gymnocarpos decander*, *Rhus tripartitus*, and *Withania adpressa*. Ephemeral species are also found in areas after wet winters (January–April) and may not appear in rainless years (Ozenda 1983; Kassas and Batanouny 1984; Zahran and Willis 1992).

Outstanding or Distinctive Biodiversity Features

The ecoregion's flora is species poor, with figures for the total number of higher plants varying between 1,150 species (Le Houérou 1990) and 1,200 species (Ozenda 1983). Three families represent around 35 percent of the flora (Asteracea, Fabaceae, Poaceae). The Zygophyllaceae is the only example of a true Saharan family, with seven genera and thirty species. Overall, the proportion of endemism is around 25 percent, or about 300 species (Ozenda 1983). Sandy habitats harbor several endemic plants, such as *Danthonia fragilis*, *Genista saharae*, *Calligonum azel*, and *C. arich (calvescens)*. The highest numbers of endemic species are found in nonsaline wadis and dayas, including *Argyrolobium saharae*, *Aristida acutiflora*, *Astragalus gombo*, *Centaurea maroccana*, *Crotalaria vialettei*, *Daucus biseriatus*, *Euphorbia calyptrata*, *Fagonia zilloides*, *Helianthemum getu-*

lum, *Lotononis dichotoma*, *Oudneya africana*, *Pituranthos battandieri*, and *Zilla macroptera*. In the saline wadis endemic plants include *Frankenia thymifolia*, *Limoniastrum guyonianum*, *Nucularia perrini*, *Tamarix pauciovulata*, *Zygophyllum cornutum*, *Z. gaetulum*, and *Z. geslini*. Jebel endemics are *Diplotaxis pitardiana*, *Limoniastrum feei*, *Lotus jolyi*, *Perralderia coronopifolia*, *Plantago akkensis*, and *Trichodesma calcaratum*. Hamadas and regs have fewer endemic species, such as *Fredolia aretioides* and *Haloxylon schmittianum* (Ozenda 1983).

The ecoregion has a fairly rich fauna, although the number of species and their density have fallen dramatically for two reasons. First, many species have been lost from this ecoregion as the climate has dried in recent centuries (Newby 1984, 1988). By the time of the Romans many large mammals (e.g., giraffe [*Giraffa camelopardalis*]) had already disappeared (Le Houérou 1991). Second, hunting has also had a major effect in recent decades. For instance, ostrich (*Struthio camelus*) was fairly common in the northern Sahara at the end of the nineteenth century but was eradicated from the area by the early twentieth century. Also, wild boar (*Sus scrofa*) and cheetah (*Acinonyx jubatus hecki*, EN), present 20 years ago in many wadis, are now rare. Other species have also declined dramatically, including striped hyena (*Hyaena hyaena barbara*, DD), dorcas gazelle (*Gazella dorcas*, EN), Cuvier's gazelle (*Gazella cuvieri*, EN), Barbary sheep or aodad (*Ammotragus lervia*, VU), and slender-horned gazelle (*Gazella leptoceros*, EN). Populations of the Houbara bustard (*Chlamydotis undulata*) and Nubian bustard (*Neotis nuba*) have been greatly reduced in recent decades.

Reduced predator populations mean that desert rodents can be extremely numerous in favorable years. These include the North African gerbil (*Gerbillus campestris*), James's gerbil (*G. jamesi*), pale gerbil (*G. perpallidus*), lesser short-tailed gerbil (*G. simoni*), sand gerbil (*G. syrticus*, CR), fat-tailed gerbil (*Pachyuromys duprasi*), Shaw's jird (*Meriones shawi*), and the threatened four-toed jerboa (*Allactaga tetradactyla*, EN). Other common mammal species include the Atlas gundi (*Ctenodactylus gundi*) and Val's gundi (*Ctenodactylus vali*).

Several important bird areas have been recognized, both as important wintering grounds and for resident species (Fishpool and Evans 2001). In the Grand Erg Occidental and Grand Erg Oriental, wintering waterbirds include gadwall (*Anas strepera*), Eurasian wigeon (*Anas penelope*), northern shoveler (*Anas clypeata*), common pochard (*Aythya ferina*), and northern lapwing (*Vanellus vanellus*) (Coulthard 2001a). In the Grand Erg Occidental (Béni Abbés), important species include spotted sandgrouse (*Pterocles senegallus*), crowned sandgrouse (*P. coronatus*), pharaoh eagle-owl (*Bubo ascalaphus*), Egyptian nightjar (*Caprimulgus aegyptius*), and pale crag-martin (*Hirundo obsoleta*) (Coulthard 2001a).

Although the diversity of reptiles is moderately high in the ecoregion, there are very few endemics. Common reptiles include the horned viper (*Cerastes cerastes*), common sand viper (*Cerastes vipera*), Peters's banded skink (*Scincopus fasciatus*), fringe-fingered lizards (*Acanthodactylus pardalis*, *A. scutellatus*), Anderson's short-fingered gecko (*Stenodactylus petrii*), Dabbs mastigure (*Uromastyx acanthinura*), and desert monitor lizard (*Varanus griseus*).

Status and Threats

Current Status

Habitats of this ecoregion are extensive and in good condition over vast areas. However, they can be quite badly degraded when close to settlements or along the coast where higher rainfall occurs, providing better pastures. People are concentrated in areas where water is available, such as oases. Various irrigation techniques (basin system, diversion weir, quanat, foggara) enabled the development of small-scale agriculture, particularly of date palm (Hannachi et al. 1998).

Legal protection through national parks and nature reserves is poor. Officially, the ecoregion has one protected area in Mauritania, the Iriki Permanent Hunting Reserve, and two national parks in Tunisia, the Jebil National Park, created in 1993 (1,500 km²), and the Sidi Toui National Park, created in 1993 (63 km²) (Chaïeb and Boukhris 1998). One proposed protected area is the Taghit oasis, which lies at the foot of the great western erg in the Algerian Sahara.

Recent conservation efforts include the establishment of new protected areas and the reintroduction of ungulates such as the dama gazelle (*Gazella dama*), addax (*Addax nasomaculatus*), and scimitar-horned oryx (*Oryx dammah*) in Tunisia and Morocco.

Types and Severity of Threats

Over the last 30 years resettlement schemes, such as those undertaken in Algeria, and the acquisition of off-road vehicles by nomads have created greater pressure on habitats in this ecoregion. Threats are concentrated in areas with more rainfall or around water sources, where the local pressure can be intense. Overgrazing by livestock is a serious problem that has resulted in severe environmental degradation in many areas. The cutting of woody vegetation for firewood is another problem in remote areas.

The remaining populations of larger animals and birds, such as dorcas gazelle, Cuvier's gazelle, slender-horned gazelle, Barbary sheep, Houbara bustard, and Nubian bustard, are threatened by hunting for sport and recreation.

Justification for Ecoregion Delineation

This ecoregion follows the boundaries for two vegetation units ("regs, hammadas and wadis" and "desert dunes with perennial vegetation") mapped by White (1983). These two vegetation units were first amalgamated and then separated into a North [93] and South [94] Saharan Steppe ecoregion because of the

significant climatic (winter rainfall in the north and summer rainfall in the south) and biological (higher rate of endemism for the northern ecoregion) differences.

Ecoregion Number:	**94**
Ecoregion Name:	**South Saharan Steppe**
Bioregion:	**African Palearctic**
Biome:	**Deserts and Xeric Shrublands**
Political Units:	**Mauritania, Mali, Algeria, Niger, Chad, Sudan**
Ecoregion Size:	**1,101,700 km²**
Biological Distinctiveness:	**Bioregionally Outstanding**
Conservation Status:	**Relatively Intact**
Conservation Assessment:	**V**
Author:	*Chris Magin*
Reviewed by:	*John Newby*

Location and General Description

The South Saharan Steppe [94] extends in a narrow band from central Mauritania, through Mali, southwestern Algeria, Niger, and Chad, and across Sudan to the Red Sea, covering the southern fringes of the Sahara Desert. Rainfall occurs mainly in the summer months of July and August but is unreliable and varies greatly in space and time from year to year. On average, annual rainfall is between 100 and 200 mm (in most years 50–100 mm would be more likely), declining along a gradient from south to north. However, droughts lasting several years often occur. Except near the coast of Sudan, mean annual temperature throughout the Sahelian portion of the ecoregion is between 26°C and 30°C. Pastoralism and domestic grazing are the foundations of the economy in this Sahelo-Saharan region. Rainfall in this ecoregion is generally insufficient for rain-fed agriculture, but some irrigated agriculture is practiced near waterpoints and along wadis (seasonally dry watercourses).

In terms of the phytogeographic classification of White (1983), the ecoregion is in both the Sahara regional transition zone and the Sahel regional transition zone. Delineated by the "regs, hamadas and wadis" vegetation type of White (1983), the northern border of the ecoregion lies several hundred kilometers north of the 100-mm rainfall isohyet, which is the northern limit of summer grassland pasture composed of the grasses *Eragrostis*, *Aristida*, and *Stipagrostis* spp., with the herbs *Tribulus*, *Heliotropium*, and *Pulicharia*. Woody species include *Acacia tortilis*, *Acacia ehrenbergiana*, *Balanites aegyptiaca*, and *Maerua crassifolia*, which grow mainly along wadis. In the south, the vegetation of the ecoregion grades into the Sahelian *Acacia* Savanna [35] and includes steppes of *Panicum turgidum* perennial tussock grass.

Outstanding or Distinctive Biodiversity Features

The South Saharan Steppe [94] has few endemic plants. The Mediterranean flora that is characteristic in the northern Sahara is almost completely absent from the southern Sahara, where a tropical flora dominates (White 1983). Endemic plant species are found on the higher massifs emerging from this ecoregion: the Termit Massif in Niger, the Adrar des Iforas in Mali, and Jebel Elba and Jebel Hadai Aweb in Sudan (White 1983; WWF and IUCN 1994). However, these have all been assigned to montane xeric woodland ecoregions [96, 97, 99].

Only one vertebrate species is strictly endemic to the South Saharan Steppe, the Dongola gerbil (*Gerbillus dongolanus*). Other near-endemic mammals found here are two more gerbil species, *Gerbillus mauritaniae* and *G. principulus*. Many larger animals that once occurred have been reduced to extremely small and scattered populations, including addax (*Addax nasomaculatus*, CR), slender-horned gazelle (*Gazella leptoceros*, EN), dorcas gazelle (*Gazella dorcas*, VU), dama gazelle (*Gazella dama*, EN), striped hyena (*Hyaena hyaena*), cheetah (*Acinonyx jubatus*, VU), wild dog (*Lycaon pictus*, EN), and ostrich (*Struthio camelus*). Small populations of the Barbary sheep or aoudad (*Ammotragus lervia*, VU) may exist on scattered rocky outcrops. The once-widespread scimitar-horned oryx (*Oryx dammah*, EW) is now believed extinct in the wild (East 1999).

Status and Threats

Current Status

The habitats of this ecoregion are heavily influenced by drought, the effects of which are locally exacerbated by the large numbers of domestic livestock and tree cutting. In wetter years, there can be an abundance of forage, and in the past large migratory ungulates used these temporary grasslands (Newby 1988). Tree cover in the south and along wadis has declined. Many trees have been cut for charcoal, firewood, and building materials. The recruitment of young trees into mature age classes is almost nonexistent in many areas because of high grazing pressure.

The Aïr and Ténéré National Nature Reserve in Niger and the Ouadi Rimé-Ouadi Achim Faunal Reserve in Chad are the two most important protected areas in the Sahelo-Saharan zone. They contain many of the last viable populations of larger ungulates in this ecoregion. However, both reserves have been plagued by political insecurity and civil unrest, and the protection of their wildlife is far from certain. Although parks and reserves have been planned, there are no managed protected areas in this ecoregion in Mauritania, Mali, or Sudan. Key areas for urgent attention include the Manga (Chad and Niger), Termit-Tin Toumma (Niger), North Tamesna (Niger and Mali), Wadi Howar (Sudan), and the Majabat al-Koubra (Mali and Mauritania).

Types and Severity of Threats

Niger, Mali, Sudan, and Chad have suffered severe degradation of their natural resources in recent decades. Prolonged droughts and overgrazing by livestock are primarily responsible. The most severe degradation has occurred in the higher-rainfall areas, or around boreholes. Threats to the larger animals are intense. The populations of many ungulate species, as well as ostriches, have been greatly reduced by subsistence and recreational hunting, particularly by military personnel and rebel forces, and through competition with humans and their livestock. Predators, including the striped hyena and cheetah, have been persecuted throughout the ecoregion by the widespread use of poisons such as strychnine because they kill livestock (MHE et al. 1996). The underrepresentation of the desert and semi-desert habitats of this ecoregion in the protected area systems of the Sahelian countries poses a threat to their long-term stability and conservation.

Justification for Ecoregion Delineation

This ecoregion is delineated from White's (1983) "regs, hamadas and wadis" and "desert dunes with perennial vegetation" units south of the Sahara Desert. Although these vegetation types surround the Sahara Desert, the southern habitats were delineated as a distinct ecoregion from the northern unit because of their different rainfall regimes and the presence of Afrotropical species.

Ecoregion Number:	**95**
Ecoregion Name:	**Sahara Desert**
Bioregion:	**African Palearctic**
Biome:	**Deserts and Xeric Shrublands**
Political Units:	**Mauritania, Mali, Algeria, Niger, Libya, Sudan, Egypt**
Ecoregion Size:	**4,639,900 km²**
Biological Distinctiveness:	**Locally Important**
Conservation Status:	**Relatively Intact**
Conservation Assessment:	**V**
Authors:	*Nora Berrahmouni, Salima Benhouhou*
Reviewed by:	*John Newby, Pedro Regato*

Location and General Description

The greater Sahara stretches across northern Africa from the Atlantic Ocean in the west to the Red Sea in the east, encompassing an area of 9,100,000 km². The Sahara Desert [95] ecoregion lies in the central part of the Sahara, between 18°N and

30°N, and covers an area of roughly 4,639,900 km². The northern and southern fringes of the Sahara, which are generally wetter and have greater vegetation cover, are described as separate ecoregions.

Precambrian basement rocks underlie much of the Sahara, but these are generally concealed beneath younger marine rock formations from the Mesozoic period and wind-blown sands, with a few emergent volcanic massifs (Menchikoff 1957; Williams 1984). The dominant landforms are stony plateaus called hamada (Tadmait Plateau, Hamada El Hamra), gravely plains known as reg or serir, and sand dunes or ergs (Erg Chech, Erg Raoui, Libyan Sand Plain). Other landforms are dry river beds (wadis), rocky mountains (djebels), and saline depressions (sebkhas or chotts) (Williams and Faure 1980; Cloudsley-Thompson 1984). Vast underground aquifers underlie much of the region and sometimes reach the surface, forming oases. A number of large mountain ranges rise from the desert (Ahaggar, Tassili N'Ajjer, Tibesti, Aïr) but are delineated as separate ecoregions. Mechanical and chemical weathering of rocks over the past 50 million years has produced the soils of the ecoregion. Over the hamadas and regs the soils are yermosols with minimal horizon development. Sandy soils are regosols, those in nonsaline valleys and basins are fluvisols, and those of saline depressions are known as solonchaks (Mitchell 1984).

The Sahara is one of the hottest regions in the world, with an annual mean temperature of around 25°C. Highest temperatures above 50°C occur in July and August, with the lowest temperatures (below freezing) in December to February. The Sahara is also extremely windy, with moderate to strong winds raising dust and sand grains. Dust and sand particles, swept up from a hot desert surface, elevate the air temperature.

The Sahara Desert is located in a climatic divide. The Intercontinental Convergence Zone moves up from the south and stops before the center of the Sahara; consequently, it hardly ever brings rain to this area. Similarly, the winter rainfall of North Africa does not reach far enough south to bring regular rain to the central Sahara. As a result, rainfall is below 25 mm everywhere and in the eastern region is less than 5 mm per year (Le Houérou 1990, 1991). The scarcity of rainfall is aggravated by its irregularity: no rain may fall for many years, followed by a single intense thunderstorm (Ozenda 1983; Smith 1984). The extreme aridity of this area is a recent feature; much larger areas of the Sahara had adequate water only 5,000–6,000 years ago (Climap 1976), and vegetation may have been semi-desert or savanna woodland.

In terms of the phytogeographic classification of White (1983), the ecoregion is classified in the Sahara regional transition zone. The flora of this region shows strictly Sahara-Arabian affinities and exceptional adaptations to aridity. Perennial vegetation is found in areas that receive runoff water, waterborne material, and wind-blown sediments, such as wadis, channels, runnels, depressions, and hill slopes (Monod 1954). Where the water table is close to the surface, wadi beds may harbor tree communities with *Acacia* species in nonsaline wadis and *Tamarix* species in saline wadis. Other habitats such as regs, hamadas, and to a lesser extent ergs are devoid of visible plant life for years (Ozenda 1983). After rain, a plant community of ephemeral annuals germinates, which can reach more than 50 percent cover of sand dunes and 20 percent on gravel plains (WWF and IUCN 1994).

Outstanding or Distinctive Biodiversity Features

The flora of the central Sahara Desert [95] is poor, supporting an estimated 500 species across a huge area (Le Houérou 1990). As many as 162 of the plant species are endemic to the Sahara (Zahran and Willis 1992). Most genera are represented by one to three species, but there are a number of isolated monotypic genera, of both wide and narrow distribution. The monotypic genera have a suggested Tertiary origin, with probable extinction of linking forms (Cloudsley-Thompson 1984).

The fauna of the central Sahara is richer than is generally believed. Arthropods are numerous, especially ants. The Sahara used to support populations of desert-adapted antelopes, many of which are now threatened if not already extirpated, such as the addax (*Addax nasomaculatus*, CR). Small numbers of other species still occur, such as the slender-horned gazelle (*Gazella leptoceros*, EN) and more widespread dorcas gazelle (*Gazella dorcas*) (Newby 1984).

At least one important bird area is located in this ecoregion (Robertson and Essghaier 2001). Situated in Libya the Zallaf is an oasis region at the eastern end of the great Awbari Erg. Among the Sahara-Sindian biome avifauna are greater hoopoe-lark (*Alaemon alaudipes*), sooty falcon (*Falco concolor*), spotted sandgrouse (*Pterocles senegallus*), and desert sparrow (*Passer simplex*) (Robertson and Essghaier 2001).

Status and Threats

Current Status

The Sahara Desert [95] contains a vast area of largely intact habitat. Because of the low nomadic population of Twaregs, Tubbus, and Moors in this part of the ecoregion, the hamadas, regs, ergs, djebels, and wadis often are undisturbed. Areas with good moisture-retaining capacities, such as wadis, harbor a permanent vegetation cover that constitutes the best pastures for pastoralist nomads. In a few oases available water forms the basis for irrigated agriculture, and there is a permanent human population.

There is little official protection of the Sahara because of the impracticality of defining borders over this vast area. Only one protected area is recorded: Zellaf Nature Reserve in Libya (1,000 km²) (IUCN 1992a). The populations of large mammals in the ecoregion have been decimated by hunting, and some species have been entirely removed, or almost so.

Types and Severity of Threats

Resettlement schemes undertaken in many countries of North Africa and the increasing use of off-road vehicles have had a negative impact on natural resources in and around the centers of population, particularly through wood cutting for fuel and the gathering of fodder. More recently the development of oil exploitation infrastructure in Algeria and Libya has contributed to the development of modern cities, creating intense local degradation.

The use of off-road cars and modern weaponry is a major threat to animals in this ecoregion, particularly the larger ungulates. The populations of desert-adapted species have been greatly reduced by hunting for food and recreation. As a result, desert-adapted antelopes have been entirely eliminated or reduced to tiny and highly threatened populations.

Justification for Ecoregion

The borders for this ecoregion follow those mapped by White (1983) and correspond approximately to the region with less than 25 mm of average rainfall per year. The two vegetation units that White distinguishes as "desert dunes with perennial vegetation" and "absolute desert" are merged here.

Ecoregion Number:	**96**
Ecoregion Name:	**West Saharan Montane Xeric Woodlands**
Bioregion:	**African Palearctic**
Biome:	**Deserts and Xeric Shrublands**
Political Units:	**Algeria, Mali, Niger, Mauritania**
Ecoregion Size:	**258,100 km²**
Biological Distinctiveness:	**Locally Important**
Conservation Status:	**Relatively Intact**
Conservation Assessment:	**V**
Author:	*Christine Burdette*
Reviewed by:	*John Newby*

Location and General Description

The West Saharan Montane Xeric Woodlands [96] occur in isolated mountain ranges and several smaller outliers in the Sahara Desert. The Ahaggar or Hoggar Mountains, found largely in south central Algeria, comprise the largest block and include peaks such as Mount Tahat (2,981 m) and Mount Assekrem (2,728 m). Four smaller mountain blocks are also included in the ecoregion: the Tassili N'Ajjer in southeast Algeria, the Aïr in northern Niger (with Mount Baguezane reaching more than 2,000 m), the Adrar in Mauritania (reaching almost 700 m), and

the Adrar des Iforas in Mali and Algeria (reaching 900 m). The Hoggar Mountains are volcanic in origin. Other parts of the ecoregion consist of wind-eroded Tassili sandstone formations (Cloudsley-Thompson 1984). The soils developed over these rocks are a mixture of bare rock and lithosols at higher elevations and yermosols at lower elevations (Cloudsley-Thompson 1984).

The climate is cold and dry in the winter and hot and dry in the summer. Rainfall is variable but averages less than 150 mm per annum, with most falling at higher elevations in June to September. The mean maximum temperature reaches 30°C at lower elevations and 18°C to 12°C at the highest elevations, whereas the mean minimum temperatures are as low as 3°C at the highest elevations. In the winter, frosts are common, and snow can be found on the higher peaks of the Hoggar. Throughout the ecoregion permanent waterholes, called gueltas, are protected from the sun in narrow gorges, which reduces evaporation and increases the permanence of the pools; it is primarily these areas that give the ecoregion its floral and faunal values.

This ecoregion is regarded as a part of the Saharan regional transition zone in White's (1983) phytogeographic classification of Africa. At lower elevations, the vegetation is mapped as regs, hamadas, and wadis, but at the highest altitudes there is a transition to Saharomontane vegetation (White 1983). This ecoregion supports an interesting relict flora, with Mediterranean, Sudano-Deccan, and Saharo-Sindien affinities. The affinities to the Mediterranean region are particularly notable (White 1983; Cloudsley-Thompson 1984).

In general, larger woody species are typically Saharan endemics with Mediterranean origin, or they also occur in the Mediterranean region (White 1983). Stands of relict forest vegetation comprising Duprey cypress (*Cupressus dupreziana*), wild olive (*Olea laperrini*), and myrtle (*Myrtus nivellei*) are characteristic and grow at the bottom of wadis, in intermittent stream valleys, or beside gueltas. Other species with a preference for moist habitats are *Trianthema pentandra, Lupinus pilosus,* and *Convolvulus fatmensis.* Species such as *Silene kiliani, Acacia laeta, A. scorpiodes,* and *Cordia rochii* also grow in wadis.

Outstanding or Distinctive Biodiversity Features

Vegetation in this ecoregion varies according to elevation and landscape features and contains a number of endemic and rare species. The most notable of these is Duprey cypress or *tarout,* wild olive, and myrtle, all of which are relict Saharan-Mediterranean species.

Migratory Palearctic birds use this ecoregion as a rest area because of the year-round water and cooler temperatures. BirdLife International has identified a number of important bird areas for conservation in this ecoregion, including Parc National de Tassili N'Ajjer, Parc National de l'Ahaggar, and the Aïr Ténéré (Brouwer et al. 2001; Coulthard 2001a). Species include pallid harrier (*Circus macrourus*), bar-tailed desert lark (*Ammomanes cincturus*), Lichtenstein's sandgrouse (*Pterocles lichtensteinii*), and greater hoopoe-lark (*Alaemon alaudipes*).

The plateaus of this ecoregion provide the last refuges for some species. These include populations of globally threatened antelope, such as dorcas gazelle (*Gazella dorcas,* VU), slender-horned gazelle (*Gazella leptoceros,* EN), and dama gazelle (*Gazella dama,* EN) (East 1999). Cheetah (*Acinonyx jubatus,* VU), Barbary sheep (*Ammotragus lervia,* VU), sand cat (*Felis margarita*), and caracal (*Caracal caracal*), species more closely associated with arid climates, also exist here (Coulthard 2001a). Locally threatened species include the rock hyrax (*Procavia capensis*) and Val's gundi (*Ctenodactylus vali*) (Boitani et al. 1999). Only one near-endemic mammal species occurs, Olga's dormouse (*Graphiurus olga*).

Many reptiles are also present, including the snakes *Telescopus obtusus, Psammophis sibilans,* and *Echis leucogaster.* Amphibians include the European green toad (*Bufo viridis*) (Schleich et al. 1996).

Status and Threats

Current Status

The largest intact blocks of habitat found in Algeria lie in the huge protected areas of Parc National de l'Ahaggar and Parc National de Tassili N'Ajjer. Tourism has become a source of income for the area, and efforts are being made to curb poaching and woodcutting. The large Aïr and Ténéré National Nature Reserve in Niger protects another significant habitat area.

Types and Severity of Threats

Although naturally fragmented by geography, much of the ecoregion is intact and protected by its inaccessible location and rugged terrain. However, pastoralism and wood collection by nomads have led to the alteration and erosion of river channels and elimination of rare vegetation.

Motorized sport hunting by people from North Africa and the Middle East, in combination with local subsistence hunting, poses a threat to most large mammal populations, such as cheetah, slender-horned gazelle, and dama gazelle.

Justification for Ecoregion Delineation

This ecoregion covers several disjunct arid montane areas of the Sahara Desert that share some plant communities and collectively act as refugia for species that had wider distributions in more mesic climates. The boundaries for the largest part of this ecoregion, which includes the Tassili N'Ajjer, l'Ahaggar Massif, follow the "regs, hamadas and wadis" and the "Saharomontane" vegetation units of White (1983) above the 1,000 m contour. Areas further south were included in this ecoregion (Aïr ou Azbine in northern Niger, Dhar Adrar in Mauritania, and Adrar des Iforas in Mali and Algeria), using the 500-m elevation contour.

Ecoregion Number:	**97**
Ecoregion Name:	**Tibesti-Jebel Uweinat Montane Xeric Woodlands**
Bioregion:	**African Palearctic**
Biome:	**Deserts and Xeric Shrublands**
Political Units:	**Chad, Libya, Egypt, Sudan**
Ecoregion Size:	**82,200 km²**
Biological Distinctiveness:	**Locally Important**
Conservation Status:	**Relatively Intact**
Conservation Assessment:	**V**
Author:	Christine Burdette
Reviewed by:	John Newby

Location and General Description

This Tibesti-Jebel Uweinat Montane Xeric Woodlands [97] ecoregion comprises two isolated montane areas in the central part of the Sahara Desert. Lying halfway between Lake Chad and the Gulf of Syrte, the larger is the Tibesti Massif, which is found in the northern portion of Chad and extends marginally into southern Libya. The Tibesti Mountains consist of seven inactive volcanoes, with the highest peak reaching 3,415 m elevation. The second, smaller portion is the Jebel Uweinat, located further to the east along the intersection of eastern Libya, southwestern Egypt, and northwestern Sudan. The Jebel Uweinat includes peaks reaching elevations just under 2,000 m. Both parts of the ecoregion are basalt outcrops in an expanse of Nubian sandstones. The soils developed over the basalts are typically thin, with lithosols predominating. Bare rock and regosols also occupy large areas of the ecoregion.

Annual average rainfall in the surrounding Sahara Desert is less than 100 mm and is extremely unpredictable, as years may pass with no rainfall followed by a single thunderstorm lasting only a few hours. In this montane ecoregion rainfall is more regular but still probably under 600 mm per annum. Lowland wadis areas receive their water from the mountains down storm channels. This water remains in the ground for a long period because these areas have natural impermeable layers and shield-like sand covers that slow evaporation (Cloudsley-Thompson 1984). The mean maximum temperature is approximately 30°C in the lowlands and falls to 20°C in the highest elevations. Mean minimum temperatures are 12°C in the lowlands but fall to 9°C over most of the ecoregion and are as low as 0°C at the highest elevations in winter.

In terms of the phytogeographic classification of White (1983), this ecoregion is regarded as a part of the Saharan regional transition zone. The vegetation is mapped at the lower elevations as regs, hamadas, and wadis and at the highest elevations as Saharomontane. The Tibesti mountain vegetation varies according to elevation and slope. Large wadis radiating from the mountains to the southwest support tree species such as the doum palm (*Hyphaene thebaica*), *Salvadora persica, Tamarix articulata, Acacia nilotica adstringens,* and *Acacia albida,* and other tropical herbs in the genera *Abutilon, Hibiscus, Rhynchosia,* and *Tephrosia* (White 1983). Doum palm and date palm (*Phoenix dactylifera*) grow along deep gorges that hold water year-round (White 1983).

The peak of the Jebel Uweinat is almost devoid of vegetation, with a sparse scattering of *Lavandula* and *Salvia* (White 1983). In lower-altitude woodlands, dominant species include *Fagonia indica, Aerva javanica, Acacia tortilis,* and *Cleome chrysantha* (Leonard 2000). Low-altitude wadis receive rainwater runoff from the mountain areas and support a diverse plant community with Mediterranean affinities (White 1983).

Outstanding or Distinctive Biodiversity Features

The Saharomontane vegetation of the higher elevations of the Tibesti Mountains supports the endemic *Ficus teloukat,* which grows on the south and southwestern slopes, *Myrtus nivellei* and *Nerium oleander* on the western slopes, *Tamarix gallica, T. nilotica,* and *Nerium oleander* on the wetter northern slopes. The northern slopes also support wetland species such as *Juncus maritimus, Typha australis, Scirpus holoschoenus, Phragmites australis,* and *Equisetum ramosissimum* (White 1983). Jebel Uweinat is less species rich, with a total of only eighty-seven plant species recorded (Leonard 2000). Remnants of tropical and Mediterranean plant communities are seen throughout this ecoregion, including species such as *Hibiscus* sp. and *Rhynchosia* sp. Other species are Saharan endemics or more widely occurring across drier parts of Africa (White 1983). These affinities indicate a wetter climate during the Pleistocene, when there was a continuous connection between Mediterranean North Africa and tropical Africa.

Populations of several important Saharan large mammals remain in this ecoregion. For example, dorcas gazelle (*Gazella dorcas,* VU), dama gazelle (*Gazella dama,* EN), Barbary sheep (*Ammotragus lervia,* VU), and cheetah (*Acinonyx jubatus,* VU) have been recorded from Tibesti (Monfort et al. 2002). Dorcas gazelle and Barbary sheep were also recently documented in the Jebel Uweinat (Leonard 2000). Small mammals and their predators are also abundant, including rock hyrax (*Procavia capensis*), Cape hare (*Lepus capensis*), spiny mouse (*Acomys* spp.), North African gerbil (*Gerbillus campestris*), bushy-tailed jird (*Sekeetamys calurus*), and three different fox species, Rueppel's fox (*Vulpes rueppelli*), pale fox (*Vulpes pallida*), and fennec (*Vulpes zerda*) (Le Berre 1990; Boitani et al. 1999).

There is a rich desert avifauna, including crowned sandgrouse (*Pterocles coronatus*), Lichtenstein's sandgrouse (*P. lichtensteinii*), bar-tailed desert lark (*Ammomanes cinturus*), desert lark (*A. deserti*), greater hoopoe lark (*Alaemon alaudipes*), pale crag-martin (*Hirundo obsoleta*), white-tailed wheatear (*Oenanthe leucopyga*), blackstart (*Cercomela melanura*), fulvus babbler (*Tur-*doides fulvus), trumpeter finch (*Rhodopechys githaginea*), and desert sparrow (*Passer simplex*) (Scholte and Robertson 2001).

The reptile and amphibian richness is poor in this area. Reptiles include the beaked blindsnake (*Leptotyphlops macrorhynchus*), braid snake (*Coluber rhodorachis*), and ringed wall gecko (*Tarentola annularis*) (Schleich et al. 1996).

Status and Threats

Current Status

A period of climatic desiccation lasting several thousand years has affected the vegetation of the area, as with other parts of the Sahara. The steep, rough terrain of this ecoregion and its location deep in the Sahara Desert make it largely intact. Almost all species, both plant and animal, can seek refuge in remote parts of the ecoregion. Currently there are almost no people living in the area, allowing vegetation to grow from previous degradation caused by grazing (Le Houérou 1998). There are no official protected areas in the ecoregion.

Types and Severity of Threats

Although habitats are largely intact, an established protected area may be needed in the long term to conserve the species and habitats in this ecoregion. Hunting continues to be a major threat to large mammals. Political and economic instability limits the opportunities and resources available for protection of the habitats and biodiversity.

Justification for Ecoregion Delineation

The high-elevation areas of the Tibesti Mountains and Jebel Uweinat are unique geological formations with distinct biota in the Sahara Desert. The boundaries of the ecoregion follow White's (1983) "Saharomontane vegetation" where it divides from the lower elevation vegetation type, "regs, hamadas, wadis." Given the wide distance between other massifs in the Sahara Desert and the presence of local endemism, the Tibesti and Jebel Uweinat were defined as a separate ecoregion.

Ecoregion Number: **98**

Ecoregion Name: **Red Sea Coastal Desert**

Bioregion: **African Palearctic**

Biome: **Deserts and Xeric Shrublands**

Political Units: **Egypt, Sudan**

Ecoregion Size: **59,300 km²**

Biological Distinctiveness: **Locally Important**

Conservation Status: **Relatively Stable**

Conservation Assessment: **V**

Author: *Christine Burdette*

Reviewed by: *Mahmoud A. Zahran*

Location and General Description

The Red Sea Coastal Desert [98] ecoregion lies on the eastern coast of Egypt along the Red Sea and Gulf of Suez. It stretches from the coastline inland to a coastal mountain chain with peaks reaching 2,187 m. Rocks are both igneous and sedimentary (limestone) and soils are mainly lithosols.

The ecoregion lies in a winter rainfall area. There is as little as 3 mm per year in the coastal desert areas, but the mountains, particularly along the eastern slopes, receive significantly more rainfall year-round. Coastal mountains receive both "fog precipitation" from clouds and heavy rainfall from occasional storms. The southern mountains, such as Jebel Elba, are closer to the Red Sea and therefore receive greater fog precipitation than the northern mountains (Kassas and Zahran 1971). In many areas a dense cover of annual vegetation develops after precipitation (Abd El-Ghani 1998). In years of heavy rains, there can be torrential flooding along wadis (intermittent streams) that drain the mountains both west (to the Nile) and east (to the Red Sea). Mean maximum temperatures in the lowlands range from 21°C to 30°C, and mean minimum temperatures range from 18°C to 3°C on the mountaintops. Frost occurs frequently at higher elevations in winter.

The human population of the ecoregion is very small, and densities are less than five people per square kilometer. Most people are not resident but either are nomadic (such as the Bischarin, who come to the area for short periods of time) or live near the Nile River, where there is a better water supply. The northernmost part of the ecoregion supports a larger population because of its proximity to Cairo.

White (1983) regarded this ecoregion as part of the Somali-Masai regional center of endemism and mapped it as the northern part of the "Red Sea coastal desert." The ecoregion consists of a number of habitat types, including mangroves, littoral salt marshes, coastal plain, and montane.

Along the coast *Avicennea marina* and *Rhizophora mucronata* characterize stands of mangroves. Halophytes in the littoral salt marshes include *Halocnemum strobilaceum, Arthrocnemum macro-*

stachyum, Zygophyllum album, Nitraria retusa, Suaeda monoica, Salicornia fruticosa, Halopeplis perfoliata, and *Tamarix nilotica* (Zahran and Willis 1992).

The coastal desert plain is characterized by xerophytic vegetation dominated by species including *Hammada elegans, Arabasis articulata, Launaea spinosa, Leptadenia pyrotechnica, Zygophyllum coccineum, Salvadora persica, Acacia ehrenbergiana, Calotropis procare, Salsola baryosma, Zilla spinosa,* and *Aerva javanica.* Woodland vegetation can also be found alongside wadis and in other wetter areas in the plain. The woodlands contain *Acacia tortilis, A. raddiana, A. ehrenbergiana, Aerva versica, Euphorbis cuneata,* and *Balanites aegyptiaca* (Goodman 1985; Zahran and Willis 1992; Fossati et al. 1999).

In the mountains, east-facing slopes support richer vegetation than the western slopes, including water-loving plants such as *Phragmites australis* and *Imperata cylindrica.* Montane wadis support a rich flora, including the tree *Moringa peregrina.* The Jebel Elba group, a transitional area between the Afrotropical and Palearctic realms, supports the richest plant life in the region, especially on its north and east sides (Kassas and Zahran 1971; WWF and IUCN 1994). On north-facing slopes, *Euphorbia cuneata* dominates the lower elevations, *E. nubica* grows at middle elevations, and unusual mist vegetation, such as dragon ombet (*Dracaena ombet*), is found in the mist zone at the highest elevations (WWF and IUCN 1994; Goodman 1985; Sheded 1998).

Outstanding or Distinctive Biodiversity Features

The Red Sea Coastal Desert [98] lies along one of the major Palearctic-Afrotropical bird migration routes, with millions of birds passing through the ecoregion every year. Migrants fly north to their breeding grounds from late February to early May and return to their wintering grounds in sub-Saharan Africa from July to October. The mountains offer a hospitable stopover area, providing water, shade, and food. Large soaring birds passing through in spring and fall migrations include the white stork (*Ciconia ciconia*), great white pelican (*Pelecanus onocrotalus*), imperial eagle (*Aquila heliaca,* VU), steppe eagle (*Aquila nipalensis*), common buzzard (*Buteo buteo*), European honey-buzzard (*Pernis apivorus*), lesser kestrel (*Falco naumanni*), and Levant sparrowhawk (*Accipiter brevipes*) (Baha el Din 1999, 2001). Several coastal areas and near-shore islands have been recognized as important bird areas, especially as breeding sites, including Gebel El Zeit, Hurghada archipelago, Wadi Gimal Island, Qulân Islands, Zabargad Island, Siyal Islands, Rawabel Islands, and Gebel Elba (Baha el Din 2001). The Hurghada archipelago hosts the largest known breeding population of white-eyed gull (*Larus leucophthalmus*) in the world.

Although the ecoregion supports a number of mammal species, increasing hunting pressures have reduced remaining population numbers. Some species occur only in the montane areas, including the Nubian ibex (*Capra nubiana,* EN), aardwolf

(*Proteles cristatus*), and Barbary sheep (*Ammontragus lervia*, VU). Other mammals found in this ecoregion include dorcas gazelle (*Gazella dorcas*), rock hyrax (*Procavia capensis*), and pale fox (*Vulpes pallida*).

This ecoregion also supports a diverse assemblage of reptiles, although much of the area has not been surveyed (Schleich et al. 1996). In some regions there are important populations of the declining ocellated mastigure (*Uromastyx ocellata*) (Baha el Din 2001).

Status and Threats

Current Status

The Red Sea Coastal Desert [98] ecoregion is fragile but largely undisturbed (Goodman 1985). The Jebel Elba National Park is the only protected area that covers this ecoregion. The habitat is being slowly denuded by human activities including grazing by domestic animals, collection of woody plants for firewood and charcoal, unregulated use of off-road vehicles, road construction, solid waste disposal, unregulated quarrying, tourist development, and land reclamation.

Types and Severity of Threats

Principal threats include overgrazing, hunting, habitat alteration, firewood collection, and charcoal production (Baha el Din 2001). Cement quarrying sites also pose a serious threat not only to vegetation but also local wildlife (Hegazy 1996). One cement plant was recorded to emit more than 40 tons/km^2 of kiln dust in 1 month, affecting vegetation within a 15-km radius of the plant. Emissions may also have an effect on bird migrations and other species (Baha el Din 1999).

Hunting of birds and mammals is a popular activity that is completely unregulated. Falconry is very popular and commonly occurs along the coast, despite being illegal (Baha el Din 1999). Other methods of hunting are also a serious threat to large mammals, many of whose population numbers are very low. In the southern portion of the ecoregion, the Jebel Elba National Park is controlled cooperatively by the Sudanese and Egyptian governments, but hunting is still a problem (Goodman 1985).

Justification for Ecoregion Delineation

This ecoregion follows the northern portion of the "Red Sea coastal desert" of White (1983) down the coast of Egypt into Sudan and encompasses the chain of mountains that run parallel to the Red Sea coast. Orographic mist and rain supplies springs and wadis in the mountains, which in turn support a rich flora, particularly along its northern and eastern slopes. The southern portion of the "Red Sea coastal desert" vegetation unit was delineated as the Eritrean Coastal Desert [100] because of its isolation and modest floral and faunal differences.

Ecoregion Number:	**99**
Ecoregion Name:	**East Saharan Montane Xeric Woodlands**
Bioregion:	**Western Africa and Sahel**
Biome:	**Deserts and Xeric Shrublands**
Political Units:	**Chad, Sudan**
Ecoregion Size:	**27,900 km^2**
Biological Distinctiveness:	**Locally Important**
Conservation Status:	**Relatively Stable**
Conservation Assessment:	**V**
Author:	*Christine Burdette*
Reviewed by:	*John Newby*

Location and General Description

The East Saharan Montane Xeric Woodlands [99] ecoregion is located in the Sahelian regional transition zone where high mountains rise from the Sahel. The largest portion of the ecoregion is located in Chad, where it encompasses the Massif de l'Ennedi and Massif du Kapka above 1,400 m elevation. A smaller outlier of this ecoregion is located in Sudan, the Jebel Marra, which reaches heights over 3,000 m.

The Jebel Marra is composed of Tertiary volcanic deposits and has a relict crater now containing two lakes. Bare rock outcrops are found in many areas, although lithosols and yermosols also occur. The Massif de l'Ennedi differs geologically by having a cap of Devonian-aged sandstone from 1,800 to 2,000 m altitude (White 1983).

Most of the annual rainfall occurs between May and September. Precipitation varies greatly with elevation, ranging from 150 to 500 mm in most areas but reaching more than 1,000 mm on the higher parts of the Jebel Marra. Temperatures are also significantly lower than in the surrounding semi-desert habitats, particularly in winter.

In terms of the phytogeographic classification by White (1983), this area is part of the Sahelian regional transition zone. The vegetation is mapped as "Saharomontane" (White 1983), one of three such areas in northern Africa. The vegetation comprises dry woodland vegetation surrounded by Sahel *Acacia* wooded grassland and deciduous bushland (White 1983). The flora has affinities to that of North Africa, the Ethiopian Highlands, the Kenyan Mountains, and Europe (Wickens 1976). Originally, the higher-elevation areas probably were dominated by *Olea laperrinei* (VU) scrub forest; however, the current vegetation is dominated by secondary grassland, with *O. laperrinei* scattered across the landscape. On the steeper, more eroded slopes, bunchgrass and small shrub species such as *Andropogon distachyos*, *Themeda triandra*, and *Hyparrhenia hirta* grow together (White 1983). Areas with better drainage harbor a second type of grassland, including *Vulpia bromoides*, *Aristida congesta*, *Festuca abyssinica*, and *Pentaschistis pictigluma*, all not more than 5 cm tall (White 1983). The Jebel Marra contains multiple habitat types: valleys with deep litter soil and wooded forest, bare rocky outcrops, tallgrass savanna, and a high-elevation plateau of remnant volcanic ash with short grasses and wind-blown bushes (Happold 1969).

Outstanding or Distinctive Biodiversity Features

A number of plant species endemic to the Sahel occur in this ecoregion, including *Ammannia gracilis*, *Chrozophora brocchiana*, *Farsetia stenoptera*, *Indigofera senegalensis*, *Nymphoides ezannoi*, *Panicum laetum*, *Tephrosia gracilipes*, *T. obcordata*, and *T. quartiniana* (White 1983). Endemic plant species of the Jebel Marra area include *Kickxia aegyptiaca*, *Pletranthus jebel-marrae*, and *Felicia dentata* (Wickens 1976).

The ecoregion houses a few endemic animal species. Among mammals, two gerbils, Burton's gerbil (*Gerbillus burtoni*, CR) and Lowe's gerbil (*Gerbillus lowei*, CR), are strictly endemic to the higher elevations of the Jebel Marra, with the latter confined to hillsides near the crater lakes (Happold 1966). The arid-thicket rat (*Grammomys aridulus*) is also near endemic to the Jebel Marra and nearby lower areas. The rusty lark (*Mirafra rufa*) is near endemic to the ecoregion, also being found in other high-altitude areas of the Sahara.

This ecoregion, particularly the Massif de l'Ennedi, is also an important location for larger mammals, including threatened antelope populations (Hilton-Taylor 2000). Dama gazelle (*Gazella dama*, EN), dorcas gazelle (*Gazella dorcas*, VU), and red-fronted gazelle (*Gazella rufifrons*, VU) are all found here (East 1999). More common mammals include Rüppell's fox (*Vulpes rueppellii*), caracal (*Caracal caracal*), Cape hare (*Lepus capensis*), rock hyrax (*Procavia capensis*), African wild cat (*Felis silvestris*), common or golden jackal (*Canis aureus*), and desert hedgehog (*Hemiechinus aethiopicus*). Only a few large predators exist, such as a highly threatened population of cheetah (*Acinonyx jubatus*, VU) and a few striped hyenas (*Hyaena hyaena*).

Two parts of this ecoregion, Fada Archei (Chad) and Jebel Marra (Sudan), have been identified as important bird areas (Fishpool and Evans 2001). Birds from the Fada Archei gorge include Egyptian nightjar (*Caprimulgus aegyptius*), pharaoh eagle-owl (*Bubo ascalaphus*), crowned sandgrouse (*Pterocles coronatus*), Lichtenstein's sandgrouse (*P. lichtensteinii*), bar-tailed desert lark (*Ammomanes cincturus*), desert lark (*A. deserti*), greater hoopoe-lark (*Alaemon alaudipes*), white-tailed wheatear (*Oenanthe leucopyga*), blackstart (*Cercomela melanura*), and fulvous babbler (*Turdoides fulvus*) (Scholte and Robertson 2001). The Jebel Marra important bird area supports lesser kestrel (*Falco naumanni*, VU) and Nubian bustard (*Neotis numa*) (Robertson 2001c).

Status and Threats

Current Status

Humans have inhabited the Jebel Marra for more than 2,000 years, with all accessible slopes up to 2,600 m elevation at one point used for agriculture (Wickens 1976). The top of the Jebel Marra was once covered in scrub forest but is now secondary grassland with scattered trees (MacKinnon and MacKinnon 1986). Today, the only habitats remaining undisturbed in this ecoregion are deep valley floors. However, these areas are inhospitable to larger vertebrates because of their small size and steep-sided valleys.

The Fada Archei Game Reserve in Chad lies in part of this ecoregion (Keith and Plowes 1997). There is also a forest reserve in Sudan that covers part of the Jebel Marra area.

Types and Severity of Threats

The human population of the ecoregion is extremely low and consists mostly of nomadic and seminomadic people, with a few farmers present in the valleys surrounding the ecoregion boundaries.

Human activities are not believed to be threatening the habitats of this ecoregion, although recent reports are not readily available. Most of the agriculture once practiced in the area has been abandoned, resulting in regeneration of the landscape into more natural habitat (although not the original *Olea laperrinei* scrub forest type). This cannot be said of the fauna: the larger animal species have been removed or reduced to very small populations by hunting or competition with livestock and by drought.

Justification for Ecoregion Delineation

The boundaries of the Massif de l'Ennedi and Massif du Kapka in eastern Sahara were drawn around the 1,000-m elevation contour, and the boundaries for the Jebel Marra were taken directly from the "Saharomontane" vegetation unit of White (1983). These high jebels, which contain moisture-dependent habitat, are biologically distinct from the surrounding deserts and share affinities with areas in North Africa, the Ethiopian Highlands, Kenyan Mountains, and Europe.

Ecoregion Number:	**100**
Ecoregion Name:	**Eritrean Coastal Desert**
Bioregion:	**Horn of Africa**
Biome:	**Deserts and Xeric Shrublands**
Political Units:	**Djibouti, Eritrea**
Ecoregion Size:	**4,600 km^2**
Biological Distinctiveness:	**Locally Important**
Conservation Status:	**Vulnerable**
Conservation Assessment:	**V**
Authors:	*Chris Magin, Christine Burdette*
Reviewed by:	*Ib Friis*

Location and General Description

The Eritrean Coastal Desert [100] runs along the southern coast of the Red Sea from Balfair Assoli in Eritrea to Ras Bir near Obock in Djibouti. It occupies the southern shore of the Bab-el-Mandeb straits, which are the entrance to the Red Sea from the Gulf of Aden, and includes several groups of offshore islands, such as the Sept Frères, belonging to Djibouti.

The climate is hot and dry. Annual rainfall averages less than 100 mm but is highly variable from year to year. The mean maximum temperature is 33°C, and the mean minimum temperature (27°C) is the highest experienced anywhere in Africa. Basement rocks consist of volcanic lavas from the Ethiopian Highlands that reach the Red Sea, overlain with regosols. The landscape consists of a flat, largely featureless sand or gravel plain lying below 200 m altitude, interspersed with rocky outcrops. The shoreline is a mixture of rocky areas (e.g., around Ras Siyan in Djibouti), old coral reefs that are exposed at low tide, and sandy beaches.

White (1983) regarded the area as part of the Somali-Masai regional center of phytogeographic endemism and mapped it as the southern part of the Red Sea Coastal Desert. The northern part of this desert has been mapped as the Red Sea Coastal Desert [98] ecoregion. Vegetation consists of sparse herbaceous, grassy steppe (typical species include *Aerva javanica, Cymbopogon schoenanthus, Panicum turgidum,* and *Lasiurus scindicus*) and scattered *Acacia tortilis–Acacia asak* steppe with some *Rhigozum somalense–Caesalpinia erianthera* shrubland (Audru et al. 1987). Along the coast there are stretches of halophytic vegetation, and the few sheltered creeks are fringed with stands of mangroves consisting of *Rhizophora mucronata, Ceriops tagal,* and *Avicenna marina*. Inland, the ecoregion grades into xeric grasslands and shrublands.

Outstanding or Distinctive Biodiversity Features

This desiccated ecoregion contains no outstanding biological features, few resident fauna, and few endemics (Ibrahim 1999; Magin 1999). Only three near-endemic species are found here, all reptiles: the Ogaden burrowing asp (*Atractaspis leucomelas*),

Ragazzi's cylindrical skink (*Chalcides ragazzii*), and Indian leaf-toed gecko (*Hemidactylus flaviviridis*). Among the mammals, dorcas gazelle (*Gazella dorcas,* VU) is still common, although the status of other antelopes such as Sömmerring's gazelle (*Gazella soemmerringii,* VU) and Salt's dik-dik (*Madoqua saltiana*) is not well known (East 1999; Hilton-Taylor 2000).

Each autumn hundreds of thousands of Asian and European birds of prey cross from the Arabian Peninsula to Africa via the Bab-el-Mandeb Straits at the mouth of the Red Sea (MacKinnon and MacKinnon 1986; Welch and Welch 1988), one of the world's largest intercontinental raptor migrations. Scientists have recorded twenty-six raptor species, the two most numerous being the steppe buzzard (*Buteo buteo vulpinus*) and the steppe eagle (*Aquila nipalensis*). In addition, the Sept Frères islands have important breeding colonies of crested tern (*Sterna bergii*) and lesser-crested tern (*Sterna bengalensis*) (Magin 2001), and green turtle (*Chelonia mydas,* EN) and hawksbill (*Eretmochelys imbricata,* CR) (Hilton-Taylor 2000) nest in sheltered sandy coastal coves.

Status and Threats

Current Status

The habitats of the ecoregion are largely intact but degraded by overgrazing and fuelwood collection, particularly near settlements. Although there are no protected areas, proposals have been made to protect the Sept Frères and the section of the coast from Kadda Guêïni to Doumêra, both in Djibouti (East 1999; Magin 1999). The inhospitable condition of this ecoregion for human settlement offers some protection to the wildlife. In Eritrea and Djibouti laws ban hunting, but enforcement is very weak. Although researchers saw little or no evidence of violation of this law in areas of Eritrea in 1995, hunting remains a concern in Eritrea (East 1999) and Djibouti (Magin 1999).

Types and Severity of Threats

Poaching of gazelles, marine turtles, and nesting seabirds (for both meat and eggs) occurs. A potential future threat would be the upgrading of the Eritrea-Djibouti coastal road, which could lead to unplanned and uncontrolled development. However, it could also dramatically improve the potential for sensitively managed ecotourism in the area. A further potential threat is climate change.

Justification for Ecoregion Delineation

This ecoregion follows the boundaries of the southern portion of the "Red Sea coastal desert" of White (1983). Although communities in this region are similar to those of the Red Sea coastal desert in Egypt, this region was elevated to ecoregion status because of its isolation from the northern unit (over 10° latitude) and because of modest floral and faunal differences.

Ecoregion Number: **101**

Ecoregion Name: **Ethiopian Xeric Grasslands
and Shrublands**

Bioregion: **Horn of Africa**

Biome: **Deserts and Xeric Shrublands**

Political Units: **Ethiopia, Eritrea, Djibouti, Somalia,
Sudan**

Ecoregion Size: **153,100 km²**

Biological Distinctiveness: **Regionally Outstanding**

Conservation Status: **Relatively Stable**

Conservation Assessment: **III**

Authors: *Chris Magin, Christine Burdette*

Reviewed by: *Ib Friis*

Location and General Description

The Ethiopian Xeric Grasslands and Shrublands [101] extend from the Sudan-Eritrea border, south through Ethiopia to Djibouti, and east into Somalia, in the Somaliland region of the country. It borders the Red Sea and the Gulf of Oman and includes the Dahlak Archipelago and some other islands. Most of the area lies between sea level and 800 m, increasing westward toward the Ethiopian and Eritrean Highlands. Arid hills and massifs reach 1,300 m in altitude, but higher massifs, such as the Goda and Mabla in Djibouti, are considered outliers of the Ethiopian Lower Montane Forests, Woodlands, and Bushlands [25]. There are also fault-induced depressions, such as the Danakil (Dallol) Depression and Lac Assal, lying as much as 160 m below sea level.

The climate is very hot and dry. Mean annual rainfall varies from less than 100 mm close to the coast to around 200 mm further inland. Mean minimum temperatures range from 21°C to 24°C, and the mean maximum temperature is around 30°C. There are few permanent watercourses, with the most notable being the Awash River of Ethiopia, which terminates in a series of saline lakes and salt flats near the border with Djibouti. The region is extremely active tectonically and experiences many earthquakes because of the continuing expansion of the Rift Valley. There are also active volcanoes. Rocks are mainly Tertiary lava flows, although there are also Quaternary basinal deposits to the north and pre-Cretaceous basinal deposits on the coast of Somalia. Soils developed over the lava deposits are mainly lithosols, whereas regosols predominate on the Quaternary and pre-Cretaceous basinal deposits.

Phytogeographically, this ecoregion is a part of the Somali-Masai regional center of endemism, and the vegetation is mapped as Somalia-Masai semi-desert grassland and shrubland (White 1983). No areas are completely devoid of vegetation. *Acacia mellifera* and *Rhigozum somalense* dominate the basaltic lava fields, and scattered *Acacia tortilis, A. nubica,* and *Balanites ae-* *gyptiaca* are found in the sandy plains. Stands of *Hyphaene thebaica* occur in depressions and along wadis (Audru et al. 1987). Along the coast, mangroves occur in muddy areas.

Outstanding or Distinctive Biodiversity Features

Largely because of political instability over the last 30 years, many elements of the fauna and flora remain poorly known. An estimated 825–950 plant species have been observed in Djibouti, although many of these have been found only in the small outlying patches of the Ethiopian montane forest at the Day Forest and Mabla Massif above 1,100 m (Magin 1999). The floral diversity of the Afar mapping region (equivalent to the Danakil Depression and its surroundings) comprises 200 plant species, of which about 25 are endemic to this region and the adjacent equally dry parts of Ethiopia and Somalia (Friis et al. 2001). These data indicate that the flora is depauperate compared with that of other dry regions in the Horn of Africa. The dragon ombet (*Dracaena ombet,* EN) and Bankoualé palm (*Livistona carinensis,* VU) are notable plant species that do occur (Magin 1999; Hilton-Taylor 2000).

Among mammals, desert ungulates are well represented. The African wild ass (*Equus africanus somalicus,* CR) survives in the Buri Peninsula area of central Eritrea, possibly the last viable population (Hilton-Taylor 2000). Beira antelope (*Dorcatragus megalotis,* VU) occur where Djibouti, Ethiopia, and Somalia meet. Dorcas gazelle (*Gazella dorcas,* VU), Sömmerring's gazelle (*Gazella soemmerringii,* VU), Salt's dik-dik (*Madoqua saltiana*), and gerenuk (*Litocranius walleri*) are also still found (East 1999; Magin 1999). Beisa oryx (*Oryx gazella beisa*) persist, but the numbers have been greatly reduced by intense hunting pressure.

The reptilian fauna is fairly rich, but the avifauna is poor, and there are very few amphibian species. In all vertebrates levels of endemism are low, with strict endemics limited to one bird, Archer's lark (*Heteromirafra archeri,* VU), a rodent (*Gerbillus acticola*), and two geckos, Arnold's leaf-toed gecko (*Hemidactylus arnoldi*) and a subspecies of the northern sand gecko (*Tropiocolotes tripolitanus somalicus*).

Status and Threats

Current Status

Human population density typically is less than ten people per square kilometer. In some areas, there is less than one person per square kilometer. The dominant ethnic groups are the nomadic pastoralist Afars and a Somali clan, the Issas. However, human density does not account for grazing animals, which are common. The ecoregion contains only a few protected areas that suffer a lack of enforcement.

Although large blocks of "natural" habitat remain, much has been degraded through overexploitation. There is only one protected area, the Mille-Serdo Wildlife Reserve (8,766 km²) in

Ethiopia, but there are two proposed Eritrean protected areas on the Buri Peninsula in the coastal lowlands and in the Danakil Depression. The Eritrean government has also enacted a system of closures for the last remaining patches of natural vegetation throughout the country that should greatly assist the conservation of vegetation and animal species of the ecoregion.

Types and Severity of Threats

The major threats are widespread overgrazing and tree cutting for fuel and timber, particularly around the increasingly permanent settlements. Clearance for agriculture along the few permanent watercourses is also a major problem. Populations of most large mammal species have been severely reduced through hunting by local people and by the government and resistance armies during 30 years of war. Some species, such as giraffe (*Giraffa camelopardis*), are believed to be locally extinct. Other large mammals have been reduced to very small populations.

Justification for Ecoregion Delineation

The boundaries of this ecoregion follow the "Somalia-Masai semi-desert grassland and shrubland" vegetation unit of White (1983), from just north of the Sudan-Eritrea border to the extremely narrow coastal strip along the northern coast of Somalia (Somaliland). This eastern border follows the range limit of some species, such as the Dorcas gazelle, and encompasses the range of others, such as the African wild ass. White's (1983) larger "Somalia-Masai semi-desert grassland and shrubland" unit, which extends further east along the Horn of Africa, was separated into distinct ecoregions based in particular on differences in the vertebrate composition.

Ecoregion Number: **102**

Ecoregion Name:	**Somali Montane Xeric Woodlands**
Bioregion:	**Horn of Africa**
Biome:	**Deserts and Xeric Shrublands**
Political Units:	**Somalia**
Ecoregion Size:	**62,600 km²**
Biological Distinctiveness:	**Regionally Outstanding**
Conservation Status:	**Relatively Stable**
Conservation Assessment:	**III**
Authors:	*Chris Magin, Christine Burdette*
Reviewed by:	*Ib Friis, Mats Thulin*

Location and General Description

The Somali Montane Xeric Woodlands [102] ecoregion stretches along the coast of Somalia through the regions of Somaliland and Puntland. It extends from the Shimbiris Mountain east of Hargeysa through the northern mountains of Somalia to Raas Caseyr, covering the very tip of the Horn of Africa and continuing some 300 km south along the Somali coastal plain. Elevations range from sea level to the summit of Shimbiris at 2,416 m, the highest point in Somalia (WWF and IUCN 1994). There are also extensive coastal plains and sizable mountain escarpments with areas higher than 1,500 m. As a result, some authorities (e.g., Friis 1992) consider these mountain areas to be biogeographic extensions of the Ethiopian Highlands and would exclude them from this ecoregion.

In general, the climate is hot and dry, with wide seasonal temperature variations. Mean temperatures range from 21°C to 30°C in the lowlands to 9°C to 21°C in the mountains. The mean rainfall of the low-lying areas is less than 200 mm annually. The mountains receive more rainfall; for example, the escarpment near Maydh receives more than 650 mm each year, mainly in winter. Most of the higher mountain areas are composed of limestone and gypsum, covered with free-draining, thin rendzina lithosols that retain little moisture outside the rainy seasons.

The vegetation of this ecoregion varies with elevation, rainfall, and soil or rock types. At lower elevations, xerosols and yermosols have developed, particularly on the lowland coastal plains bordering the Indian Ocean. Here, the landscape is a desert, and there is little to no vegetation. In subcoastal areas woody vegetation becomes denser, with dominant species from the genera *Acacia*, *Commiphora*, and *Boswellia* (WWF and IUCN 1994). Along the sides of the escarpment Macchia-like evergreen and semi-evergreen scrub occurs with species such as *Dracaena ombet*, *Cadia purpurea*, *Buxus hildebrandtii*, and *Pistacia aethiopica*. At the highest altitudes remnants of *Juniperus* forest grow (White 1983; WWF and IUCN 1994). In 1983 it was estimated that 40,000 ha of the *Juniperus* forest remained, with canopy dominant species of *Juniperus procera*, *Olea europaea* ssp. *cuspidata*, *Dodonea viscosa*, *Cadia purpurea*, and *Sideroxylon mascatense* (previously *S. buxifolium*) (Simonetta and Simonetta 1983). In the lower, eastern part of this ecoregion, the endemic tree *Mimusops angel* is found along wadis and in dry bushland where groundwater is available (Friis 1992).

Outstanding or Distinctive Biodiversity Features

The biological values of the ecoregion are high and several hundred endemic plants are represented, including relict elements of arid and semi-arid groups, such as five endemic species of *Helianthemum* and one endemic species of *Thamnosma* (Friis 1992; Thulin 1994; WWF and IUCN 1994; Lovett and Friis 1996). Also, the monotypic genus *Renschia* is a strict endemic (WWF and IUCN 1994), as is *Boswellia frereana*, an economically important frankincense-producing tree. Numerous endemic succulent plants also occur, such as *Euphorbia mitriformis*, *Aloe eminens*, and *Huernia formosa*. Some of the endemic species, such as *Cyclamen somalense* and *Anemone somaliensis*, are found in genera that have their major distribution in the Mediterranean or the temperate regions. Other examples of plant species with disjunct distributions are found in Thulin (1994), and Friis and Ryding (2001). The most endemic-rich zone is in the high mountains, but plant endemics are also found at lower elevations.

There are also endemic species in different vertebrate groups. Three strict endemic reptiles occur, the snakes *Spalerosophis josephscorteccii* and *Leptotyphlops reticulatus* and the lizard *Pseuderemias savagei*, with two other reptiles nearly endemic to the ecoregion. Three strict endemic birds are found: the Somali pigeon (*Columba oliviae*, DD), the Somali thrush (*Turdus ludoviciae*, CR), and the Warsangli linnet (*Carduelis johannis*, EN), all found in the North Somali Mountains endemic bird area (Stattersfield et al. 1998; Robertson 2001b). The Somali thrush and Warsangli linnet are confined to juniper forests at higher elevations (Robertson 2001b). Three small mammal species are also considered near-endemics: Somali hedgehog (*Atelerix sclateri*), Louise's spiny mouse (*Acomys louisae*), and Somali elephant shrew (*Elephantulus revoili*). The rare beira antelope (*Dorcatragus megalotis*, VU), Sömmerring's gazelle (*Gazella soemmerringii*, VU) and Speke's gazelle (*Gazella spekei*, VU) are also found here and in a few other ecoregions in the Horn of Africa (East 1999).

Status and Threats

Current Status

Because of the long-standing and continuing political difficulties in the former Somalia, the ecoregion has no formal protection, but the vegetation of most escarpment areas is still intact, even if small-scale logging of *Juniperus* is going on (E. Barrow, pers. comm., 2003). The forests of the adjacent plains (particularly *A. bussei*) are being heavily damaged by charcoal production.

Types and Severity of Threats

The major threats to the ecoregion are thought to be intensive grazing by goats and other livestock (including cattle in the mountains) and cutting of trees for timber, charcoal, and fuelwood. Hunting of larger mammals is also a long-standing problem. The prolonged period of political instability in the ecoregion has also resulted in the breakdown of management authorities set up to conserve forests and wildlife.

Justification for Ecoregion Delineation

This ecoregion follows the "Somalia-Masai semi-desert grassland and shrubland" vegetation unit mapped by White (1983). The Somali Montane Xeric Woodlands [102] cover approximately the same area as the North Somali Mountains endemic bird area (Stattersfield et al. 1998) and includes Cal Madow (Al Medu), a center of endemism for plants (WWF and IUCN 1994). We extend the ecoregion farther west than Berbera and include an island of montane vegetation at Gacanlibaax, east of the Harer branch of the Ethiopian Highlands.

Ecoregion Number: **103**

Ecoregion Name:	**Hobyo Grasslands and Shrublands**
Bioregion:	**Horn of Africa**
Biome:	**Deserts and Xeric Shrublands**
Political Units:	**Somalia**
Ecoregion Size:	**25,600 km²**
Biological Distinctiveness:	**Bioregionally Outstanding**
Conservation Status:	**Endangered**
Conservation Assessment:	**IV**
Authors:	*Chris Magin, Christine Burdette*
Reviewed by:	*Ib Friis, Mats Thulin*

Location and General Description

The Hobyo Grasslands and Shrublands [103] cover a long, narrow coastal strip in Somalia, extending from just south of Mogadishu near the town of Merka northward to 250 km north of Hobyo. Scattered outliers of dune habitats occur along the coast beyond the towns of Brava and Kisimaio, right to the border with Kenya. The dunes reach heights of 160 m, and the dune field is about 10–15 km wide along its entire length. There are occasional rocky outcrops, especially in the northern part of the region. Further inland, the habitat changes to dry savanna and semi-desert vegetation of the Somali *Acacia-Commiphora* Bushlands and Thickets [44]. The area is sparsely populated, with one to twenty people per square kilometer.

Recently deposited sands overlie Precambrian basement or coral limestone rocks. Where there is soil, it is shallow and recent in origin. The climate is hot and dry. Temperature varies little during the year, with maximum temperatures reaching 33°C and minimum temperatures dipping to 21°C. Rainfall averages 200 mm annually, with most rain falling in the period from April to June as the Intertropical Convergence Zone moves north and then south.

This ecoregion is part of the Somali-Masai regional center of endemism (White 1983). The vegetation is mapped as deciduous bushland and thicket, although the main components are coastal dune grasslands with scattered bushes, herbs, and shrublets (White 1983; WWF and IUCN 1994). Vegetation is adapted to the severe climate in several ways. Along the coast, bushy vegetation, which is sandblasted by powerful winds, forms a specialized community of low, dense thickets of *Aerva javanica, Indigofera sparteola, Jatropha pelargoniifolia (glandulosa),* and *Tephrosia filiflora* (Kingdon 1997). The first colonizers of the dunes are wind-tolerant grasses and sedges, such as the near-endemic *Cyperus chordorrhizus* and the endemic *Aristolochia rigida*. Thorny, endemic species, such as the 5- to 6-m-tall *Solanum arundo,* are important features of the shrub vegetation. In the southern part of this ecoregion, the vegetation intergrades with the East African coastal evergreen bushland and forest (Friis 1992).

Outstanding or Distinctive Biodiversity Features

This area is a center of endemism for plants (Friis 1992; Thulin 1994; WWF and IUCN 1994; Lovett and Friis 1996). Botanical exploration of this region started at the end of the nineteenth century, but unpredictable rainfall, political instability, and inaccessibility have hindered detailed studies. There are no reliable estimates for the number of endemic plants, although many of the cushion plants shaped by the sand-laden winds are endemic (WWF and IUCN 1994). Succulents are also common, and the monotypic genus *Puntia* occurs. Other endemics include *Amphiasma gracilicaulis* and *Gymnocarpos parvibractus.* Unusual plant communities grow in moist limestone gorges in the northern part of the region, including the woody *Buxus hildebrandtii, Maytenus undata,* and *Vepris eugeniifolia*. These species are otherwise found only in the moister Somali Montane Xeric Woodlands [102]. *Dirachma somalensis,* one of two endangered species in the Dirachmaceae family, is located in limestone gorges.

The overall number of vertebrate species in the ecoregion is low, but these include several endemic and rare species, including two strictly endemic reptiles, *Haackgreerius miopus* and *Latastia cherchii.* Two strictly endemic mammals are the silver dik-dik (*Madoqua piacentinii,* VU) and the Somali golden mole (*Chlorotalpa tytonis,* CR). A number of rare larger mammals also occur: dibatag (*Ammodorcas clarkei,* VU), Sömmerring's gazelle (*Gazella soemmerringii,* VU), and Speke's gazelle (*Gazella spekei,* VU) (East 1999; Hilton-Taylor 2000). These all have restricted ranges in the Horn of Africa. The ecoregion also supports two strictly endemic bird species: Ash's lark (*Mirafra ashi,* EN) and the Obbia lark (*Spizocorys obbiensis,* DD), which are restricted to the coastal fixed-dune grasslands (Sattersfield et al. 1998).

Status and Threats

Current Status

Because of the long-standing and continuing political instability in Somalia, it is not known how much habitat remains in this ecoregion or how fragmented it has become. The only official protected area is Lag Badana Bush-Bush National Park, but this is presumably no longer functional.

Types and Severity of Threats

No recent information on threats is available. It is known that local populations use the scrub and grassland habitats of the ecoregion to graze their animals and gather fuelwood. The recent political instability and clan warfare in Somalia may have degraded habitats by displacing people to the coastal strip from urban centers and from areas further inland.

Justification for Ecoregion Delineation

This ecoregion forms the southernmost section of the "Somalia-Masai semi-desert grassland and shrubland" vegetation unit of White (1983). It is delineated based on the boundaries of the Central Somali Coast endemic bird area and is distinct in its plant and vertebrate composition (WWF and IUCN 1994; Stattersfield et al. 1998).

Ecoregion Number:	**104**
Ecoregion Name:	**Masai Xeric Grasslands and Shrublands**
Bioregion:	**Eastern and Southern Africa**
Biome:	**Deserts and Xeric Shrublands**
Political Units:	**Ethiopia, Kenya**
Ecoregion Size:	**101,000 km²**
Biological Distinctiveness:	**Bioregionally Outstanding**
Conservation Status:	**Relatively Intact**
Conservation Assessment:	**V**
Authors:	*Chris Magin, Christine Burdette*
Reviewed by:	*Chris Magin*

Location and General Description

The Masai Xeric Grasslands and Shrublands [104] are located in the northern part of Kenya and extreme southwestern Ethiopia. The ecoregion is primarily a dry semi-desert and includes the Dida-Galgalu Desert. Also included are most of Lake Turkana and the Omo River Delta. The ecoregion grades into the savanna woodlands of the Northern *Acacia-Commiphora* Bushlands and Thickets [45] to the west and the Somali *Acacia-Commiphora* Bushlands and Thickets [44] to the east. The topography is predominantly gently undulating between 200 and 700 m in elevation. Mount Kulal and Mount Marsabit rise from this ecoregion to higher altitudes but are placed in other ecoregions.

The climate is hot and dry over most of the year, with mean maximum temperatures around 30°C and mean minimum temperatures between 18°C and 21°C. There is a short wet season between March and June as the Intercontinental Convergence Zone moves north. Mean annual rainfall is between 200 and 400 mm.

The ecoregion is located on an outlier of the Tertiary volcanic materials that make up the Ethiopian Massif. The soils of the area are complex and vary from bare rocks and lithosols to solonchaks, yermosols, and regosols, indicating the general desiccation of the area.

The ecoregion is part of White's (1983) Somali-Masai regional center of endemism, with the vegetation mapped as "Somali-Masai semi-desert grassland and shrubland." In years of high rainfall, large areas of semi-desert annual grasslands occur. *Aristida adcensionis* and *A. mutabilis* are dominant species, but during droughts these and other plant species can be absent, sometimes for many years. The next most extensive vegetation types are dwarf shrublands, dominated by *Duosperma eremophilum* on heavier, wetter sedimentary soils and *Indigofera spinosa* on stabilized dunes (White 1983). The shore of Lake Turkana is mostly rocky or sandy, with little aquatic vegetation, and around the lake there are grassy plains on which yellow spear grass and doum palms (*Hyphaene thebaica*) predominate.

Outstanding or Distinctive Biodiversity Features

The ecoregion is moderately rich in species but has a low level of endemism. The grasses, shrubs, and trees of the ecoregion are fire tolerant because fires are frequent in the dry season, and they include some species endemic to the Somali-Masai region (numbers unknown).

Strictly endemic animals include Lake Turkana toad (*Bufo turkanae*) and Turkana mud turtle (*Pelusios broadleyi*, VU) (Hilton-Taylor 2000). Lake Turkana has more than 350 species of aquatic and terrestrial birds and is also an important flyway for migrant birds, including more than 100,000 little stint (*Calidris minuta*) (Bennun and Njoroge 1999). Central Island has a breeding population of African skimmers (*Rynchops flavirostris*).

The Dida-Galgalu Desert supports two endemic birds: William's lark (*Mirafra williamsi*, DD) and the locally distributed masked lark (*Spizocorys personata*) (Bennun and Njoroge 1999). The mammalian fauna includes Burchell's zebra (*Equus burchelli*), Grevy's zebra (*Equus grevyi*, EN), Beisa oryx (*Oryx gazella beisa*), Grant's gazelle (*Gazella granti*), topi (*Damaliscus lunatus jimela*), lion (*Panthera leo*, VU), and cheetah (*Acinonyx jubatus*, VU) (Boitani et al. 1999; East 1999). These species are all found in Sibiloi National Park and other areas in the ecoregion. Other mammals include leopard (*Panthera pardus*), reticulated giraffe (*Giraffa camelopardalis reticulata*), and elephant (*Loxodonta africana*, EN). Among the small mammals there is also the near-endemic cushioned gerbil (*Gerbillus pulvinatus*).

Status and Threats

Current Status

Heavy grazing from domestic livestock has degraded most habitats of this ecoregion. A small area of intact habitat remains in the few protected areas: Sibiloi National Park on the north-eastern edge of Lake Turkana, Mount Kulal Biosphere Reserve extending north and east from the southeastern coast of Lake Turkana, and the Chew Bahr Wildlife Refuge in Ethiopia. The habitats of the ecoregion are not particularly fragmented, but the populations of large mammals are greatly reduced. In particular, the black rhino (*Diceros bicornis*, CR) used to occur here but has been exterminated through overhunting.

Types and Severity of Threats

The main threats to natural resources are the rapid increases in both human and livestock populations, which have caused widespread overgrazing and soil erosion, particularly in the north, and have led to desertification. Lawlessness and poaching are rife along the Kenya-Ethiopia border. Action is needed to combat these threats, particularly overgrazing and poaching.

Justification for Ecoregion Delineation

This ecoregion forms the southern outlier of the "Somalia-Masai semi-desert grassland and shrubland" of White (1983), including the area around Lake Turkana and the Dida-Galgalu Desert. Although the ecoregion contains a similar vegetation structure to the Ethiopian Xeric Grasslands and Shrublands [101] further north, it was elevated to ecoregion status because of its disjunct position, arid nature, and elements of savanna woodland communities.

Ecoregion Number:	**105**
Ecoregion Name:	**Kalahari Xeric Savanna**
Bioregion:	**Eastern and Southern Africa**
Biome:	**Deserts and Xeric Shrublands**
Political Units:	**Botswana, Namibia, South Africa**
Ecoregion Size:	**588,100 km²**
Biological Distinctiveness:	**Bioregionally Outstanding**
Conservation Status:	**Relatively Intact**
Conservation Assessment:	**V**
Author:	*Colleen Seymour*
Reviewed by:	*W. R. J. Dean, Phoebe Barnard, Antje Burke*

Location and General Description

The Kalahari Xeric Savanna [105] stretches across northwestern South Africa, southern Botswana, and central-southeastern Namibia. Most of it lies on the level plains of the Kalahari Basin, interrupted by long parallel sand dunes in the south (Lovegrove 1993). The Kalahari sands themselves are nutrient poor and reddish-brown in color, except where leached by water (Leistner

1967; Van der Walt and Le Riche 1999). They vary in depth and are underlain mainly by calcrete (Acocks 1988). Similar sands extend from the northern cape in South Africa to the Democratic Republic of the Congo, and there is as yet no consensus regarding their origin or age (Main 1987).

Temperature fluctuations are extreme in this ecoregion. In the southern Kalahari, the temperature on winter nights can plummet to –14°C while soaring to 30°C during the day. Similarly, a cold summer night may drop to 5°C, whereas daytime temperatures may exceed 45°C (Lovegrove 1993). Rainfall is remarkably patchy, with great differences occurring between sites only a few kilometers apart. Average annual rainfall is highest in the northeast and lowest in the southwest, ranging between 150 and 500 mm (Lovegrove 1993; Knight and Joyce 1997; Schulze 1997). Latitude, high atmospheric pressure, and the barrier created by the Drakensberg Mountains between the Kalahari and the Indian Ocean control the climate (Knight and Joyce 1997). Elevation is also influential, and although the ecoregion occurs at elevations between 600 m and 1,600 m, most lies above 1,000 m (Lovegrove 1993).

In less arid areas, the vegetation is open savanna with grasses (*Schmidtia* spp., *Stipagrostis* spp., *Aristida* spp., and *Eragrostis* spp.) interrupted by trees such as camelthorn (*Acacia erioloba*), grey camelthorn (*A. haematoxylon*), false umbrella thorn (*A. luederitzii*), blackthorn (*A. mellifera*), shepherd's tree (*Boscia albitrunca*), and silver cluster-leaf (*Terminalia sericea*). Shrubs include *Grewia flava, Ziziphus mucronata, Tarchonanthus camphoratus, Rhigozum trichotomum, Acacia hebeclada,* and *Lycium* spp. In drier areas, most large trees occur in ancient riverbeds, and the rolling red dunes are sparsely populated by smaller *Acacia erioloba, A. haematoxylon,* and *B. albitrunca* trees, broom bushes (*Crotalaria spartioides*), and dune reeds (*Stipagrostis amabilis*). The creeping tsamma melons (*Citrullus lanatus*), gemsbok cucumbers (*Acanthosicyos naudinianus*), and wild cucumbers (*Cucumis africanus*) are important sources of water and food for both humans and animals.

Outstanding or Distinctive Biodiversity Features

Plant species richness per unit area is among the lowest of all the southern African ecoregions, and it is estimated that less than 3 percent of the plants are endemic (Van Rooyen 1999). Animal endemism is also low, including such species as the strictly endemic FitzSimon's legless skink (*Typhlosaurus gariepensis*) and near-endemic Brants's whistling rat (*Parotomys brantsii*).

Despite the low rates of endemism, the diversity of large mammals is remarkable for such an oligotrophic and arid system. Mammals include lion (*Panthera leo*, VU), cheetah (*Acinonyx jubatus*, VU), leopard (*Panthera pardus*), spotted (*Crocuta crocuta*) and brown (*Hyaena brunnea*) hyena, and wild dog (*Lycaon pictus*, EN). There is also a high diversity of smaller predators, including black-footed cat (*Felis nigripes*), Cape fox (*Vulpes chama*), meerkat (*Suricata suricatta*), yellow mongoose (*Cynictis*

penicillata), aardwolf (*Proteles cristatus*), caracal (*Caracal caracal*), black-backed jackal (*Canis mesomelas*), and honey badger (*Mellivora c_pensis*). The mammals of the Kalahari also show a range of adaptations to the extremes of this arid environment. Gemsbok (*Oryx gazella*) are high adapted to arid areas. The impressive communal nests of the sociable weaver (*Philetairus socius*) are up to 6 m long and 2 m high and weigh as much as 1,000 kg, with 300 individual birds (Lovegrove 1993). They are so well insulated that they buffer the temperature extremes of the outside air (Bartholomew et al. 1976). The lions of the Kalahari also show adaptations; they live in small groups, have larger home ranges, and hunt smaller prey more often than lions of more mesic areas (Knight and Joyce 1997). They are also taller at the shoulder and lighter, and many of the males possess black manes.

The avifauna is not especially rich but includes many raptor species, including the secretary bird (*Sagittarius serpentarius*), martial eagle (*Polemaetus bellicosus*), bateleur (*Terathopius ecaudatus*), and Verreux's eagle owl (*Bubo lacteus*). As in many arid areas, the amphibian fauna is not particularly species rich, but it does include Tschudi's African bullfrog (*Pyxicephalus adspersus*), which eats small birds, rodents, reptiles, and insects.

Status and Threats

Current Status

Approximately 18 percent of the ecoregion falls in protected areas, with the largest reserve being the Central Kalahari (although not all of this reserve is in the ecoregion) and adjoining Khutse Game Reserves in Botswana. The Kgalagadi Transfrontier Park (KTP) was recently proclaimed. The KTP is the official union of the former Kalahari Gemsbok National Park (South Africa) and the Gemsbok National Park (Botswana). Common strategies for both conservation management and tourism development can now be shared across the border.

Types and Severity of Threats

Both fires and fences have had major impacts in the ecoregion. Although fire is a natural ecological process, over the last few centuries mismanagement of fire regimes has contributed to habitat degradation, lowered species richness, reduced nutrient availability, and adversely affected plant and animal communities in the Namibian portion of the ecoregion (Barnard et al. 1998).

Fences (farm, veterinary, and border) have been primarily responsible for the precipitous decline in numbers of species such as blue wildebeest (*Connochaetes taurinus*) and hartebeest (*Alcelaphus buselaphus*) in the Kalahari since the 1960s (Main 1987). Construction of veterinary fences began in 1954 and continues today (Main 1987; Albertson 1998). Animals such as the blue wildebeest and hartebeest, not truly adapted to aridity, must migrate to obtain food and water. During migration some animals become caught in the fences, and others die of thirst and starvation (Albertson 1998; Keene-Young 1999). Botswana's agreement with the European Union contains incentives that have encouraged the national herd to increase to more than 3 million animals. The European Union's strict import regulations are behind the construction of veterinary fences.

Livestock farmers often use poisoned carcasses to kill "problem" animals such as black-backed jackals (*Canis mesomelas*) and caracals (*Caracal caracal*), resulting in the poisoning of non-target raptors (Anderson 2000). Some species, such as martial and black (*Aquila verreauxii*) eagles, perceived to prey on domestic livestock and poultry, may be intentionally targeted (Anderson 2000). The African wild dog (*Lycaon pictus*, EN) is particularly threatened by eradication of their prey and shooting by farmers.

Justification for Ecoregion Delineation

The boundaries for the Kalahari Xeric Savanna are based on White's (1983) "Kalahari deciduous *Acacia* wooded grassland and bushland" and "Kalahari/Karoo-Namib transition," with a small transition area of "undifferentiated woodland to *Acacia* deciduous bushland and wooded grassland." A majority of the ecoregion encompasses the plains of the Kalahari Basin.

Ecoregion Number:	**106**
Ecoregion Name:	**Kaokoveld Desert**
Bioregion:	**Eastern and Southern Africa**
Biome:	**Deserts and Xeric Shrublands**
Political Units:	**Angola, Namibia**
Ecoregion Size:	**45,700 km²**
Biological Distinctiveness:	**Globally Outstanding**
Conservation Status:	**Relatively Intact**
Conservation Assessment:	**III**
Author:	*Amy Spriggs*
Reviewed by:	*Phoebe Barnard, Rob Simmons, Antje Burke*

Location and General Description

The Kaokoveld Desert [106] stretches along the west coast of southern Africa from the Uniab and Koigab rivers in Namibia north into the Moçâmedes Desert of southern Angola. For most of its length, it is about 100 km wide and extends from the Atlantic Coast to the foot of the Great Escarpment.

This ecoregion falls in the northern summer rainfall area of the greater Namib Desert, with most rain falling as sporadic thunderstorms between October and March (von Willert et al.

1992). However, the most important climatic feature is the highly unpredictable rainfall of less than 100 mm per year. Cool air moving inland off the Benguela current forms a stable layer of fog, which is blown inland as far as 50 km. The fog provides water to the desert plants and animals (Lovegrove 1993). Air temperatures in the Kaokoveld Desert decrease toward the coast as a result of the cool air and frequent fogs, which keep daily and seasonal temperature changes to a minimum (Barnard 1998). Temperature extremes increase inland, where they become highly variable, falling below 0°C and rising above 50°C. The Kunene River is the only permanent river in the Kaokoveld Desert. Other rivers, such as the Hoanib, Hoarusib, and Khumib rivers, run intermittently and rarely reach the ocean because of intervening dunes (Barnard 1998).

The ecoregion consists of rugged hills, valleys, sand dunes, sand, and gravel plains. A large area of shifting sand dunes, which reaches up to 40 km wide, occurs in the Iona National Park in southern Angola, between Porto Alexandre and the Kunene River (White 1983). Sand and gravel plains cover much of the remaining ecoregion. No true soils with well-defined profiles are found (von Willert et al. 1992). On the gravel plain soils are cemented into a rock-hard layer by lime and gypsum, forming a hardpan at a depth of 10 to 50 mm. Salt crusts are common close to the ocean, and the soils are brackish as far as the inland limit of coastal fogs (Barnard 1998).

The relict gymnosperm *Welwitschia mirabilis* is found throughout the ecoregion and is often the most conspicuous feature of the vegetation, scattered about the arid plains. Individual *Welwitschia* plants have been estimated to be more than 2,500 years old (White 1983; Armstrong 1990; Lovegrove 1993). In some areas shifting sand dunes are completely devoid of vegetation, although elsewhere *Acanthosicyos horridus* colonizes the sand. Close to the coast gravel deserts are devoid of vegetation except for colorful fields of foliose and crustose lichens. Species of *Parmelia* and *Usnea* and the orange lichen *Teloschistes capensis* are common. Further inland, the gravel deserts are less barren and support plants such as *Zygophyllum simplex*, *Galenia africana*, *Sesuvium portulacastrum*, and *Stipagrostis subacaulis*. Some areas of sparse grassland are found and support species such as *Salsola nollothensis*, *Indigofera cunenensis*, *Stipagrostis ramulosa*, and *Eragrostis cyperoides* (Giess 1971; Werger 1978; White 1983). Halophytic communities grow on saline beaches, characterized by *Sesuvium* spp., *Suaeda fruticosa*, *Scirpus littoralis*, and *Asthenatherum forsskalii*. Dry river beds contain a higher diversity of species than found in the adjacent desert. Dense cushion-like growths of leaf succulents such as *Salsola* spp., *Zygophyllum clavatum*, and *Z. stapffii* are common (Giess 1971; Werger 1978; White 1983).

Outstanding or Distinctive Biodiversity Features

The Kaokoveld Desert lies at the northern end of the Karoo-Namib regional center of endemism (White 1983) and represents a center of floral endemism (WWF and IUCN 1994; van Wyk and Smith 2001). Families rich in endemic species include Acanthaceae, Asclepiadaceae, Burseraceae, Fabaceae, Poaceae, and Vitaceae (van Wyk and Smith 2001).

The Namib Desert and nearby plateaus have been arid for at least 55 million years, and the area is an ancient stable center of evolution (Barnard 1998). Paleoendemics, such as *Welwitscha mirabilis* and *Kaokochloa* sp., indicate the ancient history of the area. Several plant species have disjunct distributions and may be relics of a once continuous arid belt that stretched across Africa from Somalia to Botswana, Namibia, and South Africa (Werger 1978). Such patterns can be seen in *Kalanchoe laciniata,* which occurs in both the Kaokoveld and then again in Tanzania and into northeastern Africa (van Wyk and Smith 2001).

The Kaokoveld contains a high number of endemic reptiles, including *Pedioplanis benguellensis* and *Palmatogecko vanzyli* (Branch 1998). It also contains a number of invertebrate species, such as the endemic solifuge *Ceroma inerme,* which has adapted xerophilic behavior to accommodate a nearly marine existence, living just outside the high tidal zone and foraging in the intertidal area at low tide (Griffin 1998b).

The Kaokoveld Desert is home at times to elephant (*Loxodonta africana,* EN), black rhino (*Diceros bicornis,* CR), and giraffe (*Giraffa camelopardalis*), which are found in the river beds that fringe the desert (Joubert and Mostert 1975). Other mammals found in the ecoregion are springbok (*Antidorcas marsupialis*), gemsbok (*Oryx gazella*), Damara dik-dik (*Madoqua kirkii*), and black-faced impala (*Aepyceros melampus petersi,* VU). These ungulates concentrate their activities in the vegetated river beds that transect the ecoregion. Predators include lion (*Panthera leo,* VU), brown hyena (*Hyaena brunnea*), and black-backed jackal (*Canis mesomelas*). Setzer's hairy-footed gerbil (*Gerbillurus setzeri*) is near endemic to the ecoregion.

There are many bird species in the ecoregion, with highest diversity along river beds. For example, the endemic Cinderella waxbill (*Estrilda thomensis*) is restricted to the Kunene river bed. One of the few birds permanently residing in the sand dunes is Gray's lark (*Ammomanes grayi*). The edges of the desert are occupied by the pale Ruppell's korhaan (*Eupodotis ruppelli*), which moves into the dunes after rain to take advantage of the lush vegetation (Sinclair and Hockey 1996). Both of these birds are endemic to the greater Namib Desert. One near-endemic species that has been recently described is the Benguela long-billed lark (*Certhilauda benguelensis*) (Ryan et al. 1999).

Status and Threats

Current Status

A large percentage of this ecoregion falls under formal protection in Namibia and Angola. The Skeleton Coast Park stretches 500 km along the northern coast of Namibia from the Ugab River to the Kunene River on the Angolan-Namibian border, a total of 15,800 km^2 (Stuart and Stuart 1992). The park contains the three main vegetation units found in the Kaokoveld Desert (gravel plains, sand dunes, and dry river beds) (du Plessis 1992). The northern area of the park (north of the Uniab River) is a wilderness area and is closed to the public. In Angola, the Mocamedes Partial Reserve (4,450 km^2) and the Iona National Park (15,150 km^2) cover most of the northern portion of the ecoregion (Huntley 1974a).

The southeastern Kaokoveld was once under protection in Game Reserve No. 2, which originally covered 80,000 km^2 and was the largest nature reserve in the world. The reserve stretched from the Kunene River, 200 km south to the Hoarusib River (Barnard et al. 1998). This enormous park allowed for the westward seasonal migration of elephants, lions, and other mammals as far as the Atlantic Ocean. Much of Game Reserve No. 2 was degazetted in 1963, but it is now regaining conservation status through a mosaic of informally protected conservancies that stretch from the Skeleton Coast Park and into the Kunene. Conservancies are jointly managed for resource conservation by multiple landowners, with financial and other benefits shared between them. They occur on both private and tribal (communal) land (Barnard et al. 1998).

Types and Severity of Threats

The Namibian side of the Kaokoveld Desert has remained largely intact because of its arid conditions and consequent low population density. To the east, the Ovahimba, who are nomadic pastoralists, inhabit the ecoregion. Their impact has been local, mainly through the collection of wood and medicinal, edible, and culturally important plants. The biggest threat posed by the Ovahimba is seasonal overgrazing by their cattle (Sullivan and Konstant 1997). Another major threat is the impact of off-road vehicles.

The biggest threat to the Namibian side of the Kaokoveld Desert is from wildlife poaching. This started when Game Reserve No. 2 was degazetted in 1963 and administered from afar by the Department of Bantu Affairs in Pretoria, resulting in a period of corruption and widespread poaching.

The 30-year civil war in Angola has left the Angolan side of the Kaokoveld Desert open to poachers, human settlement, and agriculture. There are several hundred Ovahimba pastoralists with more than 3,000 head of livestock in the Iona National Park. Few, if any, viable populations of larger wild mammals have survived, and populations of lion, black rhino, and giraffe have been reduced to the threshold of local extinction.

Since 2000, the Namibian and Angolan governments have been negotiating the establishment and co-management of a transfrontier conservation area between the Skeleton Coast Park and Iona National Park, a move that would greatly strengthen ties between the countries and their management, surveillance, and enforcement capacity.

Justification for Ecoregion Delineation

Both the Kaokoveld [106] and Namib Desert [107] ecoregions are part of the "Namib Desert" vegetation unit of White (1983). The southern boundary of the northern Namib, recognized here as the Kaokoveld Desert [106], follows the biogeographic division made by Giess (1971 in White 1983). The northern limit reflects that of White's "bushy Karoo-Namib shrubland" unit. This area is also distinguished as a center of plant diversity, including monotypic families and genera.

Ecoregion Number: **107**

Ecoregion Name: **Namib Desert**

Bioregion: **Eastern and Southern Africa**

Biome: **Deserts and Xeric Shrublands**

Political Units: **Namibia**

Ecoregion Size: **80,900 km²**

Biological Distinctiveness: **Globally Outstanding**

Conservation Status: **Relatively Intact**

Conservation Assessment: **III**

Author: *Amy Spriggs*

Reviewed by: *Phoebe Barnard, Rob Simmons, Antje Burke*

Location and General Description

The Namib Desert [107] ecoregion extends along the coastal plain of western Namibia, from the Uniab River in the north to the town of Lüderitz in the south. It reaches inland from the Atlantic coast to the foot of the Namib Escarpment, a distance of 80–200 km. The ecoregion can be divided into two parts: the Central Namib (from the Uniab to the Kuiseb River) and the Southern Namib (from the Kuiseb River to the town of Lüderitz) (Giess 1971). In the north, the Central Namib merges with the Northern Namib or Kaokoveld Desert [106], and in the south it merges with the Succulent Karoo [110].

This ecoregion has sparse and highly unpredictable annual rainfall. Mean annual rainfall ranges from 5 mm in the west to about 85 mm along the ecoregion's eastern limits (Lovegrove 1993). The coastal area has a mean annual rainfall of 2–20 mm and has thick fog more than 180 days of the year. Air temperatures are low because of the cool air coming off the Benguela Current, and daily and seasonal temperature changes are minimal (Barnard 1998). Up to about 50 km inland the mean annual rainfall increases from 20 to 50 mm. Though still important to desert organisms, fog occurs only about 40 days in the year. Still further inland, fog is rare, and the mean annual rainfall increases from 50 mm to a maximum of 85 mm. Daily and seasonal temperatures increase sharply and become highly vari-

able, with temperatures below 0°C and above 50°C recorded (von Willert et al. 1992). There is very little surface water in the Namib, with the Southern and Central Namib bisected by ephemeral rivers. The most important is the Kuiseb River (Jacobson et al. 1995), and others are the Swakop River and the Omaruru River (north of Swakopmund).

Soils are made of raw minerals and are sandy and sometimes calcareous or with calcareous crusts, composed of particles in a wide range of sizes (von Willert et al. 1992). Salt crusts and biological crusts are common on the soil close to the ocean, and the soils are brackish as far as the inland limit of coastal fog. Gypsum accumulations are also found (Scholz 1968).

The Southern Namib is an expansive area of large, shifting dunes that reach elevations of 300 m (Giess 1971). The vegetation comprises a few perennial grasses (*Stipagrostis sabulicola*, *Monsonia ignorata*) and the succulent *Trianthema hereroensis*. Hummocks formed by *Acanthosicyos horridus* occur between these sand dunes and the coast (Giess 1971; Werger 1978; White 1983). The dunes move north, driven by prevailing southerly winds, and are brought to an abrupt halt by vegetation of the Kuiseb River. To the north of the Kuiseb River, the dunes give way to the gravel plains of the Central Namib, which are dotted with inselbergs of granite and limestone (Lovegrove 1993; Burke et al. 2002). The Central Namib contains a narrow strip of vegetation that follows the coastline north of the Swakop River, which supports shrubs such as *Psilocaulon salicornioides*, *Zygophyllum clavatum*, *Salsola aphylla*, and *S. nollothensis*. The pencil plant (*Arthraerua leubnitziae*) and dollar bush (*Zygophyllum stapfii*) are prominent in the fog-influenced coastal belt. Inland from this strip of vegetation are vast gravel plains that are largely devoid of vegetation except for fields of colorful lichens, including *Teloschistes flavicans*, *Parmelia convoluta*, and *Usnea* spp. Some of these lichens are not attached to any substrate and are known as vagrant lichens or Wanderflechten, such as *Xanthomaculina convoluta* and *Parmelia hueana* (Wessels 1989). To the east, annuals dominate the gravel plains (mostly *Stipagrostis* species). These annuals, which lie dormant as seeds through extended drought periods, grow rapidly after heavy rainfall and transform the landscape into a sea of grass (Giess 1971; Werger 1978; White 1983).

A dense growth of *Sporobolus robustus*, or more open communities of *Eragrostis spinosa*, are found along dry river beds in the ecoregion. Trees of *Acacia erioloba* are also scattered along the river beds. Along the Swakop and Kuiseb rivers, *Acacia erioloba* forms dense stands with *Faidherbia albida*, wild tamarisk (*Tamarix usneoides*), the mustard tree (*Salvadora persica*), and the alien invasive plant *Nicotiana glauca*, native to South America (White 1983).

Outstanding or Distinctive Biodiversity Features

The Namib Desert, isolated between the ocean and the escarpment, has been arid for millions of years. This has produced a stable center for the evolution of desert species, with high levels of plant endemism and numerous adaptations to arid conditions (Louw and Seely 1982). The monotypic *Welwitschia mirabilis*, one of the most remarkable plants in the world, is endemic to the Namib and the Kaokoveld Desert [106] to the north. These plants have the longest-lived leaves of any member of the plant kingdom, with the largest plants estimated to be about 2,500 years old (White 1983; Lovegrove 1993).

The Namib Desert supports a large number of small rodent species, such as the dune hairy-footed gerbil (*Gerbillurus tytonis*) and Grant's golden mole (*Eremitalpa granti*, VU), that occur among the rocky habitats in the western deserts, in the sand dunes and in the vegetation of the gravel plains (Hilton-Taylor 2000). Larger ungulates are scarce in the Namib Desert, with only gemsbok (*Oryx gazella*) and springbok (*Antidorcas marsupialis*) normally present (Griffin 1998a). Hartmann's mountain zebra (*Equus zebra hartmannae*, EN) is found in the extreme east in the transition belt between the desert and the escarpment (Joubert and Mostert 1975). Predators include cheetah (*Acinonyx jubatus*, VU), brown hyena (*Hyaena brunnea*), spotted hyena (*Crocuta crocuta*), black-backed jackal (*Canis mesomelas*), Cape fox (*Vulpes chama*), and bat-eared fox (*Otocyon megalotis*). Many species have become locally extinct in the southern areas of the Namib Desert, including lion (*Panthera leo*, VU), elephant (*Loxodonta africana*, EN), black rhinoceros (*Diceros bicornis*, CR), white rhinoceros (*Ceratotherium simum*), and giraffe (*Giraffa camelopardalis*).

The desert has low avian richness. The most prominent bird in the desert is the ostrich (*Struthio camelus*). Most of the bird life is concentrated along the coastline. For example, Sandwich Harbor is an area of high species richness, with 113 bird species recorded to date (Berry and Berry 1975). Five birds are considered endemic or near endemic to the Namib Desert: the dune lark (*Certhilauda erythrochlamys*), Benguela long-billed lark (*C. benguelensis*) (Ryan et al. 1999), Gray's lark (*Ammomanes grayi*), tractrac chat (*Cercomela tractrac*), and Rüppell's korhaan (*Eupodotis rueppellii*). The dune lark is strictly endemic, whereas Gray's lark, Rüppell's korhaan, and Benguela long-billed lark are found in this ecoregion, the Kaokoveld Desert [106], or the Succulent Karoo [110].

Reptiles contribute much of the high faunal species richness and endemism and have evolved adaptations to survive in this harsh environment. Some of the endemic reptiles, including the wedge-snouted sand lizard (*Meroles cuneirostris*), small-scaled sand lizard (*M. micropholidotus*), barking gecko (*Ptenopus kochi*), and day gecko (*Rhoptropus bradfieldi*), dive beneath the sand to escape danger (Branch 1998).

The Namib Desert is also well known for its high species richness of beetles, particularly those belonging to the family Tenebrionidae (Lovegrove 1993). Many of these, along with other desert organisms, have developed physiologic, morphologic, and behavioral adaptations to the arid environment (Louw and Seely 1982). For example, the tenebrionid beetle *Onymacris unguicularis* has evolved a behavioral method of condensing fog as a source of water.

Status and Threats

Current Status

The present conservation status of the Namib Desert is good: most of the ecoregion is intact and is protected in extensive conservation blocks (Barnard et al. 1998). The Namib-Naukluft Park (49,768 km^2) is the largest conservation area in southern Africa and protects the central area of this ecoregion. The park runs from Walvis Bay in the north to Lüderitz in the south. The park covers gravel plains, the dune sea, the eastern semi-desert, and the Kuiseb River and is therefore a good representation of the Central and Southern Namib vegetation (du Plessis 1992), except for Succulent Karoo species.

To the north of the Namib-Naukluft Park lies the National West Coast Recreation Area. This area extends for 180 km up the coast and is under less stringent protection than the national parks. The Cape Cross Seal Reserve is located in this area and protects one of the largest colonies of Cape fur seals (*Arctocephalus pusillus*) in southern Africa.

Types and Severity of Threats

A major threat in this area is off-road driving. The impact of this activity is greatest on the gravel plains, where depressions left by vehicles remain for more than 40 years. Lichens are particularly sensitive to mechanical damage because they grow extremely slowly and cannot quickly repair damaged thalli (Lovegrove 1993).

The major threat to the Namib-Naukluft Park is the drop in the water table along the Kuiseb River. This is caused primarily by the extraction of groundwater at two sites near Walvis Bay, which supplies Walvis Bay and Swakopmund and the Rössing Uranium Mine near Swakopmund. A more modest threat to the Namib-Naukluft Park is by the Topnaar pastoralists, who graze large herds of goats and small groups of donkeys over the Kuiseb river bed and along the edge of the dunes. The livestock have overgrazed the understory plant growth and fallen seedpods of the river bed and are competing for food with wild animals, such as gemsbok.

Justification for Ecoregion Delineation

The Namib Desert [107] ecoregion follows Giess's (1971) Central and Southern Namib vegetation units. The Northern Namib was delineated separately as the Kaokoveld Desert [106] ecoregion because of its high levels of floral endemism.

Ecoregion Number: **108**

Ecoregion Name: **Nama Karoo**

Bioregion: **Eastern and Southern Africa**

Biome: **Deserts and Xeric Shrublands**

Political Units: **South Africa, Namibia**

Ecoregion Size: **351,100 km²**

Biological Distinctiveness: **Globally Outstanding**

Conservation Status: **Relatively Stable**

Conservation Assessment: **III**

Author: *Colleen Seymour*

Reviewed by: *W. R. J. Dean*

Location and General Description

The Nama Karoo ecoregion [108] is a vast, open, arid region dominated by low shrub vegetation, punctuated by rugged relief (Dean and Milton 1999). Most of the Nama Karoo occurs on the central plateau of the Western, Northern, and Eastern Cape provinces, inland of the Southern Fold Mountains in South Africa, although it extends over the Orange River into Namibia in the northwest. The Great Escarpment, which runs parallel to the coast 100 to 200 km inland, divides the ecoregion into two parts: one between 550 and 900 m in elevation, the other between 900 and 1,300 m (Palmer and Hoffman 1997).

The climate of the Nama Karoo typically is harsh. Droughts are common, and both seasonal and daily temperatures fluctuate widely. Temperature variations of 25°C between day and night frosts are common (Venter et al. 1986). Mean maximum temperatures in midsummer (January) exceed 30°C, whereas mean minimum midwinter (July) temperatures are below freezing (Palmer and Hoffman 1997). Rainfall is unseasonal but generally peaks between December and March (Palmer and Hoffman 1997). Annual rainfall ranges between 100 and 500 mm, decreasing from east to west and from north to south (Palmer and Hoffman 1997; Desmet and Cowling 1999b). Variability in interannual rainfall tends to increase with increasing aridity (Schulze 1997).

Shallow, weakly developed lime-rich soils cover much of the region (Watkeys 1999). The soils are derived principally from sediments of the Dwyka Formation, which cover rocks of the Ecca and Beaufort groups, respectively (Lloyd 1999). The Karoo dolerite dikes and sills were formed when molten rock intruded into the preexisting rocks of the Ecca and Beaufort shales (Lloyd 1999). Dolerite sills, generally more resistant to weathering than the surrounding sandstones and shales, can be seen as the flat-topped hills (Watkeys 1999).

The Orange River Basin is the region's main drainage system, although many of the watercourses that feed it flow ephemer-

ally (Lloyd 1999). The ecoregion also has a number of pan systems, the largest of which is the Grootvloer-Verneukpan complex (Lloyd 1999). When summer rainfall is high, the system also provides a link between the Orange and Sak river systems, which may enable an interchange of indigenous fish and other aquatic organisms (Lloyd 1999). The seasonal Fish River flows through a canyon that is second in size only to the Grand Canyon of the United States (Barnard 1998).

Dwarf shrubs (chamaephytes) and grasses (hemicryptophytes) dominate the current vegetation, their relative abundances dictated mainly by rainfall and soil (Palmer and Hoffman 1997). As a rule, shrubs increase and grasses decrease with increasing aridity (Palmer and Hoffman 1997). However, heavy grazing by domestic livestock can obscure this pattern by suppressing the grass component (Lovegrove 1993). Some of the more abundant shrubs include species of *Drosanthemum, Eriocephalus, Galenia, Pentzia, Pteronia,* and *Ruschia,* and the principal perennial grasses are *Aristida, Digitaria, Enneapogon,* and *Stipagrostis* spp. Trees and taller woody shrubs are restricted mostly to watercourses and include *Acacia karroo, Diospyros lycioides, Grewia robusta, Rhus lancea,* and *Tamarix usneoides* (Palmer and Hoffman 1997).

Outstanding or Distinctive Biodiversity Features

There are few published data regarding species richness or endemism for the Nama Karoo flora. Gibbs Russel (1987) calculated that 2,147 species occurred in a central area of 198,000 km², of which 377 (18 percent) are endemic. Recently, however, an archipelago of mountains in a part of the ecoregion known as Bushmanland have been found to harbor both Nama Karoo and Succulent Karoo type vegetation and a diverse assemblage of succulents endemic to the archipelago itself (Desmet 2000). A study of the invertebrate fauna of one of these mountains (the Gamsberg) also revealed a collection of Succulent Karoo species (Desmet 2000).

The fauna of the Nama Karoo is species poor, and there are few strict endemics (Vernon 1999). Three small mammal species are strictly endemic: Visagie's golden mole (*Chrysochloris visagiei,* CR), Grant's rock rat (*Aethomys granti*), and Shortridge's rat (*Thallomys shortridgei*) (Hilton-Taylor 2000). Three other small mammals are near endemic: the riverine rabbit (*Bunolagus monticularis,* EN), bushy-tailed hairy-footed gerbil (*Gerbillurus vallinus*), and Brukkaros pygmy rock mouse (*Petromyscus monticularis*). Black rhinoceros (*Diceros bicornis,* CR) have already been extirpated, and the quagga (*Equus quagga*), a Nama Karoo near-endemic zebra-like mammal, was hunted to extinction in the nineteenth century (Skinner and Smithers 1990). In the mid- to late-1800s, European travelers and colonists witnessed game migrations numbering millions across the Nama Karoo, which have long since ceased (Lovegrove 1993). Although other game species such as blue wildebeest (*Connochaetes taurinus*), blesbok (*Damaliscus pygargus*), quagga, and eland (*Taurotragus oryx*) were often involved in these migrations, springbok (*Antidorcas marsupialis*) were by far the most numerous.

Among birds, the ferruginous lark (*Certhilauda burra*, VU) (Dean et al. 1991) and Sclater's lark (*Spizocorys sclateri*) are strictly endemic to this ecoregion, whereas several others are near endemic: Karoo chat (*Cercomela schlegelii*), tractrac chat (*Cercomela tractrac*), Karoo scrub-robin (*Cercotrichas coryphaeus*), red-headed cisticola (*Cisticola subruficapillus*), and Namaqua warbler (*Phragmacia substriata*). Other characteristic bird species of the Nama Karoo that are regarded as vulnerable in South Africa are the tawny eagle (*Aquila rapax*), martial eagle (*Polemaetus bellicosus*), African marsh-harrier (*Circus ranivorus*), lesser kestrel (*Falco naumanni*), blue crane (*Grus paradisea*), kori bustard (*Ardeotis kori*), and Ludwig's bustard (*Neotis ludwigii*) (Dean et al. 1991; Barnes 2000; McCann 2000).

The reptile fauna contains a number of near-endemics, but only a few are potentially confined to the Nama Karoo, including Karoo dwarf chameleon (*Bradypodion karrooicum*) and Boulenger's padloper (*Homopus boulengeri*).

Many of the endemics, and some of the other species present, are relicts of drier epochs in which there was a continuous arid corridor along the southeastern and eastern parts of Africa. Species with interrupted distributions include the bateared fox (*Otocyon megalotis*), Garman's toad (*Bufo garmani*), fawn-colored lark (*Mirafra africanoides*), and lark species related to the southern African sabota lark (*Mirafra sabota*) (Vernon 1999).

Status and Threats

Current Status

The Nama Karoo ecosystem was formerly grazed by a variety of indigenous migratory ungulates and now by domestic sheep and goats confined within farm boundaries (Skead 1982). This change is thought to be responsible for alterations in plant species composition and cover (Roux and Theron 1987). On a smaller scale, disturbances associated with *heuweltjies* (ancient termitaria) (Moore and Picker 1991) maintain habitat heterogeneity and patchiness in the landscape (Midgley and Musil 1990; Armstrong and Siegfried 1990; Milton and Dean 1990).

Less than 1 percent of the ecoregion is protected (Cowling 1986; Barnard et al. 1998). The only large park is the Fish River Canyon Park, which has recently been enlarged to include mountains to the west as far as the Orange River. The establishment of wildlife conservancies on commercial and communal farmlands could improve the protected status of this ecoregion (Barnard et al. 1998).

The Namibian portion of the ecoregion once had high mammal species richness. As settlers moved north from South Africa, large mammals receded, leaving southern Namaland devoid of vulnerable species such as lions and Burchell's zebras (*Equus burchelli*). These two species have suffered a 95 percent range reduction over the past 200 years (Griffin 1998a).

Types and Severity of Threats

Most of the ecoregion is now rangeland for livestock grazing (Hoffman et al. 1999) and therefore still intact, although heavy grazing has left parts seriously degraded (Lloyd 1999). The issue of degradation and grazing practices is complex, however, and warrants further investigation (Hoffman and Cowling 1990; Dean and Macdonald 1994; Hoffman et al. 1999; Dean and Milton 1999). The use of poisoned carcasses by livestock farmers to kill "problem" animals such as black-backed jackal (*Canis mesomelas*) and caracal (*Caracal caracal*) often results in poisoning of nontarget raptors (Lloyd 1999; Anderson 2000). Some species, such as the martial and black (*Aquila verreauxii*) eagles, perceived to prey on domestic livestock and poultry, may be intentionally targeted (Anderson 2000).

In addition to pastoralism, alien invasive plants, mining, agriculture, and the collection of succulents and reptiles for the pet trade also threaten the ecoregion's biodiversity (Lovegrove 1993; Lloyd 1999). A number of introduced ornamental (e.g., some Cactaceae) and forage (e.g., *Opuntia*, *Prosopis*, *Atriplex*, and *Bromus* spp.) plants, together with a few accidental introductions (e.g., *Salsola kali* and *Argemone ochroleuca*), have the potential to seriously alter the region's ecology and hydrology (Milton et al. 1999). These exotics disperse efficiently, lack natural controls, and can outcompete indigenous plants for water, nutrients, and light (Lovegrove 1993). Pesticides used to control brown locust (*Locustana pardalina*) outbreaks also contaminate wildlife habitat, with high concentrations being found at the top of the food chain, particularly in raptors (Lovegrove 1993).

Justification for Ecoregion Delineation

The boundaries of the ecoregion were taken from the Nama Karoo biome of Low and Rebelo (1996) and extended north to Keetmanshoop roughly around the 900-m contour. This ecoregion is distinguished from surrounding ecoregions by a range of environmental parameters including elevation, temperature, and rainfall. The Nama Karoo [108] lies between 500 and 1,500 m elevation and has more extreme temperatures and more variable rainfall compared with the adjacent Succulent Karoo [110] ecoregion.

Ecoregion Number: **109**

Ecoregion Name: **Namib Escarpment Woodlands**

Bioregion: **Eastern and Southern Africa**

Biome: **Deserts and Xeric Shrublands**

Political Units: **Angola, Namibia, South Africa**

Ecoregion Size: **225,500 km²**

Biological Distinctiveness: **Globally Outstanding**

Conservation Status: **Relatively Intact**

Conservation Assessment: **III**

Author: *Amy Spriggs*

Reviewed by: *Phoebe Barnard, Rob Simmons,
Antje Burke*

Location and General Description

The Namib Escarpment Woodlands [109] ecoregion covers the narrow escarpment inland of the Namib Desert and broadens gradually toward the south, where it comprises extensive areas of the Nama Karoo plateau south of the town of Mariental. The ecoregion extends from near the town of Sumbe in western Angola down through Namibia, with the southern boundary located just north of Groot Karas Berg.

The rainfall is low, ranging from 60 mm in the west to 200 mm in the east (Barnard 1998). Most rain falls as thundershowers in the summer, from October to March. There is great variation between years, with the driest years having the least predictable rainfall. Low humidity results in extreme temperatures, with temperatures dropping as low as –9°C in places. Winter frosts are common. The mean maximum monthly temperature can occasionally exceed 40°C (Lovegrove 1993).

The north and central part of the ecoregion is a transition zone between the low-lying desert and the central highland plateau. Mountains such as the Baynes (2,038 m), Erongo (2,319 m), Naukluft (1,974 m), Spitzkoppe (1,759 m), and Gamsberg (2,347 m) lie along the escarpment edge, which is narrow and deeply dissected. The Brandberg, Namibia's highest mountain at 2,579 m, is an outlier within the boundaries of the Namib Desert [107] ecoregion. It is described as part of this ecoregion because it shares many affinities with the Kaoko Escarpment and is often considered part of the Kaokoveld center of endemism (van Wyk and Smith 2001). To the south, the ecoregion broadens to include the stony central plateau. This plateau lies above and east of the escarpment, between about 1,000 and 2,000 m. Soils differ markedly within the ecoregion, with litholithic and sandy loams on the escarpment and the southwestern part of the plateau (Barnard 1998).

Most of the ephemeral rivers in Namibia have their major watersheds in this highland (Barnard 1998). The Kunene River, which rises in Angola and runs along the Angola-Namibia border, is the only perennial river in the ecoregion.

Other important rivers are the Swakop, Kuiseb, and Fish rivers. The Swakop and Kuiseb rivers flow infrequently and usually are dry river beds. The Fish River, one of Namibia's longest rivers, begins in the Naukluft Mountains southwest of Windhoek and winds south for 800 km before flowing into the Orange River.

The vegetation of this ecoregion is highly varied, reflecting diverse topographic factors and related soil and microclimate characteristics (Giess 1971; Werger 1978; White 1983). According to Giess's (1971) vegetation map of Namibia, the ecoregion contains three vegetation types: mopane savanna, semi-desert and savanna transition, and dwarf shrub savanna.

Colophospermum mopane, Sesamothamnus benguellensis, and *S. guerichii* characterize mopane savanna in the north and east of the ecoregion. The semi-desert and savanna transition zone supports a great variety of species, many of which are endemic. Characteristic species include *Euphorbia guerichiana, Cyphostemma* spp., *Adenolobus* spp., the quiver tree (*Aloe dichotoma*), and *Moringa ovalifolia.* The genus *Commiphora* is characteristic of both the mopane savanna and the semi-desert and savanna transition zone. To the south, the vegetation becomes more open and is classified as dwarf shrub savanna. Characteristic species include *Rhigozum trichotomum, Parkinsonia africana, Acacia nebrownii, Boscia foetida, B. albitrunca,* and *Catophractes alexandri,* as well as smaller Karoo bushes such as *Pentzia* spp. and *Eriocephalus* spp. Trees such as *Acacia erioloba, A. karroo, Tamarix usneoides, Euclea pseudebenus,* and *Rhus lancea* are found along river beds throughout the ecoregion.

Outstanding or Distinctive Biodiversity Features

The Kaoko Escarpment is one of Namibia's two distinct hotspots, defined as areas of high endemism and high species richness (Simmons et al. 1998b). Most of the endemics are clustered around the Brandberg Mountain and rugged mountains of the escarpment, with a few around the Khomas Highlands north of Windhoek. The Brandberg Mountain supports 90 of Namibia's endemic plants, including 8 plants that are found only there (Maggs et al. 1998). The insect order Mantophasmatodea was also recently discovered here. One of the most striking plants of the Brandberg Mountain is the Brandberg acacia (*Acacia montis-usti*) (Nordenstam 1974). Many plant species are also taxonomically isolated, such as the monotypic genera *Phlyctidocarpa* and *Kaokochloa* (Poaceae) (Maggs et al. 1994). Intense speciation has occurred in the *Petalidium* genus.

This ecoregion is also a center of faunal endemism and species richness, with a high number of Namibian endemic invertebrates, amphibians, reptiles, mammals, and birds (Simmons et al. 1998b). Endemic and near-endemic mammals include mainly bats, rodents, and small carnivores. The Namaqua slender mongoose (*Galerella swalius*) and Shortridge's rock mouse (*Petromyscus shortridgei*) are restricted to the escarpment. The only large mammal endemic to Namibia is Hartmann's mountain zebra (*Equus zebra hartmannae*, EN), which is near endemic to the ecoregion.

Among the larger mammals, the ecoregion is well known for its desert-dwelling populations of elephant (*Loxodonta africana*, EN) and black rhinoceros (*Diceros bicornis*, CR) (Joubert and Mostert 1975; Berger and Cunningham 1994). Other large mammal species include greater kudu (*Tragelaphus strepsiceros*), springbok (*Antidorcas marsupialis*), gemsbok (*Oryx gazella*), Damara dik-dik (*Madoqua kirkii*), and black-faced impala (*Aepyceros melampus petersi*, VU). Predators include lion (*Panthera leo*, VU), leopard (*Panthera pardus*), cheetah (*Acinonyx jubatus*, VU), bat-eared fox (*Otocyon megalotis*), and Cape fox (*Vulpes chama*).

Endemic bird species are found in the rocky habitats of the ecoregion, at elevations between 600 and 1,200 m. Characteristic species include chatshrike (*Lanioturdus torquatus*), Monteiro's hornbill (*Tockus monteiri*), violet woodhoopoe (*Phoeniculus damarensis*), Herero chat (*Namibornis herero*), Damara rock jumper (*Achaetops pycnopygius*), Carp's tit (*Parus carpi*), Rüppell's parrot (*Poicephalus rueppellii*), Hartlaub's francolin (*Pternistis hartlaubi*), and Cinderella waxbill (*Estrilda thomensis*).

Specialist reptiles include Husaben sand lizard (*Pedioplanis husabensis*), Campbell's spinytail lizard (*Cordylus campbelli*), Herero girdled lizard (*Cordylus pustulatus*), and Brandberg thick-toed gecko (*Pachydactylus gaiasensis*) (Branch 1998).

Status and Threats

Current Status

Namibia's recent biodiversity assessment (Barnard 1998) identified the Kaoko Escarpment in the northern part of this ecoregion as an endemism hotspot, yet it is one of the most significant gaps in habitat protection in Namibia (Rebelo 1994; Barnard et al. 1998). The area was once part of "Game Reserve No. 2," which originally covered 80,000 km^2 and was the largest nature reserve in the world. In 1963, this reserve was reduced in area by 72 percent to become the present Etosha Pan National Park. Since Namibia's independence in 1990, rural communities in the Kunene and Erongo political regions have established eight communal conservancies amounting to 28,021 km^2. These conservancies are rebuilding the former expanse of Game Reserve No. 2, although with a lower legal protection status.

The southern part of the Namib Escarpment ecoregion is also poorly protected. Only a small portion of the Namib-Naukluft National Park extends into the ecoregion to include the Naukluft Mountains, southwest of Windhoek. There are also two small government recreational parks, Naute (225 km^2) and Hardap (252 km^2), on the rocky plateau (Barnard 1998).

There are also some private nature reserves and game farms in the ecoregion, and freehold reserves and conservancies are expanding in the south. Game farming is increasing in popularity, partly because much of the area is only marginally suitable for livestock.

Types and Severity of Threats

At an ecoregion scale, many species such as springbok, leopard, and gemsbok have not suffered significant range reductions, and the distributions of smaller mammals have changed little during recorded history. A few species, such as the greater kudu and the Damara dik-dik, are even thought to have benefited from the bush encroachment prevalent in some of the eastern parts of the ecoregion. Although bush encroachment may benefit some species, it reduces overall diversity and therefore is not seen as positive for conservation in the ecoregion.

Severe overhunting of game mammals on private land was a major threat to wildlife for the first half of the century, but this was significantly reversed in 1967 when legislation shifted the ownership of game from the state to the individual landowner (Griffin 1998a). The Nature Conservation Amendment Act of 1996 extends similar fundamental rights to people living in communal areas, with the hope that rural dwellers will incur benefits from the value of wildlife and will manage it sustainably. Poaching is a present-day threat to wildlife, especially to the unfenced black rhino population. In an attempt to control poaching, these rhinos were dehorned (Berger and Cunningham 1994) and have previously been translocated to the Etosha Pan National Park (Hofmeyer et al. 1975). Plant poaching by collectors of succulent species along the southern escarpment is having a negative effect on the flora. Illegal trade in spectacular succulent species is believed to be considerable (Maggs et al. 1998).

Lastly, the Namib Escarpment has recently become a popular destination for off-road enthusiasts. Although this may have advantages in bringing tourism into the area, the indiscriminate use of off-road vehicles has increased cultural tensions, soil erosion, and scarring of the landscape.

Justification for Ecoregion Delineation

This ecoregion is based largely on the "bushy Karoo-Namib shrubland" of White (1983) and also includes a small transition area of "*Colophospermum* mopane scrub woodland to Karoo-Namib shrubland." Although the northern limit follows White's vegetation unit, the southern limit stops north of Groot Karasberg, near Keetmanshoop roughly around the 900-m contour. This ecoregion is characterized by faunal and floristic elements of the Namib and Kalahari. The southern portion is a part of the Nama Karoo biome, which extends into southern South Africa, and some authors (e.g., Irish 1994 in Barnard 1998) regard the Nama Karoo as extending narrowly along the west of the Namib Escarpment ecoregion into southern Angola.

Ecoregion Number:	**110**
Ecoregion Name:	**Succulent Karoo**
Bioregion:	**Cape Floristic Region**
Biome:	**Deserts and Xeric Shrublands**
Political Units:	**South Africa, Namibia**
Ecoregion Size:	**102,700 km²**
Biological Distinctiveness:	**Globally Outstanding**
Conservation Status:	**Relatively Stable**
Conservation Assessment:	**III**
Authors:	*Shirley M. Pierce, Richard M. Cowling*

Location and General Description

The Succulent Karoo [110] extends down the western coast of Namibia from the town of Lüderitz and into South Africa. It comprises two major biogeographic domains: the Namaqualand-Namib Domain and the Southern Karoo Domain (Jürgens 1991). The former encompasses the fog-affected coastal plain and adjacent escarpment in the west and receives most of its rain in the winter. The latter is located further east, is not subject to fog, and receives most of its rain in spring and autumn.

The distinctive climatic characteristics of the Succulent Karoo make it different from all other deserts in the world (Desmet and Cowling 1999b). Rainfall is reliable and predictable, falling mostly in winter, and prolonged winter droughts are rare. The climate is mild compared with that of other arid areas, particularly in the Namaqualand-Namib Domain, where frosts are extremely rare (Rutherford 1997). This domain receives annual rainfall ranging from 20 mm in the drier northwest to more than 400 mm in the escarpment zone, but most of the area receives less than 150 mm. Precipitation is supplemented by heavy dewfalls and fog. In contrast, the inland Southern Karoo Domain experiences a more extreme climate, with frequent summer maximum temperatures greater than 40°C. Dew is an important source of moisture, and rainfall is fairly reliable, with 150–300 mm falling each year.

The Namaqualand-Namib Domain is part of the Namaqualand Metamorphic Province, comprising granites and gneisses that are 1–2 billion years old (Watkeys 1999). Tertiary and recent sands of marine and aeolian origin cover the coastal plain. In the Southern Karoo Domain, the Tanqua Karoo is a large basin between the Great Escarpment and the Cape Folded Belt underlain by Mesozoic sediments of the Karoo cycle (Watkeys 1999). Soils are mainly stony and shallow.

There are only three perennial river systems in the Succulent Karoo, all of which have their source in wet mountain areas distant from the ecoregion. The perennial rivers are important in providing corridors of productivity dominated by trees that are derived from the distant savannas, such as *Acacia karroo*. On the sandy coastal plain of the Namaqualand-Namib Domain, the numerous seasonal river courses are associated with exposed bedrock.

The vegetation of the Succulent Karoo ecoregion may be divided into six broad types (Cowling and Pierce 1999b), determined primarily by soil depth, texture, moisture, and temperature regime. The most widespread vegetation type is vygieveld, a dwarf to low shrubland ranging from less than 25 cm to almost 50 cm in height, dominated by leaf succulents, notably *Crassula* spp. and members of the Mesembryanthemaceae family. Vygieveld is invariably associated with shallow soils. Strandveld is an open shrubland 0.5 to 2.0 m high that grows on the coastal plain of the Namaqualand-Namib Domain on deep sands of marine origin. Its species complement is modest, with low succulent creeping shrubs (*Cephalophyllum* spp.), dwarf forms in rocky sites (e.g., the monotypic endemic *Wooleya farinosa*, Mesembryanthemaceae), succulent shrubs (*Stoebaria* spp., *Ruschia* spp., *Zygophyllum* spp.), and nonsucculent shrubs (e.g., *Eriocephalus* spp., *Hermannia* spp.). In autumn, amaryllid bulbs of *Brunsvigia* and *Haemanthus* produce brilliant blooms, and in spring the space between the perennial shrubs is ablaze with flowering ephemerals. Broken veld is a widespread vegetation type found on rocky terrain in the escarpment zone of the Namaqualand-Namib Domain and is the predominant vegetation of the Little Karoo. It is rich in succulents, especially those in the families Mesembryanthemaceae and Euphorbiaceae. Renosterveld grows in the uplands of the Hardeveld, Little Karoo, Richtersveld, and the Western Mountain Karoo, with rainfall varying from 250 mm to more than 400 mm per year. This is a dense and taller shrubland dominated by Asteraceae, especially *Elytropappus* spp., *Euryops* spp., *Didelta* spp., *Pteronia* spp., *Eriocephalus* spp., and *Athanasia* spp. The renosterveld of the Succulent Karoo shows strong similarities with, and grades into, the renosterveld of the adjacent Lowland and Montane Fynbos and Renosterveld [89, 90] ecoregions. Fynbos, the characteristic vegetation of the two fynbos-type ecoregions, is also found in the Namaqualand-Namib Domain, patchily distributed on infertile, wind-blown sands along the coast and in the highest and wettest reaches of the Kamiesberg.

Outstanding or Distinctive Biodiversity Features

The Succulent Karoo is the world's only plant hotspot (sensu Mittermeier et al. 1999) that is entirely arid. By far the most distinctive feature of the Succulent Karoo is the diversity of succulents, especially dwarf and contracted leaf species (Cowling et al. 1999b). Some 1,700 species of leaf succulents are present, 700 of which are stone plants and their allies (e.g., *Conophytum*, *Lithops*). The major families contributing to this group are Mesembryanthemaceae, Crassulaceae, and Aloaceae. Certain succulent genera are extremely speciose, such as *Ruschia* (Mesembryanthemaceae, 136 spp.), *Conophytum* (Mesembryanthemaceae, 85 spp.), *Euphorbia* (Euphorbiaceae, 77 spp.), *Othonna* (Asteraceae, 61 spp.), and *Drosanthemum* (Mesembryanthemaceae, 55 spp.) (Hilton-Taylor 1996). Another outstanding feature is the high diversity of geophytes or bulblike plants. Most of the 630 geophyte species are petaloid monocots in the families Hyacinthaceae (*Lachenalia, Ornithogalum*), Iridaceae (*Babiana, Lapeirousia, Moraea, Romulea*), Amaryllidaceae (*Brunsvigia, Hessea, Strumaria*), and Asphodelaceae (*Bulbine, Trachyandra*). Despite the world-renowned displays of spring annuals, this plant group comprises a low proportion (8 percent) of the Succulent Karoo flora. Tree richness is poor, comprising only thirty-five species, but this paucity is offset by presence of charismatic endemics such as bastard quiver tree (*Aloe pillansii*), quiver tree (*Aloe dichotoma*), and halfmens (*Pachypodium namaquanum*).

Levels of plant endemism are extremely high. Some sixty-seven genera and 1,940 species are endemic, concentrated in four centers of endemism (three in the Namaqualand-Namib Domain and one in the Southern Karoo Domain) (Hilton-Taylor 1994), although local and point endemics are found throughout the region (Desmet and Cowling 1999a).

The Gariep Center encompasses the Richtersveld and extends north into Namibia's Sperrgebiet. This area is home to about 355 endemic plant species and three endemic genera: *Juttadinteria, Dracophilus,* and *Arenifera* (Mesembryanthemaceae). Included in this center are lichen fields that have the highest cover, density, and diversity of lichens in the world.

The Kamiesberg Center, home to eighty-six endemics, includes the peaks and upper slopes of the Kamiesberg Massif. The predominant vegetation types are fynbos and renosterveld, and most of the endemics are geophytes, especially irids in the genera *Babiana, Moraea, Romulea,* and *Lapeirousia*. Endemic dwarf succulents include species of *Cheiridopsis, Conophytum,* and *Lithops.*

The Van Rhynsdorp Center, a large expanse of coastal plain with the Knersvlakte as its hub, includes at least 150 endemic species, most of which are dwarf succulents and geophytes (with new species discovered regularly). This center includes all ten species of *Argyroderma* and all three species of *Oophytum* (both Mesembryanthemaceae).

In the Southern Karoo Domain, the Little Karoo is a distinct center of endemism, with 200–300 endemic species. Mesembryanthemaceae make up the majority of these endemics (Cowling and Hilton-Taylor 1999). Endemic genera include *Cerochlamys, Gibbaeum,* and the monotypic *Muirii* and *Zeuktophyllum. Glottiphyllum* and *Pleiospilos* have their centers of species richness and endemism in the Little Karoo.

The explosive speciation of the Mesembryanthemaceae, comprising approximately 1,800 species and 120 genera, is centered in the Succulent Karoo. This recent evolutionary phenomenon is probably unrivaled among angiosperms (Ihlenfeldt 1994). There has also been diversification in many other succulent plant lineages, including the Crassulaceae (*Crassula* and *Tylecodon*), Aloaceae (*Haworthia* and *Aloe*), Apocynaceae (Stapelieae), and *Euphorbia.*

The fauna of the Succulent Karoo [110] has a rich comple-

ment of endemics, especially among the arachnids, hopliniid beetles, aculeate Hymenoptera, and reptiles (Vernon 1999). Of the ecoregion's fifty scorpion species, twenty-two are endemic. Monkey beetles (Rutelinae: Hoplini), largely endemic to southern Africa, are concentrated in the Succulent Karoo and are important pollinators of the flora (Goldblatt et al. 1998). So too are the Hymenoptera and masarine wasps and colletid, fideliid, and melittid bees, all of which have centers of diversity and endemism in the ecoregion. The melittid bees include species of *Rediviva,* oil collectors that exclusively pollinate species of *Nemesia* and *Diascia* (Scrophulariaceae) (Steiner and Whitehead 1990).

Among the region's reptiles, the genus *Cordylus* (spinytail lizards) includes six strict endemics. Other strict endemics include Richtersveld dwarf leaf-toed gecko (*Goggia gemmula*), Calvinia thick-toed gecko (*Pachydactylus labialis*), Namaqua thick-toed gecko (*P. namaqua*), and Meyer's legless skink (*Typhlosaurus meyeri*). Amphibian endemics include Boulenger's short-headed frog (*Breviceps macrops*), Namaqualand short-headed frog (*B. namaquensis*), and *Bufo robinsoni.*

Endemism is less pronounced among the bird and mammal faunas (Vernon 1999). Of the birds, only one is strictly endemic, the recently described Barlow's lark (*Certhilauda barlowi*) which occurs only in the Sperrgebiet region (Sinclair and Hockey 1996; Ryan et al. 1999). Strictly endemic mammals include a subspecies of the pygmy rock mouse, *Petromyscus collinus barbouri,* Van Zyl's golden mole (*Cryptochloris zylii*), and De Winton's golden mole (*Cryptochloris wintoni*).

Status and Threats

Current Status

Large tracts of the ecoregion are still fairly intact, despite general overgrazing. The Little Karoo has seen the most transformation by agriculture because of the proximity of perennial streams draining into the major basin. Large areas of the higher and hence wetter areas of the Namaqualand-Namib Domain have also been converted for agriculture.

Given its global significance as a biodiversity hotspot (Cowling and Pierce 1999b) and its long-standing recognition as a regional conservation priority (Hilton-Taylor 1996; Rebelo 1997), the number and coverage of reserves are woefully inadequate. As of 1998, only 2,352 km^2, or approximately 2 percent of its original extent, was conserved in seven statutory reserves (Cowling and Lombard 1998). Larger reserves (greater than 100 km^2) were located in only four of the Succulent Karoo's twelve subregions and conserved only 80 plant species, or 9 percent of the 851 species in the *Red Data List of Southern African Plants* (Lombard et al. 1999).

The conservation situation has improved recently with the establishment of the approximately 500-km^2 Namaqua National Park and the initial phase of the Knersvlakte Nature Reserve (74 km²) in the Namaqualand-Namib Domain. The Anysberg Nature Reserve was recently expanded by more than 200 km², and Groenefontein Nature Reserve (47 km²) was created in the Little Karoo (Southern Karoo Domain). Formal (statutory) reserves now cover 2.79 percent of the Succulent Karoo, including the multiple-use Richtersveld National Park, where nomadic herders graze livestock, and the Goegab and Vrolikheid Nature Reserves. These initiatives have been directed by the outcomes of systematic conservation plans (Cowling and Lombard 1998; Desmet et al. 1999; Lombard et al. 1999; Cowling et al. 1999a).

The Sperrgebiet is enclosed in a Protected Diamond Area of 26,000 km², which has been closed to the public since 1910, with the lease expiring in 2020. The main mining area is the narrow coastal belt covering about 1 percent of the Sperrgebiet, with impacts leading to excessive inland sand movement (Williamson 1997).

Types and Severity of Threats

Although more than 90 percent of the Succulent Karoo is in a natural or semi-natural state, much of the habitat has been severely degraded by overgrazing. Land-use practices that will further threaten the ecoregion's biodiversity are as follows, in their order of importance:

- The expansion of communally owned land and associated overgrazing and desertification.
- Overgrazing of commercial (privately owned) rangelands.
- Agriculture, especially in the valleys of perennial rivers.
- Mining for diamonds, heavy minerals, gypsum, limestone, marble, monzite, kaolin, ilmenite, and titanium. For example, 65 percent of the Namaqualand coastline is or has been mined.
- Illegal and large-scale collection of succulents and geophytes.

In addition, climate change is likely to have a major impact on the biodiversity of the Succulent Karoo, given the specialized habitat requirements of the numerous local and point plant endemics (Rutherford et al. 1999).

Justification for Ecoregion Delineation

The ecoregion's northern and eastern (outer) boundaries follow Low and Rebelo's (1996) Succulent Karoo biome, which is, in turn, based on Rutherford and Westfall's (1986) biome concepts. The inner boundary (abutting the fynbos ecoregions) is according to the delimitation of the Cape Floristic Region by Cowling and Heijnis (2001). This was established by concordant patterns of geology, topography, climate, and, in some cases, vegetation types (sensu Low and Rebelo 1996).

Ecoregion Number:	**111**
Ecoregion Name:	**Socotra Island Xeric Shrublands**
Bioregion:	**Horn of Africa**
Biome:	**Deserts and Xeric Shrublands**
Political Units:	**Yemen, Somalia**
Ecoregion Size:	**3,800 km²**
Biological Distinctiveness:	**Globally Outstanding**
Conservation Status:	**Relatively Stable**
Conservation Assessment:	**III**
Author:	*Mike Evans*
Reviewed by:	*Chris Magin*

Location and General Description

The Socotra Island Xeric Shrublands [111] covers a small archipelago of islands approximately 240 km east of the Horn of Africa and 480 km south of the Arabian Coast. Socotra is the largest and most easterly island, and the other main islands are the Brothers, Abd al Kuri, Semhah, and Darsa. Yemen administers Socotra and two of the Brothers, and Puntland (a state in northeastern Somalia) administers Abd al Kuri and some of the smaller islands. Socotra often is considered to be part of the Middle East and not Africa, but geologically and biogeographically it is a continuation of the Horn of Africa.

The dominant landscape feature of Socotra Island is an extensive plateau of Cretaceous limestone averaging 300–700 m in elevation. The plateau rises near the Hagghier Mountains in the northwest (maximum elevation 1,519 m), which are composed of Precambrian granites and metamorphic rocks. The plateau then declines abruptly at the extreme western portion of the island, falling in steep escarpments to the coastal plains or directly into the ocean. The coastal plains can be up to 5 km wide and are found around most of the island. They consist mainly of alluvial soils of stone and coarse sand. Sand dunes can occur in some areas, particularly in the Noged Plain, a 60-km stretch of unbroken plain in the south.

Climate is strongly influenced by both the southwest (April–October) and northeast (November–March) monsoons. The southwest monsoons bring extremely strong, hot, and dry winds from Africa. The winter monsoon begins in November and lasts until March (WWF and IUCN 1994). Mean annual rainfall varies from about 150 mm on the coastal plains to more than 1,000 mm in the mountains. Mist and dew are more important to the water supply than monsoonal rain, especially in the high-altitude mountain belt (Mies and Beyhl 1998). Mean temperatures range from 27°C to 37°C maximum and 17°C to 26°C minimum along the coastal plain. It is substantially cooler in the Hagghier Mountains.

White (1983) considered Socotra a local center of endemism in the Somali-Masai regional center of endemism. Pronounced local variations in climate have resulted in a broad mosaic of plant communities on the island. The coastal plains and low inland hills consist of open deciduous shrubland dominated by the endemic *Croton socotranus* and scattered trees of *Euphorbia arbuscula, Dendrosicyos socotranus,* and *Ziziphus spina-christi.* Grasses and herbs develop after sufficient rainfall. The most widespread vegetation type is distinctive, species-rich, open shrubland found on the coastal foothills and the limestone escarpments. Two endemics, *Croton socotranus* and *Jatropha unicostata,* are the main shrubs present and are the most abundant plants on Socotra. Succulent trees such as *Euphorbia arbuscula, Dendrosicyos socotranus,* and *Adenium obesum sokotranum* and emergent trees such as *Boswellia* spp., *Sterculia africana,* var. *socotrana,* and *Commiphora* spp. are also present (WWF and IUCN 1994). On the limestone plateau and upward to the middle slopes of the Hagghier Mountains there are areas of semideciduous thicket dominated by *Rhus thyrsiflora, Buxus hildebrandtii, Carphalea obovata,* and *Croton* spp. (WWF and IUCN 1994). The higher montane slopes support a mosaic of dense thickets, dominated by *Rhus thyrsiflora, Cephalocroton socotranus,* and *Allophylus rhoidiphyllus* with the emergent dragon's blood tree (*Dracaena cinnabari*), low *Hypericum* shrubland, and, in many areas, anthropogenic pastures. Lichens and low cushion plants, including an endemic monotypic genus of Umbelliferae (*Nirarathamnos asarifolius*) and several endemic *Helichrysum* species, cover open rocks.

Outstanding or Distinctive Biodiversity Features

Biologically, the Socotran Archipelago is well known for its assemblage of endemic and unusual species. There are 850 recorded plant species, of which approximately 230–260 (about 30 percent) are endemic. There are also ten endemic genera (*Ankalanthus, Ballochia, Trichocalyx, Duvaliandra, Socotranthus, Haya, Lachnocapsa, Dendrosicyos, Placoda,* and *Nirarathamnos*) and one near-endemic family (Dirachmaceae). Some of the plants on Socotra represent the last surviving members of their genus. The limestone plateau and the Hagghier Mountains are the richest areas for endemic plant species, but endemics are found throughout the island in every type of vegetation. Because of habitat fragmentation and degradation, several endemic plant species are endangered. The endemic *Dirachma socotrana* is considered vulnerable by IUCN (Hilton-Taylor 2000), and *Croton pachyclados* survives only in one location (Mies and Beyhl 1998). *Dendrosicyos* is the only representative of the cucumber family to grow in tree form. *Euphorbia abdelkuri* grows only on Abd al Kuri. This endangered plant is an unusual *Euphorbia,* known for its spineless columnar stems, all linked by a single rootstock. In total, fifty-two endemic Socotran plants are found in the Red List of Threatened Species (Hilton-Taylor 2000).

Plant species in the drier parts of this ecoregion evolved morphologic and physiologica adaptations to cope with the dry climate and fierce monsoonal winds. *Adenium socotranum* has a special cell sap cycling in the caudex that prevents overheating. The succulents display several morphologic adaptations.

Plant bodies are globular or columnar, with reduced surface areas that decrease transpiration. Glaceous wax surfaces and microanatomic epidermal emergences reflect radiation. Umbrella-shaped shrubs form dense thickets, with all plants reaching the same height, a structure that protects them from strong winds (Mies and Beyhl 1998).

There are only seven terrestrial mammals in the Socotra Island Xeric Shrublands [111], most of which are introduced, although a bat (*Rhinopoma* sp.) and a shrew (*Suncus* sp.) may be endemic. Endemic bird species include island cisticola (*Cisticola haesitatus,* VU), Socotra warbler (*Cisticola incanus*), Socotra bunting (*Emberiza socotrana,* VU), Socotra sunbird (*Nectarinia balfouri*), Socotra starling (*Onychognathus frater*), and Socotra sparrow (*Passer insularis*) (Stattersfield et al. 1998). At least thirty bird species are known to breed on Socotra, including a significant population of Egyptian vultures (*Neophron percnopterus*) (Wranik 1998). An estimated ten or eleven endemic bird subspecies have been identified, some of which may warrant full species status (Sagheir and Porter 1998). The majority of terrestrial reptiles are endemic, including *Pachycalamus brevis, Chamaeleo monachus,* and *Coluber socotrae* (Joger 2000). The burrowing, legless *Pachycalamus brevis* is considered to be a relict of an ancient and once widely distributed Afro-Arabian herpetofauna. There are no amphibians, despite adequate water and the arid-adapted species present on the nearby African and Arabian mainlands. It is possible that severe drought in the past eradicated amphibians that colonized the island (Wranik 1998).

Status and Threats

Current Status

The original climax vegetation on Socotra has been altered by a combination of grazing and the cutting of wood for fuel. Some populations of endemic animal and plant species are now extremely small and occupy only scattered parts of the island. The more densely populated coastal plains are more degraded than the interior plains or especially the mountains. The main land use is pastoralism. The island's isolation has also helped to curb degradation and preserve the natural habitats. From April to October, the violent monsoon winds prevent the approach of airplanes or ships.

There are no legally protected areas, but privately owned areas in the highlands are managed traditionally and provide habitat refuges. A list of important bird areas exists (Evans 1994), and there is a proposal to designate the whole island as a biosphere reserve and United Nations Educational, Scientific and Cultural Organization (UNESCO) World Heritage Site.

Types and Severity of Threats

The main threats to biodiversity are excessive woodcutting for timber and fuel in the vicinity of the main settlements, unplanned infrastructural development on Socotra, especially the coastal zone, and an expanding population of goats and, to a lesser extent, cattle.

The small populations of some of the endemic resident bird species and endemic plants reflect the threats to their long-term survival on the island. This is particularly true for some of the breeding seabirds, whose numbers have been reduced by rats and other introduced predators.

Justification for Ecoregion Delineation

Biogeographically an extension of the Horn of Africa, the Socotra archipelago is undeniably a unique island ecosystem. It is characterized by interesting plant assemblages and high levels of plant endemism (28 to 32 percent) at both the species and generic levels and is also an important area for endemic birds and reptiles.

Ecoregion Number: **112**

Ecoregion Name: **Aldabra Island Xeric Scrub**

Bioregion: **Madagascar–Indian Ocean**

Biome: **Deserts and Xeric Shrublands**

Political Units: **Seychelles**

Ecoregion Size: **200 km²**

Biological Distinctiveness: **Globally Outstanding**

Conservation Status: **Relatively Intact**

Conservation Assessment: **III**

Author: *Helen Crowley*

Reviewed by: *Ross Wanless*

Location and General Description

The Aldabra Island Xeric Scrub [112] ecoregion occupies the island of Aldabra. This is an isolated coral atoll approximately 400 km northwest of Madagascar, 680 km east of the African mainland, and 1,100 km southwest of the main island group of the Seychelles. The atoll is 34 km long and 14.5 km wide and comprises four main islands (from largest to smallest, Grand Terre, Malabar, Picard, and Polymnie) and numerous lagoon islets. No areas are more than 18 m above sea level.

The islands are composed of coral limestone that was last inundated around 125,000 years ago. As is characteristic of atoll island systems, the four islands form a rough circle that encloses a large, shallow lagoon and are thus separated by four channels feeding to the Indian Ocean. The lagoon has an area of 310 km² and is fringed by extensive mangrove tracts. Topographically, the island surface is rugged because prolonged weathering has eroded much of the limestone into pits and fissures. There are also areas of raised lagoon sediments, coastal beaches, sand

dunes, and undercut limestone cliffs. The maximum spring tidal range is 2.7 m, giving rise to extremely strong currents in and around the channels.

The climate is tropical, with an average annual temperature of 27°C. Rainfall is variable from year to year, and it averages about 1,200 mm per year. There is a wet season from November to April and a drier season May to October. Aldabra seldom experiences cyclones.

Biologically, Aldabra and its neighboring atolls have a greater affinity with Madagascar and the Comoros archipelago than with the other islands of the Seychelles (Mittermeier et al. 1999). There are two structurally different types of xeric vegetation. The first is dominated by dense, almost monospecific stands of *Pemphis acidula* thicket that covers areas close to the saline water table. The second is a mixed scrub region, generally very dense but quite open in some places, and covers most of the rest of the atoll. The mixed scrub is composed of low trees, shrubs, herbs, and grasses. In the dry southeast, tortoises maintain extensive grazing lawns. On Picard Island and a few other locations there are abandoned coconut plantations. Aldabra is one of the few areas in the world where the dominant grazer is a reptile.

Outstanding or Distinctive Biodiversity Features

The terrestrial flora includes about 9 fern and 178 flowering plant species, 38 percent of which are believed to be endemic (Fosberg and Renvoize 1980). There are also many endemic invertebrate species (Shah et al. 1997).

The Aldabra atoll hosts a population of 80,000–100,000 giant tortoises (*Dipsochelys dussumieri*, VU) (Bourn et al. 1999). Significant breeding populations of green turtles (*Chelonia mydas*, EN) and hawksbill turtles (*Eretmochelys imbricata*, CR) use the beaches of the atoll for nesting. The lagoon and its fringing mangroves also represent a major rookery for immature turtles of both species. Aldabra supports an endemic subspecies of gecko, Abbott's day gecko (*Phelsuma abbotti abbotti*), and a large population of Bouton's fishing skink (*Cryptoblepharus boutonii*) (Cheke 1994).

Aldabra has two endemic landbird species and ten endemic subspecies (Benson 1967; Benson and Penny 1971). Of the endemic species, the Aldabra drongo (*Dicrurus aldabranus*) numbers around 1,500 individuals, whereas the Aldabra warbler (*Nesillas aldabranus*) is considered recently extinct (Hilton-Taylor 2000). Of the subspecies, one, the Aldabra white-throated rail (*Dryolimnas cuvieri aldabranus*), with an estimated 3,000–4,000 pairs on the atoll (Wanless 2002), is the last of the flightless birds for which the western Indian Ocean islands were once famed. A recent successful reintroduction to Picard Island has significantly improved the rails' conservation status (Wanless et al. 2002). The other nine endemic subspecies are Aldabra sacred ibis (*Threskiornis aethiopicus abbotti*), Aldabra blue pigeon (*Alectroenas sganzini minor*), Aldabra turtle dove (*Streptopelia picturata coppingeri*), Aldabra coucal (*Centropus toulou insularis*), Aldabra nightjar (*Caprimulgus madagascariensis aldabrensis*), Aldabra bulbul (*Hypsipetes madagascariensis rostratus*), Souimanga sunbird (*Nectarinia sovimanga aldabrensis*), Aldabra white-eye (*Zosterops maderaspatana aldabrensis*), and Aldabra fody (*Foudia eminentissima aldabrana*) (Rocomora and Skerrett 2001).

Aldabra is an important regional breeding site for several seabird species (Diamond 1971, 1994). The greater and lesser frigate birds (*Fregata minor* and *F. ariel*) both breed on Aldabra, in the second-largest colony of frigate birds in the world. Aldabra is a regional stronghold for the red-tailed tropic bird (*Phaethon rubricauda*, ~2,000 pairs) and red-footed booby (*Sula sula*, ~20,000 pairs). There are also about 2,000 pairs of white-tailed tropic birds (*Phaethon lepturus*) and 50–100 pairs of Audubon's shearwater (*Puffinus lherminieri*). Five species of tern breed on Aldabra, including up to 500 pairs of the rare black-naped tern (*Sterna sumatrana*) (Rocomora and Skerrett 2001). Aldabra atoll is also a breeding site for the greater flamingo (*Phoenicopterus ruber*).

There is one endemic mammal subspecies, the Aldabra giant fruit bat (*Pteropus seychellensis aldabranus*). Three insectivorous bat species have also been recorded. Aldabra is an important refuge for coconut crab (*Birgus latro*, DD), which has disappeared from many other oceanic islands.

Status and Threats

Current Status

In 1981, Aldabra became a special reserve under the National Parks and Nature Conservancy Act of the Seychelles. A "Management Plan for Aldabra" has been developed and is used as a guideline for managing the atoll. In 1982, the reserve was listed as a World Heritage Site. There is a policy of minimum human interference, the research and monitoring programs on the islands continue (Shah et al. 1997).

There is a research station on Picard Island. The reserve warden and support staff live at the station, and apart from visiting scientists and tourists there are no other people on the atoll. Before it was established as a reserve, there was only limited protection of the animals, many of which were commercially exploited. Attempts to grow commercial crops were abandoned before large-scale habitat alteration occurred, although several commercially and incidentally introduced plant species still occur. Green turtle numbers are growing, making Aldabra one of the only growing populations of this species in the world (Mortimer 1988). The tortoise population has recovered to reach carrying capacity on Grande Terre, the largest island of the atoll (Bourn et al. 1999).

Types and Severity of Threats

There are no introduced birds on Aldabra. In the past, rats (*Rattus rattus*), cats (*Felis catus*), and goats (*Capra hircus*) were intro-

duced to these islands. The black rat is a significant predator of bird nests. Rats also cause substantial damage to native vegetation by stripping bark and eating leaf and flower buds. Feral domestic cats have been eliminated from all islands except Grand Terre. They have been responsible for the decline in numbers of several native bird species. The introduced goats have been eradicated from all islands except Grande Terre. An alien coccid insect (*Icerya seychellarum*) parasitizes many native plant species and does significant damage to native plants, but the introduction of a ladybird beetle appears to have controlled infestations.

The presence of feral rats, cats, and goats prevents the complete restoration of Aldabra's ecology. Should cats become established on any of the islands that now support the Aldabra rail, the latter would become locally extinct (Wanless 2002). The potential for an accidental introduction of the very aggressive brown rat, an alien species that is abundant in Mahé, from which all of Aldabra's supplies are ferried, is a source of major concern. The presence of several introduced bird species on the nearby Assumption Island has been a source of concern for many years. There is a very real possibility of these species colonizing Aldabra.

Justification for Ecoregion Delineation

Aldabra is an extremely isolated coral atoll island in the Indian Ocean with unusually high levels of floristic and faunal endemism. Although Aldabra is part of the Seychelles, it has significant geological and biological differences from the granitic Seychelles and therefore is separated into its own ecoregion. Furthermore, it is one of the only largely intact oceanic island systems of any significant size anywhere in the world.

Ecoregion Number:	**113**
Ecoregion Name:	**Madagascar Spiny Thickets**
Bioregion:	**Madagascar–Indian Ocean**
Biome:	**Deserts and Xeric Shrublands**
Political Units:	**Madagascar**
Ecoregion Size:	**43,300 km²**
Biological Distinctiveness:	**Globally Outstanding**
Conservation Status:	**Endangered**
Conservation Assessment:	**I**
Author:	*Helen Crowley*
Reviewed by:	*Steve Goodman, Achille Raselimanana, Frank Hawkins, Sue O'Brien*

Location and General Description

The Madagascar Spiny Thickets [113] extend across southern and southwestern Madagascar from the Mangoky River on the west coast to the western slopes of the Anosyennes Mountain chain in the southeast. The topography of the ecoregion is fairly flat, running from sea level to altitudes between 55 and 200 m above sea level in the north. There are two major rock types in the ecoregion: the Tertiary limestone of the Mahafaly Plateau and the unconsolidated red sands of the central south and southeast. This geology corresponds to a major division in the habitat (Du Puy and Moat 1996). The taller, dense, dry forest on the sandy soils is dominated by *Didierea madagascariensis,* and the more xeric-adapted vegetation on the calcareous plateau around Lake Tsimanampetsotsa is characterized by dwarf species.

This ecoregion falls in the extreme rainshadow of Madagascar behind the eastern chain of mountains and far from the prevailing northeastern rains. Consequently, the average annual rainfall is 500 mm or less per year. The driest areas are in the southwestern coastal region, where the annual rainfall may be less than 350 mm per year, and the dry season may last 9–11 months (Donque 1972). Rainfall can be erratic from year to year; prolonged periods of drought, lasting up to several years, occur regularly. Average annual temperatures range between a maximum of 33°C and a minimum of 15°C.

Vegetation consists of spiny bush in the south and west and a mosaic of spiny bush and secondary grassland further inland (Morat 1973; White 1983; Phillipson 1996). Many of the spiny bush plants possess extreme adaptations to aridity, such as extended root systems with large tubers, enlarged, succulent trunks and branches, succulent and small leaves, thorns, and waxy and hairy coatings. The spiny thickets usually are 3–6 m in height but sometimes include emerging trees of the Didiereaceae family that reach more than 10 m in height, such as *Alluaudia ascendens* and *A. procera.* Other emergents include *Commiphora* spp. (Burseraceae), *Tetrapterocarpon geayi* (Leguminosae), and *Gyrocarpus americanus* (Hernandiaceae). A recent survey in the eastern spiny forest reported that the following families were the most dominant and diverse: Burseraceae, Didiereaceae, Euphorbiaceae, Anacardiaceae, and Fabaceae (Rakotomalaza and Messmer 1999). Other significant plant communities include the low bushy scrub of the coastal dunes and the gallery forests on the alluvial soils bordering major rivers (the Mandrare, Onilahy, Linta, and Fiherenana). There are also important areas of transition forest, such as on the western side of the Anosyennes Mountains in the southeast, where the humid forest grades into spiny forest.

The coastal Mikea Forest between Manombo and Morombe is a unique area in this ecoregion (Seddon et al. 2000). These forests grow on sandy soils and in a semi-arid climate with annual precipitation as low as 350 mm. The canopy, which rarely exceeds 12 m, is shorter than that of the forests inland and of those further north. The plant families forming the canopy include Leguminosae, Euphorbiacae, Burseraceae, and Bombaceae. The shrub layer consists mainly of *Croton* sp. (Euphorbiaceae), *Aloe vaombe* (Aloaceae), and lianas such as *Dioscorea* sp. (Dioscoreaceae) (Nicoll and Langrand 1989).

Outstanding or Distinctive Biodiversity Features

The ecoregion has the highest percentage of plant endemism in Madagascar (Jolly et al. 1984; WWF and IUCN 1994; Phillipson 1996). Some of the dominant forest species belong to the endemic family Didiereaceae. There are eleven species and four genera (*Didierea, Alluaudia, Alluaudiopsis,* and *Decaryia*) in this family. Some of the endemic plants are extremely rare and have highly restricted ranges, such as *Aloe suzannae* (Liliaceae), the palm *Dypsis decaryi,* tiny *Euphorbia* herbs, *Pachypodium* spp., and *Hibiscus* shrubs.

The fauna of the ecoregion is also distinctive and includes three strictly endemic mammals: the white-footed sportive lemur (*Lepilemur leucopus*), Grandidier's mongoose (*Galidictis grandidieri,* EN), and gray mouse lemur (*Microcebus murinus*). Near-endemic mammals include the large-eared tenrec (*Geogale aurita*) and the lesser hedgehog tenrec (*Echinops telfairi*). Other lemurs are found only in spiny thicket and the adjacent Madagascar Succulent Woodlands [114] ecoregion, including Verreaux's sifaka (*Propithecus verreauxi verreauxi,* VU) and ringtailed lemur (*Lemur catta,* VU) (Mittermeier et al. 1994). Some mammals have very restricted ranges in the ecoregion. Grandidier's mongoose (*Galidictis grandidieri,* EN) was described as new to science in 1986 and has a restricted range around Lake Tsimanampetsotsa.

Eight bird species are endemic to the ecoregion (Nicoll and Langrand 1989; Stattersfield et al. 1998), including Verreaux's coua (*Coua verreauxi*), running coua (*Coua cursor*), Lafresnaye's vanga (*Xenopirostris xenopirostris*), red-shouldered vanga (*Calicalicus rufocarpalis,* VU), and Archbold's newtonia (*Newtonia archboldi*). Three endemic species have very restricted ranges. The subdesert mesite (*Monias benschi,* VU) and long-tailed ground-roller (*Uratelornis chimaera,* VU) are known only from a narrow coastal strip on the northwest edge of the ecoregion. All of these species belong to monospecific genera and are representatives of two of the five families endemic to Madagascar (Langrand 1990). The red-shouldered vanga is known only from the Toliara region (Goodman et al. 1997; Hawkins et al. 1998). The Thamnornis warbler (*Thamnornis chloropetoides*) extends only slightly outside this ecoregion into the Madagascar Succulent Woodlands [114] ecoregion. The Madagascar plover (*Charadrius thoracicus*) is found in this ecoregion and along the west coast into the Succulent Woodlands and the Dry Deciduous Forests [114, 33] ecoregions.

Reptile species strictly endemic to the ecoregion include the chameleons *Furcifer belalandaensis* and *F. antimena* and the radiated tortoise (*Geochelone radiata,* VU). Other endemic and near-endemic species include the spider tortoise (*Pyxis arachnoides*), the rock-dwelling iguanids *Oplurus saxicola* and *O. fierinensis,* the day geckos *Phelsuma breviceps* and *Phelsuma standingi* (VU), the nocturnal geckos *Ebenavia maintimainty* and *Matoatoa brevipes,* and the snakes *Liophidium chabaudi* and *Acrantophis dumerili.*

Status and Threats

Current Status

Blocks of largely intact forest remain in the northwestern portion of the ecoregion (Seddon et al. 2000) and the extreme southeast (Du Puy and Moat 1996). Much of the inland area has been replaced by secondary grassland and wooded grassland (Morat 1973; White 1983; WWF and IUCN 1994; Lowry et al. 1997). The rate of habitat loss and degradation is lower than in other habitats around Madagascar, in part because of the low human population density. However, recent developments, such as an irrigation pipeline (Sussman et al. 1994), have increased the movement of people into the ecoregion, and recent estimates indicate that clearance of the forest has accelerated (Seddon et al. 2000).

Several small reserves, including Tsimanampetsotsa National Park and Ramsar site, Beza-Mahafaly Special Reserve, Cap Sainte Marie Special Reserve, and Berenty Private Reserve, protect about 3 percent of remaining habitat. The protected areas exclude important habitats for endemic species of birds and reptiles, such as the coastal area around the Onilahy River, the strip of forest between Mangoky and Fiherenana rivers, and Lake Ihotry (Morris and Hawkins 1998; Seddon et al. 2000).

Types and Severity of Threats

The principal threats are firewood and charcoal production. Selective logging of forests for construction wood is also a significant threat, particularly because the spiny thicket forest type has a naturally slow rate of growth and regeneration.

The cultivation of maize and grazing of domestic species (primarily cattle and goats) are also expanding in the ecoregion and pose a serious threat where they occur (Seddon et al. 2000). Invasive plant species, such as prickly pear (*Opuntia* spp.), rubber vine (*Cissus* spp., which is a threat only in gallery forest), and sisal have increased the degradation of the habitats, especially in disturbed forest areas. As in other regions of Madagascar, the collection of endemic species of plants and animals for international trade threatens the integrity of the habitats. Illegal collection is a particularly significant threat for the populations of the two rare tortoises, *Phelsuma standingi,* and various species of endemic succulent plants.

Traditionally, hunting *fady,* or taboos, of two local tribes (Antandroy and Mahafaly) protected many animal species in this region. However, with the increased movement of people across the region, the local *fady* on certain animals is becoming less effective as a means of protection.

Justification for Ecoregion Delineation

The Madagascar Spiny Thickets [113] ecoregion extends from Morombe to just west of Tolagnaro along the southwest coast of Madagascar and is based primarily on Humbert's (1955) southern vegetation domain. The dominant vegetation is deciduous thicket, representing a center of Didiereaceae endemism. The ecoregion extends inland approximately 80 km, with the eastern limits stretching further inland along the Mandrare River where it meets the southern edge of the central highlands. There has been some debate on whether the boundary should be delineated using Cornet's (1974) bioclimatic zone or Humbert's vegetation zone, each differing in extent and ecological parameters. It was decided to use Humbert's linework for the Madagascar Spiny Thicket [113] ecoregion. The Madagascar Succulent Woodlands [114] ecoregion was delineated according to the remaining extent of Cornet's subarid bioclimate because it shares floral affinities with both the spiny thicket and western dry deciduous forest, yet contains distinct assemblages.

Ecoregion Number:	**114**
Ecoregion Name:	**Madagascar Succulent Woodlands**
Bioregion:	**Madagascar–Indian Ocean**
Biome:	**Deserts and Xeric Shrublands**
Political Units:	**Madagascar**
Ecoregion Size:	**79,800 km²**
Biological Distinctiveness:	**Globally Outstanding**
Conservation Status:	**Critical**
Conservation Assessment:	**I**
Author:	*Helen Crowley*
Reviewed by:	*Steve Goodman, Achille Raselimanana, Frank Hawkins, Sue O'Brien*

Location and General Description

The Madagascar Succulent Woodlands [114] are located in southwestern and central western Madagascar, sandwiched between the Madagascar Spiny Thickets [113] and the Madagascar Dry Deciduous Forests [33]. The eastern extent, at the 600- to 800-m contour, is contiguous with the Madagascar Subhumid Forests [30] and to a large extent coincides with the southwestern limit of the central highlands.

The geology of the western part of the ecoregion includes unconsolidated sands on the coast and Tertiary limestone and sandstone further inland. To the south, there are also metamorphic and igneous basement rocks (Du Puy and Moat 1996). The soils are generally sandy, with richer alluvial soils around rivers. The terrain is flat, but there are some notable rock outcrops and deep precipitous valleys (e.g., in the Makay region).

The ecoregion has a tropical dry climate with a distinct dry season between May and October. During the wet season, November to April, rainfall may reach 750 mm, within a yearly range of 575–1,330 mm. The annual average daily temperature for the region is between 25°C and 31°C.

Vegetation is similar to the Madagascar Dry Deciduous Forests [33] but is characterized by more xerophytic species. These species often have water storage adaptations and stem photosynthesis and remain without leaves for long periods. Forests of the ecoregion may reach 15 m in height, with the endemic baobabs (Bombaceae family) *Adansonia za* and *A. grandidieri* as distinctive emergent species. Other canopy species belong to the families Euphorbiaceae and Leguminosae, including several endemic species of *Pachypodium*. The shrub layer consists of the families Sapindaceae, Euphorbiaceae, Anacardiaceae, and Burseraceae (White 1983).

Outstanding or Distinctive Biodiversity Features

This ecoregion provides important habitat for eight lemur species and sixty to ninety bird species. There is an overlap of species between the succulent woodlands, the spiny thicket to the south, and the dry deciduous forests to the north. There are also several areas of high local endemicity, such as the forests between the Tsiribihina and Mangoky rivers and around the National Park of Zombitse-Vohibasia. Endemic mammals include narrow striped mongoose (*Mungotictis decemlineata decemlineata*, EN), giant jumping rat (*Hypogeomys antimena*, EN), and Berthe's mouse lemur (*Microcebus berthae*). Near-endemics include the red-tailed sportive lemur (*Lepilemur ruficaudatus*), large-eared tenrec (*Geogale aurita*), lesser hedgehog tenrec (*Echinops telfairi*), and Verreaux's sifaka (*Propithecus verreauxi verreauxi*, VU).

Among the birds, Appert's greenbul (*Phyllastrephus apperti*) is strictly endemic, known from the Zombitse-Vohibasia forests (Langrand 1990). The white-breasted mesite (*Mesitornis variegata*, VU) and long-tailed ground-roller (*Uratelornis chimaera*, VU) are near endemic. The red-capped coua (*Coua ruficeps*) is found throughout this ecoregion.

Some of the local endemic reptiles include *Oplurus cuvieri, Chalarodon madagascariensis*, and the geckos *Phelsuma standingi* and *Paroedura vazimba*. *Pyxis planicauda* has a narrow distribution range in the ecoregion, as do *Mabuya tandrefana, Furcifer antimena*, and *Brookesia brygooi*. The rare snake *Liophidium chabaudi* also occurs. At least two frog species are endemic: the hyperoliid *Heterixalus luteostriatus* and the mycrohylid *Dyscophus insularis*.

Status and Threats

Current Status

A number of protected areas fall in this ecoregion, including Zombitse-Vohibasia National Park, Andranomena Special Reserve, and Kirindy-Mitea National Park. Furthermore, the Kirindy Forest, north of Marofandilia, is managed as a private reserve. These protected zones and several classified forests com-

prise a small area of the remaining habitat. The degree to which the forest and wildlife is protected varies widely between reserves. The classified forests offer little protection; logging continues in these areas.

Types and Severity of Threats

As with many of the habitats of Madagascar, the major threat to the succulent woodlands is fire, both intentional burning for expansion of agricultural and pasture lands and unintentional wildfires. Although it is not known whether the ecoregion included only dry woodland or a mosaic of woodland and grassland before human settlement, the increased incidence of cultivation and fire in recent times has certainly led to increasingly fragmented and isolated patches of native vegetation.

Tree cutting for charcoal production has caused extensive deforestation. Honey collection, through the felling of trees, is a traditional activity that threatens the forests to lesser degree. The Malagasy endemic tree *Hazomalania voyroni* is at risk of becoming extinct through traditional forest exploitation for construction wood. Some attempts have been made to propagate and replant this species (Randrianasolo et al. 1996). In addition to this species, several other endemic trees are removed from the forests mainly for construction purposes, including *Givotia madagascariensis, Cedrelopsis grevei*, and *Commifora* spp.

Traditional hunting occurs throughout the ecoregion, even in protected areas. With increasing human populations and greater movement of people, these traditional activities become locally unsustainable. The main species threatened by hunting are the tailless tenrec (*Tenrec ecaudatus*), the fruit bats *Pteropus rufus* and *Eidolon helvum*, and the red-fronted brown lemur (*Eulemur fulvus rufus*). In many parts of the ecoregion, cattle and goat grazing are also degrading the forests.

Justification for Ecoregion Delineation

This ecoregion forms the northern part of the Cornet's (1974) subarid bioclimate zone. Humbert's (1955) "southern vegetation domain" was used to delineate the Madagascar Spiny Thickets [113] to the south, with the remaining northern extent of Cornet's subarid bioclimate used to define the Madagascar Succulent Woodlands [114]. Previous descriptions of phytogeographic regions of Madagascar had grouped the succulent woodlands with the dry deciduous forests as a "western domain" (Guillaumet 1984; Lowry et al. 1997). The succulent thickets receive intermediate levels of precipitation between the wetter Madagascar Dry Deciduous Forests [33] and drier Madagascar Spiny Thickets [113] and are primarily a transitional area that shares floral affinities with both these other ecoregions. For example, the area is the northern limit of Didiereaceae.

Ecoregion Number: **115**

Ecoregion Name: **Guinean Mangroves**

Bioregion: **Western Africa and Sahel**

Biome: **Mangroves**

Political Units: **Senegal, the Gambia, Guinea-Bissau, Guinea, Sierra Leone, Liberia, Côte d'Ivoire**

Ecoregion Size: **23,500 km²**

Biological Distinctiveness: **Locally Important**

Conservation Status: **Critical**

Conservation Assessment: **IV**

Author: *Sylvia Tognetti*

Reviewed by: *Samba Diallo, Edem Okon Edem*

Location and General Description

The Guinean Mangroves [115] are located in western Africa and encompass mangrove stands along the coastlines of Mauritania, Senegal, the Gambia, Guinea-Bissau, Guinea, Sierra Leone, Liberia, and Côte d'Ivoire. Temperatures range from 15°C to 28°C in the north and from 23°C to 32°C in Guinea. Annual rainfall varies, from as low as 95 mm in the Senegal River Delta to approximately 9,000 mm in Sierra Leone (Spalding et al. 1997).

Salinity levels in the water surrounding the mangroves vary dramatically between the wet season and dry season. High freshwater inputs during the wet season reduce salinity levels and create an offshore estuarine zone along the coast of Guinea and Sierra Leone. During the dry season, the marine influence is more widespread and reaches farther inland. In Mauritania and the Senegal Delta, where evaporation exceeds precipitation 9 months of the year, salinity levels limit mangrove tree growth (Diop and Bâ 1993). Maximum tidal amplitudes range from 1.6 m in Senegal to 3.7 m in Guinea (Diop and Bâ 1993).

Extensive mangroves can be found where the flat topography of rivers and estuaries combines with high tidal amplitudes to allow tidal waters to penetrate deeply into the interior. This extends the habitat range of salt-tolerant mangrove species and allows a greater degree of species zonation along the low- to high-water gradient (John and Lawson 1990; Spalding et al. 1997). The mangrove and estuarine system is particularly extensive in Guinea, where littoral subsidence has led to marine incursions into river mouths (Diallo 1993).

The West African mangrove species assemblages are more similar to those found in the western Atlantic than to those in eastern Africa. Native species include *Rhizophora racemosa, R. mangle, R. harrisonii, Avicennia germinans, Conocarpus erectus,* and *Laguncularia racemosa.* Also, there is one introduced species, *Nypa fruticans.* Principal mangrove zones are the *Avicennia*-dominated swamps in Senegal and those in which *Rhizophora*

species are the primary colonists, with *Avicennia africana* found in the interior (Chapman 1977; Ukpong 1995).

Outstanding or Distinctive Biodiversity Features

Typically low in species diversity, the Guinean Mangroves [115] are important for the diversity of habitat and ecological functions that they provide. In addition, high productivity and an extensive food web support migratory shorebird populations, important marine species, and offshore fisheries.

The vertebrate fauna of these mangroves includes the African manatee (*Trichechus senegalensis,* VU), water chevrotain (*Hyemoschus aquaticus,* DD), pygmy crocodile (*Osteolaemus tetraspis,* VU), Nile monitor (*Varanus niloticus*), Nile crocodile (*Crocodilus niloticus*), and perhaps a few pygmy hippopotamus (*Hexaprotodon liberiensis,* VU) (Hughes and Hughes 1992; Schwartz 1992a; Simao 1993). Primates include Temminick's red colobus (*Procolobus badius temmincki,* EN), Western black and white colobus (*Colobus polykomos*), and Campbell's mona monkey (*Cercopithecus mona campbelli*).

Although there no bird species are restricted to this ecoregion, it supports many species, including large numbers of migratory waterbirds that use the mangrove areas seasonally during their migrations between the Palearctic and Afrotropical realms (Simao 1993). Ornithologic values of the coastal wetlands are summarized in Altenburg (1987) for West Africa and Altenburg and van der Kamp (1991) for Guinea. Important sites for migratory birds include the Djouge in Senegal and Arquipélago dos Bijagós in Guinea-Bissau (Coulthard 2001b; Robertson 2001a).

This ecoregion also supports a variety of marine fauna, including crustaceans, polychaetes, barnacles, mollusks, gastropods, and echinoids (Uschakov 1970). There are also many crab species, such as *Uca tangeri, Lepas anserifera,* and *Balanus tintinnabulum* (Uschakov 1970). A common fish species that occurs in the mangroves is *Periophthalmus papilio,* although other fish species are also present in the estuaries. Sea turtles also use this habitat, including olive ridley turtle (*Lepidochelys olivacea,* EN), loggerhead turtle (*Caretta caretta,* EN), and green turtle (*Chelonia mydas,* EN).

Status and Threats

Current Status

The largest mangrove stands in this ecoregion are found in the Casamance Delta and the Saloum River in Senegal and the Gambia River in the Gambia (Giffard 1974; Marius 1985; Diop 1993). Farther south, there are large stands on the Bijagós Archipelago of Guinea-Bissau (around 2,484 km²) (Simao 1993), the Conakry Peninsula of Guinea (2,960 km²) (Diallo 1993), and inland of Shebro Island in Sierra Leone (Johnson and Johnson 1993). Smaller areas of mangrove are found in Liberia (Momoh 1993) and Côte d'Ivoire (Blasco et al. 1980; Mathieu 1993). Remnant mangrove areas are found in Mauritania, surviving in a semi-arid climate after developing during a period of moister climatic conditions.

Protected areas containing mangroves exist in Senegal, Côte D'Ivoire, and Guinea-Bissau. The largest is Delta du Saloum in Senegal, which is 760 km^2 in size. Also in Senegal are the Basse-Casamance (50 km^2) and the Djouge National Park, which contain small areas of mangrove. Others are Lagoa de Cufada in Guinea-Bissau (391 km^2), which is a Ramsar site, and Iles Ehotile (105 km^2) in Côte d'Ivoire. Recently, the Guinea Department of Fisheries and Aquaculture has proposed several marine protected areas that include mangroves. Many of the mangroves of Sierra Leone fall in forest reserves.

Types and Severity of Threats

Of greatest concern is the loss of mangrove areas to rice farming, urban expansion, road construction, and shrimp farming. In Guinea-Bissau it is estimated that half of the mangrove area (2,480 of 4,760 km^2) has been converted to other uses (Simao 1993). In Guinea, 1,400 km^2 of mangroves has been converted to rice paddies, and only 780 km^2 now remains. About 620 km^2 of these rice paddies have been abandoned, with 270 km^2 of abandoned land recolonized by shrub species of low productivity and the remaining 350 km^2 degraded by acidity and soil toxicity (Diallo 1993). However, small-scale fragmentation does not greatly affect mangrove-associated biodiversity because mangroves are naturally fragmented and are able to disperse over long distances. Throughout the ecoregion mangrove wood is used for construction because the trees are straight and possess termite-resistant wood. Larger trunks are also used to make dugout canoes. The trees are also cut for firewood. According to Yansané (1998), in Guinea about 18,000 tons of wood are exploited per year for construction, 58,000 tons per year are exploited for firewood, and around 93,000 tons of mangrove wood are destroyed each year for salt extraction.

The entire region has been affected by a trend of reduced rainfall that began in 1968, causing a reduction in mangrove area in Senegal and the Gambia. In addition, the latitudinal extent of the mangroves means that they are particularly vulnerable to climate change. The reduction of freshwater inputs has also reduced the diversity of ostracods, foraminifers, and macrobenthic phytoplankton. Although the consequences remain uncertain, dams also interfere with the hydrological regime, such as the Diama Dam in the Senegal estuary (Diop and Bâ 1993). In Côte d'Ivoire, the reduction of freshwater inputs allows the sea swell to deposit sand sufficient to obstruct access between the lagoons and the sea (Mathieu 1993).

Justification for Ecoregion Delineation

This ecoregion stretches from Senegal to west of the Dahomey Gap. This gap is a major ecological barrier separating the rain-

forest regions of West and Central Africa, which in the marine environment represents the end of the influence of the south-to north-flowing cold waters of the Benguela Current. Although more extensive, the West African mangroves (five mangrove species) are species poor compared with the East African mangroves (nine species).

Ecoregion Number:	**116**
Ecoregion Name:	**Central African Mangroves**
Bioregion:	**Western Africa and Sahel**
Biome:	**Mangroves**
Political Units:	**Ghana, Nigeria, Cameroon, Equatorial Guinea, Gabon, Democratic Republic of the Congo, Angola**
Ecoregion Size:	**30,900 km^2**
Biological Distinctiveness:	**Locally Important**
Conservation Status:	**Endangered**
Conservation Assessment:	**IV**
Author:	*Sylvia Tognetti*
Reviewed by:	*Edem Okon Edem*

Location and General Description

The Central African Mangroves [116] are located in western Africa and encompass mangrove areas along the coastlines of Ghana, Nigeria, Cameroon, Equatorial Guinea, Gabon, Democratic Republic of the Congo (DRC), and Angola (to 19°18′S). The structure of the mangrove areas varies widely, from the lagoon systems found in the western part of this ecoregion to systems modified by complex patterns of sediment deposition at river mouths in the central and southern portions.

The climate is primarily humid and tropical but changes to more temperate conditions toward Angola. Off the coast of DRC, mangrove development is inhibited by the presence of the cold water Benguela Current, but some stands are found where high river water temperatures locally raise seawater temperatures (Makaya 1993). Annual rainfall varies from a mean of 750 mm in Angola to 6,000 mm in Cameroon. In Ghana and the western part of Nigeria, mangroves are associated primarily with extensive lagoons. These are enclosed part of the year by sediments, when rainfall is lower and freshwater outflow is not sufficient to counteract ocean swells (Sackey et al. 1993). In the remainder of the region, mangroves are associated primarily with river mouths, the largest of which is the Niger River Delta. The sediment load flowing from the Niger River has been estimated to be 20 million m^3 a year, most of which is captured in the mangrove swamps (CEC 1992). Sediment deposition and channel erosion have created a network of river creeks, estuarine swamps, and barrier islands. Soils range from recently deposited, unconsolidated, soft dark mud containing silt, clay, and peaty clay to transitional swamps, all of which are associated with different types of vegetation.

The key factors that influence these mangrove ecosystems are river floods (Adegbehin 1993) and the tidal range. Tidal range increases from west to east, reaching a maximum of 2.8 m in eastern Nigeria. This allows flood tides to penetrate up to 40–45 km into the interior. The large inputs of freshwater create a low-salinity zone offshore where salinity fluctuations range between 0 and 0.5 percent during the rainy season and 30 to 35 percent during the dry season. Farther south in Cameroon, annual rainfall reaches 6,000 mm, but this is highly variable because of variation in topography and coastal types. These freshwater inputs, together with a convergence of the Guinea Current, the Benguela Current, and an equatorial subsurface current, create a "piling up" of water that results in an increase in mean sea level of 1.2 m and creates an unusual circulation pattern (Appolinaire 1993). It also results in the formation of sand bars and the deposition of large amounts of suspended sediment in the mouths of estuaries.

Five mangrove species are found in this region, including the red mangroves (*Rhizophora racemosa, R. mangle,* and *R. harrisonii*), the white mangroves (*Avicennia germinans* and *Laguncularia racemosa*), and an introduced species, *Nypa fruticans. Rhizophora racemosa* is the primary colonist in the open lagoon systems, whereas *Avicennia africana* is the primary colonist in closed systems. In the back swamps *Nypa fruticans* is replacing red mangroves because it colonizes more rapidly and has shallow roots that destabilize riverbanks (Isebor and Awosika 1993). *Rhizophora racemosa* is dominant in the tidal and more inundated areas of Cameroon. Farther south in DRC where mangroves are found around lagoons, the dominant species is *Rhizophora mucronata* (Makaya 1993). In Angola, dominant trees are *Rhizophora racemosa, R. mangle, R. harrisonii,* and *Avicennia africana,* the former two species reaching heights of approximately 30 m (Huntley 1974b).

Outstanding or Distinctive Biodiversity Features

The Central African Mangroves [116] contain no strictly endemic species. However, they are known for their diverse pelagic fish communities, including some narrowly distributed species, abundant avifauna, and the presence of some rare mammals and turtles.

These mangroves provide habitat for the threatened African manatee (*Trichechus senegalensis,* VU) (Hughes and Hughes 1992), the soft-skinned turtle (*Trionyx triunguis*), and, in the Niger Delta, isolated populations of pygmy hippopotamus (*Hexaprotodon liberiensis heslopi,* CR). The near-endemic Sclater's guenon (*Cercopithecus sclateri,* EN) and talapoin monkey (*Miopithecus talapoin*) may also use the brackish portion of the mangroves (NDES 1997). Hippopotamus (*Hippopotamus amphibius*) and Nile crocodile (*Crocodylus niloticus*) are also found in brackish areas.

Coastal mangroves and wetlands are important primarily for large concentrations of birds that use the areas during migration, although some wetland species also breed here, such as Caspian tern (*Sterna caspia*). Several of the coastal wetland sites are internationally important for migratory wetland birds, such as those of Ghana (Piersma and Ntiamoa-Baidu 1995; Ntiamoa-Baidu et al. 2001), Gabon (Christy 2001a), and the Niger Delta (Hughes and Hughes 1992; Ezealor 2001).

The mangroves are also important for species found primarily in adjacent habitats that also depend on mangroves for parts of their life cycle. The Niger Delta provides spawning and nursery areas for the fisheries in the Gulf of Guinea. A high diversity is found in the pelagic fish community, with forty-eight species in thirty-eight families (Ajao 1993). Pelagic families and species associated with them include Clupeidae (*Pellonula leonensis, Ilisha africana, Sardinella maderensis*), Belonidae (*Ablennes hians, Strongylura senegalensis*), Megalopidae (*Tarpon atlanticus*), Hemiramphidae (*Hyporhamphuspicarti*), Elopidae (*Elops lacerta, E. senegalensis*), and Albulidae (*Albula vulpes*) (Isebor and Awosika 1993; Shumway 1999). Five species of marine turtle are also found: leatherback (*Dermochelys coriacea,* EN), loggerhead (*Caretta caretta,* EN), olive ridley (*Lepidochelys olivacea,* EN), hawksbill (*Eretmochelys imbricata,* CR), and green turtle (*Chelonia mydas,* EN).

Status and Threats

Current Status

Current estimates of mangrove area provided by Spalding et al. (1997) range between 16,673 and 17,176 km^2, of which more than two-thirds are found in Nigeria. The most important remaining blocks are found in the Niger River Delta in Nigeria, to the east of the mouth of the Cross River in Nigeria and Cameroon, around Doula in Cameroon, and in the Muni Estuary and Como River in Gabon. Smaller habitat areas are also found in Ghana, in the Conkouati lagoons of Congo, at the mouth of the Congo River in the DRC, and in Angola. The Niger Delta has been growing for millions of years and is still expanding into the Gulf of Guinea. The delta mangroves mark the transition between swamp forest habitats to pioneer communities on the coast and can extend up to 45 km wide. In Angola, mangrove communities occur at the mouths of the Cuvo, Longa, Cuanza, Dande, and M'Bridge rivers (Huntley and Matos 1994).

Approximately 3,165 km^2, or 10.68 percent, of the Central African Mangroves [116] is protected. These are included in the Douala-Edea Faunal Reserve in Cameroon (1,600 km^2) and the Anlo-Keta Lagoon Complex and Songor Lagoon in Ghana. Draft management recommendations have been prepared for the coastal zone in Ghana, including mangrove areas (Agyepong et al. 1990).

Types and Severity of Threats

Fragmentation itself does not greatly affect mangrove biodiversity because mangroves are naturally fragmented and are able to disperse over long distances. Of greater concern is the total amount of mangrove area lost to urbanization, industrialization, and agriculture, as well as impacts from timber and petroleum exploitation (Diop 1993). Timber is used primarily for firewood and poles for housing construction. Impacts from petroleum exploitation include coastal subsidence, which may aggravate the effects of sea-level rise, and infrastructure development and oil spills, which have led to large invertebrate and fish mortalities. Exporting oil from coastal areas is an economically important activity in Nigeria, Gabon, and Cameroon that can lead to oil spills (NDES 1997). In Nigeria, in the past 30 years, seismic lines have been placed in the Niger Delta mangrove forests. Other threats include the practice of gas flaring, the use of poison and dynamite for fishing, canalization, discharge of sewage and other pollutants, siltation, sand mining, erosion, construction of embankments, and growing population pressure in the coastal zone (Isebor and Awosika 1993).

Justification for Ecoregion Delineation

The Central African Mangroves range from Ghana east of the Dahomey Gap, through the Niger Delta (the largest concentration of mangroves in Africa), and south to the mouth of the Congo River, with outlying patches in Angola. This ecoregion generally follows the part of the African coastline that is affected (at least occasionally) by the cold water Benguela current.

Ecoregion Number: **117**

Ecoregion Name: **Southern Africa Mangroves**

Bioregion: **Eastern and Southern Africa**

Biome: **Mangroves**

Political Units: **South Africa, Mozambique**

Ecoregion Size: **1,000 km^2**

Biological Distinctiveness: **Locally Important**

Conservation Status: **Endangered**

Conservation Assessment: **IV**

Author: *Sylvia Tognetti*

Reviewed by: *Rudy van der Elst,*
Anthony B. Cunningham

Location and General Description

The Southern Africa Mangroves [117] contain the southernmost mangroves on the African continent, lining parts of the eastern South African and southernmost Mozambique coastline along the Indian Ocean. Mangroves extend 20 degrees farther south on the eastern coast than on the western coast of Africa because of the warming effect of the Agulhas Current. The climate is subtropical, with mean maximum temperatures ranging from 18°C to 24°C and mean minimum temperatures ranging from 12°C to 18°C. Annual rainfall ranges from 800 to more than 1,200 mm. Tidal amplitude ranges from 0.75 to 3.5 m at Maputo (Spalding et al. 1997).

In South Africa most mangroves are found in the mouths of perennial rivers, in numerous estuaries and lagoons, where large sandbars and coastal dunes often protect them. They begin to appear just north of East London at the mouth of the Nahoon River, although the first well-developed mangrove forest is found in the Mngazana estuary (Hughes and Hughes 1992). The three largest mangrove areas are in Richards Bay/Mhlatuzi (652 ha), Saint Lucia (279 ha), and Mngazana (137 ha), totaling 1,068 ha. The southernmost sites contain only the mangrove species *Avicennia marina* in association with several salt marsh species. *Bruguiera gymnorrhiza* begins to be found at the Mbashe estuary and *Rhizophora mucronata* at the Mngazana estuary. In the northernmost sites of South Africa, at Kosi Bay, *Ceriops tagal* and *Lumnitzera racemosa* are also found, but in very limited numbers (Hughes and Hughes 1992). The distribution of *Acrostichum aureum* is not known, although it has been noted in Pondoland (Steinke 1995). In many areas the estuarine mudflats and shallows often are dominated by aquatic vegetation such as *Potamogeton, Ruppia,* and *Zostera,* especially during periods of low salinity.

Many other important plant species are associated with these mangrove complexes, including *Helichrysopsis septentrionale* (a Maputaland endemic), four regional endemic genera (*Brachychloa, Ephippiocarpa, Helichrysopsis,* and *Inhambanella*), *Wolffiella welwitschii,* and *Thalasodendron ciliata* (the only marine flowering plant found on the South African coastline). Wetland vegetation types include freshwater swamp (*Phragmites* and *Papyrus*), saline reed swamp (*Phragmites mauritianus*), *Eleocharis* swamp, and salt marsh (*Sporobolus virginicus, Paspalum vaginatum, Juncus kraussii, Sarcocornia* spp., and *Ruppia maritima*). All of these vegetation types occur in or in association with mangrove formations.

Outstanding or Distinctive Biodiversity Features

Several important bird areas are located in this ecoregion (Barnes et al. 2001). These include the Kosi Bay system (northernmost patch of mangrove in this ecoregion), Lake St. Lucia and Mkuze swamps, Umlalazi Nature Reserve, Richards Bay Sanctuary (extensive *Rhizophora* and *Avicennia* mangrove estuary), and Dwesa and Cwebe nature reserves (with extensive *Bruguiera, Avicennia,* and *Rhizophora* mangroves). The Greater St. Lucia Wetland Park, which supports more than 350 bird species, is among the most important breeding areas for waterbirds in southern Africa. At least forty-eight bird species have been reported as breeding in this system (Barnes 1998; Barnes et al. 2001). Notable species include lesser flamingo (*Phoenicopterus minor*), great white and pink-backed pelicans (*Pelecanus onocrotalus* and *P. rufescens*), grey-headed gull (*Larus cirrocephalus*), Cape shoveler (*Anas smithii*), yellow-billed duck (*A. undulata*), pied avocet (*Recurvirostra avosetta*), saddlebilled stork (*Ephippiorhynchus senegalensis*), yellow-billed stork (*Mycteria ibis*), and Caspian tern (*Sterna caspia*) (Barnes et al. 2001). Only one bird is considered near endemic to this ecoregion, the mangrove kingfisher (*Halcyon senegaloides*).

Few mammals are permanent residents of mangrove ecosystems. Around the fringes and in the adjacent wetlands, antelope may forage, such as the southern reedbuck (*Redunca arundinum*). Smaller mammals, including shrews and rats, are quite common in parts of mangrove ecosystems. Bats are important transient mammals, especially the fruit bat *Rousettus aegyptiacus*. Arboreal mammals include the vervet monkey (*Chlorocebus aethiops*), large-tailed greater bush baby (*Otolemur crassicaudatus*), and red bush squirrel (*Paraxerus palliatus*). The southern tree hyrax (*Dendrohyrax arboreus*) is closely associated with mangrove forests of the Transkei region (Rautenbach et al. 1980). Predators include the African civet (*Viverra civetta*), genet (*Genetta rubiginosa*), and water mongoose (*Atilax paludinosus*). Also found in these wetlands are two otter species, *Aonyx capensis* and *Lutra maculicollis*.

Up to 200 benthic species have been reported (Hughes and Hughes 1992), among which crabs are dominant. Marine fauna includes sesmarid crabs (e.g., *Sesarma meinerti* and *S. catenata*), fiddler crabs (*Uca annulipes* and *U. urvillei*), gastropods (e.g., *Cerithidea decollata*, *Terebralia palustris*, and *Littorina scabra*), barnacles (e.g., *Balanus amphitrite*), and numerous fish species, such as grey mullet (*Mugil cephalus*) and Cape stumpnose (*Rhabdosargus holubi*) (Steinke 1995). The St. Lucia system is an important nursery area for penaeid prawns, of which the dominant species is *Penaeus indicus,* which sustains offshore fisheries in the region. Numbers of fish species recorded range from 62 at the mouth of the Mngazana River, to 108 in the St. Lucia estuary, to 133 in the Kosi Lake system (Hughes and Hughes 1992). Of the 133 species found in the Kosi lakes, 39 reside permanently in the estuary, and 85 are marine species that periodically use the estuary. The remaining fish species are restricted to freshwater. Out of the five species of marine turtle that have been recorded from southern Africa, the loggerhead (*Caretta caretta,* EN) and leatherback (*Dermochelys coriacea,* EN) commonly nest on beaches adjacent to the mangrove ecosystems. Numerous snake species have been recorded from mangrove forests, such as the night adder (*Causus rhombeatus*) and occasionally the yellow-bellied sea snake (*Pelamis platurus*).

Status and Threats

Current Status

Historically there has been widespread destruction of mangroves, largely through harbor development in South Africa and timber harvesting in Mozambique. However, there has also been extensive rehabilitation of mangroves (e.g., Richards Bay). Some of the important remaining mangrove sites are found at the mouths of the Mngazana and Bashee rivers, Durban Bay, Sipingo and Mgeni near Durban, and the mouths of the Mlalazi, Mgoboseleni, and Mfolosi estuaries in Zululand (Hughes and Hughes 1992).

Mangrove species are specially protected in South Africa and may not be destroyed without an environmental impact assessment. Furthermore, the mangrove stands are located in areas that enjoy some form of legal protection. Some stands are rapidly expanding, such as Richards Bay and Mlalazi, and in other places rehabilitation is being planned, such as in Durban Harbor. In some cases, such as in Durban Bay and Richards Bay, the mangroves have been declared Natural Heritage Sites, affording 100 percent protection.

In Mozambique, mangroves also are protected, especially in proclaimed conservation areas of Inhaca and Bazaruto islands (Kalk 1995). However, many regions remain open to mangrove destruction, especially near urbanized regions such as Maputo Bay.

Types and Severity of Threats

The main threats to the continued health and existence of these mangroves are the physical destruction of their habitat, especially through existing harbors and new harbor construction, and the development of marinas and tourist facilities. Changes in hydrology, salinity, siltation, and sedimentation resulting from upstream agricultural practices and catchment mismanagement pose a further threat. In some cases, the discharge of waste products and sewage threatens not only the mangroves but also the associated biota. Future threats stem from proposals to drain swamps for agriculture and the increasing trend of burning vegetation close to mangroves, which severely damages the edges of mangrove forests.

Justification for Ecoregion Delineation

This ecoregion is found in the subtropics where the warm water of the Agulhas Current runs along the eastern coast of South Africa. These are the only subtropical mangrove stands in Africa. The mangrove stands are found only in the most suitable sites, are poor in species compared with the mangroves of Eastern Africa, and are structurally less well developed.

Ecoregion Number: **118**

Ecoregion Name: **East African Mangroves**

Bioregion: **Eastern and Southern Africa**

Biome: **Mangroves**

Political Units: **Kenya, Mozambique, Somalia, Tanzania**

Ecoregion Size: **16,100 km^2**

Biological Distinctiveness: **Locally Important**

Conservation Status: **Critical**

Conservation Assessment: **IV**

Author: *Sylvia Tognetti*

Reviewed by: *Paul Siegel, Alan Rodgers, Rudy van der Elst, Judy Oglethorpe*

Location and General Description

East African Mangroves [118] are found in Mozambique, Tanzania, Kenya, and southern Somalia. The dominant climatic influences on most of the region are the seasonal wind patterns associated with the northeast monsoon (NEM), the southeast monsoon (SEM), and the major coastal currents. Maximum tidal amplitudes range from 3.2 m in Tanzania to 3.5 m in Kenya and 5.6 m in Mozambique (Spalding et al. 1997). Estimates of the remaining mangrove area in the region range from 2,555 to 7,211 km^2 (Spalding et al. 1997), with the most extensive mangrove areas found in the Rufiji River Delta in Tanzania and the Zambezi River Delta and associated areas in Mozambique.

At the northern end of the ecoregion, in the Puntland region of Somalia, the lack of rivers and the seasonal upwellings of cold water substantially inhibit mangrove development. Along the Kenyan, Tanzanian, and Mozambique coasts there are two general categories of mangroves: those found in fringe communities along the open coastline and those found in estuaries and at river mouths. Fringe mangroves often indicate the presence of groundwater discharge sufficient to lower salinity levels, as are found at Mida Creek and the Lamu Archipelago in Kenya. Estuarine mangroves are found in areas of low tidal energy and muddy to sandy substrates, where there are distinct zonation patterns among mangrove tree species. Rivers supporting extensive mangroves are the Tana and Sabaki in Kenya, the Rufiji, Ruvuma, Umba, and Ruvu in Tanzania, and the Zambezi and Limpopo in Mozambique. Extensive mangrove forests are also found between the Quelimane and Save rivers in Mozambique; here the coastal substrates are gently sloping and muddy, suitable for mangroves. In the northern part of Mozambique the continental shelf is narrow, and there are fewer major rivers, which results in fewer mangroves and more coral and sea grass.

Although the mangroves of East Africa are less extensive than those of West Africa (e.g., Niger Delta), East Africa has a greater diversity of mangrove species. Ten mangrove species are found throughout the region, the distribution of which is determined primarily by salinity gradients, depth of water table, and soil pH and oxygen content. *Sonneratia alba* is a pioneer species found on open coasts, with *Heritiera littoralis* and *Bruguiera gymnorrhiza* often found behind it. Other species include *Avicennia marina*, *Rhizophora mucronata*, *Ceriops tagal*, *Lumnitzera racemosa*, and *Xylocarpus granatum* (Chapman 1977; CEC 1992; Diop 1993).

Outstanding or Distinctive Biodiversity Features

The East African Mangroves [118] are an exceptionally productive ecoregion. They function as nutrient traps for river catchments and provide shelter and refuge for the juveniles of many important species of fish, shrimps, crabs, and mollusks. The shallow intertidal mudflats often associated with mangroves scattered along the coast also provide important foraging sites for migratory birds. Mangrove forests also play an important role in preventing shoreline erosion and prevent sediment from flowing onto coral reefs and seagrass beds.

The mangrove forests and associated wetland ecosystems of the Rufiji and Zambezi deltas are internationally important sites for migratory wetland birds such as curlew sandpiper (*Calidris ferruginea*), little stint (*Calidris minuta*), roseate tern (*Sterna dougallii*), and Caspian tern (*Sterna caspia*) (Bregnballe et al. 1990; Semesi 1993). Many other mangrove areas along the coast also support important numbers of migratory wetland birds, particularly Mida Creek in Kenya (Bennun and Njoroge 1999). In more intact mangrove areas, upstream mangroves also provide habitat for the Nile crocodile (*Crocodilus niloticus*), hippopotamus (*Hippotamus amphibius*), blue monkey (*Cercopithecus mitis*), and spot-necked otter (*Lutra maculicollis*).

The Tana River Delta and the Rufiji Delta (and probably other mangrove stands) are important for breeding and feeding populations olive ridley (*Lepidochelys olivacea*, EN), hawksbill (*Eretmochelys imbricata*, CR), and green turtle (*Chelonia mydas*, EN). A few dugong (*Dugong dugong*, VU) also remain in some places, but most of these have been hunted out.

Status and Threats

Current Status

Apart from their ecological value, mangroves play an extremely important economic role in East Africa. They stabilize the coastline, protect coastal developments, absorb pollution, provide feeding and breeding grounds for fisheries, and supply many natural resources.

The largest continuous mangrove stands in East Africa are located at the Rufiji Delta (532 km^2) of Tanzania and the Zambezi Delta and associated coastline of Mozambique (2,800 km^2). Other important mangrove areas are the Lamu Archipelago (160 km^2 of intact mangroves); the Shebela Delta in Somalia, which is the northern extent of mangroves in East Africa; the Tana River Delta, which includes the highest concentration of *Heritiera littoralis;* and Pemba, Nacala-Mossuril, and Mswamweni-Tanga (Kemp et al. 2000; WWF-Tz 2001). In 1990, the national mangrove cover in Mozambique was estimated at 3,961 km^2, compared with 4,081 km^2 using satellite imagery from 1972 (a decline of 3.6 percent) (Saket and Matusse 1994). A further update on the current status of mangroves in Mozambique has been produced recently (Barbosa et al. 2001).

Several protected areas contain mangrove forest, such as Mafia Island Marine Park, Jozani–Chwaka Bay National Park, and Sadaani National Park in Tanzania; Kiunga Marine Reserve, Watamu Marine National Park, and Ras Tenewi Marine National Park in Kenya; and Bazaruto Marine National Park, Ilhas da Inhaca e dos Portugueses Faunal Reserve, Marromeu Game Reserve, Pomene Game Reserve, Quirimbas Marine Park, and Maputo Game Reserve in Mozambique. In Tanzania, large mangrove stands are also managed as forest reserves by a special mangrove unit of the Forestry Division, which has developed and is implementing management plans (Semesi 1993, 1998; Gaudian et al. 1995). In Mozambique, mangroves also are protected, especially in proclaimed conservation areas of Inhaca and Bazaruto islands. Important areas of mangrove remain unprotected, such as those in the Tana River Delta of Kenya (Hughes and Hughes 1992) and the Zambezi Delta of Mozambique.

Types and Severity of Threats

For thousands of years people have harvested mangroves for timber and fuel along the East African coasts. The trade continues today with larger towns along the coast of mainland East Africa and also with Zanzibar and the Middle East. This has resulted in the clearance or degradation of large areas of mangroves in the region (Kairo et al. 2001; Ngusaru et al. 2001; Taylor et al. 2003). Large numbers of mangroves trees are still being removed as timber and poles for housing construction (the wood is termite resistant), for fuelwood used domestically and for smoking fish, and to obtain tannins from mangrove bark to use as preservatives (Semesi 1993; Saket and Matusse 1994; Wass 1995). Medicines and mosquito repellent are also obtained from mangroves (Wass 1995).

In all eastern African countries intact mangrove stands are still being cleared for rice paddies, saltpans, aquaculture, and urbanization. Mangroves also receive untreated agricultural waste and sewage discharged to rivers and coastal waters, as well as industrial pollution, silt from erosion and dumping of dredgings, and pesticides in runoff. The widespread destruction of mangroves has led to increased siltation of coral reefs, changes in seagrass beds, and coastal erosion.

Justification for Ecoregion Delineation

The East African Mangroves [118] were delineated using the mangrove vegetation unit in White (1983). The mangroves of this ecoregion are biogeographically related to mangroves along the western coast of Madagascar and South Africa. This ecoregion comprises mangrove stands in the tropical latitudes in East Africa, influenced by the South Equatorial Current and East Africa Coastal Current (EACC), as well as currents of the Mozambique Channel. The northern boundary is where the EACC meets the Somali Current, and the southern boundary is where the Agulhas Current originates.

Ecoregion Number:	**119**
Ecoregion Name:	**Madagascar Mangroves**
Bioregion:	**Madagascar–Indian Ocean**
Biome:	**Mangroves**
Political Units:	**Madagascar**
Ecoregion Size:	**5,200 km^2**
Biological Distinctiveness:	**Locally Important**
Conservation Status:	**Endangered**
Conservation Assessment:	**IV**
Author:	*Sylvia Tognetti*
Reviewed by:	*Steve Goodman, Achille Raselimanana, Frank Hawkins, Sue O'Brien*

Location and General Description

The Madagascar Mangroves [119] ecoregion contains mangrove forests found primarily along the western coast. They occur in a wide range of environmental and climatic conditions, fostered by a low coastal platform, high tidal range, and constant freshwater supply from numerous rivers that also bring a high silt load, which is deposited along the coast (CEC 1992; Rasolofo 1993). Mangroves occupy a stretch of coastline approximately 1,000 km long, where they are often associated with coral reefs, which protect the mangroves from ocean swells. The mangroves, in turn, capture river sediments that threaten reefs and seagrass beds. The largest mangrove stands are found at Mahajamba Bay, Bombetoka, south Mahavavy, Ambavankarana estuary, Tsiribihina, Soalala, and Maintirano (Spalding et al. 1997). The southern part of Madagascar has fewer mangroves because, in addition to its longer dry season and lower rainfall, it is subject to intensive ocean swells and lacks the necessary alluvial sediments deposited by major river systems. Alluvial sediments also are lacking on the eastern side of the island, where mangrove stands are small and scattered.

Water temperatures vary little from north to south, and rainfall varies with climatic zones that range from 2,000 mm in the humid subequatorial north to 350 mm in the dry subtropical south. Madagascar has two seasons: a cool dry season from May through October and a warm humid season from November through April. Variation in salinity is greater along the northwest coast, where rainfall is higher, ranging from 32 percent at the end of the rainy season to 35 percent at the end of the dry season (Rasolofo 1993). On the western coast, the tidal range may reach up to 4 m during the equinoctial periods (Gaudian et al. 1995), compared with 0.75 m on the east coast. The Mangoky, Tsiribihina, and Betsiboka are major rivers that flow toward the west coast.

Although up to nine mangrove tree species have been recorded (Gaudian et al. 1995), most of the Madagascar mangrove stands contain six species in four families: *Rhizophora mucronata*, *Bruguiera gymnorrhiza*, and *Ceriops tagal* (Rhizophoraceae), *Avicennia marina* (Avicenniaceae), *Sonneratia alba* (Sonneratiaceae), and *Lumnitzera racemosa* (Combretaceae) (Rasolofo 1993). Other reported species are *Xylocarpus granatum* and *Heritiera littoralis*. The primary colonizers are *Sonneratia* and *Avicennia*. Species of *Rhizophora* and *Bruguiera* are found behind them or along creeks. *Bruguiera*, *Ceriops tagal*, and *Xylocarpus* are found in the tidally inundated areas. Other plant species found in the Madagascar mangroves are summarized in Koechlin et al. (1974).

Outstanding or Distinctive Biodiversity Features

Several of Madagascar's endemic birds are found in the coastal areas of western Madagascar, where they use mangrove and associated wetland habitats. These species are the Madagascar heron (*Ardea humbloti*, VU), Madagascar teal (*Anas bernieri*, EN), Madagascar plover (*Charadrius thoracicus*), and Madagascar fish-eagle (*Haliaeetus vociferoides*, CR) (Stattersfield et al. 1998). The majority of the Madagascar teal population spends most of its life around mangroves, nesting in mangrove tree holes. Mangroves are also important for migratory bird species, such as common ringed plover (*Charadrius hiaticula*), crab plover (*Dromas ardeola*), and African spoonbill (*Platalea alba*).

Some sea turtles, primarily green turtle (*Chelonia mydas*, EN) and hawksbill turtle (*Eretmochelys imbricata*, CR), nest along the western coast and are occasionally found in mangroves. Dugong (*Dugong dugong*, VU) and Nile crocodile (*Crocodylus niloticus*) are also found in the mangroves.

There is particularly high diversity among the fish populations, the families of which include Mugilidae, Serranidae, Carangidae, Gerridae, Hemiramphidae, and Elopidae (CEC 1992). The neighboring coral reefs that are associated with the mangroves have also been noted for extremely high fish diversity (Rasolofo 1993). There is also high diversity among mollusks and crustaceans, including crab species of the genera *Uca*, *Scylla*, *Macrophtalmus*, and *Sesarma* (CEC 1992).

Status and Threats

Current Status

Estimates of remaining mangrove areas in Madagascar range from 2,170 to 4,000 km², with 3,270 km² considered the most likely figure (Spalding et al. 1997). Of this, only about 50 km² is found on the eastern coast at eleven sites (Spalding et al. 1997). In contrast, twenty-nine mangrove areas are found on the west coast (Hughes and Hughes 1992), with some of the largest stands occurring around Mahajamba Bay (390 km²), South Mahavavy and Soalala (340 km²), Bombetoka (460 km²), and Besalampy (457 km²) (Rasolofo 1993). Some mangroves are found in the existing marine park, Reserve Mananara Biosphere Reserve (Gaudian et al. 1995). However, there are no extensive protected mangrove stands along the west coast.

Types and Severity of Threats

Mangroves are threatened by development of urban areas, overfishing, and erosion caused by tree cutting in the highlands. Some mangrove areas have been converted to rice farming and salt production. The Malagasy government encourages development of shrimp aquaculture. As a result, mangroves have being increasingly used by the private business sector. Because of low population densities and availability of wood from other sources, direct harvesting of the mangrove trees has been low except in some areas, particularly Mahajanga and Toliara (Rasolofo 1993). However, demographic trends suggest that this situation could change in the future (Spalding et al. 1997).

Justification for Ecoregion Delineation

Ecologically, the mangroves of Madagascar are very similar to those of the African mainland. However, they were separated because of their presence in a different biogeographic region. Nearly all of the mangroves in Madagascar occur along the low-lying western coast. Of these, only the larger stands have been delineated.

Glossary

adaptive radiation The evolution of a single species into many species that occupy diverse ways of life in the same geographic range.

afforestation Reforesting an area, generally with plantations of exotic tree species.

Afroalpine African vegetation lacking trees, where frosts are common, typically above 3,000 m altitude.

Afromontane African vegetation that can be with or without trees, where frosts are rare to absent, typically above 1,800 m and below 3,000 m altitude.

Afrotropical The region of Africa south of the Sahara Desert.

alpha diversity Species diversity in a single site.

altimontane See *Afroalpine*.

amphibian A member of the vertebrate class Amphibia (frogs and toads, salamanders, and caecilians).

anthropogenic Human induced.

aquatic Growing in, living in, or frequenting water.

aquifer A formation, group of formations, or part of a rock formation that contains sufficient saturated permeable material to yield significant quantities of water to wells and springs.

artesian spring A geologic formation in which water is under sufficient hydrostatic pressure to be discharged to the surface without pumping.

assemblage In conservation biology, a predictable and particular collection of species in a biogeographic unit (e.g., ecoregion or habitat).

avifauna Bird fauna.

basin See *catchment*.

beta diversity Species diversity between habitats (thus reflecting changes in species assemblages along environmental gradients).

biodiversity (Also called biotic or biological diversity.) The variety of organisms considered at all levels, from genetic variants belonging to the same species through arrays of species to arrays of genera, families, and still higher taxonomic levels; includes the variety of ecosystems, which comprise both communities of organisms in particular habitats and the physical conditions under which they live.

biodiversity conservation Five classes of biodiversity conservation priorities were determined in this study by integrating biological distinctiveness with conservation status. The five classes roughly reflect the concern with which we should view the loss of biodiversity in different ecoregions and the timing and sequence of response to the loss of biodiversity.

biogeographic unit A delineated area based on biogeographic parameters.

biogeography The study of the geographic distribution of organisms, both past and present.

biological distinctiveness Scale-dependent assessment of the biological importance of an ecoregion based on species richness, endemism, relative scarcity of ecoregion, and rarity of ecological phenomena. Biological distinctiveness classes are globally outstanding, regionally outstanding, bioregionally outstanding, and nationally important.

biome A global classification of natural communities in a particular region based on dominant or major vegetation types and climate.

bioregion A geographically related assemblage of ecoregions that share a similar biogeographic history and thus have strong affinities at higher taxonomic levels (e.g., genera, families).

bioregionally outstanding Biological distinctiveness category.

biota   The combined flora, fauna, and microorganisms of a given region.

biotic   Biological, especially referring to the characteristics of faunas, floras, and ecosystems.

bog   A poorly drained area rich in plant residues, and having characteristic flora.

bushmeat   Meat for human consumption obtained by killing wild animals.

Cape Floral Kingdom   A region of South Africa noted for its plant endemism.

catchment   All lands enclosed by a continuous hydrologic surface drainage divide and lying upslope from a specified point on a stream; or, in the case of closed-basin systems, all lands draining to a lake.

Centres of Plant Diversity   A set of priority sites for plant conservation identified by WWF and IUCN in the early 1990s.

charcoal   Cooking fuel used in African urban areas obtained by slowly burning wild wood vegetation.

chotts   Shallow, irregularly flooded depressions.

colonization   The process by which European countries took over parts of Africa and formed them into countries under their own management, later reversed as the countries obtained independence.

community   Collection of organisms of different species that co-occur in the same habitat or region and that interact through trophic and spatial relationships.

conifer   A tree or shrub in the phylum Gymnospermae whose seeds are borne in woody cones. There are 500–600 species of living conifers.

conservation biology   New discipline that treats the content of biodiversity, the natural processes that produce it, and the techniques used to sustain it in the face of human-caused environmental disturbance.

conservation status   Assessment of the status of ecological processes and of the viability of species populations in an ecoregion. The different status categories used are "extinct," "critical," "endangered," "vulnerable," "relatively stable," and "relatively intact." The current conservation status is based on an index derived from values of four landscape-level variables. The final conservation status is the current assessment modified by an analysis of threats to the ecoregion over the next 20 years.

conversion   Alteration of habitat by human activities to such an extent that it no longer supports most characteristic native species and ecological processes.

Cretaceous   A geological period from 144 to 65 million years ago.

critical   Conservation status category characterized by low probability of persistence of remaining intact habitat.

dayas   Silty depressions in North Africa.

degradation   The loss of native species and processes caused by human activities such that only certain components of the original biodiversity still persist, often including significantly altered natural communities.

desert   An area with low rainfall and sparse or absent vegetation cover.

disjunct   A species that is found in at least two widely separated geographic regions.

disturbance   Any discrete event in time that disrupts ecosystem, community, or population structure and changes resources, substrate availability, or the physical environment.

djebels   Desert mountains in North Africa.

drainage basin   See *catchment*.

ecological processes   Complex set of interactions between animals, plants, and their environment that ensure that an ecosystem's full range of biodiversity is adequately maintained. Examples include population and predator-prey dynamics, pollination and seed dispersal, nutrient cycling, migration, and dispersal.

ecoregion   A large area of land or water that contains a geographically distinct assemblage of natural communities that share a majority of their species and ecological dynamics, share similar environmental conditions, and interact ecologically in ways that are critical for their long-term persistence.

ecoregion conservation   Conservation strategies and activities whose efficacy is enhanced through close attention to larger-scale (landscape) spatial and temporal patterns of biodiversity, ecological dynamics, threats, and strong links of these issues to fundamental goals and targets of biodiversity conservation.

ecosystem   A system resulting from the integration of all living and nonliving factors of the environment.

ecosystem service   Service provided free by an ecosystem or by the environment, such as clean air, clean water, and flood amelioration.

endangered   Conservation status category characterized by medium to low probability of persistence of remaining intact habitat.

endemic   A species or race native to a particular place and found only there.

endemic bird area (EBA)   A geographic region that contains at least two bird species as strict endemics in a region smaller than 50,000 km^2.

endemism   Degree to which a geographically circumscribed area, such as an ecoregion or a country, contains species not naturally occurring elsewhere.

endorheic   Referring to a closed basin with no natural watercourses leading to the sea.

ergs   Sandy desert in North Africa.

ericaceous   Vegetation dominated by the family Ericaceae.

ericoid   See *ericaceous*.

estuarine   Associated with an estuary.

estuary A deepwater tidal habitat and its adjacent tidal wetlands, which are usually partially enclosed by land but have open, partly obstructed, or sporadic access to the open ocean and in which ocean water is at least occasionally diluted by freshwater runoff from the land.

evolutionary phenomenon In the context of WWF regional conservation assessments, evolutionary phenomena are patterns of community structure and taxonomic composition that are the result of extraordinary examples of evolutionary processes, such as pronounced adaptive radiations.

evolutionary radiation See *radiation*.

exotic species A species that is not native to an area and has been introduced intentionally or unintentionally by humans; not all exotics become successfully established.

extinct Describes a species or population (or any lineage) with no surviving individuals.

extinction The termination of any lineage of organisms, from subspecies to species and higher taxonomic categories from genera to phyla. Extinction can be local, in which one or more populations of a species or other unit vanish but others survive elsewhere, or total (global), in which all the populations vanish.

extirpated Status of a species or population that has completely vanished from a given area but continues to exist in some other location.

extirpation Process by which an individual, population, or species is destroyed.

family In the hierarchical classification of organisms, a group of species of common descent higher than the genus and lower than the order; a related group of genera.

fauna All the animals found in a particular place.

fire regime The characteristic frequency, intensity, and spatial distribution of natural fire events in a given ecoregion or habitat.

flooded grassland A grassland habitat that experiences regular inundation.

flora All the plants found in a particular place.

fragmentation Landscape-level variable measuring the degree to which remaining habitat is separated into smaller discrete blocks; also the process by which habitats are subdivided into smaller discrete blocks.

freshwater In the strictest sense, water that has less than 0.5 percent salt concentration; in this study, refers to rivers, streams, springs, and lakes.

fuelwood Woody material that is cut and used for cooking and heating.

fynbos A type of endemic-rich shrubby vegetation found in the region around Cape Town in South Africa.

genera The plural of *genus*.

genus A group of similar species with common descent, ranked below the family.

geology The study of rocks

geomorphology The study of landforms.

Global 200 A set of approximately 200 terrestrial, freshwater, and marine ecoregions around the world that support globally outstanding or representative biodiversity as identified through analyses by WWF.

globally outstanding Biological distinctiveness category for units of biodiversity whose features are equaled or surpassed in only a few other areas around the world.

Gondwanaland An ancient continent that included part of southern Africa and contained its own unique flora and fauna.

grassland A habitat type with landscapes dominated by grasses and with biodiversity characterized by species with wide distributions, in communities resilient to short-term disturbances but not to prolonged, intensive burning or grazing. In such systems larger vertebrates, birds, and invertebrates display extensive movement to track seasonal or patchy resources.

groundwater Water in the ground that is in the zone of saturation, from which wells, springs, and groundwater runoff are supplied.

guild Group of organisms, not necessarily taxonomically related, that are ecologically similar in characteristics such as diet, behavior, or microhabitat preference or with respect to their ecological role in general.

Guineo-Congolian A geographic region of Africa that covers the rainforests of Central and West Africa with their unique flora and fauna.

gymnosperm Any of a class or subdivision of woody vascular seed plants that produce naked seeds, not enclosed in an ovary. Conifers and cycads are examples of gymnosperms.

habitat An environment of a particular kind, often used to describe the environmental requirements of a certain species or community.

habitat blocks Landscape-level variable that assesses the number and extent of blocks of contiguous habitat, taking into account size requirements for populations and ecosystems to function naturally. It is measured here by a habitat-dependent and ecoregion size–dependent system.

habitat loss Landscape-level variable that refers to the percentage of the original land area of the ecoregion that has been lost (converted). It underscores the rapid loss of species and disruption of ecological processes predicted to occur in ecosystems when the total area of remaining habitat declines.

habitat type In this study, a habitat type is defined by the structure and processes associated with one or more natural communities. An ecoregion is classified under one major habitat type (biome) but may encompass multiple habitat types.

halophytic Saline.

hamadas Stony desert in North Africa.

headwater The source of a stream or river.

heathland Vegetation type associated with Afromontane and Afroalpine regions of Africa.

herbivore A plant-eating animal, especially ungulates.

herpetofauna All the species of amphibians and reptiles inhabiting a specified region.

hotspot A geographic region that contains at least 1,500 plants as strict endemics where at least 75 percent of the habitat has been destroyed.

hypersaline An area where the salt concentrations form a supersaturated solution, often crystallized into a crust of salts.

ice age A series of climatic cycles over the past 2 million years including cold (glacial) and warm (interglacial) periods.

important bird area (IBA) A small geographic region where the assemblage of birds meets globally defined criteria.

indigenous Native to an area.

inselbergs Outcrops of ancient rocks that rise above the normal landscape.

intact habitat Largely undisturbed areas characterized by the maintenance of most original ecological processes and by communities with most of their original native species still present.

integrated conservation and development project (ICDP) A conservation project that attempts to combine conservation and development elements into a single approach for conserving a site or broader landscape.

Intercontinental Convergence Zone (ITCZ) A climate system that moves up and down Africa according to season, bringing rains.

introduced species See *exotic species.*

invasive species Exotic species (i.e., alien or introduced) that rapidly establish themselves and spread through the natural communities into which they are introduced.

invertebrate Any animal lacking a backbone or bony segment that encloses the central nerve cord.

karst Applies to areas underlain by gypsum, anhydrite, rock salt, dolomite, quartzite (in tropical moist areas), and limestone, often highly eroded, complex landscapes with high levels of plant endemism.

keystone species Species that are critically important for maintaining ecological processes or the diversity of their ecosystems.

kopjes See *inselbergs.*

landform The physical shape of the land, reflecting geologic structure and processes of geomorphology that have sculptured it.

landscape An aggregate of landforms together with its biological communities.

landscape ecology Branch of ecology concerned with the relationship between landscape-level features, patterns, and processes and the conservation and maintenance of ecological processes and biodiversity in entire ecosystems.

life cycle The entire lifespan of an organism from the moment it is conceived to the time it reproduces.

lineage The evolutionary history of a group of species.

littoral Habitats found along the shore under the influence of marine waters.

Lower Guinea A geographic region including the lowland forests of the Congo Basin.

Macronesia A biogeographic region including the Canary and Cape Verde islands.

Malagasy A biogeographic region including the island of Madagascar and others in the Indian Ocean.

mangrove A type of tree that can tolerate saltwater conditions.

marine Living in saltwater.

Mediterranean climate Dry and hot summers with cool and moist winters, found at the northern and southern margins of Africa.

mesic Mesic habitats are moist, wet areas.

mesophytic Applying to plants that grow under conditions of abundant moisture.

Miocene A geologic period from 23.8 to 5.3 million years ago.

miombo An eastern and southern African vegetation type dominated by species of trees in the genera *Brachystegia, Julbernardia,* and *Isoberlinia.*

montane Habitats typically above 1,500 m altitude in Africa.

moorland Vegetation type associated with Afromontane and Afroalpine regions of Africa.

mopane A southern African vegetation type dominated by the tree *Colophospermum mopane.*

nationally important Biological distinctiveness category.

natural disturbance event Any natural event that significantly alters the structure, composition, or dynamics of a natural community. Floods, fire, and storms are examples.

natural range of variation A characteristic range of levels, intensities, and periodicities associated with disturbances, population levels, or frequency in undisturbed habitats or communities.

nonnative species See *exotic species.*

nonspecies biological values See *evolutionary phenomenon.*

oasis Area of permanent water in a desert.

obligate species A species that must have access to particular habitat type to persist.

outlier A unit of geology or vegetation that is isolated from the major part.

Palearctic   A biogeographic region that includes Africa north of the Sahara Desert.

pans   Dried lakes often encrusted by various salts.

pastoralists   People whose lifestyle involves the keeping of cattle and other livestock and regular movements across the landscape.

phylum   Primary classification of animals that share similar body plans and development patterns.

phytogeography   The study of the distributions of plants.

population   In biology, any group of organisms belonging to the same species at the same time and place.

population sink   An area where a species displays negative population growth, often because of insufficient resources and habitat or high mortality.

Precambrian   A geological age more than 543 million years ago.

predator-prey system   An assemblage of predators and prey species and the ecological interactions and conditions that permit their long-term coexistence.

protection   Landscape-level variable that assesses how well humans have conserved large blocks of intact habitat and the biodiversity they contain. It is measured here by the number of protected blocks and their sizes in a habitat-dependent and ecoregion size–dependent system.

Quaternary   A geological period from 2 million years ago to the present.

radiation   The diversification of a group of organisms into multiple species, caused by intense isolating mechanisms or opportunities to exploit diverse resources.

rarity   Seldom occurring either in absolute number of individuals or in space.

refugia   Habitats that have allowed the persistence of species or communities because of the stability of favorable environmental conditions over time.

regionally outstanding   Biological distinctiveness category.

regs   Stony desert in North Africa.

relatively intact   Conservation status category indicating the least possible disruption of ecosystem processes. Natural communities are largely intact, with species and ecosystem processes occurring in their natural ranges of variation.

relatively stable   Conservation status category between "vulnerable" and "relatively intact" in which extensive areas of intact habitat remain but in which local species declines and disruptions of ecological processes have occurred.

relictual taxa   A species or group of organisms largely characteristic of a past environment or ancient biota.

representation   The protection of the full range of bio-diversity of a given biogeographic unit in a system of protected areas that cover different habitat types.

restoration   Management of a disturbed or degraded habitat that results in recovery of its original state.

riparian   Referring to the interface between freshwater streams and lakes and the terrestrial landscape.

Saharomontane   Montane habitats within the Sahara desert above ~2,000 m altitude.

Sahel   Geographic region in the southern margin of the Sahara Desert.

savanna   A habitat dominated by grasslands but with woodland and gallery forest elements.

sclerophyll   Type of vegetation characterized by hard, leathery evergreen foliage that is specially adapted to prevent moisture loss; generally characteristic of regions with Mediterranean climates.

sclerophyllous   Relating to sclerophyll.

sebkhas   Irregularly flooded depressions.

seral   The stages a natural community experiences after a disturbance.

shrublands   Habitats dominated by various shrub species, often with many grass and forb elements.

Somali-Masai   A biogeographic region that covers the Horn of Africa from Somalia down to northern Kenya.

source pool   A habitat that provides individuals or propagules that disperse to and colonize adjacent or neighboring habitats.

species   The basic unit of biological classification, consisting of a population or series of populations of closely related and similar organisms.

species richness   A simple measure of species diversity calculated as the total number of species in a habitat or community.

spring   A natural discharge of water as leakage or overflow from an aquifer through a natural opening in the soil and rock onto the land surface or into a body of water.

stream   A general term for a body of flowing water, often used to describe a midsized tributary (as opposed to a river).

sub-Saharan Africa   Africa South of the Sahara Desert.

subspecies   Subdivision of a species. Usually defined as a population or series of populations occupying a discrete range and differing genetically from other geographic races of the same species.

subtropical   An area in which the mean annual temperature ranges from 13°C to 20°C.

taxon (pl. taxa)   A general term for any taxonomic category, such as a species, genus, family, or order.

temperate   An area in which the mean annual temperature ranges from 10°C to 13°C.

terrestrial   Living on land.

Tertiary   A geological period ranging between 66 and 2 million years ago.

threatened species   A set of species found on the IUCN Red List of species that are threatened with extinction.

transboundary   A protected area that ranges across national borders.

transfrontier   See *transboundary.*

tropical   An area where the mean annual temperature is over 20°C.

umbrella species   A species whose effective conservation will benefit many other species and habitats, often because of their large area requirements or sensitivity to disturbance.

ungulate   A member of the group of mammals with hooves, of which most are herbivorous.

Upper Guinea   A biogeographic region that covers the lowland forests of West Africa, west of Ghana.

vagile   Able to be transported or to move actively from one place to another.

vascular plant   A plant that possesses a specialized vascular system for supplying its tissues with water and nutrients from the roots and with food from the leaves.

vulnerable   Conservation status category characterized by good probability of persistence of remaining intact habitat (assuming adequate protection) but also by loss of some sensitive or exploited species.

wadis   Ephemeral but normally dry riverbeds in North African deserts.

watershed   See *catchment.*

wetlands   Lands transitional between terrestrial and aquatic systems, where the water table usually is at or near the surface or the land is covered by shallow water; areas inundated or saturated by surface water or groundwater at a frequency and duration sufficient to support a prevalence of vegetation typically adapted for life in saturated soil conditions.

wilderness area   A geographic area that contains intact habitats and few people.

wildfires   Fires in savanna woodland habitats that are either natural or caused by people.

xeric   Describes dryland or desert areas.

xerophilous   Thriving in or tolerant of xeric climates.

Zambezian   A biogeographic region of Africa that covers most of eastern and southern Africa and is typified by savanna woodland habitats.

Zanzibar-Inhambane   A biogeographic region of Africa that covers the coastal area of eastern Africa and is typified by a mosaic of lowland forests and woodlands.

zoogeography   The study of the distributions of animals.

Literature Cited

AAAS. 2000. *AAAS atlas of population and environment*. University of California Press, Berkeley, USA.

Abbot, J. F., N. B. Ananze, P. Burnham, E. de Merode, A. Dunn, E. Fuchi, E. Hakizumwami, C. Hesse, and R. Mwinyihali. 2000. *Promoting partnerships: managing wildlife resources in Central and West Africa*. Evaluating Eden. Series No. 4. International Institute of Environment and Development, London, UK.

Abbot, J. I. O., D. H. L. Thomas, A. A. Gardner, S. E. Neba, and M. W. Khen. 2001. Understanding the links between conservation and development in the Bamenda Highlands, Cameroon. *World Development* 29:1115–1136.

Abd El-Ghani, M. M. 1998. Environmental correlates of species distribution in arid desert ecosystems of eastern Egypt. *Journal of Arid Environments* 38:297–313.

Abell, R., D. M. Olson, E. Dinerstein, P. Hurley, S. Walters, C. Loucks, T. Allnutt, and W. W. Wettengel. 2000. *A conservation assessment of the freshwater ecoregions of North America*. Island Press, Washington, DC, USA.

Achard, F., H. D. Eva, H. J. Stibig, P. Mayaux, J. Gallego, T. Richards, and J. P. Malingreau. 2002. Determination of deforestation rates in the world's humid tropical forests. *Science* 297:999–1002.

Acocks, J. P. H. 1953. Veld types of South Africa. *Memoirs of the Botanical Society of South Africa* 28:1–192.

Acocks, J. P. H. 1988. Veld types of South Africa. *Memoirs of the Botanical Society of South Africa* 57:1–146.

Acreman, M. C. and G. E. Hollis, editors. 1996. Water management and wetlands in sub-Saharan Africa. IUCN, Gland, Switzerland.

Adam, J. G. 1958. *Elements pour l'étude de la vegetation des hauts plateaux du Fouta Djalon. Part 1*. Bureau des sols, Direction des Services Economiques, Gouvernement General de l'Afrique Occidentale Français, Dakar, Senegal.

Adams, W. M., A. S. Goudie, and A. R. Orme, editors. 1996. *The physical geography of Africa*. Oxford University Press, Oxford, UK.

Adams, W. M. and D. Hulme. 2001. If community conservation is the answer in Africa, what is the question? *Oryx* 35:193–200.

Adegbehin, J. O. 1993. Mangroves in Nigeria. Pages 135–153 in E. D. Diop, editor. *Conservation and sustainable utilization of mangrove forests in Latin America and Africa regions. Part II: Africa*. Mangrove Ecosystems Technical Reports, Volume 3. International Society for Mangrove Ecosystems and Coastal Marine Project of UNESCO, Okinawa, Japan.

Adham, K. G., I. F. Hassan, N. Taha, and T. Amin. 1999. Impact of hazardous exposure to metals in the Nile and Delta lakes on the catfish, *Clarias lazera. Environmental Monitoring and Assessment* 54:107–124.

Adjanahoun, E. J., L. Aké Assi, A. Ahmed, J. Eymé, S. Guinko, A. Kayonga, A. Keita, and M. Lebras. 1982. *Contributions aux études ethnobotaniques et floristiques aux Comores*. Agence de Coopération et Technique, Paris, France.

Adler, G. H. 1994. Avifaunal diversity and endemism on tropical Indian Ocean islands. *Journal of Biogeography* 21:85–95.

AECCG. 1991. *The African elephant conservation review*. African Elephant Conservation Coordinating Group, United Nations Environment Programme, IUCN, World Wildlife Fund, Oxford, UK.

Agyepong, G. T. K., P. W. K. Yankson, and Y. Ntiamoa-Baidu. 1990. Coastal zone indicative management plan. E.P.C., Accra, Ghana.

Ajao, E. A. 1993. Mangrove ecosystems in Nigeria. Pages 155–167 in E. D. Diop, editor. *Conservation and sustainable utilization of mangrove forests in Latin America and Africa regions. Part II: Africa*. Mangrove Ecosystems Technical Reports, Volume 3. International Society for Mangrove Ecosystems and Coastal Marine Project of UNESCO, Okinawa, Japan.

Akerele, O., V. Heywood, and H. Synge. 1991. *The conservation of medicinal plants*. Cambridge University Press, Cambridge, UK.

Albertson, A. 1998. Northern Botswana veterinary fences: critical ecological impacts. Retrieved 2000 from the World Wide

433

Web: http://www.stud.ntnu.no/~skjetnep/owls/fences/index
.html.

Alers, M. P. T., A. Blom, K. C. Sikubwabo, T. Masunda, and R. F. W.
Barnes. 1992. A preliminary assessment of the status of the
forest elephant in Zaire. *African Journal of Ecology* 30:279–
291.

Alexander, K. A. and M. J. G. Appel. 1994. African wild dogs (*Lycaon pictus*) endangered by a canine distemper epizootic
among domestic dogs near the Masai Mara National Reserve,
Kenya. *Journal of Wildlife Diseases* 30:481–485.

Allan, D. G. 1992. Distribution, relative abundance and habitat of the blue crane in the Karoo and the southwestern cape.
Pages 29–46 in D. J. Porter, editor. *Proceedings of the first
Southern African Crane Conference*. South African Crane Foundation, Durban, South Africa.

Allen, J. A. 1878. The geographic distribution of mammals. Bulletin, U.S. Geological and Geographical Survey of the Territories 4:39–343.

Allport, G. 1991. The status and conservation of threatened
birds in the Upper Guinea Forest. *Bird Conservation International* 1:53–74.

Allport, G., M. Ausden, P. V. Hayman, P. Robertson, and P. Wood.
1989. The conservation of the birds of Gola Forest, Sierra
Leone. Report of the UEA-ICBP Gola Forest Project (Bird Survey), October 1989 to February 1989. ICBP, Cambridge, UK.

Almond, S. 2000. Itigi thicket monitoring using Landsat TM imagery. M.S. dissertation. University College, London, UK.

Alpers, E. A. 1975. *Ivory and slaves in east Central Africa: changing patterns of international trade in the later nineteenth century.*
Heinemann, London, UK.

Alpert, P. 1993. Conserving biodiversity in Cameroon. *Ambio*
22:44–49.

Alpert, P. 1996. Integrated conservation and development projects. *BioScience* 46:845–855.

Altenburg, W. 1987. Waterfowl in West African coastal wetlands:
a summary of current knowledge. WIWO report 5. Zeist, The
Netherlands.

Altenburg, W. and J. van der Kamp. 1991. *Ornithological importance of coastal wetlands in Guinea.* ICBP and WIWO, Cambridge, UK.

Amiet, J.-L. 1975. Ecologie et distribution des amphibiens
anoures de la région de Nkongsamba (Cameroun). *Annales
Faculté Sciences Yaoundé* 20:33–107.

Amiet, J.-L. and F. Dowsett-Lemaire. 2000. Un nouveau Leptodactylodon de la dorsale camerounaise (Amphibia, Anura).
Alytes 18:1–14.

Amman, K. and J. Pierce. 1995. *Slaughter of the apes: how the tropical timber industry is devouring Africa's great apes.* World Society for Protection of Animals, New York, USA.

Anderson, D. and R. Grove. 1987. *Conservation in Africa: peoples,
policies and practice.* Cambridge University Press, Cambridge,
UK.

Anderson, J., P. Dutton, P. Goodman, and B. Soto. 1990. *Evaluation of the wildlife resource in the Marromeu complex with recommendations for its future use.* LOMACO, Maputo, Mozambique.

Anderson, M. D. 2000. Raptor conservation in the Northern
Cape Province, South Africa. *Ostrich* 71:25–32.

Ando, A., J. Camm, S. Polasky, and A. Solow. 1998. Species distributions, land values, and efficient conservation. *Science*
279:2126–2128.

Anonymous. 1954. Geology, water supply and minerals. *The
Nigeria handbook.* The Government Printer, Lagos, Nigeria.

Ansell, W. F. H. 1960. *Mammals of northern Rhodesia.* Government Printer, Lusaka, Zambia.

Ansell, W. F. H. and R. J. Dowsett. 1988. *Mammals of Malawi.*
Trendrine Press, Cornwall, UK.

Anstey, S. 1991. *Wildlife utilization in Liberia.* World Wildlife
Fund and Liberian Forestry Development Authority, UK.

Appleton, M. 1997. Conservation in a conflict area. *Oryx*
31:153–154.

Appolinaire, Z. 1993. Mangroves of Cameroun. Pages 193–209
in E. D. Diop, editor. *Conservation and sustainable utilization
of mangrove forests in Latin America and Africa regions. Part II:
Africa.* Mangrove Ecosystems Technical Reports, Volume 3.
International Society for Mangrove Ecosystems and Coastal
Marine Project of UNESCO, Okinawa, Japan.

Arbonnier, M. 2001. *Arbres, arbustes et lianes des zones sèches
d'Afrique de l'Ouest.* CIRAD, MNHN and IUCN, Montpellier
and Paris, France.

Arinaitwe, H., D. Pomeroy, and H. Tushabe. 2000. *The state of
Uganda's biodiversity.* MUIENR, Kampala, Uganda.

Arlidge, E. Z. and Y. Wong you Cheong. 1975. *Notes on the land
resources and agricultural suitability map of Mauritius.* Mauritius Sugar Industry Research Institute & FAO (UN), Reduit,
Mauritius.

Armstrong, A. J. and W. R. Siegfried. 1990. Selective use of
heuweltjie earthmounds by sheep in the Karoo. *South African
Journal of Ecology* 1:77–80.

Armstrong, S. 1990. Fog, wind and heat: life in the Namib
Desert. *New Scientist* 127:46–50.

Arnold, J. E. M. 1998. *Managing forests as common property.* Community Forestry Paper 13. FAO, Rome, Italy.

Atkinson, P. W., J. S. Dutton, N. Peet, and V. A. S. Sequeira, editors. 1993. *A study of the birds, small mammals, turtles and
medicinal plants of São Tomé, with notes on Príncipe.* BirdLife
International Study Report 56. Cambridge, UK.

Atkinson, P. W., N. Peet, and J. Alexander. 1991. The status and
conservation of the endemic bird species of São Tomé and
Príncipe, West Africa. *Bird Conservation International* 1:255–
282.

Atkinson, P., P. A. Turner, S. Pockwell, G. Broad, A. P. Karoma, D.
Annaly, and S. Rowe. 1992. *Land use and conservation of the
Mount Loma Reserve, Sierra Leone.* University of East Anglia/
BirdLife International, Cambridge, UK.

Attwell, C. A. M. and F. P. D. Cotterill. 2000. Postmodernism
and African conservation science. *Biodiversity and Conservation* 9:559–577.

Aubreville, A. 1937. Les forêts du Dahomey et du Togo. *Bulletin
Comite d'Etude Historique et Scientifique Occidentale Française*
20:1–112.

Audru, J., G. Cesar, G. Forgiarini, and J. Lebrun. 1987. *La végétation et les potentialités pastorales de la République de Djibouti.*
Institut d'Elevage et de Médecine Vétérinaire des Pays Tropicaux, Maisons-Alforts, France.

Auzel, P. and D. S. Wilkie. 2000. Wildlife use in northern Congo:

hunting in a commercial logging concession. Pages 413–426 in J. G. Robinson and E. L. Bennett, editors. *Hunting for sustainability in tropical forests.* Columbia University Press, New York, USA.

Aveling, C. and R. Aveling. 1989. Gorilla conservation in Zaire. *Oryx* 23:64–70.

Aveling, C. and A. H. Harcourt. 1984. A census of the Virunga gorillas. *Oryx* 18:8–13.

AWF. 2003. Program Impact Assessment (PIMA): internal document. AWF, Washington, DC, USA.

AWF. 2003. Heartland Conservation Process (HCP): internal document. AWF, Washington, DC, USA.

Axelrod, D. I. and P. H. Raven. 1978. Late Cretaceous and Tertiary vegetation history of Africa. Pages 77–130 in M. J. A. Werger, editor. *Biogeography and ecology of southern Africa.* Dr. W. Junk Publishers, The Hague, The Netherlands.

Bacallado, J. J., M. Báez, A. Brito, T. Cruz, F. Domínguez, E. Moreno, and J. M. Pérez. 1984. *Fauna (marina y terrestre) del Archipiélago Canario.* Ed. Edirca, Las Palmas, Canary Islands, Spain.

Baha el Din, S. M. 1997. A new species of *Tarentola* (Squamata: Gekkonidae) from the western desert of Egypt. *African Journal of Herpetology* 46:30–35.

Baha el Din, S. M. 1999. *Directory of important bird areas in Egypt.* BirdLife International, Palm Press, Cairo, Egypt.

Baha el Din, S. M. 2001. Egypt. Pages 241–264 in L. D. C. Fishpool and M. I. Evans, editor. *Important bird areas in Africa and associated islands: priority sites for conservation.* Pisces Publications and BirdLife International (BirdLife Conservation Series No. 11), Newbury and Cambridge, UK.

Bailey, R. G. 1998. *Ecoregions: the ecosystem geography of the oceans and continents.* Springer Verlag, New York, USA.

Bakarr, M. I., B. Bailey, D. Byler, R. Ham, S. Oliverieri, and M. Omland. 2001a. *From the forest to the sea: biodiversity connections from Guinea to Togo.* Conservation International, Washington, DC, USA.

Bakarr, M. I., B. Bailey, M. Omland, N. Myers, L. Hannah, C. G. Mittermeier, and R. A. Mittermeier. 1999. Guinean forests. Pages 239–253 in R. A. Mittermeier, N. Myers, P. R. Gil, and C. G. Mittermeier, editors. *Hotspots: Earth's biologically richest and most endangered terrestrial ecoregions.* CEMEX and Conservation International, Mexico City, Mexico and Washington, DC, USA.

Bakarr, M. I., G. A. B. da Fonseca, R. Mittermeier, A. B. Rylands, and K. W. Painemilla. 2001b. *Hunting and bushmeat utilization in the African rain forest: perspectives toward a blueprint for conservation action.* Center for Applied Biodiversity Science, Conservation International, Washington, DC, USA.

Baker, N. E. 1996. Tanzania waterbird count. *The first coordinated count on the major wetlands of Tanzania: January 1995.* Wildlife Conservation Society of Tanzania, Dar es Salaam, Tanzania.

Baker, N. E. and E. M. Baker. 2001. Tanzania. Pages 897–946 in L. D. C. Fishpool and M. I. Evans, editors. *Important bird areas in Africa and associated islands: priority sites for conservation.* Pisces Publications and BirdLife International (BirdLife Conservation Series No. 11), Newbury and Cambridge, UK.

Baker, N. E. and E. M. Baker. 2002. *Important bird areas of Tanzania: a first inventory.* The Wildlife Conservation Society of Tanzania and Royal Society for the Protection of Birds, Dar es Salaam, Tanzania and Sandy, UK.

Balfour, D. and S. Balfour. 1992. *Etosha.* Struik Publishers, Cape Town, South Africa.

Balinsky, B. I. 1962. Patterns of animal distribution on the African continent. *Annals of the Cape Provincial Museum* 2:299–310.

Balmford, A. 2002. Selecting sites for conservation. Pages 74–104 in K. Norris and D. J. Pain, editors. *Conserving bird biodiversity: general principles and their application.* Cambridge University Press, Cambridge, UK.

Balmford, A., A. Bruner, P. Cooper, R. Costanza, S. Farber, R. E. Green, M. Jenkins, P. Jefferiss, V. Jessamy, J. Madden, K. Munro, N. Myers, S. Naeem, J. Paavola, M. Rayment, S. Rosendo, J. Roughgarden, K. Trumper, R. Turner, and R. Kerry Turner. 2002. Economic reasons for conserving wild nature. *Science* 297:950–953.

Balmford, A., K. J. Gaston, S. Blyth, A. James, and V. Kapos. 2003. Global variation in conservation costs, conservation benefits, and unmet conservation needs. Proceedings of the National Academy of Sciences USA 100:1046–1050.

Balmford, A., K. J. Gaston, A. S. L. Rodrigues, and A. James. 2000. Integrating costs of conservation into international priority setting. *Conservation Biology* 14:597–605.

Balmford, A., N. Leader-Williams, and M. Green. 1992. The protected area system. Pages 69–80 in J. A. Sayer, C. S. Harcourt, and N. M. Collins, editors. *The conservation atlas of tropical forests: Africa.* IUCN, Gland, Switzerland and Cambridge, UK.

Balmford, A. and A. Long. 1994. Avian endemism and forest loss. *Nature* 372:623–624.

Balmford, A., J. L. Moore, T. Brooks, N. D. Burgess, L. A. Hansen, P. Williams, and C. Rahbek. 2001a. Conservation conflicts across Africa. *Science* 291:2616–2619.

Balmford, A., J. L. Moore, T. Brooks, N. D. Burgess, L. A. Hansen, J. C. Lovett, S. Tokumine, P. Williams, and C. Rahbek. 2001b. People and biodiversity in Africa. *Science* 293:1591–1592.

Barbéro, M. and P. Quézel. 1975. Les forêts de sapin du pourtour méditerranéen. *Anales del Instituto Botanico* A. J. Cavanilles 32:1–38.

Barbier, E. B., M. Acreman, and D. Knowler. 1996. *Economic valuation of wetlands: a guide for policy makers and planners.* Ramsar Convention Bureau, IUCN, Gland, Switzerland.

Barbosa, F. M. A., C. C. Cuambe, and S. O. Bandeira. 2001. Status and distribution of mangroves in Mozambique. *South African Journal of Botany* 67:393–398.

Barbosa, L. A. G. 1970. *Carta fitogeografica de Angola.* IICA, Luanda, Angola.

Barbour, K. M., J. S. Oguntoyinbo, J. O. C. Onyemelukwe, and J. C. Nwafor. 1982. *Nigeria in maps.* Hodder and Stoughton, London, UK.

Barnard, P., editor. 1998. *Biological diversity in Namibia: a country study.* Namibian National Biodiversity Task Force, Directorate of Environmental Affairs, Windhoek, Namibia.

Barnard, P., C. J. Brown, A. M. Jarvis, A. Robertson, and L. Van Rooyen. 1998. Extending the Namibian protected area network to safeguard hotspots of endemism and diversity. *Biodiversity and Conservation* 7:531–547.

Barnes, K. N., editor. 1998. *The important bird areas of southern Africa*. BirdLife South Africa, Johannesburg, South Africa.

Barnes, K. N. 2000. *Eskom Red Data Book of birds of South Africa, Lesotho and Swaziland*. BirdLife South Africa, Johannesburg, South Africa.

Barnes, K. N. 2001. Lesotho. Pages 465–471 in L. D. C. Fishpool and M. I. Evans, editors. *Important bird areas in Africa and associated islands: priority sites for conservation*. Pisces Publications and BirdLife International (BirdLife Conservation Series No. 11), Newbury and Cambridge, UK.

Barnes, K. N., D. J. Johnson, M. D. Anderson, and P. B. Taylor. 2001. South Africa. Pages 793–876 in L. D. C. Fishpool and M. I. Evans, editors. *Important bird areas in Africa and associated islands: priority sites for conservation*. Pisces Publications and BirdLife International (BirdLife Conservation Series No. 11), Newbury and Cambridge, UK.

Barnes, R. F. W. 1987. A review of the status of elephants in the rain forests of central Africa. Pages 41–46 in A. Burrill and I. Douglas-Hamilton, editors. *African Elephant Database Project*. Global Environment Monitoring System, United Nations Environment Programme, Nairobi.

Barnes, R. F. W. 1993. Indirect methods for counting elephants in forests. *Pachyderm* 16:24–30.

Barnes, R. F. W. 1999. Is there a future for elephants in West Africa? *Mammal Review* 29:175–199.

Barnes, R. F. W., A. Blom, and M. P. T. Alers. 1995. A review of the status of forest elephants *Loxodonta africana* in Central Africa. *Biological Conservation* 71:125–132.

Barnes, R. F. W., G. C. Craig, H. T. Dublin, G. Overton, W. Simons, and C. R. Thouless. 1999. *African Elephant Database 1998*. IUCN/Species Survival Commission African Elephant Specialist Group. IUCN, Gland, Switzerland and Cambridge, UK.

Barnett, A., M. Prangley, P. V. Hayman, D. Diawara, and J. Koman. 1994. A preliminary survey of Kounounkan Forest, Guinea, West Africa. *Oryx* 28:269–275.

Barnett, R. 2000. *Food for thought: the utilisation of wild meat in eastern and southern Africa*. TRAFFIC, Eastern and Southern Africa, Nairobi, Kenya.

Barré, N. 1988. Une avifaune menacée: les oiseaux de la Réunion. Pages 167–196 in J. C. Thibault and I. Guyot, editors. *Livre Rouge oiseaux menacés des regions françaises d'outre-mer*. International Council for Bird Preservation, Monograph No. 5, France.

Barrett, C. B. and P. Arcese. 1995. Are integrated conservation-development projects (ICDPs) sustainable? On the conservation of large mammals in sub-Saharan Africa. *World Development* 23:1073–1084.

Barrow, E., H. Gichohi, and M. Infield. 2000. *Rhetoric or reality? A review of community conservation policy and practice in East Africa*. Evaluating Eden. Series 5. International Institute of Environment and Development, London, UK.

Bartholomew, G. A., F. N. White, and T. R. Howell. 1976. The thermal significance of the nest of the sociable weaver *Philetairus socius*: summer observations. *Ibis* 118:402–410.

Bauer, A. M., W. R. Branch, and D. A. Good. 1996. A new species of rock-dwelling *Phyllodactylus* (Squamata: Gekkonidae) from the Richtersveld, South Africa. *Occasional Papers of the Museum of Natural Science Louisiana State University* 71:1–13.

Bauer, A. M., D. A. Good, and W. R. Branch. 1997. The taxonomy of the southern African leaf-toed geckos (Squamata: Gekkonidae), with a review of Old World *Phyllodactylus* and the description of five new genera. Proceedings of the California Academy of Science 49:447–497.

Beadle, L. C. 1981. *The inland waters of tropical Africa*. Longman Group Limited, London, UK.

Beard, J. S. 1992. *The proteas of tropical Africa*. Kangaroo Press, Hong Kong, China.

Bearder, S. K. 1999. Physical and social diversity among nocturnal primates: a new view based on long term research. *Primates* 40:267–282.

Bedward, M., R. L. Pressey, and D. A. Keith. 1992. A new approach for selecting fully representative reserve networks: addressing efficiency, reserve design and land suitability with an iterative analysis. *Biological Conservation* 62:115–125.

Beentje, H. 1988. An ecological and floristic study of the forests of the Taita Hills, Kenya. *Utafiti* 1:23–66.

Beilfuss, R. D. and D. G. Allan. 1996. Wattled crane and wetland surveys in the Great Zambezi Delta, Mozambique. Pages 345–353 in R. D. Beilfuss, W. R. Tarboton, and N. N. Gichuki, editors. Proceedings of 1993 African Crane and Wetland Training Workshop. International Crane Foundation, Baraboo, WI, USA.

Belcastro, C. 1986. A preliminary list of Hesperiidae (Lepidoptera) from Sierra Leone with description of a new species. Ricerche Biologiche in Sierra Leone (Parte II). *Accademia Nazionale dei Lincei* 260:165–194.

Bell, R. H. V. 1971. A grazing ecosystem in the Serengeti. *Scientific American* 224:86–93.

Bellefontaine, R., A. Gaston, and Y. Petrucci. 1997. *Aménagement des forêts naturelles des zones tropicales sèches*. Cahier Conservation Food and Agricultural Organization 32, Food and Agricultural Organization, Rome, Italy.

Belsky, A. J. 1986. Population and community processes in a mosaic grassland in the Serengeti, Tanzania. *Journal of Ecology* 74:841–856.

Belsky, J. A. and R. G. Amundson. 1992. Effects of trees on understory vegetation and soils at forest-savanna boundaries. Pages 353–366 in P. A. Furley, J. Proctor, and J. A. Ratter, editors. *Nature and dynamics of forest-savanna boundaries*. Chapman and Hall, London.

Benabid, A. 1985. Les écosystèmes forestiers, préforestiers et présteppiques du Maroc: diversité, répartition biogéographiques et problèmes posés par leur aménagement. *Forêt Méditerranéenne* 7:53–64.

Benabid, A. and M. Fennane. 1994. Connaissances sur la végétation du Maroc: phytogégraphie, phytosociologie et séries de végétation. *Lazaroa* 14:21–97.

Benhouhou, S. S., T. C. D. Dargie, and O. L. Gilbert. 2001. Vegetation associations in the Great Western Erg and the Saoura Valley, Algeria. *Phytocoenologia* 31:311–324.

Benjaminsen, T. A. 1993. Fuelwood and desertification: Sahel orthodoxies discussed on the basis of field data from the Gourma region in Mali. *Geoforum* 24:397–409.

Bennet, A. F. 2003. *Linkages in the landscape: the role of corridors and connectivity in wildlife conservation.* IUCN: The World Conservation Union, Gland, Switzerland.

Bennet, K. D. 1990. Milankovitch cycles and their effects on species in ecological and evolutionary time. *Paleobiology* 16:11–21.

Bennett, E. L. and J. G. Robinson. 2001. *Hunting of wildlife in tropical forests.* The World Bank, Washington, DC, USA.

Bennun, L. A., R. A. Aman, and S. A. Crafter, editors. 1995. *Conservation of biodiversity in Africa: local initiatives and institutional roles.* Center for Biodiversity, National Museums of Kenya, Nairobi.

Bennun, L. and P. Njoroge. 1999. *Important bird areas in Kenya.* Nature Kenya/BirdLife International, Nairobi, Kenya.

Bennun, L. and P. Njoroge. 2001. Kenya. Pages 411–464 in L. D. C. Fishpool and M. I. Evans, editors. *Important bird areas in Africa and associated islands: priority sites for conservation.* Pisces Publications and BirdLife International (BirdLife Conservation Series No. 11), Newbury and Cambridge, UK.

Ben-Shahar, R. 1993. Patterns of elephant damage to vegetation in northern Botswana. *Biological Conservation* 65:249–256.

Benson, C. W. 1967. The birds of Aldabra and their status. *Atoll Research Bulletin* 118:63–111.

Benson, C. W. and M. P. S. Irwin. 1965. The birds of *Cryptosepalum* forests, Zambia. *Arnoldia (Rhodesia)* 28:1–12.

Benson, C. W. and M. P. S. Irwin. 1966. The *Brachystegia* avifauna. *Ostrich,* Supplement 6:297–321.

Benson, C. W. and M. J. Penny. 1971. The land birds of Aldabra. *Philosophical Transactions of Royal Society of London B* 260:529–548.

Beresford, P. and J. Cracraft. 1999. *Speciation in African forest robins (Stiphronis): species limits, phylogenetic relationships, and molecular biogeography.* No. 3270. American Museum of Natural History, New York, USA.

Berger, J. and C. Cunningham. 1994. Active intervention and conservation: Africa's pachyderm problem. *Science* 263:1241–1242.

Berkes, F. 1999. *Sacred ecology: traditional ecological knowledge and resource management systems.* Taylor and Francis, Philadelphia, USA.

Berry, C. 1991. *Trees and shrubs of the Etosha National Park.* Directorate of Nature Conservation, Windhoek, Namibia.

Berry, H. H. 1972. Flamingo breeding on the Etosha Pan South-West Africa during 1971. *Madoqua* 1:5–31.

Berry, H. H. and C. U. Berry. 1975. A check list and notes on the birds of Sandvis, South West Africa. *Madoqua* 9:5–18.

Berry, H. H., H. P. Shark, and A. S. Van Vuuren. 1973. White pelicans *Pelecanus onocrotalus* breeding on the Etosha Pan South-West Africa during 1971. *Madoqua* 1:17–31.

Beuning, K. R. M., M. R. Talbot, and K. Kelts. 1997. A revised 30,000-year paleoclimatic and paleohydrologic history of Lake Albert, East Africa. *Palaeogeography, Palaeoclimatology, Palaeoecology* 136:259–279.

Bikie, H., J. G. Collomb, L. Djomo, S. Minnemeyer, R. Ngouffo, and S. Nguiffo. 2000. *An overview of logging in Cameroon: A Global Forest Watch Cameroon report.* World Resources Institute, Washington, DC, USA.

Bingham, M. 1995. Zambia's vegetation. Retrieved 2001 from the World Wide Web: http://www.africa-insites.com/zambia/info/General/vegetati.htm.

Bingham, M., J. Golding, B. Luwiika, C. Nguvulu, P. Smith, and G. Sichima. 2000. Red Data List: spotlight on Zambia. *SABONET News* 5:93–95.

Bird, M. I. and J. A. Cali. 1998. A million year old record of fire in sub-Saharan Africa. *Nature* 394:767–769.

BirdLife International. 2000. *Threatened birds of the world.* Lynx Edicions and BirdLife International, Barcelona, Spain and Cambridge, UK.

Bisby, F. A. 1995. Characterisation of biodiversity. Pages 21–106 in UNEP, editor. *Global biodiversity assessment.* UNEP, Nairobi, Kenya.

Blake, S., E. Rogers, J. M. Fay, M. Ngangoue, and G. Ebeke. 1995. Swamp gorillas in northern Congo. *African Journal of Ecology* 33:285–290.

Blasco, F., C. Caratini, A. Fredoux, P. Giresse, G. Rouguedet, C. Tissot, and H. Weiss. 1980. *Les rivages tropicaux, mangroves d'Afrique et d'Asie, travaux et documents de geographie tropicale* CEGT, CNRS, No. 39, Bordeaux.

Blom, A., M. P. T. Alers, and R. F. W. Barnes. 1990. Gabon. Pages 113–120 in R. East, editor. *Antelopes: global survey and regional action plans part 3.* West and Central Africa: IUCN/Species Survival Commission Antelope Specialist Group. IUCN, Gland, Switzerland.

Blom, A., M. P. T. Alers, A. T. C. Feistner, R. F. W. Barnes, and K. L. Jensen. 1992. Notes on the current status and distribution of primates in Gabon. *Oryx* 26:223–234.

Blom, A., A. Almasi, I. M. A. Heitkonig, J.-B. Kpanou, and H. H. T. Prins. 2001. A survey of the apes in the Dzanga-Ndoki National Park, Central African Republic: a comparison between the census and survey methods of estimating the gorilla (*Gorilla gorilla gorilla*) and chimpanzee (*Pan troglodytes*) nest group density. *African Journal of Ecology* 39:98–105.

Blondel, J. and J. Aronson. 1999. *Biology and wildlife of the Mediterranean region.* Oxford University Press, Oxford, UK.

Boahene, K. 1998. The challenge of deforestation in tropical Africa: reflections on its principal causes, consequences and solutions. *Land Degradation and Development* 9:247–258.

Bober, S. O., M. Herremans, M. Louette, J. C. Kerbis Peterhans, and J. M. Bates. 2001. Geographical and altitudinal distribution of endemic birds in the Albertine Rift. *Ostrich Supplement* 15:189–196.

Bocian, C. 1998. *Preliminary observations on the status of primates in the Etiema community forest.* Report for A. G. Leventis and the Nigerian Conservation Foundation, Lagos, Nigeria.

Boffa, J.-M. 1999. *Agroforestry parklands in sub-Saharan Africa.* Food and Agricultural Organization Conservation Guide 34. Food and Agricultural Organization, Rome, Italy.

Böhme, W. and A. Schmitz. 1996. A new lygosomine skink (Lacertilia: Scincidae: *Panaspis*) from Cameroon. *Revue Suisse de Zoologie* 103:767–774.

Boitani, L., F. Corsi, A. De Biase, I. D. Carranza, M. Ravagli, G. Reggiani, I. Sinbaldi, and P. Trapanese. 1999. *A databank for the conservation and management of the African mammals.* Istituto Ecologia Applicata, Rome, Italy.

Bonnefille, R., J. C. Roeland, and J. Guiot. 1990. Temperature and rainfall estimates for the past 40,000 years in equatorial Africa. *Nature* 346:347–349.

Booth, A. H. 1958. The Niger, the Volta and the Dahomey Gap as geographic barriers. *Evolution* 12:48–62.

Borkin, L. J. 1999. Distribution of amphibians in North Africa, Europe, western Asia, and the former Soviet Union. Pages 329–420 in W. E. Duellman, editor. *Patterns of distribution of amphibians.* Johns Hopkins University Press, Baltimore, MD, USA.

Borrini-Feyerabend, G. and D. Buchan. 1997. *Beyond fences. Seeking social sustainability in conservation. Volume 1: A process companion edition.* IUCN, Gland, Switzerland.

Borrow, N. and R. Demey. 2001. *A guide to the birds of western Africa.* Princeton University Press, Princeton, NJ, USA.

Bourn, D., C. Gibson, D. Augeri, C. J. Wilson, J. Church, and S. Hay. 1999. The rise and fall of the Aldabran giant tortoise population. Proceedings of the Royal Society Biological Sciences B266:1091–1100.

Bourquin, O. 1991. A new genus and species of snake from the Natal Drakensberg, South Africa. *Annals of the Transvaal Museum* 35:199–203.

Bourquin, O. and A. J. L. Lambiris. 1996. A new species of *Acontias cuvier* (Sauria: Scincidae) from southeastern KwaZulu-Natal, South Africa. *Annals of the Transvaal Museum* 36:223–227.

Bowden, C. G. R. and S. M. Andrews. 1994. Mount Kupe and its birds. *Bulletin African Bird Club* 1:13–16.

Bowen-Jones, E. and S. Pendry. 1999. The threat to primates and other mammals from the bushmeat trade in Africa, and how this threat could be diminished. *Oryx* 33:233–246.

Brady, N. C. and R. R. Weil. 1999. *The nature and property of soils,* 12th edition. Prentice Hall, Englewood Cliffs, NJ.

Bramwell, D. and Z. Bramwell. 1983. *Flores silvestres de las Islas Canarias,* 2nd edition. Ediciones Rueda, Madrid, Spain.

Branch, B. 1998. *Field guide to the snakes and other reptiles of southern Africa.* Struik Publishers, Capetown, South Africa.

Branch, W. R. 1988. *South African Red Data Book reptiles and amphibians.* South African National Programmes Report No 151. CSIR, Pretoria, South Africa.

Branch, W. R. 1997. A new adder (Bitis; Viperidae) from the Western Cape Province, South Africa. *South African Journal of Zoology* 32:37–42.

Branch, W. R., A. M. Bauer, and D. A. Good. 1995. Species limits in the *Phyllodactylus lineatus* complex (Reptilia: Gekkonidae), with the elevation of two to specific status and the description of two new species. *Journal of Herpetological Association of Africa* 44:33–54.

Branch, W. R., A. M. Bauer, and D. A. Good. 1996. A review of the Namaqua gecko, *Pachydactylus namaquensis* (Reptilia: Gekkonidae) from southern Africa, with the description of two new species. *South African Journal of Zoology* 31:53–69.

Branch, W. R. and M. J. Whiting. 1997. A new *Platysaurus* (Squamata: Cordylidae) from the northern Cape Province, South Africa. *African Journal of Herpetology* 46:124–136.

Brandt, C. J. and J. B. Thornes, editors. 1996. *Mediterranean desertification and land use.* Wiley, Chichester, UK.

Bredenkamp, G. J., A. F. Joubert, and H. Bezuidenhout. 1989. A reconnaissance survey of the vegetation of the plains in the Potchefstroom-Fochville-Parys area [South Africa]. *Suid-Afrikaanse Tydskrif Vir Plantkunde* 55:199–206.

Bregnballe, T., K. Halberg, L. N. Hansen, I. K. Petersen, and O. Thorup. 1990. *Ornithological winter surveys on the coast of Tanzania 1988–89.* ICBP Study Report 43. ICBP, Cambridge, UK.

Breman, H. and J. J. Kessler. 1995. *Woody plants in agro-ecosystems of semi-arid regions with an emphasis on the Sahelian countries.* Springer-Verlag, Berlin, Germany.

Brenan, J. P. M. 1953–1954. Plants collected by the Vernay Nyasaland expedition of 1946. *Memoirs of the New York Botanical Garden* 8:191–256, 9:409–510, 1–132.

Brenan, J. P. M. 1978. Some aspects of the phytogeography of tropical Africa. *Annals of Missouri Botanical Garden* 65:437–478.

Breytenbach, G. J. 1986. Impacts of alien organisms on terrestrial communities with emphasis on the South West Cape. Pages 229–238 in I. A. W. MacDonald, editor. *The ecology and management of biological invasions.* Oxford University Press, Oxford, UK.

Bridges, E. M. 1990. *World geomorphology.* Cambridge University Press, Cambridge, UK.

Bristow, M. 1996. Dog jabs to save lions. *BBC Wildlife* 14:61.

Broadley, D. 1994. Reptiles. Pages 100–107 in J. Timberlake and P. Shaw, editors. *Chirinda Forest: a visitor's guide.* Forestry Commission, Harare, Zimbabwe.

Broadley, D. G. 1971. The reptiles and amphibians of Zambia: a checklist with key distribution records and ecological data. *Puku* 6:1–143.

Broadley, D. G. 1990. The herpetofaunas of the islands off the coast of south Mozambique. *Arnoldia Zimbabwe* 9:469–493.

Broadley, D. G. 1991. A review of the Namibian snakes of the genus *Lycophidion* (Serpentes: Colubridae), with the description of a new endemic species. *Annals of the Transvaal Museum* 35:209–214.

Broadley, D. G. 1992. Reptiles and amphibians from the Bazaruto Archipelago, Mozambique. *Arnoldia Zimbabwe* 9:539–548.

Broadley, D. G. 1994a. A collection of snakes from eastern Sudan, with the description of a new species of *Telescopus* Wagler, 1830 (Reptilia: Ophidia). *Journal of African Zoology* 108:201–208.

Broadley, D. G. 1994b. The genus *Scelotes* Fitzinger (Reptilia: Scincidae) in Mozambique, Swaziland and Natal, South Africa. *Annals of the Natal Museum* 35:237–259.

Broadley, D. G. 1994c. A review of *Lygosoma* Hardwicke & Gray 1827 (Reptilia Scincidae) on the East African coast, with the description of a new species. *Tropical Zoology* 7:217–222.

Broadley, D. G. 1995a. A new species of *Prosymna* Gray (Serpentes: Colubridae) from coastal forest in northeastern Tanzania. *Arnoldia Zimbabwe* 10:29–32.

Broadley, D. G. 1995b. A new species of *Scolecoseps* (Reptilia: Scincidae) from southeastern Tanzania. *Amphibia-Reptilia* 16:241–244.

Broadley, D. G. 1996. A revision of the genus *Lycophidion* Fitzinger (Serpentes: Colubridae) in Africa south of the Equator. *Syntarsus* 3:1–33.

Broadley, D. G. 1997. A review of the *Monopeltis capensis* complex in southern Africa (Reptilia: Amphisbaenidae). *African Journal of Herpetology* 46:1–12.

Broadley, D. G. 1998. A review of the genus *Atheris* Cope (Serpentes: Viperidae), with the description of a new species from Uganda. *Journal of Herpetology* 8:117–135.

Broadley, D. G. 1999. A new species of worm snake from Ethiopia (Serpentes: Leptotyphlopidae). *Arnoldia Zimbabwe* 10:141–144.

Broadley, D. G. 2001. An annotated check list of the herpetofauna of Mulanje Mountain. *Nyala* 21:29–36.

Broadley, D. G. and S. Broadley. 1997. A revision of the African genus *Zygaspis* Cope (Reptilia: Amphisbaenia). *Syntarsus* 4:1–23.

Broadley, D. G. and S. Broadley. 1999. A review of the African worm snakes from south of latitude 12 deg. S (Serpentes: Leptotyphlopidae). *Syntarsus* 5:1–36.

Broadley, D. G. and B. Hughes. 1993. A review of the genus *Lycophidion* (Serpentes: Colubridae) in northeastern Africa. *Herpetology Journal* 3:8–18.

Broadley, D. G. and B. Schätti. 2000. A new species of *Coluber* from northern Namibia (Reptilia: Serpentes). *Madoqua* 19:171–174.

Broadley, D. G. and V. Wallach. 1996. Remarkable new worm snake (Serpentes: Leptotyphlopidae) from the East African coast. *Copeia* 162–166.

Broadley, D. G. and V. Wallach. 1997a. A review of the genus *Leptotyphlops* (Serpentes: Leptotyphlopidae) in KwaZulu-Natal, South Africa, with the description of a new forest-dwelling species. *Durban Museum Novitates* 22:37–42.

Broadley, D. G. and V. Wallach. 1997b. A review of the worm snakes of Mozambique (Serpentes: Leptotyphlopidae) with the description of a new species. *Arnoldia Zimbabwe* 10:111–119.

Bromage, T. G. and F. Schrenk. 1999. *African biogeography, climate change and human evolution*. Oxford University Press, Oxford, UK.

Brooks, T., A. Balmford, N. Burgess, J. Fjeldså, L. A. Hansen, J. Moore, C. Rahbek, and P. Williams. 2001a. Toward a blueprint for conservation in Africa. *BioScience* 54:613–624.

Brooks, T., A. Balmford, N. Burgess, L. A. Hansen, J. Moore, C. Rahbek, P. Williams, L. Bennun, A. Byaruhanga, P. Kasoma, P. Njoroge, D. Pomeroy, and M. Wondafrash. 2001b. Conservation priorities for birds and biodiversity: do East African important bird areas represent species diversity in other terrestrial vertebrate groups? *Ostrich Supplement* 15:3–12.

Brooks, T. M., R. A. Mittermeier, C. G. Mittermeier, G. A. B. da Fonseca, A. B. Rylands, W. R. Konstant, P. Flick, J. Pilgrim, S. Oldfield, G. Magin, and C. Hilton-Taylor. 2002. Habitat loss and extinction in the hotspots of biodiversity. *Conservation Biology* 16:909–923.

Brooks, T. M., S. L. Pimm, and J. O. Oyugi. 1999. Time lag between deforestation and bird extinction in tropical forest fragments. *Conservation Biology* 13:1140–1150.

Brouwer, J. S., C. François, and W. C. Mullié. 2001. Niger. Pages 661–672 in L. D. C. Fishpool and M. I. Evans, editors. *Important bird areas in Africa and associated islands: priority sites for conservation*. Pisces Publications and BirdLife International (BirdLife Conservation Series No. 11), Newbury and Cambridge, UK.

Brown, E. and W. Henry. 1993. The viewing value of elephants. Pages 146–155 in B. Barbier, editor. *Economics and ecology: new frontiers and sustainable development*. Chapman and Hall, London, UK.

Brown, L., E. K. Urban, and K. Newman. 1982. *The birds of Africa, Volume I*. Academic Press, London, UK.

Brühl, C. A. 1997. Flightless insects: a test case for historical relationships of African mountains. *Journal of Biogeography* 24:233–250.

Bruner, A. G., R. E. Gullison, R. E. Rice, and G. A. B. da Fonseca. 2001. Effectiveness of parks in protecting tropical biodiversity. *Science* 291:125–128.

Brussard, P. F., D. D. Murphy, and D. R. Tracy. 1994. Cattle and conservation biology: another view. *Conservation Biology* 8:919–921.

Bryant, D., D. Nielsen, and L. Tangley. 1997. *The last frontier forests: ecosystems and economies on the edge*. World Resources Institute, Washington, DC, USA.

Buchanan, K. M. and J. C. Pugh. 1955. *Land and people in Nigeria*. University of London Press, London, UK.

Buckle, C. 1978. *Landforms in Africa*. Longman, London, UK.

Bullock, D. J. 1986. The ecology and conservation of reptiles on Round Island and Gunner's Quoin, Mauritius. *Biological Conservation* 37:135–156.

Bullock, S. H., H. A. Mooney, and E. Medina, editors. 1996. *Seasonally dry tropical forests*. Cambridge University Press, Cambridge UK.

Burgess, N. D. and G. P. Clarke, editors. 2000. *The coastal forests of eastern Africa*. IUCN, Cambridge, UK and Gland, Switzerland.

Burgess, N. D., G. P. Clarke, and W. A. Rodgers. 1998a. Coastal forests of eastern Africa: status, endemism patterns and their potential causes. *Biological Journal of the Linnean Society* 64:337–367.

Burgess, N. D., H. de Klerk, H. Fjeldså, T. Crowe, and C. Rahbek. 2000. A preliminary assessment of congruence between biodiversity patterns in Afrotropical forest birds and forest mammals. *Ostrich* 71:286–290.

Burgess, N. D., N. Doggart, and J. C. Lovett. 2002a. The Uluguru Mountains of eastern Tanzania: the effect of forest loss on biodiversity. *Oryx* 36:140–152.

Burgess, N. D., C. Fitzgibbon, and G. P. Clarke. 1996. Coastal forests of East Africa. Pages 329–359 in T. McClanaghan and T. P. Young, editors. *Ecosystems and their conservation in East Africa*. Oxford University Press. New York, USA.

Burgess, N. D., J. Fjeldså, and R. Botterweg. 1998b. The faunal importance of the Eastern Arc Mountains. *Journal of the East African Natural History Society* 87:37–58.

Burgess, N., J. Fjeldså, and C. Rahbek. 1998c. Mapping the distributions of Afrotropical vertebrate groups. *Species* 30:16–17.

Burgess, N. D., M. Nummelin, J. Fjeldså, K. M. Howell, K. Lukumbyzya, L. Mhando, P. Phillipson, and E. Vanden Berghe, eds. 1998d. Biodiversity and conservation of the Eastern Arc Mountains of Tanzania and Kenya. *Special Issue of the Journal of the East African Natural History Society* 87:1–367.

Burgess, N. D., C. Rahbek, F. Wugt Larsen, P. Williams, and A. Balmford. 2002b. How much of the vertebrate diversity of sub-Saharan Africa is catered for by recent conservation proposals? *Biological Conservation* 107:327–339.

Burgess, N., T. S. Romdal, and M. Rahner. 2001. Forest loss in the Ulugurus, Tanzania and the status of the Uluguru bush shrike *Malconotus alius. Bulletin of the African Bird Club* 8:89–90.

Burke, A., K. Esler, E. Pienaar, and P. Barnard. 2002. Species richness and floristic relationships between mesas and their surroundings in southern African Nama Karoo. *Diversity and Distributions* 9:43–53.

Burke, K. 2001. Origin of the Cameroon line of volcano-capped swells. *Journal of Geology* 109:349–362.

Burney, D. A. 1996. Paleoecology of humans and their ancestors. Pages 19–36 in T. R. McClanahan and T. P. Young, editors. *East African ecosystems and their conservation.* Oxford University Press, New York, USA and Oxford, UK.

Burney, D. A. 1997. Theories and facts regarding Holocene environmental change before and after human colonization. Pages 75–89 in S. M. Goodman and D. B. Patterson, editors. *Natural change and human impact in Madagascar.* Smithsonian Institution Press, Washington, DC, USA.

Burtt, B. D. 1942. Some East African vegetation communities. *Journal of Ecology* 30:65–145.

Busby, J. R. 1991. BIOCLIM: a bioclimatic analysis and prediction system. Pages 64–68 in C. R. Margules and M. P. Austin, editors. *Nature conservation: cost effective biological surveys and data analysis.* CSIRO, Melbourne, Australia.

Bushmeat Crisis Task Force. 2001. Bushmeat: a wildlife crisis in West and Central Africa and around the world. Retrieved 2001 from http://www.bushmeat.org.

Butynski, T. and M. J. Kalina. 1993. Three new mountain national parks for Uganda. *Oryx* 27:214–224.

Byaruhanga, A., P. Kasoma, and D. Pomeroy. 2001. *Important bird areas in Uganda.* Nature Uganda, Kampala, Uganda.

Byers, B. 1991. Ecoregions, state sovereignty and conflict. *Bulletin of Peace Proposals* 22:65–76.

Byers, B. 2001. Miombo ecoregion reconnaissance: synthesis report. Draft report for WWF-SARPO, 28 March 2001. World Wildlife Fund–SARPO, Harare, Zimbabwe.

Bytebier, B. 2001. *Taita Hills biodiversity project final report.* National Museums of Kenya, Nairobi, Kenya.

Cable, S. and M. Cheek. 1998. *The plants of Mount Cameroon: a conservation checklist.* Royal Botanic Gardens, Kew, UK.

Cadet, L. J. T. 1977. La végétation de l'Île de la Réunion: étude phytoécologique et phytosociologique. Thesis, Université de Marseille, Imprimerie Cazal, St. Denis, Réunion.

Cahen, L., N. J. Snelling, J. Delhal, and J. R. Val. 1984. *The geochronology and evolution of Africa.* Clarendon Press, Oxford, UK.

Caldecott, J. 1998. *Designing conservation projects.* Cambridge University Press, Cambridge, UK.

Campbell, A. 1990. *The nature of Botswana: a guide to conservation and development.* IUCN, Harare, Zimbabwe.

Campbell, B. M. 1985. A classification of the mountain vege-

tation of the Fynbos biome. *Memoirs of the Botanical Survey of South Africa* 50:1–115.

Campbell, B., editor. 1996. *The miombo in transition: woodlands and welfare in Africa.* Centre for International Forestry Research, Bogor, Indonesia.

Campbell, B., P. Frost, and N. Byron. 1996. Miombo woodlands and their use: overview and key issues. Pages 1–10 in B. Campbell, editor. *The miombo in transition: woodlands and welfare in Africa.* Centre for International Forestry Research, Bogor, Indonesia.

Campbell, B. M. and M. K. Luckert. 2002. *Uncovering the hidden harvest: valuation methods for woodland and forest resources.* Earthscan, London, UK.

Campbell, K. and M. Borner. 1995. Population trends and distribution of Serengeti herbivores: implications for management. Pages 117–145 in A. R. E. Sinclair and P. Arcese, editors. *Serengeti II: dynamics, management and conservation of an ecosystem.* University of Chicago Press, Chicago, USA.

Campbell, K. and H. Hofer. 1995. People and wildlife: spatial dynamics and zones of interaction. Pages 534–570 in A. R. E. Sinclair and P. Arcese, editors. *Serengeti II: dynamics, management and conservation of an ecosystem.* University of Chicago Press, Chicago, USA.

CAMPFIRE. 1997. *CAMPFIRE's income and expenditure: the bottom line.* Africa Resources Trust, Harare, Zimbabwe.

Caputo, V. 1993. Taxonomy and evolution of the *Chalcides chalcides* complex (Reptilia, Scincidae) with description of two new species. *Bollettino del Museo Regionale di Science Naturali di Torino* 11:47–120.

Caputo, V. and J. Mellado. 1992. A new species of *Chalcides* (Reptilia: Scincidae) from northeastern Morocco. *Bollettino di Zoologia* 59:335–342.

Carleton, M. D. 1994. Systematic studies of Madagascar's endemic rodents (Muroidea: Nesomyinae): revision of the genus *Eliurus. American Museum Novitates* 3087:1–55.

Carleton, M. D. and S. M. Goodman. 1996. Systematic studies of Madagascar's endemic rodents (Muroidea: Nesomyinae): a new genus and species from the Central Highlands. *Fieldiana Zoology New series:*231–256.

Carleton, M. D. and S. M. Goodman. 1998. New taxa of nesomyine rodents (Muroidea: Muridae) from Madagascar's northern highlands, with taxonomic comments on previously described forms. *Fieldiana Zoology New series:*163–200.

Carroll, R. W. 1988. Elephants of the Dzanga-Sangha dense forest of south-western Central African Republic. *Pachyderm* 10:12–15.

Carter, J. 1987. *Malawi: wildlife parks and reserves.* Macmillan, London, UK and New York, USA.

Caspary, H. U. 1999. *Wildlife utilization in Côte d'Ivoire and West Africa: potentials and constraints for development cooperation.* Tropical Ecology Support Program Publication No. TOB F-V/10e. GTZ, Eschborn, Germany.

Castellini, G. 1990. Quattro nuovi *Euconnus* di Sierra Leone (Coleoptera, Scydmaenidae). Ricerche biologiche in Sierra Leone (parte III). *Accademia Nazionale dei Lincei* 265:185–190.

Castro, R., F. Tattenbach, L. Gámez, and N. Olson. 1998. *The Costa Rican experience with market instruments to mitigate cli-*

mate change and conserve biodiversity. Fundecor, San José, Costa Rica.

Castroviejo Bolívar, J., A. Blom, and M. P. T. Alers. 1990. Equatorial Guinea. Pages 110–113 in R. East, editor. *Antelopes: global survey and regional action plans part 3. West and Central Africa.* IUCN/Species Survival Commission Antelope Specialist Group, IUCN, Gland, Switzerland.

Castroviejo Bolívar, J., J. J. Balleste, and R. C. Alvarez. 1986. *Investigación y conservación de la naturaleza en Guinea Ecuatorial.* Secretaria de Estado para la Cooperación Internacional y para Iberoamerica, Madrid, Spain.

CEC (Commission of the European Communities). 1992. *Mangroves of Africa and Madagascar.* Office for Official Publications of the European Communities, Luxembourg.

Central Africa Regional Program for the Environment (CARPE). 1998. CARPE data CD-ROM. CARPE and USAID, Washington, DC, USA.

Chabanaud, P. 1920. *Contribution à l'étude de la faune herpétologique de l'Afrique occidentale: note préliminaire sur les résultats d'une mission scientifique en Guinée française (1919–1920).* Bulletin du Comité d'Études Historiques et Scientifiques de l'Afrique Occidentale Française, Paris 3:489–497.

Chabwela, H. N. 1992. The ecology and resource use of the Bangweulu Basin and the Kafue Flats. Pages 11–25 in R. C. V. Jeffrey, H. N. Chabwela, G. Howard, and P. J. Dugan, editors. *Managing the wetlands of Kafue Flats and Bangweulu Basin.* Kafue National Park, Zambia. IUCN, Gland, Switzerland.

Chabwela, H. N. and W. Mumba. 1998. Integrating water conservation and population strategies on the Kafue Flats. Retrieved 2001 from the World Wide Web: http://www.aaas.org/international/psd/waterpop/Zambia.htm.

Chaïeb, M. and M. Boukhris. 1998. *Flore succinte et illustrée des zones arides et sahariennes de Tunisie.* Edition l'Or du Temps, Tunis, Tunisie.

Chapin, J. P. 1932. Faunal relations and subdivisions of the Congo. *Bulletin American Museum of Natural History* LXV:83–98.

Chapman, C. A. and L. J. Chapman. 1996. Mid-elevation forests: a history of disturbance and regeneration. Pages 385–400 in T. R. McClanahan and T. P. Young, editors. *East African ecosystems and their conservation.* Oxford University Press, New York, USA.

Chapman, J. D. 1962. *The vegetation of the Mulanje Mountains, Nyasaland.* The Government Printer, Zomba, Malawi.

Chapman, J. D. 1990. *Mount Mulanje, Malawi: a plea for its future.* Braeriach, Urlar Road, Aberfeldy, Perthshire, UK.

Chapman, J. D. 1994. Mount Mulanje, Malawi. Pages 240–247 in World Wildlife Fund and IUCN, editors. *Centres of plant diversity. A guide and strategy for their conservation.* Volume 1: Europe, Africa, South West Asia and the Middle East. IUCN Publications Unit, Cambridge, UK.

Chapman, J. D. and F. White. 1970. *The evergreen forests of Malawi.* Commonwealth Forestry Institute, University of Oxford, Oxford, UK.

Chapman, V. J. 1977. Africa B. The remainder of Africa. Pages 233–240 in V. J. Chapman, editor. *Ecosystems of the world 1: wet coastal ecosystems.* Elsevier Scientific Publishing Company, New York, USA.

Charco, J. 1999. *El bosque Mediterraneo en El Norte de Africa. Biodoversita y lucha contra la desertificacion.* Agencia Espagnola de Cooperation Internacional, Madrid, Spain.

Cheek, M., S. Cable, F. N. Hepper, N. Ndam, and J. Watts. 1994. Mapping plant biodiversity on Mount Cameroon. Pages 110–210 in L. J. G. Van der Masen, X. M. van der Burgt, and J. M. van Medenbach de Rooy, editors. *The biodiversity of African plants.* Proceedings XIVth AETFAT Congress, Wageningen, The Netherlands. Kluwer Academic Press, Dordrecht, The Netherlands.

Cheek, M., J.-M. Onana, and B. J. Pollard, compilers and editors. 2000. *The plants of Mount Oku and the Ijim Ridge, Cameroon: a conservation checklist.* Royal Botanic Gardens, Kew, UK.

Cheke, A. S. 1987. The ecological history of the Mascarene Islands, with particular reference to extinction's and introductions of land vertebrates. Pages 5–89 in A. W. Diamond, editor. *Studies of Mascarene Island birds.* Cambridge University Press, Cambridge, UK.

Cheke, A. S. 1994. Lizards of the Seychelles. Pages 331–360 in D. R. Stoddart, editor. *Biogeography and ecology of the Seychelles Islands.* Dr. W. Junk Publishers, The Hague, Netherlands.

Chenje, M. and P. Johnson, editors. 1994. *State of the environment in southern Africa.* Southern African Research and Documentation Centre, Harare, Zimbabwe.

Chidumayo, E. and P. Frost. 1996. Population biology of miombo trees. Pages 59–71 in B. Campbell, editor. *The miombo in transition: woodlands and welfare in Africa.* Centre for International Forestry Research, Bogor, Indonesia.

Chidumayo, E., J. Gambiza, and I. Grundy. 1996. Managing miombo woodlands. Pages 175–194 in B. Campbell, editor. *The miombo in transition: woodlands and welfare in Africa.* Centre for International Forestry Research, Bogor, Indonesia.

Chidumayo, E. N. 1987. Woodland structure, destruction and conservation in the copperbelt area of Zambia. *Biological Conservation* 40:89–100.

Child, G. 1996. Realistic "game laws." *Oryx* 30:228–229.

Childes, S. and P. J. Mundy. 2001. Zimbabwe. Pages 1025–1041 in L. D. C. Fishpool and M. I. Evans, editors. *Important bird areas in Africa and associated islands: priority sites for conservation.* BirdLife Conservation Series No. 11. Pisces Publications and BirdLife International, Newbury and Cambridge, UK.

Chirio, L. and I. Ineich. 1991. Les genres *Rhamphiophis* Peters, 1854 et *Dipsina* Jan, 1863 (Serpentes, Colubridae): revue des taxons reconnus et description d'une espèce nouvelle. *Bulletin du Museum National d'Histoire Naturelle, Paris* 13:217–235.

Christy, P. 1999. Bird list for the Dzanga-Sangha complex. World Wildlife Fund–Central African Republic, Banguy, Central African Republic.

Christy, P. 2001a. Gabon. Pages 349–356 in L. D. C. Fishpool and M. I. Evans, editor. *Important bird areas of Africa and associated islands: priority sites for conservation.* BirdLife Conservation Series No. 11. Pisces Publications and BirdLife International, Newbury and Cambridge, UK.

Christy, P. 2001b. São Tomé and Príncipe. Pages 727–732 in L. D. C. Fishpool and M. I. Evans, editor. *Important bird areas in Africa and associated islands: priority sites for conservation.* Pisces Publications and BirdLife International (BirdLife Conservation Series No. 11), Newbury and Cambridge, UK.

Christy, P. and W. V. Clarke. 1998. *Guide des oiseaux de São Tomé et Príncipe.* ECOFAC, Libreville, Gabon.

Cibois, A., B. Slikas, T. S. Schulenberg, and E. Pasquet. 2001. An endemic radiation of Malagasy songbirds is revealed by mitochondrial DNA sequencing. *Evolution* 55:1198–1206.

CIESIN. 1995. *Gridded population data of the world.* Center for International Earth Science Information Network, Columbia University, Palisades, NY, USA.

CIESIN, Columbia University, IFPRI, and WRI. 2000. Gridded Population of the World (GPW), Version 2. Retrieved 2001 from the World Wide Web: http://sedac.ciesin.columbia.edu/plue/gpw.

Cincotta, R. P. and R. Engelman. 2000. *Nature's place: human population and the future of biological diversity.* Population Action International, Washington, DC, USA.

Cincotta, R. P., J. Wisnewski, and R. Engelman. 2000. Human population in the biodiversity hotspots. *Nature* 404:990–992.

Clark J. D., J. E. Dunn, and K. G. Smith. 1993. A multivariate model of female black bear habitat use for geographic information system. *Journal of Wildlife Management* 57:519–526.

Clarke, G. P. 1998. A new regional centre of endemism in Africa. Pages 53–65 in D. F. Cutler, C. R. Huxley, and J. M. Lock, editors. *Aspects of the ecology, taxonomy and chorology of the floras of Africa and Madagascar.* Kew Bulletin Additional Series. Royal Botanic Gardens, Kew, UK.

Clarke, T. and D. Collins. 1996. *A birdwatchers' guide to the Canary Islands.* Prion Ltd., Perry, UK.

Clausen, M. and V. Gayler. 1997. The greening of the Sahara during the mid-Holocene: results of an interactive atmosphere-biome model. *Global Ecology and Biogeography Letters* 6:369–377.

Clements, J. F. 1991. *Birds of the world. A check list.* Ibis Publishing Company, Vista, CA, USA.

CLIMAP. 1976. The surface of the ice age earth. *Science* 191:1131–1144.

Clinebell, R. R., O. L. Phillips, A. H. Gentry, N. Starks, and H. Zuuring. 1995. Prediction of neotropical tree and liana species richness from soil and climatic data. Biodiversity and *Conservation* 4:56–90.

Cline-Cole, R. A. 1987. The socio-ecology of firewood and charcoal on the Freetown Peninsula. *Africa* 57:457–497.

Cloudsley-Thompson, J. L. 1984. *Sahara Desert.* Pergamon Press, Oxford, UK and New York, USA.

CNE. 1991. *Rapport national environnement.* Secrétariat Technique du Comité National pour l'Environnement ONTA/SPSE, Djibouti.

Cobb, S. and D. Western. 1989. The ivory trade and the future of the African elephant. *Pachyderm* 12:32–37.

Coche, A. 1998. *Supporting aquaculture development in Africa: research network on integration of aquaculture and irrigation.* CIFA Occasional Paper No. 23. Food and Agriculture Organization (FAO), Accra, Ghana.

Coe, M. and K. Curry-Lindahl. 1965. Ecology of a mountain: first report on Liberian Nimba. *Oryx* 8:177–184.

Coe, M. J., N. C. McWilliam, G. N. Stone, and M. J. Packer, editors. 1999. *Mkomazi: the ecology, biodiversity and conservation of a Tanzanian savanna.* Royal Geographical Society, London, UK.

COEFOR/CI. 1993. *Répertoire et carte de distribution: domaine forestier de Madagascar.* Direction des Eaux et Forêts, Service des Ressources Forestières, Projet COEFOR (Contribution a l'étude des forêts classés), et Conservation International. Antananarivo, Madagascar.

Coetzee, J. P., G. J. Bredenkamp, and N. Van Rooyen. 1993. The sub-humid warm temperature mountain bushveld plant communities of the Pretoria-Witbank-Heidelberg area. *South African Journal of Botany* 59:623–632.

Cole, M. 1986. *The savannas: biogeography and geobotany.* Academic Press, London, UK.

Cole, M. 1992. Influence of physical factors on the nature and dynamics of forest-savanna boundaries. Pages 63–75 in P. A. Furley, J. Proctor, and J. A. Ratter, editors. *Nature and dynamics of forest-savanna boundaries.* Chapman and Hall, London, UK.

Cole, N. H. A. 1967. *Ecology of the montane community at the Tingi Hills in Sierra Leone.* Bulletin de l'IFAN 29A:904–924.

Cole, N. H. A. 1968. *The vegetation of Sierra Leone.* Njala University College Press, Freetown, Sierra Leone.

Cole, N. H. A. 1974. Climate life forms and species distribution on the Loma montane grassland Sierra-Leone. *Botanical Journal of the Linnean Society* 69:197–210.

Collar, N. J. and S. N. Stuart. 1985. *Threatened birds of Africa and related islands,* 3rd edition. The ICBP/IUCN Red Data Book, Part 1, Cambridge, UK.

Collar, N. J. and S. N. Stuart. 1988. *Key forests for threatened birds in Africa.* ICBP, Cambridge, UK.

Collomb, J. G., J. B. Mikissa, S. Minnemeyer, S. Mundunga, H. Nzao Nzao, J. Madouma, J. de Dieu Mapaga, C. Mikolo, N. Rabenkogo, S. Akagah, E. Bayani-Ngoye, and A. Mofouma. 2000. *A first look at logging in Gabon.* Global Forest Watch, World Resources Institute, Washington, DC, USA.

Colston, P. P. and K. Curry-Lindahl. 1986. *The birds of Mount Nimba, Liberia.* Publication No. 982. Bulletin of the British Museum (Natural History), Zoology, London, UK.

Colyn, M. 1991. Zoogeographical importance of the Zaire River basin for speciation. *L'importance zoogeographique du bassin du fleuve Zaire pour la speciation: le cas des primates simiens.* Koninklijk Museum voor Midden-Afrika, Tervuren, Belgium.

Colyn, M., A. Gautier-Hion, and D. Thys Van Den Audenaerde. 1991a. *Cercopitecus dryas,* Schwartz 1932 and *C. salongo,* Thys Van Den Audenaurde 1997 are the same species with an age-related coat pattern. *Folia Primatologica* 56:167–170.

Colyn, M., A. Gautier-Hion, and W. Verheyen. 1991b. A re-appraisal of the palaeoenvironmental history in Central Africa: evidence for a major fluvial refuge in the Zaire Basin. *Journal of Biogeography* 18:403–407.

Comley, P. and S. Meyer. 1994. *Traveller's guide to Botswana.* New Holland Publishers, London, UK.

Connah, G. 1987. *African civilisations: precolonial cities and states in tropical Africa an archaeological perspective.* Cambridge University Press, Cambridge, UK.

Connor, E. F. and E. D. McCoy. 1979. The statistics and biology of the species-area relationship. *American Naturalist* 113:791–833.

Cornet, A. 1974. Essai cartographique bioclimatique à Madagascar, carte à 1/2'000'000 et notice explicative no. 55. ORSTROM, Paris, France.

Corsi, F., I. De Leeuw, and A. Skidmore. 2000. Species distribution modeling with GIS. Pages 389–434 in L. Boitani and T. K. Fuller, editors. *Research techniques in animal ecology*. Columbia University Press, New York, USA.

Corsi, F., E. Duprè, and L. Boitani. 1999. A large-scale model of wolf distribution in Italy for conservation planning. *Conservation Biology* 13:150–159.

Costanza, R., R. d'Arge, R. de Groot, S. Farber, M. Grasso, B. Hannon, K. Limburg, S. Naeem, R. V. O'Neill, J. Paruelo, R. G. Raskin, P. Sutton, and M. Van den Belt. 1997. The value of the world's ecosystem services and natural capital. *Nature* 387:253–260.

Cottrell, C. B. and J. P. Loveridge. 1966. Observations on the *Cryptosepalum* forest of the Mwinilunga District of Zambia. *Proceedings and Transactions of the Rhodesia Scientific Association* 51:79–120.

Coulthard, N. D. 2001a. Algeria. Pages 51–70 in L. D. C. Fishpool and M. I. Evans, editors. *Important bird areas in Africa and associated islands: priority sites for conservation*. Pisces Publications and BirdLife International (BirdLife Conservation Series No. 11), Newbury and Cambridge, UK.

Coulthard, N. D. 2001b. Senegal. Pages 733–750 in L. D. C. Fishpool and M. I. Evans, editors. *Important bird areas in Africa and associated islands: priority sites for conservation*. Pisces Publications and BirdLife International (BirdLife Conservation Series No. 11), Newbury and Cambridge, UK.

Cowling, R. M. 1983a. Phytochorology and vegetation history in the south-Eastern Cape, South Africa. *Journal of Biogeography* 10:393–419.

Cowling, R. M. C. 1983b. Vegetation studies in the Humansdorp region of the fynbos biome. Ph.D. thesis, University of Cape Town, Cape Town, South Africa.

Cowling, R. M. 1984. A syntaxonomic and synecological study in the Humansdorp region of the fynbos biome. *Bothalia* 15:175–228.

Cowling, R. M. 1986. *A description of the Karoo Biome Project*. South African National Scientific Programmes Report No. 122. Council for Scientific and Industrial Research, Pretoria, South Africa.

Cowling, R. M. and W. J. Bond. 1991. How small can reserves be? An empirical approach in Cape Fynbos, South Africa. *Biological Conservation* 58:243–256.

Cowling, R. M., K. J. Esler, and P. W. Rundel. 1999a. Namaqualand, South Africa: an overview of a unique winter-rainfall desert ecosystem. *Plant Ecology* 142:3–21.

Cowling, R. M. and C. E. Heijnis. 2001. The identification of broad habitat units as biodiversity entities for systematic conservation planning in the Cape Floristic Region. *South African Journal of Botany* 67:15–38.

Cowling, R. M. and C. Hilton-Taylor. 1994. Patterns of plant diversity and endemism in South Africa: an overview. Pages 31–52 in B. J. Huntley, editor. *Botanical diversity in southern Africa*. National Botanical Institute, Pretoria, South Africa.

Cowling, R. M. and C. Hilton-Taylor. 1997. Phytogeography, flora and endemism. Pages 43–61 in R. M. Cowling, D. M. Richardson, and S. M. Pierce, editors. *Vegetation of southern Africa*. Cambridge University Press, Cambridge, UK.

Cowling, R. M. and C. Hilton-Taylor. 1999. Plant biogeography, endemism and diversity. Pages 42–56 in W. R. J. Dean and S. J. Milton, editors. *The Karoo: ecological patterns and processes*. Cambridge University Press, Cambridge, UK.

Cowling, R. M., P. M. Holmes, and A. G. Rebelo. 1992. Plant diversity and endemism. Pages 62–112 in R. M. Cowling, editor. *The ecology of fynbos: nutrients, fire and diversity*. Oxford University Press, Cape Town, South Africa.

Cowling, R. M. and A. T. Lombard. 1998. *A strategic and systematic framework for conserving the plant life of the Succulent Karoo*. IPC Report 9802 submitted to the Trustees of the Leslie Hill Succulent Karoo Trust. Institute for Plant Conservation, University of Cape Town, Cape Town, South Africa.

Cowling, R. M. and D. J. McDonald. 1999. Local endemism and plant conservation in the Cape Floristic Region. Pages 64–86 in P. W. Rundel, G. Montenegro, and F. Jaksic, editors. *Landscape degradation in Mediterranean-climate ecosystems*. Springer-Verlag, Heidelberg, Germany.

Cowling, R. M. and S. M. Pierce. 1999a. Cape Floristic Province. Pages 218–227 in R. A. Mittermeier, N. Myers, and C. G. Mittermeier, editors. *Hotspots: Earth's biologically richest and most threatened terrestrial ecoregions*. CEMEX and Conservation International, Mexico City, Mexico and Washington, DC, USA.

Cowling, R. M. and S. M. Pierce. 1999b. *Namaqualand: a succulent desert*. Fernwood Press, Cape Town, South Africa.

Cowling, R. M. and R. L. Pressey. 2001. Rapid plant diversification: planning for an evolutionary future. Proceedings of the National Academy of Sciences of the United States of America 98:5452–5457.

Cowling, R. M., R. L. Pressey, A. T. Lombard, P. G. Desmet, and A. G. Ellis. 1999b. *From representation to persistence: requirements for a sustainable system of conservation areas in the species-rich Mediterranean-climate desert of southern Africa*. Diversity and Distributions 5:51–71.

Cowling, R. M., R. L. Pressey, A. T. Lombard, C. E. Heijnis, D. M. Richardson, and N. Cole. 1999c. *Framework for a conservation plan for the Cape Floristic Region*. IPC Report 9902 submitted to World Wide Fund for Nature, Stellenbosch, South Africa.

Cowling, R. M., R. L. Pressey, M. Rouget, and A. T. Lombard. 2003. A conservation plan for a global biodiversity hotspot: the Cape Floristic Region, South Africa. *Biological Conservation* 112:191–216.

Cowling, R. M. and D. M. Richardson. 1995. *Fynbos: South Africa's unique floral kingdom*. Fernwood Press, Vlaeberg, South Africa.

Cowling, R. M., D. M. Richardson, and P. J. Mustart. 1997a. Fynbos. Pages 99–130 in R. M. Cowling, D. M. Richardson, and S. M. Pierce, editors. *Vegetation of southern Africa*. Cambridge University Press, Cambridge, UK.

Cowling, R. M., D. M. Richardson, and S. M. Pierce, editors. 1997b. *Vegetation of southern Africa*. Cambridge University Press, Cambridge, UK.

Cowling, R. M., P. W. Rundel, B. B. Lamont, M. K. Arroyo, and M. Arianoutsou. 1996. Plant diversity in Mediterranean-climate regions. *Trends in Ecology & Evolution* 11:362–366.

Cowlishaw, G. 1999. Predicting the pattern of decline of African primate diversity: an extinction debt from historical deforestation. *Conservation Biology* 13:1183–1193.

Cox, C. B. and P. D. Moore. 1993. *Biogeography: an ecological and evolutionary approach*. Blackwell Scientific Publications, London, UK.

Craw, R. C., J. R. Grehan, and M. J. Heads. 1999. *Panbiogeography: tacking the history of life*. Oxford University Press, New York, USA.

Cribb, P. J. and G. P. Leedal. 1982. *The mountain flowers of southern Tanzania: a field guide to the common flowers*. A.A. Balkema, Rotterdam, Netherlands.

Croizat, L. 1965. An introduction to the subgeneric classification of *Euphorbia* L. with stress on the South African and Malagasy species. *Webbia* 20:573–706.

Cronk, Q. C. B. 1992. Relict floras of Atlantic Islands: patterns assessed. *Biological Journal of the Linnean Society* 46:91–103.

Cronk, Q. C. B. 1997. Islands: stability, diversity, conservation. *Biodiversity and Conservation* 6:477–494.

Cronk, Q. C. B. and J. L. Fuller. 2002. *Plant invaders: threats to natural ecosystems*. Earthscan, London, UK.

Crowe, T. M. 1990. A quantitative analysis of patterns of distribution, species richness and endemism in southern African vertebrates. Pages 124–138 in G. Peters and R. Hutterer, editors. *Vertebrates in the tropics*. Museum Alexander Koenig, Bonn, Germany.

Crowe, T. M. and A. A. Crowe. 1982. Patterns of distribution, diversity and endemism in Afrotropical birds. *Journal of Zoology* 198:417–442.

CSIR. 2000a. *Cape Action Plan for the Environment: implementation programme report*. CSIR Report No: ENV-S-C 99130 D. Prepared for WWF-SA, Stellenbosch, South Africa.

CSIR. 2000b. *Cape Action Plan for the Environment: independently submitted projects report*. CSIR Report No. ENV-S-C 99130 E. Prepared for WWF-SA, Stellenbosch, South Africa.

Culverwell, J. 1998. *Long-term recurrent costs of protected area management in Cameroon*. World Wildlife Fund–Cameroon and MINEF, Yaoundé, Cameroon.

Cumming, D. H. M. 1999. *Study on the development of transboundary natural resource management areas in southern Africa environmental context: natural resources, land use, and conservation*. Biodiversity Support Program, Washington, DC, USA.

Cumming, D. H. M., R. F. du Toit, and S. N. Stuart. 1990. *African elephants and rhinos: status survey and conservation action plan*. IUCN, Gland, Switzerland.

Cumming, D. H. M., M. B. Fenton, I. L. Rautenbach, R. D. Taylor, G. S. Cumming, M. S. Cumming, J. M. Dunlop, A. G. Ford, M. D. Hovorka, D. S. Johnston, M. Kalcounis, Z. Mahlangu, and C. V. R. Portfors. 1997. Elephants, woodlands and biodiversity in southern Africa. *South African Journal of Science* 93:231–236.

Cumming, D. H. M. and T. J. P. Lynam. 1997. *Land use changes, wildlife conservation and utilisation, and the sustainability of agro-ecosystems in the Zambezi Valley: final technical report*. European Union Contract B7-5040/93/06. WWF Southern Africa Programme Office, Harare, Zimbabwe.

Cumming, D. H. M., C. Mackie, S. Magane, and R. D. Taylor. 1994. *Aerial census of large herbivores in the Gorongosa National Park and the Marromeu area of the Zambezi delta in Mozambique: June 1994*. Description of Gorongosa-Marromeu National Resource Management Area. IUCN ROSA, Harare, Zimbabwe.

Cunningham, A. B. 1988. *An investigation of the herbal medicine trade in Natal/Kwazulu*. Investigational Report No. 29. Institute of Natural Resources, University of Natal, Pietermaritzburg, South Africa.

Cunningham, A. B. 1991. Development of a conservation policy on commercially exploited medicinal plants: a case study from southern Africa. Pages 337–358 in O. Akerele, V. Heywood, and H. Synge, editors. *Conservation of medicinal plants*. Cambridge University Press, Cambridge, UK.

Cunningham, A. B. 2002. *Applied ethnobotany: people, wild plant use, and conservation*. Earthscan, London, UK.

Curry-Lindahl, K. 1966. Zoological aspects on the conservation of vegetation in tropical Africa. *Acta Phytogeographica Suecia* 54:25–32.

Curtis, B., K. S. Roberts, M. Griffin, S. Bethune, C. J. Hay, and H. Kolberg. 1998. Species richness and conservation of Namibian freshwater macro-invertebrates, fish and amphibians. *Biodiversity and Conservation* 7:447–466.

da Fonseca, G. A. B., A. Balmford, C. Bibby, L. Boitani, F. Corsi, T. Brooks, C. Gascon, S. Olivieri, R. A. Mittermeier, N. Burgess, E. Dinerstein, D. Olson, L. Hannah, J. Lovett, D. Moyer, C. Rahbek, S. Stuart, and P. Williams. 2000. Following Africa's lead in setting priorities. *Nature* 405:393–394.

Dale, I. R. 1954. Forest spread and climatic change in Uganda during the Christian era. *Empire Forestry Review* 33:23–29.

Dallman, P. R. 1998. *Plant life in the world's Mediterranean climates*. Oxford University Press, Oxford, UK.

Daszak, P., A. A. Cunningham, and A. D. Hyatt. 2000. Emerging infectious diseases of wildlife: threats to biodiversity and human health. *Science* 287:443–449.

Davenport, T. R. B. 1996. *An ecological monitoring programme for Bwindi Impenetrable and Mgahinga Gorilla national parks*. Report to the Institute of Tropical Forest Conservation. Kabale, Uganda.

Davenport, T. R. B. 2002a. Food for thought: orchid conservation in Tanzania. *Wildlife Conservation* February:12.

Davenport, T. R. B. 2002b. Garden of the Gods: Kitulo Plateau, a new national park for Tanzania. *Wildlife Conservation* June:15.

Davenport, T., P. Howard, and C. Dickinson, editors. 1996a. *Mount Elgon National Park biodiversity report*. Forest Department, Kampala, Uganda.

Davenport, T., P. Howard, and C. Dickinson, editors. 1996b. *Moroto, Kadam and Napak forest reserves biodiversity report*. Forest Department, Kampala, Uganda.

Davenport, T., P. Howard, and R. Matthews, editors. 1996c. *Moroto, Kadam and Napak forest reserves*. Biodiversity Report No. 6. Forest Department, Kampala, Uganda.

Davenport, T. R. B. and H. J. Ndangalasi. 2003. An escalating trade in orchid tubers across Tanzania's southern highlands: assessment, dynamics and conservation implications. *Oryx* 37:55–61.

Davies, B. and J. Day. 1998. *Vanishing waters*. University of Cape Town Press, Cape Town, South Africa.

Davies, G. 1987. *The Gola Forest Reserves, Sierra Leone*. IUCN, Gland, Switzerland and Cambridge, UK.

Davies, G. and B. Birkenhager. 1990. Jentink's duiker in Sierra Leone: evidence from the Freetown Peninsula. *Oryx* 24:143–146.

Davies, J. N. P. 1979. *Pestilence and disease in the history of Africa.* Witwatersrand University Press, Johannesburg, South Africa.

Davies, O. 1976. The older coastal dunes in Natal and Zululand and their relation to former shorelines. *Annals of the South African Museum* 71:19–32.

Deacon, H. J., M. R. Jury, and F. Ellis. 1992. Selective regime and time. Pages 6–22 in R. M. Cowling, editor. *The ecology of fynbos: nutrients, fire and diversity.* Oxford University Press, Cape Town, South Africa.

Deall, G. B., G. K. Theron, and R. H. Westfall. 1989. The vegetation ecology of the Eastern Transvaal Escarpment in the Sabie area: 2. floristic classification. *Bothalia* 19:69–90.

Dean, W. R. J. 2000. *The birds of Angola.* BOU Checklist No. 18. British Ornithologist's Union, Tring, UK.

Dean, W. R. J. and I. A. W. Macdonald. 1994. Historical changes in stocking rates of domestic livestock as a measure of semi-arid and arid rangeland degradation in the Cape Province, South Africa. *Journal of Arid Environments* 26:281–298.

Dean, W. R. J. and S. J. Milton. 1999. *The Karoo: ecological patterns and processes.* Cambridge University Press, Cambridge, UK.

Dean, W. R. J., S. J. Milton, M. K. Watkeys, and P. A. R. Hockey. 1991. Distribution, habitat preference and conservation status of the red lark *Certhilauda burra* in Cape Province, South Africa. *Biological Conservation* 58:257–274.

Decher, J. 1997. Conservation, small mammals, and the future of sacred groves in West Africa. *Biodiversity and Conservation* 6:1007–1026.

Deem, S. L., W. B. Karesh, and W. Weisman. 2001. Putting theory into practice: wildlife health in conservation. *Conservation Biology* 15:1224–1233.

Dehn, M. and L. Christiansen. 2001. Additions to the known avifauna of the Rwenzori Mountains National Park in western Uganda. *Scopus* 21:19–22.

Dejardin, J., J. L. Guillaumet, and M. Mangenot. 1973. Contribution à la connaissance de l'élément non éndemique de la flore malgache (végétaux vasculaires). *Candollea* 28:325–391.

de Klerk, H., T. Crowe, J. Fjeldså, and N. D. Burgess. 2002a. Biogeographical patterns of endemic terrestrial Afrotropical birds. *Diversity and Distributions* 8:147–162.

de Klerk, H., T. Crowe, J. Fjeldså, and N. D. Burgess. 2002b. Patterns in the distribution of Afrotropical birds. *Journal of Zoology* 256:327–342.

de Klerk, H. M., J. Fjeldså, S. Blyth, and N. D. Burgess. 2004. Gaps in the protected area network for threatened Afrotropical birds. *Biological Conservation* 117:529–537..

de Laubenfels, D. J. 1975. *Mapping the world's vegetation: regionalization of formations and flora.* Syracuse University Press, Syracuse, CA, USA.

de Menocal, P. B. 1995. Plio-Pleistocene African climate. *Science* 270:53–59.

Demissew, S. 1996. Ethiopia's natural resource base. Pages 36–53 in S. Tilahun, S. Edwards, and T. B. G. Egziabher, editors. *Important bird areas of Ethiopia.* Ethiopian Wildlife and Natural History Society, Semayata Press, Addis Ababa, Ethiopia.

Demy, R. and M. Louette. 2001. Democratic Republic of Congo. Pages 199–218 in L. D. C. Fishpool and Evans M. I., editors. *Important bird areas in Africa and associated islands: priority sites for conservation.* Pisces Publications and BirdLife International (BirdLife Conservation Series No. 11), Newbury and Cambridge, UK.

Denny, P. 1991. Africa. Pages 115–148 in M. Finlayson and M. Moser, editors. *Wetlands.* International Waterfowl and Wetlands Research Bureau. Facts on File, Oxford, UK.

Desanker, P. V., P. G. H. Frost, C. O. Justice, and J. J. Scholes. 1997. *The miombo network: framework for a terrestrial transect study of land-use and land-cover change in the miombo ecosystems of central Africa.* IGBP Report 41. The International Geosphere-Biosphere Programme (IGBP), Stockholm, Sweden.

Desegaulx de Nolet, A. 1984. *Lépidoptères de l'Océan Indien: Comores, Mascareignes, Seychelles.* ACCT, Paris, France.

de Smet, K. J. M. 1989. Distribution and habitat choice of larger mammals in Algeria, with special reference to nature protection. Ph.D. thesis. Ghent State University, Ghent, Belgium.

Desmet, P. G. 2000. The succulents of northern Bushmanland: their distribution and implications for conservation. *Aloe* 37:32–35.

Desmet, P. G., T. Barret, R. M. Cowling, A. G. Ellis, C. Heijnis, A. le Roux, A. T. Lombard, and R. L. Pressey. 1999. *A systematic plan for a protected area system in the Knersvlakte region of Namaqualand.* IPC Report 9901 submitted to the Trustees of the Leslie Hill Succulent Karoo Trust. Institute for Plant Conservation, University of Cape Town, Cape Town, South Africa.

Desmet, P. G. and R. M. Cowling. 1999a. Biodiversity, habitat and range-size aspects of a flora from a winter-rainfall desert in North-Western Namaqualand, South Africa. *Plant Ecology* 142:23–33.

Desmet, P. G. and R. M. Cowling. 1999b. The climate of the Karoo: a functional approach. Pages 3–16 in W. R. J. Dean and S. J. Milton, editors. *The Karoo: ecological patterns and processes.* Cambridge University Press, Cambridge, UK.

d'Hoore, J. L. 1964. *Soil map of Africa scale 1 to 5,000,000: explanatory monograph.* Commission for Technical Co-operation in Africa, Lagos, Nigeria.

Diallo, A. 1993. The mangroves of Guinea. Pages 47–57 in E. D. Diop, editor. *Conservation and sustainable utilization of mangrove forests in Latin America and Africa regions. Part II: Africa.* Mangrove Ecosystems Technical Reports, Volume 3. International Society for Mangrove Ecosystems and Coastal Marine Project of UNESCO, Okinawa, Japan.

Diamond, A. W. 1971. The ecology of the seabirds of Aldabra. *Philosophical Transactions of the Royal Society of London* B 260:561–571.

Diamond, A. W. 1994. Seabirds of the Seychelles, Indian Ocean. Pages 258–267 in D. N. Nettleship, J. Burger, and M. Gochfeld, editors. *Seabirds on islands: threats, case studies and action plans.* BirdLife International Conservation Series No. 1. BirdLife International, Cambridge, UK.

Diamond, A. W. and A. C. Hamilton. 1980. The distribution of forest passerine birds and Quaternary climatic change in tropical Africa. *Journal of Zoology, London* 191:379–402.

Dikobe, L. 1995. People and parks: do they mix? Pages 91–94 in K. Legget, editor. *Present status of wildlife and its future in Botswana.* Proceedings of a symposium organized by the Kalahari Conservation Society and Chobe Wildlife Trust, Gaborone, Botswana.

Dillon, T. C. and E. D. Wikramanayake. 1997. *A forum for transboundary conservation in Cambodia, Laos, and Vietnam.* WWF, Hanoi, Vietnam and Washington, DC, USA.

Dinerstein, E. 2003. *The return of the unicorns: the natural history and conservation of the greater one-horned rhinoceros.* Columbia University Press, New York, USA.

Dinerstein, E., D. Olson, D. Graham, A. Webster, S. Primm, M. Bookbinder, and G. Ledec. 1995. *A conservation assessment of the terrestrial ecoregions of Latin America and the Caribbean.* World Bank, Washington, DC, USA.

Dinerstein, E., G. Powell, D. Olson, E. Wikramanayake, R. Abell, C. Loucks, E. Underwood, T. Allnutt, W. Wettengel, T. Ricketts, H. Strand, S. O'Connor, and N. Burgess. 2000. *A workbook for conducting biological assessments and developing biodiversity visions for ecoregion-based conservation: part I, terrestrial ecoregions.* World Wildlife Fund, Washington, DC.

Dinerstein, E. and E. D. Wikramanayake. 1993. Beyond "hotspots": how to prioritize investments to conserve biodiversity in the Indo-Pacific region. Conservation Biology 7:53–65.

Dinesen, L., T. Lehmberg, J. O. Svendsen, L. A. Hansen, and J. Fjeldså. 1994. A new genus and species of perdicine bird (Phasianidae, Perdicini) from Tanzania: a relict form with Indo-Malayan affinities. *Ibis* 136:3–11.

Diop, E. D. (ed.) 1993. *Conservation and sustainable utilization of mangrove forests in Latin America and Africa regions. Part II: Africa.* Mangrove Ecosystems Technical Reports, Volume 3. International Society for Mangrove Ecosystems and Coastal Marine Project of UNESCO. Okinawa, Japan.

Diop, E. S. and M. Bâ. 1993. The mangroves of Sénégal and Gambia. Pages 19–35 in E. D. Diop, editor. *Conservation and sustainable utilization of mangrove forests in Latin America and Africa regions. Part II: Africa.* Mangrove Ecosystems Technical Reports, Volume 3. International Society for Mangrove Ecosystems and Coastal Marine Project of UNESCO. Okinawa, Japan.

Djebaili, S. 1978. Recherches phytosociologiques et écologiques sur la végétation des hautes plaines steppiques et de l'Atlas saharien Algérien. Thèse Doc. University of Montpellier, Montpellier, France.

Djebaili, S. 1990. Syntaxonomie des groupements prèforetsiers et steppiques de l'Algérie aride. *Ecologia Mediterranea* XVI:231–244.

Djellouli, Y. 1990. Flores et climats en Algérie septentrionale. Déterminismes climatiques de la répartition des plantes. Thèse Doc. ès Sc. USTHB, Algiers, Algeria.

Dobson, A. P. 1996. *Conservation and biodiversity.* Freeman and Company, New York, USA.

Dobson, J. E., E. A. Bright, P. R. Coleman, R. C. Durfee, and B. A. Worley. 2000. LandScan: a global population database for estimating populations at risk. *Photogrammetric Engineering and Remote Sensing* 66:849–857.

Dodman, T., H. Y. Béibro, E. Hubert, and E. Williams. 1999. *African waterbird census 1998.* Wetlands International, Wageningen, The Netherlands.

Dodman, T., C. de Vaan, E. Hubert, and C. Nivet. 1997. *African waterfowl census 1997.* Wetlands International, Wagingen, The Netherlands.

Dodman, T., B. Kamweneshe, D. Kamweneshe, V. Katanekwa, and L. Thole. 1996. Zambia crane and wetland action plan. In R. D. Beilfuss, W. R. Tarboton, and N. N. Gichuki, editors. *Proceedings of the African Crane and Wetland Training Workshop.* Wildlife Training Institute, Botswana. International Crane Foundation, Baraboo, WI, USA.

Donque, G. 1972. The climatology of Madagascar. Pages 87–144 in R. Battistini and G. Richard-Vindard, editors. *Biogeography and ecology of Madagascar.* Dr. W. Junk Publishers, The Hague, The Netherlands.

Dorr, L. J. and E. G. H. Oliver. 1999. New taxa, names, and combinations in *Erica* (Ericaceae-Ericoideae) from Madagascar and the Comoro Islands. *Adansonia,* sér 3 21:75–91.

Douady, C. J., F. Catzeflis, D. J. Kao, M. S. Springer, and M. J. Stanhope. 2002. Molecular evidence for the monophyly of Tenrecidae (Mammalia) and the timing of the colonization of Madagascar by Malagasy tenrecs. *Molecular Phylogenetics and Evolution* 22:357–363.

Doumenge, C. 1990. *La conservation des ecosystémes forestiers du Zaire.* IUCN, Gland, Switzerland.

Dowsett, R. 1986. Origins of the high-altitude avifaunas of tropical Africa. Pages 557–585 in F. Vuilleumier and M. Monasterio, editors. *High altitude tropical biogeography.* Oxford University Press, New York, USA.

Dowsett, R. J., editor. 1989. *A preliminary natural history survey of Mambilla Plateau and some lowland forests of eastern Nigeria.* Tauraco Research Report No. 1. Tauraco Press, Ely, UK.

Dowsett, R. J. and F. Dowsett-Lemaire. 1993. *A contribution to the distribution and taxonomy of Afrotropical and Malagasy birds.* Tauraco Press, Liège, Belgium.

Dowsett, R. J. and F. Dowsett-Lemaire, editors. 1997. *Flore et faune du Parc National d'Odzala, Congo.* Tauraco Research Report No. 6. Tauraco Press, Liège, Belgium.

Dowsett, R. J. and A. D. Forbes-Watson. 1993. *Checklist of birds of the Afrotropical and Malagasy regions. Volume 1: Species limits and distribution.* Tauraco Press, Liège, Belgium.

Dowsett-Lemaire, F. 1988. The forest vegetation of Mt. Mulanje (Malawi): a floristic and chorological study along an altitudinal gradient (650–1950 m). *Bulletin du Jardin Botanique National de Belgique* 58:77–108.

Dowsett-Lemaire, F. 1989a. Ecological and biogeographical aspects of forest bird communities in Malawi. *Scopus* 13:1–80.

Dowsett-Lemaire, F. 1989b. The flora and phytogeography of the evergreen forests of Malawi: I. Afromontane and mid-altitude forests. *Bulletin du Jardin Botanique National de Belgique* 59:3–132.

Dowsett-Lemaire, F. 1990. The flora and phytogeography of the evergreen forests of Malawi: II. Lowland forests. *Bulletin du Jardin Botanique National de Belgique* 60:9–71.

Dowsett-Lemaire, F. 1996. Composition et evolution de la végétation forestière au Parc National d'Odzala Congo. *Bulletin du Jardin Botanique National de Belgique* 65:253–292.

Dowsett-Lemaire, F. 1997. The avifauna of Odzala National Park, northern Congo. In R. J. Dowsett and F. Dowsett-Lemaire, editors. *Flore et faune du Parc National d'Odzala, Congo.* Tauraco Research Report No. 6. Tauraco Press, Liège, Belgium.

Dowsett-Lemaire, F. 2001. Congo. Pages 191–218 in L. D. C. Fishpool and M. I. Evans, editors. *Important bird areas of Africa and associated islands: priority sites for conservation.* Pisces Publications and BirdLife International (BirdLife Conservation Series No. 11), Newbury and Cambridge, UK.

Dowsett-Lemaire, F. and R. J. Dowsett. 1998. *Zoological surveys of small mammals, birds and frogs in the Bakossi and Kupe Mts., Cameroon.* World Wildlife Fund–Cameroon, Yaoundé, Cameroon.

Dowsett-Lemaire, F. and R. J. Dowsett. 2000. *Further biological surveys of Manenguba and central Bakossi in March 2000, and an evaluation of the conservation importance of Manenguba, Bakossi, Kupe and Nlonako Mts., with special reference to birds.* World Wildlife Fund–Cameroon, Yaoundé, Cameroon.

Dowsett-Lemaire, F. and R. J. Dowsett. 2001. *Birds and mammals of Mt. Cameroon: an update of the state of knowledge and further fieldwork around Mann's Spring.* World Wildlife Fund–Cameroon, Yaoundé, Cameroon.

Dowsett-Lemaire, F., R. J. Dowsett, and M. Dyer. 2001. Malawi. Pages 539–556 in L. D. C. Fishpool and M. I. Evans, editors. *Important bird areas in Africa and associated islands: priority sites for conservation.* Pisces Publications and BirdLife International (BirdLife Conservation Series No. 11), Newbury and Cambridge, UK.

DPNRF (Direction des Parcs Nationaux et Réserves de Faune). 1997. Zakouma: projet conservation de l'environnement dans le sud-est du Tchad. *Rapport d'Activité* 1996–1997. Direction des Parcs Nationaux et Réserves de Faune, Ministère de l'Environnement et de l'Eau, République du Tchad.

Dransfield, J. and H. Beentje. 1995. *The palms of Madagascar.* Royal Botanic Gardens, Kew, London.

Draulans, D. and E. Van Krunkelsven. 2002. The impact of war on forest areas in the Democratic Republic of Congo. *Oryx* 36:35–40.

Dublin, H. T. 1991. Dynamics of the Serengeti-Mara woodlands: an historical perspective. *Forest and Conservation History* 35:169–178.

Dubois, O. and J. Lowore. 2000. *The journey towards collaborative forest management in Africa: lessons learned and some navigational aids. An overview.* IIED, Forestry and Land Use Series No. 15. IIED, London, UK.

Duellman, W. E. 1993. *Amphibian species of the world: additions and corrections.* University of Kansas, Lawrence, USA.

Duellman, W. E., editor. 1999. *Patterns of distribution of amphibians: a global perspective.* Johns Hopkins University Press, Baltimore, MD, USA.

Dumont, H. J. 1992. The regulation of plant and animal species and communities in African shallow lakes and wetlands. *Revue d'Hydrobiologie Tropicale* 25:303–346.

du Plessis, M. A. 1995. The effects of fuelwood removal on the diversity of some cavity-using birds and mammals in South Africa. *Biological Conservation* 74:77–82.

du Plessis, W. 1992. In situ conservation in Namibia: the role of national parks and nature reserves. *Dinteria* 23:132–141.

du Puy, D. J. and J. Moat. 1996. A refined classification of the primary vegetation of Madagascar based on the underlying geology: using GIS to map its distribution and to assess its conservation status. Pages 205–218 + 3 maps in W. R. Lourenço, editor. *Biogéographie de Madagascar.* Editions de l'ORSTOM, Paris, France.

du Toit, A. L. 1966. *The geology of South Africa,* 3rd edition. Oliver and Boyd, Edinburgh, Scotland and London, UK.

Dutton, J. 1994. Introduced mammals in São Tomé and Príncipe: possible threats to biodiversity. *Biodiversity and Conservation* 3:927–938.

Duvall, C. S. 2001. Habitat, conservation and use of *Gilletio-dendron glandulosum* (Fabaceae-Caesalpinoideae) in south-western Mali. Pages 699–737 in E. Robbrecht, J. Degreef, and I. Friis, editors. *Plant systematics and phytogeography for the understanding of African biodiversity,* 71. Proceedings of the XVIth AETFAT Congress. Systematics and Geography of Plants, National Botanic Garden of Belgium, Meise, Belgium.

Dwasi, J. 2002a. *Findings from Kenya, Namibia, South Africa and Uganda on the direct and indirect impacts of the HIV/AIDS pandemic on management and conservation of natural resources in Africa: lessons learned on strategies for coping with the impacts.* Africa Biodiversity Collaborative Group, Washington, DC, USA.

Dwasi, J. 2002b. *Impacts of HIV/AIDS on natural resources management and conservation in Africa: case studies of Botswana, Kenya, Namibia, Tanzania and Zimbabwe.* International Resources Group, Washington, DC, USA.

Dye, C. 1996. Serengeti wild dogs: what really happened? *Trends in Ecology and Evolution* 11:188–189.

East, M. L. and H. Hofer. 1996. Wild dogs in the Serengeti ecosystem: what really happened. *Trends in Ecology and Evolution* 11:509.

East, R., editor. 1997. *Antelope survey update.* No. 4 (Feb. 1997) IUCN/Species Survival Commission Antelope Specialist Group Report. IUCN, Gland, Switzerland.

East, R., compiler. 1999. *African antelope database 1998.* Occasional Paper of the IUCN Species Survival Commission No. 21. IUCN, Gland, Switzerland.

East, R. 2000. Antelope captures in Niokola-Koba National Park, Senegal. *Gnusletter* 19:4–6.

Ebedes, H. 1976. Anthrax epizootics in Etosha National Park. *Madoqua* 10:99–118.

Eckhardt, H. C., N. Van Rooyen, G. J. Bredenkamp, and G. K. Theron. 1993. An overview of the vegetation of the Vrede-Memel-Warden Area, northeastern Orange Free State. *South African Journal of Botany* 59:391–400.

Edmonds, A. C. R. 1976. *Vegetation map (1:500000) of Zambia.* Surveyor General, Lusaka, Zambia.

Ehrlich, D. and E. F. Lambin. 1996. Broad scale land-cover classification and interannual climate variability. *International Journal of Remote Sensing* 17:845–862.

Eig, A. 1931. Les éléments et les groupes phytogéographiques auxiliaires dans la flore Palestinienne. *Fedde Repert* 63:470–496.

Eilu, G., H. Arinaitwe, and H. Bakamwesiga. 2001. *The Albertine Rift Area of Endemism (ARAE): a contribution towards its conservation.* Technical Report to World Wildlife Fund. MUIENR, Kampala, Uganda.

Elenga, H., O. Peyron, R. Bonnfille, D. Jolly, R. Cheddadi, J. Guiot, V. Andrieu, S. Bottema, G. Butchet, J. L. de Beaulieu, A. C. Hamilton, J. Maley, J. Marchant, R. Perez-Obiol, M. Reille, G. Riollet, L. Scott, H. Straka, D. Taylor, E. Van Campo, A. Vincens, F. Laarif, and H. Jonson. 2000. Pollen-based biome reconstruction for southern Europe and Africa 18,000 yr B.P. *Journal of Biogeography* 27:621–634.

El Hadidi, M. N. and H. A. Hosni. 1994. Biodiversity in the flora of Egypt. In L. J. G. Van der Masen, X. M. van der Burgt, and J. M. van Medenbach de Rooy, editors. *The biodiversity of African plants: proceedings XIVth AETFAT Congress.* Wageningen, The Netherlands. Kluwer Academic Press, Dordrecht, The Netherlands.

Elkan, P., A. Moukassa, and S. Elkan. 2002. Follow up to an approach to wildlife management in a forest concession in northern Congo. Pages 121–122 in S. Manika and M. Trivedi, editors. *Links between biodiversity conservation, livelihoods and food security.* IUCN Species Survival Commission, Occasional Paper No. 24. IUCN, Gland Switzerland and Cambridge, UK.

Ellis, C. G. 1986. Medicinal plant use: a survey. *Veld and Flora* 72:72–73.

Ellis, J. and K. A. Galvin. 1994. Climate patterns and land-use practices in the dry zones of Africa: comparative regional analysis provides insight into the effects of climate variations. *BioScience* 44:340–349.

El-Raey, M., Y. Fouda, and S. Nasr. 1997. GIS assessment of the vulnerability of the Rosetta area, Egypt to impacts of sea rise. *Environmental Monitoring and Assessment* 47:59–77.

Els, H. 1996. Game ranching and rural development. Pages 581–591 in J. D. P. Bothma, editor. *Game ranch management.* Van Schaik, Pretoria, South Africa.

Els, H. and J. D. P. Bothma. 2000. Developing partnerships in a paradigm shift to achieve conservation reality in South Africa. *Koedoe* 43:19–26.

Elsokkary, I. H. 1996. Synopsis on contamination of the agricultural ecosystem by trace elements: an emerging environmental problem. *Egyptian Journal of Soil Science* 36:1–22.

EMBL (European Molecular Biology Laboratory reptiles database). 1996. Retrieved 2000 from the World Wide Web: http://www.embl-heidelberg.de/~uetz/LivingReptiles.html.

Emslie, R. and M. Brooks, compilers. 1999. *African rhino: status survey and conservation action plan.* IUCN/Species Survival Commission African Rhino Specialist Group, IUCN, Gland, Switzerland and Cambridge, UK.

Engler, A. and L. Diels. 1936. *Syllabus der Pflanzenfamilie,* 2nd edition. G. Borntraeger, Berlin, Germany.

Entwistle, A. C. and N. Dunstone. 2000. Future priorities for mammalian conservation. Pages 369–387 in A. C. Entwistle and N. Dunstone, editors. *Priorities for the conservation of mammalian biodiversity: has the panda had its day?* Cambridge University Press, Cambridge, UK.

Epstein, P. R., H. F. Diaz, S. Elias, G. Grabherr, N. E. Graham, W. J. M. Martens, E. Mosley-Thompson, and J. Susskind. 1998. Biological and physical signs of climate change: focus on mosquito-borne diseases. *Bulletin of the American Meteorological Society* 79:409–417.

ESRI. 1993. *Digital chart of the world.* Environmental Systems Research Institute, Redlands, CA, USA.

Estes, R. D. 1991. *The behavior guide to African mammals.* University of California Press, Berkeley, USA.

Eswaran, H., P. Reich, and F. Beinroth. 2001. Global desertification tension zones. Pages 24–28 in D. E. Stott, R. H. Mohtar, and G. C. Steinhardt, editors. Sustaining the global farm: selected papers from the 10th International Soil Conservation Organization Meeting, May 24–29, 1999. International Soil Conservation Organization in cooperation with the USDA and Purdue University, West Lafayette, USA.

Evans, M., compiler. 1994. *Important bird areas of the Middle East.* BirdLife Conservation Series No. 2. BirdLife International, Cambridge, UK.

Everard, D. A. 1987. A classification of the subtropical transitional thicket in the Eastern Cape, based on syntaxonomic and structural attributes. *South African Journal of Botany* 53:329–340.

Everard, D. A. 1988. Threatened plants of the Eastern Cape: a synthesis of collection records. *Bothalia* 18:271–278.

Everard, D. A., G. F. van Wyck, and J. J. Midgley. 1994. Disturbance and the diversity of forests in Natal, South Africa: lessons for their utilization. *Strelitzia* 1:275–285.

Eves, H. E. and R. G. Ruggerio. 2002. Antelopes in Africa: bushmeat, game meat and wild meat a question of sustainability. Pages 73–84 in S. Manika and M. Trivedi, editors. *Links between biodiversity conservation, livelihoods and food security.* IUCN Species Survival Commission, Occasional Paper No. 24. IUCN, Gland, Switzerland and Cambridge, UK.

Ezealor, A. U. 2001. Nigeria. Pages 673–692 in L. D. C. Fishpool and M. I. Evans, editors. *Important bird areas in Africa and associated islands: priority sites for conservation.* Pisces Publications and BirdLife International (BirdLife Conservation Series No. 11), Newbury and Cambridge, UK.

Fa, J. E. 1991. *Conservación de los ecosistemas forestales de Guinea Ecuatorial.* IUCN, Gland, Switzerland and Cambridge, UK.

Fa, J. E. 2000. Hunted animals in Bioko Island, West Africa: sustainability and future. Pages 168–198 in J. G. Robinson and E. L. Bennett, editors. *Hunting for sustainability in tropical forests.* Columbia University Press, New York, USA.

Fairhead, J. and M. Leach. 1996. *Misreading the African landscape: society and ecology in a forest-savanna mosaic.* Cambridge University Press, Cambridge, UK.

Fairhead, J. and M. Leach. 1998a. Reconsidering the extent of deforestation in twentieth century West Africa. *Unasylva* 192:38–46.

Fairhead, J. and M. Leach. 1998b. *Reframing deforestation: global analysis and local realities: studies in West Africa.* Routledge, London, UK.

Fairhead, J. and M. Leach. 2002. After desolation, conservation—and eviction: the future of the West African forests and their peoples. *Times Literary Supplement,* May 5.

Faniran, A. and L. K. Jeje. 1983. *Humid tropical geomorphology: a study of the geomorphological processes and landforms in warm humid climates.* Longman, London, UK.

Fanshawe, D. B. 1960. Evergreen forest relics in northern Rhodesia. *Kirkia* 1:20–24.

Fanshawe, D. B. 1969. *The vegetation of Zambia.* Forest Research Bulletin No. 7. Government Printer, Lusaka, Zambia.

FAO. 1997. *Irrigation potential in Africa: a basin approach.* FAO Land and Water Bulletin 4, Rome, Italy.

FAO. 1998. Mali. *Fishery country profile.* Food and Agriculture Organization of the United Nations, Rome, Italy.

FAO. 1999. *State of the world's forests.* Food and Agriculture Organization of the United Nations, Rome, Italy.

FAO. 2000. *Forest resources of Europe, CIS, North America, Australia, Japan and New Zealand (industrialized temperate/boreal countries): UN-ECE/FAO contribution to the Global Forest Resources Assessment 2000, United Nations.* United Nations, New York, USA.

FAO. 2001a. Cattle stall-feeding in the Mandara Mountains region. Retrieved 2001 from the World Wide Web: http://www.fao.org/wairdocs/ilri/x5511e/x5511e02.htm.

FAO. 2001b. *State of the world's forests.* Food and Agriculture Organization of the United Nations, Rome, Italy.

FAOSTAT. 1998. FAOSTAT database: agriculture. Retrieved 2001 from the World Wide Web: http://apps.fao.org/page/collections?subset=forestry.

Farjon, A. and C. N. Page. 1999. *Conifers: status survey and conservation action plan.* IUCN/SSC Conifer Specialist Group, IUCN, Gland Switzerland and Cambridge, UK.

Farrell, J. A. K. 1968. Preliminary notes on the vegetation of the lower Sabi-Lundi basin, Rhodesia. *Kirkia* 6:223–248.

Fay, J. M. and M. Agnagna. 1991. Forest elephant populations in the Central African Republic and Congo. *Pachyderm* 14:3–19.

Fay, J. and M. Agnagna. 1992. Census of gorillas in northern Republic of Congo. *American Journal of Primatology* 27:275–284.

Fay, J. M., M. Agnagna, J. Moore, and R. Oko. 1989. Gorilla (*Gorilla gorilla gorilla*) in the Likouala swamp forests of North Central Congo: preliminary data on populations and ecology. *International Journal of Primatology* 10:477–486.

Feduccia, A. 1995. Explosive evolution in Tertiary birds and mammals. *Science* 267:637–638.

Feely, J. M. 1980. Did iron age man have a role in the history of Zululand's wilderness landscapes? *South African Journal of Science* 76:150–152.

Fennane, M. 1989. Esquisse des séries du thuya de Bérbérie au Maroc. Bulletin de l'Institut Scientifique, *Rabat* 13:77–83.

Ferraro, P. J. and A. Kiss. 2002. Ecology: direct payments to conserve biodiversity. *Science* 298:1718–1719.

Ferrier, S., R. L. Pressey, and T. W. Barrett. 2000. A new predictor of the irreplaceability of areas for achieving a conservation goal, its application to real-world planning, and a research agenda for further refinement. *Biological Conservation* 93:303–325.

Figueiredo, E. 1994. Diversity and endemism of angiosperms in the Gulf of Guinea islands. *Biodiversity and Conservation* 3:785–793.

Figueiredo, E. 1998. The pteridophytes of São Tomé and Príncipe (Gulf of Guinea). *Bulletin of the Natural History Museum,* London (Botany) 28:41–66.

Fimbel, C. and R. Fimbel. 1997. Rwanda: the role of local participation. *Conservation Biology* 11:309–310.

Finlayson, M. and M. Moser. 1991. *Wetlands.* International Waterfowl and Wetlands Research Bureau (IWRB). Facts on File, Oxford, UK.

Fishpool, L. D. C. and M. I. Evans, editors. 2001. *Important bird areas in Africa and associated islands: priority sites for conservation.* Pisces Publications and BirdLife International (BirdLife Conservation Series No. 11), Newbury and Cambridge, UK.

Fishpool, L. D. C., M. F. Heath, Z. Waliczky, D. C. Wege, and M. J. Crosby. 1998. Important bird areas: criteria for selecting sites of global conservation significance. *Ostrich* 69:428.

FitzGibbon, C. D., H. Mogaka, and J. H. Fanshawe. 1996. Subsistence hunting and mammal conservation in Kenyan coastal forest: resolving a conflict. Pages 147–159 in V. J. Taylor and N. Dunstone, editors. *The exploitation of mammal populations.* Chapman and Hall, London, UK.

FitzGibbon, C. D., H. Mogaka, and J. H. Fanshawe. 2000. Threatened mammals, subsistence harvesting, and high human population densities: a recipe for disaster? Pages 154–167 in J. G. Robinson and E. L. Bennett, editors. *Hunting for sustainability in tropical forests.* Columbia University Press, New York, USA.

Fjeldså, J. 1994. Geographical patterns for relict and young species in Africa and South America and the dilemma of ranking biodiversity. *Biodiversity and Conservation* 3:207–226.

Fjeldså, J. 1999. The impact of human forest disturbance on the endemic avifauna of the Udzungwa Mountains, Tanzania. *Bird Conservation International* 9:47–62.

Fjeldså, J. 2000. The relevance of systematics in choosing priority areas for global conservation. *Environmental Conservation* 27:65–75.

Fjeldså, J., M. K. Bayes, M. W. Bruford, and M. S. Roy. 2000. Biogeography and diversification of African forest faunas: implications for conservation. In C. Moritz, editor. *Rainforests past and future.* Chicago University Press, Chicago, USA.

Fjeldså, J., N. D. Burgess, S. Blyth, and H. M. de Klerk. 2004. Where are the major gaps in the reserve network for Africa's mammals? *Oryx* 38:17–25.

Fjeldså, J., D. Ehrlich, E. Lambin, and E. Prins. 1997. Are biodiversity "hotspots" correlated with current ecoclimatic stability? A pilot study using the NOAA-AVHRR remote sensing data. *Biodiversity and Conservation* 6:401–422.

Fjeldså, J. and J. C. Lovett. 1997. Geographical patterns of old and young species in African forest biota: the significance of specific montane areas as evolutionary centers. *Biodiversity and Conservation* 6:325–347.

Fjeldså, J. and C. Rahbek. 1998. Continent-wide conservation priorities and diversification processes. Pages 139–160 in G. M. Mace, A. Balmford, and J. R. Ginsberg, editors. *Conservation in a changing world.* Cambridge University Press, Cambridge, UK.

Forests Monitor. 2001. *Sold down the river: the need to control transnational forestry corporations: a European case study.* Forests Monitor, Cambridge, UK.

Forman, R. T. T. 1998. *Land mosaics: the ecology of landscapes and regions.* Cambridge University Press, Cambridge UK.

Fosberg, F. R. and S. A. Renvoize. 1980. The flora of Aldabra and neighbouring islands. *Kew Bulletin Additional Series* 7:1–358.

Fossati, J., G. Pautou, and J. Peltier. 1999. Water as resource and disturbance for Wadi vegetation in a hyperarid area (Wadi Sannur, Eastern Desert, Egypt). *Journal of Arid Environments* 43:63–77.

Fossey, D. 1983. *Gorillas in the mist.* Houghton Mifflin Co., Boston, USA.

Fotso, R., F. Dowsett-Lemaire, R. J. Dowsett, Cameroon Ornithological Club, P. Scholte, M. Languy, and C. Bowden. 2001. Cameroon. Pages 133–159 in L. D. C. Fishpool and M. I. Evans, editors. *Important bird areas of Africa and associated islands: priority sites for conservation.* BirdLife Conservation Series No. 11. Pisces Publications and BirdLife International, Newbury and Cambridge, UK.

Franciscolo, M. E. 1982. Some new records of Gyrinidae (Coleoptera) from Sierra Leone. Ricerche Biologiche in Sierra Leone. *Accademia Nazionale dei Lincei* 255:63–82.

Franciscolo, M. E. 1994. Three new *Africophilus* Guignot and new records of Gyrinidae and Dytiscidae from Sierra Leone (Coleoptera). Ricerche Biologiche in Sierra Leone (Parte IV). *Accademia Nazionale dei Lincei* 267:267–298.

Franco, J. A. 1986. Género *Abies.* Pages 164–167 in S. Castroviejo, editor. *Flora ibérica: plantas vasculares de la Península Ibérica e Islas. Real Jardín Botánico, Madrid, Volume I.* Consejo Superior de Investigaciones Cientificas, Madrid, Spain.

Frazee, S. R., R. M. Cowling, R. L. Pressey, J. K. Turpie, and N. Lindberg. 2003. Estimating the costs of conserving a biodiversity hotspot: a case study of the Cape Floristic Region, South Africa. *Biological Conservation* 112:275–290.

Frazier, S., editor. 1999. *Directory of wetlands of international importance.* Wetlands International and Ramsar Convention Bureau, Waginingen, The Netherlands.

Fredoux, A. 1994. Pollen analysis of a deep-sea core in the Gulf of Guinea: vegetation and climatic changes during the last 225,000 years B.P. *Palaeogeography, Palaeoclimatology, Palaeoecology* 109:317–330.

Freitag, S., A. O. Nicholls, and A. S. van Jaarsveld. 1996. Nature reserve selection in the Transvaal, South Africa: what data should we be using? *Biodiversity and Conservation* 5:685–698.

Freitag, S., A. S. van Jaarsveld, and H. C. Biggs. 1997. Ranking priority biodiversity areas: an iterative conservation value-based approach. *Biological Conservation* 82:263–272.

Friedman, F. 1994. Seychelles. Pages 288–292 in WWF and IUCN, editor. *Centres of plant diversity. Volume 1: Europe, Africa, South West Asia, and the Middle East.* IUCN Publications Unit, Cambridge, UK.

Friis, I. 1992. *Forests and forest trees of northeast tropical Africa: their natural habitats and distribution patterns in Ethiopia, Djibouti and Somalia.* Kew Bulletin Additional Series XV, Her Majesty's Stationery Office, London, UK.

Friis, I. 1998. Frank White and the development of African chorology. Pages 25–51 in C. R. Huxley, J. M. Lock, and D. F. Cutler, editors. *Chorology, taxonomy and ecology of the floras of Africa and Madagascar.* Royal Botanic Gardens, Kew, UK.

Friis, I. and S. Demissew. 2001. Vegetation maps of Ethiopia and Eritrea. A review of existing maps and the need for a new map for the flora of Ethiopia and Eritrea. *Kongelige Danske Videnskabernes Selskab Biologiske Skrifter* 54:399–439.

Friis, I., S. Edwards, E. Kelbessa, and S. Demissew. 2001. Diversity and endemism in the flora of Ethiopia and Eritrea: what do the published flora volumes tell us? *Kongelige Danske Videnskabernes Selskab Biologiske Skrifter* 54:173–193.

Friis, I. and O. Ryding. 2001. Biodiversity research in the Horn of Africa region. *Kongelige Danske Videnskabernes Selskab Biologiske Skrifter* 54:1–439.

Frost, D. R. 1999. Amphibian species of the world: an online reference. Version 2.1 (15 November 1999). American Museum of Natural History, New York. Retrieved 2000 from http://research.amnh.org/herpetology/amphibia/index.html.

Frost, P. 1996. The ecology of miombo woodlands. Pages 11–58 in B. Campbell, editor. *The miombo in transition: woodlands and welfare in Africa.* Centre for International Forestry Research, Bogor, Indonesia.

Frost, P. G. H. and F. Robertson. 1987. The ecological effects of fire in savannas. Pages 93–140 in B. H. Walker, editor. *Determinants of tropical savannas.* IUBS, Paris, France.

Frost, P., J. Timberlake, and E. Chidumayo. 2002. Miombo-mopane woodlands and grasslands. Pages 182–208 in R. A. Mittermeier, C. G. Mittermeier, P. R. Gil, J. Pilgrim, G. Fonseca, T. Brooks, and W. R. Konstant, editors. *Wilderness: Earth's last wild places.* CEMEX, Mexico City, Mexico.

Frumhoff, P. 1995. Conserving wildlife in tropical forests managed for timber. *BioScience* 45:456–464.

Fry, C. H., S. Keith, and E. K. Urban. 1988. *The birds of Africa, Volume 3.* Academic Press, London, UK.

Fryxell, J. M. and A. R. E. Sinclair. 1988. Seasonal migration by white-eared kob in relation to resources. *African Journal of Ecology* 26:17–32.

Fuls, E. R., G. J. Bredenkamp, V. Van Rooyen, and G. K. Theron. 1993. The physical environment and major plant communities of the Heilbron-Lindley-Warden-Villiers area, Northern Orange Free State. *South African Journal of Botany* 59:345–359.

Furley, P. A., J. Proctor, and J. A. Ratter, editors. 1992. *Nature and dynamics of forest savanna boundaries.* Chapman and Hall, London, UK.

Galaty, J. G. and P. Bonte. 1992. *Herders, warriors and traders: pastoralism in Africa.* Westview Press, Boulder, Colorado, USA.

Gallais, J. 1967. *Le delta intérieur du Niger, études de géographie régionale.* Mémoires de l'Institut Fondamental d'Afrique Noir, no. 79, IFAN, Dakar, Senegal.

Ganzhorn, J. U., J. Fietz, E. Rakotovao, D. Schwab, and D. Zinner. 1999. Lemurs and the regeneration of dry deciduous forest in Madagascar. *Conservation Biology* 13:794–804.

Ganzhorn, J. U., P. P. Lowry, G. E. Schatz, and S. Sommer. 2001. The biodiversity of Madagascar: one of the world's hottest hotspots on its way out. *Oryx* 35:346–348.

Ganzhorn, J. U., B. Rakotosamimanana, L. Hannah, J. Hough, L. Iyer, S. Olivieri, S. Rajaobelina, C. Rodstrom, and G. Tilkin. 1997. Priorities for biodiversity conservation in Madagascar. *Primate Report* 48–1:1–81.

Garba-Boyi, M., N. D. Burgess, and K. G. Smith. 1993. Ornithological significance of the Hadejia-Nguru wetlands, northern Nigeria. Proceedings VIII Pan-African Ornithological Congress, 509–514.

Garbutt, N. 1999. *Mammals of Madagascar.* Pica Press, Sussex, UK.

García, R., G. Ortega, and J. M. Pérez. 1992. Insectos de Canarias. Ed. del Cabildo Insular de Gran Canaria, Las Palmas, The Canary Islands, Spain.

Gardiner, A. 1994. Insects. Pages 108–116 in J. Timberlake and

P. Shaw, editors. *Chirinda Forest: a visitor's guide.* Forestry Commission, Harare, Zimbabwe.

Gardner, A. S. 1986. *Herpetofauna of the Seychelles.* Herpetological Society Bulletin No. 16. The British Herpetological Society, London, UK.

Garnett, T. and C. Utas. 2000. *The Upper Guinea heritage: nature conservation in Liberia and Sierra Leone.* IUCN-Netherlands, Amsterdam, The Netherlands.

Garshelis, D. L. 2000. Delusions in habitat evaluation: measuring use, selection, and importance. Pages 111–164 in L. Boitani and T. K. Fuller, editors. *Research techniques in animal ecology.* Columbia University Press, New York, USA.

Gartlan, S. 1989. *La conservation des ecosystémes forestiers du Cameroun.* IUCN, Gland, Switzerland and Cambridge, UK.

Gascoigne, A. 1994a. The biogeography of land snails in the islands of the Gulf of Guinea. *Biodiversity and Conservation* 3:794–807.

Gascoigne, A. 1994b. The dispersal of terrestrial gastropod species in the Gulf of Guinea. *Journal of Conchology* 35:1–7.

Gascon, C., G. B. Williamson, and G. A. B. da Fonseca. 2000. Receding forest edges and vanishing reserves. *Science* 288:1356–1358.

Gasse, F., R. Tehet, A. Durand, E. Gibert, and J. C. Fontes. 1990. The arid transition in the Sahara and the Sahel during the last deglaciation. *Nature* 346:141–146.

Gaston, K. J. 1991. How large is a species' geographic range? *Oikos* 61:434–437.

Gaston, K. J. 2000. Global patterns in biodiversity. *Nature* 405:220–227.

Gathaara, G. N. 1999. *Aerial survey of the destruction of Mt. Kenya, Imenti and Ngara Ndare forest reserves: February–June 1999.* Kenya Wildlife Service, Nairobi, Kenya.

Gatter, W. 1997. *Birds of Liberia.* Yale University Press, New Haven, CT, USA.

Gaudian, G., A. Koyo, and S. Wells. 1995. Marine Region 12: East Africa. Pages 71–105 in G. Kelleher, C. Bleakley, and S. Wells, editors. *A global representative system of marine protected areas. Volume III: Central Indian Ocean, Arabian Seas, East Africa and East Asian Seas.* Great Barrier Reef Marine Park Authority, The World Bank, The World Conservation Union (IUCN). The World Bank Environment Department, Washington, DC, USA.

Gautier, L. and S. M. Goodman. 2002. Inventaire floristique et faunistique de la Réserve Spéciale de Manongarivo (NW Madagascar). *Boissiera* 59:1–435.

Geerling, C. 1985. The status of the woody species of the Sudan and Sahel zones of West Africa. *Forest Ecology and Management* 13:247–256.

GEF. 1999. *Experience with conservation trust funds.* Evaluation report 1-99. Global Environment Facility, Washington, DC, USA.

Geldenhuys, C. J. 1989. *Environmental and biogeographic influences on the distribution and composition of the southern cape forests (Veld type 4).* Department of Botany, University of Cape Town, South Africa.

Geldenhuys, C. J. 1992. Richness, composition and relationships of the floras of selected forests in southern Africa. *Bothalia* 22:205–233.

Geldenhuys, C. J., P. J. Le Roux, and K. H. Cooper. 1986. Alien invasions in indigenous evergreen forest. Pages 119–131 in A. MacDonald, F. J. Kruger, and A. A. Ferrar, editors. *The ecology and management of biological invasions in southern Africa.* Oxford University Press, Cape Town, South Africa.

Geldenhuys, C. J. and D. R. MacDevette. 1989. Conservation status of coastal and montane evergreen forest. Pages 224–235 in B. J. Huntley, editor. *Biotic diversity in southern Africa: concepts and conservation.* Oxford University Press, Cape Town, South Africa.

Geldenhuys, C. J. and C. J. Van der Merwe. 1988. Population structure and growth of the fern *Rumohra adiantiformis* in relation to frond harvesting in the Southern Cape Forests. *Suid-Afrikaanse Tydskrif Vir Plantkunde* 54:351–362.

Gelderblom, C. M. and G. N. Bronner. 1995. Patterns of distribution and protection status of the endemic mammals of South Africa. *South African Journal of Zoology* 30:127–135.

Gelderblom, C. M., G. N. Bronner, A. T. Lombard, and P. J. Taylor. 1995. Patterns of the distribution and current protection status of the Carnivora, Chiroptera and Insectivora in South Africa. *South African Journal of Zoology* 30:103–114.

GEO3. 2002. *Global Environmental Outlook 3: past, present and future perspectives.* UNEP and Earthscan, London, UK.

Gerlach, J., editor. 1997. *Seychelles Red Data Book 1997.* Nature Protection Trust of the Seychelles, Mahé, The Seychelles.

Gibbons, J. W., D. E. Scott, T. J. Ryan, K. Buhlmann, T. D. Tuberville, B. S. Metts, J. L. Greene, T. Mills, Y. Leiden, S. Poppy, and C. T. Winnie. 2000. The global decline of reptiles, déjà vu amphibians. *BioScience* 50:653–666.

Gibbs Russell, G. E. 1987. Preliminary floristic analysis of the major biomes in southern Africa. *Bothalia* 17:213–227.

Gibbs Russell, G. E. and E. R. Robinson. 1981. Phytogeography and speciation in the vegetation of the Eastern Cape. *Bothalia* 13:467–472.

Gibson, C. C. and S. A. Marke. 1995. Transforming local hunters into conservationists: an assessment of community-based wildlife management programmes in Africa. *World Development* 23:941–957.

Giess, W. 1971. A preliminary vegetation map of South West Africa. *Dinteria* 4:1–114.

Giffard, P. L. 1974. *L'arbre dans le paysage sénégalais.* Centre Technique Forestier Tropical, Nogent-sur-Marne, France.

Gillis, M. 1988. West Africa: resource management policies and the tropical forest. Pages 299–351 in R. Repetto and M. Gillis, editors. *Public policies and the misuse of forest resources.* Cambridge University Press, Cambridge, UK.

Ginsberg, J. 2001. Mapping wild dogs. Canid Specialist Group survey. Retrieved 2001 from the World Wide Web: http://www.canids.org/PUBLICAT/CNDNEWS1/afrwldog.htm.

Glantz, M. H. 1994. *Drought follows the plow.* Cambridge University Press, Cambridge, UK.

Glaw, F. and M. Vences. 1992. *A fieldguide to the amphibians and reptiles of Madagascar.* Moos-Druck, Leverkusen, The Netherlands.

Glaw, F. and M. Vences. 2000. Current counts of species diversity and endemism of Malagasy amphibians and reptiles. Pages 243–248 in W. R. Lourenço and S. M. Goodman, editors. *Diversité et endemism a Madagascar.* Mémoires de la Société de Biogéographie, Paris, France.

Gleason, H. A. and A. Cronquist. 1964. *The natural geography of plants*. Columbia University Press, New York, USA.

Global Witness. 2001. *The role of Liberia's logging industry on national and regional insecurity*. Global Witness, London, UK.

GLTPJMB (Greater Limpopo Transfrontier Park Joint Management Board). 2002. *Joint policy and management guidelines for the Great Limpopo Transfrontier Park*. Development Alternatives, Inc. Nelspruit, South Africa.

Gobierno de Canarias. 1995. *Legislación Canaria del suelo y el medio ambiente*. Ed. Gobierno de Canarias, Consejería de Política Territorial, Las Palmas de Gran Canaria, The Canary Islands, Spain.

Godoy, R. A., D. S. Wilkie, H. Overman, J. Demmer, A. Cubas, K. McSweeney, and N. Brokaw. 2000. Valuation of consumption and sale of forest goods from a Central African rain forest. *Nature* 406:62–63.

Goldblatt, P. 1978. An analysis of the flora of southern Africa: its characteristics, relationships and origins. *Annals of the Missouri Botanical Garden* 65:369–436.

Goldblatt, P. 1990. *Biological relationships between Africa and South America*. Yale University Press, New Haven, CT and Boston, USA.

Goldblatt, P. 1997. Floristic diversity in the Cape Flora of South Africa. *Biodiversity and Conservation* 6:359–377.

Goldblatt, P., P. Bernhardt, and J. C. Manning. 1998. Pollination of petaloid geophytes by monkey beetles (Scarabaeidae: Rutelinae: Hopliini) in southern Africa. *Annals of the Missouri Botanical Garden* 85:215–230.

Goldblatt, P. and J. Manning. 2000. *Cape plants: a conspectus of the cape flora of South Africa*. Strelitzia 9. National Botanical Institute and Missouri Botanical Garden Press, Pretoria, South Africa and St. Louis, MO, USA.

Golder, B. (ed.). 2004. *Ecoregion action programmes: a guide for practitioners*. World Wide Fund for Nature, Gland, Switzerland.

Gómez-Campo, C., editor. 1985. *Plant conservation in the Mediterranean area*. Dr. W. Junk Publishers, Dordrecht, The Netherlands.

Gonder, K. M., J. F. Oates, T. R. Disotell, M. R. J. Forstner, J. C. Morales, and D. J. Melnick. 1997. A new West African chimpanzee subspecies? *Nature* 388:337.

González, M. N., J. D. Rodrigo, and C. Suárez. 1986. *Flora y vegetación del Archipiélago Canario*. Ed. Edirca S.L., Las Palmas de Gran Canaria.

Good, R. 1964. *The geography of flowering plants*. Longmans, Green, and Co., London, UK.

Goode, D. 1989. *Cycads of Africa*. Struik Winchester, Cape Town, South Africa.

Goodman, P. S. 1990. Soil, vegetation and large herbivore relations in Mkuzi Game Reserve, Natal. Ph.D. thesis, University of the Witwatersrand, Johannesburg, South Africa.

Goodman, P. 1992. The Zambezi Delta: an opportunity for sustainable utilization of wildlife. *International Waterfowl and Wetlands Research Bureau News* 8:12.

Goodman, S. M. 1985. *Natural resources and management considerations, Gebel Elba conservation area*. Report for the IUCN/World Wildlife Fund project No. 3612, Gland, Switzerland.

Goodman, S. M. 1996. A floral and faunal inventory of the east-ern slopes of the Réserve Naturelle Intégrale d'Andringitra, Madagascar: with reference to elevational variation. *Fieldiana: Zoology,* new series 85:1–319.

Goodman, S. M. 1998. A floral and faunal inventory of the Réserve Spéciale d'Anjanaharibe-Sud, Madagascar: with reference to elevational variation. *Fieldiana: Zoology* new series 90:1–246.

Goodman, S. M. 1999. A floral and faunal inventory of the Réserve Naturelle Intégrale d'Andohahela, Madagascar: with reference to elevational variation. *Fieldiana: Zoology* new series 94:1–297.

Goodman, S., editor. 2000. A floral and faunal inventory of the Parc National de Marojejy, Madagascar: with reference to elevational variation. *Fieldiana: Zoology* new series 97:1–286.

Goodman, S. M. and J. P. Benstead, editors. 2003. *Natural history of Madagascar*. University of Chicago Press, Chicago, USA.

Goodman, S. M., A. F. A. Hawkins, and C. A. Domergue. 1997. A new species of vanga (Vangidae, *Calicalicus*) from southwestern Madagascar. *Bulletin of the British Ornithologists' Club* 117:5–10.

Goodman, S. M. and P. D. Jenkins. 1998. The insectivores of the Reserve Speciale d'Anjanaharibe-Sud, Madagascar. *Fieldiana: Zoology* new series:139–161.

Goodman, S. M., O. Langrand, and B. Whitney. 1996. A new genus and species of passerine from the eastern rain forest of Madagascar. *Ibis* 138:153–159.

Goodwin, H. J. and N. Leader-Williams. 2000. Tourism and protected areas distorting conservation priorities towards charismatic megafauna? Pages 257–275 in A. C. Entwistle and N. Dunstone, editors. *Priorities for the conservation of mammalian biodiversity: has the panda had its day?* Cambridge University Press, Cambridge, UK.

Goudie, A. S. 1973. *Duricrusts in tropical and subtropical landscapes*. Clarendon, Oxford, UK.

Government of Tanzania. 1998. *Tanzanian forest policy*. Division of Forestry and Beekeeping, Government of Tanzania, Dar es Salaam, Tanzania.

Government of Tanzania. 2001. *Tanzania National Forest Programme*. Division of Forestry and Beekeeping, Government of Tanzania, Dar es Salaam, Tanzania.

Government of Uganda. 1967. *Atlas of Uganda*, 2nd edition. Department of Lands and Surveys, Entebbe, Uganda.

Green, G. M. and R. W. Sussman. 1990. Deforestation history of the eastern rainforest of Madagascar from satellite images. *Science* 248:212–215.

Greenway, P. J. and D. F. Vesey-Fitzgerald. 1969. The vegetation of Lake Manyara National Park Tanzania. *Journal of Ecology* 57:127–149.

Greyling, T. and B. J. Huntley. 1984. *Directory of southern African conservation areas*. South African National Scientific Programmes Report 98. CSIR, Pretoria, South Africa.

Griffin, J., D. Cumming, S. Metcalfe, M. t'Sas-Rolfes, E. Chonguica, M. Rowen, and J. Oglethorpe. 1999. *Study on the development of transboundary natural resource management areas in southern Africa*. Biodiversity Support Program, Washington, DC, USA.

Griffin, M. 1998a. The species diversity, distribution and conservation of Namibian mammals. *Biodiversity and Conservation* 7:483–494.

Griffin, R. E. 1998b. Species richness and biogeography of non-acarine arachnids in Namibia. *Biodiversity and Conservation* 7:467–481.

Griffiths, C. J. 1993. The geological evolution of East Africa. Pages 9–21 in J. C. Lovett and S. K. Wasser, editor. *Biogeography and ecology of the rain forests of eastern Africa.* Cambridge University Press, Cambridge, UK.

Griffiths, O. 1996. Summary of the land snails of the Mascarene Islands, with notes on their status. *Proceedings of the Royal Society of Arts and Sciences Mauritius* 6:37–48.

Grifo, F. and J. Rosenthal, editors. 1997. *Biodiversity and human health.* Island Press, Washington, DC, USA.

Grimmett, R. F. A. and T. A. Jones, editors. 1989. *Important bird areas in Europe.* International Council for Bird Preservation (Technical Publication No. 9), Cambridge, UK.

Gritzner, J. 1988. *The West African Sahel: human agency and environmental change.* The University of Chicago, Chicago, USA.

Groombridge, B. and M. D. Jenkins, compilers. 2000. *Global biodiversity: Earth's living resources in the 21st century.* World Conservation Monitoring Centre, Cambridge, UK.

Groombridge, B. and M. Jenkins, compilers. 2002. *World atlas of biodiversity: Earth's living resources in the 21st century.* UNEP–World Conservation Monitoring Centre, Cambridge, UK.

Groves, C. P. 2001. *Primate taxonomy.* Smithsonian Institution Press, Washington, DC, USA.

Groves, C. R. 2003. *Drafting a conservation blueprint. A practitioner's guide to planning for biodiversity.* Island Press, Washington, DC, USA.

Grubb, P. 1978. Patterns of speciation in African mammals. *Bulletin of the Carnegie Museum of Natural History* 6:152–167.

Grubb, P. 1990. Primate geography in the Afro-tropical forest biome. Pages 187–214 in G. Peters and R. Hutterer, editors. *Vertebrates in the Tropics.* Museum Alexander Koenig, Bonn, Germany.

Grubb, P. 2001. Endemism in African rain forest mammals. Pages 88–100 in W. Weber, L. J. T. White, A. Vedder, and L. Naughton-Treves, editors. *African rain forest ecology and conservation.* Yale University Press, New Haven, CT, USA.

Grubb, P., T. S. Jones, E. Edberg, E. D. Starin, and J. E. Hill. 1998. *Mammals of Ghana, Sierra Leone, and the Gambia.* Tenderine Press, St. Ives, UK.

Guest, N. J. and J. A. Stevens. 1951. *Lake Natron: its springs, rivers, brines and visible saline reserves.* Report for the Geological Survey of Tanganyika, No. 28. Dar es Salaam, Tanzania.

Guillaumet, J. L. 1967. *Recherches sur la végétation et la flore de la région du Bas-Cavally (Côte d'Ivoire).* Mémoires ORSTOM No. 20, Paris, France.

Guillaumet, J. L. 1984. The vegetation: an extraordinary diversity. Pages 27–54 in A. Jolly, P. Oberlé, and R. Albignac, editors. *Key environments: Madagascar.* IUCN, Pergamon Press, New York, USA.

Guinochet, M. 1951. *Contribution à l'étude phytosociologique du Sud-Tunisien.* Bulletin de la Société d'Histoire Naturelle d'Afrique du Nord 42:131–153.

Gumbo, D., F. Shonhiwa, H. Kojwang, J. Timberlake, E. Chidu-mayo, and N. Burgess. 2003. *Conservation in the miombo ecoregions: southern and eastern Africa.* WWF Southern Africa Regional Programme Office, Harare, Zimbabwe.

Gurr, T., M. G. Marshall, and D. Khosla. 2000. *Peace and conflict: a global survey of armed conflicts, self-determination movements and democracy.* Center for International Development and Conflict Management, College Park, USA.

Gwynne-Jones, D. R. G., P. K. Mitchell, M. E. Harvey, and K. Swindell. 1977. *A new geography of Sierra Leone.* Longman, London, UK.

Haacke, W. D. 1996. *Description of a new species of Phyllodactylus Gray (Reptilia: Gekkonidae) from the Cape Fold Mountains, South Africa.* Annals of the Transvaal Museum 36:229–237.

Haacke, W. D. 1997. Systematics and biogeography of the southern African scincine genus *Typhlacontias* (Reptilia: Scincidae). *Bonner Zoologische Beitrage* 47:139–163.

Hackel, J. D. 1999. Community conservation and the future of Africa's wildlife. *Conservation Biology* 13:726–734.

Hacker, J. E., G. Cowlishaw, and P. H. Williams. 1998. Patterns of African primate diversity and their evaluation for the selection of conservation areas. *Biological Conservation* 84:251–262.

Hahn, B. H., G. M. Shaw, K. M. De Cock, and P. M. Sharp. 2000. AIDS as a zoonosis: scientific and public health implications. *Science* 287:607–614.

Hahn, D. E. and V. Wallach. 1998. Comments on the systematics of Old World *Leptotyphlops* (Serpentes: Leptotyphlopidae) with description of a new species. *Hamadryad* 23:50–62.

Hails, A. J. 1997. *Wetlands, biodiversity and the Ramsar Convention: the role of the convention on wetlands in the conservation and wise use of biodiversity.* Ramsar Convention Bureau, Gland, Switzerland.

Hall, B. P. 1960a. *The ecology and taxonomy of some Angolan birds.* Bulletin of the British Museum of Natural History 78:1–211.

Hall, B. P. 1960b. The faunistic importance of the scarp of Angola. *Ibis* 102:420–442.

Hall, B. P. and R. E. Moreau. 1970. *An atlas of speciation in African passerine birds.* British Museum (Natural History), London, UK.

Hall, J. B. and M. D. Swaine. 1981. *Distribution and ecology of vascular plants in a tropical rain forest: forest vegetation in Ghana.* Geobotany 1. Dr. W. Junk Publishers, The Hague, Netherlands.

Hall, J. S., B. I. Inogwabini, E. A. Williamson, I. Omari, C. Sikubwabo, and L. J. T. White. 1997. A survey of elephants (*Loxodonta africana*) in the Kahuzi-Biega National Park lowland sector and adjacent forest in eastern Zaire. *African Journal of Ecology* 35:213–223.

Hall, J. S., K. Saltonstall, B. Inogwabini, and I. Omari. 1998a. Distribution, abundance and conservation status of Grauer's gorilla. *Oryx* 32:122–130.

Hall, J. S., L. J. T. White, B. I. Inogwabini, I. Omari, M. H. Simons, E. A. Williamson, E. A. Saltonstall, P. Walsh, C. Sikubwabo, D. Bonny, K. Prince Kiswele, A. Vedder, and K. Freeman. 1998b. Survey of Grauer's gorillas (*Gorilla gorilla graueri*) and eastern chimpanzees (*Pan troglodytes schweinfurthi*) in the Kahuzi-Biega National Park lowland sector and adjacent forest in eastern Democratic Republic of Congo. *International Journal of Primatology* 19:207–235.

Hall, M. 1984. Man's historical and traditional use of fire in southern Africa. Pages 40–52 in P. D. V. Booysen and N. M. Tainton, editors. *Ecological effects of fire*. Springer-Verlag, Berlin, Germany.

Hall, S. L. 1988. Archaeology and early history. Pages 371–378 in R. Lubke, F. Gess, and M. Bruton, editors. *A field guide to the Eastern Cape coast*. Grahamstown Centre of the Wildlife Society of Southern Africa, Grahamstown, South Africa.

Hall, T. 1994. *Spectrum guide to Namibia*. Struik Publishers, Cape Town, South Africa.

Hallam, A. 1994. *An outline of Phanerozoic biogeography*. Oxford University Press, Oxford, UK.

Hallerman, J. and J. O. Rödel. 1995. A new species of *Leptotyphlops* (Serpentes: Leptotyphlopidae) of the *longicaudus* group from West Africa. *Stuttgarter Beiträge zur Naturkunde* 532:1–8.

Hall-Martin, A. J. 1992. Distribution and status of the African elephant *Loxodonta africana* in South Africa, 1652–1992. *Koedoe* 35:65–88.

Haltenorth, T. and H. Diller. 1977. *Field guide to the mammals of Africa, including Madagascar*. Collins, London, UK.

Hamilton, A. C. 1974. Distribution patterns of forest trees in Uganda and their historical significance. *Vegetatio* 29:550–553.

Hamilton, A. C. 1981. The Quaternary history of African forests: its relevance to conservation. *African Journal of Ecology* 19:1–6.

Hamilton, A. C. 1982. *Environmental history of East Africa: a study of the Quaternary*. Academic Press, London, UK.

Hamilton, A. C. 1984. *Deforestation in Uganda*. Oxford University Press, Nairobi, Kenya.

Hamilton, A. C. and R. Bensted-Smith. 1989. *Forest conservation in the East Usambara Mountains, Tanzania*. IUCN, Gland, Switzerland and Cambridge, UK.

Hamilton, A. C., A. Cunningham, D. Byarugaba, and F. Kayanga. 2000. Conservation in a region of political instability: Bwindi Impenetrable Forest, Uganda. *Conservation Biology* 14:1722–1725.

Hamilton, A., D. Taylor, and P. Howard. 2001. Hotspots in African forests and Quaternary refugia. Pages 57–67 in W. Weber, L. J. T. White, A. Vedder, and L. Naughton-Treves, editors. *African rain forest ecology and conservation*. Yale University Press, New Haven, CT, USA.

Hammer, T. U. 1986. *Saline lake ecosystems of the world*. Dr. W. Junk Publishers, Dordrecht, The Netherlands.

Hamunyela, E., R. E. Simmons, and W. Moller. 1998. *Checklist of the birds of Etosha National Park*. Ministry of Environment and Tourism, Windhoek, Namibia.

Hanks, J. 2000. The role of transfrontier conservation areas in southern Africa in the conservation of mammalian biodiversity. Pages 239–256 in A. C. Entwistle and N. Dunstone, editors. *Priorities for the conservation of mammalian biodiversity: has the panda had its day?* Cambridge University Press, Cambridge, UK.

Hanks, J. 2002. Kalahari Desert. Pages 385–392 in R. A. Mittermeier, C. Goettsch Mittermeier, G. P. Robles, J. Pilgrim, G.

Fonseca, T. Brooks, and W. R. Konstant, editors. *Wilderness: Earth's last wild places*. CEMEX, Mexico City, Mexico.

Hannachi, S. D. Khitri, A. Benkhalifa, and R. A. Brac de la Perriere. 1998. *Inventaire variétal de la palmeraie Algérienne*. Document réalisé sous l'égide des Ministères de l'Agriculture et de la Pêche, de l'Enseignement Supérieur et de la Recherche Scientifique, du Commissariat au Développement de l'Agriculture des Régions Sahariennes et de l'Unité de Recherche sur les Zones Arides. Ed. CDARS-URZA, Algérie.

Hannah, L., G. F. Midgeley, T. Lovejoy, W. J. Bond, M. Bush, J. C. Lovett, D. Scott, and F. I. Woodward. 2002. Conservation of biodiversity in a changing climate. *Conservation Biology* 16:264–268.

Hanotte, O., D. G. Bradley, J. W. Ochieng, Y. Verjee, E. W. Hill, and J. E. O. Rege. 2002. African pastoralism: genetic imprints of origins and migrations. *Science* 296:336–339.

Hansen, M., R. DeFries, J. R. G. Thownsend, and M. Sohlberg. 2000. Global land cover classification at 1 km resolution using a decision tree classifier. *International Journal of Remote Sensing* 21:1331–1365.

Hansen, M. C. and B. Reed. 2000. A comparison of the IGBP Discover and University of Maryland 1 km global land cover products. *International Journal of Remote Sensing* 21:1365–1373.

Happold, D. C. D. 1966. The mammals of the Jebel Marra, Sudan. *Journal of Zoology* (London) 149:126–136.

Happold, D. C. D. 1969. The mammalian fauna of some jebels in the northern Sudan. *Journal of Zoology* (London) 157:133–145.

Happold, D. C. D. 1984. Small mammals. Pages 251–275 in J. L. Cloudsley-Thompson, editor. *Sahara Desert*. Pergamon Press, Oxford, UK and New York, USA.

Happold, D. C. D. 1987. *The mammals of Nigeria*. Oxford University Press, New York, USA.

Happold, D. C. D. 1995. The interactions between humans and mammals in Africa in relation to conservation: a review. *Biodiversity and Conservation* 4:395–414.

Happold, D. C. D. 1996. Mammals of the Guinea-Congo rain forest. Pages 243–284 in I. J. Alexander, M. D. Swaine, and R. Watling, editors. *Essays on the ecology of the Guinea-Congo rain forest*. Proceedings of the Royal Society of Edinburgh Series B 104.

Harcourt, A. H., J. Kineman, G. Campbell, J. Yamagiwa, I. Redmond, C. Aveling, and M. Condiotti. 1983. Conservation and the Virunga gorilla population. *African Journal of Ecology* 21:139–142.

Harcourt, A. H., S. A. Parks, and R. Woodroffe. 2001. Human density as an influence on species/area relationships: double jeopardy for small African reserves? *Biodiversity and Conservation* 10:1011–1026.

Harcourt, C., editor. 1990. Lemurs of Madagascar and the Comoros. IUCN Red Data Book. IUCN, Gland, Switzerland.

Harcourt, C., G. Davies, J. Waugh, J. Oates, N. Coulthard, N. Burgess, P. Wood, and P. Palmer. 1992. Sierra Leone. Pages 244–250 in J. A. Sayer, C. S. Harcourt, and N. M. Collins, editors. *The conservation atlas of tropical forests: Africa*. IUCN and Macmillan Publishers, London, UK.

Hardin, G. 1968. The tragedy of the commons. *Science* 280:682–683.

Harrison, J. A., D. G. Allan, L. G. Underhill, M. Herremans, A. J. Tree, V. Parker, and C. J. Brown, editors. 1997. *The atlas of southern African birds, Volumes 1 and 2*. BirdLife South Africa, Johannesburg, South Africa.

Hart, J. A. 2000. Impact and sustainability of indigenous hunting in the Ituri Forest, Congo-Zaire: a comparison of unhunted and hunted duiker populations. Pages 106–153 in J. G. Robinson and E. L. Bennett, editors. *Hunting for sustainability in tropical forests*. Columbia University Press, New York, USA.

Hart, J. and J. Hall. 1996. Status of eastern Zaire's forest parks and reserves. *Conservation Biology* 10:316–324.

Hart, J. and T. Hart. 1988. A summary report on the behaviour, ecology and conservation of the okapi (*Okapia johnstoni*) in Zaire. *Acta Zoologica et Pathologica Antverpiensia* 80:19–28.

Hart, J. and T. Hart. 1989. Ranging and feeding behaviour of okapi (*Okapia johnstoni*) in the Ituri Forest of Zaire: food limitation in a rain-forest herbivore? *Symposia of the Zoological Society of London* 61:31–50.

Hart, J. A. and A. Upoki. 1997. Distribution and conservation status of Congo peafowl *Afropavo congensis* in eastern Zaire. *Bird Conservation International* 7:295–316.

Hart, T. B. 1990. Monospecific dominance in tropical rain forests. *Trends in Ecology & Evolution* 5:6–11.

Hart, T. and J. Hart. 1997. Conservation and civil strife: two perspectives from Central Africa—Zaire: new models for an emerging state. *Conservation Biology* 11:308–309.

Hart, T., J. Hart, R. Dechamps, M. Fournier, and M. Ataholo. 1994. Changes in forest composition over the last 4000 years in the Ituri Basin, Zaire. Pages 545–563 in L. J. G. Van der Masen, X. M. van der Burgt, and J. M. van Medenbach de Rooy, editors. *The biodiversity of African plants*. Proceedings XIVth AETFAT Congress and Kluwer Academic Press, Wageningen and Dordrecht, The Netherlands.

Hart, T. B., J. A. Hart, and P. G. Murphy. 1989. Monodominant and species-rich forests of the humid tropics: causes for their co-occurrence. *American Naturalist* 133:613–633.

Hart, T. and R. Mwinyihali. 2001. *Armed conflict and biodiversity in sub-Saharan Africa: the case of the Democratic Republic of Congo (DRC)*. Biodiversity Support Program, Washington, DC, USA.

Hasler, R. 1999. *An overview of the social, ecological and economic achievements and challenges of Zimbabwe's CAMPFIRE programme*. Evaluation Eden Discussion Paper 3. International Institute of Environment and Development, London, UK.

Hassane, A., M. Kuper, and D. Orange. 2000. *Influence des aménagements hydrauliques et hydro-agricoles du Niger supérieure sur l'onde de la crue du delta intérieur du Niger au Mali*. ORSTOM/DNIER, Bamako, Mali.

Hastenrath, S. 1992. The dramatic retreat of Mount Kenya's glaciers between 1963 and 1987: greenhouse forcing. *Annals of Glaciology* 16:127–133.

Hathout, S. A. 1972. *Soil resources of Tanzania*. Tanzania Publishing House, Dar es Salaam, Tanzania.

Hatton, J., M. Couto, and J. Oglethorpe. 2001. *Biodiversity and war: A case study from Mozambique*. Biodiversity Support Program, Washington, DC, USA.

Hawkins, A. F. A. 1999. Altitudinal and latitudinal distribution of east Malagasy forest bird communities. *Journal of Biogeography* 26:447–458.

Hawkins, F. 1993. *An integrated biodiversity conservation project under development: the ICBP Angola Scarp Project*. Proceedings of the VIII Pan-African Ornithological Congress 279–284.

Hawkins, F., M. Rabenandrasana, M. C. Virginie, R. O. Manese, R. Mulder, E. R. Ellis, and R. Ramariason. 1998. Field observations of the red-shouldered vanga *Calicalicus rufocarpalis*: a newly described Malagasy endemic. *Bulletin of the African Bird Club* 5:30–32.

Hawthorne, W. D. 1991. *Fire damage and forest regeneration in Ghana*. Forest Inventory and Management Project. Forestry Department of Ghana, Kumas, Ghana.

Hawthorne, W. D. 1993a. East African Coastal Forest botany. Pages 57–99 in J. C. Lovett and S. K. Wasser, editors. *Biogeography and ecology of the rain forests of eastern Africa*. Cambridge University Press, Cambridge, UK.

Hawthorne, W. D. 1993b. *Forest regeneration after logging*. Findings of a study in the Bia South Game Production Forest Reserve. ODA Forestry Series No. 3. Overseas Development Administration (ODA) and Natural Resources Institute (INRI), London and Chatham, UK.

Hawthorne, W. D. and M. Abu-Juam. 1995. *Forest protection in Ghana, with particular reference to vegetation and plant species*. IUCN, Gland, Switzerland and Cambridge, UK.

Hawthorne, W. D. and M. P. E. Parren. 2000. How important are forest elephants to the survival of woody plant species in Upper Guinean Forests? *Journal of Tropical Ecology* 16:133–150.

Haydon, D. T., M. K. Laurenson, and C. Sillero-Zubiri. 2002. Integrating epidemiology into population viability analysis: managing the risk posed by rabies and canine distemper to the Ethiopian wolf. *Conservation Biology* 16:1372–1385.

Hayman, R. W. 1958. A new genus and species of West African mongoose. *Annals and Magazine of Natural History* 13:448–452.

Hazevoet, C. J. 1995. *The birds of the Cape Verde Islands*. BOU Check-list No. 13. British Ornithologists' Union, Tring, UK.

Hazevoet, C. J. 2001. Cape Verde. Pages 161–168 in L. D. C. Fishpool and M. I. Evans, editors. *Important bird areas of Africa and associated islands: priority sites for conservation*. Pisces Publications and BirdLife International (BirdLife Conservation Series No. 11), Newbury and Cambridge, UK.

Hazevoet, C. J., L. R. Monteiro, and N. Ratcliffe. 1999. Rediscovery of the Cape Verde cane warbler *Acrocephalus brevipennis* on Sao Nicolau in February 1998. *Bulletin of the British Ornithologists' Club* 119:68–71.

Heath, M. F. and M. I. Evans, editors. 2000. *Important bird areas in Europe: priority sites for conservation*. BirdLife International (Conservation Series No. 8), Cambridge, UK.

Hecketsweiler, P. 1990. *La conservation des ecosystemes forestiers du Congo*. IUCN, Gland, Switzerland.

Hecketsweiler, P., C. Doumenge, and J. I. Mokoko. 1991. *Le Parc National d'Odzala, Congo*. IUCN, Gland, Switzerland.

Hedberg, I. and O. Hedberg. 1979. Tropical-alpine life-forms of vascular plants. *Oikos* 33:297–307.

Hedberg, I., O. Hedberg, P. J. Madati, K. E. Mshigeni, E. N. Mshiu, and G. Samuelsson. 1982. Inventory of plants used in traditional medicine in Tanzania: I. Plants of the families Acanthaceae-Cucurbitaceae. *Journal of Ethnopharmacology* 6:29–60.

Hedberg, I., O. Hedberg, P. J. Madati, K. E. Mshigeni, E. N. Mshiu, and G. Samuelsson. 1983a. Inventory of plants used in traditional medicine in Tanzania: II. Plants of the families Dilleniaceae–Opilliaceae. *Journal of Ethnopharmacology* 9:105–128.

Hedberg, I., O. Hedberg, P. J. Madati, K. E. Mshigeni, E. N. Mshiu, and G. Samuelsson. 1983b. Inventory of plants used in traditional medicine in Tanzania: III. Plants of the families Papilionaceae–Vitaceae. *Journal of Ethnopharmacology* 9:237–260.

Hedberg, O. 1951. Vegetation belts of East African Mountains. *Svensk Botanishe Tidskrift* 45:140–202.

Hedberg, O. 1957. Afroalpine vascular plants: a taxonomic revision. *Symbolae Botanicae Upalienses* XV:1–411.

Hedberg, O. 1961. The phytogeographical position of the Afro-alpine flora. *Recent Advances in Botany* 1:914–919.

Hedberg, O. 1964. Features of Afro-alpine plant ecology. *Acta Phytogeographica Suecica* 49:1–147.

Hedberg, O. 1986. Origins of the Afroalpine flora. Pages 443–465 in F. Vuilleumier and M. Monasterio, editors. *High altitude tropical biogeography.* Oxford University Press, New York, USA.

Hedberg, O. 1994. Afroalpine region: east and north-east tropical Africa. Pages 253–256 in S. D. Davis, V. H. Heywood, and A. C. Hamilton, editors. *Centres of plant diversity, a guide and strategy for their conservation, Volume 1.* Information Press, Oxford, UK.

Hedberg, O. 1997. High-mountain areas of tropical Africa. Pages 185–197 in F. E. Wielgolaski, editor. *Ecosystems of the world 3: polar and alpine tundra.* Elsevier, Amsterdam, the Netherlands.

Hegazy, A. K. 1996. Effects of cement-kiln dust pollution on the vegetation and seed-bank species diversity in the Eastern Desert of Egypt. *Environmental Conservation* 23:249–258.

Heinzel, H., R. Fitter, and J. Parslow. 1992. *Manual de las aves de España y de Europa, Norte de Africa y Próximo oriente.* Ed. Omega, Barcelona, Spain.

Helbig, A. J., J. Martens, I. Seibold, F. Henning, B. Schottler, and M. Wink. 1996. Phylogeny and species limits in the Palaearctic chiffchaff *Phylloscopus collybita* complex: mitochondrial genetic differentiation and bioacoustic evidence. *Ibis* 138:650–666.

Henkel, F. W. and W. Schmidt. 1995. *Amphibien und Reptilien Madagaskars der Maskarenen, Seychellen und Komoren.* Original German edition. Eugen Ulmer GmBH and Co., Stuttgart, Germany.

Henkel, F. M. and W. Schmidt. 2000. *Amphibians and reptiles of Madagascar and the Mascarene, Seychelles, and Comoro Islands.* Original English edition. Krieger Publishing Company, Malabar, FL, USA.

Herbertson, A. J. 1905. The major natural regions: an essay in systematic geography. *Geography Journal* 25:300–312.

Heringa, A. C. 1990. Mali. Pages 8–14 in R. East, editor. *Antelopes global survey and regional action plans. Part 3.* West and Central Africa. IUCN/Species Survival Commission Antelope Specialist Group. IUCN, Gland, Switzerland and Cambridge, UK.

Hesse, C. and P. Trench. 2000. *Who managing the commons? Inclusive management for a sustainable future. Securing the Commons No. 1.* International Institute of Environment and Development, London, UK.

Heywood, V. H. and R. T. Watson. 1995. *Global biodiversity assessment.* Cambridge University Press, Cambridge, UK.

Hickley, P. and R. G. Bailey. 1987. Food and feeding relationships of fish in the Sudd Swamps (River Nile, southern Sudan). *Journal of Fish Biology* 30:147–160.

Hiernaux, P. 1982. *Les végétations et les fourragères dans les systèmes pastoraux.* Centre International pour l'Élevage en Afrique, Bamako, Mali.

Higgins, S. I., D. M. Richardson, R. M. Cowling, and T. H. Trinder-Smith. 1999. Predicting the landscape-scale distribution of alien plants and their threat to plant diversity. *Conservation Biology* 13:303–313.

Hilliard, O. M. and B. L. Burtt. 1987. *The botany of the southern Natal Drakensberg.* National Botanic Gardens, Cape Town, South Africa.

Hillman, J. C. 1986. *Bale Mountains National Park, management plan.* Wildlife Conservation Organisation, Addis Ababa, Ethiopia.

Hillman, J. C. 1990. The Bale Mountains National Park area, southeastern Ethiopia, and its management. Pages 277–286 in B. Messerli and H. Hurni, editors. *African mountains and highlands, problems and perspectives.* African Mountains Association, Walsworth Press, Marceline, MO, USA.

Hilton-Taylor, C. 1987. Phytogeography and origins of the Karoo flora. Pages 70–95 in R. M. Cowling and P. W. Roux, editors. *The Karoo biome: a preliminary synthesis. Part 2: vegetation and history.* South Africa National Scientific Programmes Report No. 142. Council for Scientific and Industrial Research, Pretoria, South Africa.

Hilton-Taylor, C. 1994. Karoo-Namib region: Western Cape Domain (Succulent Karoo). Pages 204–217 in WWF and IUCN, editors. *Centres of plant diversity: a guide and strategy for their conservation.* IUCN Publications Unit, Cambridge, UK.

Hilton-Taylor, C. 1996. *Red Data List of southern African plants.* Strelitzia 4. National Botanical Institute, Pretoria, South Africa.

Hilton-Taylor, C, compiler. 2000. *2000 IUCN Red List of threatened species.* IUCN, Gland, Switzerland and Cambridge, UK.

Hoare, R. E. 1999. *A standardized data collection and analysis protocol for human-elephant conflict situations in Africa.* IUCN/SSC African Elephant Specialist Group, Nairobi, Kenya.

Hoare, R. E. 2001. *A decision support system for managing human-elephant conflict situations in Africa.* IUCN/SSC African Elephant Specialist Group, Nairobi, Kenya.

Hoare, R. E. and J. T. Du Toit. 1999. Coexistence between people and elephants in African savannas. *Conservation Biology* 13:633–639.

Hobbs, R. J. 1993. Effects of landscape fragmentation on ecosystem processes in the Western Australian wheatbelt. *Biological Conservation* 64:193–201.

Hoeffler, A. and P. Collier. 2002. On the incidence of civil war in Africa. *Journal of Conflict Resolution* 46:13–28.

Hoelzmann, P., D. Jolly, S. P. Harrison, F. Laarif, R. Bonnefille, and H. P. Pachur. 1998. Mid-Holocene land-surface conditions in northern Africa and the Arabian Peninsula: a data set for the analysis of biophysical feedbacks in the climate system. *Global Biochemical Cycles* 12:35–51.

Hoffman, M. T., B. Cousins, T. Meyer, A. Petersen, and H. Hendricks. 1999. Historical and contemporary land use and the desertification of the Karoo. Pages 257–273 in W. R. J. Dean and S. J. Milton, editors. *The Karoo: ecological patterns and processes.* Cambridge University Press, Cambridge, UK.

Hoffman, M. T. and R. M. Cowling. 1990. Vegetation change in the semi-arid eastern Karoo over the last 200 years: an expanding Karoo—fact or fiction? *South African Journal of Science* 86:286–294.

Hoffman, R. L. 1993. Biogeography of East African montane forest millipedes. Page 103–115 in J. C. Lovett and S. K. Wasser, editors. *Biogeography and ecology of the rain forests of eastern Africa.* Cambridge University Press, Cambridge, UK.

Hofmeyer, J. M., H. Ebedes, R. E. M. Fryer, and J. R. de Bruine. 1975. The capture and translocation of the black rhinoceros *Diceros bicornis* in south west Africa. *Madoqua* 9:35–44.

Högberg, P. and G. D. Piearce. 1986. Mycorrhizas of Zambian trees in relation to host taxonomy, vegetation communities and successional patterns. *Journal of Ecology* 74:775–785.

Hollis, G. E., W. M. Adams, and M. Aminu-Kano. 1993. *The Hadejia-Nguru wetlands: environment, economy and sustainable development.* IUCN, Gland, Switzerland.

Homer-Dixon, T. F., J. H. Boutwell, and G. H. Rathjens. 1993. Environmental change and violent conflict. *Scientific American* 268:38–45.

Homewood, K. 1994. Pastoralists, environment and development in East African Rangelands. In B. Zaba and J. Clarke, editors. *Environment and population change.* Ordina Editions, Paris, France.

Homewood, K. and D. Brockington. 1999. Biodiversity, conservation and development in Mkomazi, Tanzania. *Global Ecology and Biogeography* 8:301–313.

Homewood, K. M. and W. A. Rodgers. 1991. *Maasailand ecology: pastoralist development and wildlife conservation in Ngorongoro, Tanzania.* Cambridge University Press, Cambridge, UK.

Honacki, J. H., K. E. Kinman, and J. W. Koeppl. 1982. *Mammal species of the world: a taxonomic and geographic reference.* Allen Press, Lawrence, KS, USA.

Honess, P. E. and S. K. Bearder. 1996. Descriptions of the dwarf galago species of Tanzania. *African Primates* 2:75–79.

Hopkins, B. 1992. Ecological processes at the forest-savanna boundary. Pages 21–33 in P. A. Furley, J. Proctor, and J. A. Ratter, editors. *Nature and dynamics of forest-savanna boundaries.* Chapman and Hall, London, UK.

Howard, P. C. 1991. *Nature conservation in Uganda's tropical forests.* IUCN Tropical Forest Programme, Gland, Switzerland and Cambridge, UK.

Howard, P. C. and T. R. B. Davenport, editors. 1996. *Forest biodiversity reports, Volumes 1–33.* Uganda Forest Department, Kampala, Uganda.

Howard, P. C., T. R. B. Davenport, F. W. Kigenyi, P. Viskanic, M. B. Baltzer, C. J. Dickinson, J. Lwanga, R. A. Matthews, and E. Mupada. 2000. Protected area planning in the tropics: Uganda's national system of forest nature reserves. *Conservation Biology* 14:858–875.

Howard, P. C., P. Viskanic, T. R. B. Davenport, F. W. Kigenyi, M. Baltzer, C. J. Dickinson, J. S. Lwanga, R. A. Matthews, and A. Balmford. 1998. Complementarity and the use of indicator groups for reserve selection in Uganda. *Nature* 394:472–475.

Hughes, B. 1983. African snake faunas. *Bonner Zoologishe Beitrage* 34:311–356.

Hughes, R. and F. Flintan. 2001. *Integrated conservation and development experience: a review and bibliography of the ICDP literature.* IIED, Biodiversity and Livelihoods Issues No. 3. IIED, London, UK.

Hughes, R. H. and J. S. Hughes. 1992. *A directory of African wetlands.* IUCN, United Nations Environment Programme, The World Conservation Monitoring Centre, Gland, Switzerland, Nairobi, Kenya and Cambridge, UK.

Hulme, D. and M. Murphree. 1999. Communities, wildlife and the "new conservation" in Africa. *Journal of International Development* 11:277–285.

Humbert, H. 1955. Les territoires phytogéographiques de Madagascar. Colloques internationaux du C.N.R.S., 59: les divisions écologique du monde. Moyen d'expression, nomenclature, cartographie. Paris, Juin–Juillet 1954. *Année Biologique,* 3e Série 31:439–448.

Humbert, H. 1959. Origines présumées et affinités de la flore de Madagascar. *Mémoire d'Institut Science Madagascar,* Série B (Biologie et Végétation) 9:149–187.

Humbert, H. and G. M. Cours Darne. 1965. *Carte internationale du tapis vegetal et des conditions ecologiques. 3 coupures au 1/1,000,000 de Madagascar.* Centre National de la Recherche Scientifique et l'Office de la Recherche Scientifique et Technique Outre-Mer, L'Institut Français de Pondicherry, Pondicherry, India.

Huntley, B. J. 1965. A preliminary account of the Ngoye Forest Reserve, Zululand. *South African Journal of Botany* 31:177–205.

Huntley, B. J. 1974a. *Ecosystem conservation priorities in Angola.* Ecologist's Report No. 28. Servicos de Veterinaria, Luanda, Angola.

Huntley, B. J. 1974b. Outlines of wildlife conservation in Angola. *Journal of the Southern African Wildlife Management Association* 4:157–166.

Huntley, B. J. 1978. Ecosystem conservation in southern Africa. Pages 1333–1384 in M. J. A. Werger, editor. *Biogeography and ecology of southern Africa.* Dr. W. Junk Publishers, The Hague, The Netherlands.

Huntley, B. J. 1984. Characteristics of South African biomes. *Ecological Studies Analysis and Synthesis* 48:1–18.

Huntley, B. J., editor. 1994. *Botanical diversity in southern Africa.* National Botanical Institute, Pretoria, South Africa.

Huntley, B. J. and E. M. Matos. 1992. *Biodiversity: Angolan environmental status quo assessment report.* IUCN Regional Office for Southern Africa, Harare, Zimbabwe.

Huntley, B. J. and E. M. Matos. 1994. Botanical diversity and its conservation in Angola. Pages 53–74 in *Botanical diversity in southern Africa.* National Botanical Institute, Pretoria, South Africa.

Hurni, H. 1986. *Management plan, Simien National Park and surrounding Areas.* UNESCO World Heritage Committee and Wildlife Conservation Organisation, Addis Ababa, Ethiopia.

Hutchings, A., A. Scott, G. Lewis, and A. B. Cunningham. 1996. *Zulu medicinal plants: an inventory.* Natal University Press, Pietermaritzburg, South Africa.

Hutchinson, M. F., H. A. Nix, J. P. McMahon, and K. D. Ord. 1996. *Documentation for "A Topographic and Climate Data Base for Africa, Ver. 1.1."* Centre for Resource and Environmental Studies, Canberra, Australia.

Hutterer, R. and O. Fulling. 1994. Mammal diversity in the Oku Mountains, Cameroon. In *International Symposium on Biodiversity and Systematics in Tropical Ecosystems.* Museum Alexander Koenig, Bonn, Germany.

Ibrahim, M. M. 1999. *Plantes de Djibouti.* IUCN-Ethiopian Agricultural Research Organization and Direction de l'Environnement, Nairobi, Kenya and Djibouti.

ICBP. 1992. *Putting biodiversity on the map: priority areas for global conservation.* International Council for Bird Preservation, Cambridge, UK.

Ihlenfeldt, H.-D. 1994. Diversification in an arid world: the Mesembryanthemaceae. *Annual Review of Ecology and Systematics* 25:521–546.

Ilambu, O., J. A. Hart, T. M. Butynski, N. R. Birhashirwa, U. Agnenonga, Y. M'Keyo, F. Bengana, M. Bashonga, and N. Bagurubumwe. 1999. The Itombwe Massif, Democratic Republic of Congo: biological surveys and conservation, with an emphasis on Grauer's gorilla and birds endemic to the Albertine Rift. *Oryx* 33:301–22.

Iliffe, J. 1995. *Africans: the history of a continent.* Cambridge University Press, Cambridge, UK.

ILOG. 1997–2000. *CPLEX Linear Optimiser 6.6.1 with Mixed and Barrier Solvers.* ILOG, Gentilly, France.

IMF. 1999. *Transforming the Enhanced Structural Adjustment Facility (ESAF) and the Initiative for the Heavily Indebted Poor Countries (HIPCs).* International Monetary Fund, Washington, DC, USA.

Infield, M. 1988. *Hunting, trapping and fishing in villages within and on the periphery of the Korup National Park.* World Wide Fund for Nature, Gland, Switzerland.

INRAT (Annales de l'Institut National de la Recherche Agronomique de Tunisie). 1967. *Carte phyto-ecologique de la Tunisie Septrionale.* Centre d'Etudes Phytosociologiques et Ecologiques, Montpellier, France.

IPCC. 2001. *The regional impacts of climate change: an assessment of vulnerability.* UNEP and WMO, Nairobi, Kenya.

IPG. 2000. *The IPG handbook on environmental funds: a resource book for the design and operation of environmental funds.* Pact Publications, New York, USA.

Irish, J. 1994. *The biomes of Namibia, as determined by objective categorisation.* Navorsing van die Nasionale Museum, Bloemfontein 10:549–592.

Isebor, C. E. and L. F. Awosika. 1993. Nigerian mangrove resources, status and management. Pages 169–185 in E. Diop, editor. *Conservation and sustainable utilization of mangrove forests in Latin America and Africa Regions. Part II: Africa.* Mangrove Ecosystems Technical Reports, Volume 3. International Society for Mangrove Ecosystems and Coastal Marine Project of UNESCO, Okinawa, Japan.

Ite, U. and W. Adams. 2000. Expectations, impacts and attitudes: conservation and development in Cross River National Park, Nigeria. *Journal of International Development* 12:325–342.

ITTO. 1999. *Annual review and assessment of the world timber situation.* International Tropical Timber Organization, Yokohama, Japan.

IUCN. 1987. *The IUCN directory of Afrotropical protected areas.* IUCN, Gland, Switzerland and Cambridge, UK.

IUCN. 1989. *La conservation des ecosystems forestiers d'Afrique centrale.* IUCN, Gland, Switzerland and Cambridge, UK.

IUCN. 1992a. *Protected areas of the world: a review of national systems. Volume 2: Palearctic.* IUCN, Gland, Switzerland and Cambridge, UK.

IUCN. 1992b. *Protected areas of the world: a review of national systems. Volume 3: Afrotropical.* IUCN, Gland, Switzerland and Cambridge, UK.

IUCN. 1997. *1996 IUCN Red List of threatened animals.* The IUCN Species Survival Commission, Gland, Switzerland.

IUCN. 1998. *1997 United Nations list of protected areas.* World Conservation Monitoring Center/IUCN, Gland, Switzerland and Cambridge, UK.

IUCN. 2001. *IUCN Red List categories and criteria, Version 3.1.* IUCN SSC, Gland, Switzerland.

IUCN and Peace Parks Foundation. 1997. *Parks for Peace.* Draft proceedings International Conference on Transboundary Protected Areas as a Vehicle for International Co-operation. Somerset West, South Africa.

IUCN Regional Office for Southern Africa and SASUSG (Southern Africa Sustainable Use Specialist Group). 2000. *Community wildlife management in southern Africa: a regional review.* Evaluating Eden discussion paper 11. IIED, London, UK.

IUCN Species Survival Commission. 1994. *IUCN Red List categories.* IUCN, Gland, Switzerland.

IUCN and UNEP. 2003. *2003 world database on protected areas.* IUCN and WCMC-UNEP, Gland, Switzerland and Cambridge, UK.

IUCN and WCPA. 1997. Special issue on Parks for Peace. *Parks* 7:1–56.

Iverson, J. B. 1992. *A checklist with distribution maps of the turtles of the world.* Published by the author, Richmond, Indiana.

Iwu, M. M. 1993. *Handbook of African medicinal plants.* CRC Press, Boca Raton, FL, USA.

Jacobsen, N. H. G. 1992. New *Lygodactylus* taxa (Reptilia: Gekkonidae) from the Transvaal. *Bonner Zoologishe Beitrage* 43:527–542.

Jacobsen, N. H. G. 1994. The *Platysaurus intermedius* complex (Sauria: Cordylidae) in the Transvaal, South Africa, with descriptions of three new taxa. *South Africa Journal of Zoology* 29:132–143.

Jacobson, P. J., K. M. Jacobson, and M. K. Seely. 1995. *Ephemeral rivers and their catchments: sustaining people and development in western Namibia.* Desert Research Foundation of Namibia, Windhoek, Namibia.

Jaeger, P. 1983. Le recesement des plantes vasculaires et les originalites du peuplement végétal des monts Loma en Sierra Leone (Afrique Occidentale). *Bothalia* 14:539–542.

Jaeger, P. and J. G. Adam. 1975. Les forêts de l'étage culminal du Nimba Liberien. *Adansonia* 15:177–188.

Jaeger, P., N. Halle, and J. G. Adam. 1968. Contribution à l'étude des orchidées des Monts Loma (Sierra Leone). *Adansonia* 8:265–310.

Jakobsen, A. 1996. A review of some East African members of the genus *Elapsoidea* Bocage with the description of a new species from Somalia and a key for the genus (Reptilia, Serpentes, Elapidae). *Steenstrupia* 22:59–82.

James, A. N., K. J. Gaston, and A. Balmford. 1999. Balancing the earth's accounts. *Nature* 401:323–324.

Jauro, A. B. 1998. Lake Chad Basin Commission (LCBC) perspectives. Working paper presented at workshop for International Network of Basin Organizations, March 19–21 1998. Paris, France.

Jeanrenaud, S. 2002. *People-orientated approaches in global conservation: the leopard changing its spots?* IIED, Institutionalising Participation Series. IIED, London, UK.

Jenkins, M. D., editor. 1987. *Madagascar: an environmental profile.* IUCN Conservation Monitoring Centre, IUCN, Gland, Switzerland.

Jenkins, M. and A. Hamilton. 1992. Biological diversity. Pages 26–32 in J. A. Sayer, C. S. Harcourt, and N. M. Collins, editors. *The conservation atlas of tropical forests: Africa.* IUCN and Macmillan Publishers, London, UK.

Jenkins, P. D. 1988. A new species of *Microgale* (Insectivora: Tenrecidae) from northeastern Madagascar. *American Museum Novitates* 2910:1–7.

Jenkins, P. D. 1993. A new species of *Microgale* (Insectivora: Tenrecidae) from eastern Madagascar with an unusual dentition. *American Museum Novitates* 3067:1–11.

Jenkins, P. D. and S. M. Goodman. 1999. A new species of *Microgale* (Lipotyphla, Tenrecidae) from isolated forest in southwestern Madagascar. *Bulletin of the Natural History Museum, London* 65:155–164.

Jenkins, P. D., S. M. Goodman, and C. J. Raxworthy. 1996. The shrew tenrecs (*Microgale*) (Insectivora: Tenrecidae) of the Réserve Naturelle Intégrale d'Andringitra, Madagascar. Pages 191–217 in S. M. Goodman, editor. *A floral and faunal inventory of the eastern slopes of the Réserve Naturelle Intégrale d'Andringitra, Madagascar: with reference to elevational variation.* Field Museum of Natural History, Chicago, USA.

Jenkins, P. D., C. J. Raxworthy, and R. A. Nussbaum. 1997. A new species of *Microgale* (Insectivora, Tenrecidae), with comments on the status of four other taxa of shrew tenrecs. *Bulletin of the Natural History Museum, London* (Zoology) 63:1–12.

Jepson, P. 2001. Global biodiversity plans need to convince local policy makers. *Nature* 409:12.

Jepson, P. and R. J. Whittaker. 2002. Ecoregions in context: a critique with special reference to Indonesia. *Conservation Biology* 16:42–57.

Jetz, W. and C. Rahbek. 2001. Geometric constraints explain much of the species richness pattern in African birds. *Proceedings of the National Academy of Sciences* 98:5661–5666.

Jetz, W. and C. Rahbek. 2002. Geographic range size and determinants of avian species richness. *Science* 297:1548–1551.

Joger, U. 1990. The herpetofauna of the Central African Republic, with description of a new species of *Rhinotyphlops* (Serpentes: Colubridae). Pages 85–102 in G. Peters and R. Hutterer, editors. *Vertebrates in the tropics.* Museum Alexander Koenig, Bonn, Germany.

Joger, U. 2000. The reptile fauna of the Socotra archipelago. Pages 337–350 in G. Rheinwald, editor. *Isolated vertebrate communities in the tropics.* Proceedings of the 4th International Symposium of Zoologisches Foschungsinstitut und Museum Alexander Koenig, Bonn, May 13–17, 1999. Bonner Zoologische Monographien 46, Bonn, Germany.

Joger, U. and M. R. K. Lambert. 1996. Analysis of the herpetofauna of the Republic of Mali, I. Annotated inventory, with description of a new *Uromastyx* (Sauria: Agamidae). *Journal of African Zoology* 110:21–51.

Johansson, D. 1974. Ecology of vascular epiphytes in West African rain forest. *Acta Phytogeographica Suecica* 59.

John, D. M. and G. W. Lawson. 1990. A review of mangrove and coastal ecosystems in West Africa and their possible relationships. *Mangrove Oceanography* 31:505–518.

John, D. M., C. Lévêque, and L. E. Newton. 1993. Western Africa. Pages 47–78 in D. Whigham, D. Dykjova, and S. Hejny, editors. *Wetlands of the world 1.* Kluwer Academic Publishers, Dordrecht, The Netherlands.

Johnson, A. R. and L. Bennun. 1994. *Lesser flamingo: concern over Lake Natron.* International Waterfowl Research Bureau News 11:10–11.

Johnson, N. C. 1995. *Biodiversity in the balance: setting geographical conservation priorities.* Biodiversity Support Program, Washington, DC, USA.

Johnson, R. and R. Johnson. 1993. The mangroves of Sierra Leone. Pages 59–69 in E. D. Diop, editor. *Conservation and sustainable utilization of mangrove forests in Latin America and Africa Regions. Part II: Africa.* Mangrove Ecosystems Technical Reports, Volume 3. International Society for Mangrove Ecosystems and Coastal Marine Project of UNESCO, Okinawa, Japan.

Jolly, A., P. Oberlé, and R. Albignac. 1984. *Key environments: Madagascar.* Pergamon Press, Oxford, UK.

Jolly, D., R. Bonnefille, A. Vincens, and J. Guiot. 2000. Climate of East Africa 6000 ^{14}C Yr B.P. as inferred from pollen data. *Quaternary Research* 54:90–101.

Jolly, D., I. C. Pretice, R. Bonnefille, A. Ballouche, M. Bengo, P. Brenac, G. Buchet, D. Burney, J. P. Cazet, R. Cheddadi, T. Edorh, H. Elenga, S. Elmoutaki, J. Guiot, F. Laarif, H. Lamb, A. M. Lezine, J. Maley, M. Mbenza, O. Peyron, M. Reille, I. Reynaud-Farrera, G. Riollet, J. C. Ritchie, E. Roche, L. Scott, I. Ssemmanda, H. Straka, M. Umer, E. Van Campo, S. Vilimumbalo, A. Vincens, and M. Waller. 1998. Biome reconstruction from pollen and macrofossil data for Africa and the Arabian Peninsula at 0 and 6000 years. *Journal of Biogeography* 25:1007–1027.

Jolly, D., D. Taylor, R. Marchant, A. Hamilton, R. Bonnefille, G. Buchet, and G. Riollet. 1997. Vegetation dynamics in central Africa since 18,000 yr BP: pollen records from the interlacustrine highlands of Burundi, Rwanda and western Uganda. *Journal of Biogeography* 24:495–512.

Jones, C. G. and J. Hartley. 1995. A conservation project on Mauritius and Rodriguez: an overview and bibliography. *Dodo* 31:40–65.

Jones, C. G., W. Heck, R. E. Lewis, Y. Mungroo, G. Slade, and T. Cade. 1995. The restoration of the Mauritius kestrel *Falco punctatus* population. *Ibis* 137:S173–S180.

Jones, E. W. 1955. Ecological studies of the rain forest of southern Nigeria IV. The plateau forest of the Okomu Forest Reserve, Part 1. The environment, the vegetation types of the forest, and the horizontal distribution of species. *Journal of Ecology* 43:564–594.

Jones, E. W. 1956. Ecological studies of the rain forest of southern Nigeria IV. The plateau forest of the Okomu Forest Reserve, Part 2. The reproduction and history of the forest. *Journal of Ecology* 44:83–117.

Jones, P. J. 1994. Biodiversity in the Gulf of Guinea: an overview. *Biodiversity and Conservation* 3:772–785.

Jones, P. J. and A. Tye. 1988. *A survey of the avifauna of São Tomé and Príncipe.* ICBP Study Report No. 24. ICBP, Cambridge, UK.

Joubert, E. and P. K. N. Mostert. 1975. Distribution patterns and status of some mammals in south west Africa. *Madoqua* 9:5–44.

Juo, A. S. R. and L. P. Wilding. 1994. Soils of the lowland forests of West and Central Africa. Pages 15–30 in I. J. Alexander, M. D. Swaine, and R. Watling, editors. *Essays on the ecology of the Guinea-Congo rain forest.* Proceedings of the Royal Society of Edinburgh Series B 104.

Jürgens, N. 1991. A new approach to the Namib Region. 1. Phytogeographic subdivision. *Vegetatio* 97:21–38.

Jürgens, N. 1997. Floristic biogeography and history of African arid regions. *Biodiversity and Conservation* 6:495–514.

Juste, J. 1996. Trade in the gray parrot *Psittacus erithacus* on the island of Príncipe (São Tomé and Príncipe, Central Africa): initial assessment of the activity and its impact. *Biological Conservation* 76:101–104.

Juste, J., J. Fa, J. Perez del Val, and J. Castroviejo. 1995. Market dynamics of bushmeat species in Equatorial Guinea. *Journal of Applied Ecology* 32:454–467.

Juste, J. and C. Ibanez. 1994. Bats of the Gulf of Guinea islands: faunal composition and origins. *Biodiversity and Conservation* 3:837–850.

Kaabache, M. 1993. Les forêts de pin d'Alep de l'Atlas Saharien (Algérie). Essai de synthèse sur la végétation steppique du Maghreb. Thèse Doct. University of Paris-Sud, Paris, France.

Kahurananga, J. and F. Silkiluwasha. 1997. The migration of zebra and wildebeest between Tarangire National Park and Simanjiro Plains, northern Tanzania, in 1972 and recent trends. *African Journal of Ecology* 35:179–185.

Kairo, J. G., F. Dahdouh-Guebas, J. Bosire, and N. Koedam. 2001. Restoration and management of Mangrove systems: a lesson for and from the East African Region. *South African Journal of Botany* 67:383–389.

Kaiser, J. and A. Lambert. 1996. *Debt swaps for sustainable development.* IUCN, Gland, Switzerland.

Kalapula, E. S. 1992. Settlement patterns and resource utilization in the Bangweulu Basin and Kafue Flats. Pages 25–33 in R. C. V. Jeffrey, H. N. Chabwela, G. Howard, and P. J. Dugan, editors. *Managing the wetlands of Kafue Flats and Bangweulu Basin.* Kafue National Park and IUCN, Zambia and Gland, Switzerland.

Kalk, M., editor. 1995. *A natural history of Inhaca island, Mozambique.* Witwatersrand University Press, Johannesburg, South Africa.

Kalpers, J. 2001. *Volcanoes under siege: impact of a decade of armed conflict in the Virungas.* Biodiversity Support Program, Washington, DC, USA.

Kalpers, J., E. A. Williamson, M. M. Robbins, A. McNeilage, A. Nzamurambaho, N. Lola, and G. Mugiri. 2003. Gorillas in the crossfire: assessment of population dynamics of the Virunga mountain gorillas over the past three decades. *Oryx* 37:326–337.

Kamdem-Toham, A. 2002. *Western Congo Forest ecoregion: ecoregion action plan.* WWF Gabon Country Office, Libreville, Gabon.

Kamdem-Toham, A., A. W. Adeleke, N. D. Burgess, R. Carroll, J. D'Amico, E. Dinerstein, D. M. Olson, and L. Some. 2003a. Forest conservation in the Congo Basin. *Science* 299:346.

Kamdem-Toham, A., J. D'Amico, A. D. Olson, A. Blom, L. Trowbridge, N. Burgess, M. Thieme, R. Abell, R. Carroll, S. Gartlan, O. Langrand, R. Mikala Mussavu, D. O'Hara, and H. Strand, editors. 2003b. *Biological priorities for conservation in the Guinean-Congolian forest and freshwater region.* WWF Gabon Country Office, Libreville, Gabon.

Kano, T., G. Idani, and C. Hashimoto. 1994. The present situations of bonobos at Wamba, Zaire. *Primate Research* 10:191–214.

Kanyamibwa, S. and O. Chantereau. 2000. Building regional linkages and supporting stakeholders in areas affected by conflicts: experiences from the Albertine Rift region. In E. Blom, W. Bergmans, I. Dankelman, P. Verweij, M. Voeten, and P. Wit, editors. *Nature in war: biodiversity conservation during conflicts.* Netherlands Commission for International Nature Protection, Amsterdam, the Netherlands.

Karesh, W. B., S. A. Osofsky, T. E. Rocke, and P. L. Barrows. 2002. Joining forces to improve our world. *Conservation Biology* 16:1432–1434.

Kassas, M. 1967. Die Pflanzen des Sahara. Pages 162–181 in C. Krüger, editor. *Sahara.* Scroll Publishers, Munich, Germany.

Kassas, M. and K. H. Batanouny. 1984. Plant ecology. Pages 77–90 in J. L. Cloudsley-Thompson, editor. *Sahara Desert.* Pergamon Press, Oxford, New York.

Kassas, M. and M. Zahran. 1971. Plant life on the coastal mountains of the Red Sea, Egypt. *Journal of the Indian Botanical Society* 50A:571–589.

Katende, A. B. and D. E. Pomeroy. 1997. Was Sango Bay a Pleistocene refugium? *Bulletin of the East Africa Natural History Society* 27:6–8.

Kaufmann, D., A. Kraay, and P. Zoido-Lobatón. 1999. *Governance matters.* World Bank Policy Research Working Paper 2196. The World Bank, Washington, DC, USA.

Keene-Young, R. 1999. A thin line: Botswana's cattle fences. *Africa Environment and Wildlife* 7:71–79.

Keith, J. O. and D. C. H. Plowes. 1997. *Considerations of wildlife resources and land use in Chad.* Office of Sustainable Development, Africa Bureau, USAID, Washington, DC, USA.

Keith, S., E. K. Urban, and C. H. Fry. 1992. *The birds of Africa, Volume 4.* Academic Press, London, UK.

Kellert, S. R., M. Black, C. R. Rush, and A. J. Bath. 1996. Human culture and large carnivore conservation in North America. *Conservation Biology* 10:977–990.

Kemp, J., J. C. Hatton, and H. Sosovele. 2000. *East African ma-*

rine ecoregion reconnaissance: synthesis report. World Wildlife Fund–Tanzania, Dar es Salaam, Tanzania.

Kemper, J., R. M. Cowling, and D. M. Richardson. 1999. Fragmentation of South African renosterveld shrublands: effects on plant community structure and conservation implications. *Biological Conservation* 90:103–111.

Kent, P. E., J. A. Hunt, and M. A. Johnstone. 1971. *The geology and geophysics of coastal Tanzania.* Geophysical Paper No. 6. Institute of Geological Sciences, London, UK.

Kerfoot, O. 1964a. The distribution and ecology of *Juniperus procera* Endl. in East Central Africa, and its relationship to the genus *Widdringtonia* Endl. *Kirkia* 4:75–86.

Kerfoot, O. 1964b. A preliminary account of the vegetation of the Mbeya Range, Tanganyika. *Kirkia* 4:191–206.

Kerley, G. I. H., A. F. Boshoff, and M. H. Knight. 1999. Ecosystem integrity and sustainable land-use in the thicket biome, South Africa. *Ecosystem Health* 5:104–109.

Kerley, G. I. H., M. H. Knight, and M. de Kock. 1995. Desertification of subtropical thicket in the Eastern Cape, South Africa: are there alternatives? *Environmental Monitoring and Assessment* 37:211–230.

Kershaw, M., P. H. Williams, and G. M. Mace. 1994. Conservation of Afrotropical antelopes: consequences and efficiency of using different site selection methods and diversity criteria. *Biodiversity and Conservation* 3:354–372.

Kideghesho, J. R. 2001. The status of wildlife habitats in Tanzania and its implications to biodiversity. *Tanzania Wildlife* 21:9–17.

Kielland, J. 1990. *Butterflies of Tanzania.* Hill House, Melbourne, Australia and London, UK.

Kier, G. and W. Barthlott. 2001. Measuring and mapping endemism and species richness: a new methodological approach and its application on the flora of Africa. *Biodiversity and Conservation* 10:1513–1529.

Kier, G., W. Küper, J. Mutke, D. Rafigpoor, and W. Barthlott. In press. *African vascular plant species richness: A comparison of mapping approaches.* In S. A. Ghazanfar and H. J. Beentje, editors. *Biodiversity, Ecology, Taxonomy, and Phytogeography of African plants.* Proceedings of the 17th AETFAT Congress. Royal Botanic Gardens, Kew, UK and National Herbarium, Addis Abba University, Ethiopia.

Killick, D. J. B. 1979. African mountain heathlands. Pages 97–116 in R. L. Specht, editor. *Heathlands and related shrublands.* Ecosystems of the World 9A. Elsevier, Amsterdam, The Netherlands.

King, F. W. and R. L. Burke. 1989. *Crocodilian, tuatara and turtle species of the world: a taxonomic and geographical reference (I–IV).* The Association of Systematics Collections, Lawrence, Kansas.

King, L. D. 1963. *South African scenery,* 3rd edition. Oliver & Boyd, London, UK.

Kingdon, J. 1989. *Island Africa: the evolution of Africa's rare animals and plants.* Princeton University Press, Princeton, NJ, USA.

Kingdon, J. 1997. *The Kingdon field guide to African mammals.* Academic Press, London, UK and San Diego, CA, USA.

Kirkwood, D. and J. J. Midgley. 1999. The floristics of Sand Forest in northern Kwazulu-Natal, South Africa. *Bothalia* 29:293–304.

Kistner, D. H. 1986. A new species and a new record of termitophilous Staphylinidae from Sierra Leone with a revision of the genus *Termitusodes* (Coleoptera). Ricerche Biologiche in Sierra Leone (Parte II). *Accademia Nazionale dei Lincei* 260:5–10.

Kjekshus, H. K. 1977. *Ecological control and development in East Africa.* Heinemann Educational Books, Nairobi, Kenya.

Knick, S. T. and D. L. Dyer. 1997. Distribution of black-tailed jackrabbit habitat determined by GIS in southwestern Idaho. *Journal of Wildlife Management* 61:75–85.

Knight, M. and P. Joyce. 1997. *The Kalahari: survival in a thirstland wilderness.* Struik Publishers, Cape Town, South Africa.

Knight, R. S., C. J. Geldenhuys, P. H. Masson, M. L. Jarman, and M. J. Cameron. 1987. *The role of aliens in forest edge dynamics: a workshop report.* Ecosystem Programmes Occasional Report 22. CSIR, Pretoria, South Africa.

Knox, E. B. and J. D. Palmer. 1995. Chloroplast DNA variation and the recent radiation of the giant senecios (Asteraceae) on the tall mountains of eastern Africa. *Proceedings of the National Academy of Sciences of the United States of America* 92:10349–10353.

Knox, E. B. and J. D. Palmer. 1998. Chloroplast DNA evidence on the origin and radiation of the giant lobelias in eastern Africa. *Systematic Botany* 23:109–149.

Koechlin, J., J.-L. Guillaumet, and P. Morat. 1974. *Flore et végétation de Madagascar.* J. Cramer Verlag, Vaduz, Liechtenstein.

Kok, D. J. 1987. Invertebrate inhabitants of temporary pans. *African Wildlife* 41:239.

Komdeur, J. 1994. Conserving the Seychelles warbler *Acrocephalus sechellensis* by translocation from Cousin Island to the Islands of Aride and Cousine. *Biological Conservation* 67:143–152.

Kortlandt, A. 1996. A survey of the geographical range, habitats and conservation of the pygmy chimpanzee (*Pan paniscus*): an ecological perspective. *Primate Conservation* 16:21–36.

Kramer, R., C. V. Schaik, and J. Johnson. 1997. *Last stand: protected areas and defense of tropical biodiversity.* Oxford University Press, London, UK.

Kurzweil, H. 2000. Notes on the orchids of the Nyika Plateau, Malawi/Zambia. *Orchids South Africa* 31:76–85.

La Ferla, B., J. Taplin, D. Ockwell, and J. C. Lovett. 2002. Continental scale patterns of biodiversity: can higher taxa accurately predict African plant distributions? *Botanical Journal of the Linnean Society* 138:225–235.

Lafferty, K. D. and L. R. Gerber. 2002. Good medicine for conservation biology: the intersection of epidemiology and conservation theory. *Conservation Biology* 16:593–604.

Laird, S. A., editor. 2002. *Biodiversity and traditional knowledge: equitable partnerships in practice.* Earthscan, London, UK.

Lambin, E. F. and D. Ehrlich. 1996. The surface temperature-vegetation index space for land cover and land-cover change analysis. *International Journal of Remote Sensing* 17:463–487.

Langdale-Brown, I., H. A. Osmaston, and J. G. Wilson. 1964. *The vegetation of Uganda and its bearing on land-use.* Government of Uganda, Entebbe, Uganda.

Langrand, O. 1990. *Guide to the birds of Madagascar.* Yale University Press, New Haven, CT, USA.

Langrand, O. and S. M. Goodman. 1997. Inventaire des oiseaux et des micro-mammifères des zones sommitales de la Réserve Naturelle Intégrale d'Andringitra. *Akon'ny Ala* 20:39–54.

Lanjouw, A. 1987. *Data review on the central Congo swamp and floodplain forest ecosystem*. Royal Tropical Institute, Amsterdam, The Netherlands.

Lanjouw, A., A. Kayitare, H. Rainer, E. Rutagarama, M. Sivha, S. Asuma, and J. Kalpers. 2001. *Beyond boundaries: transboundary natural resource management for mountain gorillas in the Virunga-Bwindi region*. Biodiversity Support Program, Washington, DC, USA.

Largen, M. J. 2001. Catalogue of the amphibians of Ethiopia, including a key for their identification. *Tropical Zoology* 14: 307–402.

Largen, M. J. and J. B. Rasmussen. 1993. Catalogue of the snakes of Ethiopia (Reptilia, Serpentes), including identification keys. *Tropical Zoology* 6:313–434.

Larsen, T. B. 1994. The butterflies of Ghana: their implications for conservation and sustainable use. Privately distributed, London, UK.

Larson, P. S., M. Freudenberger, and B. Wyckoff-Baird. 1998. *WWF integrated conservation and development project: ten lessons from the field 1985–1996*. WWF, Washington, DC, USA.

Laurance, W. F. 1999. Reflections on the tropical deforestation crisis. *Biological Conservation* 91:109–117.

Lawesson, J. E. 1995. Studies of woody flora and vegetation in Senegal. *Opera Botanica* 0:5–172.

Laws, R. M. 1970. Elephants as agents of habitat and landscape change to East Africa. *Oikos* 21:1–15.

Lawson, D. P. 1999. A new species of arboreal viper (Serpentes: Viperidae: Atheris) from Cameroon, Africa. *Proceedings of the Biological Society of Washington* 112:793–803.

Lawson, G. W., editor. 1986. *Plant ecology in West Africa: systems and processes*. John Wiley & Sons, Chichester, UK.

Lawson, G. W. 1996. The Guinea-Congo lowland rain forest: an overview. *Proceedings of the Royal Society of Edinburgh* 104B:5–13.

Lawton, J. H., D. E. Bignell, B. Bolton, G. F. Bloemers, P. Eggleton, P. M. Hammond, M. Hodda, R. D. Holt, T. B. Larsen, N. A. Mawdsley, and N. E. Stork. 1998. Biodiversity indicators, indicator taxa and effects of habitat modification in tropical forest. *Nature* 391:72–76.

Lay, D. M. 1991. Implications of climatic change for the distribution and evolution of mammals in the Sahara-Gobian desert complex. Pages 136–146 in J. A. McNeely and V. M. Neronov, editor. *Mammals in the Palaearctic Desert: status and trends in the Sahara-Gobian region*. The Russian Academy of Sciences, and the Russian Committee for the UNESCO Programme on Man and the Biosphere (MAB), Paris, France.

Leach, G. and R. Mearns. 1988. *Beyond the fuelwood crisis*. Earthscan, London, UK.

Leach, M. and J. Fairhead. 1998. *Reframing deforestation: global analysis and local realities studies in West Africa*. Routledge, London, UK.

Leader-Williams, N. and H. Dublin. 2000. Charismatic megafauna as "flagship species." Pages 53–81 in A. C. Entwistle and N. Dunstone, editors. *Priorities for the conserva-tion of mammalian biodiversity: has the panda had its day?* Cambridge University Press, Cambridge, UK.

Leader-Williams, N., J. A. Kayera, and G. L. Overton, editors. 1996. *Community-based conservation in Tanzania*. IUCN, Gland, Switzerland and Cambridge, UK.

Lebbie, A. R. 2001. Distribution, exploitation and valuation of non-timber forest product from a forest reserve in Sierra Leone. Ph.D. dissertation, University of Wisconsin, Madison, USA.

Le Berre, M. 1989. *Faune du Sahara: poissons–amphibiens–reptiles, Volume 1*. Lechevalier-R. Chabaud, Paris, France.

Le Berre, M. 1990. *Faune du Sahara: mammifères, Volume 2*. Lechevalier-R. Chaubaud, Paris, France.

Le Corre, M. and R. J. Safford. 2001. La Réunion and Iles Eparse. Pages 693–702 in L. D. C. Fishpool and M. I. Evans, editors. *The important bird areas of Africa and associated islands: priority sites for conservation*. Pisces Publications and BirdLife International (BirdLife Conservation Series No. 11), Newbury and Cambridge, UK.

Le Houérou, H. N. 1990. *Recherches écoclimatique et biogéographique sur les zones arides de L'Afrique du Nord*. CEPE/CNRS, Montpellier, France.

Le Houérou, H. N. 1991. Outline of a biological history of the Sahara. Pages 146–174 in J. A. McNeely and V. M. Neronov, editors. *Mammals in the Palaearctic Desert: status and trends in the Sahara-Gobian region*. The Russian Academy of Sciences, and the Russian Committee for the UNESCO Programme on Man and the Biosphere (MAB), Paris, France.

Le Houérou, H. N. 1992. Outline of a biological history of the Sahara. *Journal of Arid Environments* 22:3–20.

Le Houérou, H. N. 1998. *The grazing land ecosystems of the African Sahel*. Springer-Verlag, Berlin, Germany and New York, USA.

Leistner, O. A. 1967. The plant ecology of the southern Kalahari. *Memoirs of the Botanical Survey of South Africa* 38:1–172.

Lejoly, J. 1996. *Synthese regionale sur la biodiversité vegetale des ligneux dans les 6 sites du projet ECOFAC en Afrique Centrale*. Projet ECOFAC, Composante Congo. AGRECO-CTFT.

Leonard, J. 2000. Flora and vegetation of Jebel Uweinat (Libyan Desert: Libya, Egypt, Sudan). Fourth Part. General considerations on the flora and the vegetation. *Systematics and Geography of Plants* 70:3–73.

Le Roux, C. J. G. 1980. Vegetation classification and related studies in the Etosha National Park. D.S. thesis. Department of Plant Production, University of Pretoria, South Africa.

Leroux, M. 2001. *The meteorology and climate of tropical Africa*. Springer/Praxis, Chichester, UK.

Letouzey, R. 1968. *Etude phytogéographique du Cameroun*. Paul Lechevalier, Paris, France.

Letouzey, R. 1985. *Notice de la carte phytogéographique du Cameroun au 1: 500 000: region Afro-montagnarde et etage submontagnard*. Institute de la Carte Internationale de la Végétation. Institut de la Carte Internationale de la Végétation and Institut de la Recherche Agronomique, Toulouse, France and Yaoundé, Cameroon.

Lewis, A. and D. Pomeroy. 1989. *A bird atlas of Kenya*. A.A. Balkema, Rotterdam, The Netherlands.

Lewis, B. A., P. B. Phillipson, M. Andrianarisata, G. Rahajasoa, P. J. Rakotomalaza, M. Randriambololona, and J. F. Mcdonagh.

1996. *A study of the botanical structure, composition and diversity of the eastern slopes of the Reserve Naturelle Integrale D'Andringitra, Madagascar.* Fieldiana Zoology New series:24–75.

Lewis, O. T., R. J. Wilson, and M. C. Harper. 1998. Endemic butterflies on Grande Comore: habitat preferences and conservation priorities. *Biological Conservation* 85:113–121.

Lézine, A.-M. 1989. Late Quaternary vegetation and climate of the Sahel. *Quaternary Research* 32:317–334.

Lind, E. M. and M. E. S. Morrison. 1974. *East African vegetation.* Longman, London, UK.

Lindeque, P. M. and M. Lindeque. 1997. Special edition on the Etosha National Park. *Madoqua* 20:155.

Linder, H. P. 1998. Numerical analyses of African plant distribution patterns. Pages 67–86 in C. R. Huxley, J. M. Lock, and D. F. Cutler, editors. *Chorology, taxonomy and ecology of the floras of Africa and Madagascar.* Royal Botanic Gardens, Kew, UK.

Linder, H. P. 2001. Plant diversity and endemism in sub-Saharan tropical Africa. *Journal of Biogeography* 28:169–182.

Linnell, J. D. C., J. Anderson, T. Kvam, H. Andren, O. Liberg, J. Odden, and P. F. Moa. 2000. Conservation of biodiversity in Scandinavian boreal forests: large carnivores as flagships, umbrellas, indicators or keystones? *Biodiversity and Conservation* 9:857–868.

Livingstone, D. A. 1990. Evolution of African climate. *Biological relationships between Africa and South America.* Yale University Press, New Haven, CT, and Boston, USA.

Lloyd, J. W. 1999. Nama Karoo. Pages 84–93 in J. Knobel, editor. *The magnificent natural heritage of South Africa.* Sunbird Publishing, Cape Town, South Africa.

Lock, J. M. 1998. Aspects of fire in tropical African vegetation. In C. R. Huxley, J. M. Lock, and D. F. Cutler, editors. *Chorology, taxonomy and ecology of the floras of Africa and Madagascar.* Royal Botanical Gardens, Kew, UK.

Lombard, A. T. 1995. The problems with multi-species conservation: do hotspots, ideal reserves and existing reserves coincide? *South Africa Journal of Zoology* 30:145–163.

Lombard, A. T., C. Hilton-Taylor, A. G. Rebelo, R. L. Pressey, and R. M. Cowling. 1999. Reserve selection in the succulent Karoo, South Africa: coping with high compositional turnover. *Plant Ecology* 142:35–55.

Lombard, A. T., A. O. Nicholls, and P. V. August. 1995. Where should nature reserves be located in South Africa? A snake's perspective. *Conservation Biology* 9:363–372.

Lomborg, B. 2001. *The skeptical environmentalist: measuring the real state of the world.* Cambridge University Press, Cambridge, UK.

Longman, K. A. and J. Jenik. 1992. Forest-savanna boundaries: general considerations. Pages 3–17 in P. A. Furley, J. Proctor, and J. A. Ratter, editors. *Nature and dynamics of forest-savanna boundaries.* Chapman and Hall, London, UK.

Lorence, D. H. 1978. The pteridophytes of Mauritius (Indian Ocean): ecology and distribution. *Botanical Journal of the Linnean Society* 76:207–247.

Lott, C., editor. 1992. *The Spectrum guide to Tanzania.* Camperapix Publishers International, Nairobi, Kenya.

Louette, M. 1984. Apparent range gaps in African forest birds. Pages 275–286 in J. Ledger, editor. *Proceedings of the Fifth Pan-African Ornithological Congress.* SAOS, Johannesburg, South Africa.

Louette, M. 1988. *Les oiseaux des Comores.* Musée Royal de l'Afrique Centrale. Annales, Série IN-8 No. 255. Tervuren, Belgique.

Lourenço, W. R. 1996. *Biogeography of Madagascar.* ORSTOM, Paris, France.

Louw, G. N. and M. K. Seely. 1982. *Ecology of desert organisms.* Longman Scientific, London, UK.

Lovegrove, B. 1993. *The living deserts of southern Africa.* Fernwood Press, Cape Town, South Africa.

Loveland, T. R., B. C. Reed, J. F. Brown, D. O. Ohlen, Z. Zhu, L. Yang, and J. W. Merchant. 2000. Development of a global land cover characteristics database and IGBP discover from 1 km AVHRR data. *International Journal of Remote Sensing* 21:1303–1330.

Lovett, J. C. 1993a. Eastern Arc moist forest flora. Pages 35–55 in J. C. Lovett and S. K. Wasser, editors. *Biogeography and ecology of the rain forests of eastern Africa.* Cambridge University Press, Cambridge, UK.

Lovett, J. C. 1993b. Temperate and tropical floras in the mountains of eastern Tanzania. *Opera Botanica* 121:217–227.

Lovett, J. C. 1996. Elevational and latitudinal changes in tree associations and diversity in the Eastern Arc Mountains of Tanzania. *Journal of Tropical Ecology* 12:629–650.

Lovett, J. C. 1998a. Continuous change in Tanzanian moist forest tree communities with elevation. *Journal of Tropical Ecology* 14:719–722.

Lovett, J. C. 1998b. Eastern tropical African centre of endemism: a candidate for World Heritage status? *Journal of the East African Natural History Society* 87:359–366.

Lovett, J. C. 1998c. Importance of the Eastern Arc Mountains for vascular plants. *Journal of the East African Natural History Society* 87:59–74.

Lovett, J. C. 1999. Tanzanian forest tree plot diversity and elevation. *Journal of Tropical Ecology* 15:689–694.

Lovett, J. C., G. P. Clarke, R. Moore, and G. Morrey. 2001. Elevational distribution of restricted range forest tree taxa in eastern Tanzania. *Biodiversity and Conservation* 10:541–550.

Lovett, J. C. and I. Friis. 1996. Patterns of endemism in the woody flora of north-east and east Africa. Pages 582–601 in L. J. G. van der Maesen, X. M. van der Burgt, and J. M. van Medenbach de Rooy, editors. *The biodiversity of African Plants.* Kluwer Academic Publishers, The Netherlands.

Lovett, J. C., J. R. Hansen, and V. Hørlyck. 2000a. Comparison with Eastern Arc Forests. Pages 115–125 in N. D. Burgess and G. P. Clark, editors. *Coastal forests of eastern Africa.* IUCN, Cambridge, UK and Gland, Switzerland.

Lovett, J. C. and T. Pócs. 1993. *Assessment of the condition of the catchment forest reserves, a botanical appraisal.* Catchment Forest Project, Ministry of Tourism, Natural Resources and the Environment, Dar es Salaam, Tanzania.

Lovett, J. C. and E. Prins. 1994. Estimation of land-use changes on Kitulo Plateau, Tanzania, using satellite imagery. *Oryx* 28:173–182.

Lovett, J. C., S. Rudd, J. Taplin, and C. Frimodt-Møller. 2000b. Patterns of plant diversity in Africa south of the Sahara and their implications for conservation management. *Biodiversity and Conservation* 9:33–42.

Lovett, J. C. and S. K. Wasser, editors. 1993. *Biogeography and ecology of the rain forests of eastern Africa.* Cambridge University Press, Cambridge, UK.

Low, A. B. and A. G. Rebelo. 1996. *Vegetation of South Africa, Lesotho and Swaziland.* Department of Environmental Affairs and Tourism, Pretoria, South Africa.

Lowe-McConnell, R. H. 1985. The biology of the river systems with particular reference to the fishes. Pages 101–141 in A. T. Grove, editor. *The Niger and its neighbours: environmental history and hydrobiology, human use and hazards of the major West African rivers.* A.A. Balkema, Rotterdam, The Netherlands.

Lowry, P. P. I., G. E. Schatz, and P. B. Phillipson. 1997. The classification of natural and anthropogenic vegetation in Madagascar. Pages 93–123 in S. M. Goodman and B. D. Patterson, editors. *Natural change and human impact in Madagascar.* Smithsonian Institution Press, Washington, DC, USA.

Lubke, R. A., D. A. Everard, and S. Jackson. 1986. The biomes of the Eastern Cape with emphasis on their conservation. *Bothalia* 16:251–262.

Lubke, R. and B. McKenzie. 1996. Coastal forest. Page 11 in A. B. Low and A. G. Rebelo, editors. *Vegetation of South Africa, Lesotho and Swaziland.* Department of Environmental Affairs and Tourism, Pretoria, South Africa.

Luger, A. D. and E. J. Moll. 1993. Fire protection and Afromontane forest expansion in Cape Fynbos. *Biological Conservation* 64:51–56.

Luling, V. and J. C. Kenrick. 1998. *Forest foragers of tropical Africa.* Survival International, London, UK.

Lycklama à Nijeholt, R., S. de Bie, and C. Geerling. 2001. *Beyond boundaries: transboundary natural resource management in West Africa.* Biodiversity Support Program, Washington, DC, USA.

Mabberley, D. J. 1986. Adaptive syndromes of the Afroalpine species of *Dendrosenecio*. Pages 81–102 in F. Vuilleumier and M. Monasterio, editors. *High altitude tropical biogeography.* Oxford University Press and American Museum of Natural History, New York, USA and Oxford, UK.

MacArthur, R. A. 1972. *Geographic ecology.* Princeton University Press, Princeton, NJ, USA.

MacDevette, D. R., D. K. MacDevette, I. G. Gordon, and R. L. C. Bartholomew. 1989. Floristics of the Natal indigenous forests. Pages 124–144 in C. J. Geldenhuys, editor. *Biogeography of the mixed evergreen forests of southern Africa.* Ecosystem Programmes Occasional Report No. 45. Foundation for Research Development, Pretoria, South Africa.

MacDonald, I. A. W. 1994. Global change and alien invasions: implications for biodiversity and protected area management. In O. T. Solbrig, H. M. van Emden, and P. G. W. J. van Oordt, editors. *Biodiversity and global change.* CAB International, Wallingford, UK.

Mace, G. M., A. Balmford, L. Boitani, G. Cowlishaw, A. P. Dobson, D. P. Faith, K. J. Gaston, C. J. Humphries, J. H. Lawton, C. R. Margules, R. M. May, A. O. Nicholls, H. P. Possingham, C. Rahbek, A. S. van Jaarsveld, R. I. Vane-Wright, and P. H. Williams. 2000. It's time to work together and stop duplicating conservation efforts. *Nature* 405:393.

Mace, G. M. and R. Lande. 1991. Assessing extinction threats: towards a reassessment of IUCN endangered species categories. *Conservation Biology* 5:148–157.

Machado, A. 1998. Biodiversidad. *Un paseo por el concepto y las Islas Canarias.* Ed. Cabildo Insular de Tenerife, Santa Cruz de Tenerife, The Canary Islands, Spain.

MacKinnon, J. and K. MacKinnon. 1986. *Review of the protected areas system in the Afrotropical realm.* International Union for the Conservation of Nature and Natural Resources, Commission on National Parks and Protected Areas, Gland, Switzerland.

MacKinnon, K. 2001. Editorial: integrated conservation and development projects can they work. *Parks* 11:1–5.

MacKinnon, K. and W. Wardojo. 2001. ICDPs: imperfect solutions for imperiled forests in South-East Asia. *Parks* 11:50–59.

Maclean, G. L. 1984. Avian adaptations to the Kalahari environment: a typical continental semidesert. Pages 187–194 in G. De Graaf and D. J. Van Rensburg, editors. *Proceedings of the Symposium on the Kalahari Ecosystem,* 11–12 October 1983. Koedoe Supplement.

Macleod, H. L. 1987. *The conservation of Oku Mountain Forest, Cameroon.* Study Report No. 15. International Council for Bird Preservation, Cambridge, UK.

Madge, S. and H. Burn. 1988. *Wildfowl: an identification guide to the ducks, geese and swans of the world.* Christopher Helm, London, UK.

Madgwick, F. J. 1988. Riverine forest in the Jubba Valley: vegetation analysis and comments on forest conservation. The biogeography of Somalia. *Biogeographia* 14:67–88.

Magenot, G. 1955. Etudes sur les forêts de plaines et plateaux de Côte d'Ivoire. *Etudes Eburneennes* 4:5–61.

Maggi, A. 1995. Wild dog sighting in Laikipa, Kenya. *Canid News* 3:35.

Maggs, G. L., P. Craven, Kolberg, and H. Herta. 1998. Plant species richness, endemism, and genetic resources in Namibia. *Biodiversity and Conservation* 7:435–446.

Maggs, G. L., H. H. Kolberg, and C. J. H. Hines. 1994. Botanical diversity in Namibia: an overview. Page 93–104 in B. J. Huntley, editor. *Botanical diversity in southern Africa.* Strelitzia 1. National Botanical Institute, Pretoria, South Africa.

Magha, M. I., J. B. Kambou, and J. Koudenoukpo. 2001. Beyond boundaries: transboundary natural resource management in "W" Park. Biodiversity Support Program, editor. *Beyond boundaries: transboundary natural resource management in West Africa.* Biodiversity Support Program, Washington, DC, USA.

Magin, C., editor. 1999. *Monographie nationale de la diversité biologique de Djibouti.* Direction de l'Environnement, Ministère de l'Habitat, de l'Urbanisme, l'Environnement et de l'Aménagement du Territoire. Djibouti et l'UICN, Nairobi, Kenya.

Magin, G. 2001a. Djibouti. Pages 233–240 in L. D. C. Fishpool and M. I. Evans, editors. *Important bird areas in Africa and associated islands: priority sites for conservation.* Pisces Publications and BirdLife International (BirdLife Conservation Series No. 11), Newbury and Cambridge, UK.

Magin, G. 2001b. Morocco. Pages 603–626 in L. D. C. Fishpool and M. I. Evans, editors. *Important bird areas in African and associated islands: priority sites for conservation.* Pisces Publications and BirdLife International (BirdLife Series No. 11), Newbury and Cambridge, UK.

Main, M. 1987. *Kalahari: life's variety in dune and delta.* Southern Book Publishers, Johannesburg, South Africa.

Maisels, F. G., M. Cheek, and C. Wild. 2000. Rare plants on Mount Oku Summit, Cameroon. *Oryx* 34:136–140.

Maisels, F. and P. Forboseh. 1999. The Kilum-Ijim Forest project: biodiversity monitoring in the montane forests of Cameroon. *Bulletin of the African Bird Club* 7:110–114.

Makaya, J. F. 1993. Mangroves in Congo. Pages 187–192 in E. D. Diop, editor. *Conservation and sustainable utilization of mangrove forests in Latin America and Africa Regions. Part II: Africa.* Mangrove Ecosystems Technical Reports, Volume 3. International Society for Mangrove Ecosystems and Coastal Marine Project of UNESCO, Okinawa, Japan.

Malaisse, F. 1978. High termitaria. Pages 1279–1300 in M. J. A. Werger, editor. *Biogeography and ecology of southern Africa.* Dr. W. Junk Publishers, The Hague, Netherlands.

Maley, J. 1991. The African rain forest vegetation and paleoenvironments during late Quaternary. *Climatic Change* 19:79–98.

Maley, J. 1996. The African rain forest: main characteristics of changes in vegetation and climate from the Upper Cretaceous to the Quaternary. Pages 31–74 in I. J. Alexander, M. D. Swaine, and R. Watling, editors. *Essays on the ecology of the Guinea-Congo rain forest.* Proceedings of the Royal Society of Edinburgh Series B 104.

Maley, J. 2001. The impact of arid phases on the African rain forest through geological history. Pages 68–87 in W. Weber, L. J. T. White, A. Vedder, and L. Naughton-Treves, editors. *African rain forest ecology and conservation: an interdisciplinary perspective.* Yale University Press, New Haven, CT, USA.

Maley, J. and P. Brenac. 1998. Vegetation dynamics, palaeoenvironments and climatic changes in the forests of western Cameroon during the last 28,000 years B.P. *Review of Palaeobotany and Palynology* 99:157–187.

Manders, P. T. and D. M. Richardson. 1992. Colonization of Cape Fynbos communities by forest species. *Forest Ecology and Management* 48:277–293.

Mapaura, A. 2002. Endemic plant species in Zimbabwe. *Kirkia* 18:117–148.

Marchant, S. 1949. Aspects of the fauna of Owerri Province. *Nigerian Field* 14:47–51.

Margolius, R. and N. Salafsky. 1998. *Measures of success: designing, managing and monitoring conservation and development projects.* Island Press, Washington, DC, USA.

Margules, C. R., I. D. Creswell, and A. O. Nicholls. 1994. A scientific basis for establishing networks of protected areas. Pages 327–350 in P. L. Forey, C. J. Humphries, and R. I. Vane-Wright, editors. *Systematics and conservation evaluation.* Clarendon Press, Oxford, UK.

Margules, C. R., A. O. Nicholls, and R. L. Pressey. 1988. Selecting networks of reserves to maximise biological diversity. *Biological Conservation* 43:63–76.

Margules, C. R. and R. L. Pressey. 2000. Systematic conservation planning. *Nature* 405:243–353.

Marius, C. 1985. *Mangroves du Sénégal et de la Gambie: écologie, pédologie, géochimie, mise en valeur et aménagement.* Col Travaux et Documents No. 193. ORSTOM, Paris, France.

Marshall, N. T. 1998. *Searching for a cure: conservation of medicinal wildlife resources in East and Southern Africa.* TRAFFIC International, Cambridge, UK.

Martín A. 1987. *Atlas de las aves nidificantes en la isla de Tenerife.* Monografía XXXII. Instituto de Estudios Canarios, Santa Cruz de Tenerife, The Canary Islands, Spain.

Martin, C. 1991. *The rainforests of West Africa: ecology, threats, conservation.* Birkhauser, Verlag, Basel, Switzerland.

Martin, E. and D. Stiles. 2000. *The ivory markets of Africa.* Save the Elephants, Nairobi, Kenya.

Martin, P. S. and R. G. Klein, editors. 1984. *Quaternary extinctions: a prehistoric revolution.* University of Arizona Press, Tucson, USA.

Marzol, M. V. 1998. *El clima, en geografía de Canarias,* 2nd edition. Ed. Interinsular Canaria, Santa Cruz de Tenerife.

Mathieu, E. W. 1993. The mangroves of Côte d'Ivoire. Pages 77–91 in E. D. Diop, editor. *Conservation and sustainable utilization of mangrove forests in Latin America and Africa regions. Part II: Africa.* Mangrove Ecosystems Technical Reports, Volume 3. International Society for Mangrove Ecosystems and Coastal Marine Project of UNESCO, Okinawa, Japan.

Matola, Y. G., G. B. White, and S. A. Magayuka. 1987. The changed pattern of Malaria endemicity and transmission at Amani in the Eastern Usambara Mountains, Northeastern Tanzania. *Journal of Tropical Medicine and Hygiene* 90:127–134.

Matthews, E., R. Payne, M. Rohweder, and S. Murray. 2000. *Pilot analysis of global ecosystems: forest ecosystems.* World Resources Institute, Washington, DC, USA.

Matthews, W. S., A. E. Van Wyk, and G. J. Bredenkamp. 1993. Endemic flora of the north-eastern Transvaal escarpment, South Africa. *Biological Conservation* 63:83–94.

Matzke, G. 1996. Impacts of tsetse eradication programmes. Page 81 in B. Campbell, editor. *The miombo in transition: woodlands and welfare in Africa.* Centre for International Forestry Research, Bogor, Indonesia.

Maud, R. R. 1980. The climate and geology of Maputaland. Pages 1–7 in M. N. Bruton and K. H. Cooper, editors. *Studies on the ecology of Maputaland.* Rhodes University Press and the Natal Branch of the Wildlife Society of Southern Africa, Grahamstown, South Africa.

Mayaux, P., T. Richards, and E. Janodet. 1999. A vegetation map of Central Africa derived from satellite imagery. *Journal of Biogeography* 26:353–366.

Mayers, J., S. Anstey, and A. Peal. 1992. Liberia. Pages 214–220 in J. A. Sayer, C. S. Harcourt, and N. M. Collins, editors. *The conservation atlas of tropical forests: Africa.* IUCN and Macmillan Publishers, London, UK.

Mayr, E. and R. J. O'Hara. 1986. The biogeographic evidence supporting the Pleistocene forest refuge hypothesis. *Evolution* 40:55–67.

McCann, K. 2000. Blue crane (*Anthropoides paradiseus*). Pages 92–94 in K. N. Barnes, editor. *The Eskom Red Data Book of birds of South Africa, Lesotho and Swaziland.* BirdLife South Africa, Johannesburg, South Africa.

McCloskey, J. M. and H. Spalding. 1989. A reconnaissance-level inventory of the amount of wilderness remaining in the world. *Ambio* 18:221–227.

McKay, C. R. 1994. *Survey of important bird areas for Bannerman's turaco* Tauraco bannermani *and banded wattle-eye* Platysteira latincincta *in north-west Cameroon,* 1994. BirdLife International, Cambridge, UK.

McKay, C. and N. Coulthard. 2000. The Kilum-Ijim Forests IBA in Cameroon: monitoring biodiversity using birds as indicators. *Ostrich* 71:177–180.

McKean, M. A. 2000. Common property: what is it, what is it good for, and what makes it work? Pages 27–55 in C. G. Gibson and M. A. O. E. McKean, editors. *People and forests: communities, institutions, and governance.* The MIT Press, Cambridge, USA and London, UK.

McKone, D. 1995. A brief survey of the traditional forest reserves of Rungwe District, Mbeya Region, Tanzania. Unpublished draft report. Government of Tanzania/EEC Agroforestry, Soil and Water Conservation Project, Mbeya and District Forestry Office, Rungwe District. Retrieved 2001 from http://www.mckone.org/tfrrung.html.

McKone, D. and V. Walzem. 1994. A brief survey of Mbeya Region catchment forest reserves. Government of Tanzania/EEC Agroforestry, Soil and Water Conservation Project/Regional Natural Resources Office, Mbeya. Retrieved 2001 from http://www.mckone.org/mbeyacfr.html.

McMahon, L. and M. Fraser. 1988. *A fynbos year.* David Philip, Cape Town, South Africa.

McNaughton, S. J. 1979. Grassland-herbivore dynamics. Pages 46–81 in A. R. E. Sinclair and M. Norton-Griffiths, editors. *Serengeti: dynamics of an ecosystem.* University of Chicago Press, Chicago, Illinois, USA.

McNaughton, S. J. 1983. Serengeti grassland ecology: the role of composite environmental factors and contingency in community organization. *Ecological Monographs* 53:291–320.

McNaughton, S. J. and F. F. Banyikwa. 1995. Plant communities and herbivory. Pages 49–70 in A. R. E. Sinclair and P. Arcese, editors. *Serengeti II: dynamics, management and conservation of an ecosystem.* University of Chicago Press, Chicago, USA.

McNeeley, J. A., K. P. Miller, W. V. Reid, R. A. Mittermeier, and T. B. Werner. 1990. *Conserving the world's biological diversity.* International Union for Conservation of Nature and Natural Resources, World Resources Institute, Conservation International, World Wildlife Fund–US, and the World Bank, Gland, Switzerland and Washington, DC, USA.

McNeely, J. 2000. War and biodiversity: an assessment of impacts. Pages 353–369 in J. Austin and C. E. Burch, editors. *The environmental consequences of war: legal, economic and scientific perspectives.* Cambridge University Press, Cambridge, UK and New York, USA.

McNeilage, A. 1996. Ecotourism and mountain gorillas in the Virunga volcanoes. Pages 334–344 in V. Taylor and N. Dunstone, editors. *The exploitation of mammal populations.* Chapman and Hall, London, UK.

McPhee, R. D. E. and P. A. Marx. 1997. The 40,000-year plague. Humans, hyperdisease, and first-contact extinctions. Pages 169–217 in S. M. Goodman and B. D. Patterson, editors. *Nat-*

ural change and human impact in Madagascar. Smithsonian Institution Press, Washington, DC, USA.

Mduma, S., R. Hilborn, and A. R. E. Sinclair. 1998. Limits to exploitation of Serengeti wildebeest and implications for its management. Pages 243–265 in D. M. Newbery, H. N. T. Prins, and N. D. Brown, editors. *Dynamics of tropical communities.* British Ecological Society Symposium, Volume No. 37. Blackwell Science, Oxford, UK.

Meadows, M. E. and H. P. Linder. 1993. A palaeoecological perspective on the origin of Afromontane grasslands. *Journal of Biogeography* 20:345–355.

Meadows, M. E. and K. F. Meadows. 1988. Late Quaternary vegetation history of the Winterberg mountains, Eastern Cape, South Africa. *South African Journal of Science* 84:253–259.

Medail, F. and P. Quézel. 1997. Hot-spots analysis for conservation of plant biodiversity in the Mediterranean Basin. *Annals of the Missouri Botanical Garden* 84:112–127.

Mediouni, K. 2000. *Bilan, straégie et plan d'action d'utilisation durable de la diversité biologique algérienne.* Direction Générale de l'Environnement/PNUD-GEF.

Mediouni, K. and R. Boutemine. 1988. *Etude structurale et dynamique du peuplement de pin noir* Pinus nigra *ssp.* mauretanica *du Djurdjura.* Comm. Sem. Dehesas, Spain.

Mediouni, K. and N. Yahi. 1989. Etude structurale de la série du cèdre (*Cedrus atlantica* Manetti) d'Ait Ouabane (Djurdjura). *Forêt Méditerranéenne* 11:103–112.

Meier, B. and R. Albignac. 1991. Rediscovery of *Allocebus trichotis* Guenther 1875 primates in northeast Madagascar. *Folia Primatologica* 56:57–63.

Meier, B., R. Albignac, A. Peyrieras, Y. Rumpler, and P. Wright. 1987. A new species of *Hapalemur* primates from south east Madagascar. *Folia Primatologica* 48:211–215.

Meijer, P. T. and M. J. R. Wortel. 1999. Cenozoic dynamics of the African plate with emphasis on the Africa Eurasia collision. *Journal of Geophysical Research* 104:7405–7418.

Menaut, J.-C., M. Lepage, and L. Abbadie. 1995. Savannas, woodlands and dry forests in Africa. In S. H. Bullock, H. A. Mooney, and E. Medina, editors. *Seasonally dry tropical forests.* Cambridge University Press, Cambridge, UK and New York, USA.

Menchikoff, N. 1957. Les grandes lignes de la géologie saharienne. *Revue de Géographie Physique* 2:37–45.

Mendelsohn, J., S. el Obeid, and C. Roberts. 2000. *A profile of north-central Namibia.* Directorate of Environmental Affairs, Windhoek/Gamsberg Macmillan Publishers, Windhoek, Namibia.

Menzies, J. I. 1967. An ecological note on the frog *Pseudhymenochirus merlini* Chabanaud in Sierra Leone. *Journal of the West African Science Association* 12:23–28.

Mertens, B. and E. F. Lambin. 1997. Spatial modeling of deforestation in southern Cameroon. Spatial disaggregation of diverse deforestation processes. *Applied Geography* 17:143–162.

Metcalfe, S. 1999. *Study on the development of transboundary natural resource management areas in southern Africa: community perspectives.* Biodiversity Support Program, Washington, DC, USA.

MHE, WWF, and IUCN. 1996. *La Réserve Naturelle Nationale de l'Aïr et du Ténéré.* Ministère de l'Hydraulique et de l'Environnement, World Wide Fund for Nature, and IUCN, Gland, Switzerland.

M'Hirit, O. M., F. Benzyane, S. M. Bencherkroun, S. M. El Yousfi, and M. Bendaanoun. 1998. *L'arganier. Une espèce fruitière-forestière à usages multiples.* Edition Pierre, Belgium.

Midgley, G. F. and C. F. Musil. 1990. Substrate effects of zoogenic soil mounds on vegetation composition in the Worcester-Robertson valley, Cape Province. *South African Journal of Botany* 56:158–166.

Midgley, J. J. and W. J. Bond. 1990. Knysna fynbos "islands": origins and conservation. *South African Forestry Journal* 153:18–21.

Midgley, J. J., R. M. Cowling, A. H. W. Seydack, and G. F. van Wyk. 1997. Forest. Pages 278–299 in R. M. Cowling, D. M. Richardson, and S. M. Pierce, editors. *Vegetation of southern Africa.* Cambridge University Press, Cambridge, UK.

Miehe, S. and G. Miehe. 1994. *Ericaceous forests and heathlands in the Bale Mountains of South Ethiopia. Ecology and man's impact.* Stiftung Walderhaltung in Afrika and Bundesforschungsanstalt für Forst- und Holzwirtschaft, Hamburg, Germany.

Mies, B. A. and F. E. Beyhl. 1998. Vegetation ecology of Soqotra. Pages 35–82 in H. J. Dumont, editor. *Proceedings of the first international symposium on Soqotra Island: present and future, Aden 1996.* Soqotra Technical Series, Volume 1. UNDP, New York, USA.

Millington, A. C., P. J. Styles, and R. W. Critchley. 1992. Mapping forests and savannas in sub-Saharan Africa form advances very high resolution radiometer (AVHRR) imagery. Pages 37–62 in P. A. Furley, J. Proctor, and J. A. Ratter, editors. *Nature and dynamics of forest-savanna boundaries.* Chapman and Hall, New York, USA.

Mills, G. and L. Hes. 1997. The complete book of southern *African mammals.* Struik, Cape Town, South Africa.

Milne, G. 1937. Note on soil conditions and two East African vegetation types. *Journal of Ecology* 25:254–258.

Milner-Gulland, E. J. and J. R. Beddington. 1993. The exploitation of elephants for the ivory trade: an historical perspective. *Proceedings of the Royal Society of London,* series B 252:29–37.

Milton, S. J. 1987a. Effects of harvesting on four species of forest ferns in South Africa. *Biological Conservation* 41:133–146.

Milton, S. J. 1987b. Growth of seven-weeks fern (*Rumohra adiantiformis*) in the southern cape forests: implications for management. *South African Forestry Journal* 143:1–4.

Milton, S. J. and W. R. J. Dean. 1990. Mima-like mounds in the southern and western cape: are the origins so mysterious? *South African Journal of Science* 86:207–208.

Milton, S. J., H. G. Zimmermann, and J. H. Hoffmann. 1999. Alien plant invaders of the Karoo: attributes, impacts and control. Pages 274–287 in W. R. J. Dean and S. J. Milton, editors. *The Karoo. Ecological patterns and processes.* Cambridge University Press, Cambridge, UK.

Ministère de l'Agriculture et de la Mise en Valeur Agricole. 1992. *Plan directeur des aires protégées. Volumes I and II.* Royaume du Maroc, Morocco.

Minnemeyer, S. 2002. *An analysis of access to Central Africa's rainforests.* World Resources Institute, Washington, DC, USA.

Misana, S., C. Mung'ong'o, and B. Mukamuri. 1996. Miombo woodlands in the wider context: macro-economic and inter-sectoral influences. Pages 73–99 in B. Campbell, editor. *The miombo in transition: woodlands and welfare in Africa.* Centre for International Forestry Research, Bogor, Indonesia.

Mitchell, C. W. 1984. Soils. Pages 41–55 in J. L. Cloudsley-Thompson, editor. *Sahara Desert.* Pergamon Press, Oxford, UK and New York, USA.

Mitchelmore, F., K. Beardsley, R. F. W. Barnes, and I. Douglas-Hamilton. 1989. Elephant population estimates for the central African forests. S. Cobb, editor. *The ivory trade and the future of the African elephant.* Ivory Trade Review Group, Oxford, UK.

Mittermeier, R. A., C. G. Mittermeier, P. R. Gil, J. Pilgrim, G. Fonseca, W. R. Konstant, and T. Brooks, editors. 2002. *Wilderness: Earth's last wild places.* CEMEX, Mexico City, Mexico.

Mittermeier, R. A., N. Myers, and C. G. Mittermeier. 1999. *Hotspots: Earth's biologically richest and most endangered terrestrial ecoregions.* CEMEX Conservation International, Agrupacion Sierra Madre, Mexico City, Mexico.

Mittermeier, R. A., N. Myers, J. B. Thomsen, G. A. B. da Fonseca, and S. Olivieri. 1998. Biodiversity hotspots and major tropical wilderness areas: approaches to setting conservation priorities. *Conservation Biology* 12:516–520.

Mittermeier, R. A., I. Tattersall, W. Konstant, D. Meyers, and R. Mast. 1994. *Lemurs of Madagascar: an action plan for their conservation, 1993–1999.* IUCN, Gland, Switzerland.

Moll, E. J. 1976. *The vegetation of the Three Rivers region, Natal.* Natal Town and Regional Planning Commission, Pietermaritzburg, South Africa.

Moll, E. J. 1978. The vegetation of Maputaland: a preliminary report of the plant communities and their present and future conservation status. *Trees in South Africa* 29:31–58.

Moll, E. J. 1980. Terrestrial plant ecology. Pages 52–68 in M. N. Bruton and K. H. Cooper, editors. *Studies on the ecology of Maputaland.* Rhodes University Press and the Natal Branch of the Wildlife Society of Southern Africa, Grahamstown, South Africa.

Moll, E. and F. White. 1978. The Indian Ocean coastal belt. Pages 561–598 in M. J. A. Werger, editor. *Biogeography and ecology of southern Africa.* Dr. W. Junk Publishers, The Hague, The Netherlands.

Momoh, C. M. 1993. Liberian mangrove ecosystem. Pages 71–76 in E. D. Diop, editor. *Conservation and sustainable utilization of mangrove forests in Latin America and Africa regions. Part II: Africa.* Mangrove Ecosystems Technical Reports, Volume 3. International Society for Mangrove Ecosystems and Coastal Marine Project of UNESCO, Okinawa, Japan.

Monfort, S. L., J. Newby, and T. Wacher. 2002. *Sahelo-Saharan antelope survey: Republic of Chad.* Sahelo-Saharan Interest Group. Smithsonian Institution, Washington, DC, USA.

Monod, T. 1954. Modes "contracté" et "diffus" de la végétation saharienne. Pages 35–44 in J. L. Cloudsley-Thompson, editor. *Biology of deserts.* Proceedings of a symposium on the biology of hot and cold deserts organized by the Institute of Biology, London, UK.

Monod, T. 1957. *Les grandes divisions chorologiques de l'Afrique.* CSA-CCTA, London, UK.

Monod, T. 1963. The late Tertiary and Pleistocene in the Sahara. Pages 117–229 in F. C. Howell and F. Bouliére, editors. *African ecology and human evolution.* Aldine Publishing Co., Chicago, USA.

Montaggioni, L. and P. Nativel. 1988. *La Réunion, Ile Maurice.* Géologie et aperçus biologiques. Masson, Paris, France.

Moore, J. L., A. Balmford, T. Allnutt, and N. Burgess. 2004. Integrating costs into conservation planning across Africa. *Biological Conservation* 117:343–350.

Moore, J. L., A. Balmford, T. Brooks, N. D. Burgess, L. A. Hansen, C. Rahbek, and P. H. Williams. 2003. Performance of sub-Saharan vertebrates as indicator groups for identifying priority areas for conservation. *Conservation Biology* 17:207–218.

Moore, J. L., L. Manne, T. Brooks, N. D. Burgess, R. Davies, C. Rahbek, P. Williams, and A. Balmford. 2002. The distribution of cultural and biological diversity in Africa. *Proceedings of the Royal Society Biological Sciences* Series B 269:1645–1653.

Moore, J. M. and M. D. Picker. 1991. Heuweltjies (earth mounds) in the Clanwilliam district, Cape Province, South Africa: 4000-year-old termite nests. *Oecologia (Heidelberg)* 86:424–432.

Morat, P. 1973. Les savanes du sud-ouest de Madagascar. *Mémoires ORSTOM* 68:1–236.

Moreau, R. E. 1966. *The bird faunas of Africa and its islands.* Academic Press, London, UK.

Moreau, R. E. 1969. Climatic changes and the distribution of forest vertebrates in West Africa. *Journal of Zoology (London)* 158:39–61.

Morell, V. 1995. Dogfight erupts over animal studies in the Serengeti. *Science* 270:1302–1303.

Moreno, J. M. 1988. *Guía de las aves de las Islas Canarias.* Ed. Interinsular Canaria, Santa Cruz de Tenerife, Islas Canarias, Spain.

Morris, P. and F. Hawkins. 1998. *Birds of Madagascar: a photographic guide.* Yale University Press, New Haven, CT, USA.

Mortimer, J. A. 1988. Green turtle nesting at Aldabra Atoll: Indian Ocean population estimates and trends. *Bulletin of the Biological Society of Washington* 8:116–128.

Morton, J. K. 1972. Phytogeography of the West African mountains. Pages 221–236 in D. H. Valentine, editor. *Taxonomy, phytogeography and evolution.* Academic Press, London, UK.

Morton, J. K. 1986. Montane vegetation. Pages 247–271 in G. W. Lawson, editor. *Plant ecology in West Africa.* John Wiley and Sons Ltd., London, UK.

Moss, B. 1998. *Ecology of fresh waters: man and medium, past to future.* Blackwell Science, Oxford, UK.

Mothoagae, M. 1995. Present utilization of wildlife. Pages 69–76 in K. Leggett, editor. *Present status of wildlife and its future in Botswana.* Proceedings of a symposium organized by the Kalahari Conservation Society and Chobe Wildlife Trust. Gaberone, Botswana.

Moulaert, N. 1998. *Etude et conservation de la forêt de Mohéli (RFI Comores), massif menacé par la pression anthropique.* Communauté française de Belgique, Faculté Universitaire des Sciences Agronomiques de Gembloux, Belgium.

Mourer-Chauviré, C., R. Bour, S. Ribes, and F. Moutou. 1999. The avifauna of Réunion Island (Mascarene Islands) at the time of the arrival of the first Europeans. *Smithsonian Contributions to Paleobiology* 89:1–38.

Mouton, P. le F. N. and J. H. van Wyk. 1990. Taxonomic status of the melanistic forms of the *Cordylus cordylus* complex (Reptilia: Cordylidae) in the South-Western Cape, South Africa. *Suid-Afrikaanse Tydskrif Vir Dierkunde* 25:31–38.

Mouton, P. le F. N. and J. H. van Wyk. 1994. Taxonomic status of geographical isolates in the *Cordylus minor* complex (Reptilia: Cordylidae): a description of three new species. *Journal of the Herpetological Association of Africa* 43:6–18.

Mouton, P. le F. N. and J. H. van Wyk. 1995. A new crag lizard from the Cape Folded Mountains in South Africa. *Amphibia-Reptilia* 16:389–399.

Moye, M. 2000. *Overview of debt conversion.* Debt Relief International, London, UK.

Moye, M. and B. Carr-Dirick. 2002. *Feasibility study on financing mechanisms for conservation and sustainable management of Central African forests.* Implementation document for the Yaoundé Summit Declaration. Retrieved 2002 from http://www.worldwildlife.org/conservationfinance.

Moyo, S., P. O'Keefe, and M. Sill. 1993. *The southern African environment: profiles of the SADC countries.* The ETC Foundation Earthscan, London, UK.

Muleta, S., P. Simasiku, G. Kalyocha, C. Kasutu, M. Walusiku, and S. Mwiya. 1996. *Proposed terms of reference for the preparation of the management plan for Liuwa Plains National Park.* Report prepared for IUCN Upper Zambezi Wetlands and Natural Resources Management Project, Western Province, Zambia.

Müller, T. 1999. The distribution, classification and conservation of rainforests in Zimbabwe. Pages 221–235 in J. Timberlake and S. Kativu, editors. *African plants: biodiversity, taxonomy and uses.* Royal Botanic Gardens, Kew, UK.

Munari, L. 1994. *Sepsidae from Sierra Leone: new records and descriptions of two new species (Diptera: Sepsidae).* Ricerche Biologiche in Sierra Leone (Parte IV). Accademia Nazionale dei Lincei 267:231–242.

Mundy, P., D. Butchart, J. Ledger, and S. Piper. 1992. *The vultures of Africa.* Acorn Books and Russel Freidman, South Africa.

Munslow, B., Y. Katerere, A. Ferf, and P. O'Keefe. 1988. *The fuelwood trap: a study of the SADCC region.* Earthscan, London, UK.

Muraah, W. M. and W. N. Kiarie. 2001. *HIV and AIDS: facts that could change your life.* English Press Limited, Nairobi, Kenya.

Muriuki, J. N., H. M. De Klerk, P. H. Williams, L. A. Bennun, T. M. Crowe, and E. Vanden Berge. 1997. Using patterns of distribution and diversity of Kenyan birds to select and prioritize areas for conservation. *Biodiversity and Conservation* 6:191–210.

Murphree, M. W. 1990. *Decentralising the proprietorship of wildlife resources in Zimbabwe's communal lands.* Centre for Applied Social Sciences, University of Zimbabwe, Harare, Zimbabwe.

Murray, D. L., C. A. Kapke, J. F. Evermann, and T. K. Fuller. 1999. Infectious disease and the conservation of free-ranging large carnivores. *Animal Conservation* 2:241–254.

Mutke, J., G. Kier, G. Braun, C. Schulz, and W. Barthlott. 2001. Patterns of African vascular plant diversity: a GIS based analysis. *Systematics and Geography of Plants* 71:1125–1136.

Mwikomeke, S., T. H. Msangi, C. K. Mabula, J. Ylhäisi, and K.

C. H. Mundeme. 1998. Traditionally protected forests and nature conservation in the North Pare Mountains and Handeni District, Tanzania. *Journal of the East African Natural History Society* 87:279–290.

Myers, N. 1988. Threatened biotas: "hot spots" in tropical forests. *The Environmentalist* 8:187–208.

Myers, N. 1990. The biological challenge: extended hot-spots analysis. *The Environmentalist* 10:243–256.

Myers, N. 1993. Tropical forests: the main deforestation fronts. *Environmental Conservation* 20:9–16.

Myers, N., J. C. Lovett, and N. D. Burgess. 1999. The Eastern Arc Mountains and Coastal Lowland Forests hotspot. Pages 204–213 in R. Mittermeier, N. Myers, P. R. Gil, and C. G. Mittermeier, editors. *Hotspots: Earth's biologically richest and most endangered terrestrial ecoregions.* CEMEX Conservation International, Agrupacion Sierra Madre, Mexico City, Mexico.

Myers, N., R. A. Mittermeier, C. G. Mittermeier, G. A. B. da Fonseca, and J. Kent. 2000. Biodiversity hotspots for conservation priorities. *Nature* 403:853–858.

Myre, M. 1964. *A vegetação do extremo sul da província de Moçambique.* Number 110. Estudos, Ensaios e Documentos, Lisbon, Portugal.

Nabli, M. A. 1995. *Essai de synthèse sur la végétation et la phyto-écologie tunisienne. I. Eléments de botanique et phytoécologie.* Programme Flore et végétations tunisiennes. Volumes 5 et 6. Faculté des Sciences de Tunis, Tunis, Tunisia.

NASCOP. 1997. *AIDS in Kenya: background, projections, impact and interventions surveillance report for the year 1997.* Kenya National AIDS/STDs Control Programme (NASCOP), Nairobi, Kenya.

National Directorate of Forestry and Wildlife. 1997. *Forest and wildlife resources of the north of Sofala.* National Directorate of Forestry and Wildlife, Ministry of Agriculture and Fisheries, Maputo, Mozambique.

NBDB (National Biodiversity Data Bank). 1995. *Bird and mammal checklists for ten national parks in Uganda.* Kampala, Uganda.

NDES (Niger Delta Environment Survey). 1997. *Niger Delta Environmental Survey: final report. Volumes I–IV.* Environmental Resources Managers, Lagos, Nigeria.

Necas, P. 1994. Bemerkungen zur Chamäleon-Sammlung des Naturhistorischen Museum in Wien, mit vorläufiger Beschreibung eines neuen Chamäleons aus Kenia (Squamata: Chamaeleonidae). *Herpetozoa* 7:95–108.

NEDECO. 1961. *The Niger Delta: report on an investigation.* Netherlands Engineering Consultants, The Hague, The Netherlands.

NEDECO. 1966. *The soils of the Niger Delta.* Netherlands Engineering Consultants, The Hague, The Netherlands.

Nelson, R. and N. Horning. 1993. AVHRR-LAC estimates of forest area in Madagascar, 1990. *International Journal of Remote Sensing* 14:1463–1475.

Netting, R. M. 1968. *Hill farmers of Nigeria.* University of Washington Press, Seattle, WA, USA and London, UK.

Neumann, R. P. 1996. Dukes, earls, and ersatz Edens: aristocratic nature preservationists in colonial Africa. *Environment and Planning* 14:79–98.

Neumann, R. P. 1998. *Imposing wilderness: struggles over livelihood and nature preservation in Africa.* University of California Press, Berkley, USA.

Newby, J. E. 1984. Large mammals. Pages 277–290 in J. L. Cloudsley-Thomson, editor. *Sahara Desert.* Pergamon Press, Oxford, UK and New York, USA.

Newby, J. E. 1988. Aridland wildlife in decline: the case of the scimitar-horned oryx. Page 146–166 in A. Dixon and D. Jones, editors. *Conservation and biology of desert antelopes.* Christopher Helm, London, UK.

Newmark, W. 1991a. *The conservation of Mount Kilimanjaro.* IUCN, Gland, Switzerland and Cambridge, UK.

Newmark, W. D. 1991b. Tropical forest fragmentation and the local extinction of understorey birds in the Eastern Usambara Mountains, Tanzania. *Conservation Biology* 5:6778.

Newmark, W. D. 1993. The role and design of wildlife corridors with examples from Tanzania. *Ambio* 22:500–504.

Newmark, W. D. 1996. Insularization of Tanzanian parks and the local extinction of large mammals. *Conservation Biology* 10:1549–1556.

Newmark, W. D. 1998. Forest area, fragmentation, and loss in the Eastern Arc Mountains: implications for the conservation of biological diversity. *Journal of the East African Natural History Society* 87:29–36.

Newmark, W. D. 2002. *Conserving biodiversity in East African forests: a study of the Eastern Arc Mountains.* Ecological Studies, Volume 155. Springer, Berlin, Germany.

Newmark, W. D. and J. L. Hough. 2000. Conserving wildlife in Africa: integrated conservation and development projects and beyond. *BioScience* 50:585–592.

Newton, S. F., B. Haddane, E. H. Ali, G. Atta, M. Y. Al Salam, L. I. Ojok, and E. Yohannes. 1996. North and northeast African crane and wetland action plan. Pages 619–622 in R. D. Beilfuss, W. R. Tarboton, and N. N. Gichuki, editors. *Proceedings of the African crane and wetland training workshop.* Wildlife Training Institute, Botswana. International Crane Foundation, Baraboo, WI, USA.

Ngjelé, M. 1988. Principales distributions obtenus par l'analyse factorielle des éléments phytogeographiques présumés endémiques dans la flore du Zaïre. Monograph of Systematic *Botany of the Missouri Botanical Garden* 25:631–638.

Ngoile, M., Y. Kohi, and W. A. Rodgers, editors. 2001. *Planning for the long term conservation of the Mount Kilimanjaro ecosystem.* NEMC, Costech, IUCN, UNESCO, Dar es Salaam, Tanzania.

Ngusaru, A. S., J. Tobey, and G. Luhikula. 2001. *Tanzania state of the coast 2001: People and the environment.* TCMP-STWG, Dar es Salaam, Tanzania.

Nichols, R. K., P. Phillips, C. G. Jones, and L. G. Woolaver. 2002. Status of the critically endangered Mauritius fody *Foudia rubra* in 2001. *Bulletin of the African Bird Club* 9:95–100.

Nicholson, S. E. 1994. Recent rainfall fluctuations in Africa and their relationships to past conditions over the continent. *The Holocene* 4:121–131.

Nicholson, S. E. and H. Flohn. 1980. African environmental and climatic changes and the general atmospheric circulation in Late Pleistocene and Holocene. *Climatic Change* 2:313–348.

Nicoll, M. E. and O. Langrand. 1989. *Madagascar: revue de la conservation et des aires protégées.* World Wide Fund for Nature, Gland, Switzerland.

Nievergelt, B., T. Good, and R. Güttinger. 1998. A survey of the flora and fauna of the Simien Mountains National Park, Ethiopia. Special Issue of *Walia: Journal of the Ethiopian Wildlife and Natural History Society.* Addis Ababa, Ethiopia.

Nilsson, E. 1940. Ancient changes of climate in British East Africa. *Geografiska Annaler* 22:1–79.

Nix, H. A. 1983. Climate of tropical savannas. Pages 37–62 in F. Bourliere, editor. *Tropical savannas.* Elsevier, Amsterdam, The Netherlands.

Nix, H. 1986. A biogeographic analysis of Australian elapid snakes. R. Longmore, editor. *Atlas of elapid snakes of Australia.* Australian Flora and Fauna Series No. 7. Australian Government Publishing Service, Canberra, Australia.

Nketiah, K. S., E. B. Hagan, and S. T. Addo. 1988. *The charcoal cycle in Ghana: a baseline study.* Report Prepared for UNDP, Accra, Ghana.

Nordenstam, B. 1974. Flora of the Brandberg. Dinteria 11:1–67.

Norris, S. 2001. A new voice in conservation: conservation medicine seeks to bring ecologists, veterinarians, and doctors together around a simple unifying concept: health. *BioScience* 51:7–12.

Norton-Griffiths, M., D. Herlocker, and L. Pennycuick. 1975. The patterns of rainfall in the Serengeti ecosystem Tanzania. *East African Wildlife Journal* 13:347–374.

Noss, A. 2000. Cable snares and nets in the Central African Republic. Pages 282–304 in J. G. Robinson and E. L. Bennett, editors. *Hunting for sustainability in tropical forests.* Columbia University Press, New York, USA.

Noss, R. F. 1992. The Wildlands Project land conservation strategy. *Wild Earth Special Issue:*10–25.

Noss, R. F. 1996. Ecosystems as conservation targets. *Trends in Ecology and Evolution* 10:921–922.

Noss, R. F., H. B. Quigley, M. G. Hornocker, T. Merrill, and P. C. Paquet. 1996. Conservation biology and carnivore conservation in the Rocky Mountains. *Conservation Biology* 10:949–963.

Nowak, R. M. and J. L. Paradiso. 1999. *Walker's mammals of the world,* 6th edition. John Hopkins University Press, Baltimore, USA.

Nowell, K., P. Jackson, and IUCN/Species Survival Commission Cat Specialist Group. 1996. *Wild cats: status survey and conservation action plan.* IUCN, Gland, Switzerland.

Ntiamoa-Baidu, Y., E. H. Owusu, D. T. Daramani, and A. A. Nuoh. 2001. Ghana. Pages 367–389 in L. D. C. Fishpool and M. I. Evans, editors. *Important bird areas in Africa and associated islands: priority sites for conservation.* Pisces Publications and BirdLife International (BirdLife Conservation Series No. 11), Newbury and Cambridge, UK.

Nyerges, A. E. 1989. Coppice swidden fallows in tropical deciduous forest: biological, technological, and sociocultural determinants of secondary forest successions. *Human Ecology* 17:379–400.

Oates, J., compiler. 1986. *Action plan for African primate conservation 1986–1990.* IUCN/SSC Primate Specialist Group, Gland, Switzerland.

Oates, J. F. 1989. *A survey of primates and other forest wildlife in Anambra, Imo and Rivers states, Nigeria.* Report to the National Geographic Society, the Nigerian Conservation Foundation,

the Nigerian Federal Department of Forestry, and the Governments of Anambra, Imo and Rivers States, Nigeria.

Oates, J. F. 1995. The dangers of conservation by rural development: a case-study from the forests of Nigeria. *Oryx* 29:115–122.

Oates, J. F. 1996. *African primates: status survey and conservation action plan, revised edition.* IUCN, Gland, Switzerland.

Oates, J., editor. 1998. *The 1998 IUCN primate action plan.* IUCN, Gland, Switzerland and Cambridge, UK.

Oates, J. F. 1999. *Myth and reality in the rain forest: how conservation strategies are failing in West Africa.* University of California Press, Berkeley, USA.

Oates, J. F., M. Abedi-Lartey, W. S. Mcgraw, T. T. Struhsaker, G. H. Whitesides, and H. George. 2000. Extinction of a West African red colobus monkey. *Conservation Biology* 14:1526–1532.

Oates, J. F. and P. Anadu. 1982. *The status of wildlife in Bendel State, Nigeria, with recommendations for its conservation.* Report to Bendel State Ministry of Agriculture and Natural Resources, Benin City, Nigeria.

Oates, J. F., P. A. Anadu, E. L. Gadsby, I. Inahoro, and J. L. R. Werre. 1992. Sclater's guenon: a rare Nigerian monkey threatened by deforestation. *National Geographic Research and Exploration* 8:476–491.

Oatley, T. B. 1969. Bird ecology in the evergreen forests of northwestern Zambia. *Puku* 5:141–181.

O'Brien, E. M. 1993. Climatic gradients in woody plant species richness: towards an explanation based on an analysis of southern Africa's woody flora. *Journal of Biogeography* 20:181–198.

O'Brien, E. M. 1998. Water-energy dynamics, climate, and predicting of woody plant species richness: an interim general model. *Journal of Biogeography* 25:379–398.

Oldfield, S., C. Lusty, and A. MacKinven. 1998. *World list of threatened trees.* World Conservation Press, Cambridge, UK.

Oliver, W. L., editor. 1993. *Pigs, peccaries, and hippos: status survey and conservation plan.* IUCN, Gland, Switzerland.

Olson, D. M. and E. Dinerstein. 1998. The Global 200: a representation approach to conserving the earth's most biologically valuable ecoregions. *Conservation Biology* 12:502–515.

Olson, D. M. and E. Dinerstein. 2002. The Global 200: priority ecoregions for global conservation. *Annals of the Missouri Botanical Garden* 89:199–224.

Olson, D. M., E. Dinerstein, E. D. Wikramanayake, N. D. Burgess, G. V. N. Powell, E. C. Underwood, J. A. D'Amico, I. Itoua, H. E. Strand, J. C. Morrison, C. J. Loucks, T. F. Allnutt, T. H. Ricketts, Y. Kura, J. F. Lamoreux, W. W. Wettengel, P. Hedao, and K. R. Kassem. 2001. Terrestrial ecoregions of the world: a new map of life on Earth. *BioScience* 51:933–938.

Omari, I., J. A. Hart, T. M. Butynski, N. R. Birhashirwa, A. Upoki, Y. M'keyo, F. Bengana, M. Bashonga, and N. Bagurubumwe. 1999. The Itombwe Massif, Democratic Republic of Congo: biological surveys and conservation, with an emphasis on Grauer's gorilla and birds endemic to the Albertine Rift. *Oryx* 33:301–322.

Osmaston, H. A. and G. Kaser. 2001. *Tropical glaciers.* Cambridge University Press, Cambridge, UK.

Osmaston, H. A., J. Tukahirwa, C. Basalirwa, and J. Nyakaana, editors. 1998. *The Rwenzori Mountains National Park, Uganda.* Department of Geography, Makerere University, Kampala, Uganda.

Osofsky, S. A., W. B. Karesh, and S. L. Deem. 2000. Conservation medicine: a veterinary perspective. *Conservation Biology* 14:336–337.

Ottichilo, W. K. 1987. The causes of the recent heavy elephant mortality in the Tsavo Ecosystem, Kenya, 1975–80. *Biological Conservation* 41:279–290.

Owens, M. and D. Owens. 1980. The fences of death. *African Wildlife* 34:25–77.

Owusu, J. G. K., C. K. Manu, G. K. Ofosu, and Y. Ntiamoa-Baidu. 1989. *Report of the Working Group on Forestry and Wildlife (revised version).* Report Prepared for the Environmental Protection Council, Accra, Ghana.

Ozenda, P. 1975. Sur les étages de végétation dans les montagnes du bassin méditerranéen. *Document de Cartographie Ecologique* 16:1–32.

Ozenda, P. 1983. *Flore du Sahara.* Centre National de la Recherche Scientifique, Paris, France.

Packer, C. 1996. Who rules the park? *Wildlife Conservation* 99:36–39.

Padya, B. M. 1989. *Weather and climate of Mauritius.* Mahatma Gandhi Institute, Moka, Mauritius.

Pakenham, T. 1991. *The scramble for Africa 1876–1912.* Weidenfeld & Nicolson, London, UK.

Palmer, A. R. and M. T. Hoffman. 1997. Nama Karoo. Pages 167–186 in R. M. Cowling, D. M. Richardson, and S. M. Pierce, editors. *Vegetation of southern Africa.* Cambridge University Press, Cambridge, UK.

Parker, I. S. C. and A. D. Graham. 1989. Men elephants and competition. *Symposia of the Zoological Society of London* 61:241–252.

Parren, M. P. E., B. M. de Leede, and F. Bongers. 2002. A proposal for a transnational forest network area for elephants in Côte d'Ivoire and Ghana. *Oryx* 36:249–256.

Passmore, N. and V. Carruthers. 1995. *South African frogs: a complete guide.* Halfway House: Southern Book Publishers and Witwatersrand University Press, Johannesburg, South Africa.

Pasteur, G. 1995. Biodiversité et reptiles: diagnoses de sept nouvelles espèces fossiles et actuelles du genre de lézards *Lygodactylus* (Sauria: Gekkonidae). *Dumerilia* 2:1–21.

Paulian, R., J. M. Betsch, J. L. Guillaumet, C. Blanc, and P. Griveaud. 1971. RCP 225. Etudes des écosystèmes montagnards dans la région malgache. I. Le massif de l'Andringitra. 1970–1971. Géomorphologie, climatologie et groupements végétaux. *Bulletin de la Société d'Ecologie II:*198–226.

Pauw, C. A. 1998. Will a new name save Malawi's cedars? *Sabonet News* 3:33–34.

Payne, R. B. 1998. A new species of firefinch *Lagonosticta* from northern Nigeria and its association with the Jos Plateau indigobird *Vidua maryae*. *Ibis* 140:368–381.

Pearce, D. W. 1993. Developing Botswana's savannas. Pages 205–220 in M. D. Young and O. T. Solbrig, editors. *The world's savannas. Economic driving forces, ecological constraints and policy options for sustainable land use.* Man & the Biosphere Series, Volume 12. UNESCO Paris and Parthenon Publishing Group, London, UK.

Peel, M. J. S. and M. Stalmans. 1999. The systematic reconnaissance flight (SRF) as a tool in assessing the ecological impact of a rural development programme in an extensive area of the lowveld of South Africa. *African Journal of Ecology* 37:449–456.

Peet, N. B. and P. W. Atkinson. 1994. The biodiversity and conservation of the birds of São Tomé and Príncipe. *Biodiversity and Conservation* 3:851–867.

Peltier, J. P. 1983. Les séries de l'arganeraie steppique dans le souss (Maroc). *Ecologia Mediterranea* 9:77–88.

Pence, G. Q. K. 2003. Evaluating combinations of on- and off-reserve conservation strategies for the Agulhas Plain, South Africa: a financial perspective. *Biological Conservation* 112:253–273.

Pennington, K. M. 1978. *Pennington's butterflies of southern Africa.* Ad Donker, Johannesburg, South Africa.

Penny, M. 1974. *The birds of Seychelles and outlying islands.* Taplinger Publishing Co., New York, USA.

Penry, H. 1994. *Bird atlas of Botswana.* University of Natal Press, Pietermaritzberg, South Africa.

Pérez del Val, J. 2001. Equatorial Guinea. Pages 265–272 in L. D. C. Fishpool and M. I. Evans, editor. *Important bird areas of Africa and associated islands: priority sites for conservation.* Pisces Publications and BirdLife International (BirdLife Conservation Series No. 11), Newbury and Cambridge, UK.

Pérez del Val, J., J. E. Fa, J. Castroviejo, and F. J. Purroy. 1994. Species richness and endemism of birds in Bioko. *Biodiversity and Conservation* 9:886–892.

Perrier de la Bathie, H. 1936. *Biogéographie des plantes de Madagascar.* Société d'Éditions Géographiques, Maritimes et Coloniales, Paris, France.

Perring, F. H. and S. M. Walters, editors. 1962. *Atlas of the British flora.* Published for the Botanical Society of the British Isles by T. Nelson, London, UK.

Perrings, C. and J. C. Lovett. 2000. Policies for biodiversity conservation in Sub-Saharan Africa. Pages 309–342 in C. Perrings, editor. *The economics of biodiversity conservation in sub-Saharan Africa: mending the ark.* Edward Elgar, Cheltenham, UK.

Peters, C. R. 1990. *African wild plants with rootstocks reported to be eaten raw: the monocotyledons, Part 1.* Proceedings 12th Plenary Meeting AETFAT. Mitteilungen aus dem Institut fur Allgemeine Botanik in Hamburg 23:935–952.

Peterson, A. T., J. Soberon, and V. Sanchez-Cordero. 1999. Conservatism of ecological niches in evolutionary time. *Science* 285:1265–1267.

Petit-Maire, N., G. Delibrias, and C. Gave. 1980. Pleistocene lakes in the Shati area, Fezzan (2730'N). *Paleoecology of Africa* 12:289–295.

Petit-Maire, N. and J. Riser, editors. 1983. *Sahara ou Sahel? Quaternaire recent du bassin du Taoudenni (Mali).* Librairie du Muséum, Paris, France.

Phillips, A., editor. 2000. *Financing protected areas: guidelines for protected area managers.* IUCN, Gland, Switzerland and Cambridge, UK.

Phillips, J. F. V. 1931. Forest succession and ecology in the Knysna region. *Memoirs of the Botanical Survey of South Africa* 14:13–27.

Phillips, J. 1963. *The forests of George, Knysna and the Zitzikama: a brief history of their management 1778–1939*. Government Printer, Pretoria, South Africa.

Phillips, J. 1965. Fire as master and servant: its influence in the bioclimatic regions of trans-Saharan Africa. *Proceedings of the Tall Timbers Fire Ecology Conference* 4:7–109.

Phillipson, P. B. 1996. Endemism and non-endemism in the flora of south-west Madagascar. Pages 125–136 in W. R. Lourenço, editor. *Biogéographie de Madagascar*. Editions de l'ORSTOM, Paris, France.

Phipps, J. B. and R. Goodier. 1962. A preliminary account of the plant ecology of the Chimanimani Mountains. *Journal of Ecology* 50:291–319.

Piersma, T. and Y. Ntiamoa-Baidu. 1995. *Waterbird ecology and the management of coastal wetlands in Ghana*. NIOZ Report 1995-96. Netherlands Institute for Sea Research, Texel, The Netherlands.

Pimbert, M. C. and J. M. Pretty. 1995. *Parks, people and professionals: putting "participation" into protected area management*. UNRISD Discussion Paper DP57. United Nations Research Institute for Social Development, Geneva, Switzerland.

Pimm, S. L., M. Ayers, A. Balmford, G. Branch, K. Brandon, T. Brooks, R. Bustamante, R. Costanza, R. Cowling, L. M. Curran, A. Dobson, S. Farber, G. A. B. da Fonseca, D. Gascon, R. Kitching, J. McNeely, T. Lovejoy, R. A. Mittermeier, N. Myers, J. A. Patz, B. Raffle, D. Rapport, P. Raven, C. Roberts, J. P. Rodriguez, A. B. Rylands, C. Tucker, C. Safina, C. Samper, M. L. J. Stiassny, J. Supriatna, D. H. Wall, and D. Wilcove. 2001. Can we defy nature's end? *Science* 293:2207–2208.

Pimm, S. L. and T. M. Brooks. 1999. The sixth extinction: how large, how soon, and where? Pages 46–62 in P. H. Raven and T. Williams, editors. *Nature and human society: the quest for a sustainable world*. National Academy Press, Washington, DC, USA.

Pimm, S. L. and P. Raven. 2000. Extinction by numbers. *Nature* 403:843–845.

Pimm, S. L., G. J. Russell, J. L. Gittleman, and T. M. Brooks. 1995. The future of biodiversity. *Science* 269:347–350.

Pinhey, E. 1978. Lepidoptera. Pages 763–774 in M. J. A. Werger, editor. *Biogeography and ecology of southern Africa*. Dr. W. Junk Publishers, The Hague, The Netherlands.

Plowright, W. 1982. The effect of rinderpest and rinderpest control on wildlife in Africa. *Symposium of the Zoological Society of London* 50:1–28.

Plummer, P. S. and E. R. Belle. 1995. Mesozoic tectonostratigraphic evolution of the Seychelles microcontinent. *Sedimentary Geology* 96:73–91.

Plumptre, A. J. 1996. Changes following sixty years of selective timber harvesting in Budongo forest reserve, Uganda. *Forest Ecology and Management* 89:101–113.

Plumptre, A. J., M. Behangana, T. R. B. Davenport, C. Kahindo, R. Kityo, E. Ndomba, D. Nkuutu, I. Owiunji, P. Ssegawa, and G. Eliu. 2003. *The biodiversity of the Albertine Rift*. Albertine Rift Technical Reports No. 3. Wildlife Conservation Society, Kampala, Uganda.

Plumptre, A., M. Masozera, and A. Vedder. 2001. *The impact of civil war on the conservation of protected areas in Rwanda*. Biodiversity Support Program, Washington, DC, USA.

Pócs, T. 1976. Vegetation mapping in the Uluguru Mountains (Tanzania, East Africa). *Boissera* 24:477–498.

Pócs, T. 1998. Bryophyte diversity along the Eastern Arc. *Journal of East African Natural History* 87:75–84.

Poiani, K. A., B. D. Richter, M. G. Anderson, and H. E. Richter. 2000. Biodiversity at multiple scales: functional sites, landscapes and networks. *BioScience* 50:133–146.

Polet, G. 2000. Waterfowl and flood extent in the Hadejia-Nguru wetlands of north-east Nigeria. *Bird Conservation International* 10:203–209.

Pomeroy, D. 1993. Centres of high biodiversity in Africa. *Conservation Biology* 7:901–907.

Pomeroy, D. and C. Dranzoa. 1998. Do tropical plantations of exotic trees in Uganda and Kenya have conservation value for birds? *Bird Populations* 4:23–36.

Pomeroy, D. and A. Lewis. 1987. Bird species richness in tropical Africa: some comparisons. *Biological Conservation* 40:11–28.

Pomeroy, D. and D. Ssekabiira. 1990. An analysis of the distributions of terrestrial birds in Africa. *African Journal of Ecology* 28:1–14.

Pomeroy, D. and H. Tushabe. 1996. *Biodiversity of Karamoja*. MUIENR, Kampala, Uganda.

Poore, D. 1989. *No timber without trees: sustainability in the tropical forest*. Earthscan, London, UK.

Posey, D. A. 1999. *Cultural and spiritual values of biodiversity*. UNEP and Intermediate Technology Publications, London, UK.

Possingham, H. P., S. J. Andelman, M. A. Burgman, R. A. Medellin, L. L. Master, and D. A. Keith. 2002. Limits to the use of threatened species lists. *Trends in Ecology & Evolution* 17:503–507.

Powell, C. B. 1993. *Sites and species of conservation interesting the central axis of the Niger Delta (Yenegoa, Sagbama, Ekeremor and southern Ijo local government areas)*. A report of recommendations to the Natural Resources Conservation Council (NARESCON), Abuja, Nigeria.

Powell, C. B. 1995. *Wildlife study I*. Final report (Contract E-00019). submitted to Environmental Affairs Department, Shell Petroleum Development Company of Nigeria Ltd., Port Harcourt, Nigeria.

Powell, C. B. 1997. Discoveries and priorities for mammals in the freshwater forests of the Niger Delta. *Oryx* 31:83–85.

Poynton, J. C. 1961. Biogeography of south-east Africa. *Nature* 189:801–803.

Poynton, J. C. 1995. The "arid corridor" distribution in Africa: a search for instances among amphibians. *Madoqua* 19:45–48.

Poynton, J. C. 1999. Distribution of amphibians in sub-Saharan Africa, Madagascar, and Seychelles. Pages 483–540 in W. E. Duellman, editor. *Patterns of distributions of amphibians: a global perspective*. Johns Hopkins University Press, Baltimore, USA and London, UK.

Poynton, J. C. and D. G. Broadley. 1978. The herpetofauna. Pages 927–948 in M. J. A. Werger, editor. *Biogeography and ecology of southern Africa*. Dr. W. Junk Publishers, The Hague, The Netherlands.

Poynton, J. C. and D. G. Broadley. 1991. Amphibia Zambesiaca 5. Zoogeography. *Annals of the Natal Museum* 32:221–277.

Poynton, J. C., K. M. Howell, B. T. Clarke, and J. C. Lovett. 1998. A critically endangered new species of *Nectophrynoides*

(Anura: Bufonida) from the Kihansi Gorge, Udzungwa Mountains, Tanzania. *African Journal of Herpetology* 47:59–67.

Pratt, R. M. and M. D. Gwynne. 1977. *Rangeland management in East Africa*. Krieger Publishing Company, Malabar, FL, USA.

Prell, W. L., W. H. Hutson, D. F. Williams, A. Be, K. Geitzenauer, and B. Molfino. 1980. Surface circulation of the Indian Ocean during the Last Glacial Maximum, approximately 18,000 yr BP. *Quaternary Research* 14:309–336.

Pressey, R. L. 1998. Algorithms, politics, and timber: an example of the role of science in a public political negotiation process over new conservation areas in production forests. Pages 73–87 in R. Wills and R. Hobbs, editors. *Ecology for everyone: communicating ecology to scientists, the public and the politicians*. Surrey Beatty and Sons, Sidney, Australia.

Pressey, R. L. and R. M. Cowling. 2001. Reserve selection algorithms and the real world. *Conservation Biology* 15:275–277.

Pressey, R. L., I. R. Johnson, and P. D. Wilson. 1994. Shades of irreplaceability: towards a measure of the contribution of sites to a reservation goal. *Biodiversity and Conservation* 3:242–262.

Pressey, R. L., H. P. Possingham, and J. R. Day. 1997. Effectiveness of alternative heuristic algorithms for identifying indicative minimum requirements for conservation reserves. *Biological Conservation* 80:207–219.

Preston, F. W. 1962. The canonical distribution of commonness and rarity. Parts 1 and 2. *Ecology* 43:185–215, 410–432.

Preston-Mafham, K. 1991. *Madagascar: a natural history*. Facts on File Inc., New York, USA.

Prigogine, A. 1985. Conservation of the avifauna of the forests of the Albertine Rift. Pages 277–295 in A. W. Diamond and T. E. Lovejoy, editors. *Conservation of tropical forest birds*. International Council on Bird Preservation, Cambridge, UK.

Prigogine, A. 1986. Speciation pattern of birds in the Central African forest refugia and their relationship with other refugia. Pages 2537–2546 in H. Ouellet, editor. *Acta XIX Congressus Internationalis Ornithologici*, Volume II. University of Ottawa Press, Ontario, Canada.

Procter, J. 1984. Vegetation of the granitic islands of the Seychelles. Pages 193–208 in D. R. Stoddart, editor. *Biogeography and ecology of the Seychelles Islands*. Dr. W. Junk Publishers, The Hague, Netherlands.

Putz, F. E., D. P. Dykstra, and R. Heinrich. 2000. Why poor logging practices persist in the tropics. *Conservation Biology* 14:951–956.

Pyrcz, T. 1992. Provisional checklist of the butterflies of São Tomé and Príncipe islands. *Lambillionea* 92:48–52.

Quensière, J., editor. 1994. *La pêche dans le Delta Central du Niger*. Karthala, Paris, France.

Quézel, P. 1965. *La végétation du Sahara, du Tchad à la Mauritanie*. Fisher Verlag, Stuttgart, Germany.

Quézel, P. 1971. *La haute montagne méditerranéenne: signification phytosociologique et bioclimatique génèrale*. Collection Interdisciplinaire sur les Milieux Naturales Supra-Foresties des Montagnes du Bassin Occidental de la Méditerranée, Perpignan, France.

Quézel, P. 1981. Les hautes montagnes du Maghreb et du Proche Orient: essai de mise en parallèle des caractères phytogé-

graphiques. Actas III CONGR: OPTIMA: *Anales del Jardin Botanico de Madrid* 37:353–372.

Quézel, P. 1983. Flore et végétation actuelles de l'Afrique du Nord, leur signification en fonction de l'origine, de l'évolution et des migrations des flores et structures de végétation passées. *Bothalia* 14:3–4.

Rabinowitz, P. D., M. F. Coffin, and D. Falvey. 1983. The separation of Madagascar and Africa. *Science* 220:304–324.

Rakotomalaza, P. J. and N. Messmer. 1999. Structure and floristic composition of the vegetation in the Réserve Naturelle Intégrale d'Andohahela, Madagascar. A floral and faunal inventory of the Réserve Naturelle Intégrale d'Andohahela, Madagascar: with reference to elevational variation. *Fieldiana: Zoology, New series* 94:51–96.

Ramsar. 1998. Botswana rolls back fences for wildlife. Retrieved 2000 from the World Wide Web: http://www.ramsar.org/w.n.botswana_fences.htm.

Ramsar Convention Bureau. 2002. Directory of wetlands of international importance. Retrieved 2002 from the World Wide Web: http://www.ramsar.org/.

Randrianasolo, J., P. Rakotovao, P. Deleporte, C. Rarivoson, J.-P. Sorg, and U. Rohner. 1996. Local tree species in the tree nursery. Pages 117–132 in J. U. Ganzhorn and J. P. Sorg, editors. *Ecology and economy of a tropical dry forest in Madagascar*. Primate Report 46-1. Göttingen, Germany.

Raselimanana, A. P., C. J. Raxworthy, and R. A. Nussbaum. 2000. A revision of the dwarf *Zonosaurus* Boulenger (Reptilia: Squamata: Cordylidae) from Madagascar, including descriptions of three new species. Scientific Papers, The Natural History Museum, University of Kansas 18:1–16.

Rasmussen, J. B. 1993. A taxonomic review of the *Dipsadoboa unicolor* complex, including a phylogenetic analysis of the genus (Serpentes, Dipsadidae, Boiginae). *Steenstrupia* 19:129–196.

Rasmussen, J. B. and M. J. Largen. 1992. A review of *Pseudoboodon peracca* with the description of a new species from southwest Ethiopia (Serpentes, Dipsadidae, Lycodontinae, Boadontini). *Steenstrupia* 18:65–80.

Rasoloarison, R., S. M. Goodman, and J. U. Ganzhorn. 2000. Taxonomic revision of mouse lemurs (*Microcebus*) in the western portions of Madagascar. *International Journal of Primatology* 21:963–1019.

Rasolofo, V. M. 1993. Mangroves of Madagascar. Pages 225–260 in E. Diop, editor. *Conservation and sustainable utilization of mangrove forests in Latin America and Africa regions. Part II: Africa*. Mangrove Ecosystems Technical Reports, Volume 3. International Society for Mangrove Ecosystems and Coastal Marine Project of UNESCO, Okinawa, Japan.

Rauh, W. 1995. *Succulent and xerophytic plants of Madagascar*, Volume I. Strawberry Press, Stewart, OH, USA.

Rautenbach, I. L., J. D. Skinner, and J. A. J. Nel. 1980. The past and present status of the mammals of Maputaland. Pages 322–345 in M. Bruton and K. H. Cooper, editors. *Studies on the ecology of Maputaland*. Wildlife Society of Southern Africa, Durban, South Africa.

Raxworthy, C. J. and R. A. Nussbaum. 1996. Amphibians and reptiles of the Reserve Naturelle Integrale D'Andringitra, Madagascar: a study of elevational distribution and local endemicity. *Fieldiana: Zoology, New series*:158–170.

Ray, J. C. and R. Hutterer. 1996. Structure of a shrew community in the Central African Republic based on the analysis of carnivore scats, with the description of a new *Sylvisorex* (Mammmalia; Soricidae). *Ecotropica* 1:85–97.

Raynaut, C., editor. 1997. Sahels: diversité et dynamiques des relations sociétés-nature. Karthala, Paris, France.

Reader, J. 1998. *Africa: a biography of a continent.* Penguin, London, UK.

Rebelo, A. G. 1994. Iterative selection procedures: centres of endemism and optimal placement of reserves. Pages 231–254 in B. J. Huntley, editor. *Botanical diversity in southern Africa.* Strelitzia 1. National Botanical Institute, Pretoria, South Africa.

Rebelo, A. G. 1997. Conservation. Pages 571–590 in R. M. Cowling, D. M. Richardson, and S. M. Pierce, editors. *Vegetation of southern Africa.* Cambridge University Press, Cambridge, UK.

Redford, K. H. 1992. The empty forest. *BioScience* 42:412–422.

Redford, K. H., P. Coppolillo, E. W. Sanderson, G. A. B. da Fonseca, E. Dinerstein, C. Groves, G. Mace, S. Maginnis, R. Mittermeier, R. Noss, D. Olson, J. G. Robinson, A. Vedder, and M. Wright. 2003. Mapping the conservation landscape. *Conservation Biology* 17:116–131.

Reid, W. V., J. A. McNeely, D. B. Tunstall, D. A. Bryant, and M. Winograd. 1993. *Biodiversity indicators for policy-makers.* World Resources Institute and IUCN, Washington, DC, USA.

Republic of Kenya. 1997. Sessional paper no. 4 of 1997 on AIDS in Kenya. Kenya Government Printer, Nairobi, Kenya.

Restor, J. P. M. 1997. Debt-for-nature swaps: a decade of experience and new directions for the future. *Unasylva* 48. FAO, Rome, Italy.

Richards, P. W. 1939. Ecological studies on the rain forest of southern Nigeria. I. The structure and floristic composition of the primary forest. *Journal of Ecology* 27:1–61.

Richards, P. W. 1973. Africa, the "odd man out." Pages 21–26 in B. J. Meggers, E. S. Ayensu, and W. D. Duckworth, editors. *Tropical forest ecosystems in Africa and South America: a comparative review.* Smithsonian Institution Press, Washington, DC, USA.

Richardson, J. E., F. M. Weitz, M. F. Fay, Q. C. B. Cronk, H. P. Linder, G. Reeves, and M. W. Chase. 2001. Rapid and recent origin of species richness in the cape flora of South Africa. *Nature* 412:181–183.

Ricketts, T. H., E. Dinerstein, D. Olson, C. J. Loucks, W. Eichbaum, D. DellaSala, K. Kavanagh, P. Hedao, P. T. Hurley, K. M. Carney, R. Abell, and S. Walters. 1999. *Terrestrial ecoregions of North America: a conservation assessment.* Island Press, Washington, DC, USA.

Riddell, J. C. and D. J. Campbell. 1986. Agricultural intensification and rural development: the Mandara Mountains of north Cameroon. *African Studies Review* 29:89–106.

Ripley, S. D. and G. M. Bond. 1966. *The birds of Socotra and Abd-El-Kuri.* Smithsonian Institution, Washington, DC, USA.

Robertson, A., A. M. Jarvis, C. J. Brown, and R. E. Simmons. 1998. Avian diversity and endemism in Namibia: patterns from the Southern African Bird Atlas Project. *Biodiversity and Conservation* 7:495–511.

Robertson, P. 2001a. Guinea-Bissau. Pages 403–409 in L. D. C. Fishpool and M. I. Evans, editors. *Important bird areas in Africa and associated islands: priority sites for conservation.* Pisces Publications and BirdLife International (BirdLife Conservation Series No. 11), Newbury and Cambridge, UK.

Robertson, P. 2001b. Somalia. Pages 779–792 in L. D. C. Fishpool and M. I. Evans, editors. *Important bird areas in Africa and associated islands: priority sites for conservation.* Pisces Publications and BirdLife International (BirdLife Conservation Series No. 11), Newbury and Cambridge, UK.

Robertson, P. 2001c. Sudan. Pages 877–890 in L. D. C. Fishpool and M. I. Evans, editors. *Important bird areas in Africa and associated islands: priority sites for conservation.* Pisces Publications and BirdLife International (BirdLife Conservation Series No. 11), Newbury and Cambridge, UK.

Robertson, P. and M. Essghaier. 2001. Libyan Arab Jamahiriya. Pages 481–487 in L. D. C. Fishpool and M. I. Evans, editors. *Important bird areas in Africa and associated islands: priority sites for conservation.* Pisces Publications and BirdLife International (BirdLife Conservation Series No. 11), Newbury and Cambridge, UK.

Robertson, S. A. 1989. *Flowering plants of Seychelles.* Royal Botanic Gardens, Kew, UK.

Robertson, S. A. and W. R. Q. Luke. 1993. *Kenya coastal forests: the report of the NMK/World Wildlife Fund Coast Forest Survey.* National Museums of Kenya, Nairobi, Kenya.

Robinson, J. G. and E. L. Bennett. 2000a. Carrying capacity limits to sustainable hunting in tropical forests. Pages 13–30 in J. G. Robinson and E. L. Bennett, editors. *Hunting for sustainability in tropical forests.* Columbia University Press, New York, USA.

Robinson, J. G. and E. L. Bennett, editors. 2000b. *Hunting for sustainability in tropical forests.* Columbia University Press, New York, USA.

Robinson, J. G., K. H. Redford, and E. L. Bennett. 1999. Wildlife harvested in logged tropical forests. *Science* 284:594–596.

Rocamora, G. and A. Skerrett. 2001. Seychelles. Pages 751–768 in L. D. C. Fishpool and M. I. Evans, editors. *The important bird areas of Africa and associated islands: priority sites for conservation.* Pisces Publications and BirdLife International (BirdLife Conservation Series No. 11), Newbury and Cambridge, UK.

Rodgers, W. A., R. Nabanyumya, and J. Salehe. 2001. Beyond boundaries: transboundary natural resource management in the Minziro–Sango Bay forest ecosystem. In Biodiversity Support Program, editor. *Beyond boundaries: transboundary natural resource management in eastern Africa.* Biodiversity Support Program, Washington, DC, USA.

Rodgers, W. A., J. Salehe, and G. Howard. 1996. The biodiversity of miombo woodlands. Page 12 in B. Campbell, editor. *The miombo in transition: woodlands and welfare in Africa.* Centre for International Forestry Research, Bogor, Indonesia.

Rodgers, W. A. 1993. The conservation of the forest resources of eastern Africa: past influences, present practices and future needs. Pages 283–331 in J. C. Lovett and S. Wasser, editors. *Biogeography and ecology of the rain forests of eastern Africa.* Cambridge University Press, Cambridge, UK.

Rodgers, W. A. 1996. The miombo woodlands. Pages 299–326 in T. R. McClanahan and T. P. Young, editors. *East African ecosystems and their management.* Oxford University Press, Oxford, UK and New York, USA.

Rodgers, W. A. and K. M. Homewood. 1982. Species richness and endemism in the Usambara Mountain Forests, Tanzania. *Biological Journal of the Linnean Society* 18:197–242.

Rodríguez, L. and J. Urioste. 2000. *Fauna exótica en Canarias.* Makaronesia no. 2. Museo de Ciencias Naturales de Tenerife, Cabildo de Tenerife, The Canary Islands, Spain.

Roe, D. 2001. *Community-based wildlife management: improved livelihoods and wildlife conservation?* Bio-brief 1. IIED Biodiversity and Livelihoods Group, London, UK.

Roe, D. and M. Jack. 2001. *Stories from Eden: case studies of community-based wildlife management.* Evaluating Eden Series 9, IIED, London, UK.

Roelke-Parker, M. E., L. Munson, C. Packer, R. Kock, S. Cleaveland, M. Carpenter, S. J. O'Brien, A. Pospischil, R. Hofmann-Lehmann, H. Lutz, G. L. M. Mwamengele, M. N. Mgasa, G. A. Machange, B. A. Summers, and M. J. G. Appel. 1996. A canine distemper virus epidemic in Serengeti lions (*Panthera leo*). *Nature* 379:441–445.

Rogers, D. J. 1990. A general model for tsetse populations. *Insect Science and Its Application* 11:331–346.

Rogers, D. J. 2000. Satellites, space, time and the African trypanosomiases. *Advances in Parasitology* 47:129–171.

Rogers, D. J., S. E. Randolph, R. W. Snow, and S. I. Hay. 2002. Satellite imagery in the study and forecast of Malaria. *Nature* 415:710–715.

Rosenzweig, M. L. 1995. *Species diversity in space and time.* Cambridge University Press, Cambridge, UK. and New York, USA.

Rosevear, D. R. 1954. *Vegetation and forestry. The Nigeria handbook.* The Government Printer, Lagos, Nigeria.

Rosevear, D. R. 1974. *The carnivores of West Africa.* British Museum (Natural History), London, UK.

Ross, D., editor. 1996. *Structural adjustment, the environment, and sustainable development.* Earthscan, London, UK.

Roth, H. H. and H. I. Douglas-Hamilton. 1991. Distribution and status of elephants in West Africa. *Mammalia* 55:489–527.

Roux, F. and G. Jarry. 1984. Numbers, composition and distribution of populations of Anatidae wintering in West Africa. *Wildfowl* 35:48–60.

Roux, P. W. and G. K. Theron. 1987. Vegetation change in the Karoo biome. Pages 50–69 in R. M. Cowling and P. W. Roux, editors. *The Karoo biome: a preliminary synthesis. Part 2: Vegetation and history.* South African National Scientific Programmes Report No. 142. CSIR, Pretoria, South Africa.

Roy, M. S. 1997. Recent diversification in African greenbuls (Pycnonotidae: Andropadus) supports a montane speciation model. *Proceedings of the Royal Society of London* B 264:1337–1344.

Roy, M. S., R. Sponer, and J. Fjeldså. 2001. Molecular systematics and evolutionary history of akalats (genus *Shepardia*): a pre-Pleistocene radiation in a group of African forest birds. *Molecular Phylogenetics and Evolution* 18:74–83.

Roy, S. M., J. M. C. da Silva, P. Arctander, J. García-Moreno, and J. Fjeldså. 1997. The role of montane regions in the speciation of South American and African birds. Pages 325–343 in D. P. Mindell, editor. *Avian molecular evolution and systematics.* Academic Press, London, UK and New York, USA.

Rudel, T. and J. Roper. 1997. Forest fragmentation in the humid tropics: a cross-national analysis. *Singapore Journal of Tropical Geography* 18:99–109.

Rugalema, G., S. Weigang, and J. Mbwika. 1999. *HIV/AIDS and the commercial agricultural sector in Kenya.* Food and Agriculture Organization of the United Nations, Rome, Italy.

Rundel, P. W. 1994. Tropical alpine climates. Pages 21–44 in P. W. Rundel, A. P. Smith, and F. C. Meinzer, editors. *Tropical alpine environments: plant form and function.* Cambridge University Press, Cambridge, UK.

Rutherford, M. C. 1997. Categorization of biomes. Pages 91–98 in R. M. Cowling, D. M. Richardson, and S. M. Pierce, editors. *Vegetation of southern Africa.* Cambridge University Press, Cambridge, UK.

Rutherford, M. C., G. F. Midgley, W. J. Bond, L. W. Powrie, R. Roberts, and J. Allsopp. 1999. *South African country study on climate change. Plant biodiversity: vulnerability and adaptation assessment.* National Botanical Institute, Claremont, South Africa.

Rutherford, M. C. and R. H. Westfall. 1986. Biomes of southern Africa: an objective categorization. *Memoirs van die Botaniese Opname van Suid-Afrika* I–VI:1–98.

Ryan, P. G., I. Hood, P. Bloomer, J. Komen, and T. Crowe. 1999. Barlow's lark: a new species in the Karoo lark *Certhilauda albescens* complex. *Ibis* 140:605–619.

Saboureau, M. 1962. Note sur quelques températures relevées dans les réserves naturelles. *Bulletin de l'Académie Malgache,* nouvelle série 40:12–22.

Sachs, J. and P. Malaney. 2002. The economic and social burden of malaria. *Nature* 415:680–685.

Sackey, I., E. Laing, and J. K. Adomako. 1993. Status of the mangroves of Ghana. Pages 93–101 in E. D. Diop, editor. *Conservation and sustainable utilization of mangrove forests in Latin America and Africa Regions. Part II: Africa.* Mangrove Ecosystems Technical Reports, Volume 3. International Society for Mangrove Ecosystems and Coastal Marine Project of UNESCO, Okinawa, Japan.

SADC. 1998. *Tourism protocol.* Southern Africa Development Community. SADC, Gaborone, Botswana.

SADC. 1999. *Protocol on wildlife conservation and law enforcement in the Southern African Development Community.* SADC Wildlife Sector TCU, Lilongwe, Malawi.

Sadki, N. 1995. Etude des groupments à olivier et lentisque de la région d'Annaba (nord-est Algérien). Essai phytosociologique. *Documents Phytosociologiques* XV:253–271.

Safford, R. J. 1997a. Distribution studies on the forest-living native passerines of Mauritius. *Biological Conservation* 80:189–198.

Safford, R. J. 1997b. A survey of the occurrence of native vegetation remnants of Mauritius in 1993. *Biological Conservation* 80:181–188.

Safford, R. J. 2001a. The Comoros. Pages 185–190 in L. D. C. Fishpool and M. I. Evans, editors. *The important bird areas of Africa and associated islands: priority sites for conservation.* Pisces Publications and BirdLife International (BirdLife Conservation Series No. 11), Newbury and Cambridge, UK.

Safford, R. J. 2001b. Mauritius. Pages 583–596 in L. D. C. Fishpool and M. I. Evans, editors. *The important bird areas of Africa and associated islands: priority sites for conservation.* BirdLife Conservation Series No. 11. Pisces Publications and BirdLife International, Newbury and Cambridge, UK.

Safford, R. J. 2001c. Mayotte. Pages 597–601 in L. D. C. Fishpool and M. I. Evans, editors. *The important bird areas of Africa and associated islands: priority sites for conservation.* Pisces Publications and BirdLife International (BirdLife Conservation Series No. 11), Newbury and Cambridge, UK.

Safford, R. J. and C. G. Jones. 1998. Strategies of land-bird conservation on Mauritius. *Conservation Biology* 12:169–176.

Sagheir, O. A. and R. F. Porter. 1998. The bird biodiversity of Soqotra. Pages 199–212 in H. J. Dumont, editor. *Proceedings of the first international symposium on Soqotra Island: present and future,* Aden 1996. Soqotra Technical Series, Volume 1. UNDP, New York, USA.

Said, M. Y., R. N. Chunge, C. G. Craig, C. R. Thouless, R. F. W. Barnes, and H. T. Dublin. 1995. *African elephant database 1995.* Occasional Paper of the IUCN Species Survival Commission No. 11. IUCN, Gland, Switzerland.

Sakai, I. 1989. *A report on the Mulanje Cedar resources and the present crisis.* Forestry Research Record No. 65. F.R.I.M., Zomba, Malawi.

Saket, M. and R. Matusse. 1994. *Estudos de determinação da taxa de deflorestamento da vegetação de Mangal em Moçambique.* FAO/PNUD MOZ/92/013. Unidade de Inventário Florestal, Departamento Florestal, Direcção Nacional de Florestas e Fauna Bravia, Ministério da Agricultura, Maputo, Mozambique.

Salafsky, N. and R. Margolius. 1999. Threat reduction assessment: a practical and cost-effective approach to evaluating conservation and development projects. *Conservation Biology* 13:830–841.

Salt, G. 1954. A contribution to the ecology of upper Kilimanjaro. *Journal of Ecology* 42:375–423.

Salt, G. 1987. Insects and other invertebrate animals collected at high altitudes in the Ruwenzori and on Mount Kenya. *African Journal of Ecology* 25:95–106.

Sandalow, D. B. and I. A. Bowles. 2001. Fundamentals of treaty-making on climate change. *Science* 292:1839–1840.

Sanderson, E. W., M. Jaiteh, M. A. Levy, K. H. Redford, A. V. Wannebo, and G. Woolmer. 2002a. The human footprint and the last of the wild. *BioScience* 52:891–904.

Sanderson, E. W., K. H. Redford, A. Vedder, P. B. Coppolillo, and S. E. Ward. 2002b. A conceptual model for conservation planning based on landscape species requirements. *Landscape and Urban Planning* 58:41–56.

Sandwith, T. S., C. Shine, L. S. Hamilton, and D. A. Sheppard. 2001. *Transboundary protected areas for peace and cooperation.* IUCN, Gland, Switzerland and Cambridge, UK.

Sanjayan, M. A., S. Shen, and M. Jansen. 1997. *Experiences with conservation-development projects in Asia.* World Bank Technical Paper No. 388. The World Bank, Washington, DC, USA.

Sarch, M. T. and C. Birkett. 2000. Fishing and farming at Lake Chad: responses to lake-level fluctuations. *Geographic Journal* 166:156–172.

Sarmiento, E. E. and J. F. Oates. 2000. The Cross River gorillas: a distinct subspecies, *Gorilla gorilla diehli* Matchie 1904. *American Museum Novitates* 3304:1–55.

Sauer, J. D. 1967. *Plants and man on the Seychelles coast.* University of Wisconsin Press, Madison, WI, USA and London, UK.

Saunders, D. A., R. J. Hobbs, and C. R. Margules. 1991. Biological consequences of ecosystem fragmentation: a review. *Conservation Biology* 5:18–32.

Sayer, J. A., C. S. Harcourt, and N. M. Collins. 1992. *The conservation atlas of tropical forests: Africa.* Macmillan Publishers Ltd., Basingstoke, UK.

Schaller, G. B. 1963. *The mountain gorilla: ecology and behavior.* University of Chicago Press, Chicago, UK.

Schaller, G. B. 1972. *The Serengeti lion: study of predator-prey relations.* University of Chicago Press, Chicago, Illinois, USA.

Scharff, N. 1992. The linyphiid fauna of eastern Africa (Araneae: Linyphiidae): distribution patterns, diversity and endemism. *Biological Journal of the Linnean Society* 45:117–154.

Schatz, G. E. 2000. Endemism in the Malagasy tree flora. Pages 1–9 in W. R. Lourenço and S. M. Goodman, editors. *Diversity and endemism in Madagascar.* Memoires de la Société de Biogeographie, Paris, France.

Schatz, G. E., C. Birkinshaw, and P. P. I. Lowry. 2000. The Endemic Plant Families of Madagascar project: integrating taxonomy and conservation. Pages 11–24 in W. R. Lourenço and S. M. Goodman, editors. *Diversité et endemism à Madagascar.* Mémoires de la Société de Biogéographie, Paris, France.

Schiøtz, A. 1964. Preliminary list of amphibians collected in Sierra Leone. *Videnskabelige Meddelelser fra Dansk Naturhistorisk Forening I Kobenhavn* 127:19–33.

Schiøtz, A. 1967. The treefrogs (Rhacophoridae) of West Africa. *Spolia Zoologica Musei Hauniensis* 25:1–346.

Schiøtz, A. 1999. *Treefrogs of Africa.* Edition Chimaira, Frankfurt am Main, Germany.

Schlage, C., C. Mabula, R. L. A. Mahunnah, and M. Heinrich. 1999. Medicinal plants of the Washambaa (Tanzania): Documentation and ethnopharmacological evaluation. *Plant Biology* 2:83–92.

Schleich, H. H., W. Kastle, and K. Kabisch. 1996. *Amphibians and reptiles of North Africa: biology, systematics, field guide.* Koeltz Scientific Books, Koenigstein, Germany.

Schlitter, D. A. 1974. Notes on the Liberian mongoose, *Liberiictis kuhni* Hayman 1958. *Journal of Mammalogy* 55:438–442.

Schmidl, D. 1982. *The birds of the Serengeti National Park, Tanzania.* Serengeti Research Institute No. 225. BOU Checklist No. 5. British Ornithologists' Union, London, UK.

Schnell, R. 1952. Végétation et flore de la region montagneuse du Nimba. *Memoires l'IFAN* 22:1–604.

Scholte, P. 1998. Status of vultures in the Lake Chad Basin, with special reference to northern Cameroon and western Chad. *Vulture News* 39:3–19.

Scholte, P. and P. Robertson. 2001. Chad. Pages 177–184 in L. D. C. Fishpool and M. I. Evans, editors. *Important bird areas in Africa and associated islands: priority sites for conservation.* Pisces Publications and BirdLife International (BirdLife Conservation Series No. 11), Newbury and Cambridge, UK.

Scholtz, A. 1983. Houtskool weerspieël die geskiedenis van inheemse woude. *Bosbounuus* 3:18–19.

Scholz, H. 1968. Die Böden der Wüste Namib/Südwestafrika. *Zeitschrift für Pflanzenern und Bodenkunde* 119:91–107.

Schrijver, N. 1997. *Sovereignty over natural resources: balancing rights and duties.* Cambridge University Press, Cambridge, UK.

Schulze, R. E. 1997. Climate. Pages 21–42 in R. M. Cowling, D. M. Richardson, and S. M. Pierce, editors. *Vegetation of southern Africa.* Cambridge University Press, Cambridge, UK.

Schulze, R. E. and O. S. McGee. 1978. Climatic indices and classification in relation to the biogeography of southern Africa. Pages 19–54 in M. J. A. Werger, editor. *Biogeography and ecology of southern Africa.* Dr. W. Junk Publishers, The Hague, The Netherlands.

Schwartz, B. 1992a. *Identification, establishment and management of specially protected areas in the WACAF Region.* IUCN Wetlands Programme, Gland, Switzerland.

Schwartz, D. 1992b. Assèchement climatique vers 3000 B.P. et expansion Bantu en Afrique Centrale Atlantique: quelques réflections. *Bulletin Société Géologique de France* 163:353–361.

Schwartz, J. H. 1996. *Pseudopotto martini:* a new genus and species of extant lorisform primate. Anthropological Papers of the American Museum of Natural History, New York, USA.

Sclater, P. L. 1858. On the general geographical distribution of the class Aves. *Journal of the Proceedings of the Linnean Society (London) Zoology* 2:130–145.

Scott, D. A. and P. M. Rose. 1996. *Atlas of Anitidae populations in Africa and western Eurasia.* Wetlands International, Wagingingen, The Netherlands.

Scott, J. M., H. Anderson, F. Davis, S. Caicoo, B. Csuti, T. C. Edwards, R. Noss, J. Ulliman, C. Groves, and R. G. Wright. 1993. Gap analysis: a geographic approach to protection of biological diversity. *Wildlife Monographs* 123:1–41.

Scott-Shaw, R. 1999. *Rare and threatened plants of KwaZulu-Natal and neighbouring regions.* KwaZulu-Natal Nature Conservation Service, Pietermaritzburg, South Africa.

Seddon, N., J. Tobias, J. W. Yount, J. R. Ramanampamonjy, S. Butchart, and H. Randrianizahana. 2000. Conservation issues and priorities in the Mikea Forest of south-west Madagascar. *Oryx* 34:287–304.

Semesi, A. 1993. Mangrove ecosystems of Tanzania. Pages 211–225 in E. Diop, editor. *Conservation and sustainable utilization of mangrove forests in Latin America and Africa regions. Part II: Africa.* Mangrove Ecosystems Technical Reports, Volume 3. International Society for Mangrove Ecosystems and Coastal Marine Project of UNESCO, Okinawa, Japan.

Semesi, A. 1998. Mangrove management and utilization in eastern Africa. *Ambio* 27:620–626.

Serag, M. S. 2000. The rediscovery of papyrus (*Cyperus papyrus*) on the bank of Damietta Branch, Nile Delta Egypt. *Taeckholmia* 20:195–198.

Servant, M., J. Maley, B. Turcq, M. L. Absy, P. Brenac, M. Fournier, and M. P. Ledru. 1993. Tropical forest changes during the late Quaternary in African and South American lowlands. *Global and Planetary Change* 7:25–40.

Seydack, A. H. W. 1995. An unconventional approach to timber yield regulation for multi-aged, multispecies forests. I. Fundamental considerations. *Forest Ecology and Management* 77:139–153.

Seydack, A. H. W., W. J. Vermeulen, H. E. Heyns, G. P. Durrheim, C. Vermeulen, D. Willems, M. A. Ferguson, J. Huisamen, and

J. Roth. 1995. An unconventional approach to timber yield regulation for multi-aged, multispecies forests. II. Application to a South African forest. *Forest Ecology and Management* 77:155–168.

Shah, N. J., R. Payet, and K. Henri. 1997. *Seychelles national biodiversity strategy and action plan.* Republic of Seychelles.

Shambaugh, J., J. Oglethorpe, and R. Ham. 2001. *The trampled grass: mitigating the impacts of armed conflict on the environment.* Biodiversity Support Program, Washington, DC, USA.

Shaw, P. A., S. Stokes, D. S. G. Thomas, F. B. M. Davies, and K. Holmgren. 1997. Palaeoecology and age of a quaternary high take level in the Makgadikgadi Basin of the Middle Kalahari, Botswana. *South African Journal of Science* 93:273–276.

Shaw, T. 1977. *Unearthing Igbo-Ukwu.* Oxford University Press, Idaban, Nigeria.

Sheded, M. G. 1998. Vegetation pattern along an edaphic and climatic gradient in the Southern Eastern Desert of Egypt. *Feddes Repertorium* 109:329–335.

Shine, T., P. Robertson, and B. Lamarche. 2001. Mauritania. Pages 567–581 in L. D. C. Fishpool and M. I. Evans, editors. *Important bird areas in Africa and associated islands: priority sites for conservation.* Pisces Publications and BirdLife International (BirdLife Conservation Series No. 11), Newbury and Cambridge, UK.

Shmueli, M., I. Izhaki, A. Arieli, and Z. Arad. 2000. Energy requirements of migrating great white pelicans *Pelecanus onocrotalus. Ibis* 142:208–216.

Showler, D. 1994. Reptiles on the Island of Socotra, Republic of Yemen. *British Herpetological Society Bulletin* No. 487-13.

Shumway, C. A. 1999. *Forgotten waters: freshwater and marine ecosystems in Africa: strategies for biodiversity conservation and sustainable development.* Boston University, Boston, USA.

Sibley, C. G. and B. L. Jr. Monroe. 1990. *Distribution and taxonomy of birds of the world.* Yale University Press, New Haven, CT, USA.

Sibley, C. G. and J. E. Monroe. 1993. *A supplement to "Distribution and taxonomy of birds of the world."* Yale University Press, New Haven, CT, USA.

Sillero-Zubiri, C., A. A. King, and D. W. Macdonald. 1996. Rabies and mortality in Ethiopian wolves (*Canis simensis*). *Journal of Wildlife Diseases* 32:80–86.

Sillero-Zubiri, C., D. W. Macdonald, and Species Survival Commission/IUCN Canid Specialist Group. 1997. *The Ethiopian wolf: status survey and conservation action plan.* IUCN, Gland, Switzerland.

Sillero-Zubiri, C., F. H. Tattersall, and D. W. Macdonald. 1995. Bale Mountains rodent communities and their relevance to the Ethiopian wolf (*Canis simensis*). *African Journal of Ecology* 33:301–320.

Simao, A. D. S. 1993. The mangroves of Guinea Bissau. Pages 37–46 in E. D. Diop, editor. *Conservation and sustainable utilization of mangrove forests in Latin America and Africa regions. Part II: Africa.* Mangrove Ecosystems Technical Reports, Volume 3. International Society for Mangrove Ecosystems and Coastal Marine Project of UNESCO, Okinawa, Japan.

Simmons, R. E. 1996. Population declines, viable breeding areas and management options for flamingos in southern Africa. *Conservation Biology* 10:504–514.

Simmons, R. E. 2000. Declines and movements of lesser flamingos in Africa. *Waterbirds* 23:40–46.

Simmons, R. E., C. Boix-Hinzen, K. N. Barnes, A. M. Jarvis, and A. Robertson. 1998a. Important bird areas of Namibia. Pages 295–332 in K. Barnes, editor. *The important bird areas of southern Africa.* BirdLife South Africa, Johannesburg, South Africa.

Simmons, R. E., M. Griffin, R. E. Griffin, E. Marais, and H. Kolberg. 1998b. Endemism in Namibia: patterns, processes and predictions. *Biodiversity and Conservation* 7:513–530.

Simon, E. L. 1988. A new species of *Propithecus* (Primates) from northeast Madagascar. *Folia Primatologica* 50:143–151.

Simonetta, A. M. and J. Simonetta. 1983. An outline of the status of the Somali fauna and of its conservation and management problems. *Rivista di Agricoltura Subtropicale e Tropicale* 73:456–483.

Simwinji, N. 1997. *Summary of existing relevant socio-economic and ecological information.* Report to IUCN on Zambia's Western Province and Barotseland.

Sinclair, A. R. E. 1977. *The African buffalo.* University of Chicago Press, Chicago, Illinois, USA.

Sinclair, A. R. E. 1979. The Serengeti environment. Pages 31–45 in A. R. E. Sinclair and M. Norton-Griffiths, editors. *Serengeti: dynamics of an ecosystem.* University of Chicago Press, Chicago, Illinois, USA.

Sinclair, A. R. E., S. A. R. Mduma, and P. Arcese. 2000. What determines phenology and synchrony of ungulate breeding in Serengeti? *Ecology* 81:2100–2111.

Sinclair, I. and P. Hockey. 1996. *The larger field guide to birds of southern Africa.* Struik, Cape Town, South Africa.

Sinclair, I. and O. Langrand. 1998. *Birds of the Indian Ocean Islands.* Struik Publishers, Cape Town, South Africa.

Singh, A., A. Dieye, and M. Finco. 1999. Assessing environmental conditions of major river basins in Africa as surrogates for watershed health. *Ecosystem Health* 5:265–274.

Singini, P. J. T. 1996. The Marromeu complex of the Zambezi Delta: Mozambique's unique wetland. Pages 341–344 in R. D. Beilfuss, W. R. Tarboton, and N. N. Gichuki, editors. *Proceedings of the African crane and wetland training workshop.* Wildlife Training Institute, Botswana. International Crane Foundation, Baraboo, WI, USA.

Skead, C. J. 1982. *Historical mammal incidence in the Cape Province. Volume 1: The western and northern Cape.* Department Nature and Environmental Conservation, Cape Town, South Africa.

Skead, C. J. 1987. *Historical mammal incidence in the Cape Province. Volume 2: The eastern half of the Cape Province,* including the Ciskei, Transkei and East Griqualand. Department Nature and Environmental Conservation, Cape Town, South Africa.

Skelton, P. H., J. A. Cambray, A. Lombard, and G. A. Benn. 1995. Patterns of distribution and conservation status of freshwater fishes in South Africa. *South African Journal of Zoology* 30:71–81.

Skidmore, A. K., A. Gauld, and P. W. Walker. 1996. A comparison of GIS predictive models for mapping kangaroo habitat. *International Journal of Geographical Information Systems* 10:441–454.

Skinner, J. D. and R. H. N. Smithers. 1990. *The mammals of the southern African subregion.* University of Pretoria, Pretoria, South Africa.

Skinner, J. R., P. J. Wallace, W. Altenburg, and B. Fofana. 1987. The status of heron colonies in the Niger Delta, Mali. *Malimbus* 9:65–82.

Smith, A. P. and T. P. Young. 1987. Tropical alpine plant ecology. *Annual Review of Ecology and Systematics* 18:137–158.

Smith, G. 1984. Climate. Pages 17–30 in J. L. Cloudsley-Thompson, editor. *Sahara Desert.* Pergamon Press, Oxford, UK and New York, USA.

Smith, G. F., P. Chesselet, E. J. van Jaarsveld, H. Hartmann, S. Hammer, B. E. van Wyk, P. Burgoyne, C. Klak, and H. Kurzweil. 1998. *Mesembs of the world.* Briza Publications, Pretoria, South Africa.

Smith, P. P. 1998. A reconnaissance survey of the vegetation of the North Luangwa National Park, Zambia. *Bothalia* 28:197–211.

Smith, T. B., M. W. Bruford, and R. K. Wayne. 1993. The preservation of process: the missing element of conservation programmes. *Biodiversity Letters* 1:164–167.

Smith, T. B., R. K. Wayne, D. J. Girman, and M. W. Bruford. 1997. A role for ecotones in generating rainforest biodiversity. *Science* 276:1855–1857.

Smithers, R. H. N. 1983. *The mammals of the southern African subregion.* University of Pretoria, Pretoria, South Africa.

Smithers, R. H. N. 1986. *South African Red Data Book: terrestrial mammals.* South African National Scientific Programmes Report No. 125. Council for Scientific and Industrial Research, Pretoria, South Africa.

Snow, D. W., editor. 1978. *An atlas of speciation in African non-passerine birds.* Trustees of the British Museum (Natural History), London, UK.

Snow, R. W., M. Craig, U. Deichmann, and K. Marsh. 1999. Estimating mortality, morbidity and disability due to Malaria among Africa's non-pregnant population. *Bulletin of the World Health Organization* 77:624–640.

Soler, O. 1997. *Atlas climatique de la Réunion.* Ste. Clothilde, Meteo-France, Réunion, France.

Sollenberg, M. and P. Wallensteen. 2001. Patterns of major armed conflicts, 1990–2000. SIPRI Yearbook 2001: armaments, disarmament and international security. Press release on Stockholm International Peace Research Institute (SIPRI). Web site: http://editors.sipri.se/pubs/yb02/pr02.html, accessed on July 18, 2002.

Some, E. S. 1994. Effects and control of highland malaria epidemic in Uasin Gishu District, Kenya. *East African Medical Journal* 71:2–8.

Soulé, M. E., B. A. Wilcox, and C. Holtby. 1979. Benign neglect: a model of faunal collapse in the game reserves of East Africa. *Biological Conservation* 15:259–272.

Sowunmi, M. A. 1986. Change of vegetation with time. Pages 273–307 in G. W. Lawson, editor. *Plant ecology in West Africa.* John Wiley and Sons, New York, USA.

Spalding, M. D., F. Blasco, and C. D. Field, editors. 1997. *World mangrove atlas.* The International Society for Mangrove Ecosystems, Okinawa, Japan.

Spear, T. 1978. *The Kaya complex: a history of the Mijikenda peoples of Kenya to 1900.* Kenya Literature Bureau, Nairobi, Kenya.

Spergel, B. 2001. *Raising revenues for protected areas: a menu of options*. World Wildlife Fund, Washington, DC, USA.

Spergel, B. 2002. Financing protected areas. Pages 364–382 in J. Terborgh, C. van Schaik, L. Davenport, and M. Rao, editors. *Making parks work: strategies for preserving tropical nature*. Island Press, Washington, DC, USA.

Spinage, C. A. 1996. The rule of law and African game a review of some recent trends and concerns. *Oryx* 30:178–186.

Spinage, C. 1998. Social change and conservation misrepresentation in Africa. *Oryx* 32:265–276.

Sponsel, L. E., T. N. Headland, and R. Bailey, editors. 1996. *Tropical deforestation: the human dimension*. Columbia University Press, New York, USA.

Sprawls, S. and B. Branch. 1996. *A guide to the poisonous snakes of Africa*. Struik Publishers, Cape Town, South Africa.

Stattersfield, A. J., M. J. Crosby, A. J. Long, and D. C. Wege. 1998. *Endemic bird areas of the world. Priorities for biodiversity conservation*. BirdLife Conservation Series No. 7. BirdLife International, Cambridge, UK.

Steel, L. and B. Curran. 2001. Beyond boundaries: transboundary natural resource management in the Sangha River Trinational Initiative. In Biodiversity Support Program, editor. *Beyond boundaries: transboundary natural resource management in Central Africa*. Biodiversity Support Program, Washington, DC, USA.

Stein, B. A., L. S. Kutner, and J. S. Adams, editors. 2000. *Precious heritage: the status of biodiversity in the United States*. Oxford University Press, New York, USA.

Steiner, K. E. and V. B. Whitehead. 1990. Pollinator adaptation to oil-secreting flowers *Rediviva* and *Diascia*. *Evolution* 44:1701–1707.

Steinke, T. D. 1995. A general review of the mangroves of South Africa. Pages 53–74 in G. Cowan, editor. *Wetlands of South Africa*. Department of Environmental Affairs and Tourism, Pretoria, South Africa.

Stephenson, P. J. 2002. *Elephant update number 2*. World Wildlife Fund, Gland, Switzerland.

Stephenson, P. J. and J. E. Newby. 1997. Conservation of the Okapi Wildlife Reserve, Zaire. *Oryx* 31:49–58.

Stewart, W. N. 1983. *Paleobotany and the evolution of plants*. Cambridge University Press, New York, USA and Cambridge, UK.

Steyn, P. 1990. Soda Ash Botswana, the paradox of Sua. *African Wildlife* 44:244–247.

Stiles, D. 1994. Tribals and trade: a strategy for cultural and ecological survival. *Ambio* 23:106–111.

Stockwell, H. M. and P. H. Noble. 1991. Induction of sets of rules from animal distribution data: a robust and informative method of data analysis. *Mathematics and Computer Simulation* 32:249–254.

Stoddart, J. W. 1984. Impact of man in the Seychelles. Pages 641–654 in D. R. Stoddart, editor. *Biogeography and ecology of the Seychelles Islands*. Dr. W. Junk Publishers, The Hague, The Netherlands.

Stoms, D. M. 1994. Scale dependence of species richness maps. *Professional Geographer* 46:346–358.

Stoms, D. M., F. W. Davis, and C. B. Cogan. 1992. Sensitivity of wildlife habitat models to uncertainties in GIS data. *Photogrammetric Engineering & Remote Sensing* 58:843–850.

Strahm, W. A. 1989. *Plant Red Data Book for Rodriguez*. IUCN, Gland, Switzerland.

Strahm, W. A. 1996a. Mascarene Islands: an introduction. *Curtis's Botanical Magazine* 13:182–185.

Strahm, W. A. 1996b. The vegetation of the Mascarene Islands. *Curtis's Botanical Magazine* 13:214–217.

Strahm, W. A. 1999. Invasive species in Mauritius: examining the past and charting the future. Pages 325–347 in O. T. Sandlund, P. J. Schei, and L. Viken, editors. *Invasive species and biodiversity management*. Kluwer Academic Press, Dordrecht, The Netherlands.

Strasburger, E., F. Noll, H. Schenck, and A. F. W. Schimper. 1986. *Tratado de botánica*, 7th edition. Ed. Marin, Barcelona, Spain.

Struhsaker, T. T. 1997. *Ecology of an African rainforest: logging in Kibale and the conflicts between conservation and development*. University Press of Florida, Gainesville, USA.

Stuart, C. and T. Stuart. 1992. Guide to southern African game and nature reserves. Struik Publishers, Cape Town, South Africa.

Stuart, C. and T. Stuart. 1995. *Field guide to the mammals of southern Africa*. Struik, Cape Town, South Africa.

Stuart, C. and T. Stuart. 1996. *Africa's vanishing wildlife*. Smithsonian Institution Press, Washington, DC, USA.

Stuart, S. N., editor. 1986. *Conservation of Cameroon montane forests*. International Council for Bird Preservation, Cambridge, UK.

Stuart, S. N., R. J. Adams, and M. D. Jenkins, editors. 1990. *Biodiversity in sub-Saharan Africa and its islands: conservation management and sustainable use*. Occasional papers of the IUCN Species Survival Commission. IUCN, Gland, Switzerland.

Stuckenberg, B. R. 1962. The distribution of the montane palaeogenic element in the South African invertebrate fauna. *Annals of the Cape Provincial Museum* 11:119–158.

Suliman, M., editor. 1999. *Ecology, politics and violent conflict*. Zed Books, London, UK.

Sullivan, S. and T. L. Konstant. 1997. Human impacts on woody vegetation, and multivariate analysis: a case study based on data from Khowarib settlement, Kunene Region. *Dinteria* 25:87–120.

Sussman, R. W., G. M. Green, and L. K. Sussman. 1994. Satellite imagery, human ecology, anthropology, and deforestation in Madagascar. *Human Ecology* 22:333–354.

Sutton, J. E. G. 1990. *A thousand years of East Africa*. British Institute in Eastern Africa, Nairobi, Kenya.

Swaine, M. D. 1992. Characteristics of dry forest in West Africa and the influence of fire. *Journal of Vegetation Science* 3:365–374.

Swaine, M. D., W. D. Hawthorne, and T. K. Orgle. 1992. The effects of fire exclusion on Savanna vegetation at Kpong, Ghana. *Biotropica* 24:166–172.

Swift, J. 1996. Desertification: narratives, winners and losers. Pages 73–90 in M. Leach and R. Mearns, editor. *The lie of the land: challenging received wisdom on the African environment*. James Currey, Oxford, UK.

Swynnerton, C. F. M. 1917. Some factors in the replacement of the ancient East African forest by wooded pasture land. *South African Journal of Science* 14:493–518.

Taylor, M. E. 1989. New records of two species of rare viverrids from Liberia. *Mammalia* 53:122–125.

Taylor, M. E. 1992. The Liberian mongoose. *Oryx* 26:103–106.

Taylor, M., C. Ravilious, and E. P. Green. 2003. *Mangroves of East Africa*. United Nations Environment Program–World Conservation Monitoring Centre, Cambridge, UK.

Tello, J. L. P. L. and P. Dutton. 1979. *Programa de Operação Bufalo*. Departamento do Fauna Bravia, Maputo, Mozambique.

Terborgh, J. 1999. *Requiem for nature*. Island Press, Washington, DC, USA.

Texeira, J. B. 1968. Angola. Pages 193–197 in I. Hedberg and O. Hedberg, editors. *Conservation of vegetation in Africa south of the Sahara*. Proceedings of the AETFAT Congress, Uppsala. Acta Phytogeographica Suecica 54. Uppsala, Almqvist and Wiksells, Sweden.

Thayer's Birder's Diary v. 2.5 CD-ROM. 1997. Thayer Birding Software Ltd., Florida, USA.

The Nature Conservancy. 2000. *Site conservation planning. A framework for developing and measuring the impact of effective biodiversity conservation strategies*. TNC, Arlington, Virginia.

Thibault, J. C. and I. Guyot, editors. 1988. *Livre Rouge des oiseaux menacés des regions françaises d'outre-mer*. International Council for Bird Preservation, France.

Thieme, M. L., R. A. Abell, B. Lehner, E. Dinerstein, M. L. J. Stiassny, P. H. Skelton, G. G. Teugels, A. Kamdem-Toham, and D. M. Olson, editors. 2005. *Freshwater ecoregions of Africa and Madagascar: A conservation assessment*. Island Press, Washington, DC, USA.

Thirgood, J. V. 1981. *Man and the Mediterranean forest*. Academic Press, New York, USA.

Thomas, D. 1987. Vegetation of Mount Oku. Pages 54–56 in H. MacLeod, editor. *Conservation of Oku Mountain Forest, Cameroon*. ICBP Report No. 15. ICBP, Cambridge, UK.

Thomas, D. S. G. and P. A. Shaw. 1991. *The Kalahari environment*. Cambridge University Press, Cambridge, UK.

Thomas, D. and J. Thomas. 1996. *Tchabal Mbabo botanical survey*. Consultants' Report to World Wildlife Fund Cameroon Program Office. Yaoundé, Cameroon.

Thompson, H. S. S. 1993. Status of white-necked picathartes: another reason for the conservation of the Peninsula Forest, Sierra Leone. *Oryx* 27:155–158.

Thompson, J. A. M. 1996. New information about the presence of the Congo peafowl (*Afropavo congensis*). *World Pheasant Association News* 50:3–8.

Thompson, J. A. M. 2001. The status of bonobos in their southernmost geographic range. Pages 75–82 in B. M. F. Galdikas, N. Briggs, L. K. Sheeran, G. L. Shapiro, and J. Goodall, editors. *All apes great and small. Volume 1: African apes*. Kluwer Academic/Plenum Publishers, New York, USA.

Thompson, J. A. M. 2003. *A model of the biogeographical journey from proto-pan to Pan paniscus: the westside story. Special edition of Primates*. Primate Research Institute of Kyoto University, Kyoto, Japan.

Thompson, P. J. 1975. The role of elephants, fire and other agents in the decline of a *Brachystegia boehmii* woodland. *Journal of the South African Wildlife Management Association* 5:11–18.

Thompson-Handler, N., R. Malenky, and G. Reinartz. 1995. *Action plan for* Pan paniscus: *report on free ranging populations and proposals for their preservation*. The Zoological Society of Milwaukee County, Milwaukee, WI, USA.

Thomsen, J. B. 1988. Recent U.S. imports of certain products from the African elephant. *Pachyderm* 10:1–5, 21.

Thorbjarnarson, J. 1992. *Crocodiles: an action plan for their conservation*. IUCN/Species Survival Commission Crocodile Specialist Group, Gland, Switzerland.

Thorbjarnarson, J. 1999. Crocodile tears and skins: international trade, economic constraints, and limits to the sustainable use of crocodilians. *Conservation Biology* 13:465–470.

Thorstrom, R. and L. A. René de Roland. 1997. First nest record and nesting behaviour of the Madagascar red owl *Tyto soumangei*. *Ostrich* 68:42–43.

Thorstrom, R. and L. A. René de Roland. 2000. First nest description, breeding behaviour and distribution of the Madagascar serpent-eagle *Eutriorchis astur*. *Ibis* 142:217–224.

Thulin, M. 1994. Aspects of disjunct distributions and endemism in the arid parts of the Horn of Africa, particularly Somalia. Pages 1105–1119 in J. H. Seyani and A. C. Chikuni, editors. *Proceedings of the XIII Plenary Meeting AETFAT, 2*. Zomba, Malawi.

Ticheler, H. 2000. *Fish biodiversity in West African wetlands*. Wetlands International, Wageningen, The Netherlands.

Tiffen, M., M. Mortimore, and F. Gichuki. 1994. *More people, less erosion. Environmental recovery in Kenya*. John Wiley and Sons, Chichester, UK.

Tilahun, S., S. Edwards, and T. B. G. Egziabher, editors. 1996. *Important bird areas of Ethiopia*. Ethiopian Wildlife and Natural History Society. Semayata Press, Addis Ababa, Ethiopia.

Tilbury, C. R. 1991. A new species of *Chamaeleo* Laurenti 1768 (Reptilia, Chamaeleonidae) from a relict montane forest in northern Kenya. *Tropical Zoology* 4:159–166.

Tilbury, C. R. 1992. A new dwarf forest chameleon (Sauria: Rhampholeon Guenther 1874) from Malawi, Central Africa. *Tropical Zoology* 5:1–9.

Tilbury, C. R. 1998. Two new chameleons (Sauria: Chamaeleonidae) from isolated Afromontane forests in Sudan and Ethiopia. *Bonner Zoologische Beiträge* 47:293–299.

Tilbury, C. R. and D. Emmrich. 1996. A new dwarf forest chameleon (Squamata: Rhampholeon, Guenther 1874) from Tanzania, East Africa with notes on its infrageneric and zoogeographic relationships. *Tropical Zoology* 9:61–71.

Tilman, D., J. Fargione, B. Wolff, C. D'Antonio, A. Dobson, R. Howarth, D. Schindler, W. H. Schlesinger, D. Simberloff, and D. Swackhamer. 2001. Forecasting agriculturally driven global environmental change. *Science* 292:281–284.

Timberlake, J. R. 1995. *Colophospermum mopane*: annotated bibliography and review. *Zimbabwe Bulletin of Forestry Research* No. 11. Forest Research Centre, Harare, Zimbabwe.

Timberlake, J. 1998. *Biodiversity of the Zambezi Basin wetlands. Phase 1: Review and preliminary assessment of available information. Volume 1: Main report*. Consultancy report for IUCN Regional Office for Southern Africa, Harare, Zimbabwe.

Timberlake, J. R. 1999. *Colophospermum mopane*: an overview of current knowledge. Pages 565–571 in J. R. Timberlake and S. Kativu, editors. *African plants: biodiversity, taxonomy and uses*. Royal Botanic Gardens, Kew, UK.

Timberlake, J. R. 2000. *Biodiversity of the Zambezi Basin wetlands.* Consultancy report for IUCN Regional Office for Southern Africa. Biodiversity Foundation for Africa, Bulawayo/Zambezi Society, Harare, Zimbabwe.

Timberlake, J. R., R. B. Drummond, P. Smith, and M. Bingham. 2000. Wetland plants of the Zambezi Basin. Pages 31–93 in J. R. Timberlake, editor. *Biodiversity of the Zambezi Basin wetlands.* BFA/Zambezi Society, Harare, Zimbabwe.

Timberlake, J. R. and T. Müller. 1994. Identifying and describing areas for vegetation conservation in Zimbabwe. Pages 125–139 in B. J. Huntley, editor. *Botanical diversity in southern Africa.* National Botanical Institute, Pretoria, South Africa.

Timberlake, J., T. Müller, and I. Mapaure. 1994. Vegetation. Pages 34–47 in J. Timberlake and P. Shaw, editors. *Chirinda Forest: a visitor's guide.* Forestry Commission, Harare, Zimbabwe.

Timberlake, J. and C. Musokonyi. 1994. Forest conservation and utilization. Pages 117–125 in J. Timberlake and P. Shaw, editors. *Chirinda Forest: a visitor's guide.* Forestry Commission, Harare, Zimbabwe.

Timberlake, J. and P. Shaw. 1994. Introduction. Pages 1–3 in J. Timberlake and P. Shaw, editors. *Chirinda Forest: a visitor's guide.* Forestry Commission, Harare, Zimbabwe.

Tinley, K. L. 1977. *Framework of the Gorongosa ecosystem.* D.S. thesis. University of Pretoria, Pretoria, South Africa.

Tinley, K. L. 1985. *Coastal dunes of South Africa.* South Africa National Scientific Programmes Report No. 109. CSIR, Pretoria, South Africa.

Tonge, S. J. 1989. *The past, present and future of the herpetofauna of Mauritius.* Chicago Herpetological Society Bulletin 25:220–226.

Townsend Peterson, A. and D. A. Vieglais. 2001. Predicting species invasions using ecological niche modeling: new approaches from bioinformatics attack a pressing problem. *BioScience* 51:363–371.

TRAFFIC East and Southern Africa. 2000. *Food for thought: the utilization of wild meat in eastern and southern Africa.* TRAFFIC East and Southern Africa, Nairobi, Kenya.

Transparency International. Transparency International corruption perceptions index. Retrieved 2002 from the World Wide Web: http://www.transparency.org.

Trape, J. F. and R. Roux-Estève. 1990. Note sur une collection des serpents du Congo avec description d'une espèce nouvelle. *Journal of African Zoology* 104:375–383.

Trapnell, C. G. 1959. Ecological results of woodland burning in northern Rhodesia. *Journal of Ecology* 47:129–168.

Trollope, W. S. W., L. A. Trollope, H. C. Biggs, D. Pienaar, and A. L. F. Potgieter. 1998. Long-term changes in the woody vegetation of the Kruger National Park, with special reference to the effects of elephants and fire. *Koedoe* 41:103–112.

Trujillo, D. 1991. *Murciélagos de las Islas Canarias.* Colección técnica. Ministerio de Agricultura, Pesca y Alimentación. Ed. Icona, Madrid, Spain.

Tucker, C. J., H. E. Dregne, and W. W. Newcomb. 1991. Expansion and contraction of the Sahara Desert from 1980 to 1990. *Science* 253:299–301.

Turlin, B. 1995. Faune lépidoptérologique de l'Archipel des Comores (Océan Indien): rhopalocères, Sphingidae, Attacidae (7). *Lambillionea* 95:443–452.

Turpie, J. K. and T. M. Crowe. 1994. Patterns of distribution, diversity and endemism of larger African mammals. *South African Journal of Zoology* 29:19–32.

Turpie, J., B. Smith, L. Emerton, and J. Barnes. 1999. *Economic value of the Zambezi Basin wetlands.* Report prepared for IUCN Zambezi Basin Wetlands Conservation and Resource Utilization Project. IUCN Regional Office for Southern Africa, Cape Town, South Africa.

Tutin, C. E. G. and M. Fernandez. 1984. Nation wide census of gorilla (*Gorilla g. gorilla*) and chimpanzee (*Pan t. troglodytes*) populations in Gabon. *American Journal of Primatology* 6:313–336.

Tyler, S. J. and D. R. Bishop. 2001. Botswana. Pages 99–112 in L. D. C. Fishpool and M. I. Evans, editors. *Important bird areas in African and associated islands: priority sites for conservation.* Pisces Publications and BirdLife International (BirdLife Series No. 11), Newbury and Cambridge, UK.

Uetz, P. 2001. The EMBL reptile database. Retrieved 2001 from the World Wide Web: http://www.embl-heidelberg.de/~uetz/LivingReptiles.html.

Uganda Ministry of Health. 2001. *HIV/AIDS surveillance report, June 2001.* STD/AIDS Control Programme, Uganda Ministry of Health, Kampala, Uganda.

Ukpong, I. E. 1995. An ordination study of mangrove swamp communities in West Africa. *Vegetatio* 116:147–159.

UNAIDS. 2000. Report on the global HIV/AIDS epidemic. Retrieved 2001 from the World Wide Web: http://www.unaids.org/worldaidsday/2001/EPIgraphic2_en.

UNAIDS. 2001. Report on the global HIV/AIDS epidemic, December 2001. Retrieved 2001 from the World Wide Web: http://www.unaids.org/worldaidsday/2001/EPIgraphic2_en.

UNAIDS and WHO. 2000. Epidemiological fact sheets on HIV/AIDS and sexually transmitted infections, update (revised). Joint United Nations Programme on HIV/AIDS, Geneva, Switzerland.

UNDP. 1999. *World population prospects: the 1998 revision,* Volume 1. United Nations, New York, USA.

UNDP. 2001. *World development report.* UNDP, New York, USA.

UNEP/CMS, editors. 1998. *Proceedings of the Seminar on the Conservation and Restoration of Sahelo-Saharan Antelopes.* CMS Technical Series Publication No. 3. UNEP/CMS, Bonn, Germany.

UNEP/CMS, editors. 1999. *Conservation measures for Sahelo-Saharan antelopes.* Action Plan and Status Reports, CMS Technical Series Publication No. 4. UNEP/CMS, Bonn, Germany.

UNEP, KWS, UNF, and WCST. 2002. *Aerial survey of the threats to Mount Kilimanjaro forest.* United Nations Environmental Program, Nairobi, Kenya.

UNESCO. 2000. MAB Seville 5+ recommendations for the establishment and functioning of transboundary biosphere reserves. Retrieved 2002 from the World Wide Web: http://www.unesco.org/mad/mabicc/2000/eng/TBREng.htm.

UNESCO. 2003a. Biosphere and world heritages sites. http://www.unesco.rg/mab.

UNESCO. 2003b. The World Heritage list. http://www.unesco.rg/heritage.htm.

Urban, E. K., C. H. Fry, and S. Keith, editors. 1986. *The birds of Africa,* Volume II. Academic Press, London, UK.

Urban, E. K., C. H. Fry, and S. Keith, editors. 1997. *The birds of Africa,* Volume V. Academic Press, San Diego, USA and London, UK.

Urioste, J. 1999. CD-ROM base de datos de especies introducidas en Canarias. Viceconsejería de Medio Ambiente del Gobierno de Canarias. Sección de Flora y Fauna/GESPLAN S.A. http://www.gobiernodecanarias.org/medioambiente/biodiversidad/introducidas/especiesinvasoras.html.

USAID. 2000a. *AIDS brief for sectoral planners and managers: community-based natural resource management.* Development Alternatives, Inc., Bethesda, MD, USA.

USAID. 2000b. *AIDS toolkits: HIV/AIDS and community-based natural resource management.* Development Alternatives, Inc. Bethesda, MD, USA.

Uschakov, P. V. 1970. Observation sur la repartition de la faune benthique du littoral Guinéen. *Cahier de Biologie Marine* XI:435–457.

van der Kamp, J. and M. Diallo. 1999. Suivi écologique du Delta Intérieur du Niger: les oiseaux d'eau comme bio-indicateurs. Recensements crue 1998–1999. *Malipin publication* 99-02. Wetlands International, Waginingen, The Netherlands.

van der Kamp, J., M. Diallo, B. Fofana, and B. Kone. 2001. Bio-indicateurs dans le Delta Intérieur du fleuve Niger: suivi de populations d'oiseaux d'eau par dénombrements aériens 1999–2001. *Mali-PIN publication* 01-03. Wetlands International, Waginingen, The Netherlands.

van der Linde, H., J. Oglethorpe, T. Sandwith, D. Snelson, and Y. Tessema. 2001. *Beyond boundaries: transboundary natural resource management in sub-Saharan Africa.* Biodiversity Support Program, Washington, DC, USA.

Van der Merwe, I. 1998. *The Knysna and Tsitsikamma forests: their history, ecology and management.* Department of Water Affairs and Forestry, Knysna, South Africa.

Van der Meulen, F. 1979. Plant sociology of the western Transvaal bushveld, South Africa: a taxonomic and synecological study. Dissertations Botanicae. Cramer, Vaduz, Liechtenstein.

Van der Straeten, E. 1980. A new species of *Lemniscomys* (Muridae) from Zambia. *Annals of the Cape Provincial Museums, Natural History* 13:55–62.

Van der Walt, P. and E. Le Riche. 1999. *The Kalahari and its plants.* Published by the authors, Pretoria, South Africa.

Vande weghe, J. P. 1988a. Distribution of central African montane forest birds: preliminary observations. Pages 195–204 in G. C. Bakchurst, editor. *Proceedings Sixth Pan-African Ornithological Congress.* Nairobi, Kenya.

Vande weghe, J. P. 1988b. Distributional ecology of montane forest birds: ideas for further research. Pages 469–474 in L. Bennun, editor. *Proceedings Seventh Pan-African Ornithological Congress.* Nairobi, Kenya.

Van Dijk, D. 1987. Management of indigenous evergreen high forests. Pages 454–464 in K. Von Gadow, D. W. Van der Zel, A. Van Laar, A. P. G. Schönau, H. W. Kassier, P. W. Warkotsch, H. F. Vermaas, D. L. Owen, and J. V. Jordaan, editors. *South African forestry handbook.* Southern African Institute of Forestry, Pretoria, South Africa.

Vane-Wright, R. I., C. J. Humphries, and P. H. Williams. 1991. What to protect? Systematics and the agony of choice. *Biological Conservation* 55:235–254.

Van Gils, H. 1988. *Environmental profile of Western Province, Zambia.* ITC report to Provincial Planning Unit, Mongu, Zambia.

Van Jaarsveld, A. S., S. Freitag, S. L. Chown, C. Muller, S. Koch, H. Hull, C. Bellamy, M. Kruger, S. Endrody-Younga, M. W. Mansell, and C. H. Scholtz. 1998a. Biodiversity assessment and conservation strategies. *Science* 279:2106–2108.

Van Jaarsveld, A. S., K. J. Gaston, S. L. Chown, and S. Freitag. 1998b. Throwing out biodiversity with the binary data. *South African Journal of Science* 94:210–214.

Van Keulen, H. and H. Breman. 1990. Agricultural development in the West African Sahelian region: a cure against land hunger? *Agriculture Ecosystems and Environment* 32:177–198.

Van Krunkelsven, E. 2001. Density estimation of bonobos (*Pan paniscus*) in Salonga National Park, Congo. *Biological Conservation* 99:387–391.

Van Krunkelsven, E., I. Bila-Isia, and D. Draulans. 2000. A survey of bonobos and other large mammals in the Salonga National Park, Democratic Republic of Congo. *Oryx* 34:180–187.

Van Perlo, B. 1995. *Collins illustrated checklist: birds of eastern Africa.* Harper Collins Publishers Ltd., London, UK.

Van Rompaey, H. and M. Colyn. 1998. A new servaline genet (Carnivora, Viverridae) from Zanzibar Island. *South African Journal of Zoology* 33:42–46.

Van Rooyen, N. 1999. Kalahari. Pages 10–25 in J. Knobel, editor. *The magnificent natural heritage of South Africa.* Sunbird Publishing, Cape Town, South Africa.

Van Rooyen, N. 2001. *Flowering plants of the Kalahari dunes.* Ekotrust, Lynwood, Pretoria, South Africa.

van Strien, N. J. 1989. Notable vertebrates of Mulanje Mountain. *Nyala* 14:75–79.

Van Wilgen, B. W., M. O. Andreae, J. G. Goldammer, and J. A. Lindesa. 1997. *Fire in southern African savannas.* Witwatersrand University Press, Johannesburg, South Africa.

van Wyk, A. E. 1990. Floristics of the Natal/Pondoland sandstone forests. Pages 145–158 in C. J. Geldenhuys, editor. *Biogeography of the mixed evergreen forests of southern Africa.* Occasional Report No. 45, FDR. CSIR, Pretoria, South Africa.

van Wyk, A. E. 1994a. Biodiversity of the Maputaland centre. Pages 198–209 in L. J. G. Van der Masen, X. M. van der Burgt, and J. M. van Medenbach de Rooy, editors. *Biodiversity in African savannas.* XIVth AETFAT Congress, 22–27 August 1994, International Conference Centre, Waginingen, The Netherlands. Kluwer Academic Press, Dordrecht, The Netherlands.

van Wyk, A. E. 1994b. Maputaland-Pondoland region. Pages 227–235 in WWF and IUCN, editors. *Centres of plant diversity,* 1. IUCN Publications Unit, Cambridge, UK.

van Wyk, A. E. and G. F. Smith. 2001. *Regions of floristic endemism in southern Africa: a review with emphasis on succulents.* Umdaus Press, Hatfield, South Africa.

van Zinderen Bakker, E. M. 1979. Quaternary paleoenvironments of the Sahara region. *Paleoecology of Africa* 11:83–104.

Vaughan, R. E. and P. O. Wiehé. 1937. Studies of the vegetation of Mauritius I. A preliminary survey of the plant communities. *Journal of Ecology* 25:289–343.

Venter, J. M., C. Mocke, and J. M. De Jager. 1986. Climate. In

R. M. Cowling, P. R. Roux, and A. J. H. Pieterse, editors. *The Karoo biome: a preliminary synthesis. Part 1. Physical environment*. South African National Scientific Progress Report 124. CSIR, Pretoria, South Africa.

Verboom, G. A. 1992. A report on the invasive status of *Pinus patula* on Mount Mulanje, Malawi. Honors thesis. University of Cape Town, Cape Town, South Africa.

Vernon, C. J. 1999. Biogeography, endemism and diversity of animals in the Karoo. Pages 57–78 in W. R. J. Dean and S. J. Milton, editors. *The Karoo: ecological patterns and processes*. Cambridge University Press, Cambridge, UK.

Villiers, A. 1965. Coleopteres languriides des monts Loma (Sierra Leone). *Bulletin de l'IFAN* 27A:1015–1021.

Vitousek, P. M., C. M. D'Antonio, L. L. Loope, M. Rejmanek, and R. Westbrooks. 1997a. Introduced species: a significant component of human-caused global change. *New Zealand Journal of Ecology* 21:1–16.

Vitousek, P. M., H. A. Mooney, J. Lubchenco, and J. M. Melillo. 1997b. Human domination of the earth's ecosystems. *Science* 277:494–499.

Von Breitenbach, F. 1974. *Southern cape forests and trees: a guide*. Government Printer, Pretoria, South Africa.

Von Gadow, K. 1973. Observations on the utilization of indigenous trees by the Knysna elephants. *Forestry in South Africa* 14:13–17.

Von Maltitz, G. P. and G. Fleming. 2000. Status of conservation of indigenous forests in South Africa. Pages 93–99 in A. H. W. Seydack, W. J. Vermeulen, and C. Vermeulen, editors. *Towards sustainable management based on scientific understanding of natural forests and woodlands*. Department of Water Affairs and Forestry, Knysna, South Africa.

Von Willert, D. J., B. M. Eller, M. J. A. Werger, E. Brinckmann, and H. D. Ihlenfeldt. 1992. *Life strategies of succulents in deserts*. Cambridge University Press, Cambridge, UK.

Vooren, F. and J. Sayer. 1992. Côte d'Ivoire. Pages 133–142 in J. A. Sayer, C. S. Harcourt, and N. M. Collins, editors. *The conservation atlas of tropical forests: Africa*. IUCN and Macmillan Publishers, London, UK.

Voorhoeve, A. G. 1965. *Liberian high forest trees*. Centre for Agricultural Publications and Documentation, Wageningen, The Netherlands.

Vuilleumier, F. and M. Monasterio. 1986. *High altitude tropical biogeography*. Oxford University Press, New York, USA.

Walker, P. A. 1990. Modelling wildlife distributions using geographic information system: kangaroos in relation to climate. *Journal of Biogeography* 17:279–289.

Wallace, A. R. 1876. *The geographical distribution of animals*, Volume 1. Macmillan & Co, London, UK.

Wallach, V. and D. E. Hahn. 1997. *Leptotyphlops broadleyi*, a new species of worm snake from Côte d'Ivoire (Serpentes: Leptotyphlopidae). *African Journal of Herpetology* 46:103–109.

Walter, K. S. and H. J. Gillett. 1998. *1997 IUCN Red List of threatened plants*. IUCN and WCMC, Gland, Switzerland and Cambridge, UK.

Wanless, R. M. 2002. The reintroduction of the Aldabra rail *Dryolimnas cuvieri aldabranus* to Picard Island, Aldabra Atoll. Master's thesis. University of Cape Town, Cape Town, South Africa.

Wanless, R. M., J. Cunningham, P. A. R. Hockey, J. Wanless, R. W. White, and R. Wiseman. 2002. The success of a soft-release reintroduction of the flightless Aldabra rail (*Dryolimnas (cuvieri) aldabranus*) on Aldabra Atoll, Seychelles. *Biological Conservation* 107:203–210.

Wass, P., editor. 1995. *Kenya's indigenous forests: status, management and conservation*. IUCN Forest Conservation Programme, Gland, Switzerland and Cambridge, UK.

Watkeys, M. K. 1999. Soils of the arid south-western zone of Africa. Pages 17–26 in W. R. J. Dean and S. J. Milton, editors. *The Karoo: ecological patterns and processes*. Cambridge University Press, Cambridge, UK.

Watkeys, M. K., T. R. Mason, and P. S. Goodman. 1993. The role of geology in the development of Maputaland, South Africa. *Journal of African Earth Sciences* 16:205–221.

Watts, D. P. 1998a. Long-term habitat use by mountain gorillas (*Gorilla gorilla beringei*). 1. Consistency, variation, and home range size and stability. *International Journal of Primatology* 19:651–680.

Watts, D. P. 1998b. Long-term habitat use by mountain gorillas (*Gorilla gorilla beringei*). 2. Reuse of foraging areas in relation to resource abundance, quality, and depletion. *International Journal of Primatology* 19:681–702.

Watts, D. P. 1998c. Seasonality in the ecology and life histories of mountain gorillas (*Gorilla gorilla berengei*). *International Journal of Primatology* 19:929–948.

WCMC. 1992. *Global biodiversity: status of the earth's living resources*. Chapman and Hall, London, UK.

WCMC. 1993. *Ecologically sensitive sites in Africa*, Volumes I–V. World Bank, Washington, DC, USA.

WCMC. 2002. Protected areas database. UNEP-WCMC, Cambridge, UK.

Weber, W., L. J. T. White, A. Vedder, and L. Naughton-Treves, editors. 2001. *African rain forest ecology and conservation: an interdisciplinary perspective*. Yale University Press, New Haven, CT, USA and London, UK.

Welch, G. and H. Welch. 1988. The autumn migration of raptors and other soaring birds across the Bab-el-Mandeb straits. *Sandgrouse* 10:26–50.

Welch, K. R. G. 1982. *Herpetology of Africa: a checklist and bibliography of the orders Amphisbaenia, Sauria and Serpentes*. Robert E. Krieger Publishing Company, Malabar, FL, USA.

Welcomme, R. L. 1979. *Fisheries ecology of floodplain rivers*. Longman, London, UK and New York, USA.

Welcomme, R. L. 1986. *The ecology of river systems*. Dr. W. Junk Publishers, Dordrecht, The Netherlands.

Wellington, J. E. 1938. *The Kunene River and the Etosha plain*. South African Geography Journal 20:150.

Wells, M., S. Guggenheim, A. Khan, W. Wardojo, and P. Jepson. 1999. *Investing in biodiversity: a review of Indonesia's integrated conservation and development projects*. The World Bank, Washington, DC, USA.

Werger, M. J. A., editor. 1978. *Biogeography and ecology of southern Africa*. Dr. W. Junk Publishers, The Hague, The Netherlands.

Werger, M. J. A. and B. J. Coetzee. 1978. The Sudano-Zambezian region. Pages 301–462 in M. J. A. Werger, editor. *Biogeography and ecology of southern Africa*. Dr. W. Junk Publishers, The Hague, The Netherlands.

Werre, J. L. R. 1991. *A survey of the Taylor Creek forest area, Rivers State, Nigeria*. Report to the Shell Petroleum Development Company of Nigeria, Ltd. and the Nigerian Conservation Foundation, Lagos, Nigeria.

Werre, J. L. R. 2000. Ecology and behavior of the Niger Delta red colobus (*Procolobus badius epieni*). Ph.D. thesis. City University of New York, New York, USA.

Wessels, D. C. J. 1989. Lichens of the Namib Desert, Namibia. *Dinteria* 20:3–28.

Western, D. 1989. Ivory trade under scrutiny. *Pachyderm* 12:2–3.

Wheatley, N. 1996. *Where to watch birds in Africa*. Princeton University Press, Princeton, NJ, USA.

White, F. 1965. The savanna woodlands of the Zambezian and Sudanian domains: an ecological and phytogeographical comparison. *Webbia* 19:651–681.

White, F. 1976. The underground forests of Africa: a preliminary review. *Gardens' Bulletin* (Singapore) 29:57–72.

White, F. 1978. The Afromontane region. Pages 559–620 in M. J. A. Werger, editor. *Biogeography of ecology of southern Africa*. Dr. W. Junk Publishers, The Hague, The Netherlands.

White, F. 1979. The Guineo-Congolian region and its relationships to other phytochoria. *Bulletin du Jardin Botanique National de Belgique* 49:11–56.

White, F. 1983. *The vegetation of Africa, a descriptive memoir to accompany the UNESCO/AETFAT/UNSO vegetation map of Africa (3 plates, northwestern Africa, northeastern Africa, and southern Africa, 1:5,000,000)*. United Nations Educational, Scientific and Cultural Organization, Paris, France.

White, F. 1993. Refuge theory, ice-age aridity and the history of tropical biotas: an essay in plant geography. *Fragmenta Floristica et Geobotanica* 2:385–409.

White, F., F. Dowsett-Lemaire, and J. D. Chapman. 2001. *Evergreen forest flora of Malawi*. Royal Botanic Gardens, Kew, London, UK.

White, F. and M. J. A. Werger. 1978. The Guineo-Congolian transition to southern Africa. Pages 599–620 in M. J. A. Werger, editor. *Biogeography and ecology of southern Africa*. Dr. W. Junk Publishers, The Hague, The Netherlands.

White, L. 2001. Forest-savanna dynamics in central Gabon. Pages 165–182 in W. Weber, L. J. T. White, A. Vedder, and L. Naughton-Treves, editors. *African rain forest ecology and conservation: an interdisciplinary perspective*. Yale University Press, New Haven, CT, USA.

White, L. and K. Abernethy. 1997. *A guide to the vegetation of the Lopé Reserve Gabon*. Wildlife Conservation Society, New York, USA.

White, L. J. T. 1995. *Etude de la végétation*. Réserve de a Lopé. ECOFAC, AGRECO CTFT, Libreville, Gabon.

White, L. J. T., G. McPherson, C. E. G. Tutin, E. A. Williamson, K. A. Abernethy, J. M. Reitsma, J. J. Wieringa, A. Blom, M. J. S. Harrison, and M. Leal. 2000. *Plant species of the Lopé reserve, Gabon, with emphasis on the northern half*. Special publication of the Missouri Botanical Gardens, St. Louis, USA.

White, L. J. T. and J. F. Oates. 1999. New data on the history of the plateau forest of Okomu, southern Nigeria: an insight into how human disturbance has shaped the African rain forest. *Global Ecology and Biogeography Letters* 8:355–361.

Whiteside, A. and C. Sunter. 2000. *AIDS: the challenge for South Africa*. Human & Rosseau (pty) Ltd and Tafelberg Publishers Ltd, West Cape, South Africa.

Whitmore, T. C. and J. A. Sayer, editors. 1992. *Tropical deforestation and species extinction*. IUCN Forest Conservation Programme, Gland, Switzerland and Cambridge, UK.

Wickens, G. E. 1976. The flora of the Jebel Marra (Sudan Republic) and its geographical affinities. *Kew Bulletin Additional Series* V:1–368.

Wickens, G. E. 1984. Flora. Pages 67–75 in J. L. Cloudsley-Thompson, editor. *Sahara Desert*. Pergamon Press, Oxford, UK and New York, USA.

Wikramanayake, E., E. Dinerstein, C. Loucks, D. Olson, J. Morrison, J. Lamoreux, M. McKnight, and P. Hedao. 2002. *Terrestrial ecoregions of the Indo-Pacific*. Island Press, Washington, DC, USA.

Wild, C. 1994. The status and ecology of the montane herpetology of Mount Oku, Cameroon, Africa. *ASRA Journal* 73–91.

Wild, H. and L. A. G. Barbosa. 1967. *Vegetation map (1:2 500 000 in colour), of the Flora Zambesiaca area*. Descriptive memoir. Supplement to Flora Zambesiaca. M.O. Collins, Salisbury, Rhodesia.

Wildlife Survey Team. 2001. *Report on the impact of conflict on the Boma National Park*. New Sudan Wildlife Society, Nairobi, Kenya.

Wilkie, D. S. and J. Carpenter. 1998. *The impact of bushmeat hunting on forest fauna and local economies in the Congo Basin*. Wildlife Conservation Society, Bronx, USA.

Wilkie, D. S., J. F. Carpenter, and Q. Zhang. 2001. The under-financing of protected areas in the Congo Basin: so many parks and so little willingness-to-pay. *Biodiversity and Conservation* 10:691–709.

Wilkie, D. S. and B. Curran. 1993. Historical trends in forager and farmer exchange in the Ituri Rain Forest of northeastern Zaire. *Human Ecology* 21:389–417.

Wilkie, D. S., B. Curran, R. Tshombe, and G. A. Morelli. 1998a. Managing bushmeat hunting in Okapi Wildlife Reserve, Democratic Republic of Congo. *Oryx* 32:131–144.

Wilkie, D. S., B. Curran, R. Tshombe, and G. A. Morelli. 1998b. Modeling the stainability of subsistence farming and hunting in the Ituri Forest of Zaire. *Conservation Biology* 12:137–147.

Wilkie, D. S. and J. T. Finn. 1990. Slash-burn cultivation and mammal abundance in the Ituri Forest, Zaire. *Biotropica* 22:90–99.

Wilkie, D. S. and N. Laporte. 2001. Forest area and deforestation in Central Africa: current knowledge and future directions in African rain forest ecology and conservation. Pages 119–139 in W. Weber, L. J. T. White, A. Vedder, and L. Naughton, editors. *African rain forest ecology and conservation*. Yale University Press, New Haven, CT, USA.

Wilkie, D. S., G. A. Morelli, F. Rotberg, and E. Shaw. 1999. Wetter isn't better: global warming and food security in the Congo Basin. *Global Environmental Change* 9:323–328.

Wilkie, D., E. Shaw, F. Rotberg, G. Morelli, and P. Auzel. 2000. Roads, development, and conservation in the Congo Basin. *Conservation Biology* 14:1614–1622.

Wilkie, D. S., J. G. Sidle, and G. C. Boundzanga. 1992. Mechanized logging, market hunting, and a bank loan in Congo. *Conservation Biology* 6:570–580.

Wilks, C. 1990. *La conservation des ecosystemes forestiers du Gabon.* IUCN, Gland, Switzerland.

Williams, M. 1984. Geology. Pages 31–39 in J. L. Cloudsley-Thompson, editor. *Sahara Desert.* Pergamon Press, Oxford, UK and New York, USA.

Williams, M. A. J. and H. Faure. 1980. *The Sahara and the Nile.* Balkema, Rotterdam, The Netherlands.

Williams, P. H. 1996. Measuring biodiversity value. *World Conservation* 1:12–14.

Williams, P. H. 1998. Key sites for conservation: area-selection methods for biodiversity. Pages 211–249 in G. M. Mace, A. Balmford, and J. Ginsberg, editors. *Conservation in a changing world.* Cambridge University Press, Cambridge, UK.

Williams, P. H., N. D. Burgess, and C. Rahbek. 2000. Flagship species, ecological complementarity, and conserving the diversity of mammals and birds in sub-Saharan Africa. *Animal Conservation* 3:249–260.

Williams, P. H., H. M. de Klerk, and T. M. Crowe. 1999. Interpreting biogeographical boundaries among Afrotropical birds: spatial patterns in richness gradients and species replacement. *Journal of Biogeography* 26:459–474.

Williams, P. H., J. L. Moore, T. M. Brooks, H. Strand, J. D'Amico, J. Oglethorpe, M. Wisz, N. D. Burgess, A. Balmford, and C. Rahbek. 2003. Integrating biodiversity priorities with conflicting socio-economic values in the Guinean–Congolian forest region. *Biodiversity and Conservation* 12:1297–1320.

Williamson, D. 1994. *Botswana: environmental policies and practices under scrutiny.* The Lomba Archives. Lindlife, Cape Town, South Africa.

Williamson, G. 1979. The orchid flora in the Nyika Plateau, Africa. *Journal of South African Botany* 45:459–468.

Williamson, G. 1997. Preliminary account of the floristic zones of the Sperrgebiet (protected diamond area) in southwest Namibia. *Dintera* 25:1–68.

Willis, C. K., J. E. Burrows, M. Koekemoer, and L. Fish. 2000. *Plants of the Nyika. Volume I: Preliminary checklist.* National Botanical Institute (SABONET), Pretoria, South Africa.

Wilson, D. E. and F. R. Cole. 2000. *Common names of mammals of the world.* Smithsonian Institution Press, Washington, DC, USA.

Wilson, D. E. and D. M. Reeder. 1993. *Mammal species of the world: a taxonomic and geographic reference.* Smithsonian Institution, Washington, DC, USA.

Wily, L. A. and S. Mbaya. 2001. *Land, people and forests in eastern and southern Africa at the beginning of the 21st century: the impact of land relations on the role of communities in forest future.* IUCN Eastern Africa Programme Office, Nairobi, Kenya.

Winterbottom, J. M. 1978. Birds. In M. J. A. Werger, editor. *Biogeography and ecology of southern Africa.* Dr. W. Junk Publishers, The Hague, The Netherlands.

Witte, J. 1995. Deforestation in Zaire: logging and the landlessness. Pages 179–197 in M. Colchester and L. Lohmann, editors. *The struggle for land and the fate of the forests.* World Rainforest Movement and Zed Books, London, UK.

Wojtusiak, J. and T. Pyrcz. 1997. Taxonomy of three butterfly species from the islands of São Tomé and Príncipe in the Gulf of Guinea: *Neptis larseni* n. sp., *Charaxes thomasius* n. stat., *Papilio furvus* n. stat. (Lepidoptera: Papilionoidaea). *Lambillionea* 97:53–63.

Wolfire, D., J. Brunner, and N. Sizer. 1998. *Forests and the Democratic Republic of Congo: opportunity in a time of crisis.* World Resources Institute, Washington, DC, USA.

Wood, A., P. Stedman-Edwards, and J. Mang, editors. 2000a. *The root causes of biodiversity loss.* Earthscan, London, UK.

Wood, S., K. Sebastian, and S. J. Scherr. 2000b. *Pilot analysis of global ecosystems: agroecosystems.* International Food Policy Research Institute and World Resources Institute, Washington, DC, USA.

Woodroffe, R. 1999. Managing disease threats to wild mammals. *Animal Conservation* 2:185–193.

Woodroffe, R. 2000. Predators and people: using human densities to interpret declines of large carnivores. *Animal Conservation* 3:165–173.

Woodroffe, R. and J. R. Ginsberg. 1998. Edge effects and extinctions of populations inside protected areas. *Science* 280:2126–2128.

Woodroffe, R., J. R. Ginsberg, D. W. Macdonald, and IUCN/Species Survival Commission Canid Specialist Group. 1997. *The African wild dog: status survey and conservation action plan.* IUCN, Gland, Switzerland.

World Bank. 1994. *Adjustment in Africa.* Oxford University Press, New York, USA.

World Bank. 2000. *Public information document for reversal of land and water degradation trends in the Lake Chad Basin ecosystem.* Project ID 3APE70252.

World Bank. 2002. World development report. http://www.worldbank.org.

Wozencraft, W. C. 1986. A new species of striped mongoose from Madagascar. *Journal of Mammalogy* 67:561–571.

Wranik W. 1998. Faunistic notes on Soqotra Island. In H. J. Dumont, editor. *Proceedings of the first international symposium on Soqotra Island: present and future, Aden 1996.* Soqotra Technical Series, Volume 1. UNDP, New York, USA.

WRI. 1995. African data sampler. World Resources Institute, Washington, DC, USA.

WRI. 2001. *World resources 2000–2001. People and ecosystems: the fraying web of life.* World Resources Institute, Washington, DC, USA.

Wright, R. G., M. Murray, and T. Merill. 1998. Ecoregions as a level of ecological analysis. *Biological Conservation* 86:207–213.

WWF. 1994. Parc National des Virunga, Zaire. *Stratégies de la conservation à long terme des écosystèmes et ébauche d'un plan directeur du Parc National.* World Wildlife Fund, Gland, Switzerland.

WWF. 1997. *Conserving Africa's elephants: current issues and priorities for action.* WWF, Gland, Switzerland.

WWF. 1998. *Elephants in the balance: conserving Africa's elephants.* WWF, Gland, Switzerland.

WWF. 2000. *Vision de la biodiversité: région ecologique des forêts d'epineux.* WWF Madagascar and West Indian Ocean Programme Office, Antananarivo, Madagascar.

WWF. 2001. *Plan d'action pour la Région Ecologique de Forêt Epineuse.* WWF Madagascar and West Indian Ocean Programme Office, Antananarivo, Madagascar.

WWF and IUCN. 1994. *Centres of plant diversity. A guide and strategy for their conservation. Volume 1: Europe, Africa, South West Asia and the Middle East.* IUCN Publications Unit, Cambridge, UK.

WWF MedPO. 2001. *The Mediterranean forests. A new conservation strategy* (plus map). WWF MedPO, Rome, Italy.

WWF and Terralingua. 2000. *Indigenous and traditional peoples of the world and ecoregion conservation.* WWF International and Terralingua, Gland, Switzerland.

WWF-Tz. 2001. *The eastern African marine ecoregion.* World Wildlife Fund–Tanzania, Dar es Salaam, Tanzania.

WWF-US. 2002. Ecoregion conservation: securing living landscapes through science-based planning and action, an interactive users guide. CD-ROM. World Wildlife Fund–US, Washington, DC, USA.

Wymenga, E., B. Kone, and L. Zwarts, editors. 2002. *Le Delta Intérieur du fleuve Niger: écologie et gestion durable des resources naturelles.* Wetlands International, Waginingen, The Netherlands.

Yalden, D. W. and M. J. Largen. 1992. The endemic mammals of Ethiopia. *Mammal Review* 22:115–150.

Yalden, D. W., M. J. Largen, D. Kock, and J. C. Hillman. 1996. Catalogue of the mammals of Ethiopia and Eritrea: 7. Revised checklist, zoogeography and conservation. *Tropical Zoology* 9:73–164.

Yana Njabo, K. and M. Languy. 2000. *Surveys of selected montane and submontane areas of the Bamenda Highlands in March 2000.* Cameroon Ornithological Club, Yaoundé, Cameroon.

Yansané, A. 1998. Le Schema Directeur d'Amenagement de la Mangrove de Guineé (SDAM) et sa mise en oeuvre. *Bulletin du Centre de Rogbané* 12:210–223.

Yeoman, G. 1989. *Africa's Mountains of the Moon.* Elm Tree Books, London, UK.

Yoder, A. D., M. M. Burns, S. Zehr, T. Delefosse, G. Veron, S. M. Goodman, and J. J. Flynn. 2003. Single origin of Malagasy carnivora from an African ancestor. *Nature* 421:734–737.

Yoder, A. D., M. Cartmill, M. Ruvolo, K. Smith, and R. Vilgalys. 1996. Ancient single origin for Malagasy primates. *PNAS USA* 93:5122–5126.

Young, T. P. 1996. High montane forest and Afroalpine ecosystems. Pages 401–424 in T. R. McClanahan and T. P. Young, editors. *East African ecosystems and their conservation.* Oxford University Press, New York, USA and Oxford, UK.

Younge, A. and S. Fowkes. 2003. The Cape Action Plan for the Environment: overview of an ecological planning process. *Biological Conservation* 112:15–28.

Younge, A. G. N., G. Negussie, and N. D. Burgess. 2002. Eastern Africa Coastal Forest Programme. Regional workshop report, Nairobi, February 4–7 2002. World Wildlife Fund-EARPO, Nairobi, Kenya.

Zahran, M. A. and A. J. Willis. 1992. *The vegetation of Egypt.* Chapman and Hall, London, UK.

Zahran, M. A. and A. J. Willis. 2002. *Plant life in the River Nile in Egypt.* Mars Publishers, Cairo, Egypt.

Zheng, X. and E. A. B. Eltahir. 1998. The role of vegetation in the dynamics of West African monsoons. *Journal of Climate* 11:2078–2096.

Zimmermann, E., S. Cepok, N. Rakotoarison, V. Zietemann, and U. Radespiel. 1998. Sympatric mouse lemurs in north-west Madagascar: a new rufous mouse lemur species (*Microcebus ravelobensis*). *Folia Primatologica* 69:106–114.

Zohary, M. 1973. *Geobotanical foundations of the Middle East,* 2 volumes. Fisher, Stuttgart, Germany.

Authors

Neil Burgess, Ph.D.
Senior Conservation Scientist
Conservation Science Program
World Wildlife Fund–United States
Zoology Department
Cambridge University
Downing Street
Cambridge CB2 3EJ, UK

Jennifer D'Amico Hales, M.S.
Senior Conservation Specialist
Conservation Science Program
World Wildlife Fund–United States

Emma Underwood, M.S.
Senior Conservation Specialist
Conservation Science Program
World Wildlife Fund–United States

Eric Dinerstein, Ph.D.
Chief Scientist
Conservation Science Program
World Wildlife Fund–United States

David Olson, Ph.D.
Director
Conservation Science Program
World Wildlife Fund–United States

Illanga Itoua, B.A.
Conservation Specialist
Conservation Science Program
World Wildlife Fund–United States

Jan Schipper, M.S.
Conservation Specialist
Conservation Science Program
World Wildlife Fund–United States

Taylor Ricketts, Ph.D.
Director
Conservation Science Program
World Wildlife Fund–United States

Kate Newman, M.S.
Director
East and Southern Africa
Current Director of Marine Ecoregions
Endangered Spaces Program
World Wildlife Fund–United States

Contributing Essay Authors

Tom Allnutt
Conservation Science Program
World Wildlife Fund–US
1250 24th Street NW
Washington, DC 20037, USA

Andrew Balmford
Zoology Department
Cambridge University
Downing Street
Cambridge CB2 3EJ, UK

Elizabeth L. Bennett
The Wildlife Conservation Society
2300 Southern Boulevard
Bronx, NY 10460, USA

Luigi Boitani
Department of Animal and Human Biology
University of Rome
Viale Università 32
Rome, Italy

Donald G. Broadley
Biodiversity Foundation for Africa
P.O. Box FM 730
Famona, Bulawayo, Zimbabwe

Thomas Brooks
Center for Applied Biodiversity Science
Conservation International
1919 M Street NW, Suite 600
Washington, DC 20036, USA

Brigitte Carr-Dirick
Consultant
c/o WWF Central Africa Program Office
B.P. 6776 Yaoundé
Cameroon

Alan Channing
Zoology Department
University of the Western Cape
Private Bag X17
Bellville 7535, South Africa

Fabio Corsi
International Institute for Geo-Information
Science and Earth Observation
P.O. Box 6
7500 AA Enschede, The Netherlands

Richard Cowling
University Way, Summerstrand
University of Port Elizabeth 6031
South Africa

David H. M. Cumming
Tropical Resource Ecology Programme
Department of Biological Sciences
University of Zimbabwe
P.O. Box MP 167
Mount Pleasant, Harare, Zimbabwe

Jane Dwasi
University of Nairobi
P.O. Box 30197
Nairobi, Kenya

Lincoln Fishpool
BirdLife International
Wellbrook Court, Girton Road
Cambridge CB3 0NA, England

Tommy Garnett
Environmental Foundation for Africa
PMB 34, 1 Beach Road
Lakka, Freetown, Sierra Leone

Steven M. Goodman
Field Museum of Natural History
Roosevelt Road at Lake Shore Drive
Chicago, IL 60605
and
WWF-Madagascar
B.P. 738
Antananarivo 101, Madagascar

Rachel Gylee
Zoology Department
Cambridge University
Downing Street
Cambridge CB2 3EJ, UK

Nina Jackson
Zoology Department
Cambridge University
Downing Street
Cambridge CB2 3EJ, UK

Sam Kanyamibwa
WWF East Africa Regional Office
P.O. Box 62440
Nairobi, Kenya

Rebecca Kormos
Conservation International
1919 M Street NW, Suite 600
Washington, DC 20036, USA

Mandy Lombard
Lombard, Wolf & Cole CC
P.O. Box 1465
Sedgfield 6573
South Africa

Colby Loucks
Conservation Science Program
World Wildlife Fund–US
1250 24th Street NW
Washington, DC 20037, USA

Jon C. Lovett
Centre for Ecology Law and Policy
Environment Department
University of York
York YO10 5DD, UK

Chris Margules
P.O. Box 780
Atherton, QLD 4883, Australia

Susan Minnemeyer
Global Forest Watch
World Resources Institute
10 G Street NE, Suite 800
Washington, DC 20002, USA

Joslin Moore
CSIRO Entomology
GPO Box 1700
Canberra ACT 2601, Australia

Melissa Moye
Center for Conservation Finance
World Wildlife Fund
1250 24th Street NW
Washington, DC 20037, USA

Malik Doka Morjan
New Sudan Wildlife Service
P.O. Box 8874-00100 GPO
Nairobi, Kenya

Philip Muruthi
Chief Scientist
African Wildlife Foundation
1400 Sixteenth Street NW, Suite 120
Washington, DC 20036, USA

Judy Oglethorpe
World Wildlife Fund–US
1250 24th Street NW
Washington, DC 20037, USA

Steven A. Osofsky
Wildlife Conservation Society
2300 Southern Boulevard
Bronx, NY 10460, USA

Andrew Perkin
Oxford Brookes University
Gipsy Lane Campus
Headington
Oxford, OX3 0BP, UK

Bob Pressey
New South Wales National Parks and Wildlife Service
87 Faulkner Street
P.O. Box 402
Armidale NSW 2350, Australia

Gabriella Reggiani
University of Rome
Viale Università 32
Rome, Italy

John G. Robinson
Wildlife Conservation Society
2300 Southern Boulevard
Bronx, NY 10460, USA

Liz Selig
Global Forest Watch
World Resources Institute
10 G Street NE, Suite 800
Washington, DC 20002, USA

James Shambaugh
Peace Corps
1111 20th Street NW
Washington, DC 20526, USA

Peter J. Stephenson
WWF International
Avenue du Mont Blanc
CH-1196 Gland, Switzerland

James Taplin
Centre for Ecology Law and Policy
Environment Department
University of York
York YO10 5DD, UK

David Thomas
BirdLife International
Wellbrook Court, Girton Road
Cambridge, UK

Anada Tiéga
The Ramsar Convention Secretariat
Rue Mauverney 28
CH-1196 Gland, Switzerland

Harry van der Linde
African Wildlife Foundation
1400 Sixteenth Street NW, Suite 120
Washington, DC 20036, USA

Paul Williams
Biogeography and Conservation Lab
The Natural History Museum
Cromwell Road
London, SW7 5BD, UK

Index

Island Press Board of Directors